Proceedings of the 28th International Symposium on
Lepton Photon Interactions at High Energies

Proceedings of the 28th International Symposium on

Lepton Photon Interactions at High Energies

Sun Yat-sen University, Guangzhou, China, 7 – 12 August 2017

Edited by

Wei Wang
Sun Yat-sen University, China

Zhi-Zhong Xing
Chinese Academy of Sciences, China

World Scientific

NEW JERSEY · LONDON · SINGAPORE · BEIJING · SHANGHAI · HONG KONG · TAIPEI · CHENNAI · TOKYO

Published by

World Scientific Publishing Co. Pte. Ltd.

5 Toh Tuck Link, Singapore 596224

USA office: 27 Warren Street, Suite 401-402, Hackensack, NJ 07601

UK office: 57 Shelton Street, Covent Garden, London WC2H 9HE

Library of Congress Cataloging-in-Publication Data

Names: International Symposium on Lepton and Photon Interactions at High Energies (28th : 2017 :
   Guangzhou, China), author. | Wang, Wei, (Professor of physics), editor. | Xing, Zhi-Zhong, editor.
Title: Lepton photon interactions at high energies (Lepton Photon 2017) :
   proceedings of the 28th International Symposium, 7–12 August 2017,
   Sun Yat-Sen University, Guangzhou, China / edited by Wei Wang, Zhi-Zhong Xing.
Description: New Jersey : World Scientific, [2019] | Includes bibliographical references.
Identifiers: LCCN 2019020530 | ISBN 9789811204609 (hardcover)
Subjects: LCSH: Lepton interactions--Congresses. | Photonuclear reactions--Congresses.
Classification: LCC QC793.9 .I7 2017 | DDC 503--dc23
LC record available at https://lccn.loc.gov/2019020530

British Library Cataloguing-in-Publication Data
A catalogue record for this book is available from the British Library.

For any available supplementary material, please visit
https://www.worldscientific.com/worldscibooks/10.1142/11393#t=suppl

The XXVIII 28th International Symposium on
Lepton Photon Interactions at High Energies
7-12 August 2017, Guangzhou, China

The International Symposium on Lepton Photon Interactions at High Energies (Lepton Photon), as designated by the International Union of Pure and Applied Physics (IUPAP), is a conference series held every other year as the premiere conference for high-energy physics and related fields. Physicists from around the world gather at Lepton Photon to discuss the latest advancements in particle physics, nuclear physics, astrophysics, cosmology and plans for future facilities.

The XXVIII International Symposium on Lepton Photon Interactions at High Energies was held on the South Campus of Sun Yat-sen University (SYSU), Guangzhou, China from Aug 7 to Aug 12, 2017, co-hosted by SYSU and IHEP (Institute of High Energy Physics). The conference followed the tradition of Lepton Photon: the program features one plenary session covering topics of major interest to the extended HEP community; in addition, initiated by Lepton Photon 2015, a poster session is also organized for individual researchers, working groups and collaborations to have more opportunities to present their work.

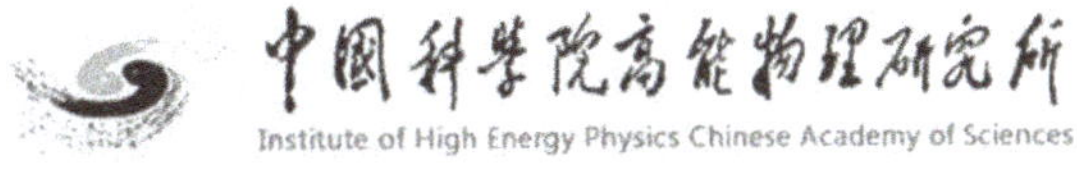

International Advisory Committee (IAC)

Halina Abramowicz (Tel Aviv, Israel)
Nima Arkani-Hamed (IAS, U.S.)
Laura Baudis (Univ. of Zurich, Switzerland)
William Brooks (UTF-SM, Chile)
Tiziano Camporesi (CERN, Europe)
Ariella Cattai (CERN, Europe)
Dave Charlton (Birmingham, U.K.)
John Ellis (KCL, U.K.)
Lyn Evans (CERN, Europe)
Fernando Ferroni (INFN, Italy)
Juan Fuster (IFIC, Spain)
Gian Francesco Giudice (CERN, Europe)
Fabiola Gianotti (CERN, Europe)
Francis Halzen (UW-Madison, U.S.)
Beate Heinemann (DESY and Freiburg University)
Takaaki Kajita (ICRR, Japan)
Soo-Bong Kim (SNU, Korea)
Young-Kee Kim (FNAL, U.S.)
Joseph D. Lykken (FNAL, U.S.)
Bob McKeown (JLab, U.S.)
Naba K. Mondal (TIFR, India)
Joachim Mnich (DESY, Germany)
Michael Peskin (SLAC, U.S.)
Stefano Ragazzi (Gran Sasso, Italy)
Nigel Smith (SNO Lab, Canada)
Yifang Wang (IHEP, China)
Masanori Yamauchi (KEK, Japan)

Local Organizing Committee (LOC)

Wei Wang (Co-chair, SYSU)
Zhi-Zhong Xing (Co-chair, IHEP)
Miao He (IHEP)
Yan-Ping Huang (IHEP)
Xiaonan Li (IHEP Neutrino Center Kaiping)
Zhi-Bing Li (SYSU, Guangzhou)
Jiajie Ling (SYSU, Guangzhou)
Jian Tang (SYSU, Guangzhou)
Liangjian Wen (IHEP, Guangzhou)
Fan-Rong Xu (JNU, Guangzhou)
Zhengyun You (SYSU, Guangzhou)
Hong-Hao Zhang (SYSU, Guangzhou)
Shun Zhou (IHEP, Guangzhou)
Xuai Zhuang (IHEP)

XXVIII International Symposium on Lepton Photon Interactions at High Energies
第２８届轻子光子国际会议
2017.8.7 中国·广州

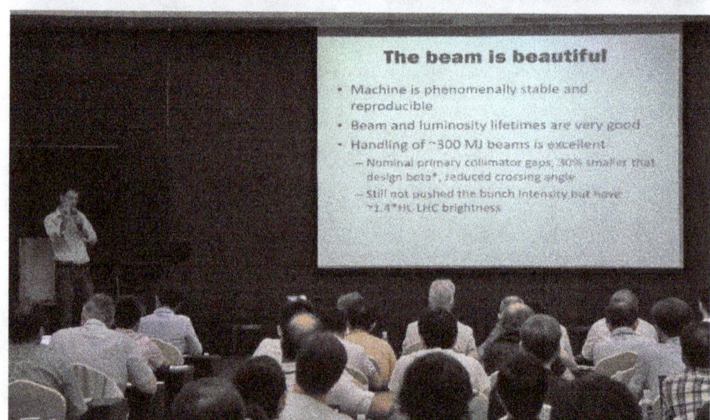

The beam is beautiful
• Machine is phenomenally stable and reproducible
• Beam and luminosity lifetimes are very good
• Handling of ~300 MJ beams is excellent
– Nominal primary collimator gaps, 30% smaller that design beta*, reduced crossing angle
– Still not pushed the bunch intensity but have ~1.4*HL LHC brightness

Neutrino oscillation experiments in Japan
Intense Neutrino Beam for study
Super-K
Tokai
295 km
Hyper-K
470 kW (today)
1 MW (2020)
1.3 MW (2026)
Seamless program
physics results
kton (Super-K, ~2020)
kton (Hyper-K, 2026~)

Top Quark Physics example

Particle Physics in China:
Past, Present and Future
Yifang Wang
Institute of High Energy Physics, Beijing
Sun Yat-Sen University, Aug.10, 2017

大亚湾反应堆中微子实验奠基
Daya Bay Reactor Neutrino Experiment Foundation Stone

Contents

PLENARY SESSIONS

Recent Highlights from the CMS Experiment

Christian Autermann

on behalf of the CMS Collaboration

I. Phys. Inst. B, RWTH Aachen University
Sommerfeldstr. 14, 52074 Aachen, Germany
E-mail: christian.autermann@cern.ch

This article summarizes the latest highlights from the CMS experiment as presented at the Lepton Photon conference 2017 in Guangzhou, China. A selection of the latest physics results, the latest detector upgrades, and the current detector status was shown. CMS has analyzed the full dataset of proton-proton collision data delivered by the LHC in 2016 at a center-of-mass energy of 13 TeV corresponding to an integrated luminosity of 40 fb^{-1}. The leap in center-of-mass energy and in luminosity with respect to the 7 and 8 TeV runs enabled interesting and relevant new physics results. A new silicon pixel tracking detector has been installed during the LHC shutdown 2016/17 and has successfully started operation.

Keywords: CMS; LHC; highlights; upgrades; detector; 13 TeV.

1. Introduction

The CMS detector[1] is a multipurpose detector located at the LHC accelerator at CERN. The LHC started operation in 2010 and has delivered until the end of 2012 proton-proton collisions at center-of-mass energies of 7 TeV and 8 TeV corresponding to a total integrated luminosity of 5 fb^{-1} and 20 fb^{-1}, respectively. In 2015 the operations at 13 TeV started and delivered about 4 fb^{-1}. The dataset delivered in 2016 at 13 TeV corresponding to a luminosity of 40 fb^{-1} is used for the majority of physics results discussed here.

CMS is characterized by the large superconducting solenoid magnet with 6 m inner diameter providing a magnetic field of 3.8 T. Inside this magnet the silicon pixel and strip tracking detectors and the calorimeter systems are installed. The magnet itself is located within the iron return yoke, which houses also the muon system of drift tube and resistive plate chambers. The tracking system is the largest silicon detector of any high energy detector and allows excellent particle momentum measurement at high efficiency and resolution. In particular the innermost pixel vertex detector enables precise b jet and τ tagging. The calorimeters are divided into the electromagnetic (ECAL) and hadronic calorimeters (HCAL). The ECAL is made out of dense but clear and scintillating lead tungstate crystals and silicon avalanche photodiodes for readout. The HCAL is a sampling calorimeter using brass and steel plates as absorbers and plastic scintillators as active material. A particle-flow (PF) event reconstruction algorithm is used to identify and reconstruct any particle of a proton−proton collision event, using the best possible combination of all detector subsystems. The PF event reconstruction leads to an improved performance for the reconstruction of jets and missing transverse momentum $p_{\mathrm{T}}^{\mathrm{miss}}$, and for the identification of electrons, muons, and taus.

2. Status of the CMS Detector

The CMS status[2] during the successful data-taking period 2016, the upgrades during the extended end-of-year technical stop 2016/17, and the startup in spring 2017 will be discussed briefly in the following.

2.1. *Detector upgrades*

Several upgrades were installed during the technical stop 2016/17. The main changes were the improvements of the hadronic forward calorimeter readout, the installation of a muon-endcap GEM prototype, and most importantly, of the new pixel vertex detector.

The readout electronics of the hadronic forward calorimeter were upgraded in order to make full use of the photomultiplier tubes (PMTs) replaced in the long shutdown of 2013. The PMTs read out the Cherenkov light produced in quartz fibers embedded in the iron absorbers of the hadronic calorimeter. However, occasionally charged particles produce Cherenkov light directly in the PMT window mimicking high-energetic particles. The new PMTs have 1/6 windows thickness, twice the quantum efficiency and gain, and dual readout that required the electronics to be upgraded.

Gas electron multiplier (GEM) detectors are being considered for an upgrade of the CMS muon system. A GEM demonstrator was installed in the forward high-eta region to test the system during LHC run conditions. GEMs are micro-pattern gaseous detectors that are expected to improve muon tracking and triggering utilizing a good spatial and time resolution, high efficiency, and radiation hardness.

The silicon pixel tracker is the innermost CMS sub-detector and crucial for precise track reconstruction and for vertex reconstruction with high resolution. The pixel tracker was replaced during the end-of-year shutdown 2016/17 due to radiation damage and significant data losses at instantaneous luminosity larger than 10^{34} cm^{-2}s^{-1}. The maximum LHC instantaneous luminosity is routinely exceeding this threshold by a factor of two, corresponding to a mean number of primary vertices of 50, causing too high occupancy in the old pixel readout chip. The upgraded pixel detector uses a digital readout chip with higher rate capability. The main new features are four instead of three barrel layers and three instead of two end-cap disks. The mechanical structure is designed to be very light and a new powering scheme and CO_2 cooling system is used in order to further reduce the amount of material.

Operation of the newly installed pixel detector has started. Layer 1 and 2 share a common programmable time delay. The time alignment was difficult to achieve due a faster Layer 1 readout chip. An optimal common plateau of efficiency with values

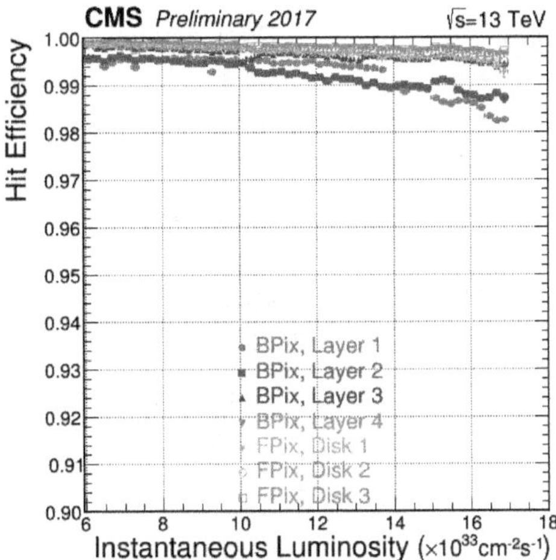

Fig. 1.   Hit efficiencies per layer of the upgraded CMS pixel detector in 2017.

close to 99% for all pixel layers and disks at luminosities $\mathcal{L} = 1.6 \cdot 10^{34}\,\mathrm{cm}^{-2}\mathrm{s}^{-1}$ as shown in Fig. 1 has been established. The timing is chosen to favor the Layer 1 performance. Although not yet at the ultimate detector performance, more complex functions like vertexing, b-tagging, and high-level trigger electron reconstruction are significantly better than with the old detector, which would not have been able to cope with the rates in the first place.

2.2. *Detector performance*

The LHC restarted collisions after the end-of-year technical stop in 2017 very successfully, delivering fast peak instantaneous luminosities near the design goal of $\mathcal{L} = 2 \cdot 10^{34}\,\mathrm{cm}^{-2}\mathrm{s}^{-1}$ as shown in Fig. 2. Until the Lepton-Photon conference already 9 fb^{-1} of data at a center-of-mass energy of 13 TeV had been delivered[3]. While this 2017 data set is not yet used for physics analyses, the detector performance studies discussed in this section, in particular of the recently installed upgrades make use of this data.

The CMS Level 1 trigger system reduces the event rate from the proton bunch crossings rate at 40 MHz down to 100 kHz without detrimental impact on the physics acceptance, in particular for measurements and searches at the electroweak scale. The trigger system was upgraded during the long shutdown 1 (LS1) in 2013/14 to cope with the increased rates due to the larger instantaneous luminosity and the increase in center-of-mass energy. A significant increase in trigger threshold with negative impact on the physics program was avoided. The performance of the level 1 electron or photon and muon trigger in $\sqrt{s} = 13$ TeV data was measured[4] and is shown exemplary in Fig. 3 (left) for isolated electron or photon candidates. The efficiency was measured using the tag&probe method on $Z \to ee$ events. The

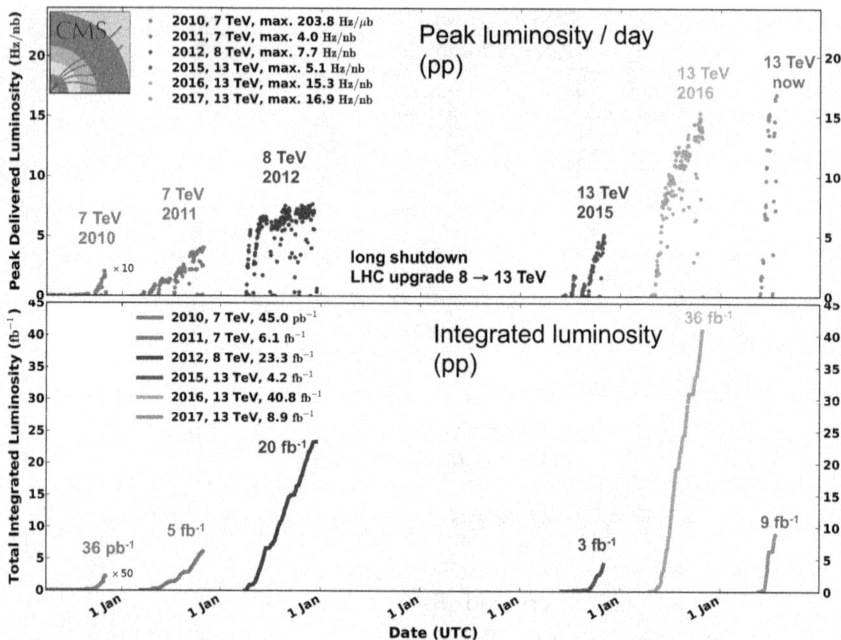

Fig. 2. Delivered and recorded instantaneous (top) and integrated luminosity (bottom) to CMS for the proton-proton runs since 2010 [3].

higher 2017 trigger efficiency is due to an improved relaxation scheme of the isolation selection as a function of E_T of the e/γ candidate. By recalibrating the calorimeter level-1 object also the transverse energy resolution of L1 e/γ candidates with respect to the offline reconstructed E_T could be improved in 2017, as shown in Fig. 3 (right).

To maintain good quality track reconstruction with high resolution for charged particles the position and the orientation of the silicon pixel and strip modules need to be known with a precision of several micrometers. The tracker alignment follows a global fit approach using Millepede-II and a local fit approach using the HipPy fit [5]. For the 2016 data taking, the modul-level alignment was obtained from cosmic-ray data collected before and a small amount of pp collision data. Time-dependent movements of the pixel detector large-scale structures were monitored. The tracker alignment in data taking, shown as red histogram in Fig. 4 (left), was corrected if necessary by an automated prompt calibration loop, as shown in Fig. 4 (right).

The missing transverse momentum p_T^{miss} is a measure of the momentum imbalance caused by undetectable particles such as neutrinos and is crucial for high precision measurements and in particular for searches for new physics with undetectable dark matter candidate particles. On the other hand, because the missing transverse momentum is calculated using the vectorial sum of all energy depositions reconstructed in the detector, it is a quantity that is highly sensitive to the precision of the energy measurement in all sub-detectors, in particular of the

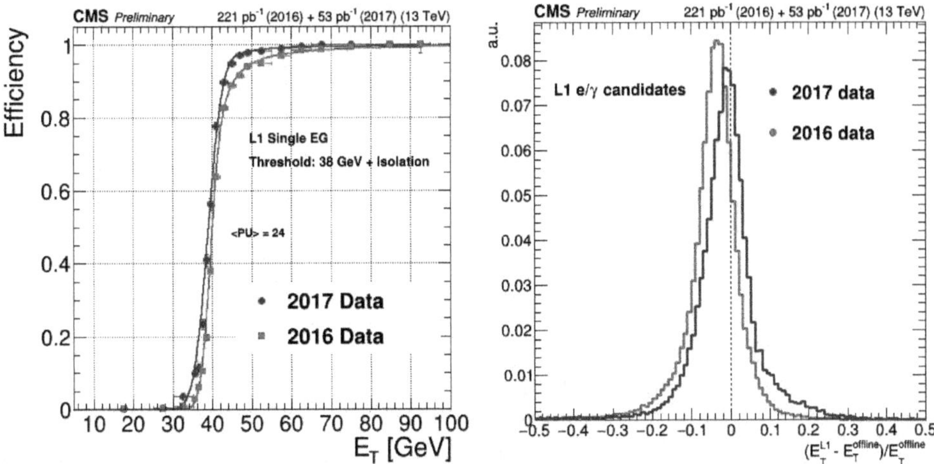

Fig. 3. L1 electron or photon candidate trigger efficiency versus the transverse energy for an average of 24 simultaneous interactions (left) and the L1 E_T resolution with respect to the offline reconstruction[4] (right).

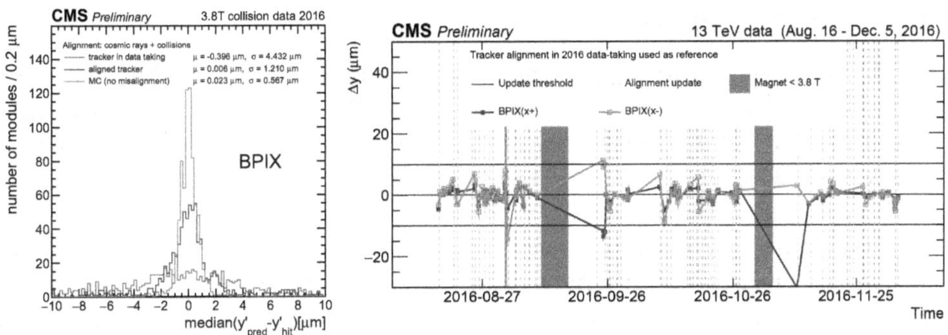

Fig. 4. Alignment of the pixel detector. The median residuals of the y-positions of all pixel barrel detector modules are shown for simulation, during data taking, and for the end-of-year alignment data (left). The movements of the pixel barrel detector in y-direction for the second half of the 2016 data taking are shown (right). The grey bands indicate runs during which the CMS magnet was not at 3.8 T. Vertical dashed lines illustrate updates of the pixel reference geometry, after which mis-alignments are cured[5].

calorimeter energy resolution. The missing transverse momentum performance and its dependence on the simultaneous luminosity, i.e., the amount of simultaneous proton-proton interactions called pileup, is studied using $Z \to ll$ events[6]. The well reconstructed leptonically decaying Z boson allows to probe the detector response of the hadronic system, and therefore to measure the scale and the resolution of the missing transverse momentum. The hadronic recoil is projected onto the axis of the well measured Z boson transverse momentum and its parallel and perpendicular components are used to study the performance of p_T^{miss} as shown in Fig. 5 for particle-flow and "pileup per particle identification" (PUPPI) corrected p_T^{miss}.

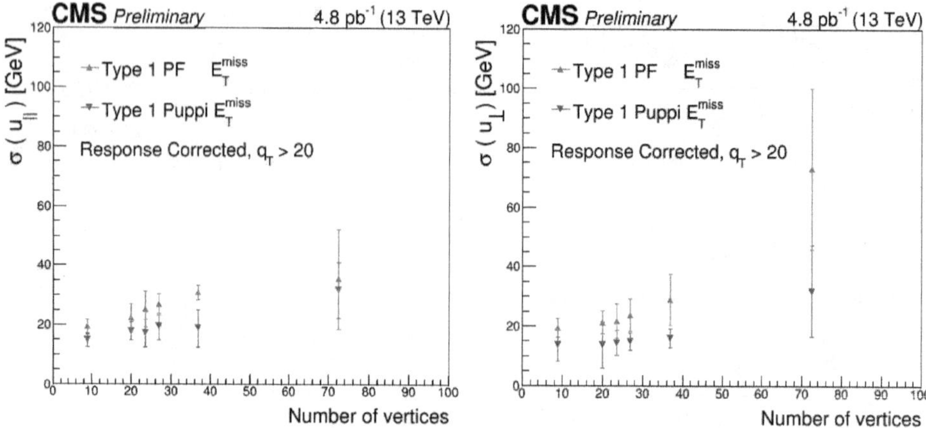

Fig. 5. Comparison of resolution of parallel (left) and perpendicular (right) components of hadronic recoil measured using jet-energy-scale corrected PF p_T^{miss} (red markers) and corrected PUPPI (blue markers) as a function of number of vertices[6] (color online).

3. Recent CMS Physics Highlights

The LHC started operations at a center-of-mass energy of 13 TeV in 2015 delivering an integrated luminosity of 3 fb^{-1}. In 2016 the 13 TeV data run continued, CMS recorded 36 fb^{-1} of data, almost twice the amount of data collected before the long shutdown in 2012 at 8 TeV corresponding to an integrated luminosity of 20 fb^{-1}, as shown in Fig. 2. In combination with the leap in center-of-mass energy the 2016 data set therefore provides the basis for significant updates of physics analyses in terms of precision, sensitivity, and reach. The physics results discussed in the following are selected highlights of recent publications typically based on this data set, significantly extending and exceeding previous CMS results.

3.1. QCD multijet measurements

QCD multijet production processes dominate the spectrum of physics produced at the LHC and are therefore substantial standard model backgrounds to many precision measurements and searches for new physics. Direct measurements of QCD multijet production can provide valuable information for physics at higher energy scales, for example by extracting the energy scale depending strong coupling α_s[7] allowing for more precise fits to higher scales, as shown in Fig. 6.

The differential QCD cross-section measurement in dijet event topologies as a function of jet masses and transverse momenta is sensitive to QCD parton showering and in particular relevant for searches for new physics with boosted objects. The dijet topology was unfolded to correct for the effect of the detector reconstruction using 2.3 fb^{-1} of 13 TeV data[7]. Jet grooming algorithms were used to remove low

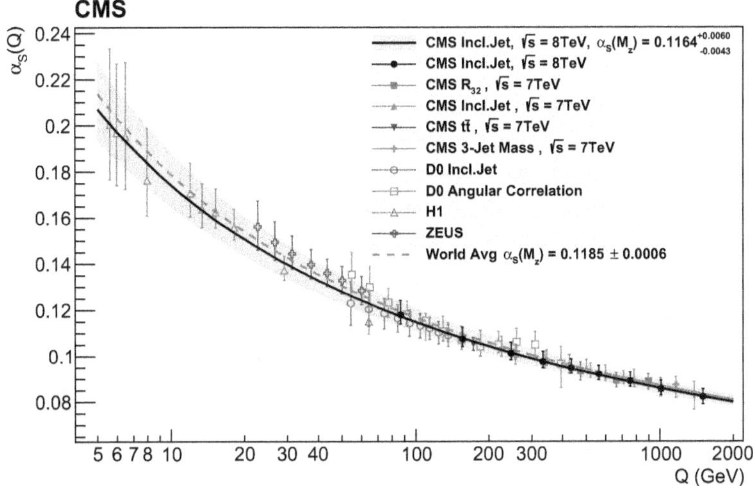

Fig. 6. The running of the strong coupling parameter α_s and its uncertainty[7]. Previous measurements from the H1, ZEUS, and D0 experiments at the HERA and Tevatron colliders are shown together with recent CMS measurements.

energy portions from the jets arising from soft radiations that are difficult to model. The Monte Carlo simulation of the jet mass spectrum was found to be improved for groomed jets, as shown in Fig. 7.

3.2. *Electroweak production*

The effective weak mixing angle was measured in 18.8 fb^{-1} of 8 TeV data using the forward-backward asymmetry of Drell-Yan events in the dielectron and dimuon final state[8]. The forward-backward asymmetry A_{FB} is defined as $A_{\mathrm{FB}} = (\sigma_{\mathrm{F}} - \sigma_{\mathrm{B}})/(\sigma_{\mathrm{F}} + \sigma_{\mathrm{B}})$, where σ_{F} and σ_{B} are the cross sections in the forward and backward hemispheres in the Collins-Soper frame of the dilepton system. The measured asymmetry A_{FB}, as shown in Fig. 8 (left), allows to extract the most precise measurement of leptonic effective weak mixing angle $\sin^2 \theta_{\mathrm{eff}}^{\mathrm{lept}} = 0.23101 \pm 0.00052$ at the LHC. Uncertainties of the proton density functions (PDFs) translate into sizable variations of A_{FB}. Since the PDF uncertainties affect A_{FB} and $\sin^2 \theta_{\mathrm{eff}}^{\mathrm{lept}}$ differently, a Baysian χ^2 reweighting method can be applied to constrain the PDF uncertainties and to reduce the uncertainty of the extracted value of $\sin^2 \theta_{\mathrm{eff}}^{\mathrm{lept}}$. The CMS measurement precision is comparable to Tevatron measurements and is compared in Fig. 8 (right) to other LHC, Tevatron, and LEP results.

The CMS collaboration has measured a wide range of standard model production cross sections, branching ratios, particle masses, and couplings. In Fig. 9 a selection of electroweak, top quark, and Higgs boson production cross section measurements is summarized. Even without QCD multijet processes cross sections over nine orders of magnitude have been measured and compared to theory predictions

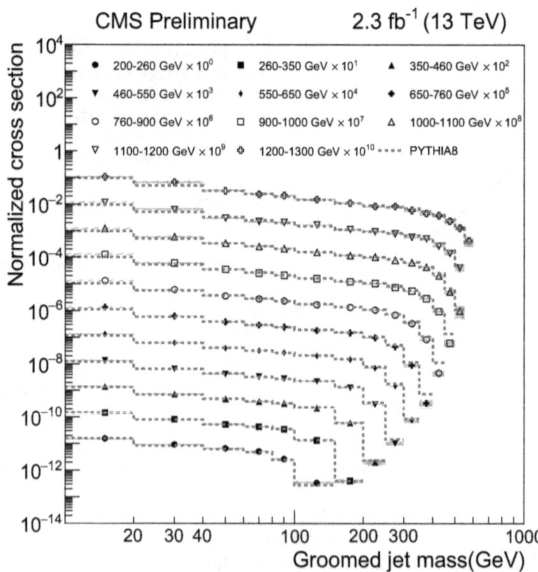

Fig. 7. Differential dijet cross section unfolded to particle level as a function of the jet mass[7].

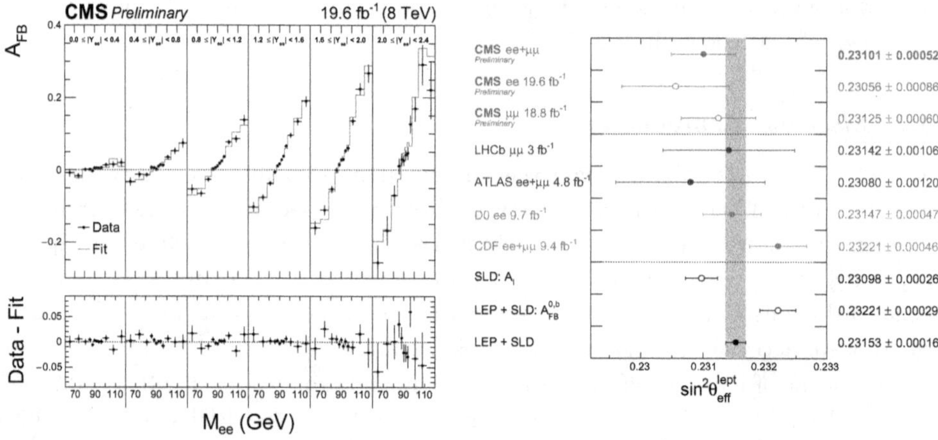

Fig. 8. Measurement of the forward-backward asymmetry of Drell-Yan events (left) and the extracted electroweak mixing angle fitted to several previous and current analyses (right)[8].

for the $\sqrt{s} = 7$ TeV, 8 TeV, and 13 TeV runs. A remarkable agreement is observed. The consistency of experimental observations and theoretical predictions over this wide range of cross sections, production mechanisms, and final states illustrates nicely the excellent control over the data reconstruction, the precision of theoretical calculations, and in result the yet unchallenged validity of the standard model at collider experiments.

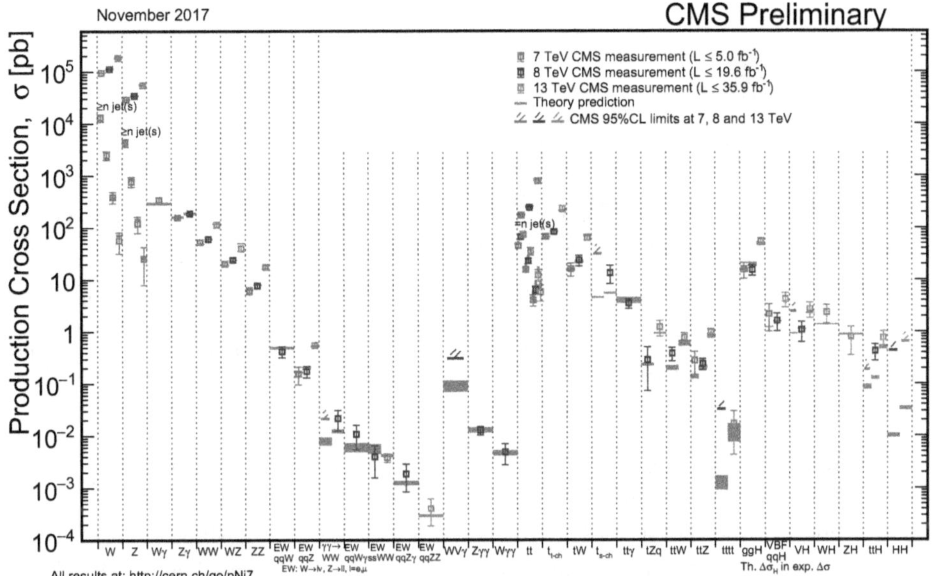

Fig. 9. Selection of inclusive cross sections measurements of electroweak processes, top quark, and Higgs physics[2].

In the following some exemplary measurements and searches for physics beyond the standard model are discussed in more detail.

3.3. *Top quark physics*

The top quark t, discovered in 1995 by the Tevatron experiments CDF and D0, is the heaviest standard model particle with the largest Yukawa coupling to the Higgs. As physics beyond the standard model is generally expected at higher scales, the top quark is the particle with the smallest mass gap to any other not yet discovered new particle and is often expected to be produced in association with new physics, such as supersymmetry. The precise knowledge of standard model top processes is essential, as it is an important background for search analyses but also because it provides through standard model precision fits direct constraints on new physics.

The top quark pair and single top quark production cross sections at the LHC are summarized in Fig. 10 depending on the center of mass energy. While the Tevatron proton−anti-proton collider enabled the top quark discovery, the succeeding LHC pp collider can be considered a top-quark-factory. At average instantaneous luminosity $\mathcal{L} = 10^{34}$ cm^{-2}s^{-1} about 50 top pairs are created every second.

The large cross section and therefore the abundance of top quarks at the LHC enabled the study of rare top processes. Four examples are briefly summarized on page 12.

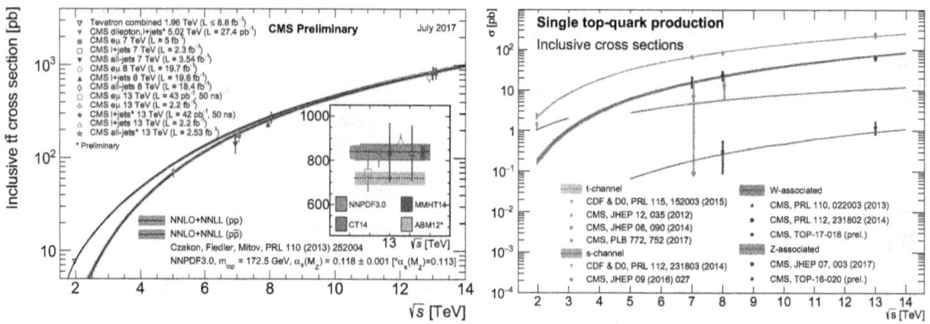

Fig. 10. Top production cross sections depending on the center of mass energy for $t\bar{t}$ pair production (left) and single-t (right)[2].

- Recently the cross section of top quark pair production in association with a W or Z boson has been measured in the 13 TeV data set collected in 2016[9]. The measurement was performed in the same-sign dilepton, three- and four-lepton final states where the jet and b-tagged jet multiplicities were exploited to enhance the signal-to-background ratio. The resulting cross section is included in the summary plot in Fig. 9.

- One of the most promising channels for the direct measurement of the Higgs-top quark Yukawa coupling strength is the standard model production of the Higgs boson in association with a $t\bar{t}$ pair, where the top quarks decay leptonically (e or μ) and the Higgs boson as $H \rightarrow b\bar{b}$. The inclusive top quark pair plus dijet $t\bar{t} + b\bar{b}$ and $t\bar{t} + jj$ production cross sections as well as their ratio $\sigma_{t\bar{t}b\bar{b}}/\sigma_{t\bar{t}jj}$ have been measured in dileptonic events using the first 2.3 fb^{-1} of 13 TeV data[10]. The ratio in the full phase space was measured to be $\sigma_{t\bar{t}b\bar{b}}/\sigma_{t\bar{t}jj} = 0.022 \pm 0.007$ in agreement with the prediction from the standard model. The contribution from $Ht\bar{t}$ was found to be negligibly small.

- The measurement of the semileptonic $t\bar{t}\gamma$ production cross section[11] is directly probing the coupling of the top quark and the photon. A deviation between the measurement and the standard model prediction could hint at exotic quarks with anomalous electric charge or electric dipole moment. The fiducial cross section for associated $t\bar{t}$ and photon production per semileptonic final state was found to be 127 ± 27 fb in agreement with standard model predictions.

- The single top quark production cross section in association with a Z boson in final states with three leptons (e or μ) was measured to 10^{+8}_{-7} fb at a significance of 2.4 σ in the 8 TeV data set corresponding to 19.7 fb^{-1} of integrated luminosity. This measurement allows to constrain flavor-changing neutral current interactions to $\mathcal{B}(t \rightarrow Zu) < 0.022\%$ and $\mathcal{B}(t \rightarrow Zc) < 0.049\%$ at 95% confidence level[12].

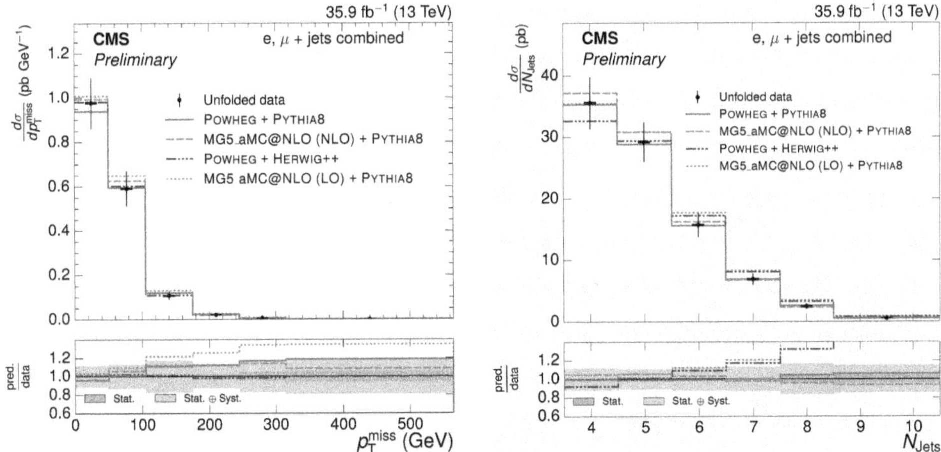

Fig. 11. Differential $t\bar{t}$ cross section measurements in missing transverse momentum (left) and in the jet multiplicity variable (right) in 35.9 fb^{-1} of 13 TeV data compared to different Monte Carlo simulations[13].

Differential cross section measurements can provide highly sensitive tests to verify theoretical models of standard model processes. These precision tests allow to tune Monte Carlo generators and to improve for example the standard model background estimation relevant for search analysis. The top differential cross section in the semileptonic channel with one electron or muon has been measured with respect to variables that do not require to reconstruct the $t\bar{t}$ system[13]. The 13 TeV data collected in 2016 corresponds to an integrated luminosity of 35.9 fb^{-1} and to about $3 \cdot 10^7$ $t\bar{t}$ pairs that are used for this analysis. The resulting differential cross sections are unfolded to particle level in a phase space resembling the fiducial volume of CMS and are compared to state-of-the-art leading order and next-to-leading order $t\bar{t}$ simulations, as shown in Fig. 11.

One important example for top quark property measurements is the measurement of the top mass. Plots summarizing the various top quark mass measurement analyses at CMS in the traditional and in alternative channels using data recorded at $\sqrt{(s)} = 7$, 8 and 13 TeV are shown in Fig. 12. It is remarkable that the top mass measurement in the single semileptonic decay channel using 19.7 fb^{-1} of 8 TeV data exceeds the precision of the Tevatron combination. Worth mentioning is also the compatible precision of the various alternative channels on the top quark mass.

3.4. *Higgs physics*

The Higgs boson is the last particle discovered by the ATLAS and CMS experiments in 2012 and completes the standard model of particle physics. The discovery channels are the fully reconstructible $H \to \gamma\gamma$ diphoton and $H \to l^+l^-l^+l^-$ four lepton (electron or muon) decay channels, where the Higgs boson decays via a charged

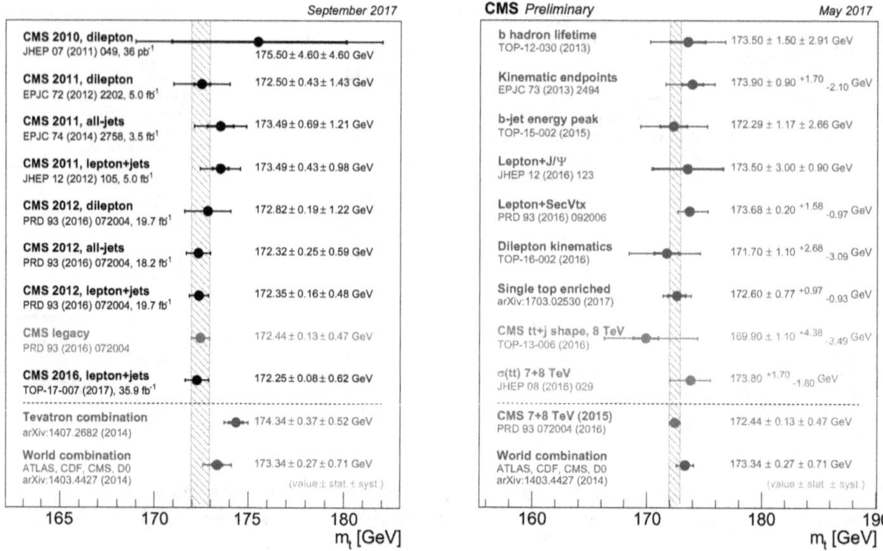

Fig. 12. Summary plot of the top mass measurements in the classic channels (left) and in alternative channels with compatible sensitivity (right) [14].

particle loop or two Z bosons, respectively. Despite the small branching fractions of $\mathrm{BF}_{\gamma\gamma} = 0.23\%$ and $\mathrm{BF}_{4l} = 0.013\%$, these channels offer a clear fully reconstructible invariant mass resonance signal over a continuous background and allow for measurements of the Higgs boson properties, such as the mass, spin, CP-parity, and couplings.

The analysis of the diphoton final state using 35.9 fb^{-1} of 13 TeV data has measured the Higgs signal strength relative to the standard model prediction for different production modes and effective couplings to fermions and bosons as well as to photons and gluons [15]. The results are in agreement with the expectations from the standard model, in particular the $Ht\bar{t}$ production was measured relative to the expectation to $\mu_{Ht\bar{t}} = 2.2^{+0.9}_{-0.8}$, corresponding to a 3.3 σ excess with respect to the absence of $Ht\bar{t}$ production. The analysis of the four-lepton final state in the same data set lead to similar measurements of cross section and coupling strengths that were also found to be consistent with standard model expectations [16]. The decay width of the Higgs boson was measured to be $\Gamma_H < 1.10$ GeV at 95% confidence level.

A combined ATLAS and CMS analysis of the LHC pp collision data at $\sqrt{s} = 7$ and 8 TeV [17] has allowed to measure the Higgs boson production and decay rates and to constrain the Higgs couplings. In particular the couplings to the top and bottom quarks, to the Z and W vector bosons, and the tau lepton have been measured and the coupling to muons has been constrained. Recently, individual analyses of the 13 TeV data in the $H \to W^+W^-$, $H \to \tau\tau$, and $H \to b\bar{b}$ channels have been published, as will be briefly discussed in the following.

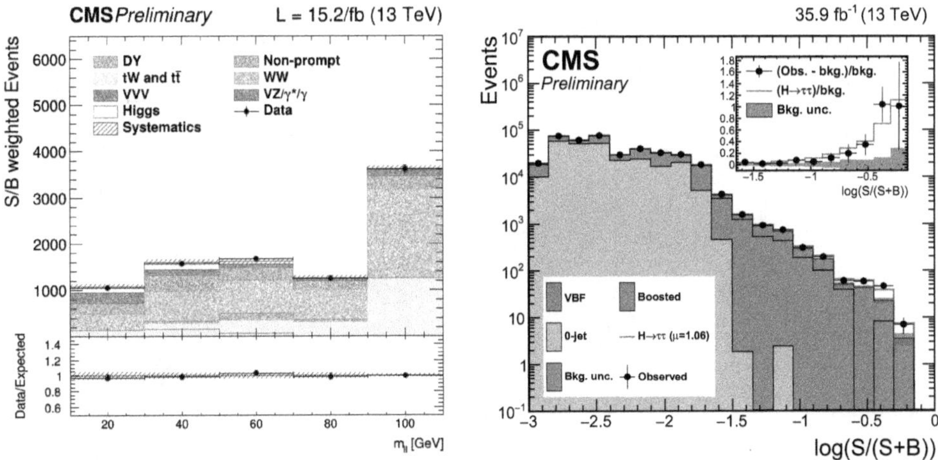

Fig. 13. Invariant dilepton mass for the $H \to W^+W^-$ analysis[18] using 15.2 fb^{-1} of 13 TeV data (left) and the distribution of the decimal logarithm of the ratio between the expected signal and the sum of expected signal and expected background in each bin of the mass distributions to extract the results, in all signal regions for the $H \to \tau\tau$ analysis[19] using 35.9 fb^{-1} of data (right).

The measurement of the Higgs to WW decay channel uses 15.2 fb^{-1} of 13 TeV data recorded in 2015 and 2016[18]. The W^+W^- candidate events are selected by requiring an opposite charged $e\mu$ pair and large missing transverse momentum, as shown in Fig. 13 (left). The analysis categorized the selection depending on the Higgs production mode, i.e. gluon or vector boson (meaning in the following W or Z) fusion and vector boson associated production, distinguished by the jet and lepton multiplicity. The observed cross section times branching fraction is measured to be 1.05 ± 0.26 times the standard model prediction, corresponding to an observed significance of 4.3 σ.

The first observation of the Higgs boson decay to a pair of tau leptons by a single experiment has been reported[19]. The 13 TeV data set corresponding to 35.9 fb^{-1} has been analyzed simultaneously for gluon and vector boson fusion production categories as shown in Fig. 13 (right), using maximum likelihood fits in two-dimensional planes to extract the results. The production cross section times branching fraction was measured $1.09^{+0.27}_{-0.26}$ times the standard model expectation. Combining this measurement with previously published CMS results using the 7 and 8 TeV data sets leads to an observed significance of 5.9 σ for the observation of the $H \to \tau\tau$ decay channel by the CMS experiment.

A recent CMS highlight result is the observed evidence for the Higgs boson decay to a bottom quark-antiquark pair using 35.9 fb^{-1} of data recorded at 13 TeV[20]. The $H \to b\bar{b}$ decay directly tests the coupling to fermions, i.e. to down-type quarks. In spite of the large Higgs branching fraction of 58% into b quarks, this channels has not yet been established with observation level significance. The combined analysis

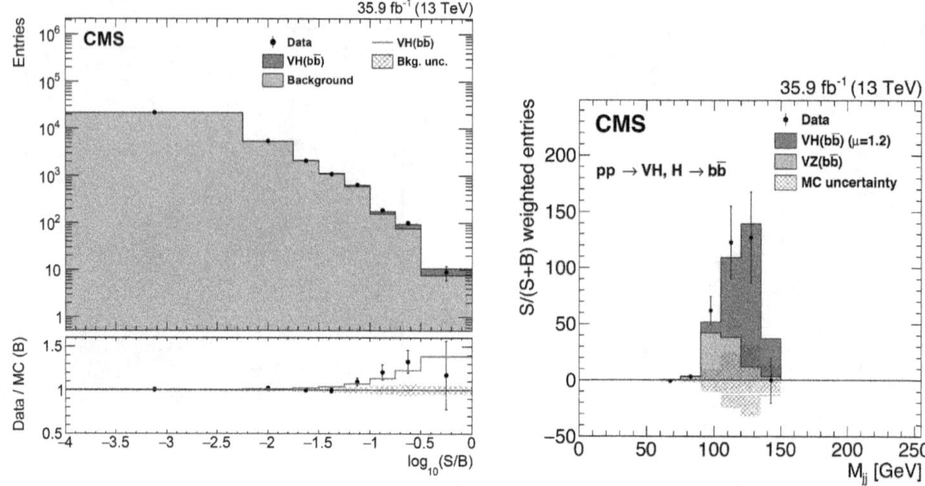

Fig. 14. Combination of all channels of the VH search into a single BDT distribution (left) and weighted invariant dijet mass distribution (right) for 35.9 fb^{-1} of data recorded at 13 TeV[20].

of the 7 and 8 TeV data by the ATLAS and the CMS experiments observed this channel at a significance of 2.6 σ. The Tevatron experiments CDF and D0 reported an excess of events at $m(H) = 125$ GeV corresponding to a local significance of 3.0 σ for the combined searches, that were dominated by VH production and the $H \to b\bar{b}$ decay channel. Here, the same VH production channel is studied, where the Higgs boson decays to bottom quarks and the vector bosons W or Z decays fall into one of the five categories $Z \to \nu\nu$, ee, $\mu\mu$, and $W \to e\nu$, $\mu\nu$. The category with two charged leptons is further divided into a low and a high boson p_T region. For each of the seven signal selections boosted decision tree (BDT) discriminators are trained using a large number of kinematic variables, such as transverse momenta, invariant masses, angular distances, and b-tagging discriminants. Figure 14 (left) shows the gathered BDT output values for these seven channels and Fig. 14 (right) shows the dijet invariant mass distribution, combined for all channels, with all background processes other than VH and VZ subtracted and weighted by $s/(s + b)$, where s and b are the numbers of expected signal and background events in the bin of the BDT output distribution. The observed signal strength for $m(H) = 125.09$ GeV is $\mu = 1.2 \pm 0.4$ corresponding to an observed significance of 3.3 σ. Combining this measurement with the 7 and 8 TeV data recorded by CMS yields a signal strength of $\mu = 1.06^{+0.31}_{-0.29}$ and observed and expected significances of 3.8 σ.

3.5. *Search for exotic new physics*

The standard model of particle physics is until today successfully describing all physics processes studied at particle accelerator experiments. Also most recent results as discussed before are in good agreement with the standard model predictions.

And yet, astrophysical observations like dark mass and dark energy, the observation of non-negligible neutrino masses, and theoretical considerations imply that the standard model cannot be complete or valid up to arbitrary high energy scales. The standard model does not offer a candidate dark matter particle and cannot describe quantum gravity effects that become relevant at the Planck scale Λ_P. An important objective of the LHC experiments is to search for physics beyond the standard model (BSM physics). Analyses of the 8 TeV data have been found to be in agreement with standard model expectations[21]. The sensitivity of searches for new physics profits greatly from the increased center-of-mass energy to 13 TeV and the large integrated luminosity of data taken during 2016.

A popular BSM theory is the scenario of extra dimensions that would lower the effective Λ_P scale, as predicted for example by the Arkani-Hamed–Dimopoulos–Dvali model. A non-resonant enhancement of dijet production through virtual graviton exchange leads to signatures differing from the predictions of QCD. The analysis of dijet angular distributions using 35.9 fb^{-1} of 13 TeV data is sensitive to the graviton exchange, as well as to other BSM scenarios such as the production of quantum black holes, dark matter, and quark contact interaction models[22]. The differential dijet production cross section dominated by QCD Rutherford scattering is flat in $\chi_{\mathrm{dijet}} = \exp|y_1 - y_2| = (1 + |\cos\theta^*|)/(1 - |\cos\theta^*|)$, where θ^* is the jet angle to the beam axis in the dijet rest frame. The χ_{dijet} analysis is complementary to dijet narrow bump searches and allows to set for the first time limits on the universal quark coupling to large dark matter mediator masses $2.5 < M_{\mathrm{Med}} < 5$ TeV, as shown in Fig. 15.

3.6. *Searches for supersymmetry*

Supersymmetry (SUSY) is probably the most popular model of physics beyond the standard model. SUSY is the last possible space-time symmetry that can provide a dark matter candidate particle, if the R-parity is conserved and the lightest SUSY particle (LSP) is neutral. It provides a solution to the Higgs hierarchy problem by virtue of the symmetry between fermions and bosons. For each known standard model particle a new supersymmetric partner is predicted differing in spin by half a unit, and in mass. Since no SUSY particle has been discovered so far[23], the mass differences between the supersymmetric partners must be sufficiently large to escape detection by previous searches. The doubled luminosity of the data taken in 2016 and in particular the increased center-of-mass energy have favored the sensitivity of searches for supersymmetry. The jump in sensitivity and reach within a short time period will be unmatched in future, until a collider with significantly more energy will be build, as the LHC experiments will now only gradually increase their sensitivity to BSM physics with the amount of integrated luminosity. In the following, the most recent result of three different flavors of supersymmetry will be discussed.

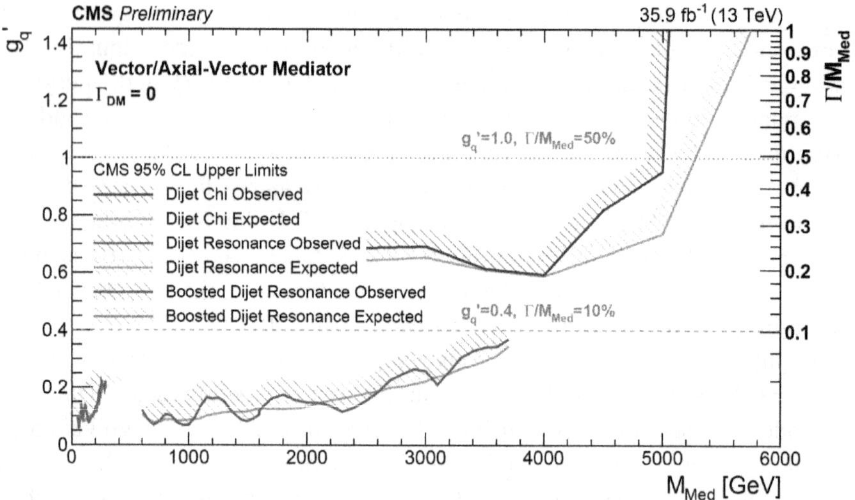

Fig. 15. The 95% CL upper limits on the universal quark coupling g'_q as a function of the mass of a DM mediator that only decays to quarks from the CMS dijet narrow resonance search, boosted dijet resonance search, and the χ_{dijet} analysis[22].

A search for supersymmetry, with gauge mediated SUSY breaking (GMSB), has been carried out using events with at least one photon, missing transverse momentum, and large hadronic event activity (H_T) in the transverse plane[24]. R-parity conservation is assumed, so that the LSP, the almost massless Gravitino $\tilde{G}$, is stable. The final state signature in signal events is determined by the nature of the next-to-lightest SUSY particle, the NLSP. In the studied signal scenarios the NLSP is a gaugino that decays promptly as $\tilde{\chi}_1^{0/\pm} \to V\tilde{G}$ depending on its mixing, where V is a SM boson $\gamma, Z^0, W^\pm$, or H. Different regions in the p_T^{miss} spectrum are used to normalize and validate the SM background estimation and to extract the signal acceptance, as shown in Fig. 16 (left). The analyses excludes squark masses up to 1.65 TeV and gluino masses up to 2.0 TeV, as compared in Fig. 16 (right) to the exclusion contours of a search for GMSB supersymmetry in events with at least one photon and missing transverse momentum[25] that is specialized for electroweak production of GMSB SUSY with little hadronic event activity.

CMS searches for electroweak production of charginos and neutralinos in several previously published final states using the full 13 TeV data set recorded in 2016 and corresponding to 35.9 fb^{-1} have been statistically combined, including a new analysis of the final state with three or more charged leptons (e or μ)[26]. Strongly coupling SUSY particles might be too massive to be directly produced at the LHC, motivating direct searches for the electroweakly produced gauginos. Assuming R-parity conservation and depending on the production channel, an abundance of SM bosons is produced through cascade decays of the heavy gauginos into the LSP. The analyses have statistically independent signal regions $1l2b$, $4b$, $2l$ on-Z, $2l$ soft, $\geq 3l$,

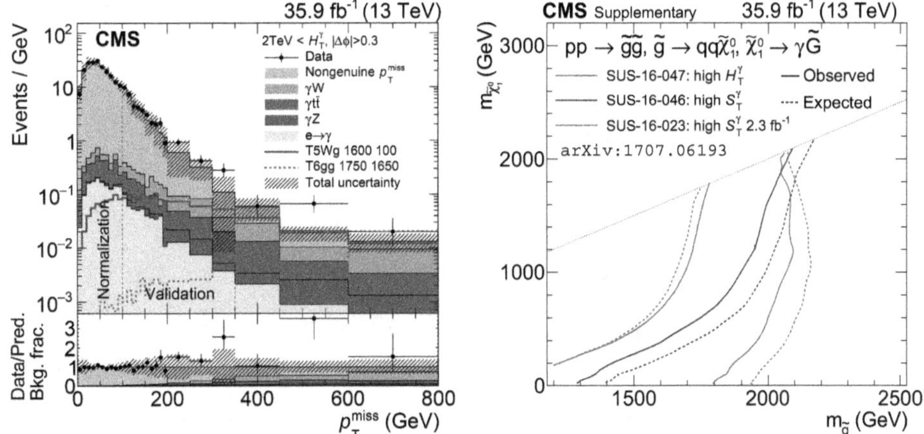

Fig. 16. Missing transverse momentum distribution as observed in the search for gauge mediated supersymmetry breaking in final states with at least one photon and jets[24] (left). The resulting exclusion contour for in a simplified signal scenario of gluino pair production and bino-like neutralino decays to photons and gravitinos is compared to a contour from the analysis with at least one photon and p_T^{miss} but without jet requirements[25] (right).

and $H(\gamma\gamma)$ targeting final states with two SM bosons WZ, WH, ZZ, ZH, or HH. The 4b search drives the exclusion at large branching fraction values of $\tilde{\chi}_1^0 \to H\tilde{G}$ while the on-Z dilepton search and the multilepton search are competing at lower branching fraction values. The search with three or more charged leptons targets $\tilde{\chi}_1^\pm \tilde{\chi}_2^0$ production, where the mass difference between $\tilde{\chi}_1^0$ and $\tilde{\chi}_2^0$ is approximately equal to the Z boson mass. The statistical combination of all final states is able to exclude for chargino-neutralino production $m(\tilde{\chi}_1^\pm) = m(\tilde{\chi}_2^0)$ up to 650 GeV. Similar limits are derived for electroweak production in the GMSB scenario[26].

Most searches for supersymmetry assume the R-parity to be conserved, causing the lightest SUSY particle to be stable. Typically the LSP is the neutral neutralino $\tilde{\chi}_1^0$, the sneutrino $\tilde{\nu}$, or the Gravitino $\tilde{G}$ that are dark matter candidate particles. A common feature of R-parity conserving SUSY signal events is missing transverse momentum, as the dark matter candidate particles do not interact with the detector. However, there is no fundamental theoretical reason for R-parity conservation. In particular in the light of the most recent increasingly stringent limits on SUSY particle masses that challenge "natural SUSY" scenarios, R-parity violation (RPV) can ease this tension, as the limits obtained from R-parity conserving analyses do not generally apply here, because depending on the RPV coupling, little or no p_T^{miss} is created in the signal events. Baryon and meson lifetime measurements place strong indirect limits on the combination of two RPV couplings. Therefore, it is typically assumed that all RPV couplings except one are negligibly small or zero.

Minimal flavor-violating SUSY motivates the search for R-parity violating supersymmetry involving the third generation via the $U_i D_j D_k$ coupling λ''_{ijk}[27], where the U_i and D_j are the up-type and down-type quark $SU(2)$ singlet fields. Gluinos

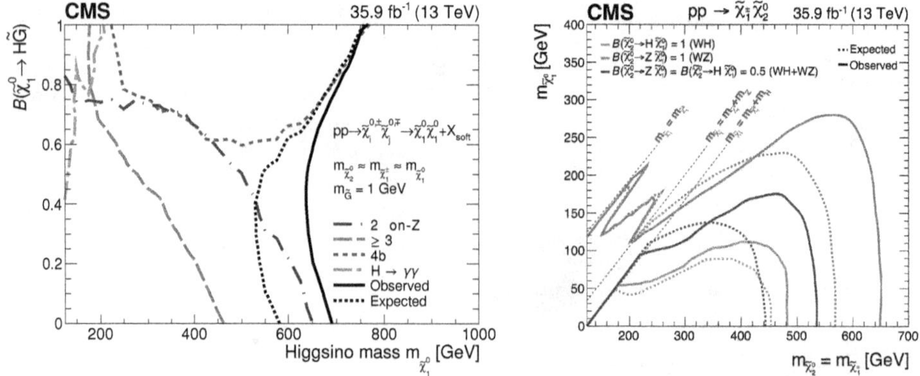

Fig. 17. Observed exclusion contours at the 95% CL in the plane of the higgsino mass and the branching fraction $\tilde{\chi}_1^0 \to H\tilde{G}$ (left) and exclusion contours at the 95% CL in the plane of $m(\tilde{\chi}_1^0)$ and $m(\tilde{\chi}_1^{\pm})$ for different assumed decay modes for the statistical combination of electroweak searches[26] (right).

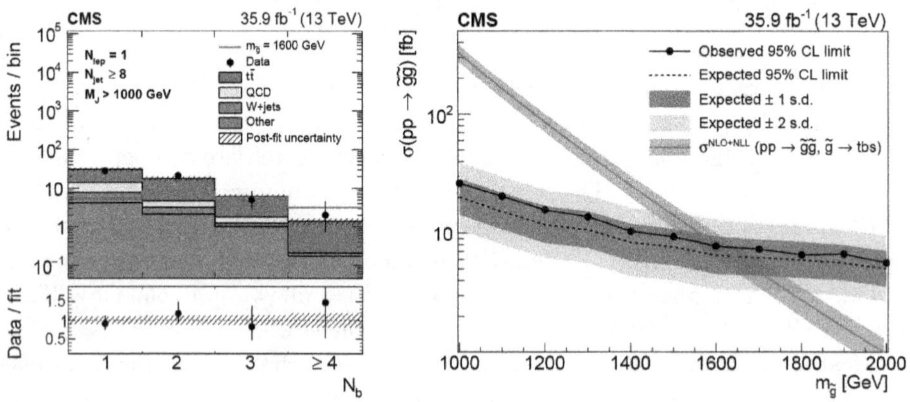

Fig. 18. The b-tagged jet multiplicity of the R-parity violating search for SUSY in gluino pair production and prompt decays $\tilde{g} \to tbs$ using 35.9 fb^{-1} of 13 TeV data (left) and the corresponding cross section and resulting limit (right)[27].

can decay through λ_{332}'' into third generation SM quarks, i.e. $\tilde{g} \to t\tilde{t} \to tbs$. The single-lepton ($e$ or μ) final state depending on the jet and b-tagged jet multiplicity, as shown in Fig. 18 (left), and the sum of masses of large-radius jets is analyzed for gluino pair-production that both decay promptly as $\tilde{g} \to tbs$. No p_T^{miss} requirement is applied, enabling the re-interpretation of this result in the context of other BSM scenarios with large jet or b-tagged jet multiplicities and little or no missing transverse momentum. An upper limit of approximately 10 fb at 95% confidence level is determined for gluino pair production, allowing to exclude gluinos below $m(\tilde{g}) < 1.61$ TeV, as shown in Fig. 18 (right).

4. Conclusion and Outlook

The outstanding performance of the Large Hadron Collider and the CMS detector have allowed to collect a 13 TeV data set corresponding to an integrated luminosity of 40 fb^{-1} in 2016. The most recent result highlights derived using this data have been presented here. The leap in center-of-mass energy and the increase in integrated luminosity by a factor of two compared to the data recorded during the Run I have resulted in standard model measurements of excellent precision. For the first time strong evidence for the $H \to b\bar{b}$ decay channel has been observed at a single experiment. Searches for physics beyond the standard model are one of the main objectives of the LHC experiments that likewise have profited from the increase in center-of-mass energy in the new data set. The large amount of available data as well as the excellent performance of the detector make more complicated or specialized search analyzes possible and worthwhile.

During the end-of-the-year shutdown 2016/17 different detector upgrades were installed, in particular a new silicon pixel detector that has since then successfully started operation. The new pixel tracking and vertexing detector is able to cope with the large number of simultaneous interactions expected at the maximum design instantaneous luminosities of $\mathcal{L} = 2 \cdot 10^{34}\,\mathrm{cm}^{-2}\mathrm{s}^{-1}$.

So far, only of the order of one percent of the total expected luminosity during the lifetime of the LHC has been recorded. The data that is yet to come will allow for many more precision measurements and direct or indirect searches for new physics.

References

1. CMS Collaboration, The CMS Experiment at the CERN LHC, *JINST* **3**, p. S08004 (2008).
2. CMS Collaboration, The CMS Public Physics Results and Detector Performance Webpages, *http://cms.web.cern.ch/news/cms-physics-results* (2017).
3. CMS Collaboration, The CMS Public Luminosity Results and Measurement Webpages, *https://twiki.cern.ch/twiki/bin/view/CMSPublic/LumiPublicResults* (2017).
4. CMS Collaboration, Level-1 E/Gamma and Muon performance on 2017 data, Tech. Rep. CMS-DP-2017-024 (2017), cds.cern.ch/record/2273270.
5. CMS Collaboration, CMS Tracker Alignment Performance Results 2016, Tech. Rep. CMS-DP-2017-021 (2017), cds.cern.ch/record/2273267.
6. CMS Collaboration, Missing transverse energy performance in high pileup data collected by CMS in 2016, Tech. Rep. CMS-DP-2017-028 (2017), cds.cern.ch/record/2275227.
7. CMS Collaboration, Measurement of the differential jet production cross section with respect to jet mass and transverse momentum in dijet events from pp collisions at $\sqrt{s} = 13$ TeV, Tech. Rep. CMS-PAS-SMP-16-010 (2017), cds.cern.ch/record/2273393.
8. CMS Collaboration, Measurement of the weak mixing angle with the forward-backward asymmetry of Drell-Yan events at 8 TeV, Tech. Rep. CMS-PAS-SMP-16-007, CERN (Geneva, 2017), cds.cern.ch/record/2273392.
9. A. M. Sirunyan *et al.*, Measurement of the cross section for top quark pair production in association with a W or Z boson in proton-proton collisions at $\sqrt{s} = 13$ TeV, *hep-ex:1711.02547*, subm. to JHEP (2017).

10. CMS Collaboration, Measurements of ttbar cross sections in association with b jets and inclusive jets and their ratio using dilepton final states in pp collisions at sqrt(s) = 13 TeV (2017).

11. CMS Collaboration, Measurement of the semileptonic $t\bar{t} + \gamma$ production cross section in pp collisions at $\sqrt{s} = 8$ TeV, *JHEP* **10**, p. 006 (2017).

12. CMS Collaboration, Search for associated production of a Z boson with a single top quark and for tZ flavour-changing interactions in pp collisions at $\sqrt{s} = 8$ TeV, *JHEP* **07**, p. 003 (2017).

13. CMS Collaboration, Measurement of the differential cross sections of top quark pair production as a function of kinematic event variables in pp collisions at $\sqrt{s} = 13$ TeV, Tech. Rep. CMS-PAS-TOP-16-014, CERN (Geneva, 2017), cds.cern.ch/record/2273459.

14. CMS Collaboration, Measurement of the top quark mass using proton-proton data at $\sqrt{(s)} = 7$ and 8 TeV, *Phys. Rev.* **D93**, p. 072004 (2016).

15. CMS Collaboration, Measurements of properties of the Higgs boson in the diphoton decay channel with the full 2016 data set, Tech. Rep. CMS-PAS-HIG-16-040, CERN (Geneva, 2017), cds.cern.ch/record/2264515.

16. CMS Collaboration, Measurements of properties of the Higgs boson decaying into the four-lepton final state in pp collisions at sqrt(s) = 13 TeV (2017).

17. ATLAS and CMS collaborations, Measurements of the Higgs boson production and decay rates and constraints on its couplings from a combined ATLAS and CMS analysis of the LHC pp collision data at $\sqrt{s} = 7$ and 8 TeV, *JHEP* **08**, p. 045 (2016).

18. CMS Collaboration, Higgs to WW measurements with 15.2 fb^{-1} of 13 TeV proton-proton collisions, Tech. Rep. CMS-PAS-HIG-16-021, CERN (Geneva, 2017), cds.cern.ch/record/2273908.

19. CMS Collaboration, Observation of the Higgs boson decay to a pair of tau leptons, *hep-ex:1708.00373*, subm. to PLB (2017).

20. CMS Collaboration, Evidence for the Higgs boson decay to a bottom quark-antiquark pair, *hep-ex:1709.07497*, subm. to PLB (2017).

21. T. Golling, LHC searches for exotic new particles, *Prog. Part. Nucl. Phys.* **90**, 156 (2016).

22. CMS Collaboration, Search for new physics with dijet angular distributions in proton-proton collisions at $\sqrt{s} = 13$ TeV and constraints on dark matter and other models, Tech. Rep. CMS-PAS-EXO-16-046, CERN (Geneva, 2017), cds.cern.ch/record/2273455.

23. C. Autermann, Experimental status of supersymmetry after the LHC Run-I, *Prog. Part. Nucl. Phys.* **90**, 125 (2016).

24. CMS Collaboration, Search for supersymmetry in events with at least one photon, missing transverse momentum, and large transverse event activity in proton-proton collisions at sqrt(s) = 13 TeV (2017).

25. A. M. Sirunyan *et al.*, Search for gauge-mediated supersymmetry in events with at least one photon and missing transverse momentum in pp collisions at $\sqrt{s} = 13$ TeV, *hep-ex:1711.08008*, subm. to PLB (2017).

26. CMS Collaboration, Combined search for electroweak production of charginos and neutralinos in proton-proton collisions at $\sqrt{s} = 13$ TeV (2018).

27. CMS Collaboration, Search for R-parity violating supersymmetry in pp collisions at $\sqrt{s} = 13$ TeV using b jets in a final state with a single lepton, many jets, and high sum of large-radius jet masses, *hep-ex:1712.08920*, subm. to PLB (2017).

Search for Beyond SM Higgs Bosons

S. Tsuno

on behalf of the ATLAS and CMS Collaborations

High Energy Accelerator Research Organization, KEK
1-1 Oho, Tsukuba, Ibaraki 305-0801, Japan
E-mail: Soshi.Tsuno@cern.ch

This article summarizes the latest results within the context of searches for Higgs boson production and properties beyond the Standard Model obtained with ATLAS and CMS experiments in summer 2017.

Keywords: Higgs; LHC; ATLAS; CMS.

1. Introduction

Recent studies of the Higgs boson[1,2] with the ATLAS[3] and CMS[4] experiments suggest the existence of a spontaneous symmetry breaking phenomena in the electroweak sector. The observation of couplings of Higgs boson to fermions[5,6] indeed proved the existence of the mass generation mechanism through the scalar Higgs field. However, the nature of this mechanism has not yet been precisely established. Several underlying questions like the hierarchy problem, unification of the forces, mystery of the asymmetry of the matter and anti-matter in universe, dark matter and the origin of the dark energy in universe are not explained by the Standard Model (SM). To answer such fundamental questions a new theory, beyond the SM, is required. The quest to discover new particles beyond those predicted by the SM is therefore a primary task of the LHC experiments in addition to performing precise measurements of the Higgs boson.

This article summarizes the latest results within the context of the searches for additional Higgs bosons, beyond that predicted by the SM, presented by the ATLAS and CMS experiments in summer 2017. An overview of the searches for MSSM Higgs bosons is discussed in Sec. 2, a summary of the use of the Higgs boson as a tool in searches for new physics is presented in Sec. 3, searches for anomalous Higgs couplings and rare decays are presented in Sec. 4, and the summary is given in Sec. 5.

2. Searches for MSSM Higgs bosons

2.1. *Overview*

The Minimal Supersymmetric Standard Model (MSSM)[7–9] is an extension of the SM inducing the supersymmetric partners of the SM particles. The MSSM requires two Higgs doublets with opposite hypercharge which results in one CP-odd (A), two CP-even (h, H) neutral Higgs bosons and two charged Higgs bosons $(H^\pm)$ with two additional mixing parameters, α the mixing angle of h and H, and $\tan\beta$ the ratio of the vacuum expectation values of the up-type and down-type fermions in the two Higgs doublets.

Extensive searches for these Higgs bosons have been carried out over the years. For example, Fig. 1 summarizes the 95% CL exclusion contours[10] in the MSSM $m_h^{\mathrm{mod}+}$ scenario[11] (with $\mu=+200$ GeV), obtained from selected CMS analyses that have been performed with the LHC Run 1 dataset. The $m_h^{\mathrm{mod}+}$ scenario is chosen to be the mass of the lightest CP-even Higgs boson, m_h, is close to the measured mass of the Higgs boson that was discovered at the LHC. The colored filled areas correspond to the excluded regions in m_A and $\tan\beta$. In the Fig. 1 on the right in addition to the direct exclusion contours, the constraint that is obtained from the compatibility of the scenario with the couplings of the SM Higgs boson when interpreted as the h is also displayed.

The couplings of the Higgs bosons to down-type fermions are enhanced for large $\tan\beta$, which results in a large production rate for the Higgs bosons with couplings to τ leptons and b-quarks. Therefore, direct searches for the heavy H/A bosons decaying to τ leptons[12,13] and b-quarks[14] are appropriate approaches to high $\tan\beta$

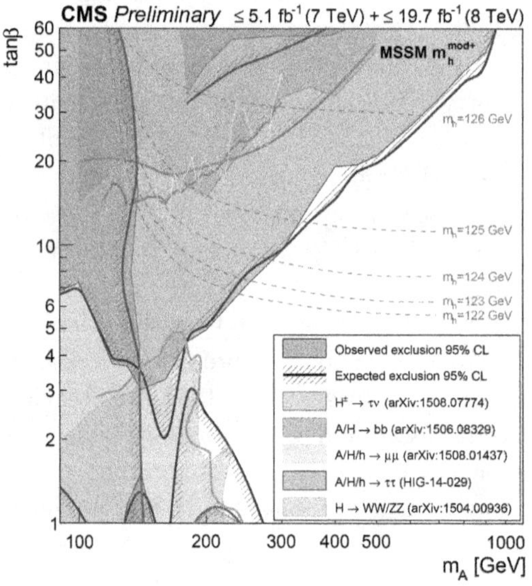

Fig. 1. 95% CL exclusion contours, in the MSSM $m_h^{\mathrm{mod}+}$ scenario (with $\mu=+200$ GeV), which is obtained by selected CMS analyses that have been performed with the LHC Run 1 dataset. The colored filled areas correspond to the excluded regions in m_A and $\tan\beta$. The colored (slightly darker shaded) lines with indicated hatches to the regions that were expected to be excluded, based on the null-hypothesis assumption of a SM-like Higgs sector. In the figure on the right in addition to the direct exclusion contours, the constraint is also displayed that is obtained from the compatibility of the scenario with the couplings of the SM Higgs boson when interpreted as the h (color online).

region, while searches for Higgs bosons coupled to the up-type fermions explore the low $\tan\beta$ region. Since almost all Higgs boson measurements so far indicate the obtained parameters are consistent with the SM, that is, low $\tan\beta$ region is preferred, the number of analyses searching in the low $\tan\beta$ region has rapidly grown during Run 2 of the LHC. The charged Higgs boson $H^\pm$ with couplings to the t- and b-quark[15,16] is one channel which can explore the low $\tan\beta$ region. Searches for the decays of a heavy Higgs boson, A, to a light Higgs boson, h, and the SM weak boson Z[17,18] also provides indirect constraints. Analyses may also search for interference effects in the top-pair production process caused by the decay $H \to t\bar{t}$ in high mass region[19]. Finally, searches for charged Higgs bosons decaying to $\tau\nu$[16,20] provide global constraints ranging from the low to high m_A regions.

In the following, only the selected physics analyses, $H/A \to \tau\tau$ and $H^\pm \to \tau\nu/tb$, which represent the most powerful channels in constraining the MSSM parameter space of $\tan\beta$ as a function of Higgs mass, are presented as benchmark searches.

2.2. *Search for $H/A \to \tau\tau$*

The analysis[12] is performed with 36.1 fb^{-1} data at a centre-of-mass energy of 13 TeV collected with the ATLAS detector in 2015 and 2016. Figure 2 shows the typical lowest order Feynman diagrams for gluon-gluon fusion production (a), b-quark associated production with 4 flavor scheme (b) and 5 flavor scheme (c) of a neutral MSSM Higgs boson. The b-quark associated production mode is enhanced in the high $\tan\beta$ region. The search is therefore split according to the presence or absence of jets originating from b-quarks in the final state as well as the decay mode of the τ leptons. The final discriminant is $m_{\mathrm{T}}^{\mathrm{tot}}$, defined as

$$m_{\mathrm{T}}^{\mathrm{tot}} = \sqrt{(p_{\mathrm{T}}^{\tau_1} + p_{\mathrm{T}}^{\tau_2} + E_{\mathrm{T}}^{\mathrm{miss}})^2 - (\vec{p}_{\mathrm{T}}^{\,\tau_1} + \vec{p}_{\mathrm{T}}^{\,\tau_2} + \vec{E}_{\mathrm{T}}^{\mathrm{miss}})^2} \tag{1}$$

where $p_{\mathrm{T}}^{\tau_{1/2}}$ is the transverse momentum of the visible component of the τ decay products, $E_{\mathrm{T}}^{\mathrm{miss}}$ is the opposite direction of the transverse vector sum of all detected particles from the collision events, which represents the direction in transverse plane of the sum of the momenta of all neutrinos produced in τ decays. The dominant backgrounds are W+jets, $t\bar{t}$ and multi-jet processes where a jet is mis-identified as a hadronically decaying τ-lepton. They are estimated from data. The dominant

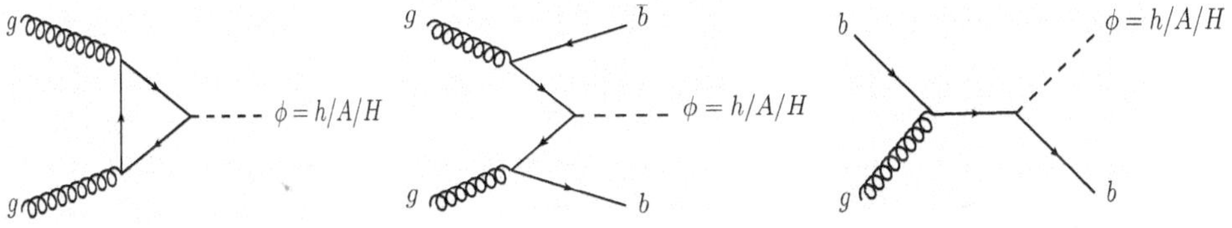

(a) gluon-gluon fusion　　(b) 4FS associated production　　(c) 5FS associated production

Fig. 2.　Lowest order Feynman diagrams for gluon-gluon fusion production (a), b-quark associated production with 4 flavor scheme (b) and 5 flavor scheme (c) of a neutral MSSM Higgs boson.

26

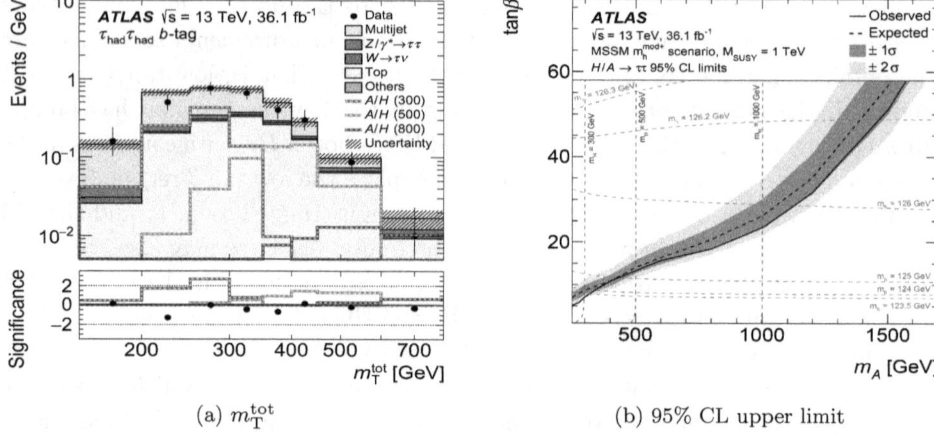

(a) $m_{\mathrm{T}}^{\mathrm{tot}}$ (b) 95% CL upper limit

Fig. 3. (a) $m_{\mathrm{T}}^{\mathrm{tot}}$ for the b-tag category in the hadronic τ channel. The hypothetical Higgs boson signals with masses of 300, 500 and 800 GeV and $\tan\beta = 10$ are also overlaid. (b) The observed and expected 95% CL upper limits on $\tan\beta$ as a function of m_A in the MSSM $m_h^{\mathrm{mod}+}$ scenario. Dashed lines of constant m_h and m_H are shown in red and blue, respectively (color online).

experimental uncertainty is associated with the modeling of high-p_{T} hadronic τ decays which cannot be verified in data due to the lack of a sufficiently large control sample of $Z \to \tau\tau$ events[21]. To address this, hadronic τ identification is studied in a sample of high-p_{T} di-jet events, where the modeling of the detector response to such events is assumed to be the same as for high-p_{T} hadronic τ events. An additional uncertainty of $\sim 20-25\%$ is assigned for high-p_{T} hadronic τ events.

Figure 3 presents the $m_{\mathrm{T}}^{\mathrm{tot}}$ distribution in the b-tagged category (a) and the observed and expected 95% CL upper limits on $\tan\beta$ as a function of m_A in the MSSM $m_h^{\mathrm{mod}+}$ scenario (b). Comparing with Fig. 1, the exclusion limit is largely expanded in Run 2, mostly owing to the increased production cross section from 8 TeV to 13 TeV. The region $\tan\beta > 25$ at $m_A = 1$ TeV is excluded. Model independent limits of $\sigma \times \mathrm{Br}(\phi \to \tau\tau) = 0.0058-0.85$ pb for gluon-gluon fusion and $0.0041-0.95$ pb for b-quark associated processes in the mass range of 0.2-2.25 TeV are also obtained.

2.3. Search for $H^{\pm} \to \tau\nu$ and tb

The main production mode of the charged Higgs boson at high mass ($m_{H^{\pm}} > m_{\mathrm{top}}(=172.5\ \mathrm{GeV})$) is through the associated production of tbH, as shown in Fig. 4. The $H^{\pm}$ predominantly decays into $\tau\nu$ or tb final state depending on $m_{H^{\pm}}$ and $\tan\beta$. In both case, the t- and b-quarks in the final state are associated with the $H^{\pm}$, so that the analyses[15,16,20] require at least one b-quark jet and high jet multiplicity. The dominant background is from the $t\bar{t}$+jets process as well as a jet faking the hadronic τ-lepton from W+jets and multi-jet events. For fully hadronic events without containing e or μ in the final state, the $E_{\mathrm{T}}^{\mathrm{miss}}$ trigger is used in combination with the hadronic τ trigger. The final discriminant is the distribution of the transverse mass between hadronic τ and missing transverse energy for the $H \to \tau\nu$ mode,

$$m_{\mathrm{T}} = \sqrt{2 \cdot p_{\mathrm{T}}^{\tau} |E_{\mathrm{T}}^{\mathrm{miss}}|(1 - \cos \Delta\phi(\vec{p}_{\mathrm{T}}^{\,\tau}, \vec{E}_{\mathrm{T}}^{\mathrm{miss}}))}. \tag{2}$$

On the other hand, for the $H \to tb$ decay mode, the reconstruction of the invariant mass of the final state is non-trivial and the analysis relies instead on a fit to a multivariate discriminant. The dominant experimental uncertainties are the jet energy scale uncertainty and $E_{\mathrm{T}}^{\mathrm{miss}}$ as well as the modeling of the hadronic τ events. For the multivariate analysis, the theory uncertainty is large, at around 50%, due to the uncertainty in estimating the number of additional jets produced in $t\bar{t}$ events. Figure 5 shows the observed and expected 95% CL upper limits on $\tan\beta$ as a function of $m_{H^\pm}$ of the $H^\pm \to \tau\nu$ (a) and $H^\pm \to tb$ (b) analyses, respectively. The $H^\pm \to \tau\nu$ analysis is sensitive to the high $\tan\beta$ region while the $H^\pm \to tb$ is sensitive to the low $\tan\beta$ region. These analyses extend the constraints on the MSSM parameter space using Run 2 data.

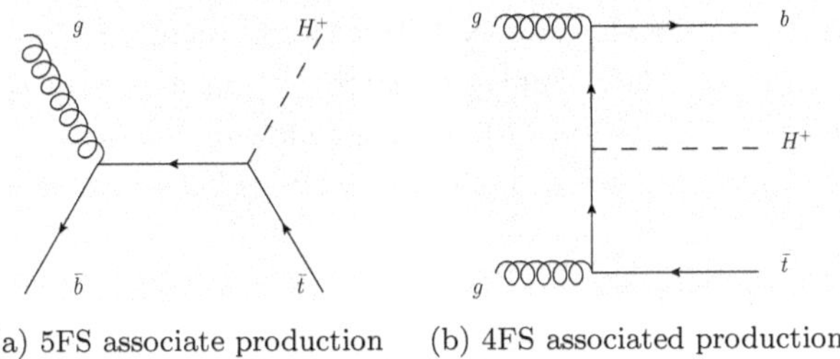

(a) 5FS associate production   (b) 4FS associated production

Fig. 4.   Lowest order Feynman diagrams for associated production with 4 flavor scheme (a) and 5 flavor scheme (b) of a charged MSSM Higgs boson.

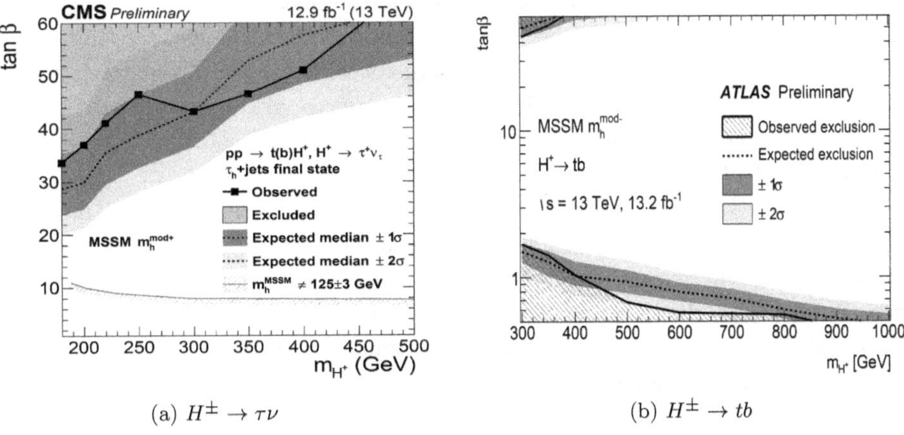

(a) $H^\pm \to \tau\nu$ (b) $H^\pm \to tb$

Fig. 5.   The observed and expected 95% CL upper limits on $\tan\beta$ as a function of $m_{H^\pm}$ of the $H^\pm \to \tau\nu$ (a) and $H^\pm \to tb$ (b) analyses, respectively.

3. The Higgs boson as a probe of new physics

3.1. *Overview*

The observation of the Higgs boson[22,23] opens up the possibility of its use as a tool to probe new physics. Many theories beyond the SM predict new particles decaying into Higgs bosons. If such a particle has a mass larger than that of the Higgs boson (125 GeV), the decay products can become highly boosted leading a collimated shower of final state particles. The SM Higgs boson decays into a b-quark pair around 60% of the time. In this case, the two b-quark jets are reconstructed as a merged single jet object with a mass close to the Higgs boson mass. Other decay modes can be considered in addition to the b-quark pair decay. Therefore, such analyses are typically split into two categories, a merged category and a resolved category. The merged category targets events where the decay products, typically jets, are reconstructed as a single object and exploit the sub-structure within the merged object to maximize the sensitivity. The resolved category attempts to resolve two distinct objects identified with, for example, the standard b-quark jet or hadronic τ reconstruction algorithms. This categorization depends on the mass of the new particle and typically $m_X = 1$ TeV is the boundary between the merged and resolved categories. Many analyses search for a heavy resonance decaying into bosons (including the Higgs boson). In the following sections, only the decay of heavy resonances into final states including a Higgs boson are considered.

3.2. *Search for $X \to Vh$*

A new particle X can decay into a Higgs boson, h, and SM W or Z boson, where the h is required to decay to a b-quark pair and the W or Z bosons are required to decay leptonically or hadronically into $\ell\ell$, $\ell\nu$, $\nu\nu$, jj[24]. The resonance mass m_{Vh} is well-reconstructed from the 4-vectors from the measured objects in both the merged and resolved categories. When a neutrino is in the final state, the $m_{\mathrm{T}}^{\mathrm{tot}}$ variable is used instead. Figure 6 shows the m_{Vh} distribution in the resolved (a) and merged (b) categories for a particular analysis channel with leptons and b-quark jets in the final state. The signal events with $m_X = 1.5$ TeV are also overlaid as 10 times of the predicted cross section for a benchmark model. The resolution for the $m_{\mathrm{T}}^{\mathrm{tot}}$ is about 30-40% at $m_X = 1.5$ TeV.

The 95% CL upper limits on the production cross section of the A multiplied by its branching ratio into the Zh final state and the branching ratio of $h \to b\bar{b}$ are obtained as a function of resonance mass in Fig. 7 for gluon-fusion (a) and b-quark associated (b) production processes, respectively. The data are also interpreted in terms of limits at 95% CL on the 2HDM parameters of $\tan\beta$ and $\cos(\beta - \alpha)$ for several benchmark models.

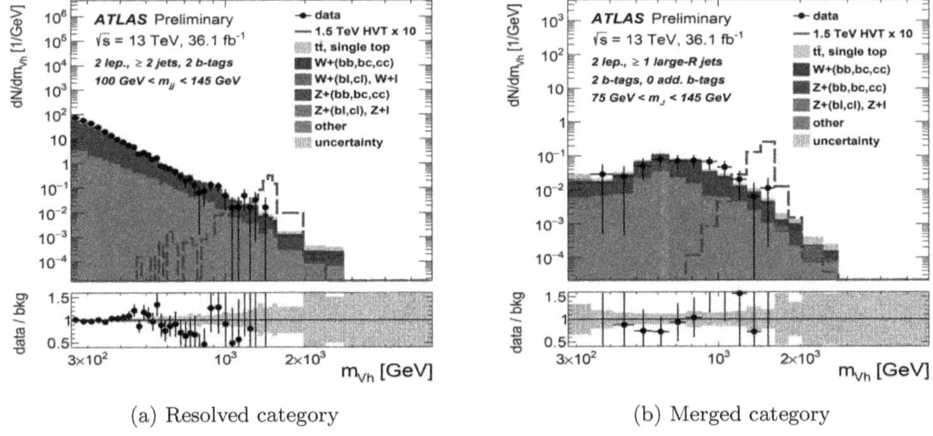

(a) Resolved category   (b) Merged category

Fig. 6. The resonance mass m_{Vh} in the resolved (a) and merged (b) categories for a particular analysis channel with leptons and b-quark jets in the final state. The signal events with $m_X = 1.5$ TeV are also overlaid as 10 times of the predicted cross section for a benchmark model.

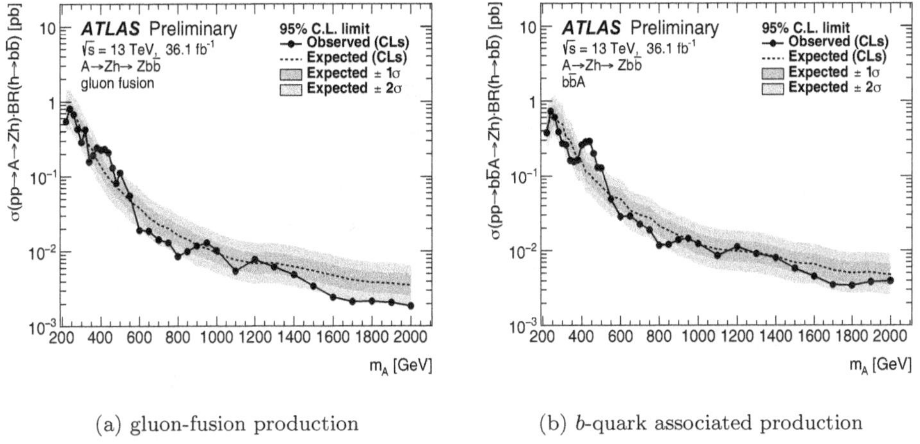

(a) gluon-fusion production   (b) b-quark associated production

Fig. 7. Upper limits at the 95% CL on the product of the production cross-section for $pp \to A$ and the branching ratios for $A \to Zh$ and $h \to b\bar{b}$. The possible signal components of the data are interpreted assuming (a) pure gluon-fusion production, and (b) pure b-quark associated production.

3.3. Search for $X \to hh$

Given the unique topology of the h decay, backgrounds can be well suppressed when a new resonance decays into hh. The considered decay modes are $b\bar{b}b\bar{b}$[25,26], $b\bar{b}\ell\nu\ell\nu$[27], $b\bar{b}\tau\tau$[28], and $b\bar{b}\gamma\gamma$[29]. Figure 8(a) shows the di-jet invariant mass distribution m_{jj}^{red} in the $b\bar{b}b\bar{b}$ channel, where the invariant mass of two small-jets inside the merged object (jet) is required to be $105 < m_{jj}^{\mathrm{small}} < 135$ GeV, the soft component inside jets is trimmed and selection of the N-subjettiness[25] is required. The signal cross section

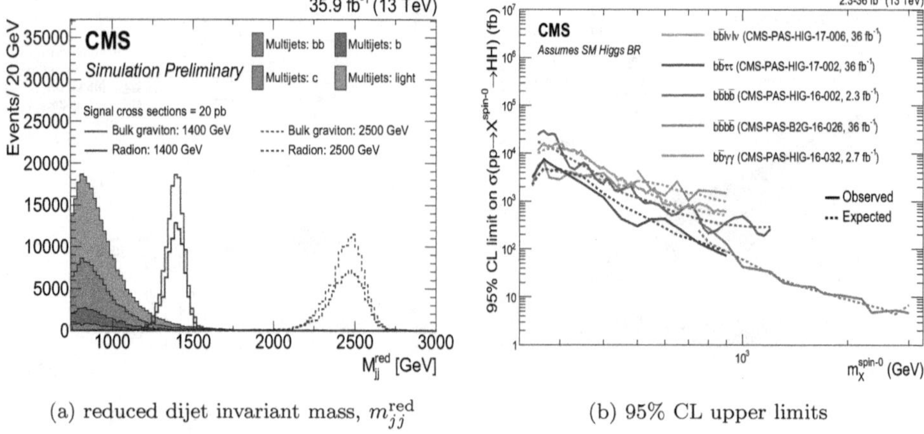

(a) reduced dijet invariant mass, m_{jj}^{red} (b) 95% CL upper limits

Fig. 8. (a) The reduced di-jet invariant mass m_{jj}^{red} in $b\bar{b}b\bar{b}$ channel, where the invariant mass of two small-jets inside the merged object (jet) is required to be $105 < m_{jj}^{\mathrm{small}} < 135$ GeV. The signal cross section is assumed to be 20 pb for all the mass hypotheses at 1400, 1800 and 2500 GeV. (b) Observed and expected 95% CL upper limits on the product of cross section and the branching ratio $\sigma(gg \to X) \times B(X \to hh)$ obtained by different analyses assuming spin-0 hypothesis in an extended mass range beyond 1 TeV.

is assumed to be 20 pb for all the mass hypotheses at 1400, 1800 and 2500 GeV for illustration purposes. Only a very small background contribution is present in the large resonance mass region ($m_X > 1.5$ TeV). Figure 8(b) presents the observed and expected 95% CL upper limits on the product of cross section and the branching ratio $\sigma(gg \to X) \times B(X \to hh)$ obtained by different analyses assuming spin-0 hypothesis in an extended mass range beyond 1 TeV.

4. Anomalous Higgs couplings and rare decays

4.1. *Overview*

Unlike the case of the SM gauge bosons, the fermion couplings in the Higgs sector is not universal across fermion flavors. Although the Yukawa couplings are predicted to be proportional to the mass of the fermion, the detailed structure of the Yukawa couplings has not yet been determined. This inspires the possibility of flavor violating decays through off-diagonal elements of the Yukawa mass matrix or rare decays involving new particles beyond the SM.

In this section, searches for rare and exotic decays of the Higgs boson are discussed.

4.2. *Search for flavor-violating decay modes*

The flavor-changing neutral current process is strictly forbidden in the framework of the SM, however searches for such processes in the decay of the Higgs boson rep-

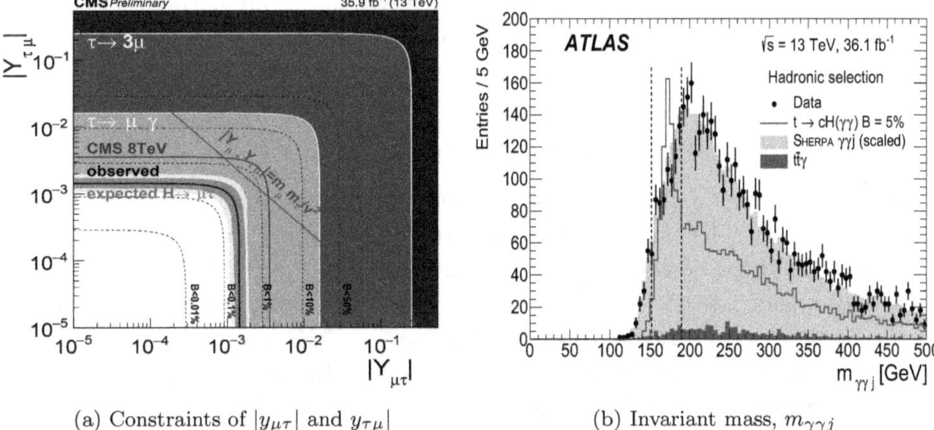

(a) Constraints of $|y_{\mu\tau}|$ and $y_{\tau\mu}|$ (b) Invariant mass, $m_{\gamma\gamma j}$

Fig. 9. (a) Constraints on the flavor violating Yukawa couplings, $|y_{\mu\tau}|$, $y_{\tau\mu}|$. The expected (red solid line) and observed (black solid line) limits are derived from the limit on Br($H \to \mu\tau$). The flavor diagonal Yukawa couplings are approximated by their SM values. The green (yellow) band indicates the range that is expected to contain 68% (95%) of all observed limit excursions from the expected limit. The shaded regions are derived constraints from null searches for $\tau \to 3\mu$ (dark green) and $\tau \to \mu\gamma$ (lighter green). The purple diagonal line is the theoretical naturalness limit $y_{ij}y_{ji} \leq m_i m_j/v^2$. (b) Invariant mass distribution of the two selected photons and a jet from t-quark. The signal distributions are normalized assuming a branching ratio of 5% (color online).

resent important tests of flavor violation in the Higgs sector. High energy colliders provide many opportunities to study the Yukawa couplings of the third generation fermions. A search for lepton flavor violating decays of the 125 GeV Higgs boson in the $\mu\tau$ and $e\tau$ decay mode, $H \to \mu\tau, e\tau$, has been carried out[30,31]. No significant excess over the SM background expectation is observed. The observed (expected) upper limits on the branching ratio of the Higgs boson are set to be Br($H \to \mu\tau$) < 0.25(0.25)% and Br($H \to e\tau$) < 0.61(0.37)% at 95% CL. Constraints on the flavor violating Yukawa couplings, $|y_{\mu\tau}|$, $|y_{\tau\mu}|$, are obtained from these results as shown in Fig. 9(a) for $\mu\tau$ mode. The flavor diagonal Yukawa couplings are approximated by their SM values. The green (yellow) band indicates the range that is expected to contain 68% (95%) of all observed limit excursions from the expected limit. The shaded regions are constraints derived from null searches for $\tau \to 3\mu$ (dark green) and $\tau \to \mu\gamma$ (lighter green). The purple diagonal line is the theoretical naturalness limit $y_{ij}y_{ji} \leq m_i m_j/v^2$, where v is the vacuum expectation value 246 GeV. Upper limits on the off-diagonal $\mu\tau$ and $e\tau$ Yukawa couplings are derived as $\sqrt{|y_{\mu\tau}|^2 + |y_{\tau\mu}|^2} < 1.43 \times 10^{-3}$ and $\sqrt{|y_{e\tau}|^2 + |y_{\tau e}|^2} < 2.26 \times 10^{-3}$ at 95% CL, respectively.

Flavor violation in the quark sector is also tested using the t-quark decay, $t \to qH$[32], where q is an up-type quark, c or u. The $H \to \gamma\gamma$ decay mode is used since it offers optimal sensitivity. The SM prediction is $\mathrm{Br}_{\mathrm{SM}}(t \to qH) = 3 \times 10^{-15}$. Figure 9(b) shows the invariant mass distribution of the two selected photons and a jet from the decay of the t-quark. The signal distributions are normalized assuming a branching ratio of 5%. The limit is obtained to be $\mathrm{Br}(t \to qH) < 2.4 \times 10^{-3}$ at 95% CL.

4.3. *Search for exclusive decay mode*

The Higgs boson can also decay into hadrons exclusively through loop diagrams as illustrated in Fig. 10. According to the SM, the branching ratios for the exclusive decays of the Higgs boson to a meson and photon, $J/\psi\gamma$, $\phi\gamma$ and $\rho\gamma$, are 2.95×10^{-6}, 2.31×10^{-6}, and 1.68×10^{-7}, respectively[33]. The mesons are experimentally identified using the decay modes of $J/\psi \to \mu^+\mu^-$, $\phi \to K^+K^-$ and $\rho \to \pi^+\pi^-$.

Figure 11 shows the invariant masses $m_{K^+K^-}$ (a) and $m_{K^+K^-\gamma}$ (b) for the selected $\phi\gamma$ candidates[34]. The reconstructed mass resolution of $m_{K^+K^-}$ is about 4 MeV. The $m_{K^+K^-}$ distribution exhibits a clear phi mass peak over the combinatoric

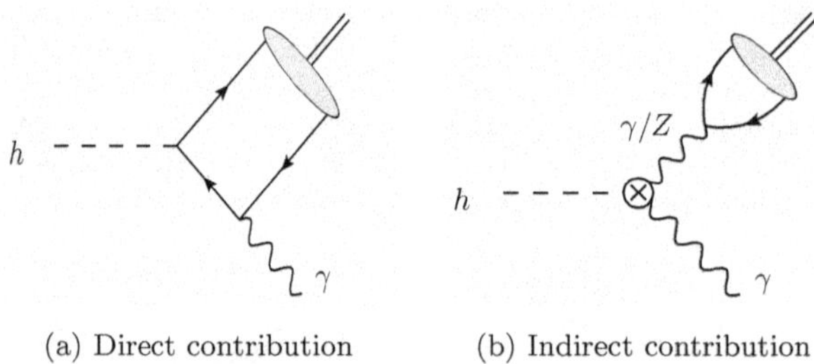

(a) Direct contribution   (b) Indirect contribution

Fig. 10. Feynman diagrams for direct (a) and indirect (b) contributions in the exclusive decay of $H \to V\gamma$. The crossed circle denotes the off-shell $H \to \gamma\gamma^*$ and $H \to \gamma Z^*$ amplitudes, which in the SM arise first at one-loop order.

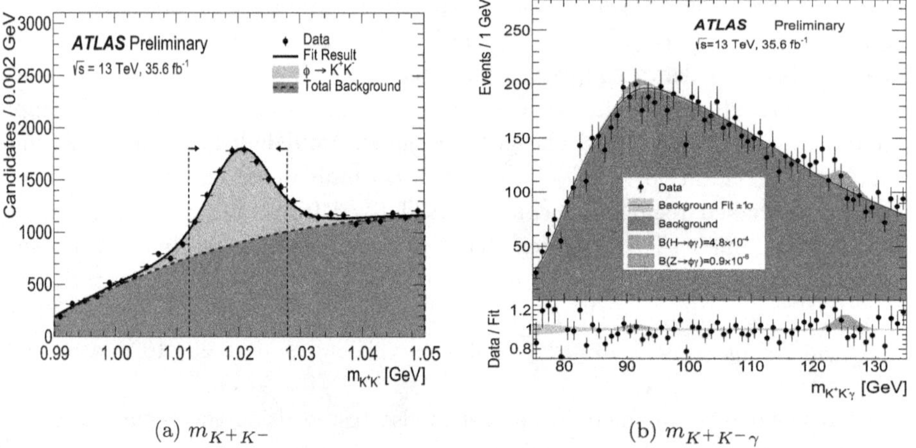

(a) $m_{K^+K^-}$ (b) $m_{K^+K^-\gamma}$

Fig. 11. The $m_{K^+K^-}$ (a) and $m_{K^+K^-\gamma}$ (b) distributions for the selected $\phi\gamma$ candidates. The Higgs and Z boson contributions for the branching ratio values corresponding to the observed 95% CL upper limits are also shown.

background. Combining with the γ, the Higgs (and Z) boson mass is reconstructed with high precision. No significant excess of events is observed above the background, in agreement with the SM expectations. Upper limits at 95% CL on the branching ratios of the exclusive radiative Higgs boson decays to mesons are set as $\mathrm{Br}(H \to \phi\gamma)$ $< 4.8\times10^{-4}$ and $\mathrm{Br}(H \to \rho\gamma) < 8.8\times10^{-4}$, respectively. The current limits on the branching fractions for such decays are two orders of magnitude higher than the SM prediction. However, assuming a dataset of 3000 fb^{-1} at $\sqrt{s} = 14$ TeV, a limit of $\mathrm{Br}(H \to J/\psi\gamma) = 5.5\times10^{-7}$ is expected[35].

4.4. *Search for exotic decay mode*

Several theories beyond the SM predict the existence of a dark sector that is weakly coupled with the SM particles. Due to the weakness of the coupling, the particles of the dark sector can have non-negligible lifetime which could decay beyond the beam-pipe or even outside the detector. The CMS experiment have searched for long lived particles which could decay inside the tracking detector volume ($c\tau \leq$ 100 mm)[36], while the ATLAS experiment searches for decays within the region between the beam-pipe and the muon detectors ($3.8 \leq c\tau \leq 600$ mm)[37]. Limits on models predicting Higgs boson decays to neutral long-lived particles (dark photons γ_D) are derived as a function of the decay length, $c\tau_{\gamma_D}$ as shown in Fig. 12.

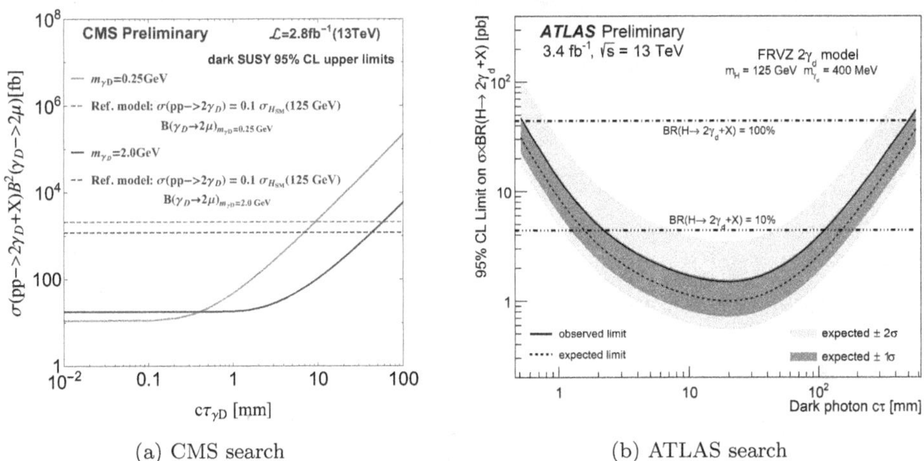

(a) CMS search   (b) ATLAS search

Fig. 12.   The 95% CL upper Limit on $\sigma(pp \to 2\gamma_D + X) \times Br^2(\gamma_D \to \mu\mu)$ as a function of $c\tau_{\gamma_D}$ for two dark photon masses of 0.25 and 2.0 GeV in CMS (b), and 0.4 GeV in ATLAS (a).

5. Summary and prospects

In this article, three different topics are discussed within the context of the latest searches for Higgs boson production and decays beyond that predicted in the SM.

Searches for MSSM Higgs bosons explore both the low and high $\tan\beta$ regions in term of the mass of H/A boson. With current dataset of 36 fb^{-1}, an MSSM Higgs bosons are excluded up to 1 TeV at $\tan\beta \sim 20$ with $m_h^{\mathrm{mod}+}$ scenario. The HL-LHC will be sensitive to the region up to $m_{H/A} = 400$ GeV for the whole parameter space, but the region around $\tan\beta \sim 10$ at high mass region is not excluded, and may require a future linear or circular collider to investigate.

The Higgs boson can be used as a tool to probe a new physics, where a new heavy particle may decay into Higgs bosons. When the mass of a particle is sufficiently larger than that of the 125 GeV Higgs boson, its decay products can be largely boosted and reconstructed as single object. By resolving the individual decay products or reconstructing them as a merged object, the good rejection of background processes can be achieved, leading to almost model independent exclusion limits which extend up to 1 TeV.

The search for anomalous Higgs couplings and rare decays also represent important tests of the nature of the Higgs boson, since the detailed structure of the Yukawa coupling is not yet well established. Searches for flavor violating Higgs boson decays have been performed by the ATLAS and CMS experiments and already lead to very stringent limits. The exclusive radiative decays of the Higgs boson to light mesons are also searched for. To date, the limits on these decays are far from the SM prediction, but in HL-LHC era, such decays may offer the possibility to study the Yukawa couplings of the first and second generation quarks. Searches for exotic Higgs boson decay modes have also been performed, including decays to dark matter candidates. The results from the collider experiments are very complementary to the programme of underground direct detection experiments.

In conclusion, it should be emphasized that both the ATLAS and CMS experiments are very active in the area of Higgs boson studies, both in terms of production and coupling measurements and searches for phenomena beyond the SM.

References

1. ATLAS and CMS Collaborations, *Measurements of the Higgs boson production and decay rates and constraints on its couplings from a combined ATLAS and CMS analysis of the LHC pp collision data at $\sqrt{s} = 7$ and 8 TeV*, JHEP **08**, p. 045 (2016).
2. CMS Collaboration, *Precise determination of the mass of the Higgs boson and tests of compatibility of its couplings with the standard model predictions using proton collisions at 7 and 8 TeV*, Eur. Phys. J. **C75**, p. 212 (2015).
3. ATLAS Collaboration, *The ATLAS Experiment at the CERN Large Hadron Collider*, JINST **3**, p. S08003 (2008).
4. CMS Collaboration, *The CMS experiment at the CERN LHC*, JINST **3**, p. S08004 (2008).
5. ATLAS Collaboration, *Evidence for the Higgs-boson Yukawa coupling to tau leptons with the ATLAS detector*, JHEP **04**, p. 117 (2015).
6. CMS Collaboration, *Evidence for the direct decay of the 125 GeV Higgs boson to fermions*, Nature Phys. **10**, 557 (2014).
7. A. Djouadi, *The Anatomy of electro-weak symmetry breaking. II. The Higgs bosons in the minimal supersymmetric model*, Phys. Rept. **459**, 1 (2008).

8. G. C. Branco, P. M. Ferreira, L. Lavoura, M. N. Rebelo, M. Sher and J. P. Silva, *Theory and phenomenology of two-Higgs-doublet models*, Phys. Rept. **516**, 1 (2012).

9. P. Fayet, *Supersymmetry and weak, electromagnetic and strong interactions*, Physics Letters B **64**, 159 (1976).

10. CMS Collaboration, *Summary results of high mass BSM Higgs searches using CMS run-I data*, CMS-PAS-HIG-16-007 (2016).

11. M. Carena, S. Heinemeyer, O. Stål, C. E. M. Wagner and G. Weiglein, *MSSM Higgs Boson Searches at the LHC: Benchmark Scenarios after the Discovery of a Higgs-like Particle*, Eur. Phys. J. **C73**, p. 2552 (2013).

12. ATLAS Collaboration, *Search for additional heavy neutral Higgs and gauge bosons in the ditau final state produced in 36 fb^{-1} of pp collisions at $\sqrt{s} = 13$ TeV with the ATLAS detector*, arXiv:1709.07242 (2017).

13. CMS Collaboration, *Search for neutral MSSM Higgs bosons decaying to a pair of tau leptons in pp collisions*, JHEP **10**, p. 160 (2014).

14. CMS Collaboration, *Search for neutral MSSM Higgs bosons decaying into a pair of bottom quarks*, JHEP **11**, p. 071 (2015).

15. ATLAS Collaboration, *Search for charged Higgs bosons in the $H^{\pm} \to tb$ decay channel in pp collisions at $\sqrt{s} = 8$ TeV using the ATLAS detector*, JHEP **03**, p. 127 (2016).

16. CMS Collaboration, *Search for a charged Higgs boson in pp collisions at $\sqrt{s} = 8$ TeV*, JHEP **11**, p. 018 (2015).

17. ATLAS Collaboration, *Search for a CP-odd Higgs boson decaying to Zh in pp collisions at $\sqrt{s} = 8$ TeV with the ATLAS detector*, Phys. Lett. **B744**, 163 (2015).

18. CMS Collaboration, *Searches for a heavy scalar boson H decaying to a pair of 125 GeV Higgs bosons hh or for a heavy pseudoscalar boson A decaying to Zh, in the final states with $h \to \tau\tau$*, Phys. Lett. **B755**, 217 (2016).

19. ATLAS Collaboration, *Search for heavy Higgs bosons A/H decaying to a top quark pair in pp collisions at $\sqrt{s} = 8$ TeV with the ATLAS detector*, arXiv:1707.06025 (2017).

20. ATLAS Collaboration, *Search for charged Higgs bosons produced in association with a top quark and decaying via $H^{\pm} \to \tau\nu$ using pp collision data recorded at $\sqrt{s} = 13$ TeV by the ATLAS detector*, Phys. Lett. **B759**, 555 (2016).

21. ATLAS Collaboration, *Measurement of the tau lepton reconstruction and identification performance in the ATLAS experiment using pp collisions at $\sqrt{s} = 13$ TeV*, ATLAS-CONF-2017-029 (2017).

22. ATLAS Collaboration, *Observation of a new particle in the search for the Standard Model Higgs boson with the ATLAS detector at the LHC*, Phys. Lett. **B716**, 1 (2012).

23. CMS Collaboration, *Observation of a new boson at a mass of 125 GeV with the CMS experiment at the LHC*, Phys. Lett. **B716**, 30 (2012).

24. ATLAS Collaboration, *Search for heavy resonances decaying to a W or Z boson and a Higgs boson in final states with leptons and b-jets in 36.1 fb^{-1} of pp collision data at $\sqrt{s} = 13$ TeV with the ATLAS detector*, ATLAS-CONF-2017-055 (2017).

25. CMS Collaboration, *Search for heavy resonances decaying to a pair of Higgs bosons in the four b quark final state in proton-proton collisions at $\sqrt{s} = 13$ TeV*, CMS-PAS-B2G-16-026 (2017).

26. CMS Collaboration, *Search for resonant pair production of Higgs bosons decaying to two bottom quark-antiquark pairs in proton-proton collisions at 13 TeV*, CMS-PAS-HIG-16-002 (2016).

27. CMS Collaboration, *Search for resonant and nonresonant Higgs boson pair production in the bblnulnu final state in proton-proton collisions at $\sqrt{s} = 13$ TeV*, arXiv:1708.04188 (2017).

28. CMS Collaboration, *Search for Higgs boson pair production in events with two bottom quarks and two tau leptons in proton-proton collisions at $\sqrt{s} = 13$ TeV*, arXiv:1707.02909 (2017).

29. CMS Collaboration, *Search for Higgs boson pair production in the final state containing two photons and two bottom quarks in proton-proton collisions at $\sqrt{s} = 13$ TeV*, CMS-PAS-HIG-17-008 (2017).

30. CMS Collaboration, *Search for lepton flavour violating decays of the Higgs boson to $\mu\tau$ and $e\tau$ in proton-proton collisions at $\sqrt{s} = 13$ TeV*, CMS-PAS-HIG-17-001 (2017).

31. ATLAS Collaboration, *Search for lepton-flavour-violating decays of the Higgs and Z bosons with the ATLAS detector, Eur. Phys. J.* **C77**, p. 70 (2017).

32. ATLAS Collaboration, *Search for top quark decays $t \to qH$, with $H \to \gamma\gamma$, in $\sqrt{s} = 13$ TeV pp collisions using the ATLAS detector, JHEP* **10**, p. 129 (2017).

33. D. de Florian *et al.*, *Handbook of LHC Higgs Cross Sections: 4. Deciphering the Nature of the Higgs Sector*, arXiv:1610.07922 (2016).

34. ATLAS Collaboration, *Search for exclusive Higgs and Z boson decays to $\phi\gamma$ and $\rho\gamma$ with the ATLAS Detector*, ATLAS-CONF-2017-057 (2017).

35. ATLAS Collaboration, *Search for the Standard Model Higgs and Z Boson decays to $J/\psi\gamma$: HL-LHC projections*, ATL-PHYS-PUB-2015-043 (2015).

36. CMS Collaboration, *A Search for Beyond Standard Model Light Bosons Decaying into Muon Pairs*, CMS-PAS-HIG-16-035 (2016).

37. ATLAS Collaboration, *Search for long-lived neutral particles decaying into displaced lepton jets in proton–proton collisions at $\sqrt{s} = 13$ TeV with the ATLAS detector*, ATLAS-CONF-2016-042 (2016).

Determination of Top-Quark Properties

Yuji Yamazaki

on behalf of ATLAS, CMS and LHCb Collaborations

Graduate School of Science, Kobe University,
Kobe, Hyogo 657-8501, Japan
E-mail: yamazaki@phys.sci.kobe-u.ac.jp

Top quarks are copiously produced at the LHC. The measurements are in stage for precision. A review on top-quark properties measured by the ATLAS, CMS and LHCb collaborations is presented, including the mass of the top quark, coupling, production cross sections and decay properties.

Keywords: Top quark; electroweak; QCD; Standard Model; BSM Physics.

1. Introduction

The discovery of top quarks in 1994 has opened up completely new possibilities of testing the Standard Model (SM). The top quark has large coupling to the Higgs boson and has been having large sensitivity to SM and Beyond-Standard-Model (BSM) physics by looking at top quark radiative corrections in SM electroweak measurements. After the discovery of the Higgs boson, the top quark properties became even more important since it provides unique opportunities for consistency checks between $m_W, m_{\mathrm{top}}, m_H$ and other EW measurements. Deviation among these masses is an implication for presence of physics beyond the Standard Model. In addition, couplings, spin properties and high-p_T cross section behaviour are the important probe for new physics strongly coupled to the top quark at TeV scale.

A top quark pair, $t\bar{t}$, is mostly produced through gluon-gluon fusion diagrams at the LHC energies (7–13 TeV). The production cross section ($\simeq 820\,\mathrm{pb}$) is much larger at the LHC than that at the Tevatron: cross section increases rapidly with $\sqrt{s}$. This comes mainly from parton densities, which also increases rapidly with $1/x$, where x here represents the longitudinal momentum fraction of the parton in the proton. The large production rate allows to investigate the properties of the top quarks with very high precision. More than 40 millions of the $t\bar{t}$ pairs had already been produced for each experiment at the time of this symposium.

So-called "single-top" production cross sections are also very large, where a top quark is produced from a bottom quark through flavour excitation by a W boson. The single-top cross sections amount to about 1/4 of the pair production. Therefore, the singly-produced top quarks are not only sensitive to its coupling to weak bosons but also give another possibility to measure top quark properties.

The decay of the top quarks is in two steps. First it decays to a bottom quark and a W boson, $t \to b + W$, for almost 100% because of the $|V_{tb}|$ element in the CKM matrix being very close to unity. Since it is a weak decay, the helicity of the top quark is transferred to W. One very important feature of this decay is that the lifetime is very short ($\Gamma \simeq 1.3\,\mathrm{GeV}$), unlike other heavy quark decays, because of the large Q value. This makes the top quark very unique: it decays before hadronisation i.e. the decay occurs as a bare quark and produces a bottom quark before hadronisation. The top decay, therefore, is expected to be free from soft QCD effects.

The top decay can further be classified by the second step, the decay of the W boson: semi-leptonic decay ($t \to \ell\nu_\ell b$) and hadronic decay ($t \to q\bar{q}b$). Three combinations are possible In case of the top pair production. They are called dilepton (both semi-leptonic), $\ell + \mathrm{jet}$ (semi-leptonic and hadronic) and all-hadronic (both hadronic) channels. The $\ell + \mathrm{jet}$ channel has one neutrinos while dilepton decay contains two neutrinos from the top decay. The advantage of the hadronic decay is absence of neutrinos while the leptons are much more precisely reconstructed in other channels. Repeating measurements in the three decay patterns is often beneficial because of different experimental systematics.

The top quarks, especially those from the pair production, are quite pure source of the bottom quarks. The background contribution is practically negligible for dilepton channel. The jets in these events are a good source of bottom quarks. This fact is used for calibrating detectors, understanding b-tagging algorithms and jet energy scale of the b-jets and other purposes.

This article reviews recent measurements on top properties, such as top mass, top production cross sections, top couplings and single-top production, with an emphasis on new results from the ATLAS[1], CMS[2] and LHCb[3] collaborations.

2. Top Quark Mass

Since a top quark decays before hadronisation, one would assume that the invariant mass of the top decay products corresponds to the bare mass of the top quark with good precision, of order of $\Lambda_{\mathrm{QCD}} \simeq 200\,\mathrm{MeV}$. Measurements of m_{top} through the reconstruction of the top decay products is called "direct method". Experimentally, the top quarks mass could in principle be reconstructed by using hadronic decay, where all the decay products are visible. For the semi-leptonic decay, the invariant mass of a part of the decay product, such as the invariant mass $M_{\ell b}$ of the b-jet and the charged lepton ℓ, still strongly reflects assumed m_{top} values.

Figure 1 shows the current status of top quark measurement values and their uncertainties from the Tevatron and the LHC experiments. The precision from the LHC experiments now surpasses those from Tevatron. The combined result from the ATLAS experiment is that $m_{\mathrm{top}} = 172.84 \pm 0.70\,\mathrm{GeV}$, where the uncertainty is decomposed into $\pm 0.34(\mathrm{stat.}) \pm 0.67(\mathrm{syst.})$: the statistical uncertainty is not the dominant source. The best single measurement is from CMS[4], $m_{\mathrm{top}} = 172.35 \pm 0.51(\pm 0.36 \pm 0.67)$, using the $\ell + \mathrm{jet}$ channel. The CMS combined results is with the uncertainty of $0.48\,\mathrm{GeV}$, which corresponds to 0.28% precision.

The best measurement from CMS make use of the kinematic fit[6]. For each event, a likelihood is calculated for assumed m_{top} and values of kinematical variables. The

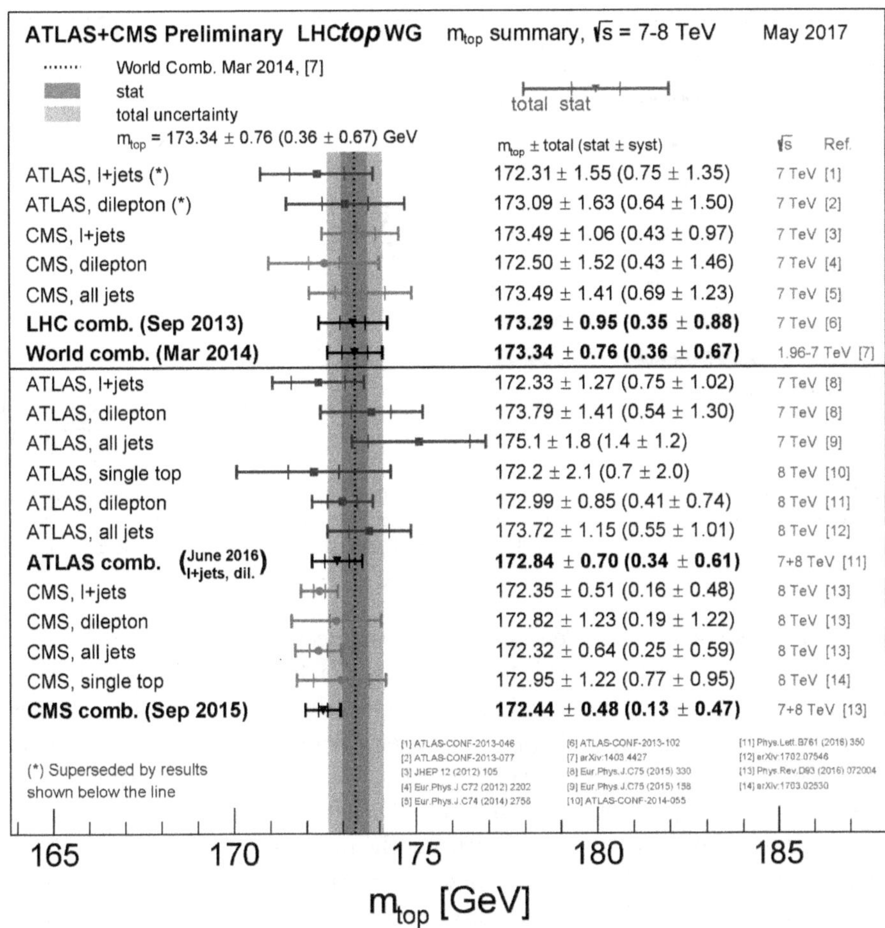

Fig. 1.   Summary of the ATLAS and CMS direct $m_{\rm top}$ measurements, in comparison to the LHC and Tevatron+LHC combinations. The figure was prepared by LHC Top Working Group[5].

$m_{\rm top}$ value is determined by maximising the product of the likelihood for all the events, scanning over assumed $m_{\rm top}$ values. In order to further reduce the experimental uncertainties, the likelihood is also combined with the likelihood for the jet energy scale (JES) determined from the W mass spectra in the hadronic decays of the top quarks, or the JES scale parameter extracted from other events containing hadronic decays of W's is included. An example of the reconstructed mass spectrum is shown in Fig. 2(left).

The benefit of using kinematic fit is demonstrated in Fig. 2(right), where the reconstructed mass as a function of p_T of hadronically decayed top quarks is shown. No strong dependence of the mass on p_T is observed both in the data and various models simulating the top quark production. This reduces the model dependence, especially on the distributions of kinematic variables in the models.

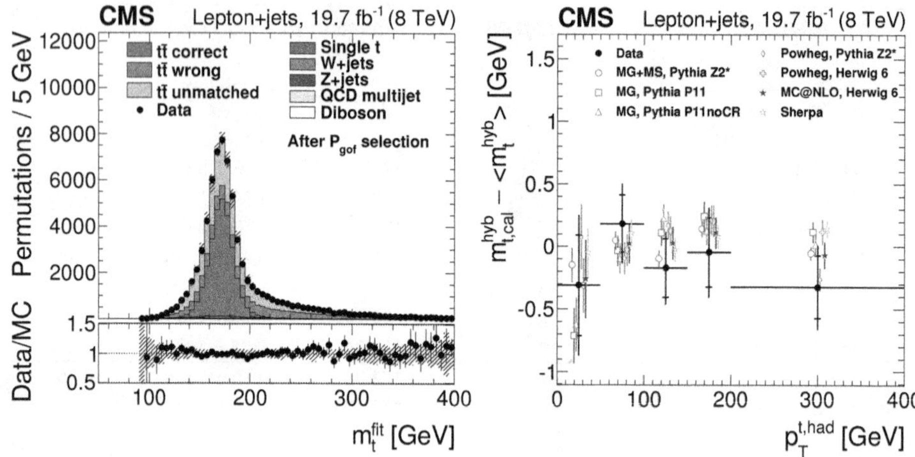

Fig. 2.   An example of the performance in the m_{top} measurement using $\ell+$jet channel by CMS[4]. (left) Mass spectrum of the hadronically decayed top quark candidates after selecting events with good likelihood in kinematic fit (right) dependence of reconstructed m_{top} on the top quark p_T.

The invariant mass of the end product of the decay, however, does not exactly correspond to the quark mass of the top quark. For lighter quarks, e.g, a b-quark, the mass of a B meson is considered to be significantly heavier than the bare mass of the heavy quark ($m_{B_0} - m_b > \Lambda_{\mathrm{QCD}}$). This effect should be smaller for the top quark but may still be important. Moreover, several issues arise when comparing the direct mass and the bare quark mass, as illustrated in Fig. 3. First is that jets are used instead of the decay partons when measuring the momenta. A significant part of the energy of the corresponding parton leaks out of the jets through final-state QCD radiation, in particular if the jet area is chosen to be small, while a sizeable amount of particles enters into the jets through soft effects (initial state QCD radiation, particles from multi-parton interactions etc.), especially in case the jet area is large. In addition, since the decay life time is short, the partons after the decay may still interact with other partons i.e. some colour reconnection occurs, which may change the way for partons to fragment. These points are specific problem for the top quarks since the partons after the decay remains coloured objects. Furthermore, the life time of top quark between the production and decay is so short that the top quark may be off-shell, distorting resonance mass shape.

Many of these effects, which may lead to bias in reconstructed mass peak of the top decay products, are simulated differently in every event generator. This means that the difference between the bare top quark mass and the reconstructed mass is strongly model-dependent. In fact, for many of the direct mass measurements, one of the dominant source of systematic uncertainties is model dependence, such as the matrix element simulation, parton shower simulation or matching between the matrix element calculation and the fragmentation model. They often have similar magnitudes to the experimental uncertainties, e.g. jet energy scale, b-jet energy

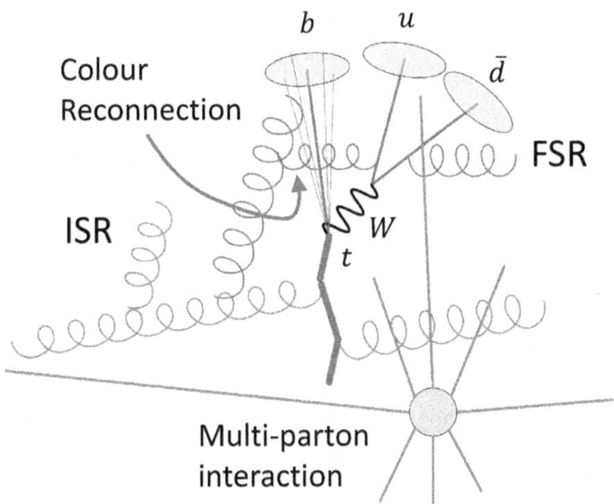

Fig. 3. An illustration on various effects affecting the top mass reconstructed from the decay particles.

scale or background estimation. It is, therefore, desired to cross-check the method of direct mass reconstruction using alternative methods.

Many of such alternatives use leptons to calculate partial masses, even though the mass reconstructed from kinematical variables from leptons can avoid only the theoretical uncertainties related to jets. One example is to utilise leptonic decays of J/ψ[7], which is originated from the b-quark from the top decay. The invariant mass of J/ψ and the charged lepton from the decay of W is compared to templates of various m_{top} values. Although the peak of the three-lepton mass is much smaller than m_{top} since it misses the neutrino from the W decay as well as the fragmented light hadrons from the b-quark. The peak of the three-lepton mass is modelled with small theoretical uncertainty. In fact, the uncertainty on the top mass determination from b-quark fragmentation ($0.37\,\text{GeV}$), governing the fraction of the b-quark momentum to J/ψ, is much smaller than the largest uncertainty, which is top p_T modelling ($0.64\,\text{GeV}$). Although this measurement is statistically not competitive for the moment, the method would have potential to become one of the most precise measurement with larger integrated luminosity, given the small theoretical uncertainties.

Such alternative methods are still not free from uncertainties related to internal lines in the top production diagram such as colour reconnection or off-shell effect in the top quark decay. The production cross section or differential cross sections, on the other hand, are much less affected by these effects. The mass of the top quark can be obtained by using the measured cross sections, since the cross section reflects the mass in the propagator. The mass obtained through cross section behaviour is called "pole mass".

The CMS collaboration has determined the top mass by comparing the total pair production cross section with NNLO+NNLL calculation[9], obtaining $m_{\text{top}} = 173.8^{+1.7}_{-1.8}$ GeV. Another example is to use kinematical distributions of leptons in $e\mu$ dilepton channel of the top-pair production[8], such as $p_T^\ell, p_T^{e\mu}, m^{e\mu}$ etc. Figure 4 (left) shows an example of how such kinematical spectrum changes in shape according to input m_{top}. It is shown that the $p_T^{e\mu}$ gives smaller tail with decreasing top mass. By using seven other variables and performing combined fit, a m_{top} value is obtained with decent precision: $m_{\text{top}} = 173.2 \pm 0.9(\text{stat.}) \pm 0.8(\text{exp.}) \pm 1.2(\text{theo.})$. The uncertainty is dominated by renormalisation and factorisation scales; soft QCD uncertainties are suppressed.

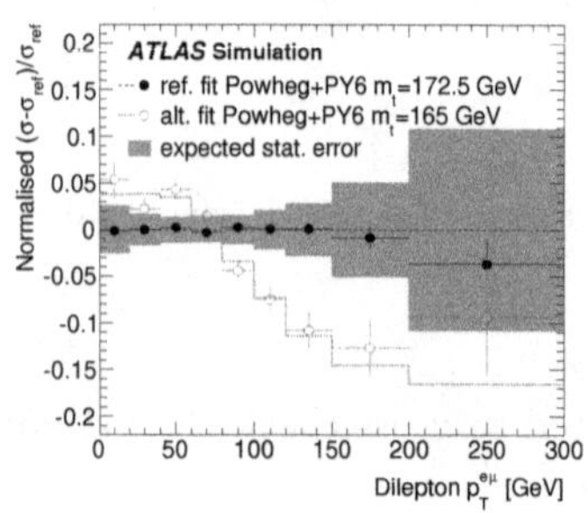
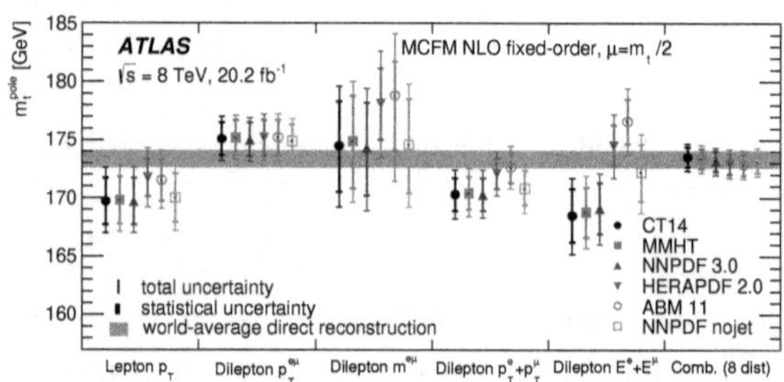

Fig. 4. Top mass determination by ATLAS using dilepton kiematics[8]. (left) Relative deviations in cross sections $(\sigma - \sigma_{\text{ref}})/\sigma$ from the reference cross-section values as a function of dilepton $p_T^{e\mu}$, when $m_{\text{top}} = 172.5$ GeV is changed to 165 GeV. (right) Measurement of m_{top} using lepton kinematic distributions in comparison to MCFM prediction with various PDF sets.

In these measurements reviewed here, there are three common major sources of systematic uncertainties: the jet and b-jet energy scale, modelling of top kinematics and decay, and cross section prediction such as scale uncertainties in QCD calculations. While the first one is experimental, the latter two are related to theory, which could be improved by giving feedback from measurements. It should, therefore, be possible to improve the mass measurement further from current precision ($\simeq 0.5$ GeV) significantly.

3. Differential Cross Sections

Cross section measurements are primary mean to understand the theoretical modelling on production and decay mechanism. The measurement should be as differential as possible so that the models can be tested in each small area of the phase space. The observables should also have good correlation to the measured quantities on the detector level in order to minimise the uncertainties arising from extrapolation from the detector to the truth level. For that reason, the top quarks in cross section measurements are defined as "pseudo-top" quarks using variables related to their decay products after hadronisation, such as momentum and flavour of particle-level (truth hadron) jets. In addition, the variables to be used in measurements should be "well-defined" so that the observables on particle level have good correlation to the partonic variables.

Thus obtained particle-level cross sections can directly be compared to various generator models. Depending on the variables used, the differential cross sections would be sensitive to various aspects of models, such as matrix element, parton shower, hadronisation or parton-density functions (PDFs) used in the simulation. Once the behaviour of the cross sections with respect to the models is understood, the systematic errors due to modelling can be reduced by choosing models describing the data better or tuning the generator models.

The differential cross sections, in particular at high-p_T, are also sensitive to BSM physics by comparing with SM predictions. Top quarks are also important source of background in many BSM searches because of their complicated final state with leptons, jets, b-jets and missing E_T. Understanding normalisation of the top quark cross sections as well as the behaviour of events at the edge of phase space would help in reducing the uncertainties in background, and thus increasing the sensitivity on BSM searches.

The hadronically decayed top quarks at high p_T, $p_T^{\mathrm{top}} \gg m_{\mathrm{top}}$, has three jets in collimated $\eta - \phi$ space, while at $p_T^{\mathrm{top}} \lesssim m_{\mathrm{top}}$ three jets are well separated. For such collimated decay, the entire objects from a top quark decay may be contained in a single cone. The momentum of the top quark then can be obtained from large-radius jets, e.g. the jet radius $R = 1.0$. Such "fat jets" can be identified as a top quark decay product if clear structure of three jets is observed, one of them being b-tagged, other two being within a window of M_W etc. This technique is called boosted-top tagging, which is widely used to improve the tagging efficiency at high-p_T, complimenting top tagging that requests three well-separated jets (called "resolved top"). An example of measurements using boosted tagging at $13\,\mathrm{TeV}$ LHC run with $14.7\,\mathrm{fb}^{-1}$ of the data[10] is shown in Fig. 5. The highest p_T^{top} is now beyond $1\,\mathrm{TeV}$, or above $2\,\mathrm{TeV}$ in $m_{t\bar{t}}$.

Large statistics from $13\,\mathrm{TeV}$ collisions also allow to have high precision measurements in other channels. In order to avoid theoretical uncertainties, cross sections are measured as a function of kinematical variables of decay objects, such as p_T^{ℓ}, or event kinematics like H_T, instead of properties related to top quark momentum. Figure 6 shows results from CMS $\ell +$ jet measurement using full statistics of 2015 and 2016 data[11] in comparison to various models: different combination of matrix element (POWHEG, MadGraph5) and parton showers (PYTHIA8, HERWIG++) simulation, with different matching scale between the matrix element and parton shower and various parameters on parton shower suppression. The data are enough precise to discriminate models causing shifts of the order of 10%.

Many of these models show a tendency that the p_T^{ℓ} spectrum is harder than the data, which is one of features common in many measurements. Figure 7 shows examples from other cross section measurements vs. p_T^{top} from ATLAS[12,13] and

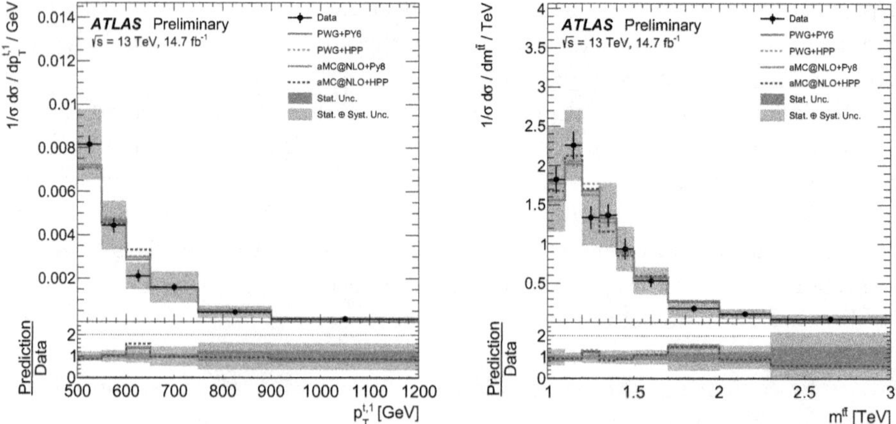

Fig. 5. Normalised top quark differential cross sections in fiducial phase-space measured by ATLAS[10] as a function of (left) transverse momentum of the leading top-quark jet (right) invariant mass of the $t\bar{t}$ system. The top quarks are tagged by boosted top technique.

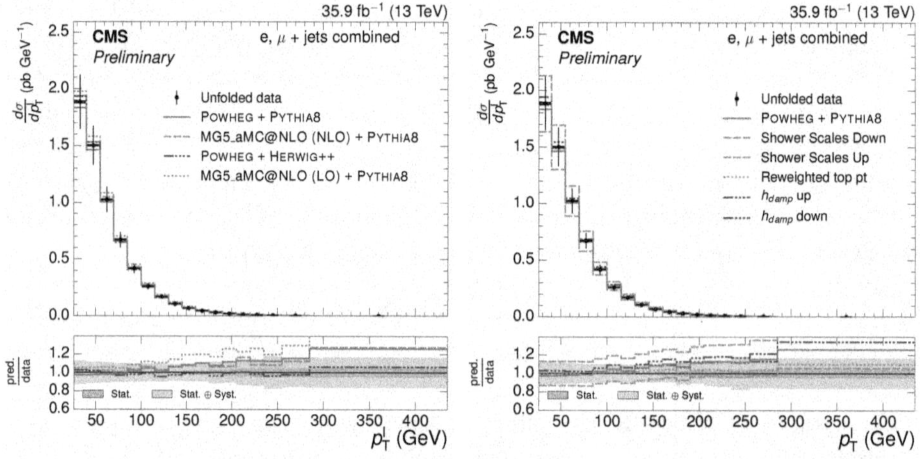

Fig. 6. Absolute differential cross sections as a function of p_T^ℓ using ℓ + jet channel of 13 TeV data[11]. (left) Comparison to models with different matrix element calculation and parton shower and (right) to models with various parton shower parameters.

CMS[14,15]. Many of the models predict harder spectrum than the data. This may imply that the models do not simulate some energies not given to the system of the top quark pair but to additional jets, either inside or the outside the fiducial acceptance of the detector ($\eta_{\rm jet} \lesssim 2.5$). In fact, recent calculation[16] shows that the spectrum becomes softer by increasing the order of approximation from NLO to NNLO. Models reproducing such behaviour of the data, in particular event generators, are awaited for reducing the systematic effects due to models in precision measurements.

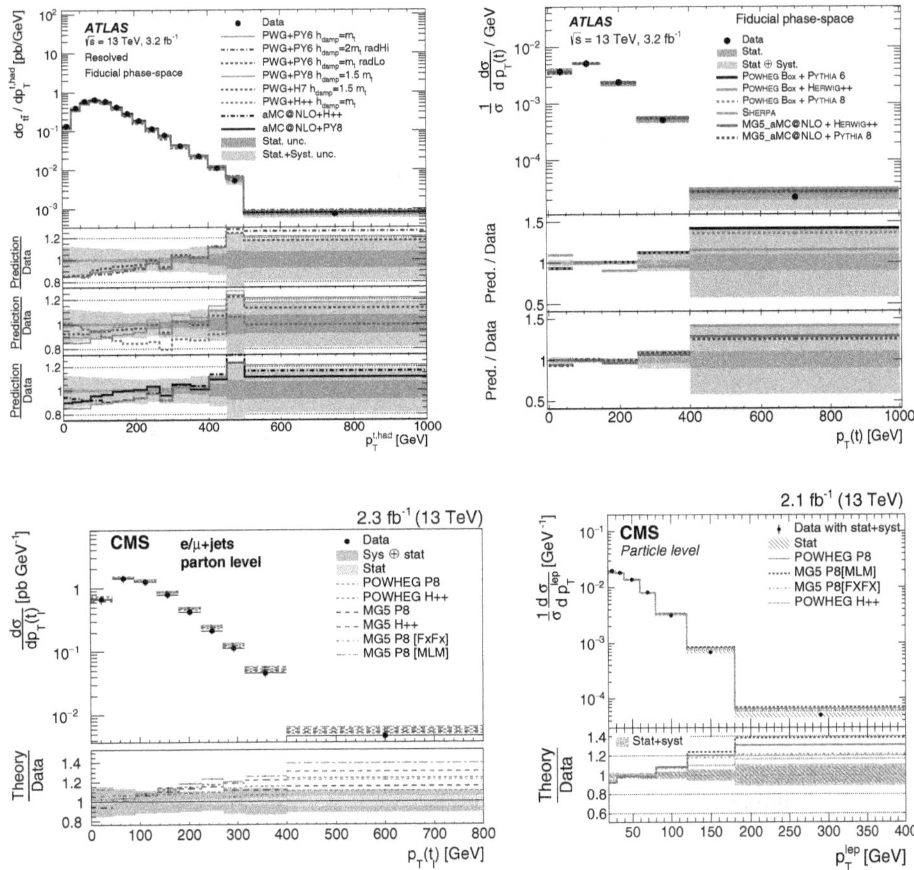

Fig. 7. Differential cross sections as a function of p_T^{top} for (upper left) ATLAS $\ell + \text{jet}$ [12], (upper right) ATLAS dilepton [13], (lower left) CMS $\ell + \text{jet}$ [14] and (lower right) CMS dilepton [15] measurements using 13 TeV collision data.

The differences between top production models are often more clearly seen in the properties related to additional QCD radiation, such as number of jets or p_T imbalance between the two top quarks. Figure 8 shows the cross section for each number of jets (left) or additional jets to those originated from the top quark (right). CMS [17] (Fig. 8(left)) used previous measurements at 8 TeV to tune the parameters related to parton shower simulation in the model. The result of the tuned model is indicated as solid line in Fig. 8(left), giving very good description of the data. For ATLAS [18] (Fig. 8(right)), the comparison is made with models not tuned to this particular data. The deviation between models show high sensitivity of the measurement.

The LHCb collaboration also have measured top production cross sections for the first time [19] with 4.9σ significance. Since the LHCb detector covers forward rapidity

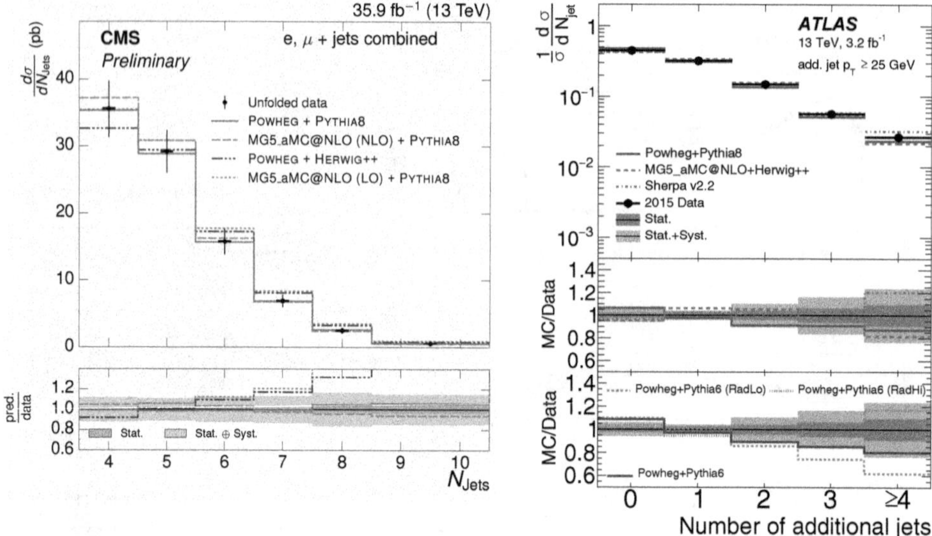

Fig. 8. Differential cross sections for top-pair production as a function of the number of jets from CMS[17] (left) and the number of additional jets to the top quark jets from ATLAS[18] (right).

region ($2 < \eta \lesssim 4.5$), the events are originated from parton-parton collisions with largely asymmetric configuration. The cross section, therefore, is sensitive to the parton densities of both high and low-x gluons. The measured fiducial cross section, $0.05^{+0.02}_{-0.01}$(stat.)$+0.02_{-0.01}$(syst.) pb, is consistent with prediction of 0.045 pb.

4. Top Quark Couplings and Spin

Although the top quark QCD coupling at around the scale of top mass is well tested through production cross section measurements, its electroweak couplings are not strongly constrained yet. Additional boson production with $t\bar{t}$ is sensitive to possible deviation of these couplings from the SM values. Top decay width is measured in order to find possible deviation in top couplings and couplings to unknown particle. The spin properties, such as spin correlation between the top and anti-top quarks in the $t\bar{t}$ production, the $V - A$ vertex structure of the top decay, can be measured by decay angles (not reviewed in this article).

The first observation on $t\bar{t}W$ and $t\bar{t}Z$ production by CMS[20] with full 2015 and 2016 statistics at 13 TeV is shown in Fig. 9. The Z^0 coupling to third-generation quarks is not experimentally well constrained. A good part of Z^0s are radiated from the top quarks in the $t\bar{t}Z$ production, an example of which is shown in 9 (left). Such diagrams are sensitive to the top-Z^0 coupling and also important for understanding $t\bar{t}H$ process, since all the $t\bar{t}Z$ diagrams, especially those where a Z^0 is coupled to a top quark. has corresponding $t\bar{t}H$ diagrams. The cross sections are constrained at about 10% level (Fig. 9 (right)). The result is interpreted in terms of effective

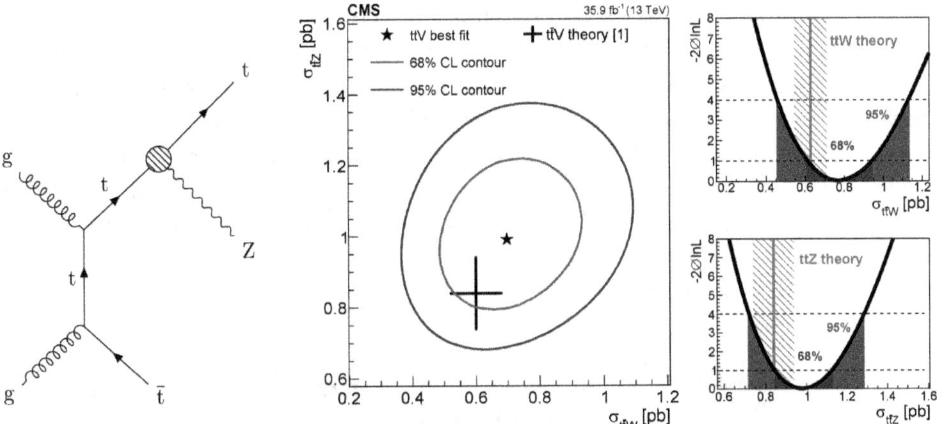

Fig. 9. (left) An example of LO diagram with top-Z^0 coupling through SM or new physics. (right) Result of simultaneous fit for $t\bar{t}W$ and $t\bar{t}Z$ cross sections [20]. The right panels show the individual cross sections along with 68 and 95% CL intervals.

Lagrangian. Assuming that the coefficients to the effective terms are about unity, a presence of new physics through the effective operators in sub-TeV range is already excluded.

The corresponding measurements with a photon in the final state, $t\bar{t}\gamma$, are performed by both ATLAS[21] and CMS[22] using full statistics of 8 TeV data. The measurements show good agreement to the theoretical prediction. An example of kinematical distribution for signal and background is shown in Fig. 10 (left). The differential cross section vs photon p_T and $|\eta|$ are also both well reproduced by NLO prediction, as shown in Fig. 10 (right).

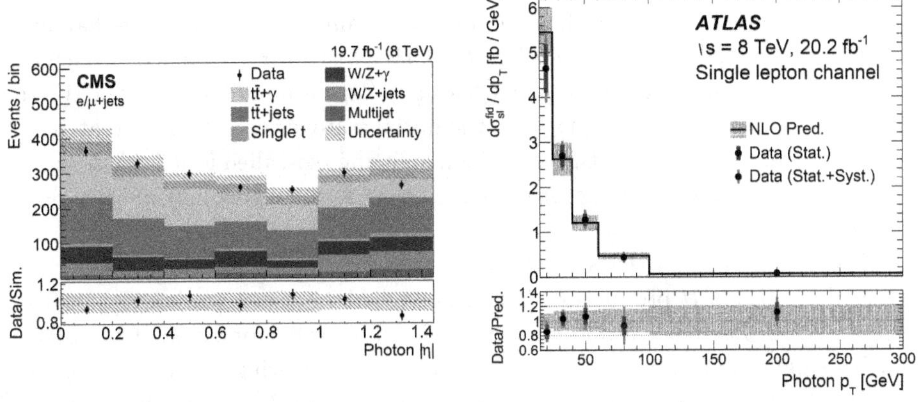

Fig. 10. (left) Photon $|\eta|$ distribution in $t\bar{t}\gamma$ event candidates with its decomposition to signal and various background sources CMS[22]. (right) Differential $t\bar{t}\gamma$ cross sections in photon p_T [21].

The SM prediction on the top width is $1.322\,\text{GeV}$ (for $m_{\text{top}} = 172.5\,\text{GeV}$). This is much smaller than experimentally reconstructed mass width due to detector resolution and QCD radiation effect as discussed in section 2. The tail of the distribution in reconstructed mass spectrum still shows shifts by changing the input top widths, as is shown in Fig. 11 (right). The most recent measurements show that top decay width is consistent with the SM prediction: $0.6 < \Gamma_t < 2.5\,\text{GeV}$ (95% CL) by CMS[23] and $\Gamma = 1.76 \pm 0.33(\text{stat.})^{+0.79}_{-0.68}(\text{syst.})$ by ATLAS[24]. These results indicate that the partial widths of top decay through non-SM particles, if any, is not much larger than that of the SM particles.

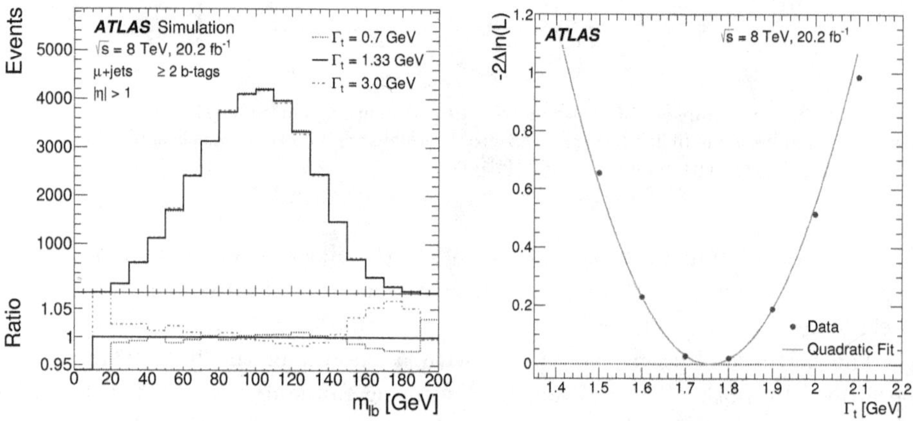

Fig. 11.   Results from top width measurement by ATLAS[24]. (left) Reconstructed m_{lb}, invariant mass of the b-jet and the lepton in the semi-leptonic decay simulated with different Γ_t values. (right) Likelihood of the data as a function of the assumed Γ_t in the templates.

5. Single-top Production

Single-top production is classified in three diagrams: $qb \rightarrow q't$ via t-channel W boson exchange (called "t-channel"), $gb \rightarrow tW$ with top exchanged in t-channel (Wt), and the s-channel production where a $q\bar{q}'$ state annihilates into a W, then goes to $t\bar{b}$ (s-channel). The b-quarks in the initial state is theoretically treated to be either dynamically produced through gluon splitting (so-called four-flavour scheme) or directly originated from the incoming protons as an active flavour (five-flavour scheme).

The mass of the particles in the final state of the t-channel process is much lighter than that of $t\bar{t}$ production. This leads to a large cross section ($\gtrsim 200\,\text{fb}$), as seen in Fig. 12. The background for t-channel can be well suppressed at the LHC, so that the single-top production can also be used for measuring top quark properties, in particular for top angular distribution studies, where the leptons from the semi-leptonic decays has been used. Only one neutrino is involved in semi-leptonic decays of for single-top production similarly to the $\ell + \text{jet}$ channel of pair production.

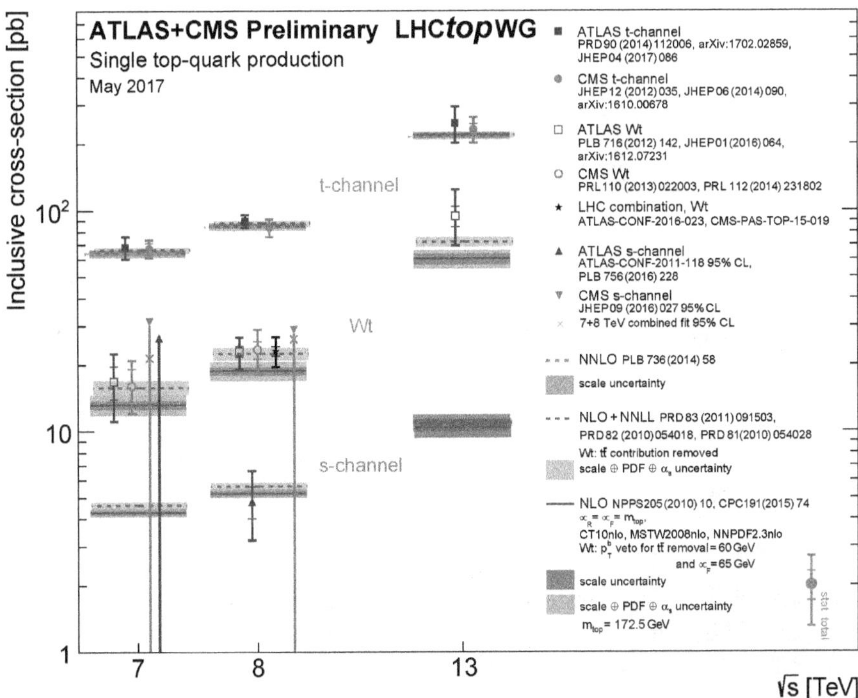

Fig. 12. Summary of ATLAS and CMS measurements of the single top production cross sections in various channels as a function of the centre-of-mass energy. The figure was prepared by LHC Top Working Group [25].

The observation on s-channel events suffers from large background since the production cross section is very small. The study at high energy collisions, i.e. 13 TeV instead of 7 or 8 TeV, remains to be difficult despite larger cross section, since the background rate increases proportionally.

The large number of events of t-channel production allows to measure differential cross sections for single-top production. Figure 13 (left) shows p_T of the top quark from single-top production at 13 TeV [26]. In this figure the prediction by the aMC@NLO model shows softer spectrum for the five-flavour scheme than the four-flavour scheme, demonstrating sensitivity to the choice of the flavour scheme. Figure 13(right) shows the tq production cross section at 8 TeV [27] in comparison with models using four- or five-flavour schemes. Although there is a tendency that the five-flavour scheme predicts higher cross sections, both the theoretical and experimental uncertainties are much larger. Differential cross sections with higher statistics would help to disentangle various theoretical uncertainties, such as the scale uncertainty, PDF uncertainty and the flavour scheme.

Another example benefiting from high purity and abundant statistics in t-channel single top production is measurements on $V-A$ top decay helicity structure.

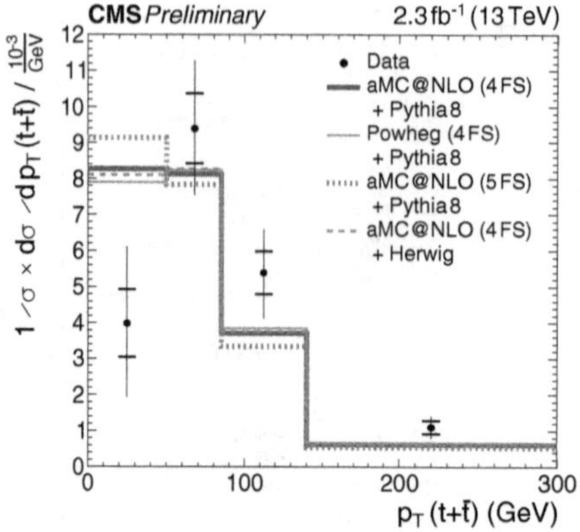
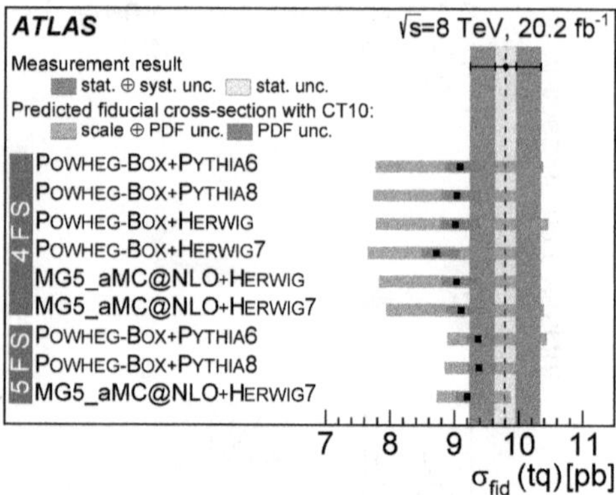

Fig. 13. Single top cross sections of the t-channel process ($t(\bar{t})q$ final state). (left) Differential cross section of t-channel single top quark production as a function of the top quark transverse momentum [26]. (right) Measured t-channel single-top fiducial cross section compared to predictions by the NLO MC generators [27].

ATLAS analysis [28] selected events by simple cut-based method and achieved signal-to-background ratio of about 1:1. This gave comparable result to the multi-variate analysis [29]: both analyses give upper limit of better than 10% on tensor coupling normalised by the corresponding vector coupling.

The first evidence of tZq production in pp collisions, an associated Z^0 production in single-top events, is reported [30] recently. The diagrams for tZq production is not only sensitive to the top coupling to Z^0 but also containing diagrams with a WWZ vertex (see Fig. 14 upper left and right). This process, therefore, allows to test the EW sector in the standard model. The cross section was measured by selecting events with three leptons and two jets, one of which being from the top decay and b-tagged, and additional jet scattered against the top quark system through W exchange. A neural-network-based analysis has shown significant excess in the signal region (Fig. 14 bottom left).

The process without any jet in the final state apart from those from a top quark and a Z^0 is, as indicated in Fig. 14(bottom right), sensitive to the flavour-changing neutral current couplings since the top quark may be produced by flavour excitation through a Z^0 boson. The analysis from CMS [31] gave best upper limit from the LHC, $\mathrm{Br}(t \to Zu) < 0.022\%$ and $\mathrm{Br}(t \to Zc) < 0.049\%$ at 95% CL.

6. Summary

Top quark measurements are now in the phase for precision thanks to the high statistics of LHC collisions and mature detectors there. The mass is obtained with the uncertainty of better than 0.5 GeV. Other top properties are also measured with precision. In these measurements, the dominant source of the systematic sources is not only the detector understanding but also theoretical origin. The experiments

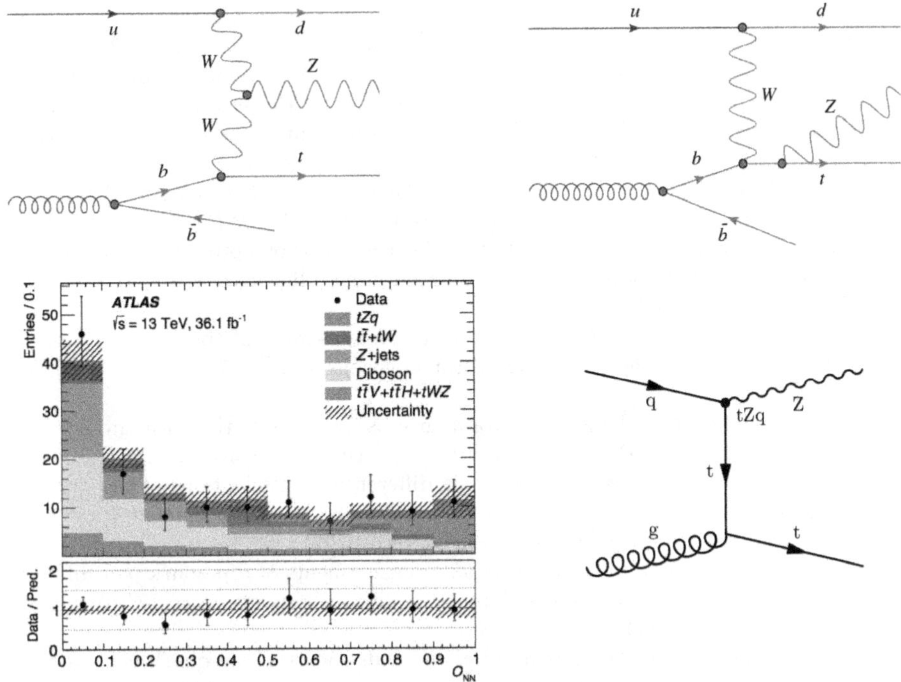

Fig. 14. (top) Two examples of tZq production in pp collisions. (bottom left) The output of the neural network-based multi-variate analysis for tZq production[30]. (bottom right) An example of the tZ final state where the top quark is produced through flavour excitation via Z^0 radiation.

should give feedback through cross section measurements for further improving QCD model description of the top-quark production.

New processes, such as $t\bar{t}Z, tZq$ etc., are observed recently, again thanks to the large data sets. This opens up possibility to precisely determine the EW couplings of top and weak bosons as well as possible new physics in these vertices.

The LHC is planning to obtain about three times more luminosity by the end of 2018 than those used for the data analysis presented in this symposium. This opens up further improvement on precision regime, provides more statistics for rare processes for electroweak and BSM coupling studies and increases discovery potential for new physics in high p_T regime.

References

1. ATLAS Collaboration, G. Aad *et al.*, The ATLAS Experiment at the CERN Large Hadron Collider, *JINST* **3**, p. S08003 (2008).
2. CMS Collaboration, S. Chatrchyan *et al.*, The CMS Experiment at the CERN LHC, *JINST* **3**, p. S08004 (2008).
3. LHCb Collaboration, A. A. Alves, Jr. *et al.*, The LHCb Detector at the LHC, *JINST* **3**, p. S08005 (2008).

4. CMS Collaboration, V. Khachatryan *et al.*, Measurement of the top quark mass using proton-proton data at $\sqrt{s} = 7$ and 8 TeV, *Phys. Rev.* **D93**, p. 072004 (2016).

5. LHC Top Working Group, Summary plot top mass measurements, May 2017. Available under: https://twiki.cern.ch/twiki/bin/view/LHCPhysics/TopMassHistory.

6. D0 Collaboration, B. Abbott *et al.*, Direct measurement of the top quark mass at D0, *Phys. Rev.* **D58**, p. 052001 (1998).

7. CMS Collaboration, V. Khachatryan *et al.*, Measurement of the mass of the top quark in decays with a J/ψ meson in pp collisions at 8 TeV, *JHEP* **12**, p. 123 (2016).

8. ATLAS Collaboration, M. Aaboud *et al.*, Measurement of lepton differential distributions and the top quark mass in $t\bar{t}$ production in pp collisions at $\sqrt{s} = 8$ TeV with the ATLAS detector (2017).

9. CMS Collaboration, V. Khachatryan *et al.*, Measurement of the $t\bar{t}$ production cross section in the $e\mu$ channel in proton-proton collisions at $\sqrt{s} = 7$ and 8 TeV, *JHEP* **08**, p. 029 (2016).

10. ATLAS Collaboration, Differential top pair cross-sections in the all-hadronic channel at 13 TeV, ATLAS-CONF-2016-100, http://cds.cern.ch/record/2217231.

11. CMS Collaboration, Measurement of the differential cross sections of top quark pair production as a function of kinematic event variables in pp collisions at $\sqrt{s} = 13$ TeV, CMS-PAS-TOP-16-014, http://cds.cern.ch/record/2273459.

12. ATLAS Collaboration, M. Aaboud *et al.*, Measurements of top-quark pair differential cross-sections in the lepton+jets channel in pp collisions at $\sqrt{s} = 13$ TeV using the ATLAS detector (2017).

13. ATLAS Collaboration, M. Aaboud *et al.*, Measurements of top-quark pair differential cross-sections in the $e\mu$ channel in pp collisions at $\sqrt{s} = 13$ TeV using the ATLAS detector, *Eur. Phys. J.* **C77**, p. 292 (2017).

14. CMS Collaboration, V. Khachatryan *et al.*, Measurement of differential cross sections for top quark pair production using the lepton+jets final state in proton-proton collisions at 13 TeV, *Phys. Rev.* **D95**, p. 092001 (2017).

15. CMS Collaboration, A. M. Sirunyan *et al.*, Measurement of normalized differential $t\bar{t}$ cross sections in the dilepton channel from pp collisions at $\sqrt{s} = 13$ TeV (2017).

16. M. Czakon, D. Heymes and A. Mitov, High-precision differential predictions for top-quark pairs at the LHC, *Phys. Rev. Lett.* **116**, p. 082003 (2016).

17. CMS Collaboration, A. M. Sirunyan *et al.*, Measurements of ttbar cross sections in association with b jets and inclusive jets and their ratio using dilepton final states in pp collisions at sqrt(s) = 13 TeV (2017).

18. ATLAS Collaboration, M. Aaboud *et al.*, Measurement of jet activity produced in top-quark events with an electron, a muon and two *b*-tagged jets in the final state in pp collisions at $\sqrt{s} = 13$ TeV with the ATLAS detector, *Eur. Phys. J.* **C77**, p. 220 (2017).

19. LHCb Collaboration, R. Aaij *et al.*, First observation of top quark production in the forward region, *Phys. Rev. Lett.* **115**, p. 112001 (2015).

20. CMS Collaboration, A. M. Sirunyan *et al.*, Measurement of the cross section for top quark pair production in association with a W or Z boson in proton-proton collisions at $\sqrt{s} = 13$ TeV (2017).

21. ATLAS Collaboration, M. Aaboud *et al.*, Measurement of the $t\bar{t}\gamma$ production cross section in proton-proton collisions at $\sqrt{s} = 8$ TeV with the ATLAS detector (2017).

22. CMS Collaboration, A. M. Sirunyan *et al.*, Measurement of the semileptonic $t\bar{t} + \gamma$ production cross section in pp collisions at $\sqrt{s} = 8$ TeV, *JHEP* **10**, p. 006 (2017).

23. ATLAS Collaboration, M. Aaboud *et al.*, Direct top-quark decay width measurement in the $t\bar{t}$ lepton+jets channel at $\sqrt{s} = 8$TeV with the ATLAS experiment (2017).

24. CMS Collaboration, Bounding the top quark width using final states with two charged leptons and two jets at $\sqrt{s} = 13\,\text{TeV}$, CMS-PAS-TOP-16-019, http://cds.cern.ch/record/2218019.

25. LHC Top Working Group, Single Top summary of all channels at 7, 8 and 13 TeV, May 2017. Available under: https://twiki.cern.ch/twiki/bin/view/LHCPhysics/SingleTopAllChannelsHistory.

26. CMS Collaboration, Measurement of the differential cross section for t-channel single-top-quark production at $\sqrt{s} = 13\,\text{TeV}$, CMS-PAS-TOP-16-004, http://cds.cern.ch/record/2151074.

27. ATLAS Collaboration, M. Aaboud *et al.*, Fiducial, total and differential cross-section measurements of t-channel single top-quark production in pp collisions at 8 TeV using data collected by the ATLAS detector, *Eur. Phys. J.* **C77**, p. 531 (2017).

28. ATLAS Collaboration, M. Aaboud *et al.*, Analysis of the Wtb vertex from the measurement of triple-differential angular decay rates of single top quarks produced in the t-channel at $\sqrt{s} = 8$ TeV with the ATLAS detector (2017).

29. CMS Collaboration, V. Khachatryan *et al.*, Search for anomalous Wtb couplings and flavour-changing neutral currents in t-channel single top quark production in pp collisions at $\sqrt{s} = 7$ and 8 TeV, *JHEP* **02**, p. 028 (2017).

30. ATLAS Collaboration, M. Aaboud *et al.*, Measurement of the production cross-section of a single top quark in association with a Z boson in proton–proton collisions at 13 TeV with the ATLAS detector (2017).

31. CMS Collaboration, A. M. Sirunyan *et al.*, Search for associated production of a Z boson with a single top quark and for tZ flavour-changing interactions in pp collisions at $\sqrt{s} = 8$ TeV, *JHEP* **07**, p. 003 (2017).

The Top Quark:
Past, Present, and Future

R. Sekhar Chivukula

Department of Physics and Astronomy, Michigan State University,
East Lansing, MI 48824, USA
E-mail: sekhar@msu.edu
http://www.pa.msu.edu/~sekhar

In this talk I discuss the widespread impact of the top quark on phenomena in elementary particle physics as codified through the Standard Model (SM), its important role in motivating the possibility of physics beyond the Standard Model (BSM), and its use as a signal for detailed studies of the SM and searches for BSM physics.

1. Past — Top Matters

In the Standard Model the masses of all of the elementary particles (and, in particular, the "current" masses of the quarks) arise through couplings between the particles and the Higgs boson.[a] With a mass of approximately 173 GeV/c^2, the top quark is the heaviest elementary particle discovered to date — and therefore has the strongest (Yukawa) coupling to the Higgs boson of all known particles. Furthermore, the top quark mass violates two different kinds of symmetries that the Standard Model (SM) would possess were the quarks all massless — the "custodial" $SU(2)$ symmetry[1] which insures that the weak-interaction ρ-parameter

$$\rho = \frac{M_W^2}{M_Z^2 \cos^2 \theta_W},\tag{1}$$

is equal to one at tree-level, and quark flavor symmetries[2] which would prevent the appearance of flavor-changing neutral currents.

The presence of non-degenerate and non-zero quark masses (and, in the case of the ρ-parameter, the hypercharge interactions) explain, through radiative corrections, the observed deviations in the ρ-parameter from one and the presence of the flavor-changing neutral currents observed in neutral meson mixing. Because of its large coupling to the Higgs sector, the largest of these radiative corrections typically arise from loop processes involving the top quark. Hence, the presence of the top quark is essential to ability of the Standard Model to account for observed particle phenomenon.

The consistency of the Standard Model is illustrated[3] in Fig. 1, by the agreement of direct measurements of the top quark and W boson masses (green horizontal and vertical regions) and the best-fit indirect determination of these masses obtained by

---

[a]This is true of the Higgs boson itself, whose mass arises through its *self-interactions*.

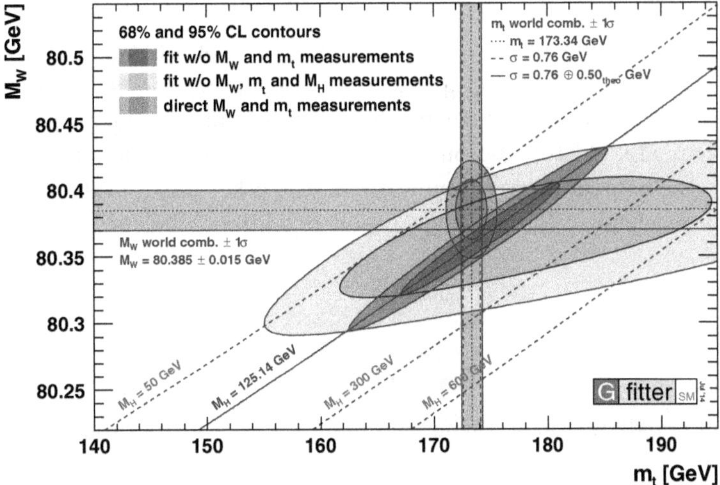

Fig. 1.   Consistency [3] of the Standard Model. The influence of the top quark on precisely measured electroweak quantities is illustrated by the "slope" of the diagonal, blue, fit-region — showing the tight correlation needed between M_W and m_t to be consistent with experimental results (color online).

using all other electroweak data (blue, oval, diagonal regions). The influence of the top quark is illustrated by the "slope" of the diagonal blue region — demonstrating how a change in the top quark mass would only be consistent with experimental data by changing the W boson mass as well.[b]

The effect of the top quark on meson mixing is illustrated through the properties of $B_d - \bar{B}_d$ mixing. The dominant contribution to this process from the Standard Model comes through the box diagram,[c] illustrated below,

$$
\begin{array}{c}
\bar{b} \quad\longleftarrow\quad W^+ \quad\longleftarrow\quad \bar{d} \\
\bar{u},\bar{c},\bar{t} \qquad\qquad u,c,t \\
d \quad\longrightarrow\quad W^- \quad\longrightarrow\quad b
\end{array}
\qquad , \tag{2}
$$

which leads to a prediction for the splitting of B-meson CP eigenstates (and, hence, to the oscillation frequency of flavor eigenstates) of order

$$
\Delta m_B \propto m_t^2 |V_{tb}V_{td}^*|^2 \approx 0.00002 \left(\frac{m_t}{\mathrm{GeV}/c^2} \right)^2 \mathrm{ps}^{-1} , \tag{3}
$$

where $V_{tb,td}$ are the relevant CKM quark-mixing elements. The initial observation of B_d-meson mixing by the ARGUS collaboration[4] found $\Delta m_B \simeq 0.5$ ps^{-1}, and

---

[b] Indeed, prior to the discovery of the Higgs boson, the measured values of M_W and m_t favored a Higgs mass in the 100 GeV range.
[c] Additional "crossed" diagrams are not shown, but have the same dependence on the top quark.

therefore concluded (on the basis of what was then known about the quark-mixing matrix) that $m_t > 50$ GeV/c^2 — the first indication of a heavy top quark!

In the cases above, the top quark Yukawa coupling enters via the coupling of the massive W boson, whose longitudinal component is (gauge-equivalent to) an "eaten" charged Goldstone boson from the Higgs doublet. The top quark coupling directly to the neutral Higgs boson itself, on the other hand, leads to the dominant production mechanism for the Higgs boson at hadron colliders,

$$\tag{4}$$

production via gluon fusion,[5] which occurs through a virtual top quark loop.

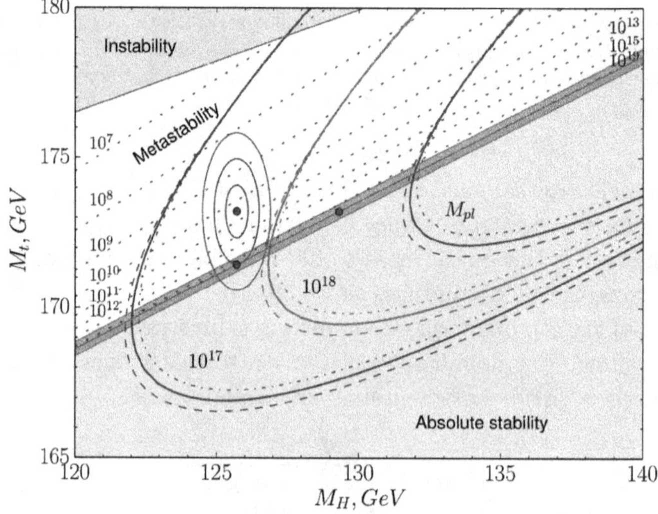

Fig. 2.   Properties of the (two-)loop corrected Higgs potential in the Standard Model.[6] Note that, for the central current values of top quark and Higgs boson masses, the Standard Model electroweak potential is metastable, a property which is exquisitely sensitive to the top quark mass.

Finally, the large top quark Yukawa coupling alters — at loop-level — the properties of the electroweak potential. As shown[6] in Fig. 2, the Standard Model electroweak potential is metastable. This property depends very senstively on the top quark mass, with the potential becoming stable if the top quark had a mass of around 170 GeV/c^2 or lower.

These examples show that *the top quark matters* — in fact, the agreement of the Standard Model with observed particle phenomenology occurs only because of the virtual contributions arising from top quark exchange.

2. Present — Top Couplings

The LHC, and the Tevatron before it, perform as veritable "top-factories" —
allowing for detailed measurements of top quark couplings. For example, Fig. 3
summarizes[7] the cross section measurements for top-pair production at the LHC
and Tevatron at various energies. The close agreement between measurement and
the theoretical QCD cross section serves as a sensitive check that the top quark is
a color-triplet quark, just like the light quarks.

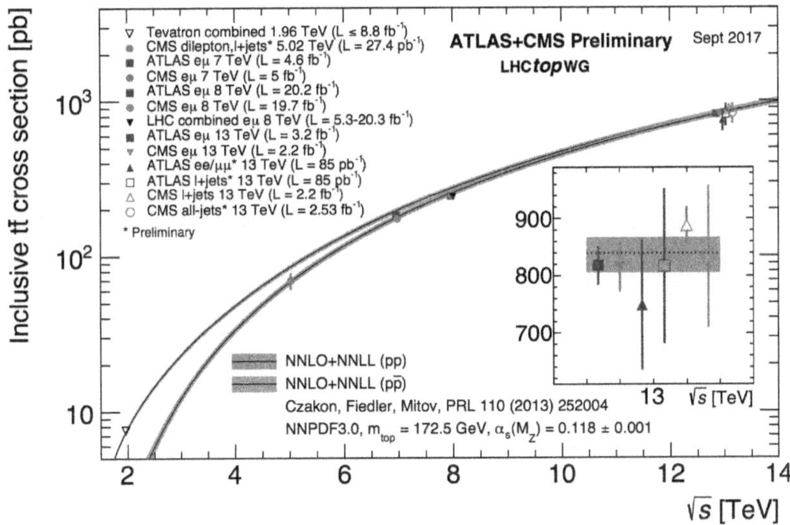

Fig. 3.   Summary[7] of top-pair production cross section measurements from the Tevatron and the
LHC. The agreement between measurement and theory is a confirmation that the top quark is a
color-triplet.

Similarly, initial direct measurements of the top quark width[8,9] are consistent
with the top quark coupling to the W boson with the strength predicted by the
Standard Model. More incisively, Fig. 4 shows that the helicity of the W boson
produced in the decay $t \to Wb$ is consistent with the standard "V-A" structure of
the weak coupling. Note, in particular, that the roughly 40% longitudinal W boson
fraction in this decay is again a measure of the top quark's large Yukawa coupling.

Furthermore, as summarized in Fig. 5, measurements of the single-top production cross section (through various channels) yields a measurement of the CKM
matrix element V_{tb} — which is very close to one.

As noted in the previous section, the top quark's large Yukawa coupling makes
the gluon-fusion cross section the dominant production mode for the Higgs boson.
Using the observed properties of the Higgs boson we can, in the context of the
Standard Model, extract measurements (or, in some cases, limits) on the coupling

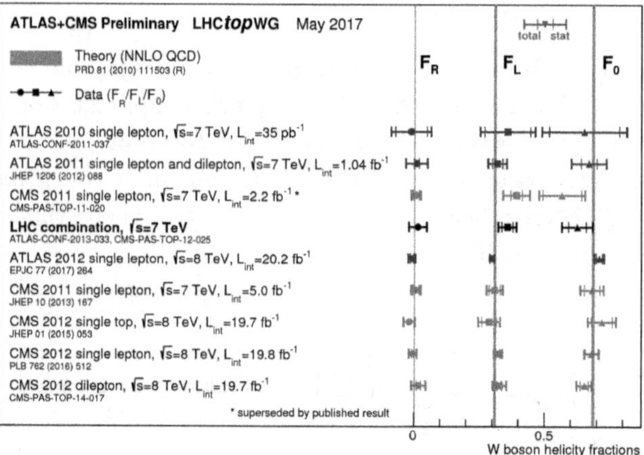

Fig. 4. Summary[10] of measured W helicity fractions by ATLAS and CMS at 7 and 8 TeV, compared to the respective theory predictions. The results are consistent with the "V-A" structure of the weak-interactions, and the large fraction of longitudinally polarized W bosons serves as an indirect check of the top quark Yukawa coupling to the Higgs doublet.

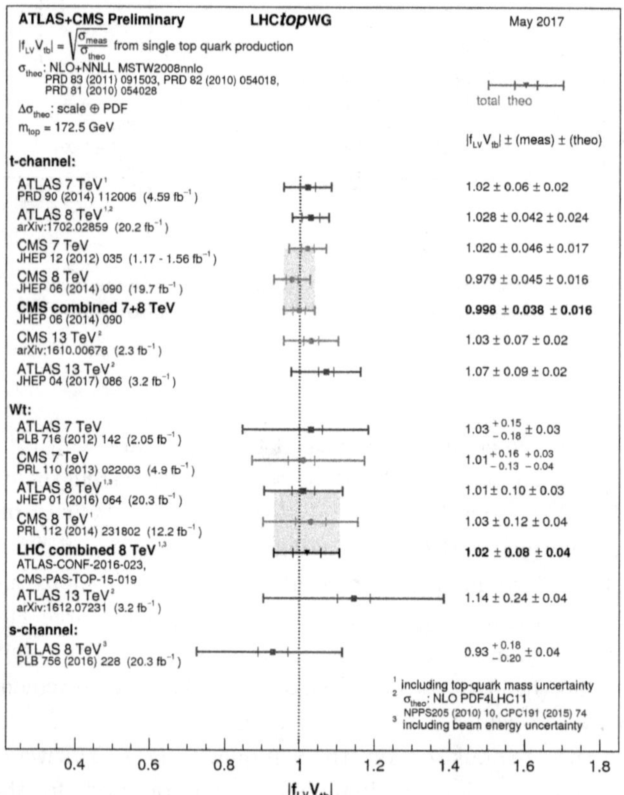

Fig. 5. Summary[11] of the ATLAS and CMS extractions of the CKM matrix element V_{tb} from single top quark production measurements.

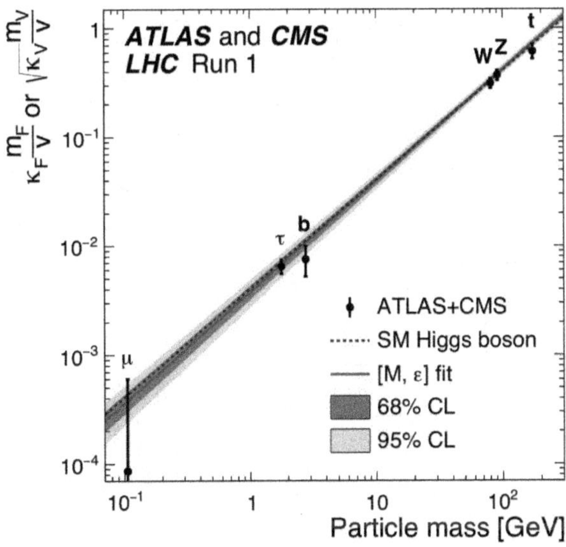

Fig. 6. Fit to Higgs boson couplings based on a combined Run 1 ATLAS and CMS analysis.[12]

of the Higgs to various particles. Current measurements are summarized in Fig. 6, which again demonstrates the consistency of the Standard Model, and graphically shows that the Higgs boson couples most strongly to the top quark.

With the large number of top quark pairs produced at the LHC, ATLAS and CMS can now begin to set meaningful bounds on flavor-changing top quark decays. In Fig. 7 we see a summary of bounds on decays of the form $t \to X + q$, where $X = g$, Z, γ or H, and $q = u$ or c. We see that, while the bounds are not near the Standard Model predictions, they are beginning to be sensitive to various models of new physics.

Presently, we are beginning to study top quark properties in detail and with precision, and to set limits on new physics from the results.

3. Future — Naturalness and the Hierarchy Problem

As foreshadowed by our discussion of the effect of the top quark on vacuum stability illustrated in Fig. 2, the top quark strongly affects the electroweak symmetry breaking sector (EWSB). The most direct effect of the top on EWSB is in terms of the "little hierarchy problem".[14] As illustrated in the diagram below, the contribution from top quark loops to the dimension-two Higgs mass term, calculated using a "cutoff" scale Λ, is

$$\Rightarrow \delta m_H^2 \propto \frac{2\lambda_t^2 \Lambda^2}{16\pi^2} \, . \tag{5}$$

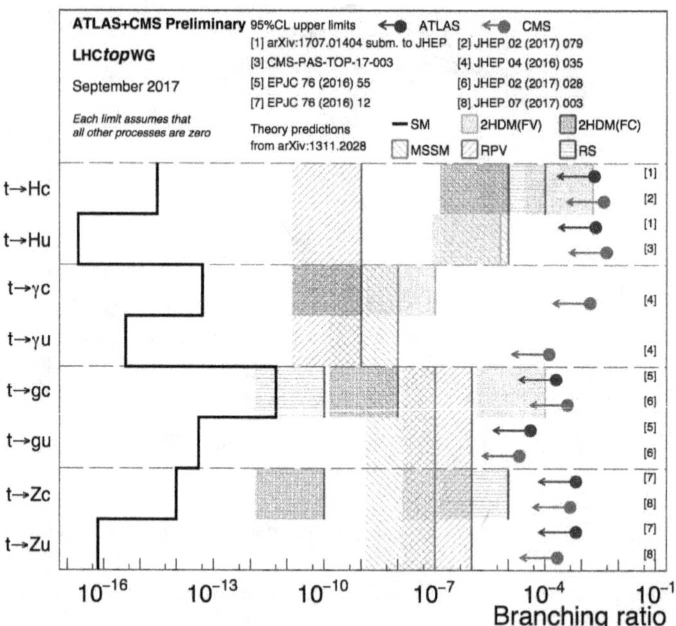

Fig. 7. Summary[13] of the 95% confidence level observed limits on the branching ratios of the top quark decays via flavour changing neutral currents to a quark and a neutral boson $t \to Xq$ ($X = g$, Z, γ or H; $q = u$ or c) by the ATLAS and CMS Collaborations compared to several new physics models.

By interpreting Λ as the scale of physics underlying the Standard Model, and requiring that $\delta m_H^2 < (100\,\text{GeV})^2$, we see that the scale of new physics cannot be larger than one or a few TeV. The Standard Model is therefore *unnatural* — it is, generically, very sensitive to any new physics occuring at scales of order a TeV or higher.

Since the Standard Model does not encompass gravity, we know that it is a low-energy effective theory description which must break down at some higher scale — presumably at a scale lower than the Planck mass, $M_{pl} = \mathcal{O}(10^{16}\,\text{TeV})$. Therefore the low-energy Standard Model effective theory suffers from the *hiearchy* problem: *if* there is no new physics between the electroweak scale and the Planck scale, then the properties of that underlying theory must be exquisitely *adjusted* ("tuned") in order for the Higgs mass, and hence the electroweak scale, to be of order 1 TeV. For example, if the scale of new physics is indeed of order the Planck Scale, we must explain how m_H^2 arises in the low-energy Standard Model (effective theory) and is of order 10^{-32} times smaller!

The naturalness and hierarchy problems motivate the search for new physics beyond the Standard Model at the TeV scale. While it is, in principle, possible that no new physics occurs until very high energies, and that the tuning required to keep the weak scale of order a TeV occurs by *accident*, it is important to explore the possiblity that the TeV scale arises dynamically.[d] One possiblity is that the elecroweak scale is associated with the dynamical breaking of a symmetry that would, if unbroken, leave the Higgs boson massless in the low-enery effective theory. The dynamically breaking of that symmetry would "generate" EWSB scale and, typically, give rise to new states at the TeV scale. The consistency of the low-energy theory would then be reflected in a "cancellation" of contributions to the Higgs boson mass — eliminating the quadratic divergence implicit in the diagram in Eq. (5).

The most popular new symmetry introduced to protect the electroweak scale is supersymmetry.[e] Supersymmetry pairs bosonic and fermionic states, and therefore the Higgs mass term(s) become related to the chiral symmetry which would be present if the corresponding fermionic states were massless. Diagramatically, for the top quark contribution in Eq. (5), there are contributions from the squarks $(\widetilde{t}_{R,L})$, the scalar partners of the top quark

$$\Rightarrow \delta m_H^2 \propto m_{\widetilde{t}}^2 \,. \tag{6}$$

As noted, the residual contributions to the Higgs boson mass are proportional to the masses of the SUSY partners in the loop (assuming $m_{\widetilde{t}} \gg m_t$) — and hence, to avoid the hierarchy problem, we would expect new supersymmetric partner states in the TeV mass range. Of particular note, however, is that supersymmetry protects the Higgs boson mass from potentially larger contributions from higher scale physics (*e.g.* quantum gravity) so long as those additional sectors are supersymmetric. Fig. 8 summarizes recent searches for top-squarks — and, although now beginning to be sensitive to top-squarks in the TeV mass range, there is as yet no sign of the top-squark.

An alternative to supersymmetry is to construct a theory in which the Higgs boson is itself a Goldstone boson of a spontaneously broken global symmetry.[21] In this case, the Higgs mass-term is protected by a "shift symmetry" in the low-energy effective lagrangian.[22] The global symmetries generally used involve chiral symmetries of fermions which feel a new (so-far undiscovered) strong interaction, which dynamically breaks this chiral symmetry in a way analogous to chiral symmetry breaking in QCD. The Higgs is then a Goldstone boson bound-state of these underlying fermionic degrees of freedom, analogous to how the pions of QCD are bound

---

[d]For example, we are not concerned with the occurrence of the QCD scale of order a GeV since this scale occurs via dimensional transmutation and the asymptotic freedom of non-abelian gauge-theories.[15–17]

[e]See review articles in [20], and references therein.

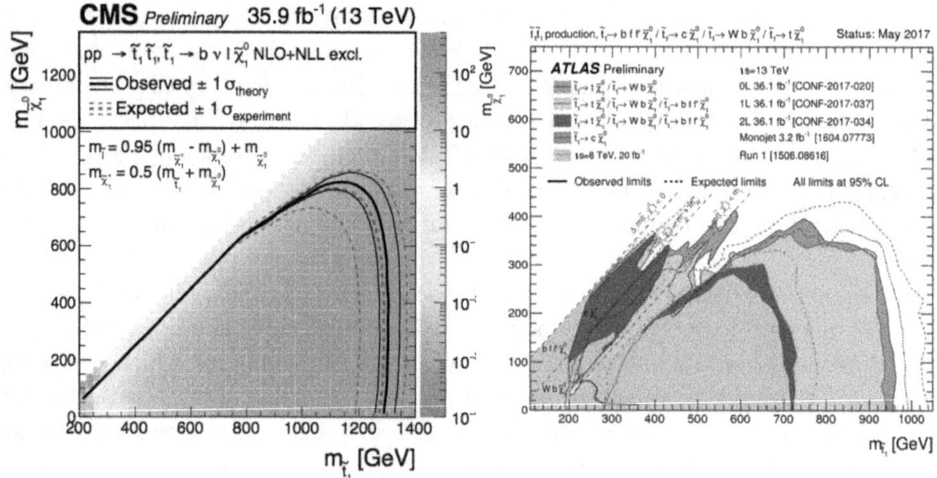

Fig. 8.   Top squark searches. CMS left[18] and ATLAS right[19].

states of the quarks. This scenario therefore realizes the possiblity that the Higgs boson is *not* a fundamental scalar, but rather a bound state of fermions interacting via some strong (gauge theory) dynamics.

Small global symmetry-breaking interactions are then added to produce the Higgs boson mass terms and to produce the "vacuum misalignment" necessary to break the electroweak symmetry. Among the symmetry-breaking terms needed are flavor-dependent ones. The Yukawa couplings necessary to generate the top quark mass, in particular, explicitly break the shift-symmetry, and the flavor symmetries must then be extended to the new fermionic degrees of freedom which are the consitutents of the Higgs boson. Diagramatically, just as in supersymmetry, additional TeV-scale states arise which cancel the potentially dangerous quadratically divergent top-loop contributions to the Higgs boson mass shown in Eq. (5). In this case, the additional states are typically "vectorial" partners of the top quark (and potentially of the other fermions as well), new heavy color-triplet fermions whose right- and left-handed components both couple to the weak interaction $SU(2)$. Interestingly, in the low-energy theory, the diagrams involving the vector-like partners of top (χ) typically interact with the Higgs through a pair-wise dimension-five interaction

$$\Rightarrow \delta m_H^2 \propto m_\chi^2\,, \qquad (7)$$

whose strength is related via the global symmetries of the theory to the top quark Yukawa coupling. Limits from LHC searches for vector-like top quark partners are illustrated in Fig. 9 and, as in the case of supersymmetry, although we are becoming

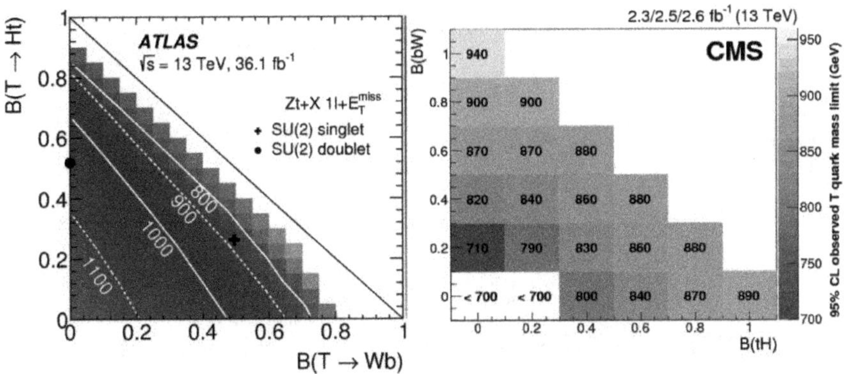

Fig. 9. Vector-quark searches. ATLAS left[23] and CMS right[24].

sensitive to these states in the TeV region there is no evidence for these additional states.

The absence (so far) of new TeV-scale states described above suggests that perhaps we need to consider new kinds of physics beyond the Standard Model. One intriguing possiblity is "neutral naturalness."[25,26] The motivation for these models is based on observation that the additional TeV states above (the top squark or vectorial top-partners) are copiously produced at the LHC because they are colored. In addition, they are colored because the additional global symmetry introduced to protect the Higgs boson mass commutes with QCD — hence the states related to the top must themselves be color triplets. Instead, if the global symmetries introduced include QCD (or some symmetry related to QCD) as a subgroup then, potentially, the states related to the top quark that are introduced to cancel the quadratically divergent contributions to the Higgs boson mass (e.g. Eq. (5) can be color neutral. The phenomenology of these models is quite different, including potentially long-lived particles and exotic signals, and they are far less constrained.[27]

For the *future*, the presence of the top quark — and, in particular, its large Yukawa-coupling to the Higgs boson — is a primary motivation for the presence of new physics at the TeV scale. The LHC was specifically built to explore the TeV region, and time will tell if the kinds of theories we discuss here are realized in nature.

4. Future — Top as a Signal

It is often said that "yesterday's signal is tomorrow's background." With the ability top to be (pair) produced via the strong interactions, and its decay to Wb, the top quark is an important background in many searches for physics beyond the Standard Model. Given the fact that the top quark couples strongly to the EWSB sector, however, *the top is also an important part of the signal in many kinds of standard and non-standard physics.* Some examples are given here.

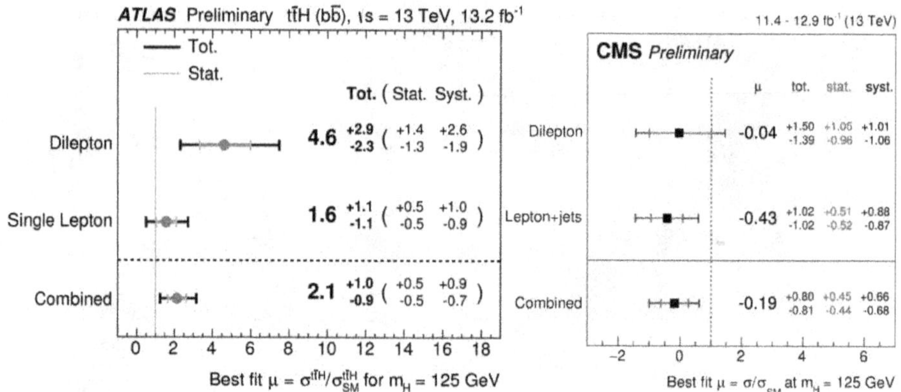

Fig. 10.   ttH searches. ATLAS left [28] and ATLAS right [29].

First, one crucial signal involving the top quark in the Standard Model is the in $t\bar{t}H$ production,

$$ \text{(8)} $$

a process which would yield a direct measurement of the top quark Yukawa coupling. Current limits on this process are shown in Fig. 10. As we see there, the measurements are only beginning to be sensitive to this process — but with additional data, the signal should become clearly visible, and we will be able to directly test whether the Standard Model mass-production mechanism applies in the case of the top quark.[f]

Second, the important contribution of the top-squark to protecting the Higgs boson mass in SUSY theories suggests that the top-squark may be substantially lighter than the scalar partners of the other quarks.[g] This possiblity, along with the large gluino (the fermionic partner of the gluon) pair-production cross section at the LHC, motivate the search shown in Fig. 11 — where the gluino decays to a top quark and the (assumed neutral and stable) lightest supersymmetric partner (LSP, $\tilde{\chi}^0$). Displayed in the $(m_{\tilde{g}}, m_{\tilde{\chi}^0})$ plane, we see that current limits extend to gluino masses of order 2 TeV, for LSP masses up to 1200 GeV.

---

[f]Similarly, observations of the Higgs decay $H \to ZZ^* \to 4\ell$ gives us direct evidence of the Standard Model mechanism for generating the masses of the electroweak gauge bosons.

[g]Just as quark mass-splittings and mixings give rise to flavor-changing neutral-currents, squark mass-splittings and mixings can as well ... so a theory with a large hierarchy of masses between the top-squark and the other scalar partners will need to have relatively small mixing between the third-generation of squarks with the other two in order to avoid dangerously large flavor-changing neutral currents.

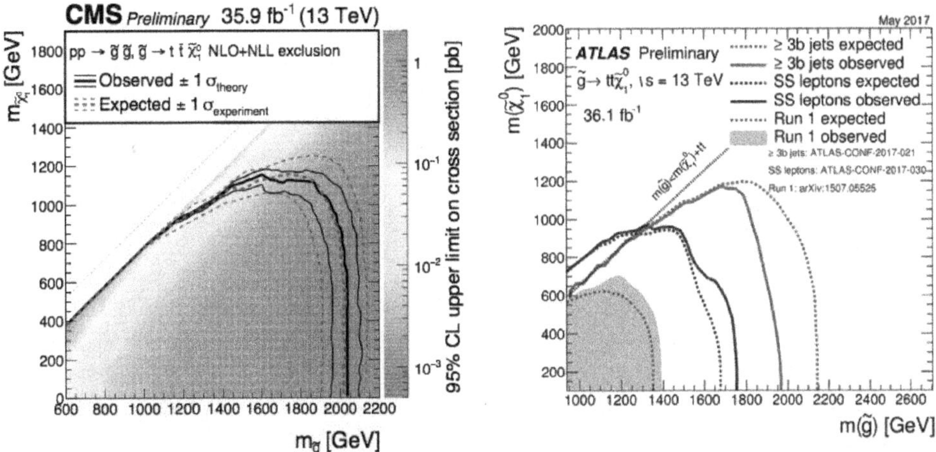

Fig. 11.   Searches for pair-production of gluinos, which subsequently decay to top plus neutralino. CMS left [30] and ATLAS right [31].

Finally, the large coupling of the top quark to the EWSB sector suggests that the top quark could play a special role in theories where the elecroweak sector is extended. Considerations of this kind motivate the search for W' and Z' decaying to tb and tt as shown in Fig. 12. We see here that, though model-dependent, these limits are now sensitive to new gauge-bosons decaying preferentially to third-generation quarks up to 2-3 TeV.

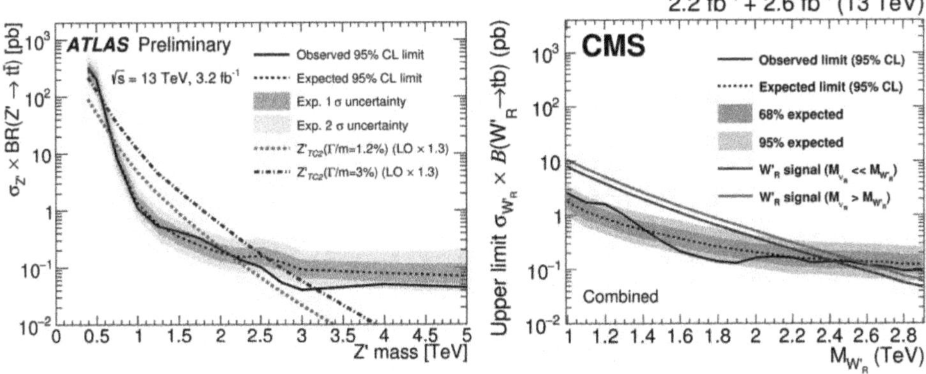

Fig. 12.   Searches for resonances decaying to top. ATLAS left [32] and CMS right [33].

In short, *the top quark provides an important signal of new physics for the LHC now, and for the future.*

5. Conclusions

To summarize, to a very good approximation, we know that the top quark

- is a $(3,2)_{+1/6}$ (left) and $(3,1)_{+2/3}$ (right) under the $SU(3)_C \times SU(2)_W \times U(1)_Y$ guage interactions of the SM,
- mixes relatively little with the other quark-generations, with $V_{tb} \simeq 1$, $|V_{ts}| \simeq 0.04$, and $|V_{td}| \simeq 0.009$,
- and has a large Yukawa coupling, $y_t = \sqrt{2} m_t / v \simeq 1$.

The top quark casts a large shadow over particle phenomenology, generating the largest electroweak and flavor radiative-corrections, providing the dominant production mechanism for the Higgs boson at the LHC, motivating the presence of TeV scale new-physics, and playing an important role in signals for physics both in the Standard Model and beyond.

Experiments at the LHC are now directly sensitive to new physics at the TeV scale, and the top quark will play a central role in experimental and theoretical advances in particle physics for the foreseeable future.

Acknowledgements

I would like to thank the conference organizers for their hospitality, and Dennis Foren for comments on the manuscript. This material is based in part upon work supported by the National Science Foundation under Grant No. 1519045.

References

1. P. Sikivie, L. Susskind, M. B. Voloshin and V. I. Zakharov, Isospin Breaking in Technicolor Models, *Nucl. Phys.* **B173**, 189 (1980).
2. R. S. Chivukula and H. Georgi, Composite Technicolor Standard Model, *Phys. Lett.* **B188**, 99 (1987).
3. M. Baak, J. Cuth, J. Haller, A. Hoecker, R. Kogler, K. Moenig, M. Schott and J. Stelzer, The global electroweak fit at NNLO and prospects for the LHC and ILC, *Eur. Phys. J.* **C74**, p. 3046 (2014).
4. H. Albrecht *et al.*, Observation of B0 - anti-B0 Mixing, *Phys. Lett.* **B192**, 245 (1987), [,51(1987)].
5. H. M. Georgi, S. L. Glashow, M. E. Machacek and D. V. Nanopoulos, Higgs Bosons from Two Gluon Annihilation in Proton Proton Collisions, *Phys. Rev. Lett.* **40**, p. 692 (1978).
6. A. V. Bednyakov, B. A. Kniehl, A. F. Pikelner and O. L. Veretin, Stability of the Electroweak Vacuum: Gauge Independence and Advanced Precision, *Phys. Rev. Lett.* **115**, p. 201802 (2015).
7. LHC Top Quark Working Group, Summary of LHC and Tevatron Top-Pair Production Cross-Section Measurements, `https://twiki.cern.ch/twiki/pub/LHCPhysics/LHCTopWGSummaryPlots/tt_curve_toplhcwg_sep17.png`, Accessed: 2017-11-09.
8. C. Collaboration, Bounding the top quark width using final states with two charged leptons and two jets at $\sqrt{s} = 13$ TeV (2016).
9. T. A. collaboration, Direct top-quark decay width measurement in the $t\bar{t}$ lepton+jets channel at $\sqrt{s} = 8$ TeV with the ATLAS experiment (2017).

10. LHC Top Quark Working Group, Summary of W Boson Helicity Fraction Measurements in Top Decay, https://twiki.cern.ch/twiki/pub/LHCPhysics/ LHCTopWGSummaryPlots/whelicity_LHC_may2017.png, Accessed: 2017-11-09.

11. LHC Top Quark Working Group, Summary of V_{tb} Measurements as Inferred from Single Top Production, https://twiki.cern.ch/twiki/pub/LHCPhysics/ LHCTopWGSummaryPlots/singletop_Vtb_may2017.png, Accessed: 2017-11-09.

12. G. Aad *et al.*, Measurements of the Higgs boson production and decay rates and constraints on its couplings from a combined ATLAS and CMS analysis of the LHC pp collision data at $\sqrt{s} = 7$ and 8 TeV, *JHEP* **08**, p. 045 (2016).

13. LHC Top Quark Working Group, Bounds on Flavor-Changing Top Quark Decays, https://twiki.cern.ch/twiki/pub/LHCPhysics/LHCTopWGSummaryPlots/fcnc_ summarybsm_sep17.png, Accessed: 2017-11-09.

14. R. Barbieri and A. Strumia, The 'LEP paradox', in *4th Rencontres du Vietnam: Physics at Extreme Energies (Particle Physics and Astrophysics) Hanoi, Vietnam, July 19-25, 2000*, 2000.

15. S. R. Coleman and E. J. Weinberg, Radiative Corrections as the Origin of Spontaneous Symmetry Breaking, *Phys. Rev.* **D7**, 1888 (1973).

16. H. D. Politzer, Reliable Perturbative Results for Strong Interactions?, *Phys. Rev. Lett.* **30**, 1346 (1973).

17. D. J. Gross and F. Wilczek, Ultraviolet Behavior of Nonabelian Gauge Theories, *Phys. Rev. Lett.* **30**, 1343 (1973).

18. A. M. Sirunyan *et al.*, Search for top squarks and dark matter particles in opposite-charge dilepton final states at $\sqrt{s} = 13$ TeV (2017).

19. ATLAS Supersymmetry Working Group, ATLAS Top Squark Pair Production Search Summary, https://atlas.web.cern.ch/Atlas/GROUPS/PHYSICS/ CombinedSummaryPlots/SUSY/ATLAS_SUSY_Stop_tLSP/ATLAS_SUSY_Stop_tLSP.png, Accessed: 2017-11-10.

20. C. Patrignani *et al.*, Review of Particle Physics, *Chin. Phys.* **C40**, p. 100001 (2016).

21. D. B. Kaplan and H. Georgi, SU(2) x U(1) Breaking by Vacuum Misalignment, *Phys. Lett.* **136B**, 183 (1984).

22. S. R. Coleman, J. Wess and B. Zumino, Structure of phenomenological Lagrangians. 1., *Phys. Rev.* **177**, 2239 (1969).

23. M. Aaboud *et al.*, Search for pair production of vector-like top quarks in events with one lepton, jets, and missing transverse momentum in $\sqrt{s} = 13$ TeV pp collisions with the ATLAS detector, *JHEP* **08**, p. 052 (2017).

24. A. M. Sirunyan *et al.*, Search for pair production of vector-like T and B quarks in single-lepton final states using boosted jet substructure techniques at sqrt(s) = 13 TeV (2017).

25. Z. Chacko, H.-S. Goh and R. Harnik, The Twin Higgs: Natural electroweak breaking from mirror symmetry, *Phys. Rev. Lett.* **96**, p. 231802 (2006).

26. G. Burdman, Z. Chacko, H.-S. Goh and R. Harnik, Folded supersymmetry and the LEP paradox, *JHEP* **02**, p. 009 (2007).

27. D. Curtin and P. Saraswat, Towards a No-Lose Theorem for Naturalness, *Phys. Rev.* **D93**, p. 055044 (2016).

28. *Search for the Standard Model Higgs boson produced in association with top quarks and decaying into $b\bar{b}$ in pp collisions at $\sqrt{s} = 13$ TeV with the ATLAS detector*, Tech. Rep. ATLAS-CONF-2016-080, CERN (Geneva, 2016).

29. C. Collaboration, Search for associated production of Higgs bosons and top quarks in multilepton final states at $\sqrt{s} = 13$ TeV (2016).

30. *Search for supersymmetry using hadronic top quark tagging in 13 TeV pp collisions*, Tech. Rep. CMS-PAS-SUS-16-050, CERN (Geneva, 2017).

31. ATLAS Supersymmetry Working Group, ATLAS Gluino Decay to Top in Simplified Model, `https://atlas.web.cern.ch/Atlas/GROUPS/PHYSICS/CombinedSummaryPlots/SUSY/ATLAS_SUSY_Gtt/ATLAS_SUSY_Gtt.png`, Accessed: 2017-11-10.

32. *Search for heavy particles decaying to pairs of highly-boosted top quarks using lepton-plus-jets events in proton–proton collisions at $\sqrt{s} = 13$ TeV with the ATLAS detector*, Tech. Rep. ATLAS-CONF-2016-014, CERN (Geneva, 2016).

33. A. M. Sirunyan *et al.*, Searches for W' bosons decaying to a top quark and a bottom quark in proton-proton collisions at 13 TeV, *JHEP* **08**, p. 029 (2017).

Results and Opportunities in Atmospheric Neutrino Measurement and Search of Proton Decays

K. Okumura

Institute for Cosmic Ray Research, University of Tokyo
Kashiwa, Chiba 2778582, Japan
E-mail: okumura@icrr.u-tokyo.ac.jp
http://www-rccn.icrr.u-tokyo.ac.jp

In this paper we discuss the current status and the future prospect of the atmospheric neutrino measurement and the proton decay search. The atmospheric neutrino has been playing an important role in studying neutrino oscillation physics in the mass range of $\Delta m_{32}^2 \simeq 2.5 \times 10^{-3}$ eV^2, and still there are many notable physics outcomes from the atmospheric neutrino observations. We review the recent results on the determination of neutrino mass hierarchy, the direct detection of tau neutrino, and the search of sterile neutrino. Proton decay is one of the unique methods to explore Grand Unified Theory (GUT). The search results of the proton decay such as $p \to e^+ \pi^0$ and $p \to \nu K^+$ are reviewed. Finally the future prospects on the atmospheric neutrino and the proton decay search by the proposed projects will be discussed.

1. Introduction

Cosmic rays, such as proton and Helium, interact with the atmosphere when they enter into the Earth, producing many secondary particles. The atmospheric neutrinos are produced as their decay products. Fig. 1 shows the measured spectrum by several experiments. The energy spectrum of the atmospheric neutrino follows the power-law shape and extends up to several 100 TeV while the flux below several GeV are reduced by the cutoff effect due to the geomagnetic field. Since the atmospheric neutrino can easily penetrate the Earth, the surface detector observes the atmospheric neutrinos in all directions. The path length to the detector from the production point depends on the zenith angle direction, ranging from several 10 km for downward-going direction to $\sim$13,000 km for upward-going direction. The atmospheric neutrinos are affected by the neutrino oscillation driven dominantly by the atmospheric mass scale ($\Delta m_{32}^2 = m_3^2 - m_2^2 \sim 2.4 \times 10^{-3}$ eV^2) below several 10 GeV depending on the energy and the path length.

Proton decay has been predicted by Grand Unified Theory (GUT), which is beyond Standard Model physics, providing the method for baryon asymmetry universe. The major channels of the proton decay are $p \to e^+ \pi^0$ and $p \to \nu K^+$, which are predicted by several GUT models, e.g. SU(5), SO(10), SUSY GUT.

The measurement of Atmospheric neutrino and the search of proton decay are complementary each other since the observed energy range is similar. Several proton decay experiments were carried out in 1980's, and the atmospheric neutrinos

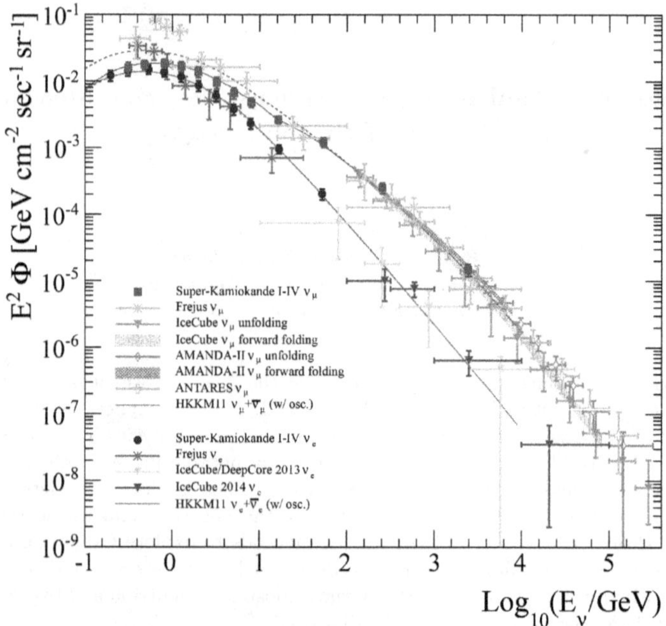

Fig. 1. Measured energy spectra of the atmospheric ν_e and ν_μ fluxes by Super-K[1], Frejus[2], AMANDA-II[3,4], IceCube[5-8], and ANTARES[9]. The solid (dashed) lines are the HKKM11 flux[10] predicted for the Kamioka site with (without) neutrino oscillation. Note that the neutrino flux below several GeV is expected to be relatively higher for the Frejus site, due to the lower geomagnetic cutoff of the primary cosmic rays. Figure was taken from[1].

were measured as a background distribution. But there was a hint of neutrino oscillation: kamiokande and IMB reported smaller ν_μ/ν_e flavor ratio than expected. In 1998 Super-Kamiokande (Super-K) discovered the deficit of upward-going muon neutrino, which showed the evidence of neutrino oscillation, leading to the Nobel Prize Physics in 2015. The atmospheric neutrino oscillation has been measured in several other experiments; MINOS experiments, which is designed to measure accelerator neutrino, also measure atmospheric neutrino with magnetized iron tracking detector. IceCube and ANTARES, which were originally designed to observe astrophysical neutrino using natural ice and sea water, also measure atmospheric neutrinos in higher energies above 10 GeV.

2. Atmospheric Neutrino Oscillation Results

2.1. *Overview*

Since the discovery of the neutrino oscillation in 1998, many efforts have been performed for measuring the neutrino mixing by several neutrino experiments such as solar, atmospheric, long baseline, and reactor neutrino experiments. PMNS matrix, which describes the mixing of three flavor neutrinos, is shown below:

$$U_{\mathrm{PMNS}} = \begin{pmatrix} 1 & 0 & 0 \\ 0 & \cos\theta_{23} & \sin\theta_{23} \\ 0 & -\sin\theta_{23} & \cos\theta_{23} \end{pmatrix} \begin{pmatrix} \cos\theta_{13} & 0 & \sin\theta_{13}e^{-i\delta_{CP}} \\ 0 & 1 & 0 \\ -\sin\theta_{13}e^{i\delta_{CP}} & 0 & \cos\theta_{13} \end{pmatrix}$$

$$\times \begin{pmatrix} \cos\theta_{12} & \sin\theta_{12} & 0 \\ -\sin\theta_{12} & \cos\theta_{12} & 0 \\ 0 & 1 & 0 \end{pmatrix}$$

Currently all three mixing angle (θ_{12}, θ_{23}, θ_{13}) and two mass square difference (Δm_{32}^2, Δm_{21}^2) have been measured. However still important pieces are missing;

- leptonic CP phase (δ_{CP}) which produces the asymmetry in neutrino and anti-neutrino oscillation is not measured.
- Mass hierarchy between 2nd and 3rd mass state is unknown. There are two state, normal ($\Delta m_{32}^2 > 0$) and inverted hierarchy ($\Delta m_{32}^2 < 0$).
- θ_{23} octant, which means e.g. θ_{23} is larger than 45 degrees or not, is unknown.

Also the issue of the sterile neutrino should be also addressed. Several indications on the existence of sterile neutrino for sterile mass of $\Delta m^2 > 0.1$ eV^2 have been pointed out from the results of short baseline neutrino experiments. In principle the atmospheric neutrino has the sensitivity to these unresolved issues of the neutrino oscillation.

2.2. *Sub-dominant Oscillation and Mass Hierarchy*

There are many opportunities to study neutrino oscillation especially three flavor mixing in the atmospheric neutrinos. The disappearance of muon neutrino is the dominant channel of the neutrino oscillation in the atmospheric neutrino, and is explained by $\nu_\mu \to \nu_\tau$ oscillation driven by Δm_{32}^2 and θ_{23}. In addition, due to non-zero θ_{13} mixing angle, which is discovered by accelerator and reactor neutrino experiments, the sub-dominant $\nu_\mu \to \nu_e$ oscillation is expected, which contains many interesting information such as neutrino mass hierarchy, leptonic CP phase (δ_{CP}), and θ_{23} octant, as shown in Fig. 2.

Mass hierarchy is the most promising physics proved by the atmospheric neutrino observation. When neutrino propagates through the matter, the effective neutrino mass is known to be changed by the effect of the additional potential caused by the forward scattering with electrons in matter, so called as *matter effect*. Accordingly, neutrino oscillation probability is changed depending on the neutrino energy and electron density. Especially in multi-GeV region, the resonant-like enhancement of $\nu_\mu \to \nu_e$ oscillation is expected. The neutrino mass hierarchy could be proved by the atmospheric neutrino via such enhancement.

2.3. *Three-flavor Oscillation Analysis*

The Super-Kamiokande (Super-K) detector[12] has been observing atmospheric neutrinos from the beginning of the experiment in 1996. Super-K is a water Cherenkov

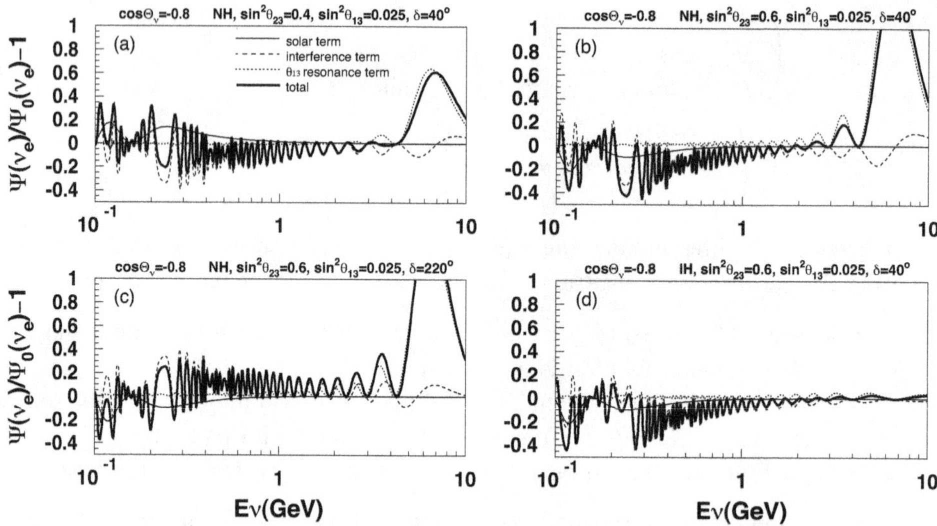

Fig. 2. Oscillated ν_e flux relative to the non-oscillated flux as a function of neutrino energy for the upward-going neutrinos with zenith angle $\cos\theta_\nu = -0.8$. $\bar{\nu}_e$ is not included in the plots. Thin solid lines, dashed lines, and dotted lines correspond to the solar term, the interference term, and the θ_{13} resonance term, respectively. Thick solid lines are total fluxes. Parameters are set as $(\sin^2\theta_{12},\ \sin^2\theta_{13},\ \sin^2\theta_{23},\ \delta_{CP},\ \Delta m_{21}^2,\ \Delta m_{32}^2) = (0.31,\ 0.025,\ 0.6,\ 40°,\ 7.6\times10^{-5}\ \text{eV}^2,\ +2.4\times10^{-3}\ \text{eV}^2)$ unless otherwise noted. The θ_{23} octant effect can be seen by comparing (a) $(\sin^2\theta_{23} = 0.4)$ and (b) $(\sin^2\theta_{23} = 0.6)$. δ_{CP} value is changed to 220° in (c) to be compared with 40° in (b). The mass hierarchy is inverted only in (d) so θ_{13} resonance (MSW) effect disappears in this plot. For the inverted hierarchy the MSW effect should appear in the $\bar{\nu}_e$ flux, which is not shown in the plot. Figure was taken from [11].

imaging detector located in Gifu prefecture of Japan. The detector size is about 40 meter in the height and the diameter, respectively. The total (fiducial) volume is 50 (22.5) kiloton. Approximately 11000 20-inch photo-multipliers (PMTs) are placed in grid on the inner detector wall, and about 1800 8″ PMTs in outer detector for the veto. Super-K has been running almost 20 years from 1996. So far it has been accomplished several remarkable achievements especially in neutrino physics. The observation of the atmospheric neutrino by Super-K covers the wide energy rage from $O(100\ \text{MeV})$ to $O(1\ \text{TeV})$. The neutrino flavor (ν_e or ν_μ) is well reconstructed by the particle identification algorithm using the Cherenkov image pattern.

To study mass hierarchy and other oscillation phenomenon, a fitting analysis is performed to the atmospheric neutrino data with three-flavor oscillation framework. Main motivation of the fitting analysis is to discriminate the state of the mass hierarchy taking the uncertainties of neutrino oscillation parameters into account. To reduce these uncertainties and maximize mass hierarchy sensitivity, several constraints from other experiments are employed in the fit; $\sin^2\theta_{13}$ value is constrained from reactor neutrino experiments, and the constraints from T2K public data are included in a form of additional χ^2; The χ^2 is calculated using the published T2K data and the modeled expectation based on Super-K analysis. It is relatively easy to construct the expectation model since T2K also uses Super-K as a far detector. The fit is done with both normal hierarchy and inverted hierarchy assumptions, respectively, and the minimum χ^2 in each fit are compared. Fig. 3 shows χ^2 difference between two hierarchy cases as a function of oscillation parameter, Δm_{32}^2, $\sin^2\theta_{23}$, δ_{CP}. The χ^2 value becomes slightly smaller for normal hierarchy case which means that the normal hierarchy is preferred. The difference of the χ^2 values is $\Delta\chi^2 = 5.2$, corresponding to the significance of about 2 σ level. The best-fit values of $\sin^2(\theta_{23})$ and δ_{CP} are 0.55 and 4.89, respectively.

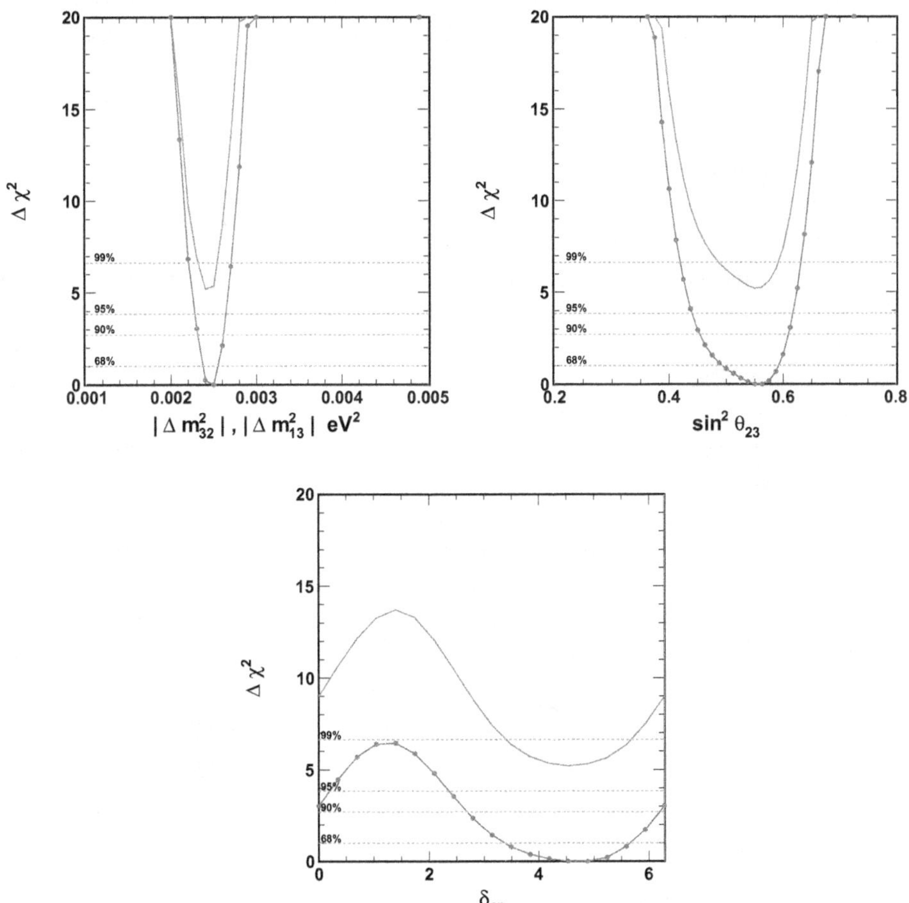

Fig. 3. χ^2 difference ($\Delta\chi^2$) from the global best fit point in the oscillation fits to Super-Kamiokande atmospheric neutrino sample with the constraints of reactor θ_{13} and T2K result. The fit is performed under the assumptions of normal (blue) and inverted (orange) mass hierarchies. The figures show $\Delta\chi^2$ value as a function of $|\Delta m_{32}^2|$ or $|\Delta m_{13}^2|$ (top left), $\sin^2\theta_{23}$ (top right), δ_{CP} (bottom), respectively (color online).

2.4. *Tau Appearance*

Though the disappearance of the upward-going muons is considered to be caused by $\nu_\mu \to \nu_\tau$ oscillation, the direct detection of the oscillation-driven ν_τ is crucial to verify three-flavor oscillation scheme. However the detection of tau is challenging in Super-K because tau production rate is much lower than ν_μ and ν_e interactions due to higher energy threshold of the tau production (3.5 GeV). Also the signal events are affected by many background; the background rate is about two order higher before the selection cut. Though Super-K detected the tau signal in 3σ level[13], they have updated the analysis by increasing the data statistics for further confirmation of the tau appearance.

The strategy for tau appearance is to discriminate signal events which decay into hadrons since the event topology are relatively different from background. A sophisticated discrimination method using neural net (NN) algorithm has been employed with the inputs of several reconstruction variables of signal and background events. The discrimination of the signal events are performed effectively using two dimensional PDFs of NN output and the zenith angle direction; tau signal appears in the area of the upward direction and the large NN output while the background events distributes in all directions with small NN value. The number of the tau events are estimated by the fit using the combined PDFs of signal and background with normalization parameter (α); $PDF_{BG} + \alpha \times PDF_{SIG}$, where $\alpha = 1$ means the number of the observed tau events is consistent with the expectation while $\alpha = 0$ no tau appearance. Fig. 4 shows the distributions of the zenith angle and NN output for tau-like (NN > 0.5) and non-tau-like (NN < 0.5) events. According to the fit result, the observed ν_τ events is estimated to be 338.1 ± 72.7 events (expectation is 224.5 events), which corresponds to 4.6σ significance with normal hierarchy (NH) assumption. The significance is increased to 5.0σ for inverted hierarchy case.

The large ν_τ sample offers the opportunity to measure the charged current (CC) ν_τ cross section in the energies between 3.5 and 70 GeV. The flux averaged cross section is measured to be $(0.94 \pm 0.20) \times 10^{-38}$ cm^2 while the expectation is 0.64×10^{-38} cm^2 (Fig. 5), which is consistent with prediction within 1.5σ. Also the expectation based on the measurement of DONUT experiment, which is scaled down to Super-K sensitive energy by using the theoretical energy dependence of the cross section, is compared and estimated to be smaller than our measurement.

2.5. *Search of Sterile Neutrino*

The three-flavor neutrino oscillation scheme is consistent with a wide range of solar, reactor, accelerator, and atmospheric neutrino experiments. However, some anomalous results, such as those from LSND and MiniBooNE motivate searches beyond three flavor mixing. The possibility of $3 + 1$ neutrino flavors, in which an additional sterile neutrino (ν_s) with $\Delta m_{41}^2 \sim 1$ eV^2 appears, is often discussed in the context of these anomalies.

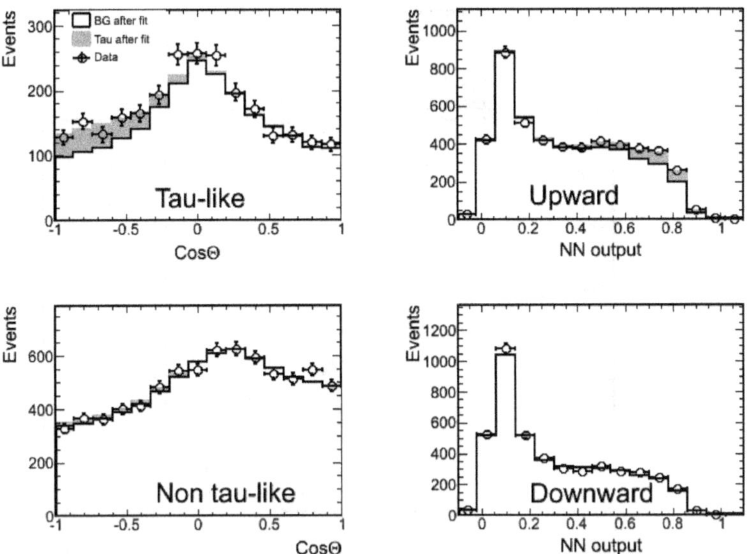

Fig. 4. Zenith angle distributions (left) and neural network (NN) classifier output (right) in tau appearance analysis. The left upper (lower) figure shows the zenith angle distribution of tau-like (non-tau-like) events, with the fitted expectation of the tau signal (shaded) overlaid on the background events. The right upper (lower) figure shows the neutral network output of the selected events in the upward-going (downward-going) sample.

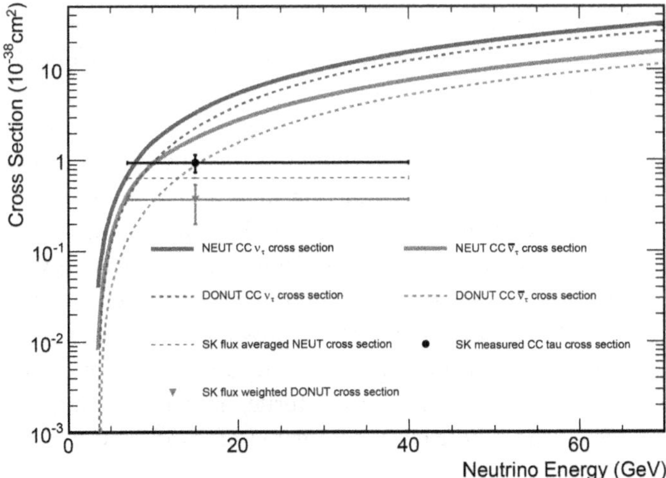

Fig. 5. Measured flux-averaged charged-current tau neutrino cross section (black point), scaled DONUT measurement (magenta), together with theoretical differential cross sections (ν_τ in red and $\bar{\nu}_\tau$ in blue), flux-averaged theoretical cross section (dashed gray). The horizontal bar of the measurement point shows the 90% range of tau neutrino energies in the simulation (color online).

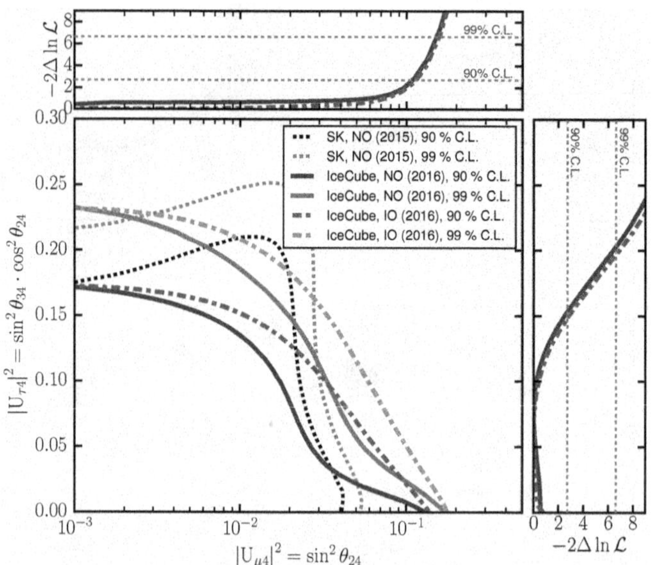

Fig. 6. Exclusion limits set by IceCube/DeepCore sterile study at 90% (dark blue) and 99% C.L. (light blue). The dashed lines show the exclusion from the Super-Kamiokande experiment [14]. The solid lines on top and right axes show the projection of the likelihood on the mixing matrix elements $|U_{\mu 4}|^2$ and $|U_{\tau 4}|^2$, respectively. Figure was taken from [15] (color online).

In the mass region of $\Delta m^2_{41} > 0.1$ eV^2, where the oscillation is rapidly averaged for atmospheric neutrinos, the limits on the mixing matrix element between the sterile and ν_μ/ν_τ neutrinos are given by Super-K [14] and IceCube/DeepCore [15] as shown in Fig. 6, where $U_{\mu 4}$ and $U_{\tau 4}$ correspond to the elements of the extended PMNS matrix between ν_μ/ν_τ and ν_s, respectively.

IceCube also searched for sterile neutrino in the energy range between 320 GeV to 20 TeV [16]. In this energy regime, mixing between active and sterile neutrinos could be enhanced by resonance-like oscillations as they propagate through the Earth, which is caused by MSW mechanism, producing distinctive, energy-dependent, distortions of the measured zenith angle distributions for sterile mass in the range $0.01 < \Delta m^2_{41} < 10$ eV^2. By examining the energy and zenith angle distributions, no such signature of sterile neutrinos was seen in the above mentioned energy range. As a result, a constraint on the parameters of sterile mass and mixing angle between ν_2 and ν_s, $(\Delta m^2_{41}, \theta_{24})$ was produced in the region of $\Delta m^2_{41} \simeq 0.1 \sim 1$ eV^2, as shown in Fig. 7.

3. Proton Decay Results

3.1. *Overview*

Super-K is the world leading experiment in proton decay, and has been explored many decay modes such as anti-lepton plus meson [17,18], ν plus K [19], di-nucleon

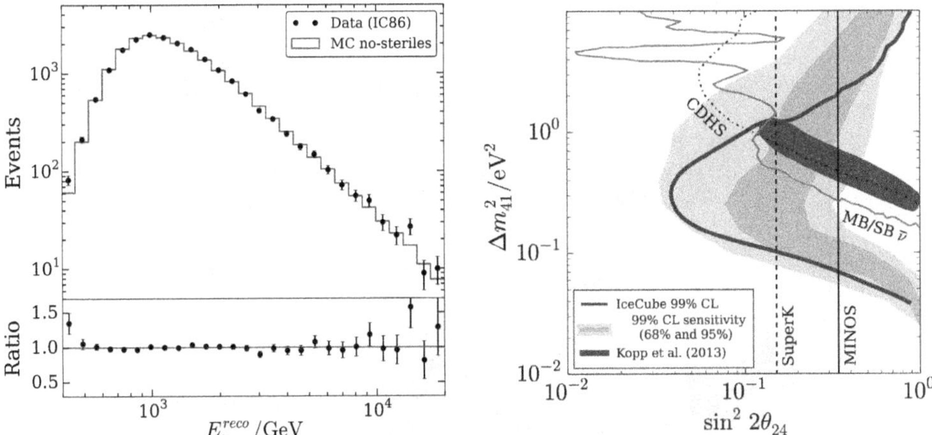

Fig. 7. Left figure: Reconstructed energy distribution in data and Monte Carlo for the no-sterile hypothesis in the analysis. Right figure: 99% C.L. contours for sterile neutrino search in IceCube with bands containing 68% (green) and 95% yellow of 99% contours in sensitivity with the assumption of 3+1 sterile oscillation. 90% C.L. allowed region (blue shaded) from the global analysis and 90% C.L. exclusion from other experiments are also shown. Figures were taken from [16] with modification (color online).

decay, n-n bar oscillation, etc. The lifetime limits set by them exceed by 1-2 orders compared to those of other experiments. Recently the updated result of $p \to \nu K^+$ search was released with the increased data based on the analysis described in [19].

3.2. *Searches of $p \to e^+\pi^0$ and $p \to \mu^+\pi^0$ Decays*

Super-K searched for proton decay via $p \to e^+\pi^0$ and $p \to \mu^+\pi^0$ with 0.306 megaton years exposure [17]. The signal events of these decay channels are possible to be discriminated by reconstructing the total momentum and the invariant mass of e/μ and π^0 particles, as shown in Fig. 8. Also they utilizes the background rejection using neutron tagging in the data of SK-IV phase and reduce to almost half compared to the previous Super-K phases. Lower limits on the proton lifetime are set at $\tau/B(p \to e^+\pi^0) > 1.6 \times 10^{34}$ years and $\tau/B(p \to \mu^+\pi^0) > 7.7 \times 10^{33}$ years at 90% confidence level.

3.3. *Search of $p \to \nu K^+$ Decay*

Two analyses are performed depending on K^+ decay channel produced by $p \to \nu K^+$; one is that K^+ decays into $\nu\mu^+$ (single muon channel), and another is K^+ decays into $\pi^0\pi^+$. The basic strategy of the event selection for single muon channel is to select single-ring muon events in the main timing window, accompanied with two sub events; the μ-e decay signal and the prompt 6 MeV gamma signal from excited oxygen nuclei. This μ-e-γ three-hold coincidence enables the significant reduction of the background events while keeping the moderate signal efficiency. Fig. 9 shows

78

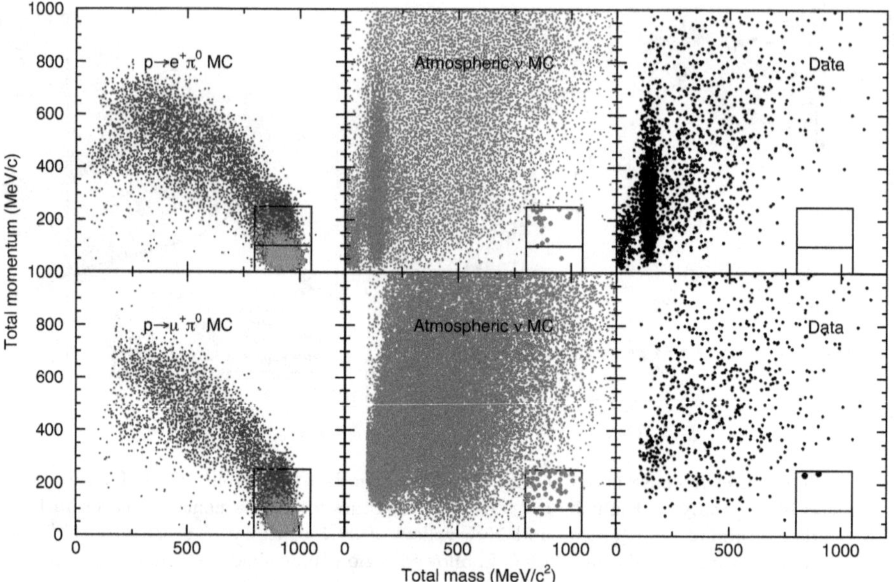

Fig. 8. Reconstructed proton mass vs. total momentum for $p \to e^+\pi^0$ (top) and $p \to \mu^+\pi^0$ (bottom). The left panels show signal MC, where light blue corresponds to free protons and dark blue is bound protons. The middle panels show atmospheric neutrino MC corresponding to 500 years live time of Super-K, and the right panels show SK-I to SK-IV data. The dot size is enlarged in the signal box. Figure was taken from [17] (color online).

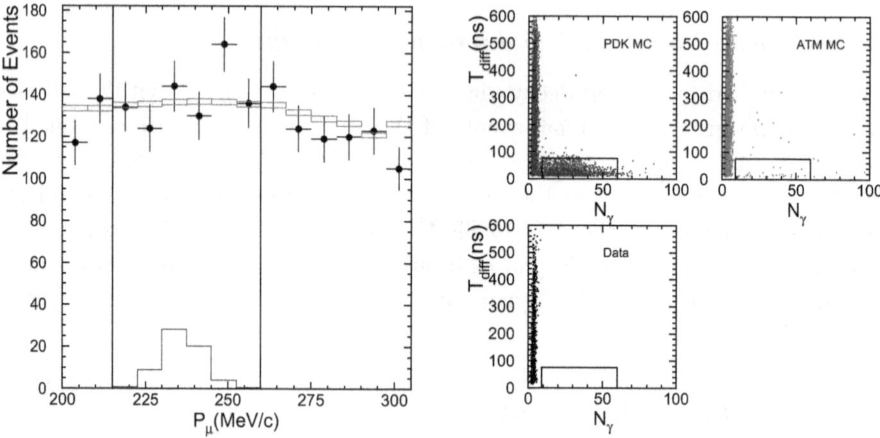

Fig. 9. Left figure: momentum distributions of single muon events in the search of $p \to \nu K^+$ via K^+ decaying into ν plus μ^+ for proton decay MC (blue histogram), data (points with error bar), atmospheric MC (red histogram). Right figure: distributions of timing difference versus number of hits for prompt gamma in $p \to \nu K^+$ analysis for proton decay MC (blue), atmospheric MC (red), and data (black) (color online).

the distributions of selection cuts by the reconstruct muon momentum and the prompt gamma observable. In $\pi^0\pi^+$ channel, $\pi^0 \to 2\gamma$ decay events are selected by the reconstructed mass, and the faint activity due to π^+ in the opposite side of π^0 direction is required (Fig. 10). As a result, no candidate events are observed for both decay channels for 349 kton-year exposure of Super-K I-IV data, corresponding to 8×10^{33} years lower limit of lifetime limit in 90% CL.

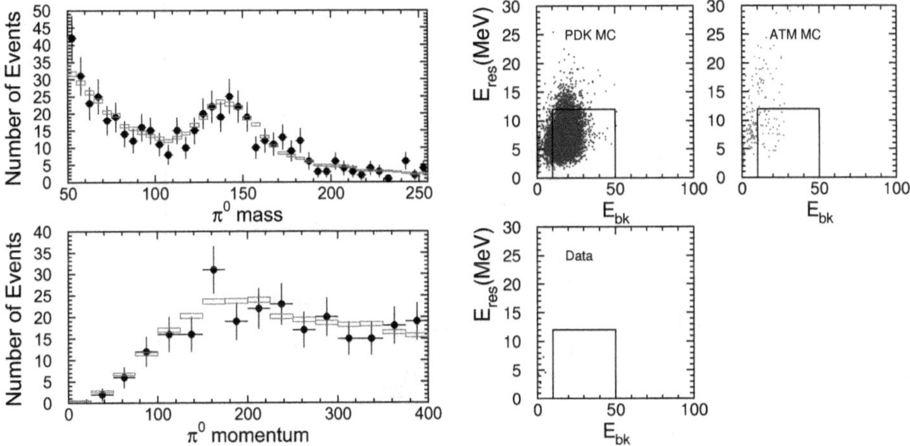

Fig. 10. Left figure: reconstructed π^0 mass and momentum distributions for data (points with error bar) and atmospheric MC (red histogram). Right figure: E_{res} versus E_{bk} distributions for proton decay MC (blue), atmospheric MC (red), and data (black), where E_{res} corresponds to the residual visible energy defined as energy not associated with the π^0 nor the π^+, and E_{bk} the visible energy backward of reconstructed π^0 direction. In the analysis selection criteria of $E_{res} < 12$ MeV and $10 < E_{bk} < 50$ MeV are required (color online).

4. Future Prospects

4.1. *Future of Atmospheric Neutrino Measurement*

Hyper-Kamiokande (Hyper-K) is a proposed water Cherenkov detector designed to study neutrino CP violation using the J-PARC neutrino beam, as well as nucleon decays and atmospheric neutrinos. With a 560 kton fiducial volume it will accumulate atmospheric neutrinos at 25 times the rate of Super-K, and will therefore offer unprecedented access to atmospheric neutrino physics. The expected sensitivity to the mass hierarchy exceeds 3σ for both hierarchy assumptions, when $\theta_{23} > 0.45$, using ten years of atmospheric neutrino data.

DUNE (Deep Underground Neutrino Experiment) is a proposed experiment that will observe both beam and atmospheric neutrinos using a liquid argon time projection chamber (TPC). Though comparatively smaller than Hyper-K at 34 kton, its high-resolution imaging offers possibilities for tagging event features that provide statistical discrimination between neutrinos and antineutrinos. According to simulation studies, taking into account the event rate of neutrino energy and zenith angle, depending on the value of θ_{23}, DUNE will have $3.4{\sim}5\sigma$ sensitivity to the mass hierarchy in ten years of running. The Hyper-K and DUNE sensitivities to the mass hierarchy are shown in Fig. 11.

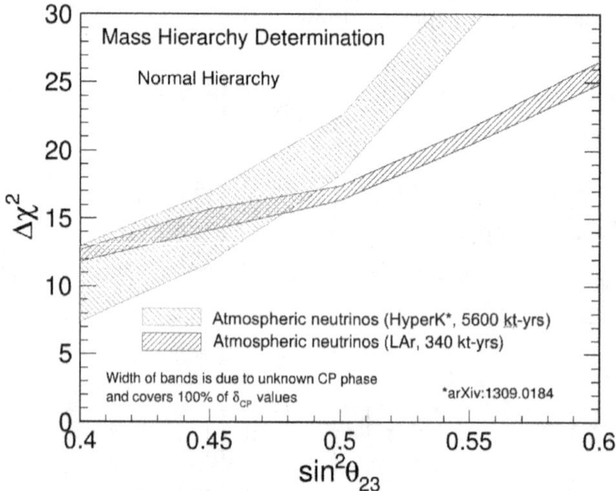

Fig. 11.   Sensitivity of the mass hierarchy determination using atmospheric neutrinos as a function of the true value of θ_{23} for Hyper-K (green) and DUNE (blue), with the assumption of true normal hierarchy. Approximately the significance of the mass hierarchy sensitivity can be estimated by the square of χ^2. Figure was taken from [20] (color online).

The Precision IceCube Next Generation Upgrade (PINGU), which is a new inner detector configuration of IceCube/DeepCore, is currently proposed in order to make precise measurements of atmospheric neutrino oscillation. PINGU will have a higher photodetector density than IceCube/DeepCore in order to lower the energy threshold below 10 GeV. A simulation based on the current detector design suggests the ability to identify the mass hierarchy with 3σ median significance within four years of operation.

ORCA (Oscillations Research with Cosmics in the Abyss) is the planned low-energy branch of KM3NeT, an underwater Cherenkov neutrino detector in the Mediterranean Sea. The detector will be a dense array of multi-PMT digital optical modules (DOMs) using technology developed for KM3NeT. According to a Monte Carlo study, the mass hierarchy is expected to be measured at more than 3σ in three years, assuming at the current world-best-fit values of θ_{23}. The expected sensitivities of mass hierarchy by PINGU and ORCA are shown in Fig. 12.

A 50 kton magnetized iron calorimeter (ICAL) at the India-based Neutrino Observatory (INO) is aiming to use charge discrimination to distinguish between neutrinos and antineutrinos, and could provide 2.2σ sensitivity to the normal mass hierarchy, assuming $\sin^2\theta_{23} = 0.5$, with 10 years of data and a 50 kton detector.

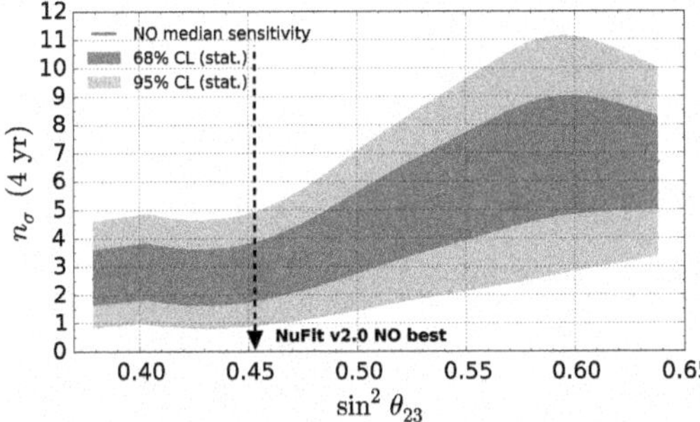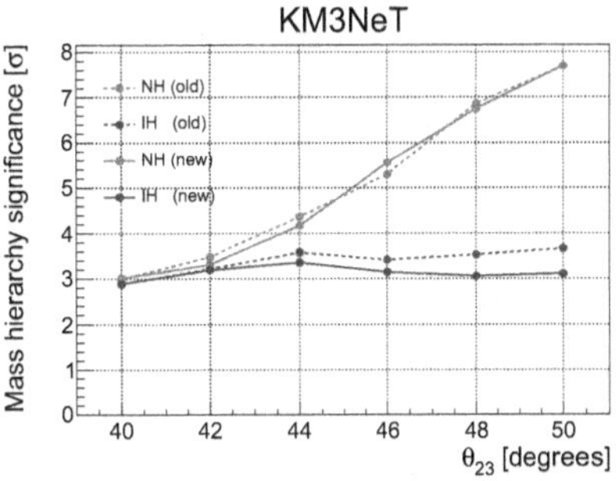

Fig. 12. Expected significance for mass hierarchy detection as a function of the true value of $\sin^2(\theta_{23})$ for PINGU (left) and ORCA (right). In left figure, four years of PINGU data are assumed. Solid red (normal) and blue (inverted) lines show median significances, while the green and yellow bands indicate the range of significances obtained in 68% and 95% of hypothetical experiments. In right figure, Right: Comparison of the mass hierarchy sensitivity calculated using the old method (dashed lines) and the new method (solid lines). In the former, the significance is calculated to reject the other hierarchy at the same θ_{23}, whereas in the latter the alternative hypothesis has a different θ_{23}. The differences are rather small, but there is a noticeable decrease in the second octant IH mass hierarchy sensitivity. This is for the 9 m spacing and three years of operation time, using the default settings (particularly, $\delta_{CP,true} = 0°$) (color online).

4.2. *Future of Proton Decay Search*

The Hyper-K is expected to improve the proton decay sensitivity compared to Super-K's generally by an order of magnitude or more mainly due to the increased fiducial volume. The expected sensitivity of $p \to e^+\pi^0$ channel, will reach to 10^{35} years by ten years of the exposure. The search of $p \to \nu K^+$ decay channel is interesting for DUNE since the liquid argon TPC detector could identify the K^+ track with high efficiency. In addition, many final states of K^+ decay would be fully reconstructible in the detector.

The Jiangmen Underground Neutrino Observatory (JUNO), a 20 kton multi-purpose underground liquid scintillator detector, was proposed for the determination of the neutrino mass hierarchy via the reactor neutrino oscillation measurement as the primary physics goal. The detector will have the good energy resolution due to the large amount of the scintillation lights and the large fiducial volume. The JUNO detector is also sensitive to $p \to \nu K^+$ proton decay in which K^+ decays into μ^+ plus ν_μ. Thanks to the lower threshold of the scintillaton lights, K^+ is visible for JUNO detector. By taking three-hold coincidence of K^+, μ^+, and e^+, as shown in Fig. 13, they can detect $p \to \nu K^+$ events with high efficiency of 64% while the background event rate, which is mainly due to neutrino induced K^+ production, is very low ($\sim$0.05 events per year).

The expected sensitivities of the proton decay detection for $p \to e^+\pi^0$ and $p \to \bar{\nu}K^+$ decay channels are compared in Fig. 14. Hyper-K's sensitivity exceeds DUNE for $p \to e^+\pi^0$ channel. The sensitivity of $p \to \nu K^+$ channel could be reached more than 10^{34} years by Hyper-K, JUNO, and DUNE.

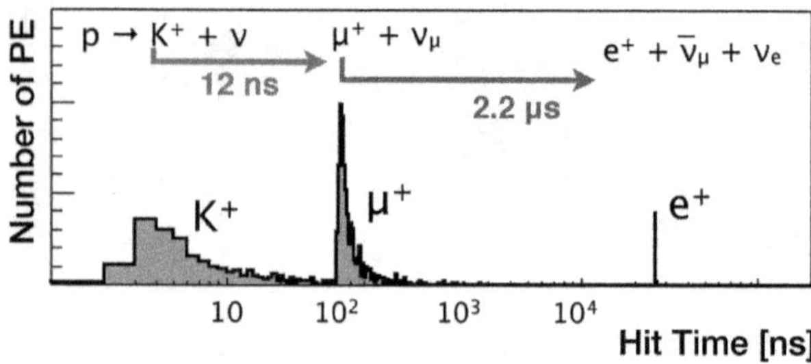

Fig. 13. The simulated hit time distribution of photoelectrons (PEs) from a $K^+ \to \mu^+\nu_\mu$ event at JUNO.

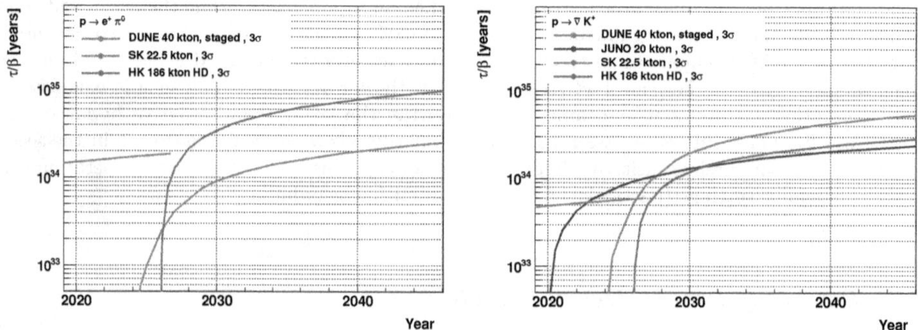

Fig. 14. Three sigma detection sensitivities for $p \to e^+\pi^0$ (left) and $p \to \bar{\nu}K^+$ (right) decay channels as a function of the year.

5. Concluding Remarks

We have reviewed the recent results and the future prospects of atmospheric neutrino measurement and proton decay searches in this paper. Atmospheric neutrino provides us the opportunity to investigate three flavor mixing scheme in various ways. The oscillation probability of $\nu_\mu \to \nu_e$ channel is expected to be enhanced by the matter effect for neutrinos which transverse inside the Earth in multi-GeV region. Thanks to mass hierarchy dependence of the resonance oscillation, atmospheric neutrino neutrino is sensitive to mass hierarchy. According to the oscillation fit result to the observed data with reactor and T2K constraints, Super-K data weakly prefers normal hierarchy. The significance of the tau appearance signal has been improved to 4.6σ mainly due to the increase of the exposure time. The larger ν_τ cross section than the prediction is measured though still consistent with theoretical prediction. The proton decay is an unique method to probe GUT theory. Super-K has been contributing to explore the possibility of the beyond Standard Model physics. The search of $p \to \nu K^+$ decay has been updated, and there are no candidate events observed from the searches of two K^+ decay channels. The lower lifetime limit has been improved to 8.0×10^{33} years. In future several experiments such as Hyper-K, DUNE, PINGU, ORCA, INO, have detection potential for neutrino mass hierarchy with more than 3σ. The search for proton decay is promising to provide about 10 times improvement by Hyper-K, DUNE, and JUNO detectors.

References

1. E. Richard *et al.*, Measurements of the atmospheric neutrino flux by Super-Kamiokande: energy spectra, geomagnetic effects, and solar modulation, *Phys. Rev.* **D94**, p. 052001 (2016).
2. K. Daum, Determination of the atmospheric neutrino spectra with the frejus detector, *Z. Phys.* **C66**, 417 (1995).
3. R. Abbasi *et al.*, Determination of the Atmospheric Neutrino Flux and Searches for New Physics with AMANDA-II, *Phys. Rev.* **D79**, p. 102005 (2009).
4. R. Abbasi *et al.*, The Energy Spectrum of Atmospheric Neutrinos between 2 and 200 TeV with the AMANDA-II Detector, *Astropart. Phys.* **34**, 48 (2010).
5. R. Abbasi *et al.*, Measurement of the atmospheric neutrino energy spectrum from 100 gev to 400 tev with icecube, *Phys. Rev.* **D83**, p. 012001 (2011).
6. R. Abbasi *et al.*, A Search for a Diffuse Flux of Astrophysical Muon Neutrinos with the IceCube 40-String Detector, *Phys. Rev.* **D84**, p. 082001 (2011).
7. M. Aartsen *et al.*, Measurement of the atmospheric flux in icecube, *Phys. Rev. Lett.* **110**, p. 151105 (Apr 2013).
8. M. G. Aartsen *et al.*, Measurement of the Atmospheric ν_e Spectrum with IceCube, *Phys. Rev.* **D91**, p. 122004 (2015).
9. S. Adrian-Martinez *et al.*, Measurement of the atmospheric ν_μ energy spectrum from 100 GeV to 200 TeV with the ANTARES telescope, *Eur. Phys. J.* **C73**, p. 2606 (2013), [Eur. Phys. J. C73, 2606(2013)].
10. M. Honda, T. Kajita, K. Kasahara and S. Midorikawa, Improvement of low energy atmospheric neutrino flux calculation using the jam nuclear interaction model, *Phys. Rev. D* **83**, p. 123001 (Jun 2011).
11. K. Abe *et al.*, Letter of Intent: The Hyper-Kamiokande Experiment — Detector Design and Physics Potential — (2011).
12. Y. Fukuda *et al.*, The Super-Kamiokande detector, *Nucl. Instrum. Meth.* **A501**, 418 (2003).
13. K. Abe *et al.*, Evidence for the Appearance of Atmospheric Tau Neutrinos in Super-Kamiokande, *Phys. Rev. Lett.* **110**, p. 181802 (2013).
14. K. Abe *et al.*, Limits on sterile neutrino mixing using atmospheric neutrinos in Super-Kamiokande, *Phys. Rev.* **D91**, p. 052019 (2015).
15. M. G. Aartsen *et al.*, Search for sterile neutrino mixing using three years of IceCube DeepCore data, *Phys. Rev.* **D95**, p. 112002 (2017).
16. M. G. Aartsen *et al.*, Searches for Sterile Neutrinos with the IceCube Detector, *Phys. Rev. Lett.* **117**, p. 071801 (2016).
17. K. Abe *et al.*, Search for proton decay via $p \rightarrow e^+\pi^0$ and $p \rightarrow \mu^+\pi^0$ in 0.31 megaton·years exposure of the Super-Kamiokande water Cherenkov detector, *Phys. Rev.* **D95**, p. 012004 (2017).
18. K. Abe *et al.*, Search for nucleon decay into charged antilepton plus meson in 0.316 megaton·years exposure of the Super-Kamiokande water Cherenkov detector, *Phys. Rev.* **D96**, p. 012003 (2017).
19. K. Abe *et al.*, Search for proton decay via $p \rightarrow \nu K^+$ using 260 kiloton·year data of Super-Kamiokande, *Phys. Rev.* **D90**, p. 072005 (2014).
20. R. Acciarri *et al.*, Long-Baseline Neutrino Facility (LBNF) and Deep Underground Neutrino Experiment (DUNE) (2015).

Short and Medium Baseline Reactor Neutrino Experiments

Marcos Dracos

IPHC, Université de Strasbourg, CNRS/IN2P3,
F-67037 Strasbourg, France
E-mail: marcos.dracos@in2p3.fr

Reactor neutrino experiments allowed during the last years to observe that the last unmeasured neutrino oscillation mixing angle θ_{13} not only was different from zero but also this angle was relatively large and just below the previous limits. Some anomalies have then been observed related to the expectations on reactor neutrino flux and to the shape of the neutrino spectrum. These anomalies are still under investigation by very short baseline reactor neutrino experiments. The observed large value of θ_{13} allows now the observation of an eventual CP violation in the leptonic sector, but also to be sensitive to the neutrino mass hierarchy using medium baseline reactor neutrino experiments.

Keywords: Neutrino oscillations; mass hierarchy; reactor neutrinos; short baseline; medium baseline.

1. Introduction

Nuclear reactors are copious neutrino sources used to study the neutrino properties. The neutrino detection method used since the first observation of neutrinos by F. Reines and C. Cowan in 1956 remains the same up to today. This method uses the inverse beta decay (IBD) of an antineutrino interacting with a proton of the detector target producing a positron annihilating with an electron (prompt signal) and a neutron captured after (delayed signal).

Up to 2012, reactor neutrino experiments were using only one "far" detector to extract their results. The unoscillated neutrino spectrum was extracted using the reactor predictions taking into account the reactor operation conditions. In order to reduce the systematic errors, new generation experiments as Double Chooz, Daya Bay and RENO, used near detectors to monitor the reactor neutrino flux. This allowed to measure for the first time the last neutrino oscillation mixing angle θ_{13} [1-3]. The measurement accuracy achieved up to now by these experiments allow to perform high precision measurements.

2. Short Baseline Neutrino Experiments

By 2011, three short baseline reactor neutrino experiments started taking data with the main objective to measure the mixing angle θ_{13} or at least set a severe limit on this parameter. These experiments were Double Chooz in France [1], Daya Bay [2] in China and RENO [3] in South Korea. The configuration of detectors and reactors for the three projects is depicted by Fig. 1. Double Chooz disposes of two detectors and two reactors, Daya Bay of three reactors and three detector sides, and RENO of six reactors and two detectors clearly identified as near and far detectors. Table 1 gives the present reactor power, the detector volume, overburden and starting date of the experiments.

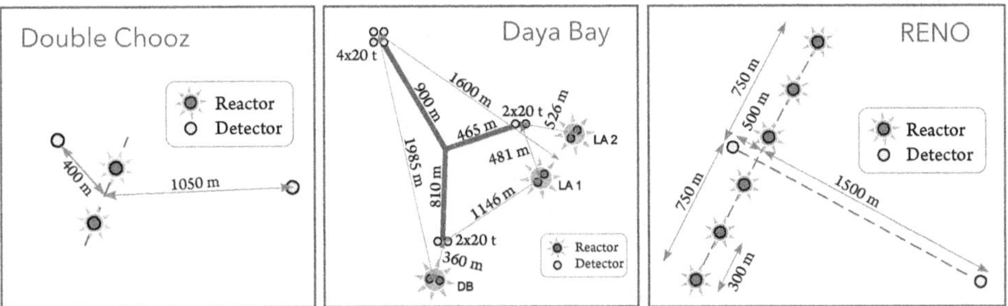

Fig. 1.   Reactor and detector configuration for Double Chooz, Daya Bay and RENO experiments.

Table 1.   Reactor power, detector volume, overburden and starting date of the experiment.

Experiment	Reactor power (GWth)	Volume (tons)	Overburden (near/far-mwe)	Start of data taking
Double Chooz	8.5	2x6	80/300	April 2011
Daya Bay	17.4	8x20	270/950	August 2011
RENO	16.8	2xyy	90/440	August 2011

The detectors of all three experiments are similar. They mainly include the liquid scintillator target doped with Gadolinium to reduce the neutron lifetime to about 30 μs, a gamma catcher filled with liquid scintillator without Gadolinium to detect escaping gammas from the neutrino target, a buffer protecting from the PMT radioactivity and a veto for cosmic ray rejection. For each experiment, the detectors on all locations are identical in order to reduce the systematic errors. Fig. 2 presents the far Double Chooz detector.

3. θ_{13} Measurement

In 2011 all three experiments, Double Chooz, Daya Bay and RENO, have started taking data. Double Chooz has started with only the far detector (the near detector was only ready beginning of 2015), while Daya Bay was disposing of six detectors over 8. RENO had started with a complete configuration. End of 2011, Double Chooz has presented the first results[1] with a significance of excluding a zero θ_{13} value of 1.7σ. It has been followed by Daya Bay (March 2012)[2] and RENO (April 2012)[3] reporting a non-zero significance of 5.2σ and 4.9σ, respectively. Fig. 3 shows the neutrino energy spectrum of the three experiments at that time.

86

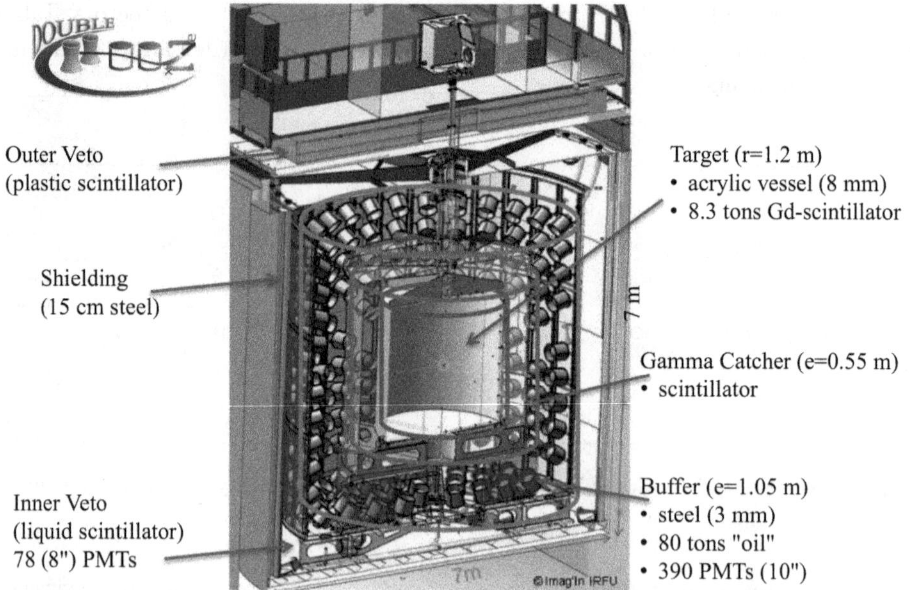

Fig. 2.   The Far Double Chooz detector.

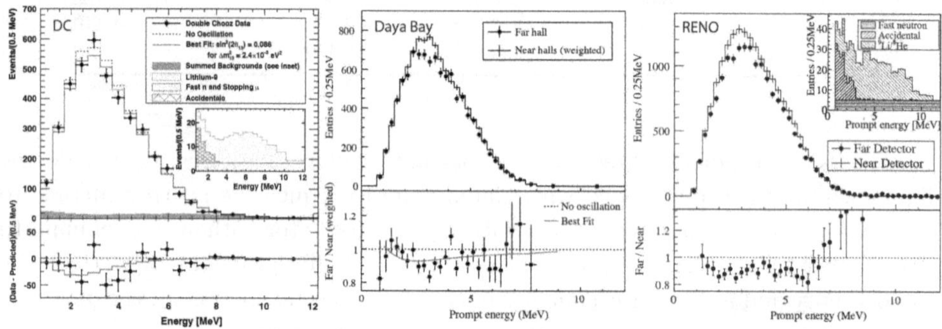

Fig. 3.   Observed energy spectrum of the prompt signal for Double Chooz, Daya Bay and RENO experiments as published in 2012.

The latest results for the three experiments for $\sin^2 2\theta_{13}$ give:

- Double Chooz: 0.119 ± 0.016 (*stat.* + *syst.*) (using data collected up to 2016)[4]
- Daya Bay: 0.0841 ± 0.0027 (*stat.*) ± 0.0019 (*syst.*) (using data collected up to July 2015)[5]
- RENO: 0.086 ± 0.006 (*stat.*) ± 0.005 (*syst.*) (using data collected up to September 2015)[6]

As it can be seen, Daya Bay already reached a 3.9% accuracy on θ_{13} while RENO and Double Chooz remain at 9.1% and 13.4%, respectively. Fig. 4 shows the dependence of the $\bar{\nu}$ disappearance probability on L/E for Daya Bay and RENO. The observed shape is as expected by the neutrino oscillation formalism. By fitting this distribution the parameter $\Delta m_{ee}^2 = \cos^2\theta_{12}|\Delta m_{21}^2| + \sin^2\theta_{12}|\Delta m_{32}^2|$ can directly be extracted from the probability distribution:

$$P(\bar{\nu}_e \to \bar{\nu}_e) \approx 1 - \underbrace{\cos^4\theta_{13}\sin^2\theta_{12}\sin^2\left(\frac{\Delta m_{21}^2 L}{4E_\nu}\right)}_{\text{small for L=few km}} - \sin^2 2\theta_{13}\sin^2\left(\frac{\Delta m_{ee}^2 L}{4E_\nu}\right)$$

The extracted values for Δm_{ee}^2 are[5,6]:

- Daya Bay: $[2.50 \pm 0.06 \; (stat.) \pm 0.06 \; (syst.)] \times 10^{-3} \text{eV}^2$
- RENO: $[2.61 \pm^{+0.06}_{-0.16} \; (stat.) \pm^{+0.09}_{-0.09} \; (syst.)] \times 10^{-3} \text{eV}^2$

Using the Daya Bay value one can get:

$$\Delta m_{32}^2 = [2.45 \pm +0.06 \; (stat.) \pm +0.06 \; (syst.)] \times 10^{-3} \text{eV}^2 \text{ for Normal Hierarchy}$$
$$= [2.56 \pm +0.06 \; (stat.) \pm +0.06 \; (syst.)] \times 10^{-3} \text{eV}^2 \text{ for Inverted Hierarchy}$$

with a world average of $[2.52 \pm +0.04] \times 10^{-3}\text{eV}^2$ for Normal Hierarchy. It has to be noted that for Daya Bay the systematic error is of the order of the statistical one not leaving room for improvement without further reduction of the systematic errors.

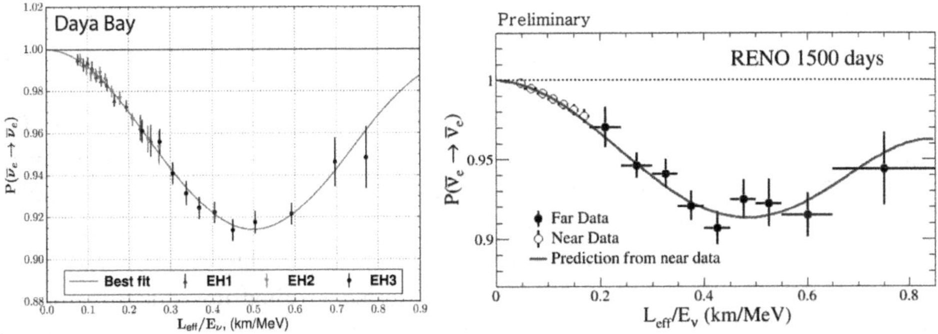

Fig. 4.   Reactor antineutrino survival probability as a function of L_{eff}/E_ν for Daya Bay and RENO experiments.

Fig. 5 presents $\sin^2 2\theta_{13}$ versus Δm_{ee}^2 for Daya Bay[5] and RENO[6] experiments.

All these precision results have been obtained using an extensive calibration program mainly based on radioactive sources deployed inside the detectors. A better background reduction and understanding of the systematic errors also helped. For

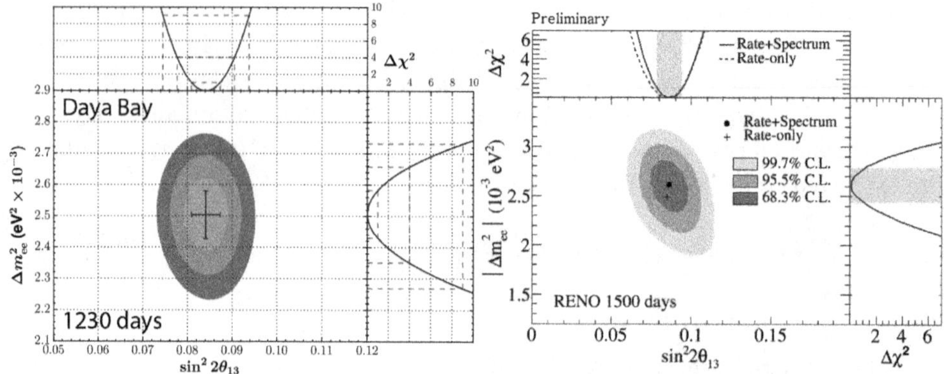

Fig. 5. $\sin^2 2\theta_{13}$ versus Δm_{ee}^2 for Daya Bay and RENO experiments.

Daya Bay, the remaining uncertainty on energy reconstruction is kept at the level of 1%. For this experiment, the energy resolution is given by:

$$\frac{\sigma_E}{E_{rec}} = \sqrt{0.016^2 + \frac{0.081^2}{E_{rec}} + \frac{0.026^2}{E_{rec}^2}}$$

giving an energy resolution of 8.7% at 1 MeV.

4. Reactor Neutrino Anomalies

With the achieved energy resolution Double Chooz, Daya Bay and RENO are able to better measure the reactor neutrino energy spectrum. All three experiments reported an excess around 5 MeV compared to predictions[6-8]. Fig. 6 presents the neutrino energy distribution compared to predictions for the three experiments. The excess around 5 MeV depends on the reactor thermal power excluding an external origine. Some possible explanations already have been presented in several publications (see e.g.[9]) based on β–decay branch contribution (thousands of individual beta–decay branches) of the main isotopes contributing to the neutrino production. This bump could also be explained by considering residual non-linearities of the experiments[10]. Further investigations are ongoing but this shows the fragility of the predicted spectrum.

On reactor antineutrino anomaly, the latest results from Daya Bay[7] and RENO[11] give for the antineutrino flux compared to the Huber[13]/Mueller[12] model (Fig. 7):

- Daya Bay: $R = 0.946 \pm 0.020$
- RENO: $R = 0.946 \pm 0.021$

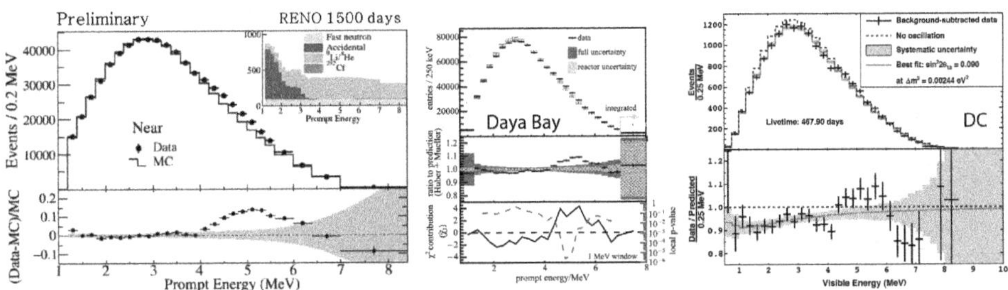

Fig. 6.   Measured neutrino energy spectrum compared to predictions for RENO, Daya Bay and Double Chooz.

while the world average including Daya Bay[7] results becomes $R = 0.9430 \pm 0.008(\text{exp.}) \pm 0.023(\text{model})$. The deficit of observed reactor neutrino flux relative to the prediction of about 6% indicates an overestimated flux or possible oscillations to sterile neutrinos.

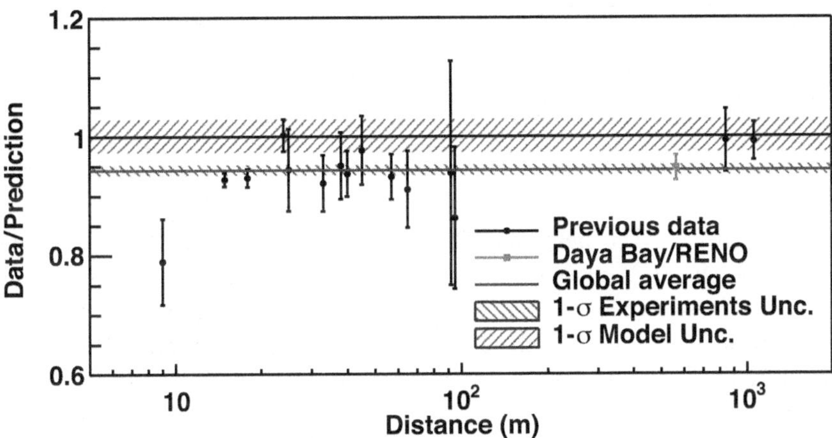

Fig. 7.   Neutrino flux measurements versus the baseline compared to predictions.

Daya Bay has studied the reactor neutrino flux and spectrum evolution with the reactor fuel isotope composition[14] using 2.2 million IBDs detected by four of its detectors. Fig. 8 shows the relative IDB yield per fission as a function of ^{239}Pu and ^{235}U fractions, for four neutrino energy ranges. A clear energy–dependent evolution is observed that induce a change in spectral shape as fuel composition evolves.

Fig. 9 presents the neutrino cross-section/fission as a function of the effective fission fraction of ^{239}Pu and ^{235}U, compared to the Huber–Mueller model prediction (rescaled by 5.1% just for better slope comparison). The hypothesis of a constant antineutrino flux as a function of the ^{239}Pu fission fraction is rejected with a significance corresponding to 10σ. The measured evolution of the reactor neutrino flux is by 3.1σ incompatible with the predictions. This indicates that the uncertainties of model calculations are underestimated. For this study, only the four primary fission isotopes, ^{235}U, ^{238}U ^{239}Pu and ^{241}Pu, are taken into account representing more than 99.7% of the total fissions.

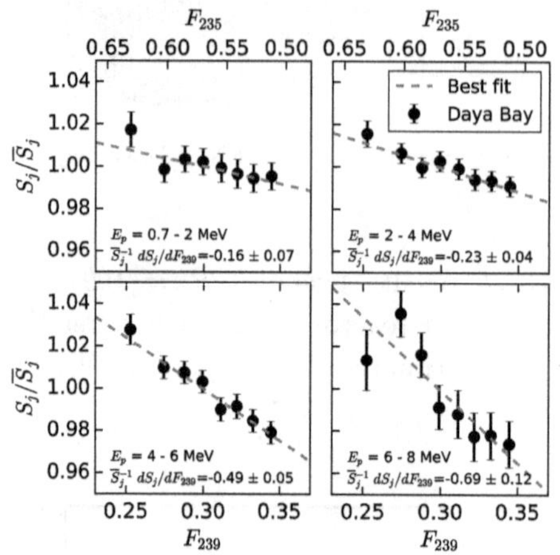

Fig. 8. Relative neutrino flux measurements versus the ^{235}U and ^{239}Pu fission fractions for four energy ranges.

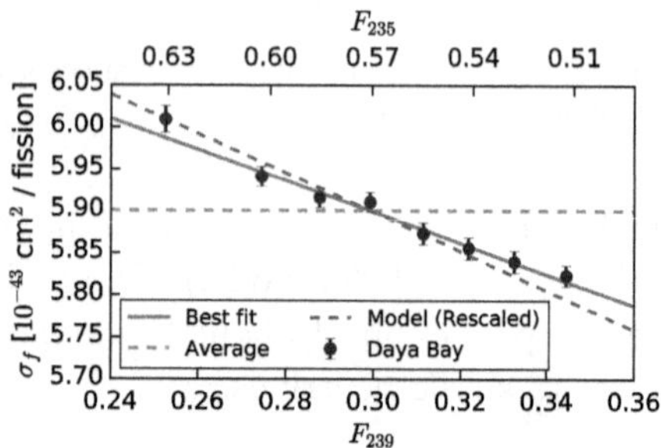

Fig. 9. Measured IBD yield versus the ^{235}U and ^{239}Pu fission fractions compared to predictions.

Fig. 10 presents the neutrino IBD yield per fission of ^{239}Pu versus the one of ^{235}U. It is clearly seen that the disagreement with the model comes from the ^{235}U evolution. A possible explanation, with 2.8σ significance, is that there is an overestimation of the antineutrino flux from ^{235}U in reactor models, instead of an anomaly produced by sterile neutrinos. Tentatives by independent analyses to combine these new Daya Bay data with all previous data coming from many other experiments have the tendency to decrease the importance of ^{235}U isotope effect[15]. However, this new Daya Bay observation constitutes an unaccounted systematic error by the reactor models.

Daya Bay doesn't report indications that the 5 MeV "bump" could come from a particular isotope. Improvements in systematic uncertainties and increasing statistics could help to better understand this particular problem.

Using the spectrum of L/E of Fig. 4 and the probability of having a 4th sterile neutrino family:

$$P(\bar{\nu}_e \to \bar{\nu}_e) \approx 1 - \cos^4\theta_{14}\sin^2\theta_{13}\sin^2\left(\frac{\Delta m_{ee}^2 L}{4E_\nu}\right) - \sin^2 2\theta_{14}\sin^2\left(\frac{\Delta m_{41}^2 L}{4E_\nu}\right)$$

exclusion limits can be established. Fig. 11 shows the exclusion limits obtained by RENO experiment[11] and the one obtained combining Daya Bay[16] results with those of MINOS and Bugey–3.

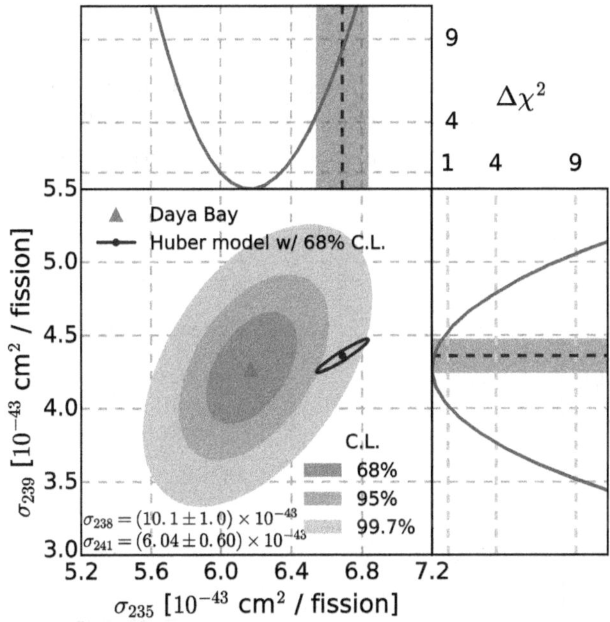

Fig. 10. IBD yield/fission for ^{235}U and ^{239}Pu compared to predictions.

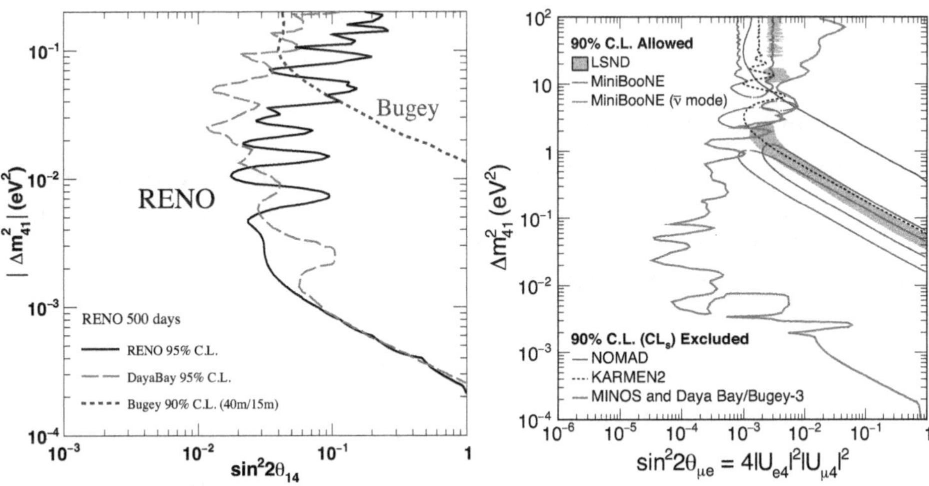

Fig. 11. Exclusion plots for sterile neutrino searches.

5. Plans and Prospects

Double Chooz will continue data taking up to beginning of 2018 after which the detector decommissioning will start. The experiment will profite of this phase to well measure the number of protons in each part of its detectors in order to decrease the systematic errors. RENO will take data up to end of 2018 with a possible extension up to 2021. The goal will be to reach a precision of 6% on θ_{13}. Daya Bay plans to take data up to 2020. The goal is to reach a 3% systematic error on θ_{13} by better understanding of the systematic errors (non–linearities, calibration,...) and better background reduction. Fig. 12[17] summarises the uncertainty evolution on θ_{13} and $|\Delta_{ee}^2|$.

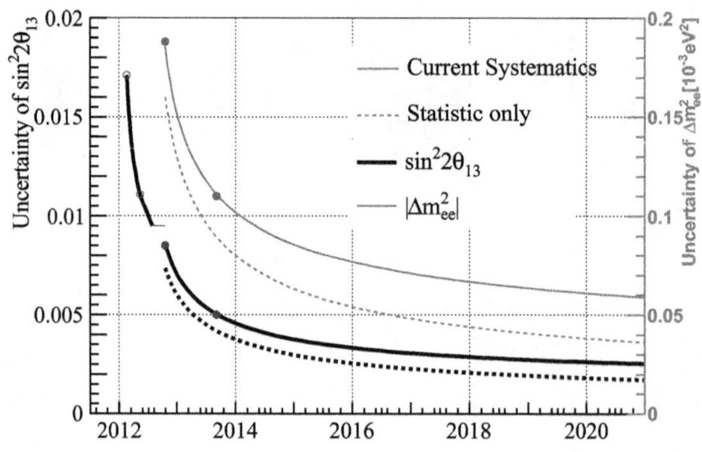

Fig. 12.   Daya Bay uncertainty evolution.

6. Medium Baseline Experiments

Within the context of relatively large θ_{13}, medium baseline reactor experiments could determine the neutrino mass hierarchy, provided that they will have a sufficient (and challenging) energy resolution[18–22]. Indeed, the survival probability of reactor neutrinos can be written as:

$$P\left(\bar{\nu}_e \rightarrow \bar{\nu}_e\right) \approx 1 - \cos^4\theta_{13}\sin^2 2\theta_{12}\sin^2\left(\frac{\Delta m_{21}^2 L}{4E_\nu}\right)$$

$$- \sin^2 2\theta_{13}\sin^2\left(\frac{\Delta m_{31}^2 L}{4E_\nu}\right)$$

$$- \sin^2\theta_{12}\sin^2 2\theta_{13}\sin^2\left(\frac{\Delta m_{21}^2 L}{4E_\nu}\right)\cos\left(\frac{2\left|\Delta m_{31}^2\right| L}{4E_\nu}\right)$$

$$\pm \frac{\sin^2\theta_{12}}{2}\sin^2 2\theta_{13}\sin\left(\frac{2\Delta m_{21}^2 L}{4E_\nu}\right)\sin\left(\frac{2\left|\Delta m_{31}^2\right| L}{4E_\nu}\right)$$

The $\pm$ at the fourth term is for Normal and Inverted hierarchy. It has to be noted that this probability doesn't depend on the CP violation parameter δ_{CP}, also the matter effect for the considered baselines is negligible.

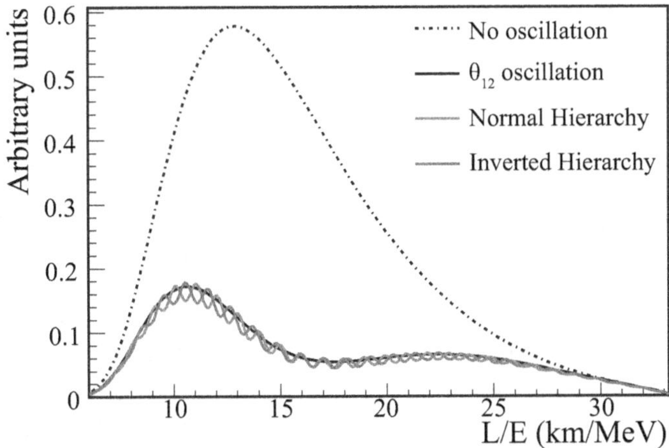

Fig. 13.   Neutrino energy spectrum for both heirarchies compared to the unoscillated spectrum.

The experiments must be placed at a distance to maximise the contribution of this last term in order to be significantly sensitive to the neutrino mass hierarchy. It comes out that this distance must be around 50 km. Fig. 13 presents the oscillated neutrino energy spectrum for the two hierarchies compared to the unoscillated one. It can be clearly seen that a very good energy resolution is needed to discriminate the two cases.

Two projects propose to exploit this effect, JUNO[23] in China and RENO50[24] in South Korea. Both experiments proposes to use a large ($\sim$20 kt) liquid scintillator detectors. Thanks to the size of these detectors, other physics subjects can be treated together with the reactor neutrino program, as the detection of supernova explosions and relic neutrinos, solar and atmospheric neutrinos, geoneutrinos, proton lifetime and exotic searches.

6.1. *JUNO*

The JUNO detector will be located in the South of China near Jiangmen at a distance of 53 km from the nuclear power plants of Yangjiang and Taishan. The thermal power of these two nuclear power plants is expected to be between 26 GWth and 35 GWth by the beginning of next decade. JUNO plans to collect 100 kIBDs over 6 years to determine the neutrino mass hierarchy.

Fig. 14 illustrates the observed neutrino spectrum which could be observed by JUNO together with the main background contributions. Thanks to the ability to observe the "solar" and "atmospheric" oscillation contributions in the same detector, JUNO will be able to precisely measure the oscillation parameters Δm^2_{21}, Δm^2_{ee} and $\sin^2 \theta_{12}$. Table 2 presents the JUNO performance for these parameters compared to the current accuracy. A sub–percent precision on all three parameters is expected.

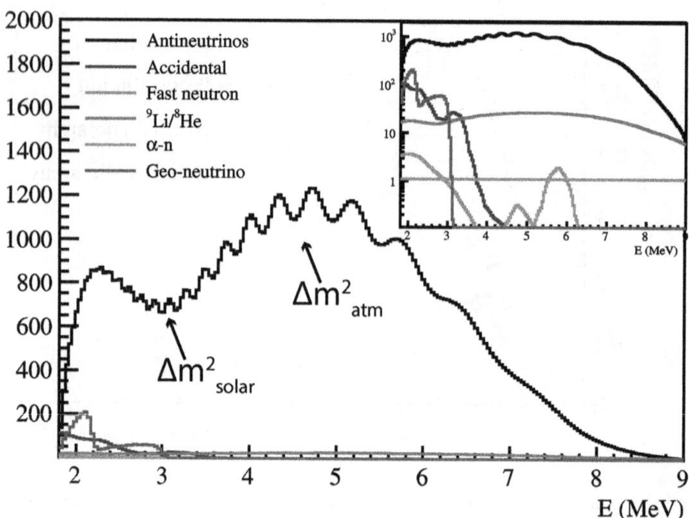

Fig. 14. Neutrino energy spectrum with main background contributions.

Table 2. Precision measurements with the JUNO detector.

	Current precision	JUNO precision
Δm^2_{21}	2.3%	0.6%
Δm^2_{ee}	4%	0.5%
$\sin^2 \theta_{12}$	6%	0.7%

For the mass hierarchy determination, JUNO has to reach an energy resolution of at leat 3% at 1 MeV. For this, JUNO detector must be able to detect more than 1200 photoelectrons par deposited MeV. To achieve this, JUNO, compared to previous liquid scintillator detectors, plans to optimise the liquid scintillator quality and composition[25]. The photocathode coverage will also considerably increase to reach a value of nearly 80%. The two last points to be improved is the photomultiplier quantum efficiency and the calibration of the detector.

6.1.1. *The JUNO detector*

The JUNO central detector will be spherical compared to the cylindrical shape of SuperKamiokande. A schematic view of the JUNO detector is depicted by Fig. 15. The neutrino interactions will occur inside an acrylic sphere (12 cm thick) immersed in water (35 kt) and containing the liquid scintillator and having a diameter of 35.4 m. Outside this sphere a mechanical structure will support all 20″ PMTs (about 17000). In between the 20″ PMTs small PMTs (3″) will be placed to mainly increase the energy dynamic range of the whole system. All the surrounding water

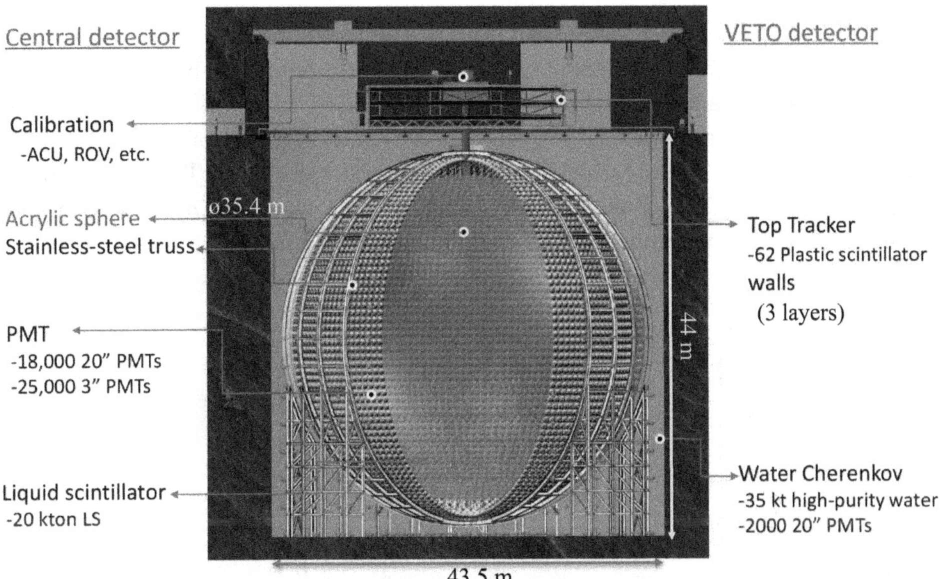

Fig. 15.   Schematic view of the JUNO detector.

will be used as Water Cherenkov veto detector with about 2000 20″ PMTs. In order to compensate the earth magnetic field significantly degrading the PMT detection performance, two magnetic coils, a vertical and horizontal ones, will be placed inside the veto detector. On top of the central detector, a Top Tracker will be placed in order to well study the cosmogenic background. This Top Tracker based on plastic scintillator strips consists of a recycling of the OPERA Target Tracker[26].

All parts of the detector are in an R&D period or in a construction phase. The 20″ PMTs will be provided by NNVT (15000) and Hamamatsu (5000) companies. The HZC company will provide all the 25000 3″ PMTs. The delivery and tests of the 20″ PMTs have already started. A distillation plant has been installed in the Daya Bay site for tests in order to find the best way to produce the JUNO liquid scintillation. Very promising results have already been obtained up to now with an attenuation length higher than 22 m (20 m required).

Taking as example Daya Bay, JUNO has elaborated a very extensive calibration program to well calibrate all the detector positions at various energies. This includes displacement of radioactive sources with ultrasonic positioning system for energy calibration, and fast 1 ns laser system for timing calibration.

The excavations of the underground laboratory to host the JUNO detector have started beginning of 2015. A 1300 m length slope tunnel to reach the detector (overburden of 700 m of rock) and a vertical shaft are already excavated. The civil engineering will finish by beginning of 2019 when the installation of the detector will start. The data taking period is supposed to start beginning of next decade.

6.1.2. *Mass hierarchy performance*

The JUNO performance to solve the neutrino mass hierarchy problem using the above described detector is shown by Fig. 16. This Fig. shows the χ^2 difference between the two hierarchies versus the parameter $|\Delta_{ee}^2|$ in case of normal hierarchy (similar performance is obtained in case of inverted hierarchy). This $\Delta\chi^2$ is of the order of 11 ($\sigma \sim 3.3$) in real conditions and could go up to 16 ($\sigma \sim 4$) in case the dispersion of the reactors in both nuclear plants is ignored.

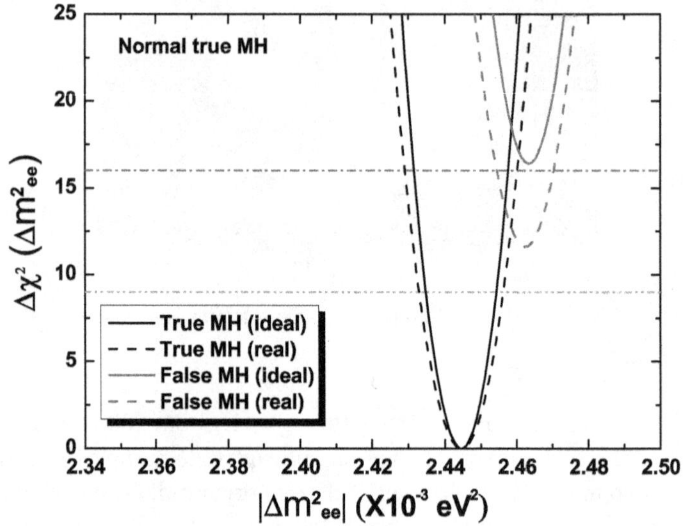

Fig. 16.   Mass hierarchy JUNO performance.

6.2. *RENO50*

RENO50 has a similar physics program and proposal than JUNO with a liquid scintillator cylindrical detector (18 kt) instead of a spherical one. This project proposes to use the same nuclear plants than those used by the RENO experiment with the detector placed at a distance of 50 km away. An extensive R&D period took place up to summer 2017 in many directions as in liquid scintillator purification and reduction of radioactivity. Unfortunately, efforts to obtain a full construction fund failed. For this reason, the Collaboration decided to stop the project.

7. Conclusions

Short baseline reactor experiments are greatly contributing in the understanding of neutrino oscillations and in measuring the oscillation parameters. They allowed in 2012 to measure the last unmeasured oscillation angle θ_{13} and thus opened the way to observe a possible CP violation in the neutrino sector and determine the neutrino mass hierarchy as well. These experiments also contribute to better understand the so called neutrino anomalies and are setting limits on sterile neutrino existence.

Thanks to the relatively large θ_{13} value, medium baseline reactor experiments, as JUNO, could determine the neutrino mass hierarchy provided that they will have a very good energy resolution better than 3% at 1 MeV. These experiments can also mesure precisely three of the oscillation parameters allowing to start unitarity tests of the PMNS mixing matrix. The very large volume ($\sim$20 kt) of these experiments will allow them to also have a very rich astroparticle physics program.

References

1. Y. Abe *et al.* [DOUBLE-CHOOZ Collaboration], "Indication for the disappearance of reactor electron antineutrinos in the Double Chooz experiment," Phys. Rev. Lett. **108**, 131801 (2012) [arXiv:1112.6353 [hep-ex]].
2. F. P. An *et al.* [DAYA-BAY Collaboration], "Observation of electron-antineutrino disappearance at Daya Bay," Phys. Rev. Lett. **108**, 171803 (2012) [arXiv:1203.1669 [hep-ex]].
3. J. K. Ahn *et al.* [RENO Collaboration], "Observation of Reactor Electron Antineutrino Disappearance in the RENO Experiment," Phys. Rev. Lett. **108**, 191802 (2012) [arXiv:1204.0626 [hep-ex]].
4. DOUBLE-CHOOZ results presented in Conferences but not yet published.
5. F. P. An *et al.* [Daya Bay Collaboration], "Measurement of electron antineutrino oscillation based on 1230 days of operation of the Daya Bay experiment," Phys. Rev. D **95** (2017) no.7, 072006 doi:10.1103/PhysRevD.95.072006 [arXiv:1610.04802 [hep-ex]].
6. S. H. Seo [RENO Collaboration], "New Results from RENO using 1500 Days of Data," arXiv:1710.08204 [hep-ex].
7. F. P. An *et al.* [Daya Bay Collaboration], "Improved Measurement of the Reactor Antineutrino Flux and Spectrum at Daya Bay," Chin. Phys. C **41** (2017) no.1, 013002 doi:10.1088/1674-1137/41/1/013002 [arXiv:1607.05378 [hep-ex]].
8. J. I. Crespo-Anadón [Double Chooz Collaboration], "Double Chooz: Latest results," Nucl. Part. Phys. Proc. **265-266** (2015) 99 doi:10.1016/j.nuclphysbps.2015.06.025 [arXiv:1412.3698 [hep-ex]].
9. D. A. Dwyer and T. J. Langford, "Spectral Structure of Electron Antineutrinos from Nuclear Reactors," Phys. Rev. Lett. **114** (2015) no.1, 012502 doi:10.1103/PhysRevLett.114.012502 [arXiv:1407.1281 [nucl-ex]].
10. G. Mention, M. Vivier, J. Gaffiot, T. Lasserre, A. Letourneau and T. Materna, "Reactor antineutrino shoulder explained by energy scale nonlinearities?," Phys. Lett. B **773** (2017) 307 doi:10.1016/j.physletb.2017.08.035 [arXiv:1705.09434 [hep-ex]].
11. S. H. Seo, "Short-baseline reactor neutrino oscillations," PoS NOW **2016** (2017) 002 [arXiv:1701.06843 [hep-ex]].
12. T. A. Mueller *et al.*, "Improved Predictions of Reactor Antineutrino Spectra," Phys. Rev. C **83** (2011) 054615 doi:10.1103/PhysRevC.83.054615 [arXiv:1101.2663 [hep-ex]].
13. P. Huber, "On the determination of anti-neutrino spectra from nuclear reactors," Phys. Rev. C **84** (2011) 024617 Erratum: [Phys. Rev. C **85** (2012) 029901] doi:10.1103/PhysRevC.85.029901, 10.1103/PhysRevC.84.024617 [arXiv:1106.0687 [hep-ph]].
14. F. P. An *et al.* [Daya Bay Collaboration], "Evolution of the Reactor Antineutrino Flux and Spectrum at Daya Bay," Phys. Rev. Lett. **118** (2017) no.25, 251801 doi:10.1103/PhysRevLett.118.251801 [arXiv:1704.01082 [hep-ex]].

15. C. Giunti, X. P. Ji, M. Laveder, Y. F. Li and B. R. Littlejohn, "Reactor Fuel Fraction Information on the Antineutrino Anomaly," JHEP **1710** (2017) 143 doi:10.1007/JHEP10(2017)143 [arXiv:1708.01133 [hep-ph]].

16. P. Adamson *et al.* [Daya Bay and MINOS Collaborations], "Limits on Active to Sterile Neutrino Oscillations from Disappearance Searches in the MINOS, Daya Bay, and Bugey-3 Experiments," Phys. Rev. Lett. **117** (2016) no.15, 151801 Addendum: [Phys. Rev. Lett. **117** (2016) no.20, 209901] doi:10.1103/PhysRevLett.117.151801, 10.1103/PhysRevLett.117.209901 [arXiv:1607.01177 [hep-ex]].

17. J. Cao and K. B. Luk, "An overview of the Daya Bay Reactor Neutrino Experiment," Nucl. Phys. B **908** (2016) 62 doi:10.1016/j.nuclphysb.2016.04.034 [arXiv:1605.01502 [hep-ex]].

18. S. T. Petcov and M. Piai, "The LMA MSW solution of the solar neutrino problem, inverted neutrino mass hierarchy and reactor neutrino experiments," Phys. Lett. B **533** (2002) 94 doi:10.1016/S0370-2693(02)01591-5 [hep-ph/0112074].

19. S. Choubey, S. T. Petcov and M. Piai, "Precision neutrino oscillation physics with an intermediate baseline reactor neutrino experiment," Phys. Rev. D **68** (2003) 113006 doi:10.1103/PhysRevD.68.113006 [hep-ph/0306017].

20. J. Learned, S. T. Dye, S. Pakvasa and R. C. Svoboda, "Determination of neutrino mass hierarchy and theta(13) with a remote detector of reactor antineutrinos," Phys. Rev. D **78** (2008) 071302 doi:10.1103/PhysRevD.78.071302 [hep-ex/0612022].

21. L. Zhan, Y. Wang, J. Cao and L. Wen, "Determination of the Neutrino Mass Hierarchy at an Intermediate Baseline," Phys. Rev. D **78** (2008) 111103 doi:10.1103/PhysRevD.78.111103 [arXiv:0807.3203 [hep-ex]].

22. L. Zhan, Y. Wang, J. Cao and L. Wen, "Experimental Requirements to Determine the Neutrino Mass Hierarchy Using Reactor Neutrinos," Phys. Rev. D **79** (2009) 073007 doi:10.1103/PhysRevD.79.073007 [arXiv:0901.2976 [hep-ex]].

23. F. An *et al.* [JUNO Collaboration], "Neutrino Physics with JUNO," J. Phys. G **43** (2016) no.3, 030401 doi:10.1088/0954-3899/43/3/030401 [arXiv:1507.05613 [physics.ins-det]].

24. See a talk given by Soo-Bong Kim at International Workshop on RENO-50 toward Neutrino Mass Hierarchy, Seoul, South Korea, 13-14 June, 2013.

25. Z. Djurcic *et al.* [JUNO Collaboration], "JUNO Conceptual Design Report," arXiv:1508.07166 [physics.ins-det].

26. T. Adam *et al.*, "The OPERA experiment Target Tracker", Nucl. Instrum. Meth. **A577** (2007) 523, arXiv:physics/0701153.

Low-Energy Neutrino Experiments (Solar Neutrinos)

G. Ranucci[*]

Istituto Nazionale di Fisica Nucleare
Milano, 20133, Italy
E-mail: gioacchino.ranucci@mi.infn.it

Low energy solar neutrino investigation has been one of the most active fields of particle physics research over the past decades, accumulating important and sometimes unexpected achievements. In this work, the characteristics of the experiments, which have done the history of solar neutrinos, are reviewed, together with the most significant physics outputs that they provided.

Keywords: Solar neutrinos.

1. Introduction

Solar neutrinos research is a mature field of investigation, which over the past five decades has accumulated fundamental insights in the particle physics arena. Originally conceived as a powerful tool to deeply investigate the core of our star, solar neutrino studies underwent a very successful detour to particle physics, crucially concurring to the successful demonstration of the neutrino oscillation phenomenon, according to the MSW mechanism [1][2]. The solar experiments were also paramount to determine with high accuracy the parameters of the PMNS mixing matrix governing the oscillation in the solar sector [3]. Hence, this extremely rich and successful chapter of particle physics produced a large part of the solid foundations upon which the next era of precision measurements of the neutrino oscillation parameters will be built. Furthermore, once completed the mission of unveiling the oscillating nature of neutrinos, solar experiments are now back to the original concept of testing the functioning mechanism of the Sun; in this context important insights are awaited from the running and future experiments, of special relevance to address the current issue regarding the surface metallic content of the Sun [4].

2. The long quest for the understanding of the Sun

One of the basic humankind's question since the dawn of civilization has surely been why and how does the Sun shine. In the era of the myths, the answer of the Ancient Greeks was the so-called "Chariot of the Sun", towed by four horses and driven by the titan Helios, originating with its circular path in the sky the night-day alternate. With modern parlance, we can somehow say that this was a first example of "theory" developed to explain the working mechanism of our star!

In the scientific era, the Sun fueling mechanism was the center of the hot debate between Lord Kelvin and Darwin triggered by the evolution theory of Darwin himself,

---

[*]Work partially supported by the INFN Borexino grant.

which Kelvin fiercely opposed. The argument of Kelvin was that no known source of energy was able to last long enough to sustain the existence of the Sun for the huge period of time implied by the proposed Darwinian evolution hypothesis.

The answer to the argument of Kelvin came after the development of the theory of relativity, and specifically after the experiment of Aston who demonstrated that the mass of an atom of helium is less than four times the mass of hydrogen. Eddington then argued in his 1920 presidential address to the British Association for the Advancement of Science that Aston's measurement meant that the Sun could shine by converting hydrogen atoms to helium, putting forward thus the first glimpse of the nuclear hypothesis as engine of the Sun and stars.

About 20 years later, the nuclear hypothesis was elaborated in its modern form in 1938, when Von Weizsacker postulated the CNO cycle and Bethe the p-p chain as the two possible sources of energies.

In both cases, the nuclear production mechanism can be summarized through the overall equation

$$4 {}^1H \rightarrow {}^4He + 2e^+ + 2\nu_e + energy$$

How can it be proved experimentally? Well, neutrinos coming from the reactions are the smoking gun! They pass undisturbed through the solar matter and if detected at Earth they would prove unambiguously the nuclear hypothesis. This argument was put forward as a concrete possibility during the debate about neutrino detection just after the World War II, among other also by Pontecorvo.

From this long prologue, the Solar Neutrino Saga emerged, whose fascinating evolution over almost 50 years represents one of the pillars of the modern understanding of the neutrino properties.

3. Solar neutrino saga: The global picture

It is now well known, after several decades of theoretical and experimental investigations, that the nuclear hypothesis for the Sun (and the stars) is fully valid, and that hence neutrinos are abundantly produced in its core. They originate from the nuclear reactions that power our star, producing the energy required to sustain it over the billions of years of its life. At the temperatures characteristic of the core of the Sun, both reactions postulated in 1938 could occur, being however the pp chain overwhelming with respect to the CNO cycle: the former indeed produces the vast majority of the energy (>99%) of the Sun, while the CNO contribution is estimated to less than 1%.

Starting from this cornerstone, the effort to produce a solar model able to reproduce fairly accurately the solar physical characteristics, as well as the spectra and fluxes of the several produced neutrino components (named after the elements involved in the producing reactions), was led for more than forty years by John Bahcall [5]. This effort culminated in the synthesis of the so-called Standard Solar Model (SSM), which represents a true triumph of the physics of XXth century, leading to extraordinary agreements between predictions and observables, in particular for the helioseismology characteristics (like the sound speed) of the Sun. Such a beautiful concordance, however, has been somehow recently spoiled, as a consequence of the controversy arisen regarding the surface metallic content of the Sun, stemming from a more accurate 3D modeling of the Sun photosphere. Therefore, there are now two versions of the SSM, according to the adoption of the old (high) or revised (low) metallicity of the surface [6]. The paradox here is that the more recent evaluation of the (low) metallicity, supposed to be more accurate, leads to a clear disagreement with the helioseismology data, while the older estimate pointing to a higher metallicity continues to represent the optimal interpretation of the acoustic wave propagation in the Sun.

Remarkably, solar neutrinos are endowed with the intrinsic capability to shed light to this astrophysical conundrum, since the high or low metallicity variants of the SSM originate different predictions for the neutrino fluxes, which become very significant for the neutrinos stemming from the CNO cycle.

On the experimental side, solar neutrino experiments constitute a successful 50 years long plot, commenced with the pioneering radiochemical experiments, i.e. Homestake, Gallex/GNO and Sage, continued with the Cerenkov detectors Kamiokande/Super-Kamiokande in Japan, and SNO in Canada, and with the last player which entered the scene, Borexino at the Gran Sasso Laboratory, which introduced in this field the liquid scintillation detection approach.

For more than 30 years, the persisting discrepancy between the experimental results and the theoretical predictions of the Solar Model formed the basis of the so-called Solar Neutrino Problem, which in the end culminated with a crystal clear proof of the occurrence of the neutrino oscillation phenomenon, via the MSW effect. In particular, the joint analysis of the results from the solar experiments and from the KamLAND antineutrino reactor experiment pin points with high accuracy the values of the oscillations parameter within the LMA (large mixing angle) region of the MSW solution [3].

So, with this spectacular conclusion in background, which surely makes the solar neutrino study one of the more productive contemporary particle physics area, in the following I will illustrate the peculiarity and the specific results of the experiments which have shaped this exciting field of research.

4. Radiochemical method

The radiochemical technique is the basis of the procedure exploited by the Homestake [7], Gallex/GNO [8] and Sage [9] experiments to detect solar neutrinos. The principle is very simple and elegant: the detection medium is a material, which, upon absorption of a neutrino, is converted into a radioactive element whose decay is afterwards revealed and counted.

The pioneering Homestake experiment used a chlorine solution as a target for the inverse β-interaction,

$$\nu_e + {}^{37}\mathrm{Cl} \rightarrow {}^{37}\mathrm{Ar} + e^-$$

characterized by a threshold of 0.814 MeV. It is worth to remind that such a technique was proposed independently by two giants of modern physics, Bruno Pontecorvo and Louis

102

Alvarez. Not only this was the first solar neutrino experiment, but it deserves also the historical merit of having established the Solar Neutrino Problem.

The Gallium experiments instead, adopted gallium as target, which allows neutrino interaction via

$$\nu_e + {}^{71}\text{Ga} \rightarrow {}^{71}\text{Ge} + e^-.$$

The threshold of this reaction is 233 keV, low enough to detect neutrinos from the initial proton fusion chain, which instead cannot be revealed with the chlorine reaction due to the higher threshold. Specific Standard Solar Model estimates gives the following breakdown for the expected contributions to the gallium signal: pp-neutrinos 53%, ${}^{7}\text{Be}$ 27%, ${}^{8}\text{B}$ 12% and CNO neutrinos 8%.

As said before, two experiments have been implemented employing Gallium, Gallex/GNO at Gran Sasso and Sage at Baksan. In Gallex/GNO the target consisted of 101 tons of a $GaCl_3$ solution in water and HCl, containing 30.3 tons of natural gallium; this amount corresponds to about 10^{29} ${}^{71}\text{Ga}$ nuclei. The solution was contained in a large tank hosted in the Hall A of the underground Gran Sasso Laboratory.

${}^{71}\text{Ge}$ produced by neutrinos is radioactive and decay back by electron capture into ${}^{71}\text{Ga}$. The mean life of a ${}^{71}\text{Ge}$ nucleus is about 16 days: thus the ${}^{71}\text{Ge}$ accumulates in the solution, reaching equilibrium when the number of ${}^{71}\text{Ge}$ atoms produced by neutrino interactions is just the same as the number of the decaying ones. When this equilibrium condition is reached, about a dozen ${}^{71}\text{Ge}$ atoms are present inside the whole gallium chloride solution. Therefore, the solar neutrino flux above threshold is deduced from the number of ${}^{71}\text{Ge}$ produced atoms, using the theoretically calculated cross section. The challenging experimental task is thus to identify the feeble amount of ${}^{71}\text{Ge}$ atoms.

This is accomplished through a complex procedure, which contemplates several steps:

a) the solution is exposed to solar neutrinos for about four weeks;

b) the ${}^{71}\text{Ge}$ atoms present at the end of the four week period in the solution are in the form of volatile $GeCl_4$, which is extracted into water by pumping about 3000 m^3 of Nitrogen trough the solution;

c) the extracted ${}^{71}\text{Ge}$ is converted into gaseous GeH_4 and introduced into miniaturized proportional counters mixed with Xenon as counting gas. At the end, a quantity variable between 95 and 98% of the ${}^{71}\text{Ge}$ present in the solution at the time of extraction is in the counter; extraction and conversion efficiencies are under constant control using non radioactive germanium isotopes as carriers;

d) decays and interactions in the counter are observed for a period of 6 months, allowing the complete decay of ${}^{71}\text{Ge}$ and a good determination of the counter background. The charge pulses produced in the counters by decays are recorded by means of fast transient digitizers;

e) the data, after application of suitable cuts, are then analyzed with a maximum likelihood algorithm to obtain the most probable number of ${}^{71}\text{Ge}$ introduced in the counter, with some final corrections applied to take into account the so called "side reaction", i.e.

interactions in the solution generated by high energy muons from cosmic rays and by natural radioactivity.

Key issue in the overall procedure is the minimization of the possible sources of backgrounds. This is performed through a triple strategy, whose first element is the rigorous application of low-level radioactivity technology in the design and construction of the counters. The second element is the use in the analysis of sophisticated pattern recognition techniques able to perform energy and shape discrimination of the signal and background events. The third and final element is the precise calibration of the counters via an external Gd/Ce X-ray source, to enhance the accuracy of the signal/background discrimination ensured by the pattern recognition method.

Thanks to this sophisticated technique, the Gallium experiments were able to provide very important results in the studies of solar neutrinos, in particularly confirming unambiguously and beyond any doubt the discrepancy between the measured and predicted flux, first hinted by Homestake.

To conclude, it is worth to mention another important ingredient of the radiochemical solar neutrino program, i.e. the source calibration efforts which were performed to prove the validity of the entire neutrino detection concept implied by this technique. In particular, Gallex and Sage underwent twice through the calibration procedure, Gallex exploiting in both cases a ^{51}Cr source, while Sage adopted two different isotopes, ^{51}Cr in the first instance and ^{37}Ar in the second test. The outcome of the source tests was the definitive validation of the radiochemical method as an effective mean to detect neutrinos, though the ratio R between the detected and predicted flux stemming from the measures is significantly less than 1: indeed taking the four tests together the global result is $R = 0.86 \pm 0.05$. This discrepancy is one of the several anomalies in the in the neutrino sector, which taken together hint to a possible additional sterile state, a very hot topic in the current debate in the neutrino field.

5. Cerenkov detectors

The widespread diffusion of the Cerenkov technique in the field of neutrino physics is witnessed by the many experimental set-ups based on this methodology, which have been employed to investigate a large portion of the neutrino spectrum, from low to high energies.

The Cerenkov radiation is produced in a material with refractive index n by a charged particle if its velocity is greater than the local phase velocity of the light. The charged particle polarizes the atoms along its trajectory, generating time dependent dipoles, which in turn generate electromagnetic radiation. If $v < c/n$ the dipole distribution is symmetric around the particle position, and the sum of all dipoles vanishes. If $v > c/n$ the distribution is asymmetric and the total time dependent dipole is different from zero, and thus radiates.

The resulting light wavefront is conical, characterized by an opening angle whose cosine is equal to $1/(\beta n)$; the spectrum of the radiation is ultraviolet-divergent, being proportional to $1/\lambda^2$. The propagation properties of the Cerenkov light are therefore fully equivalent to that of the acoustic Mach cone.

By observing the Cherenkov light in a large detector with an array of photomultipliers, the light cone is mapped into a very characteristic ring, like that shown as example in Fig. 1 (from the Super-Kamiokande detector, see next § 5.2). From the topological and timing features of the ring, the properties of the incoming particle can be inferred rather precisely.

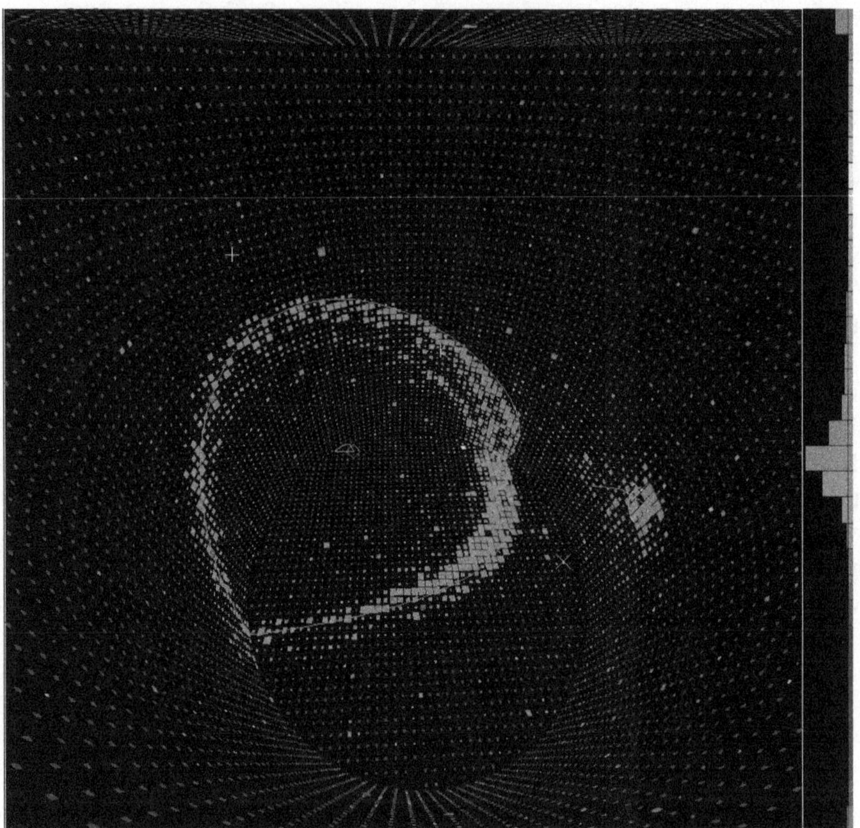

Fig. 1.   A typical Cherenkov ring as seen in a large water detector like Super-Kamiokande.

5.1. *SNO*

A paradigmatic example of how the Cerenkov light can be used as basis to build a very effective neutrino detector is constituted by the SNO detector [10]. Located underground in Canada, in the Inco mine at Sudbury, this detector employed heavy water, which acted both as target medium for the solar neutrinos and as light generating material. As shown in the scheme of Fig. 2, the heavy water is contained in an acrylic vessel, surrounded by a lattice structure holding the PMT's. The entire set-up is immersed in normal water used to shield the external radiation from the rock.

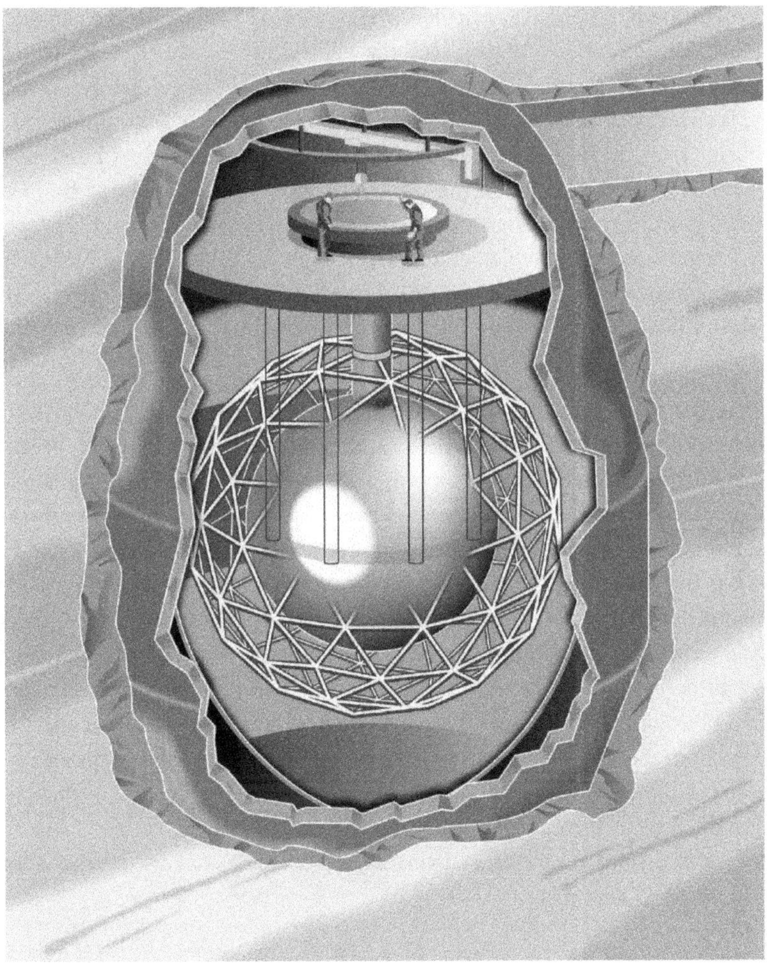

Fig. 2.   Conceptual architectural scheme of the SNO detector.

The basic idea beyond the choice of heavy water is to perform two independent solar neutrino measurements based on the deuterium target: the first is aimed to detect specifically the electron neutrino component, while the second is sensitive to the all flavour flux. Thus, the comparison of the two results can permit to discern clearly if neutrinos, generated only as electron neutrinos in the core of the Sun, undergo flavour conversion during the path Sun-Earth.

Heavy water makes this possible providing both flavour-specific and flavour-independent neutrino reactions. The first, flavour-specific reaction is the *charged current reaction*

$$\nu_e + d \rightarrow p + p + e^-$$

sensitive only to electron neutrinos.

As the neutrino approaches the deuterium nucleus, a W boson is exchanged, changing the neutron in deuterium to a proton, and the neutrino to an electron. Essentially all the neutrino energy is transferred to the electron. Due to the large energy of the incident neutrinos, the electron will be so energetic that it will be ejected at light speed, which is actually faster than the speed of light in water, therefore creating a burst of Cerenkov photons; after traveling throughout the water volume, they are revealed by the spherical array of photomultipliers instrumenting the detector. The amount of light is proportional to the incident neutrino energy, which can be inferred from the number of hits on the PMTs. From the hit pattern, also the angle of propagation of the light can be determined.

The second flavour-independent reaction is the so called *neutral current reaction*

$$\nu_x + d \rightarrow p + n + \nu_x.$$

In this reaction a neutral Z boson is exchanged. The net effect is just to break apart the deuterium nucleus; the liberated neutron is then thermalized in the heavy water as it scatters around. The reaction can eventually be observed due to gamma rays, which are emitted when the neutron is finally captured by another nucleus. The gamma rays will scatter electrons, which produce detectable light via the Cerenkov process, as discussed before.

The neutral current reaction is equally sensitive to all neutrino types; the detection efficiency depends on the neutron capture efficiency and the resulting gamma cascades. Neutrons can be captured directly on deuterium, but this is not very efficient. For this reason, SNO has employed two separated neutral current systems to enhance the neutral current detection, the first based on the ^{35}Cl which has been added to the heavy water in form of NaCl, and the second constituted by proportional helium counters which have been deployed in the core of the detector.

There is also a third reaction occurring in the detector, flavour-independent as well, which is the electron scattering

$$e + \nu_x \rightarrow e + \nu_x.$$

This reaction is not unique to heavy water, being instead the primary mechanism in other light water detectors, like Kamiokande/Super-Kamiokande. Although this reaction is sensitive to all neutrino flavors, due to the different cross sections involved the electron neutrino dominates by a factor of six. The final state energy is shared between the electron and the neutrino, thus there is very little spectral information from this reaction. On the other hand, good directional information can be obtained.

The general drawback affecting the Cerenkov technique is that, due to the feeble amount of light produced by the Cerenkov mechanism, the effective neutrino threshold is around 4-5 MeV, thus allowing the detection only of the high energy component of the solar flux, essentially the ^{8}B neutrinos.

The SNO experiment is now completed, with the heavy water returned to the owner until the very last drop. As a result, the multiple, clean and almost background-free, CC, NC and elastic scattering detection of solar neutrinos provided the unambiguous and model independent proof that neutrinos from the Sun undergo flavour conversion, which was the key to unlock the mystery of the 30 years old solar neutrino problem. This achievement is the collective output of the three phases in which the experiment evolved, characterized by the different detection procedures of the neutrons signalling the occurrence of the neutral current reactions: a pure heavy water phase, a salt phase and the final ^{3}He counters stage [11].

The estimated neutral current, charged current and elastic scattering fluxes over the three phase are statistically in agreement, with the exception of the elastic scattering measure of the ^{3}He stage, lower than the previous results, but consistent with being a downward statistical fluctuation. Also, in the latest ^{3}He measurement procedure the ratio NC/CC resulted equal to 0.301 ± 0.033, slightly lower than the previous phases. While including these updated data in the global solar + KamLAND analysis, the SNO collaboration finds $\Delta m^2 = 7.46^{+0.20}_{-0.19} \times 10^{-5} eV^2$, and $\tan^2 \theta_{12} = 0.427^{+0.027}_{-0.024}$ deg [3]. It has to be noted, in particular, that the error on the θ_{12} angle has been reduced of almost a factor two by the ^{3}He measurement, as effect of the different systematic and minimal correlation of the NC and CC measures in this phase.

The more recent LETA (Low Energy Threshold Analysis) re-processing of the data has further pushed down the analysis threshold, pursuing (similarly to Super-Kamiokande, see next sub-paragraph) the investigation of the survival probability in the 4-5 MeV region, with the aim to unravel the up-turn, if any, of the ^{8}B spectrum. Such an up-turn is expected as the imprint of the MSW effect in that specific energy region. The result stemming from this analysis is definitively puzzling, since it incredibly seems to point even to a down-turn of the spectrum.

This outcome, coupled to the absence of the up-turn in the Super-Kamiokande results, points to something new lurking in that energy region. Additional high statistic data from current and future experiments, however, will be needed to understand what is happening there.

5.2. *Super-Kamiokande*

As anticipated before, Super-Kamiokande [12], like its predecessor Kamiokande, is conceptually very similar to SNO, the major difference being the use of normal water instead of heavy water. Hence the neutrino detection occurs only via the scattering reaction off the electrons; the afterwards mechanism of Cerenkov light production and detection via an array of PMT's is equal to that already described for SNO.

Another major difference is the quantity of water employed, in total 50 ktons, which makes this detector the most massive among the solar neutrino experiments built so far, together with the geometry of the target, cylindrical instead spherical, as can be appreciated in Fig. 3. The sufficiently high statistics implied by this huge volume has made possible a fairly precise reconstruction of the spectrum of the scattered electrons, which plays an important role in the subsequent analysis for the interpretation of the data.

108

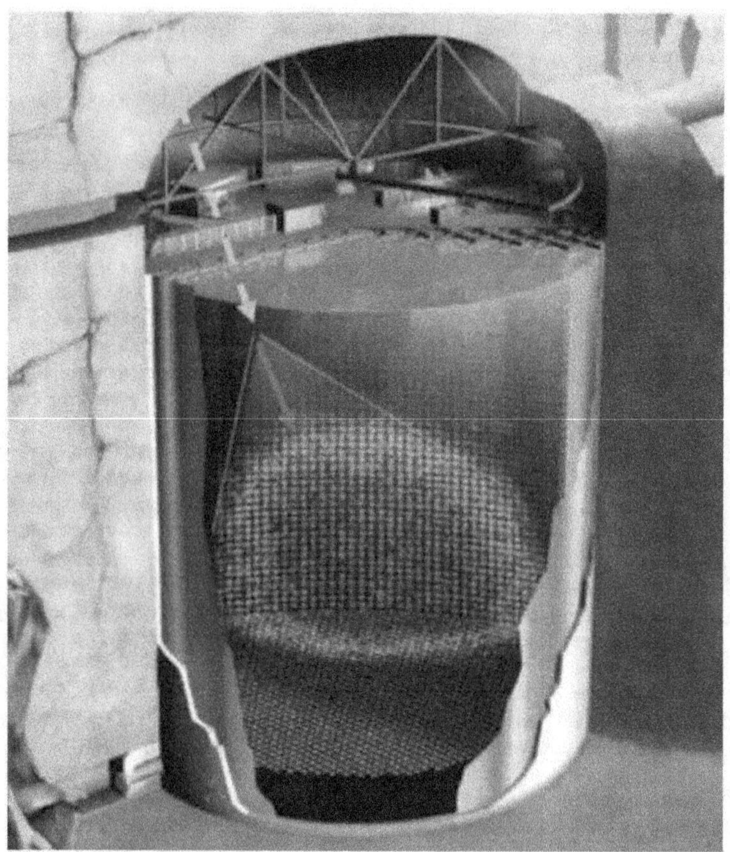

Fig. 3.   Conceptual architectural scheme of the Super-Kamiokande detector.

The intrinsic high directionality of the scattering reaction, coupled to the directionality of the Cerenkov light, provides this experiment with a powerful tool to fight the background due to trace impurities of natural radioactivity dissolved in water, by associating the reconstructed direction of the Cerenkov photons with the angular position of the Sun. Clearly, this is done on top of the purification procedure of the light water, which as for SNO was focused generally on the whole natural radioactivity, but with special emphasis on the radon, which is the factor limiting the threshold at low energy.

Contrary to SNO, Super-Kamiokande is still currently taking data. The long history of this detector started in 1996 and evolved through four phases: the first phase lasted until a major PMT incident in November 2001 and produced the first accurate measure of the ^{8}B flux via the ES detection reaction. The phase II with reduced number of PMTs, from the end of 2002 to the end 2005, confirmed with larger error the phase I measurement. After the refurbishment of the detector back to the original number of PMTs, the third phase lasted from the middle of 2006 up to the middle of 2008. After that, an upgrade of the electronics brought the detector into its fourth phase, which contemplates also an accelerator neutrino beam experiment, dubbed T2K. It is important to highlight the evolution of the energy threshold (total electron energy) in all the phases: 5 MeV in phase I, 7 MeV in phase II, 4.5 MeV in phase III and 3.5 MeV in phase IV, thanks to the continuously on-going effort to reduce the radon content in water.

The important result provided by Super-Kamiokande is the value of the equivalent ^{8}B flux, obtained converting the all flavor ES measure into an effective neutrino flux without correcting for the oscillation probability; the Super-Kamiokande IV precise measurement amounts to 2.308 ± 0.02 (stat.) ± 0.04 (sys.) $\times 10^6$ cm^{-2} s^{-1} [13] representing a high accuracy confirmation that the electron neutrino flux, which contributes mostly to the result of the experiment, is drastically reduced (about 60%) with respect to the SSM prediction.

The recent release of low threshold data had also the goal to unravel the existence of the low energy up-turn in the ^{8}B spectrum, which is expected as the imprint of the LMA-MSW electron survival probability. However, the upturn did show up only mildly, the experimental spectrum above 3.5 MeV showing only a slight preference for the distortion predicted by standard neutrino flavor oscillation parameters over a flat suppression. In this respect, therefore, the poor definition of the data around 3.5-4 MeV and the contradictory output of SNO call for additional evaluations. More data, hence, are needed to further shed light about this issue.

Two interesting ancillary results provided by the experiment over recent years are the time variation of the flux contrasted with the solar cycle [14], showing the lack of any kind of reciprocal association, and the measurement of the day night effect in the ^{8}B flux, amounting to $-3.3 \pm 1 \pm 0.5\%$ [13].

Before concluding this section, it must be stressed that historically the role of Super-Kamiokande has been that of confirming with very high statistics, accuracy and precision the existence of the Solar Neutrino Problem by measuring the ^{8}B neutrino flux, and also that its predecessor Kamiokande was the first detector to confirm the ^{8}B neutrino suppression hint coming from Homestake.

6. Scintillation method: Borexino

Borexino at the Gran Sasso Laboratory [15] is a scintillator detector, which employs as active detection medium about 300 tons of pseudocumene-based scintillator. The intrinsic high luminosity of the liquid scintillation technology is the key toward the goal of Borexino, the real time observation of sub-MeV solar neutrinos through νe^- elastic scattering, being the ^{7}Be component the main envisaged target at the time of the design. However, the lack of directionality of the method makes it impossible to distinguish neutrino-scattered electrons from electrons due to natural radioactivity, thus leading to the other crucial requirement of the Borexino technology, e.g. an extremely low radioactive contamination of the detection medium, at fantastic unprecedented levels.

The active scintillating volume is observed by 2212 PMTs located on a 13.7 m diameter sphere (see Fig. 4) and is shielded from the external radiation by more than 2500 tons of water and by 1000 tons of hydrocarbon equal to the main compound of the scintillator

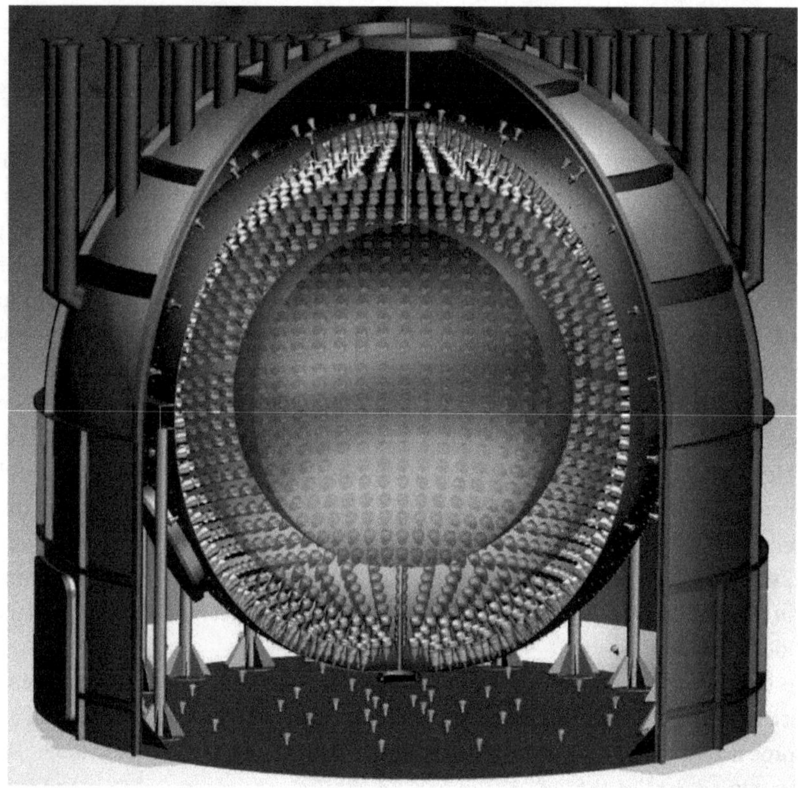

Fig. 4. Sketch of the Borexino experiment, highlighting its major components arranged according to a graded shielding design.

(pseudocumene), to ensure zero buoyancy on the Inner Vessel containing the scintillator itself. Of paramount importance for the success of the experiment are also the many purification and handling systems, which were designed and installed to ensure the proper manipulation of the fluids at the extraordinary purity level demanded by Borexino.

When data taking started in May 2007, it appeared immediately that the daunting task of the ultralow radioactivity was successfully achieved, representing *per se* a major technological breakthrough, opening a new era in the field of ultrapure detectors for rare events search.

The exceptional purity obtained implies that, once selected by software analysis the design fiducial volume of 100 tons and upon removal of the muon and muon-induced signals, the recorded experimental spectrum is so clean to show spectacularly the striking feature of the ^{7}Be scattering edge, i.e. the unambiguous signature of the occurrence of solar neutrino detection.

For the quantitative extraction of the ^{7}Be flux, the spectrum is fitted to a global signal-plus-background model. A recent ^{7}Be result has been published in [16]: taking into accounts the systematic errors, stemming essentially from the uncertainty in the energy scale and in the fiducial volume selection, the ^{7}Be evaluation is 46 ± 1.5stat$(+1.5\text{-}1.6)$sys counts/day/100 tons, hence, summing quadratically the two errors, a remarkable 5% global precision has been achieved in this critical measurement.

By assuming the MSW-LMA solar neutrino oscillations, the Borexino result can be used to infer the ^{7}Be solar neutrino flux. Using the oscillation parameters from [17], the detected ^{7}Be count rate corresponds to a total flux of $(4.84 \pm 0.24) \cdot 10^9$ cm^{-2} s^{-1}, very well in agreement with the prediction of the Standard Solar Model [18]. For comparison, the measured count rate in case of absence of oscillations would have been 74 ± 5.2 counts/day/100 tons. The resulting electrons survival probability at the ^{7}Be energy is $P_{ee} = 0.51 \pm 0.07$.

In the low energy regime the unprecedented Borexino background has allowed also the experimental study of the pep and CNO components, a possibility unanticipated during the design of the detector. By far the most important residual background for these solar neutrino fluxes is the ^{11}C decay, a radionuclide continuously produced in the scintillator by the cosmic muons surviving through the rock overburden and interacting in the liquid scintillator. The beta plus decay of ^{11}C originates a continuous spectrum which sits exactly in the middle of the energy region between 1 and 2 MeV, which is just the window for the pep and CNO investigation. Actually, to a less extent also the external background induced by the gammas from the photomultipliers is an obstacle, especially above 1.7 MeV.

In [19] a threefold coincidence strategy encompassing the parent muon, the neutron(s) emitted in the spallation of the muon on a ^{12}C nucleus, and the final ^{11}C signal has been devised and described in detail. Such a strategy applied to the Borexino data led to a pep rate of 3.13 ± 0.23 (stat.) ± 0.23 (syst.) counts per day/100ton [20], from which the corresponding flux can be calculated, assuming the current MSW-LMA parameters, as $\Phi(\text{pep}) = (1.6 \pm 0.3)\,10^8$ cm^{-2} s^{-1}, in agreement with the SSM: indeed the ratio of this result to the SSM predicted value is $f_{\text{pep}} = 1.1 \pm 0.2$. The resulting electrons survival probability at the pep energy is $P_{ee} = 0.51 \pm 0.07$; it should be underlined that the significance of the pep detection is at the 97% C.L.

The same analysis, keeping the pep flux fixed at the SSM value, originates a tight upper limit on the CNO flux, i.e. $\Phi(\text{CNO}) \leq 7.4\,10^8$ cm^{-2} s^{-1}, corresponding to a ratio with the SSM prediction less than 1.4.

The experiment has been in condition, as well, to provide important, additional insights for the higher energy ^{8}B component. The distinctive feature of the ^{8}B neutrino flux measurement performed by Borexino [21] is the very low 3 MeV threshold attained, lower than the previous measurements from the Cerenkov experiments.

The measurement is very difficult, since the total background, both of radioactive and cosmogenic origin, in the raw data is overwhelming if compared to the expected signal. The specific background suppression strategy adopted in this case is based on two ingredients: on one hand a careful MC evaluation of the main radioactive contaminants of relevance for this measure, i.e. ^{214}Bi from Radon and the external ^{208}Tl from the nylon wall of the Inner Vessel, and on the other the "in-situ" identification and suppression of the muon and associated cosmogenic signals.

The observed ^{8}B rate in the detector is $0.217 \pm 0.038(\text{stat}) \pm 0.008(\text{sys})$ cpd/100ton, corresponding to an equivalent flux $\Phi(^8\text{B}) = (2.4 \pm 0.4 \pm 0.1) \times 10^6$ cm^{-2} s^{-1}; if, as for the other components, we take into account the oscillation probability, then the ratio with the flux foreseen by the SSM is 0.88 ± 0.19.

Towards the completion of the full spectroscopy of the solar neutrino flux in a single experiment, a crucial milestones achieved by Borexino in 2014 has been the first real time spectroscopic detection of the fundamental, low energy, pp flux [22], coming from the main pp reaction providing the vast majority of the energy produced in the core of our star. This epochal result, which enabled the first direct observation of the main engine of the Sun at work, has been made possible first of all by the careful understanding of the detector response in the very low energy regime around 100 keV. Moreover, other two crucial ingredients for this success has been the reduction of the background in the same range (mainly Krypton) due to the effective purification cycle of the scintillator carried out in the years 2010–2011, and the extremely low ^{14}C level in the liquid scintillator. These two factors, coupled to the very good energy resolution, open indeed an exploration window between 200 and 240 keV in which the observation of the fundamental pp flux has been successfully performed.

The measured solar pp neutrino flux is $(6.66 \pm 0.7) \times 10^{10}$ cm^{-2} s^{-1}, in good agreement with the prediction of the standard solar model (SSM) $(5.98 \times (1 \pm 0.006)) \times 10^{10}$ cm^{-2}s^{-1}.

With more accumulated data in hand, the experiment will release soon a new complete set of measurements with improved precisions of the neutrino fluxes coming from the pp chain, reinforcing its unique status of detector able to perform solar neutrino spectroscopy over the whole relevant energy range. Meanwhile, Borexino has recently published an interesting limit on the neutrino magnetic moment [23] and a precise determination of the time variation of the ^{7}Be flux due to the eccentricity of the Earth's orbit. [24].

7. Physics implications and emerging anomalies

In Fig. 5 [22] the MSW predicted P_{ee} (electron neutrino survival probability) is shown, together with several experimental points, all from Borexino. ^{7}Be and *pep* neutrinos are mono-energetic. *pp* and ^{8}B are emitted with a continuum of energy, and the reported experimental values refer to the energy range contributing to the corresponding measurement. The violet band of the P_{ee} corresponds to the prediction of the MSW-LMA solution. It is calculated for the ^{8}B solar neutrinos, considering their production region in the Sun, and represents fairly good the other components, as well.

The vertical error bars of each data point amount to the 1σ interval; the horizontal uncertainty shows the neutrino energy range used in the measurement.

Altogether from this figure we can conclude that the solar data accumulated by Borexino so far spectacularly confirm over the entire energy range of interest the MSW-LMA solar neutrino oscillation scenario, providing the clear evidence of the transition from the high energy, more suppressed "matter" regime, to the low energy, less suppressed "vacuum" regime.

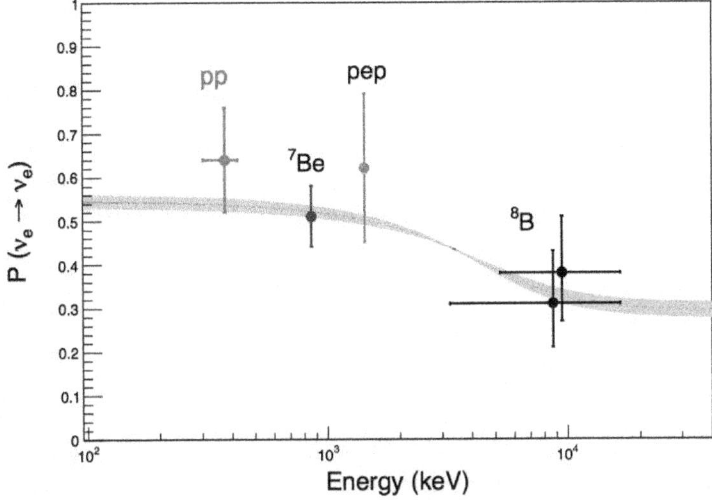

Fig. 5. Electron neutrino survival probability (MSW prediction) contrasted with the experimental points from Borexino, see description in the text.

Furthermore, all the solar and KamLAND data taken together constrain now very precisely the oscillation parameters. Fig. 6, from reference [25], shows the allowed region in the parameter space stemming from a two flavor oscillation analysis, with θ_{13} fixed. In a good approximation, the strongest constraint on the Δm^2 parameter comes from KamLAND, while the limit on the mixing angle derives from the solar data.

In this beautiful and well understood overall scenario, there are however some emerging anomalies whose status is still unclear, whether they are originated by statistical fluctuations or point to some intrinsic, profound effect.

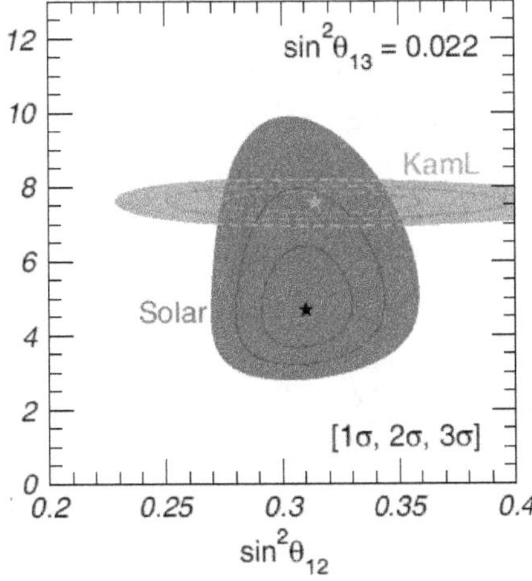

Fig. 6. Allowed region of the oscillation parameter plane.

One is the already mentioned absence of up-turn in the ^{8}B spectrum of Super-Kamiokande. The other is the persistent 2 sigma discrepancy between the Δm^2 coming from KamLAND alone and that from the solar experiments, highlighted in Fig. 6.

While this difference is compatible with being a pure fluctuation, the occurrence that it is now stably present while more and more data are accumulated, may suggest also some true intrinsic physics effect of unknown nature. Connected to this anomaly, there is also the interesting circumstance that the day-night effect detected in the ^{8}B flux by Super-Kamiokande is more aligned with the Δm^2 from the solar experiments, only, and less consistent with that of KamLAND alone.

8. Conclusions and outlook

There is no doubt that Solar neutrino investigation has been one of the most successful and prolific of results area of experimental particle physics over the past 5 decades, triggering and motivating the major developments in neutrino physics in this period, providing:

– fundamental contributions to establish the 3ν oscillation paradigm, implying oscillations in vacuum and matter with resonance flavor conversion;

– crucial data concurring to firmly demonstrate the LMA-MSW scenario as the true solution of the solar neutrino problem and to pin point the "solar" oscillation parameters $\sin^2\theta_{12}$ and Δm^2_{21}, measured with good precision together with the KamLAND data;

– the definite and conclusive quantitative assessment of the pp burning mechanism fueling the Sun.

With Borexino and Super-Kamiokande, the two still running experiments, having entered the precision measurement era, there are still important open questions, which deserve answers from future data:

– which is the nature of the hinted anomalies? Are they flukes or signatures of more profound facts?

– Are there sub-leading effects beyond the LMA MSW solution? Can the imprinting of new physics be detected in the transition region of the survival probability between the vacuum dominated and mass dominated regimes?

– Further tests for the solar model enabled by new data are definitively needed, in particular to shed light on the metallicity puzzle, since the solution can come from a future hint on the CNO flux, whose measurement *per se* represents the ultimate frontier of the solar neutrino field.

The solar neutrino saga is far from being over. New sophisticated experiments with solar neutrinos among their capabilities will enter the scene in the near-medium term, like SNO+ and JUNO (liquid scintillator), Hyper-Kamiokande (water Cherenkov), DUNE (liquid Argon). Therefore, we expect more data and further exciting results from both the running and the future detectors.

Solar neutrinos proved to be a gold mine of opportunities for physics and astrophysics, and will continue to be so still for many more years.

References

1. L. Wolfenstein, Phys. Rev. D 17, 2369 (1978).
2. S.P. Mikheyev and A.Y. Smirnov, Sov. J. Nucl. Phys. 42, 913 (1985).
3. B. Aharmim et al., arXiv:1109.0763 [nucl-ex] (2008).
4. M. Asplund et al., Annu. Rev. Astron. Astrophys. 47, 481 (2009).
5. http://www.sns.ias.edu/~jnb/ John Bahcall home page.
6. A.M. Serenelli, W.C. Haxton and C. Pena-Garay, Astrophys. J. 743, 24 (2011).
7. B.T. Cleveland et al., Ap. J. 496, 505 (1998).
8. W. Hampel et al. (GALLEX Collaboration), Phys. Lett. B 447, 127 (1999).
9. J.N. Abdurashitov et al. (SAGE Collaboration), Phys. Rev. C 80, 015807 (2009).
10. J. Boger et al., Nucl. Instr. and Meth. A449, 172 (2000).
11. B. Aharmim et al. (SNO Collaboration), arXiv:1109.0763.
12. J. Hosaka et al., Phys. Rev. D 73, 112001 (2006).
13. K. Abe et al., Phys. Rev. D 94, 052010 (2016).
14. Yasuo Takeuchi talk at the 13th Rencontres du Vietnam Quy Nhon, July 16–22, 2017 http://vietnam.in2p3.fr/2017/neutrinos/transparencies/2_tuesday/1_morning/2_takeuc hi.pdf.
15. G. Alimonti et al., Nucl. Instr. and Meth. A600, 568 (2009).
16. G. Bellini et al. (Borexino Collaboration), Phys. Rev. Lett. 107, 141302 (2011).
17. Review of Particle Physics, K. Nakamura et al. (Particle Data Group), J. Phys. G 37, 075021 (2010).
18. A.M. Serenelli, W.C. Haxton and C. Pena-Garay, Astrophys. J. 743, 24 (2011).
19. H. Back et al. (Borexino Collaboration), Phys. Rev. C 74, 045805 (2006).
20. G. Bellini et al. (Borexino Collaboration), Phys. Rev. Lett. 108, 051302 (2012).
21. G. Bellini et al. (Borexino Collaboration), Phys. Rev. D 82, 033006 (2010).
22. G. Bellini et al. (Borexino Collaboration), Nature, Volume 512, Issue 7515, pp. 383–386 (2014).
23. M. Agostini et al. (Borexino Collaboration), Phys. Rev. D 96, 091103 (2017).
24. M. Agostini et al. (Borexino Collaboration), Astroparticle Physics, Volume 92, pp. 21–29 (2017).
25. M. Maltoni and A. Yu. Smirnov, Eur. Phys. J. A 52, 87 (2016).

Theoretical Aspects of the Quantum Neutrino

Stephen Parke

Chicago, USA
E-mail: neutrinoguy@gmail.com

In this summary of my talk I will review the following the following three theoretical aspects of the quantum neutrino: current status, why we need precision measurements and neutrino oscillations amplitudes.

Keywords: Neutrino; Theory; Quantum.

1. Current Status

Circa 2017, it is now well established that neutrinos have mass and the that the flavor or interactions states ν_e, ν_μ and ν_τ are mixtures of the the mass eigenstates or propagations states, unimaginatively labelled ν_1, ν_2 and ν_3. The interaction and propagation states are related by an unitary matrix, the PMNS matrix, as follows:

$$\begin{pmatrix} \nu_e \\ \nu_\mu \\ \nu_\tau \end{pmatrix} = U_{23}(\theta_{23}, 0)\; U_{13}(\theta_{13}, -\delta)\; U_{12}(\theta_{12}, 0) \begin{pmatrix} \nu_1 \\ \nu_2 \\ \nu_3 \end{pmatrix}. \tag{1}$$

where the U's are the usual complex rotation matrices given by

$$[U_{mn}(\xi, \eta)]_{ij} = [1 + (c_\xi - 1)(\delta_{im} + \delta_{in})]\delta_{ij} + (s_\xi e^{i\eta})\,\delta_{im}\delta_{jn} - (s_\xi e^{-i\eta})\,\delta_{in}\delta_{jm}\,.$$

Placing the CP violating phase in the U_{13} matrix is customary, however oscillation physics is unaffected by placing this phase in U_{23} or U_{12} since

$$U_{23}(\theta_{23}, 0)\; U_{13}(\theta_{13}, -\delta)\; U_{12}(\theta_{12}, 0) :=: U_{23}(\theta_{23}, \delta)\; U_{13}(\theta_{13}, 0)\; U_{12}(\theta_{12}, 0)$$
$$:=: U_{23}(\theta_{23}, 0)\; U_{13}(\theta_{13}, 0)\; U_{12}(\theta_{12}, \delta)$$

where $:=:$ means equal after multiplying by a diagonal phase matrix on the left and/or right hand side.

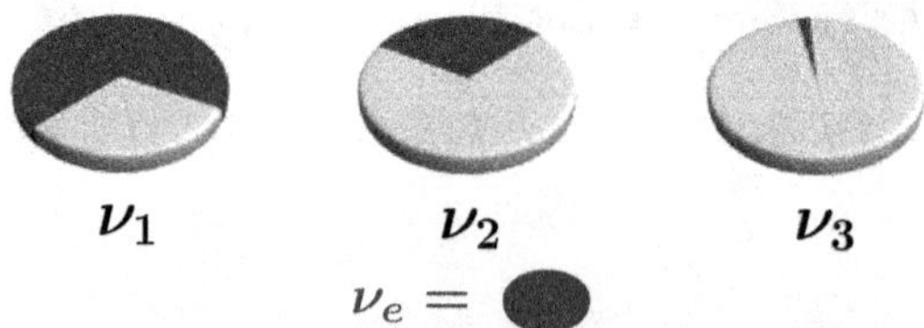

Fig. 1. The identity of the neutrino mass eigenstates or propagation states is determined by their ν_e content: ν_1 has the largest ($\sim$68%), ν_2 is in the middle ($\sim$30%) and ν_3 has the least ($\sim$2%). The remaining components are a combination of ν_μ and ν_τ but it's the ν_e fraction that defines the identity of these states. Some, use the masses to label these states, but since we don't know both the mass orderings at this stage, using the electron flavor content is simpler.

At the current time, it is most convenient to label the neutrino mass eigenstates according to the size of ν_e fraction as is shown in Fig. 1. With this choice for the mass eigenstates, it is natural to choose the order of the $U_{\alpha j}$ matrices as in Eq. (1) so that the first row and third column are simply, since these elements are most easily measured.

Since the neutrino oscillations with a $\Delta m^2 \approx 2.5 \times 10^{-3} \text{eV}^2$ has only small amounts of ν_e, it is $\nu_{1,2}$ that is has a $|\Delta m_{21}^2| \approx 7.5 \times 10^{-5} \text{eV}^2$. The SNO experiment determined the mass ordering of ν_1 and ν_2, $m_2 > m_1$, see Fig. 2.

The remaining mass ordering, whether m_3 is larger or smaller than m_1 and m_2, is shown in Fig. 3. Current and future experiments such as NOνA, JUNO, DUNE, T2HKK are designed to determine this mass ordering.

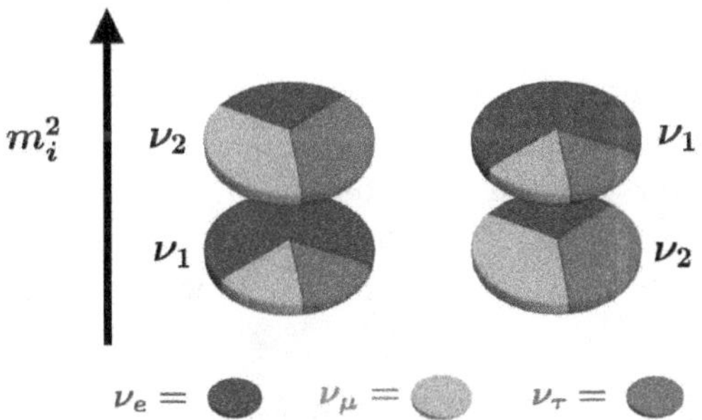

Fig. 2.   The solar mass ordering of ν_1 and ν_2 was determined by the SNO experiment using the matter effects in the solar interior. The mass of ν_2 is larger than the mass of ν_1 with $\Delta m_{21}^2 \equiv m_2^2 - m_1^2 \approx 7.5 \times 10^{-5} \text{eV}^2$. Here, I have assumed that the ν_μ fraction is equal to ν_τ fraction for both mass eigenstates.

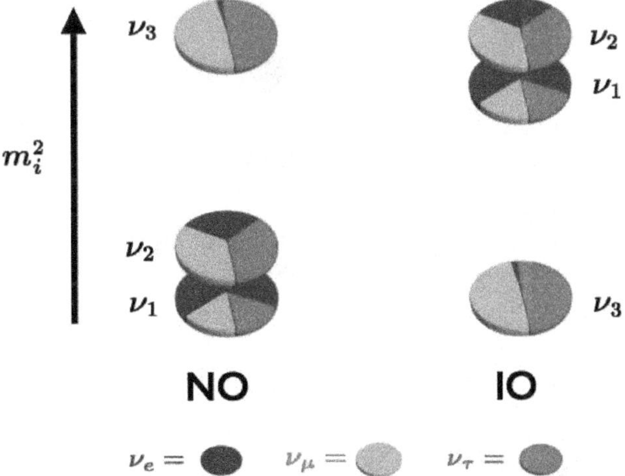

Fig. 3.   The atmospheric mass ordering, sometimes referred to as mass hierarchy, is the question of whether the mass of ν_3 is larger or smaller than ν_2, ν_1. $|\Delta m_{31}^2| \approx |\Delta m_{32}^2| \approx 2.5 \times 10^{-3} \text{eV}^2$. Here, I have assumed that the ν_μ fraction is equal to ν_τ fraction for all three mass eigenstates.

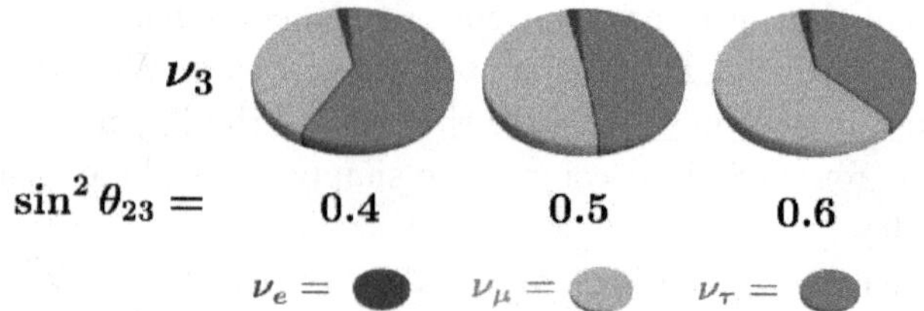

Fig. 4.   Is the dominant flavor of the mass eigenstate ν_3, ν_μ or ν_τ? If ν_τ dominates then the parameter $\theta_{23} < \pi/4$, whereas if ν_μ dominates then $\theta_{23} > \pi/4$. This is often referred to as the octant of θ_{23} puzzle.

The octant of θ_{23}, determines which flavor state dominates ν_3, see Fig. 4. Whereas the range of ν_μ and ν_τ components of ν_1 and ν_2 are determined not only by the allowed range of θ_{23} but also by the CP violating phase δ ($\cos\delta$ to be precise). There is significant depends on δ for the magnitude of $U_{\mu 1}$, $U_{\mu 2}$, $U_{\tau 1}$ and $U_{\tau 2}$ elements of the PMNS matrix. This fact is quite different than in the quark sector!

In the quark sector only the magnitude of U_{td}^{CKM} has any significant depends on δ_{CKM}. This occurs because of the hierarchy in the sizes of the mixing angles in the CKM matrix: $\theta_{12} \sim \lambda$, $\theta_{23} \sim \lambda^2$ and $\theta_{13} \sim \lambda^3$ where $\lambda \approx 0.2$.

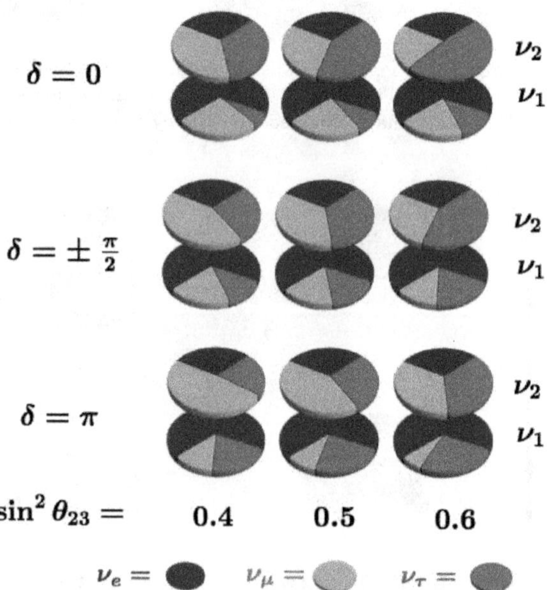

Fig. 5.   The ν_μ and ν_τ fractions of ν_1 and ν_2 depends sensitively on both the value of θ_{23} and the CP phase δ. The flavor fractions vary the most for ν_1, between $\delta = 0$, $\sin^2\theta_{23} = 0.4$ (upper left) and $\delta = \pi$, $\sin^2\theta_{23} = 0.6$ (lower right). Whereas for ν_2, the most variation is between $\delta = \pi$, $\sin^2\theta_{23} = 0.4$ (lower left) and $\delta = 0$, $\sin^2\theta_{23} = 0.6$ (upper right). Factors of 2 to 3 differences are still allowed by the data. Note that is $\delta = \pm\pi/2$, $\sin^2\theta_{23} = 0.5$ (middle middle), then the ν_μ and ν_τ content is the same for all three mass eigenstates.

2. Why We Need Precision Measurements

Four reasons for performing precision measurements:

2.1. *For Discovery of New Physics*

An experiment like ICECUBE can discover new physics in the flavor ratios of their PeV neutrinos, if precise values of the predictions for the νSM are known, see Fig. 6, from [1].

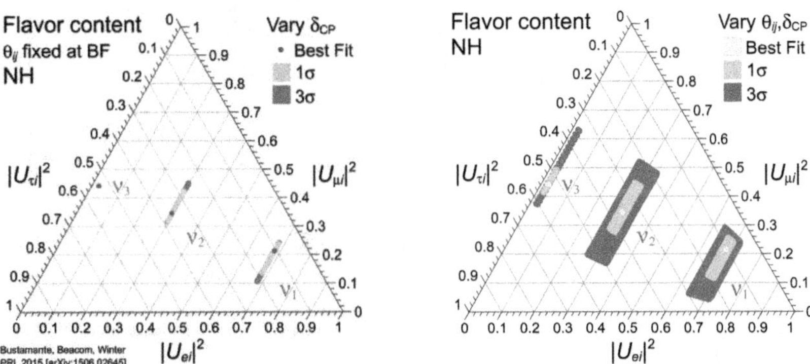

Fig. 6. Variation of flavor ratios using our current uncertainty on δ (left panel) and θ_{ij}, δ (right panel), from Bustamante, Beacom and Winter.

2.2. *Stress Test Three Neutrino Paradigm*

Compared to the Quark sector the unitarity of the PMNS matrix has only been tested at the 10% level [2].

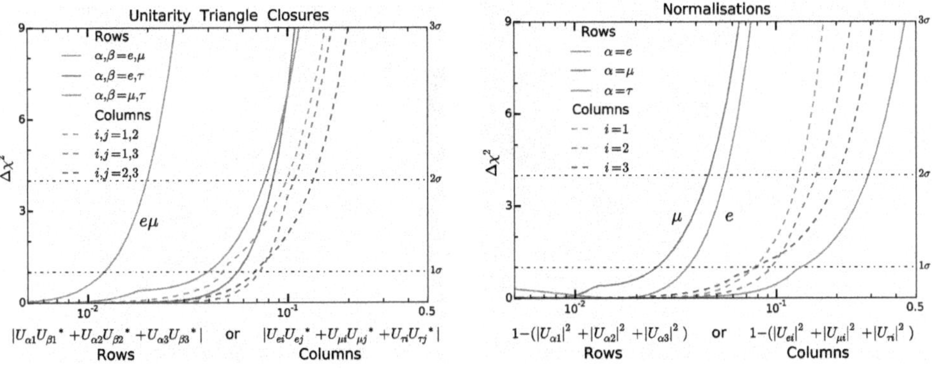

Fig. 7. The current triangle (left) and normalization (right) unitarity constraints on the elements of the PMNS matrix form Parke and Ross-Lonergan. There is still plenty of room for new physics, such as a light sterile neutrino.

120

2.3. *Test Theoretical Neutrino Models*

Fig. 8 shows how improvements on the measurements of the mixing angles and CP violating phase can used to distinguish various models that could possibly explain the mixings of the neutrinos.

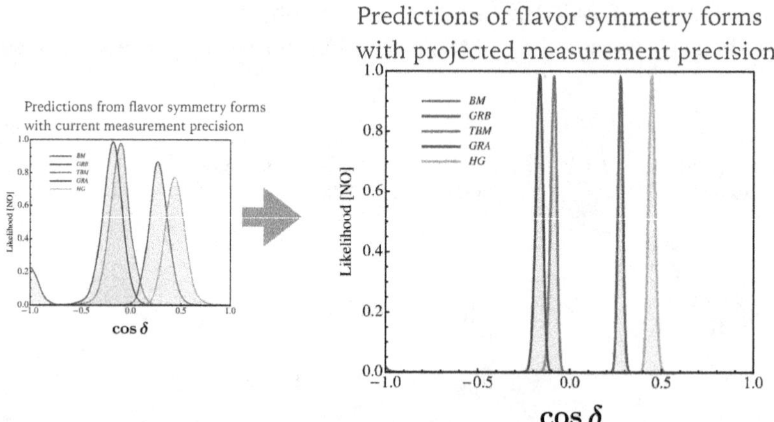

Fig. 8. Current status (left) and future status (right) of the labelled models predictions for $\cos\delta$ from Girardi, Petcov and Titov[3].

2.4. *Connection to Leptogenesis Understanding Universe*

Fig. 9 gives the allowed region for a model[4] with the measured value of the Baryon asymmetry of the universe on the neutrino parameters.

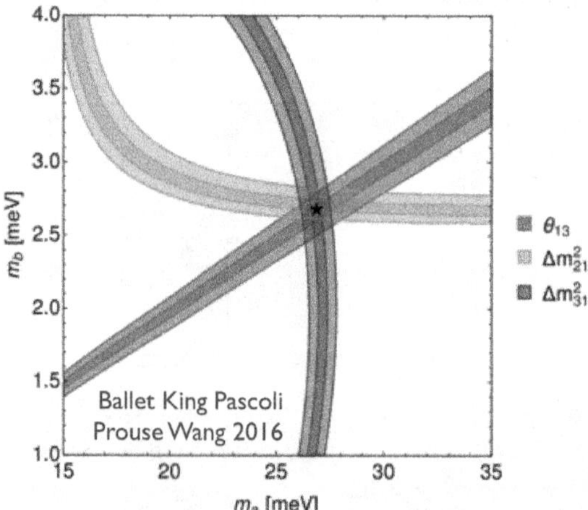

Fig. 9. Consistency of the model by Ballet, King, Pascoli, Prouse and Wang which has the measured value of the Baryon asymmetry, through Leptogenesis, with the measured values of the parameters θ_{13}, Δm^2_{21} and Δm^2_{31}.

3. Neutrino Amplitudes

The neutrino oscillation probability can be written as

$$P(\nu_\alpha \to \nu_\beta) = |\mathcal{A}_{\alpha\beta}|^2 \quad \text{where} \quad \mathcal{A}_{\alpha\beta} = \sum_j U^*_{\alpha j} U_{\beta j} e^{-im_j^2 L/2E} . \tag{2}$$

For two flavors we simple obtain

$$\mathcal{A}_{\alpha\alpha} = 1 + 2i\, s_\theta^2\, e^{+i\Delta}\, \sin\Delta \quad \text{and} \quad \mathcal{A}_{\alpha\beta} = (2s_\theta c_\theta)\, \sin\Delta$$

where $\Delta_{jk} = \Delta m_{jk}^2 L/4E$.

For three flavors, there are an infinite number of ways of writing these amplitudes which all give the same result. So we need an organizational principle. It is known that the ν_τ oscillation amplitudes can be obtained from the ν_μ oscillation amplitudes by making the following replacements $s_{23} \leftrightarrow c_{23}$ and $\delta \to \delta + \pi$. This follows from the form of the U_{23} matrix on the left hand side of the PMNS matrix, Eq. (1).

Similarly, if one considers the rotation matrix on the right hand side of the PMNS matrix, U_{12}, there is a symmetry that leaves the amplitudes invariant: the symmetry is $m_1^2 \leftrightarrow m_2^2$, $s_{12} \leftrightarrow c_{12}$ and $\delta \to \delta + \pi$ which follows from

$$U_{12}(\theta_{12},\delta)\begin{pmatrix}\nu_1\\\nu_2\end{pmatrix} = U_{12}(\pi/2+\theta_{12},\delta)\begin{pmatrix}-e^{i\delta} & \nu_2\\e^{-i\delta} & \nu_1\end{pmatrix} = U_{12}(\pi/2-\theta_{12},\delta\pm\pi)\begin{pmatrix}e^{i\delta} & \nu_2\\-e^{-i\delta} & \nu_1\end{pmatrix}.$$

To maintain this symmetry, use unitarity to remove the $U^*_{\alpha3}U_{\beta3}$ term in Eq. (2), giving

$$\mathcal{A}_{\alpha\beta} = \delta_{\alpha\beta} + (2i) \sum_{j=(1,2)} U^*_{\alpha j} U_{\beta j}\, e^{i\Delta_{3j}} \sin\Delta_{3j}, \tag{3}$$

then[a]

$$\mathcal{A}_{ee} = 1 + (2i)\, c_{13}^2 \left[c_{12}^2\, e^{i\Delta_{31}} \sin\Delta_{31} + s_{12}^2\, e^{i\Delta_{32}} \sin\Delta_{32} \right]$$

$$\begin{aligned}
\mathcal{A}_{\mu\mu} = 1 + (2i)\, \big[& (c_{23}^2 c_{12}^2 + s_{13}^2 s_{12}^2 s_{23}^2)\, e^{i\Delta_{32}} \sin\Delta_{32} \\
& + (c_{23}^2 s_{12}^2 + s_{13}^2 c_{12}^2 s_{23}^2)\, e^{i\Delta_{31}} \sin\Delta_{31} \\
& + (s_{13} s_{12} c_{12} s_{23} c_{23} \cos\delta)\, e^{i(\Delta_{31}+\Delta_{32})} \sin\Delta_{21} \big]
\end{aligned} \tag{4}$$

$$\begin{aligned}
\mathcal{A}_{\mu\tau} = (2c_{23}s_{23})\, & \left[(s_{12}^2 - s_{13}^2 c_{12}^2)e^{i\Delta_{31}} \sin\Delta_{31} + (c_{12}^2 - s_{13}^2 s_{12}^2)e^{i\Delta_{32}} \sin\Delta_{32} \right] \\
& - (2s_{13}s_{12}c_{12}) \left[c_{23}^2 e^{i\delta} - s_{23}^2 e^{-i\delta} \right] e^{i(\Delta_{31}+\Delta_{32})} \sin\Delta_{21}
\end{aligned}$$

$$\begin{aligned}
\mathcal{A}_{\mu e} = (2s_{23}s_{13}c_{13})\, & \left[c_{12}^2 e^{i\Delta_{31}} \sin\Delta_{31} + s_{12}^2 e^{i\Delta_{32}} \sin\Delta_{32} \right] \\
& + (2c_{23}c_{13}s_{12}c_{12})\, e^{i(\Delta_{31}+\Delta_{32}+\delta)} \sin\Delta_{21}
\end{aligned}$$

---

[a]Since the overall phase of an amplitude is arbitrary, there are arbitrary choices for the overall phase of each amplitude.

No approximation has been used to obtain these amplitudes and all of these amplitudes explicitly satisfy the $1 \leftrightarrow 2$ symmetry mentioned earlier. Again as an organizing principle, I have separated only the terms which involve $e^{\pm i\delta}$ as they always appear multiplied by $\sin\Delta_{21}$ (note $\sin\Delta_{21} = e^{-i\Delta_{32}}\sin\Delta_{31} - e^{-i\Delta_{31}}\sin\Delta_{32}$).

If one uses the approximation that $\Delta_{32} \approx \Delta_{31}$ then we can rewrite

$$\mathcal{A}_{\mu e} \approx (2s_{23}s_{13}c_{13})\ \sin\Delta_{31} + (2c_{23}c_{13}s_{12}c_{12})\ e^{i(\delta+\Delta_{32})}\ \sin\Delta_{21}$$

and then it's simple to see that the CP violating term is given by

$$\Delta P_{CP} = 8\ (s_{23}s_{13}c_{13})\ (c_{23}c_{13}s_{12}c_{12})\ \sin\delta\ \sin\Delta_{21}\ \sin\Delta_{31}\ \sin\Delta_{32}. \tag{5}$$

In Fig. 10 we give a graphical representation of the amplitude for $\nu_\mu \to \nu_e$ and the associated bi-probability plot.

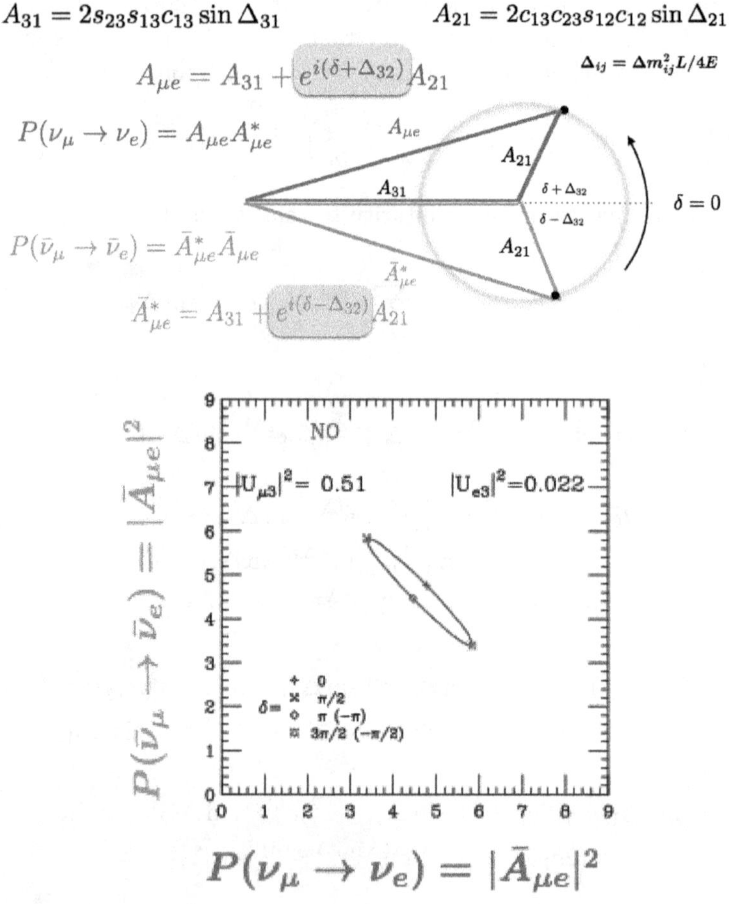

Fig. 10.   The amplitude for $\nu_\mu \to \nu_e$ as well as $\bar\nu_\mu \to \bar\nu_e$ and the associated bi-probability plot.

4. Perturbative Approximation to the Neutrino Oscillation Probabilites in Matter

In this section, a simple and accurate way to evaluate oscillation probabilities, recently shown Denton, Minakata and Parke[5], is given. Details as to the why's and how's of this method are contained in these papers.

The mixing angles in matter, which we denote by a $\widetilde{\theta}_{13}$ and $\widetilde{\theta}_{12}$ here, can also be calculated in the following way, using $\Delta m^2_{ee} \equiv \cos^2\theta_{12}\Delta m^2_{31} + \sin^2\theta_{12}\Delta m^2_{32}$, as follows, see Addendum[5]:

$$\cos 2\widetilde{\theta}_{13} = \frac{(\cos 2\theta_{13} - a/\Delta m^2_{ee})}{\sqrt{(\cos 2\theta_{13} - a/\Delta m^2_{ee})^2 + \sin^2 2\theta_{13}}}, \tag{6}$$

where $a \equiv 2\sqrt{2}G_F N_e E_\nu$ is the standard matter potential, and

$$\cos 2\widetilde{\theta}_{12} = \frac{(\cos 2\theta_{12} - a'/\Delta m^2_{21})}{\sqrt{(\cos 2\theta_{12} - a'/\Delta m^2_{21})^2 + \sin^2 2\theta_{12}\cos^2(\widetilde{\theta}_{13} - \theta_{13})}}, \tag{7}$$

where $a' \equiv a\cos^2\widetilde{\theta}_{13} + \Delta m^2_{ee}\sin^2(\widetilde{\theta}_{13} - \theta_{13})$ is the θ_{13}-modified matter potential for the 1-2 sector. In these two flavor rotations, both $\widetilde{\theta}_{13}$ and $\widetilde{\theta}_{12}$ are in range $[0, \pi/2]$.

θ_{23} and δ are unchanged in matter for this approximation.

From the neutrino mass squared eigenvalues in matter, given by

$$\widetilde{m}^2_3 = \Delta m^2_{31} + (a - a'),$$
$$\widetilde{m}^2_2 = \frac{1}{2}(\Delta m^2_{21} + \widetilde{\Delta m^2}_{21} + a'), \tag{8}$$
$$\widetilde{m}^2_1 = \frac{1}{2}(\Delta m^2_{21} - \widetilde{\Delta m^2}_{21} + a'),$$

it is simple to obtain the neutrino mass squared differences in matter, i.e. the Δm^2_{jk} in matter, which we denote by $\Delta\widetilde{m}^2_{jk}$, which are given by

$$\Delta\widetilde{m}^2_{21} = \Delta m^2_{21}\sqrt{(\cos 2\theta_{12} - a'/\Delta m^2_{21})^2 + \sin^2 2\theta_{12}\cos^2(\widetilde{\theta}_{13} - \theta_{13})},$$
$$\Delta\widetilde{m}^2_{31} = \Delta m^2_{31} + (a - \frac{3}{2}a') + \frac{1}{2}\left(\Delta\widetilde{m}^2_{21} - \Delta m^2_{21}\right), \tag{9}$$
$$\Delta\widetilde{m}^2_{32} = \Delta\widetilde{m}^2_{31} - \Delta\widetilde{m}^2_{21}.$$

To see these expressions have the correct asymptotic forms, use the fact that $(\Delta\widetilde{m}^2_{21} - \Delta m^2_{21}) = |a'| + \mathcal{O}(\Delta m^2_{21})$, for $|a| \gg \Delta m^2_{21}$. Plots of the matter mixing angles and mass squared differences are given in Fig. 11.

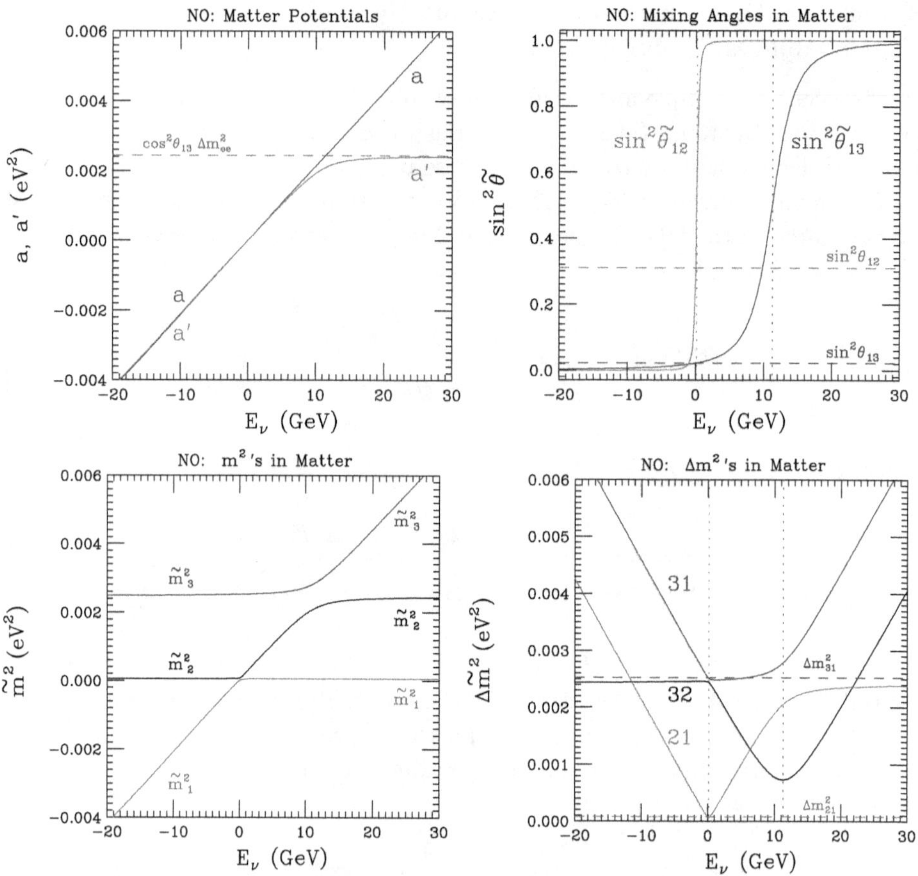

Fig. 11.   In the normal ordering (NO): Top left, the matter potentials, a and a', top right, sine squared of mixing angles in matter, $\sin^2 \widetilde{\theta}_{jk}$, bottom left, the mass squared eigenvalues in matter, $\widetilde{m^2}_j$, and bottom right, the mass squared differences in matter, $\Delta \widetilde{m^2}_{jk}$. $E_\nu \geq 0$ ($E_\nu \leq 0$) is for neutrinos (anti-neutrinos). $E_\nu = 0$ is the vacuum values for both neutrinos and anti-neutrinos.

To calculate the oscillation probabilities, to 0th order, use the above $\Delta \widetilde{m^2}_{jk}$ instead of Δm^2_{jk} and replace the vacuum MNS matrix as follows

$$U^0_{MNS} \equiv U_{23}(\theta_{23})\, U_{13}(\theta_{13}, \delta)\, U_{12}(\theta_{12}) \Rightarrow U^M_{MNS} \equiv U_{23}(\theta_{23})\, U_{13}(\widetilde{\theta}_{13}, \delta)\, U_{12}(\widetilde{\theta}_{12}).$$

That is, replace

$$\Delta m^2_{jk} \to \Delta \widetilde{m^2}_{jk}$$
$$\theta_{13} \to \widetilde{\theta}_{13}$$
$$\theta_{12} \to \widetilde{\theta}_{12}, \tag{10}$$

θ_{23} and δ remain unchanged, it is that simple. We call this the 0th order DMP approximation.

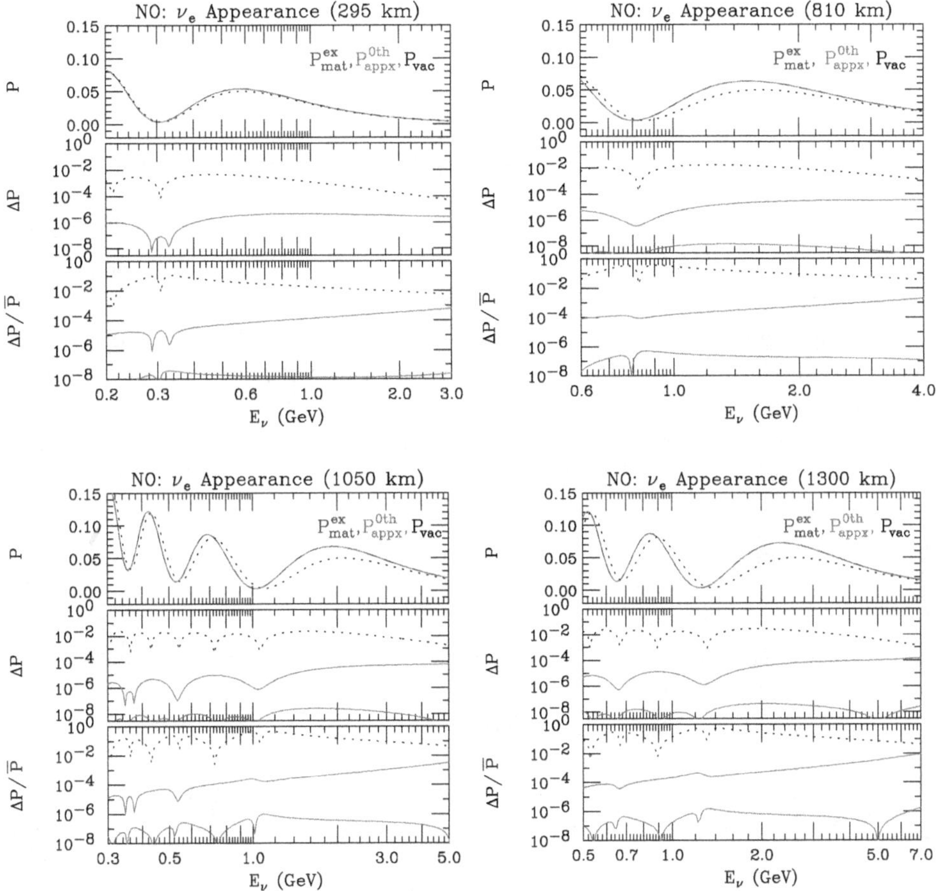

Top panel: P^{ex}_{mat}, P^{0th}_{appx} and P_{vac}

Middle and bottom panels:

black dotted lines	red solid lines	magenta solid lines
$\Delta P = \lvert P^{ex}_{mat} - P_{vac}\rvert$	$\Delta P = \lvert P^{ex}_{mat} - P^{0th}_{appx}\rvert$	$\Delta P = \lvert P^{ex}_{mat} - P^{1st}_{appx}\rvert$
$\overline{P} = \frac{1}{2}(P^{ex}_{mat} + P_{vac})$	$\overline{P} = \frac{1}{2}(P^{ex}_{mat} + P^{0th}_{appx})$	$\overline{P} = \frac{1}{2}(P^{ex}_{mat} + P^{1st}_{appx})$

Fig. 12. For normal ordering (NO), $\nu_\mu \to \nu_e$ appearance: Top Left figure is for T2K, Top Right figure is NOvA, Bottom Left figure is T2HKK, and Bottom Right is DUNE. In each figure, the top panel is exact oscillation probability in matter, P^{ex}_{mat} (blue dashes) from [6], the zeroth order DMP approximation, P^{0th}_{appx} (red dashes) from [5] and the vacuum oscillation probability, P_{vac} (black dots). The Middle panel is difference between exact oscillation probabilities in matter and vacuum (black dots), and the difference between exact and 0th DMP approximation (solid red) and exact and 1st DMP approximation (solid magenta) approximations. Bottom panel is similar to middle panel but plotting the fractional differences, $\Delta P/\overline{P}$ (color online).

These expressions are valid for both NO, $\Delta m_{ee}^2 > 0$ and IO, $\Delta m_{ee}^2 < 0$. For anti-neutrinos, just change the sign of a and δ. Our expansion parameter is

$$\left| \sin(\widetilde{\theta}_{13} - \theta_{13}) \ \sin\theta_{12}\cos\theta_{12} \ \frac{\Delta m_{21}^2}{\Delta m_{ee}^2} \right| \leq 0.015, \tag{11}$$

which is small and vanishes in vacuum, so that our perturbation theory reproduces the vacuum oscillation probabilities exactly.

If $\mathcal{A}_{\nu_\alpha \to \nu_\beta}(\Delta m_{31}^2, \Delta m_{21}^2, \theta_{13}, \theta_{12}, \theta_{23}, \delta)$ is the oscillation amplitude in vacuum, see Eq. (4), then $\mathcal{A}_{\nu_\alpha \to \nu_\beta}(\Delta \widetilde{m^2}_{31}, \Delta \widetilde{m^2}_{21}, \widetilde{\theta}_{13}, \widetilde{\theta}_{12}, \theta_{23}, \delta)$ is the oscillation probability in matter, i.e. use the same function but replace the mass squared differences and mixing angles with the matter values given in Eq. (6)–(9). The resulting oscillation probabilities are identical to the zeroth order approximation given in Denton, Minakata and Parke[5].

In Fig. 12, I have given the exact and approximate oscillation probabilities for the ν_e appearance channel for T2K[7] and T2HK[8], NOvA[9], T2HKK[10] and DUNE[11].

5. Conclusions

To summarize:

- from Nu1998 to now, tremendous experimental progress on Neutrino SM: more at Nu2018!
- LSND Sterile Nu's neither confirmed or ruled out at acceptable CL: – ultra short baseline reactor experiments.
- Great Theoretical progress on understand many aspects of Quantum Neutrino Physics: Oscillations, Decoherence, Oscillations Probabilities in Matter, Leptogenesis.
- Still searching for convincing model of Neutrino masses and mixings: with testable and confirmed predictions!

Acknowledgements

I would like to thank the organizers, and especially Prof. Wang Wei, for the wonderful hospitality so that I could attend this great conference.

This project has received funding/support from the European Union's Horizon 2020 research and innovation programme under the Marie Sklodowska-Curie grant agreement No 690575. This project has received funding/support from the European Union's Horizon 2020 research and innovation programme under the Marie Sklodowska-Curie grant agreement No 674896.

Finally, I would like to thank my collaborator Prof. Xu Zhan from Tsinghua University for a wonderful collaboration and friendship. We meet at the International Symposium on Particle and Nuclear Physics, Beijing, China, Sep 2-7, 1985 and wrote a beautiful and significant paper[12] that was a pioneering paper in the field of "Amplitudes".

References

1. M. Bustamante, J. F. Beacom and W. Winter, "Theoretically palatable flavor combinations of astrophysical neutrinos," Phys. Rev. Lett. **115**, no. 16, 161302 (2015) doi:10.1103/PhysRevLett.115.161302 [arXiv:1506.02645 [astro-ph.HE]].
2. S. Parke and M. Ross-Lonergan, "Unitarity and the three flavor neutrino mixing matrix," Phys. Rev. D **93**, no. 11, 113009 (2016) doi:10.1103/PhysRevD.93.113009 [arXiv:1508.05095 [hep-ph]].
3. I. Girardi, S. T. Petcov and A. V. Titov, "Determining the Dirac CP Violation Phase in the Neutrino Mixing Matrix from Sum Rules," Nucl. Phys. B **894**, 733 (2015) doi:10.1016/j.nuclphysb.2015.03.026 [arXiv:1410.8056 [hep-ph]].
4. P. Ballett, S. F. King, S. Pascoli, N. W. Prouse and T. Wang, "Precision neutrino experiments vs the Littlest Seesaw," JHEP **1703**, 110 (2017) doi:10.1007/JHEP03(2017)110 [arXiv:1612.01999 [hep-ph]].
5. P. B. Denton, H. Minakata and S. J. Parke, "Compact Perturbative Expressions For Neutrino Oscillations in Matter," JHEP **1606**, 051 (2016) doi:10.1007/JHEP06(2016)051 [arXiv:1604.08167 [hep-ph]]. P. B. Denton, H. Minakata and S. J. Parke, "Addendum to 'Compact Perturbative Expressions for Neutrino Oscillations in Matter'" arXiv:1801.06514 [hep-ph].
6. H. W. Zaglauer and K. H. Schwarzer, "The Mixing Angles in Matter for Three Generations of Neutrinos and the MSW Mechanism," Z. Phys. C **40** (1988) 273.
7. K. Abe *et al.* [T2K Collaboration], "The T2K Experiment," Nucl. Instrum. Meth. A **659**, 106 (2011) doi:10.1016/j.nima.2011.06.067 [arXiv:1106.1238 [physics.ins-det]].
8. K. Abe *et al.* [Hyper-Kamiokande Proto-Collaboration], "Physics potential of a long-baseline neutrino oscillation experiment using a J-PARC neutrino beam and Hyper-Kamiokande," PTEP **2015**, 053C02 (2015) doi:10.1093/ptep/ptv061 [arXiv:1502.05199 [hep-ex]].
9. D. S. Ayres *et al.* [NOvA Collaboration], "NOvA: Proposal to Build a 30 Kiloton Off-Axis Detector to Study $\nu_\mu \to \nu_e$ Oscillations in the NuMI Beamline," hep-ex/0503053.
10. K. Abe *et al.* [Hyper-Kamiokande proto-Collaboration], "Physics Potentials with the Second Hyper-Kamiokande Detector in Korea," arXiv:1611.06118 [hep-ex].
11. R. Acciarri *et al.* [DUNE Collaboration], "Long-Baseline Neutrino Facility (LBNF) and Deep Underground Neutrino Experiment (DUNE) : Volume 2: The Physics Program for DUNE at LBNF," arXiv:1512.06148 [physics.ins-det].
12. M. L. Mangano, S. J. Parke and Z. Xu, "Duality and Multi-Gluon Scattering," Nucl. Phys. B **298**, 653 (1988). doi:10.1016/0550-3213(88)90001-6

The Charged Cosmic Rays Measured by AMS on the International Space Station

W. Xu

on behalf of the AMS Collaboration

Massachusetts Institute of Technology (MIT),
Cambridge, Massachusetts 02139, USA
E-mail: Weiwei.Xu@cern.ch

The Alpha Magnetic Spectrometer, AMS, is successfully operating on the International Space Station for more than 6 years and has collected over 100 billion cosmic rays. We present the new physics results from AMS on the precision measurements of elementary particles and nuclei in the cosmic rays. Surprisingly, the spectra of both the primary cosmic rays (including proton, helium, carbon, and oxygen) and the secondary cosmic rays (including lithium, beryllium, and boron) all progressively harden above ~ 200 GV. Remarkably, the boron-to-carbon flux ratio is well described by a single power law above 65 GV and consistent with the Kolmogorov turbulence model of magnetized plasma. Unexpectedly, of the four cosmic elementary particles, protons, antiprotons and positrons have nearly identical rigidity (momentum/charge) dependence from ~ 60 GV to ~ 500 GV, electrons have distinctly different rigidity dependence. Most importantly, AMS continues studies of complex antimatter candidates with stringent detector verification and constantly collection of data.

Keywords: Primary cosmic ray; secondary cosmic ray; elementary particles; antimatter.

1. Introduction

There are two kinds of cosmic rays in space. The neutral cosmic rays, light rays and neutrinos, have been measured by many satellites and large ground-based and underground detectors. The studies of neutral cosmic rays have provided fundamental information about the universe. In additional to neutral cosmic rays, there are charged cosmic rays which are absorbed in the Earth's atmosphere and therefore their original properties can only be studied in space.

AMS is the first large acceptance precision magnetic spectrometer in space to measure the original properties of the charged cosmic rays. With its precision detectors, AMS measures the sign and value of the charge, the momentum, the energy and the arrival direction of elementary particles (electrons, positrons, protons, antiprotons), nuclei, and antinuclei.

AMS was installed on the International Space Station in May 2011, and has been taking data continuously for more than 6 years. To date, AMS has collected over 100 billion cosmic rays. With the precision, large acceptance and long exposure time, AMS is conducting the precision search for the origin of dark matter, antimatter, and cosmic rays as well as to directly explore the universe for new phenomena at high energies.

2. AMS Detector

The layout of the AMS detector[1] is shown in Fig. 1. It consists of 9 planes of precision silicon Tracker; a Transition Radiation Detector, TRD; four planes of Time of Flight counters, TOF; a Magnet; an array of anti-coincidence counters, ACC, surrounding the inner Tracker; a Ring Imaging Čerenkov detector, RICH; and an Electromagnetic Calorimeter, ECAL. The figure also shows a high energy positron of 868 GeV recorded by AMS.

Together, the tracker and the magnet measure charged cosmic rays with momentum p, charge Z and rigidity $R = pc/Ze$. The tracker[2] has nine layers, the first at the top of the detector, the second just above the magnet, six within the bore of the magnet, and the last just above the ECAL. Each layer contains double-sided silicon microstrip detectors that independently measure the x and y coordinates. The tracker accurately determines the particle trajectory by multiple measurements of the coordinates with a resolution in each layer of 10 μm in the bending (y) direction. The maximum detectable rigidity, MDR, is 2 TV for $|Z| = 1$, 3.2 TV for helium and 3.7 TV for carbon nuclei over the 3 m lever arm. Signal amplitude from each layer of the tracker also provides an independent measurement of charge $|Z|$. The overall charge resolution of the tracker is $\Delta Z/Z$ is 5% for $|Z| = 1$, 3.5% for helium and 2% for carbon nuclei.

The TOF counters measure the velocity $\beta = v/c$ and $|Z|$ of cosmic rays. Two TOF planes are located above and two planes are located below the magnet[3]. For

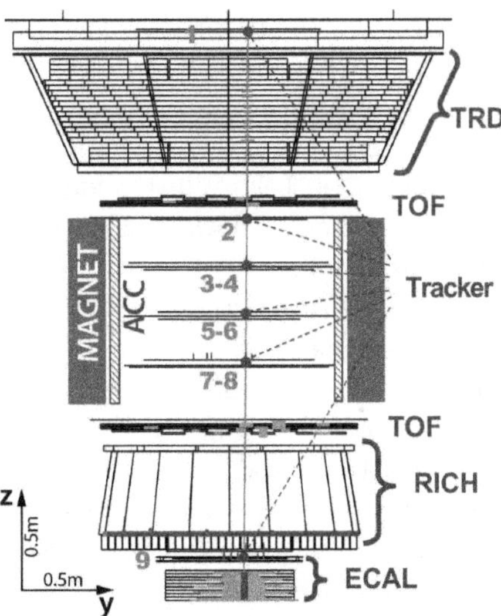

Fig. 1. A 868 GeV positron event as measured by the AMS detector on the ISS in the (y-z) plane. Tracker planes 1-9 measure the particle charge, sign and momentum. The TRD identifies the particle as an electron/positron. The TOF measures the charge and ensures that the particle is downward-going. The RICH measures the charge and velocity. The ECAL independently identifies the particle as an electron/positron and measures its energy.

$|Z| = 1$ particles, the average time resolution of each counter has been measured to be 160 ps and the overall velocity resolution to be $\Delta\beta/\beta^2 = 4\%$. The time resolution improves for higher charges to reach 50 ps for oxygen nuclei with the corresponding improvement in the velocity resolution of $\Delta\beta/\beta^2 = 1\%$. This discriminates between downward- and upward-going particles. The coincidence of signals from the four TOF planes provides a basis for charged particle trigger, whereas the coincidence of 3 out of the 4 TOF layers provides an unbiased trigger. The unbiased trigger, prescaled to 1%, is used to measure the trigger efficiency from the data.

To distinguish antiprotons and positrons from protons and electrons which are reconstructed in the tracker with wrong rigidity sign due to the finite tracker resolution or due to interactions with the detector materials, a charge confusion estimator Λ_{CC} is defined[4]. The estimator combines information from the tracker such as the track $\chi^2/d.f.$, rigidities reconstructed with different combination of tracker layers, the number of hits in the vicinity of the track, and the charge measurements in the TOF and the tracker. With this method, antiprotons/positrons have $\Lambda_{CC} \sim +1$ whereas charge confusion protons/electrons have $\Lambda_{CC} \sim -1$. This ensures efficient separation of the signal from the charge confusion events.

The TRD uses transition radiation to distinguish between $\bar{p}(p)$ and $e^-(e^+)$ and dE/dx to independently identify nuclei[5]. It consists of 5248 proportional tubes of 6 mm diameter with a maximum length of 2 m assembled in 16-tube modules. The 328 modules are mounted in 20 layers. There are 12 layers of proportional tubes along the y axis located in the middle of the TRD and, along the x axis, four layers located on top and four on the bottom. To differentiate between $\bar{p}(p)$ and $e^-(e^+)$ in the TRD, signals from the 20 layers of proportional tubes are combined into a TRD estimator Λ_{TRD} formed from the ratio of the log–likelihood probability of the $e^\pm$ hypothesis to that of the $\bar{p}(p)$ hypothesis in each layer[6]. Antiprotons and protons have $\Lambda_{TRD} \sim 1$ whereas electrons and positrons have $\Lambda_{TRD} \sim 0.5$. This provides efficient separation of $\bar{p}(p)$ and $e^-(e^+)$. For illustration, the proton rejection power of the estimator at 90% $e^\pm$ efficiency is shown in Fig. 2(a) as a function of energy. It reaches up to 10^4 in the energy range of interest.

The ring imaging Čerenkov detector[7] measures velocity and $|Z|$. It consists of two radiators, an expansion volume, and a photodetection plane. The dielectric radiators induce the emission of a cone of Čerenkov photons when traversed by charged particles with a velocity greater than the velocity of light in the material. The central radiator is formed by 16 sodium fluoride, NaF, tiles, each $85 \times 85 \times 5$ mm^3, with a refractive index $n = 1.33$. These are surrounded by 92 tiles, each $115 \times 115 \times 25$ mm^3, of silica aerogel with a refractive index $n = 1.05$. This allows the detection of particles with velocities $\beta > 0.75$ with the NaF radiator and $\beta > 0.953$ with the aerogel radiator. The expansion volume has a distance along z of 470 mm and is surrounded by a high reflectivity mirror to increase detection efficiency. The photodetection plane is an array of 10 880 photosensors with an effective spatial granularity of 8.5×8.5 mm^2. The sum of the signal amplitudes is proportional to

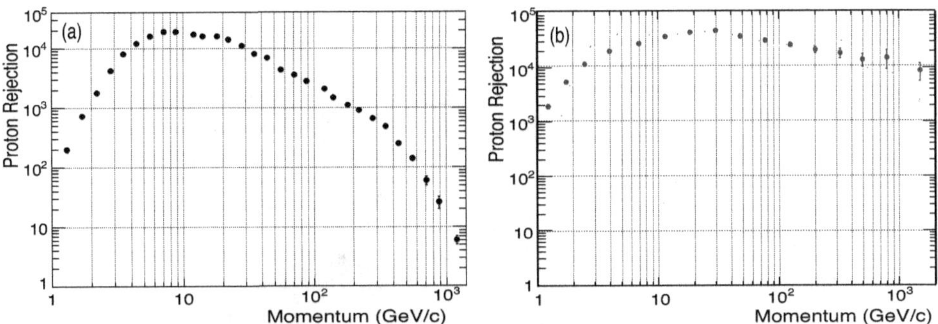

Fig. 2. (a) The proton rejection measured by the TRD as a function of track momentum at 90% selection efficiency for e$^\pm$. (b) The measured proton rejection using the ECAL and the Tracker. For 90% e$^\pm$ ECAL selection efficiency, the measured proton rejection is $\sim 10{,}000$ for the combination of the ECAL and the Tracker in the momentum range 3–500 GeV/c, independent of the TRD.

Z^2. The opening angle of the Čerenkov radiation cone is a measure of the velocity of the incoming charged particle. Typical velocity resolution is $\Delta\beta/\beta = 0.1\%$ for $|Z| = 1$.

The three-dimensional imaging capability of the 17 radiation length ECAL[8] allows for an accurate measurement of the $e^\pm$ energy E and shower shape. It consists of a multilayer sandwich of 98 lead foils and $\sim 50{,}000$ scintillating fibers with an active area of $648{\times}648\,\mathrm{mm}^2$ and a thickness of $166.5\,\mathrm{mm}$. The calorimeter is composed of 9 superlayers, with the fibers running in one direction only in each superlayer. The 3–D imaging capability of the detector is obtained by stacking alternate superlayers with fibers parallel to the x- and y-axes (5 and 4 superlayers, respectively). The energy resolution has been measured to be $\sigma(E)/E = \sqrt{(0.104)^2/E + (0.014)^2}$ (E in GeV). The uncertainty of the absolute $e^\pm$ energy scale has been verified to be 2% in the range 10–290 GeV. Below 10 GeV it increases to 5% at 0.5 GeV and above 290 GeV to 4% at 700 GeV. To cleanly separate protons from electrons and positrons, an ECAL estimator[8] is constructed using the 3–D shower shape in the ECAL. The proton rejection power of the ECAL estimator when combined with the energy-momentum matching requirement $E/p > 0.75$ is better than $\sim 10^4$ (see Fig. 2(b)).

Before launch, AMS was extensively calibrated at the CERN SPS with 180 and 400 GeV/c proton beams and positron, electron, and pion beams of 10 to 290 GeV/c. In total, calibrations with 18 different energies and particles at 2000 positions were performed. These data allow the determination of the tracker rigidity resolution function with high precision and the verification of the absolute rigidity scale. Since launch, the detector has been monitored and controlled around the clock. Its performance has been steady over time.

Monte Carlo simulated events are produced using a dedicated program developed by the collaboration from the GEANT 4.10.1 package[9]. This program simulates

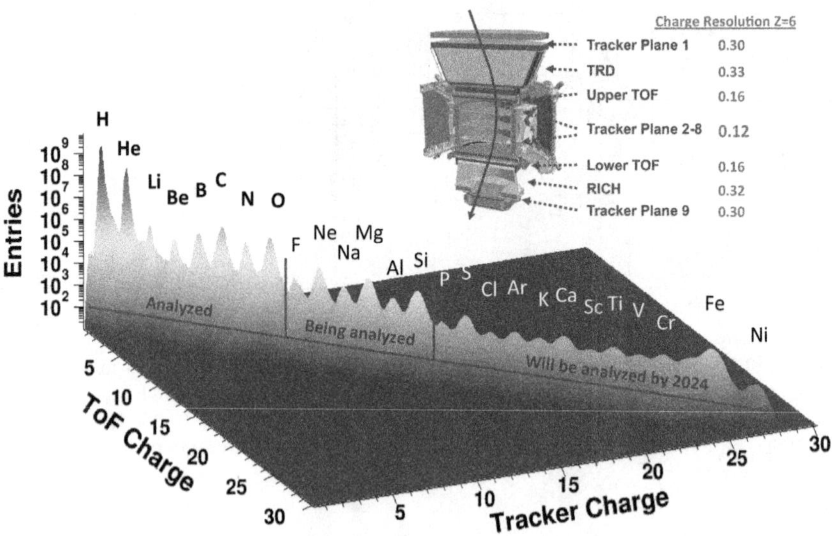

Fig. 3.   Cosmic ray nuclei studied by AMS.

electromagnetic and hadronic interactions of particles in the material of AMS and generates detector responses. The digitization of the signals is simulated precisely according to the measured characteristics of the electronics. The simulated events then undergo the same reconstruction as used for the data. In all the analysis, the MC sample is sufficient such that it does not contribute to the statistics errors.

3. Primary Cosmic Rays

The primary cosmic rays, like proton, Helium, Carbon, and Oxygen, are thought to be produced in their sources, then propagate through the interstellar medium and reach the earth. They carry information about their original spectra and propagation process. The secondary cosmic rays, like Lithium, Beryllium, and Boron, are thought to be produced by the collision of the primary cosmic rays with the interstellar medium and then propagate to reach the earth. They carry information about the propagation of primaries, secondaries and the properties of the interstellar medium. Therefore, the knowledge of the rigidity dependence of the primary and secondary cosmic rays is important in understanding the origin, acceleration and propagation of cosmic rays.

The AMS detector comprises seven instruments, which independently identify different elementary particles as well as nuclei. Protons, helium, lithium, carbon, oxygen and heavier nuclei up to iron are intensively studied by AMS (see Fig. 3).

Protons and Helium are the two most abundant cosmic rays. It has been traditionally assumed for decades that the spectra of cosmic ray protons and Helium can be described by a single power law function (see for instance the current PDG Chapter "29. COSMIC RAYS"). It was the CREAM experiment to point out first

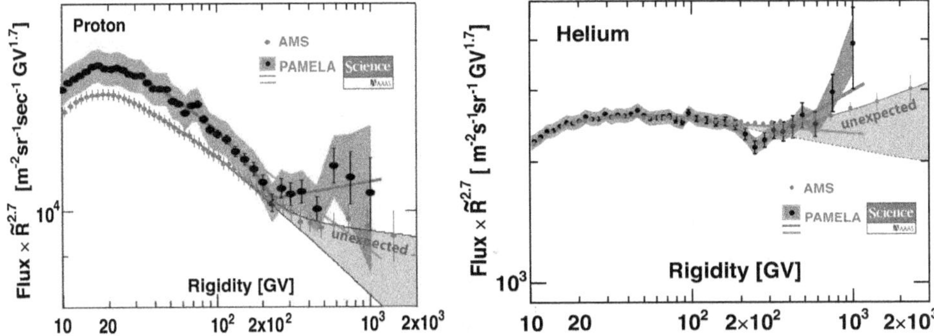

Fig. 4. The AMS proton (left) and Helium (right) spectra hardens progressively above 100 GV over the range of few hundred GV as indicated by the blue shaded area on the plot. Abrupt spectral hardening reported by the PAMELA experiment[14] (also shown on the plot) is not supported by the precision AMS measurements (color online).

that to reconcile the AMS-01 proton spectrum measurements below 200 GeV[10] (well described by a single power law function) with their own measurement in the range from 2.5 TeV to 250 TeV[11] (also well described with a single power law function) a change of the spectral index at few hundred GeV is required.

However, to study the detailed behavior of this transition required AMS precision[12,13]. Fig. 4 shows the AMS results on proton (left figure) and Helium (right figure) flux as a function of rigidity together with the earlier results of PAMELA. The proton spectrum measured by AMS hardens progressively above 100 GV over the range of few hundred GV. This is in direct contradiction with the earlier conclusion by the PAMELA experiment[14] that "At 230 to 240 GV, the proton and helium data exhibit an abrupt spectral hardening". Similarly to the proton spectrum, the helium spectrum measured by AMS hardens progressively above 100 GV over the range of few hundred GV, and not very abruptly at 230-240 GV as show PAMELA results[14].

In 2010 studying behavior of high energy primary cosmic nuclei (a combination of C, O, Ne, Mg, Si, Fe) CREAM noted that "A broken power law gives a better fit to our data."[11] (see also Fig. 5 (left plot)). The individual nuclei spectra measured by CREAM exhibit less conclusive behavior as illustrated in Fig. 5 (right plot) for the oxygen spectrum.

AMS performs unique studies in this rigidity range to understand the dynamics of progressive hardening of the individual primary nuclei spectra. For instance, protons and Helium are both primary cosmic rays and assumed to be produced in the same sources, therefore, the proton-to-Helium flux ratio is expected to be constant at high energy. As seen in Fig. 6, the precision AMS results show that the proton-to-Helium flux ratio is not a constant but decreasing with rigidity.

The carbon and oxygen spectra measured by AMS also cannot be described by current cosmic rays model GALPROP[15] based on pre-AMS data, as seen from Fig. 7. The precision AMS results provide new parameters for cosmic rays models.

134

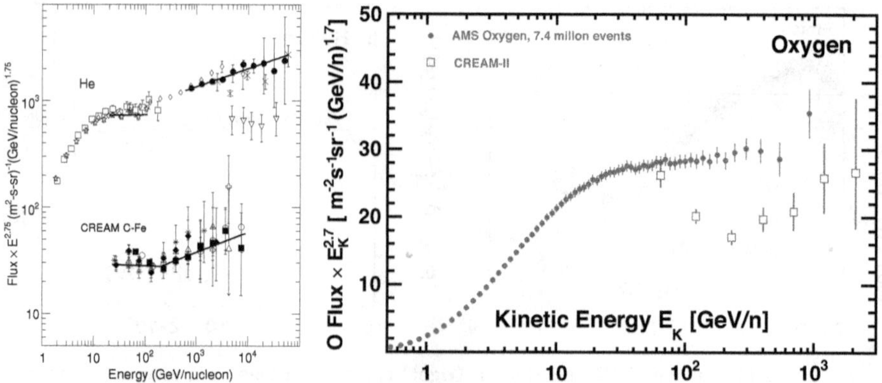

Fig. 5. (left plot, taken from [11]) A fit of "broken power law" to combined CREAM data on C, O, Ne, Mg, Fe nuclei. (right plot) Comparison of the AMS oxygen spectrum with the CREAM measurements [11] shows that conclusions on the dynamics of the oxygen spectrum hardening from these two measurements are different.

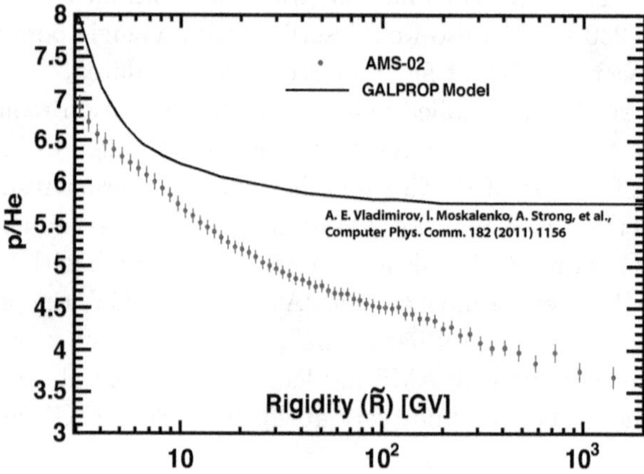

Fig. 6. The proton-to-helium flux ratio showing a steep decease at high rigidities in contrast to flat theoretical predictions.

4. Secondary Cosmic Rays

The secondary cosmic rays, like Lithium, Beryllium, and Boron, are thought to be produced by the collision of the primary cosmic rays with the interstellar medium and then propagate to reach the earth. They carry information about the propagation of primaries, secondaries and the properties of the interstellar medium. Therefore, the knowledge of the rigidity dependence of the secondary cosmic rays is important in understanding the origin, acceleration and propagation of cosmic rays.

As seen in Fig. 8, although lithium is assumed to be a secondary cosmic ray, surprisingly, its spectrum changes behavior at ~ 100 GeV/n, which is nearly identical with the energy where protons and Helium spectra start to harden.

With 1 million Beryllium nuclei and 2.9 million Boron nuclei, AMS measures the Beryllium and Boron spectra to unprecedented precision. As seen in the two top panels of Fig. 9, the AMS results cannot be explained by current GALPROP model. The unstable isotope of Beryllium, ^{10}Be, has a half-life of 1.5 million years and decays into B. The Be/B ratio therefore increases with energy due to time dilation when the Be approaches the speed of light. Hence, the ratio of beryllium to boron provides information on the age of the cosmic rays in the galaxy. As shown in Fig. 9 bottom, from this AMS has determined that the age of cosmic rays in the galaxy is 12 million years.

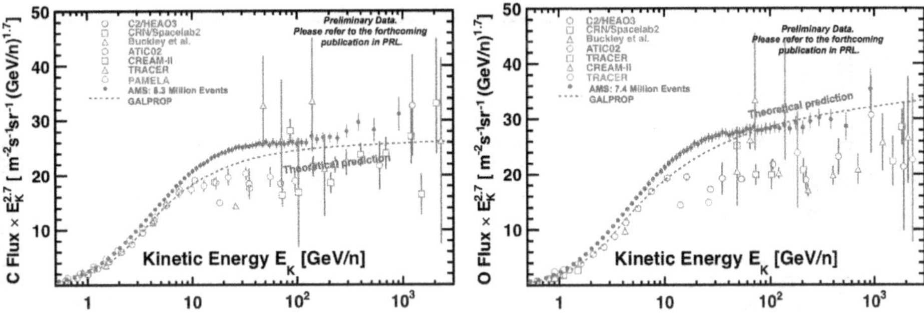

Fig. 7.   The comparison of Carbon (left) and Oxygen (right) spectra measured by AMS and earlier experiments. AMS results show that both the Carbon and Oxygen spectra progressively harden at high energy above ~ 100 GeV/n.

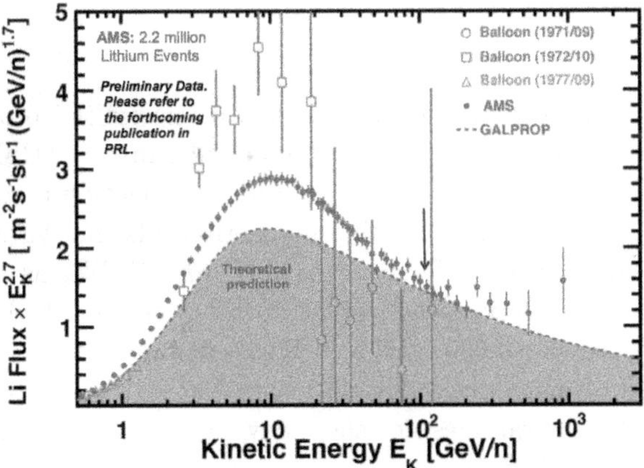

Fig. 8.   The comparison of Lithium spectrum measured by AMS and earlier experiments. AMS results show that the lithium spectrum cannot be described by a single power law and starts to harden at the similar energy with protons and Helium spectra.

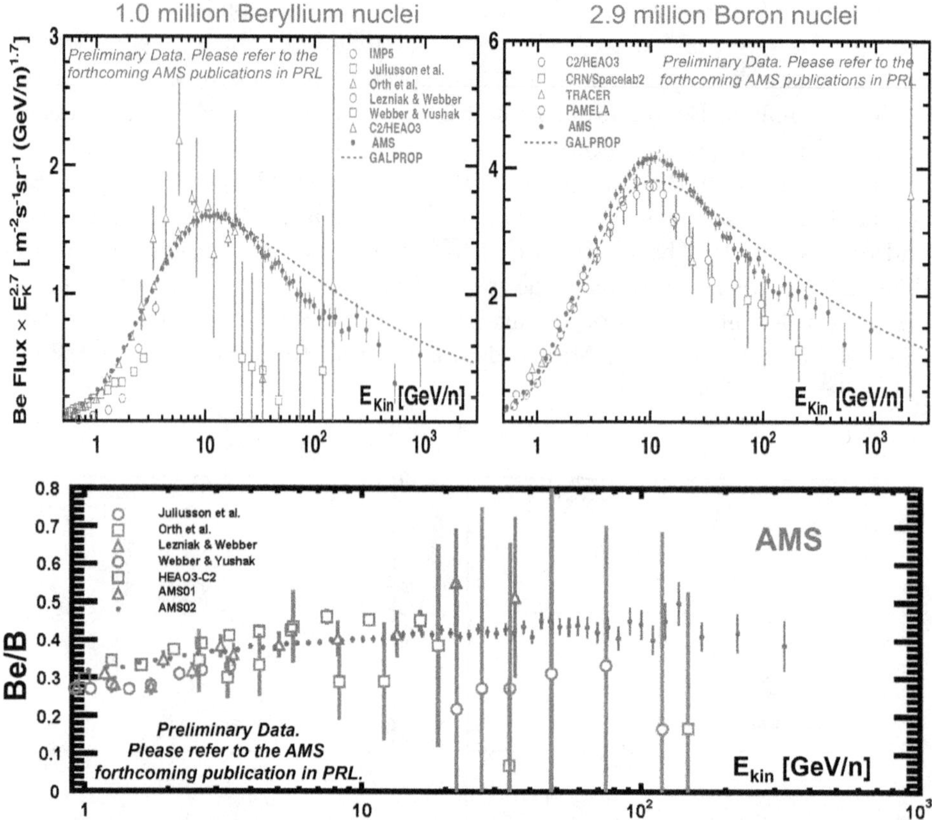

Fig. 9. Beryllium (top left plot) and Boron (top right plot) fluxes show significant improvements in the accuracy and energy reach in comparison to the previous measurements. (bottom plot) The beryllium-to-boron (Be/B) flux ratio increases with energy due to time dilation of the decaying Be. The measurement yields the age of cosmic rays in the galaxy of ~ 12 million years.

The flux ratio between secondaries (B) and primaries (C) provides information on propagation and the average amount of interstellar material (ISM) through which the cosmic rays travel in the galaxy. Cosmic ray propagation is commonly modeled as a fast moving gas diffusing through a magnetized plasma. Various models of the magnetized plasma predict different behavior. Remarkable, as shown in Fig. 10, above 65 GeV, the B/C ratio measured by AMS is well described by a single power law $B/C = R^\delta$ with $\delta = -0.333\pm0.015$.[16] This is in agreement with the Kolmogorov turbulence model of magnetized plasma where $\delta = -1/3$ asymptotically. Of equal importance, the B/C ratio does not show any significant structures in contrast to many cosmic ray models.

Cosmic ray Nitrogen is of particular importance, because it contains both primary and secondary components. With 2.1 million Nitrogen nuclei, AMS measures the cosmic ray Nitrogen flux to unprecedented precision and shows distinctly

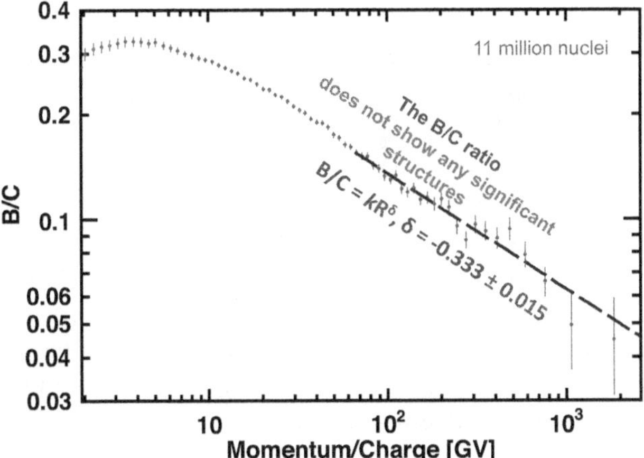

Fig. 10. The B/C ratio does not show significant structures in contrast to many cosmic ray models requiring such structures at high rigidities. Above 65 GV, the B/C ratio is well described by a single power law.

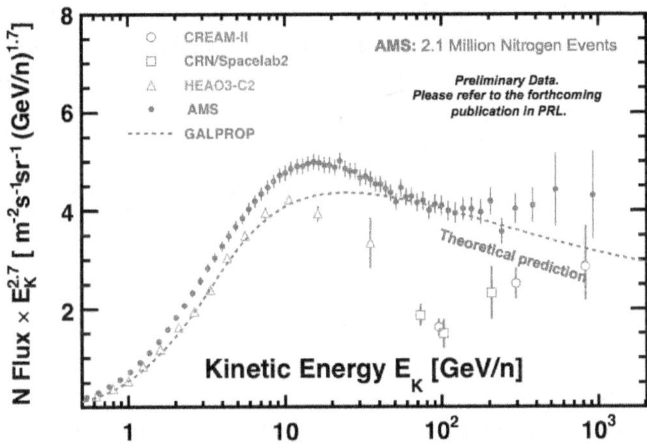

Fig. 11. The Nitrogen spectrum measured by AMS. The earlier experimental results are also shown for comparison. AMS results show that the Nitrogen spectrum cannot be described by a single power law and starts to harden at ~ 100 GeV/n.

different characteristics comparing with previous measurements. The Nitrogen spectrum shows a significant progressive hardening at ~ 100 GeV/n.

In summary, the simultaneous measurements of the cosmic ray nuclei up to Iron ($Z = 26$) and beyond, including the primary and cosmic rays, carry enormous amount of information on the origin, acceleration and propagation of cosmic rays. The current models based on pre-AMS data cannot explain our observations. The accuracy and characteristics of AMS data requires a comprehensive model to be developed.

5. Elementary Particles

Of all the charged elementary particles in the cosmos, only electrons (e^-), positrons (e^+), protons (p), and antiprotons $(\bar{p})$ have an infinite lifetime and thus can be studied accurately by AMS.

Positrons and antiprotons are rare components in the cosmic rays. They are sensitive to the new physics like Dark Matter or astrophysics phenomena, such as Supernova Remnants and Pulsars. AMS measured cosmic antiproton flux based on 3.49×10^5 antiproton events, with over 2200 events above 100 GeV. The latest positron measurement by AMS is based on 1.08 million positron events. These precision results generated widespread interests. The details on the measurements of the positron spectrum, positron fraction, antiproton spectrum, and antiproton-to-proton flux ratio and their implications in the Dark Matter searches have been reported in another talk in this Lepton-Photon Conference[17].

It is interesting to compare behavior of the spectra of all elementary particles among themselves. Traditionally, electrons and protons are assumed to be primary cosmic rays, i.e. particles produced directly at sources of cosmic rays like exploding supernovae. On the contrary, positrons and antiprotons are assumed to be secondary cosmic rays, i.e. coming from the interaction of primary cosmic rays with the interstellar media. In addition, electrons and positrons have much smaller mass than protons and antiprotons so they lose much more energy in the galactic magnetic fields. This is illustrated in Fig. 12, where the behavior of the electron spectrum is compared with that of the proton spectrum. As expected, in the rigidity region free of solar modulation effects, above 20 GeV, the proton spectrum is much harder than the electron spectrum (see Fig. 12(a)). However, further studies by AMS bring a lot of surprises. As seen in Fig. 12(b), the behavior of antiprotons (assumed to be secondaries) and protons (assumed to be primaries) at high rigidities are very similar, while the behavior of electrons (i.e. primaries) and positrons (secondaries) are very different as illustrated by the AMS data in Fig. 12(c). Most surprising is that above 60 GeV, positrons, protons and antiprotons display identical an energy dependence whereas electrons exhibit a totally different energy dependence as shown in Fig. 12(d). The physics explanation for this AMS observation is yet to be found.

More importantly, the measurement of positrons, electrons and antiprotons are currently limited by the statistics. AMS will extend the measurement to higher energy with much improved accuracy. This will provide more information to understand the origin of this unexpected observation.

6. Complex Antimatter in Cosmic Rays

The Big Bang origin of the Universe requires that matter and antimatter be equally abundant at the very hot beginning of the universe. The search for the explanation for the absence of antimatter in a complex form is known as Baryogenesis. Baryogenesis requires both a strong symmetry breaking and a finite proton lifetime. Despite the outstanding experimental efforts over the last half a century, no evidence of

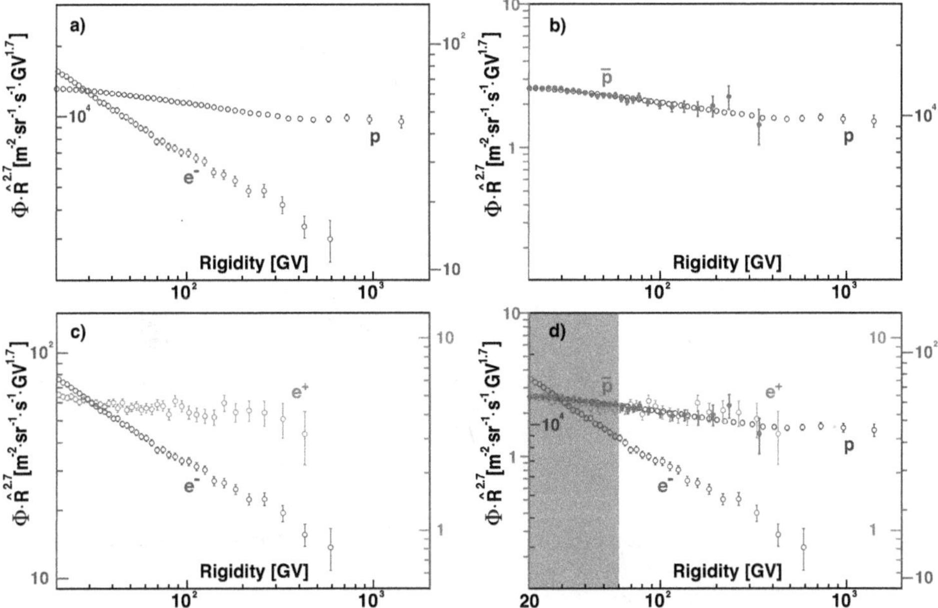

Fig. 12. AMS measurements of the fluxes of elementary particles: (a) comparison of the functional behavior of the proton and electron fluxes; (b) comparison of the functional behavior of the proton and antiproton fluxes; (c) comparison of the functional behavior of the positron and electron fluxes; (d) the positron, antiproton and proton fluxes show the very same functional behavior in the absolute rigidity range 60–500 GV, whereas electrons exhibit distinctly different dependence.

strong symmetry breaking nor of proton decay have been found. Therefore, the observation of a single anti-helium event in cosmic rays is of great importance.

In six years, AMS has collected 100 billion events. From this unparalleled sample of charged cosmic rays 700 million events with charge $|Z| = 2$ are selected. All these events have been identified as Helium nuclei, except few that have negative rigidity or $Z = -2$. One of these anti-He candidates is presented in Fig. 13. All of the $Z = -2$ candidates have one common and peculiar feature — mass around ^{3}He.

At a rate of approximately one antihelium candidate per year and a required signal to background rejection of one in a billion, a detailed understanding of the instrument is required but is exceedingly difficult and time consuming. In the coming years one of our main efforts is to perform stringent detector verification and to collect more data in order to ensure that these $Z = -2$ events are indeed antihelium.

7. Conclusion

In six years on the ISS, AMS has recorded more than 100 billion cosmic ray events. The precision AMS results reveal many unexpected phenomena.

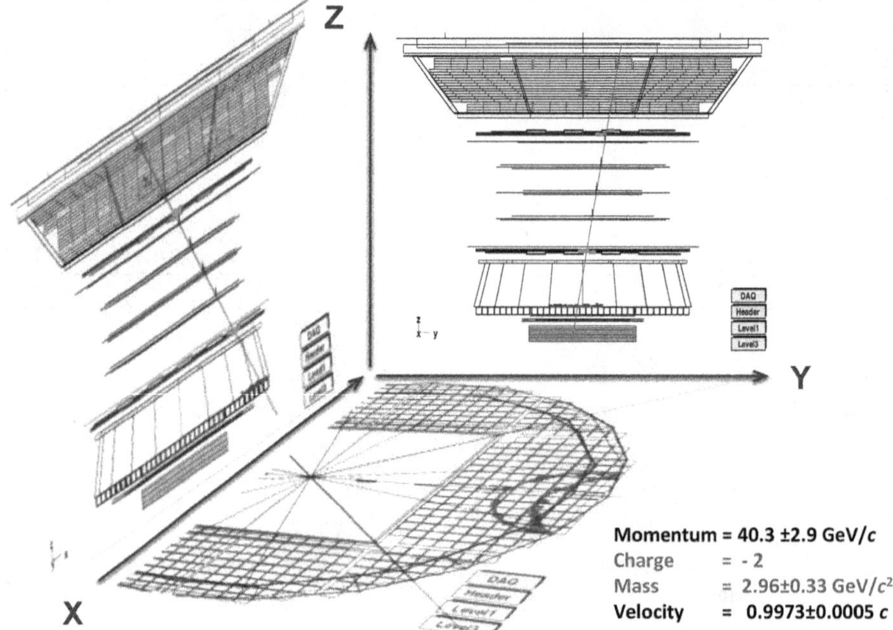

Fig. 13.   An antihelium candidate.

The primary cosmic rays (proton, Helium, Carbon and Oxygen) and secondary cosmic rays (Lithium, Beryllium and Boron) all progressively harden at high rigidity above ~ 200 GV. The proton-to-Helium flux ratio is decreasing with rigidity in contrast to the theoretical prediction. The Boron-to-Carbon flux ratio is well described by a single power law above 65 GV, in good agreement with the prediction of the Komorogov turbulence model. The positrons, antiprotons and protons have nearly identical rigidity dependence from ~ 60 to ~ 500 GV, while electrons have distinctly different rigidity dependence.

The AMS measurements provide precise and unexpected information on the production, acceleration and propagation of cosmic rays. The accuracy and characteristics of the data, simultaneously from many different types of cosmic rays require the development of a comprehensive model of cosmic rays.

Of particular significance is our study of complex antimatter in the cosmos. Through stringent detector verification, collecting additional data and anti-deuteron analysis we will ensure that the observed $Z = -2$ events are indeed anti-helium.

As a magnetic spectrometer studying cosmic rays, AMS is unique in its precision and energy reach. For the foreseeable future this is the only magnetic spectrometer in space to perform precision measurements and to explore the unknown with high expectations for exciting discoveries.

Acknowledgements

This work has been supported by acknowledged person and institutions in Ref. 16.

References

1. A. Kounine, Int. J. Mod. Phys. **E 21** (2012) 123005; S.C.C. Ting, Nucl. Phys. B, Proc. Suppl. **243-244** (2013) 12.
2. B. Alpat *et al.*, Nucl. Instrum. Meth. **A 613** (2010) 207; G. Ambrosi *et al.*, Nucl. Instrum. Meth. **A 869** (2017) 29.
3. V. Bindi *et al.*, Nucl. Instrum. Meth. **A 743** (2014) 22.
4. M. Aguilar *et al.*, Phys. Rev. Lett. **117**, 091103 (2016).
5. F. Hauler *et al.*, IEEE Trans. Nucl. Sci. **51** (2004) 1365; P. Doetinchem *et al.*, Nucl. Instrum. Meth. **A 558** (2006) 526; Th. Kirn, Nucl. Instrum. Meth. **A 706** (2013) 43.
6. M. Aguilar *et al.*, Phys. Rev. Lett. **110** (2013) 141102; L. Accardo *et al.*, Phys. Rev. Lett. **113** (2014) 121101.
7. M. Aguilar-Benitez *et al.*, Nucl. Instrum. Meth. **A 614** (2010) 237; F. Giovacchini, Nucl. Instrum. Meth. **A 766** (2014) 57.
8. C. Adloff *et al.*, Nucl. Instrum. Meth. **A 714** (2013) 147; A. Kounine, Z. Weng, W. Xu and C. Zhang, Nucl. Instrum. Meth. **A 869** (2017) 110.
9. S. Agostinelli *et al.*, Nucl. Instrum. Meth. **A 506** (2003) 250; J. Allison *et al.*, IEEE Trans. Nucl. Sci. **53** (2006) 270.
10. J. Alcaraz *et al.*, Phys. Lett. **B 472** (2000) 215; J. Alcaraz *et al.*, Phys. Lett. **B 494** (2000) 193.
11. H.S. Ahn *et al.*, Astro. J. Lett. **714** (2010) L89.
12. M. Aguilar *et al.*, Phys. Rev. Lett. **113** (2014) 221102.
13. M. Aguilar *et al.*, Phys. Rev. Lett. **115** (2015) 201101.
14. O. Adriani *et al.*, Science **332** (2011) 69.
15. We used GALPROP WebRun, A. E. Vladimirov *et al.*, Comput. Phys. Comm. 182, 5 (2011), with parametrization from R. Trotta *et al.*, Astrophys. J. 239, 2 (2011). We thank Dr. I. Moskalenko for useful discussions.
16. M. Aguilar *et al.*, Phys. Rev. Lett. **117** (2016) 231102.
17. W. Xu on behalf of AMS Collaboration, The Latest Results from AMS on the Searches for Dark Matter, 28th *Lepton Photon*, 2017, to be published.

Charged Lepton Flavor Violations (CLFV)

A. M. Baldini

Istituto Nazionale di Fisica Nucleare
Largo B. Pontecorvo 3,56127 Pisa, Italy
E-mail: alessandro.baldini@pi.infn.it

While neutrino oscillation is an experimentally verified lepton flavor violating (LFV) phenomenon, which however does not necessarily require drastic changes in the standard model (SM) of particle physics, since it could for instance be taken into account by inserting non zero dirac neutrino masses in the model, any experimental observation of a LFV process in the charged lepton sector (CLFV) would only be imputable to new physics (NP) beyond the SM (BSM). I present here a review of the current experimental limits on CLFV processes, of the compelling theoretical reasons for actively searching BSM physics in this way and of the experimental activities going on and planned for the future.

1. General overview of CLFV decays

Table 1 contains a selection of the most stringent upper limits[a] on CLFV processes and of the possible future improvements.

Table 1. A selection of the most stringent CLFV upper limits

Process	Current Limit 90%C.L.	Experiment/ Laboratory	Future Experiment
$\mu^+ \to e^+\gamma$	4.2×10^{-13}	MEG[1] at PSI	6×10^{-14} - MEGII[4]
$\mu^+ \to e^+e^+e^-$	1.0×10^{-12}	SINDRUM I[2] at PSI	$O(10^{-16})$ - Mu3e[5] at PSI
$\mu N \to eN$	7.0×10^{-13}	SINDRUM II[3]	7×10^{-17} - COMET[6], Mu2e[7], DeeMe[8]
$\tau \to e\gamma$	3.3×10^{-8}	BaBar[9] at SLAC	
$\tau \to \mu\gamma$	4.4×10^{-8}	BaBar[9]	Belle II[11]
$\tau \to \mu\mu\mu$	2.1×10^{-8}	Belle[10] at KEK	
$\tau \to eee$	2.7×10^{-8}	Belle[10]	
$K_L \to e\mu$	4.7×10^{-12}	E871[12] at BNL	
$K^+ \to \pi^+e^-\mu^+$	1.3×10^{-11}	E865[13] at BNL	NA62[16] at CERN
$B^0 \to e\mu$	2.8×10^{-9}	LHCb[14]	
$B \to Ke\mu$	3.8×10^{-8}	BaBar[15]	

[a]In the paper I usually refer to 90% Confidence Level (CL) upper limits (UL) when comparing different experimental sensitivities. However since background evaluation for future experiments is hard to evaluate it is sometimes preferred to quote the Single Event Sensitivity (SES) which is the branching ratio for which the experiment would see one event of the searched process. In the absence of background, SES multiplied by 2.3 is equal to the 90% CL UL.

Mesons decays

Improvements of several orders of magnitudes are foreseen for the sensitivities of upcoming experiments looking for CLFV processes using μ or τ beams while the situation for mesons decays is less clear, though an improvement of one order of magnitude in sensitivity to the $K^+ \to \pi^+ e^- \mu^+$ is considered for instance to be possible by the NA62 experiment at CERN[16].

Apart from technical experimental reasons, a big role was, and maybe still is, played by theoretical guidance, or prejudice, concerning CLFV processes in mesons decays since R-parity conserved supersymmetric theories predict meson CLFV decays to be strongly suppressed by muon constraints[17].

It is however probably wiser to search for BSM physics not by limiting ourselves to just one direction, on the basis of theoretical prejudice, but preferably by looking at all possible scenarios. For instance the recent LHCb results on possible Lepton Flavor Unitarity (LFU) violations in the decay of B-mesons which were reported also at this Conference[18] could be a sign of new physics which, if interpreted in the framework of specific models, for instance the leptoquark model of 19, could give rise also to CLFV. It must be noted that, as shown in 19, the parameter space region of that particular model is in any case severely constrained by the non observation of CLFV processes in muon decays.

Higgs decays

Recently new eperimental results on CLFV Higgs decays were obtained by LHC experiments[20]:

$$B_{H \to \mu\tau} < 0.25\%$$

$$B_{H \to e\tau} < 0.61\%$$

$$B_{H \to e\mu} < 0.035\%$$

at 95% confidence level. These results can be compared with those from CLFV leptonic decay limits reported in Table 1, as it is done in 21, by using an effective lagrangian:

$$\mathcal{L} \supset \sum_i (m_e)_i \bar{e}_{Li} e_{Ri} + \sum_{ij} (Y_e^h)_{ij} \bar{e}_{Li} e_{Rj} h + h.c.$$

where i and j are the lepton family indices. The limits on CLFV decays may be therefore converted to limits on the non diagonal $(Y_e^h)_{ij}$ Yukawa couplings to the physical Higgs (h). The higgs decay limits on Yukawa couplings turn out[21] to be one order of magnitude more stringent than those obtained by τ decays but two orders of magnitude less stringent than those obtained by μ decays.

144

Leptons decays

CLFV τ decays searches were performed in the past at B-factories (see Table 1) were τ's are copiously produced in pairs. One of the produced τ's is used for tagging the pair by means of its known decay channels while CLFV decays are searched by using the other τ of the pair. Invariant mass and missing energy of the τ decay products are the variables used to select good τ decays. The invariant mass distribution is obviously much narrower when all the particles in the final state are charged as in $\tau \to 3\mu$, which is therefore a "golden channel" where to search for CLF decays. Were this decay channel background free one could expect that, at the luminosity of 50 ab^{-1}, reachable by 2025, the sensitivity of Belle II to $\tau \to 3\mu$ would increase by more than one order of magnitude with respect to previous searches. An increase of roughly one order of magnitude in sensitivity is foreseen for all the other possible CLFV τ decays.

The expected ratio in sensitivity of the τ channels with respect to the μ ones is again very much dependend on the particular model considered: it may range from the unity[22] to being larger by four order of magnitudes[23].

Figure 1 shows the huge sensitivity increase of CLFV searches with muons as function of time in the past seventy years and the perspectives for the next ten years.

The first experiments with cosmic-rays and pion beams established the familiy structure content of elementary particles. At the end of the 60's the advent of muon beams enabled to further increase the sensitivity to CLFV decays therefore

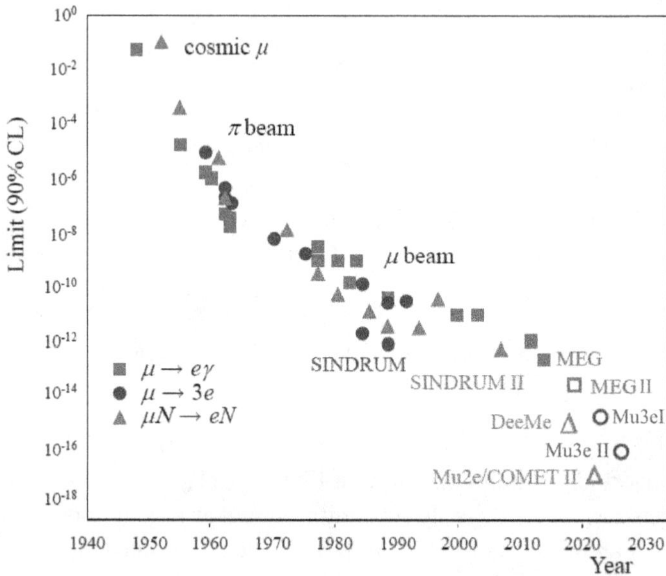

Fig. 1.   Sensitivity increase of CLFV searches as a function of time.

making possible to test new ideas such as the possible existence of new heavy leptons within the framework of the newly born gauge-symmetries[24,25]. These hypotheses essentially reflected the belief that lepton family conservation be not a fundamental symmetry of nature.

Present and future experiments are essentially following this tradition, intending to explore the possibility that new physics beyond the standard model may introduce, as we shall see in the next section, small non-diagonal terms in the lepton mass matrix leading to visible CLFV effects. The three main processes that are searched are $\mu \to e\gamma$, $\mu \to 3e$ and the conversion of the muon to an electron in the field of a nucleus ($\mu \to e$). While high intensity conventional muon beams are perfectly suited for the search of the first two processes increasing the sensitivity of $\mu \to e$ requires a completely different muon beam approach as we shall see in section 3.

2. Compelling reasons for searching CLFV

Why are CLFV processes so sensitive to BSM physics?

First of all the CLFV rates predicted by the SM, even with the introduction of conventional neutrino masses to account for oscillations, give totally negligible rates, experimentally unobservable. For instance the rate predicted[26] for $\mu \to e\gamma$ is of the order of 10^{-54}. There is therefore no need of precise calculations of SM processes to subtract: if a signal is observed "that" is a convincing, clear signal of new BSM physics.

Secondly, any modification of the SM lagrangian due to theoretically appealing necessities such as accounting for the vanishing neutrino masses (see-saw models), SM stabilization (SUperSYmmetries, grand unified or not), compositness, leptoquarks etc., makes CLFV diagrams arise. In some cases, such as for instance in a SUSY-GUT SO(10) scenario with slepton mass matrix mixing angles derived from the PMNS neutrino mass matrix, large $\mu \to e\gamma$ decay rates are predicted[21] which were excluded by the most recent experimental results[1].

As already stated above the rate of the different CLFV processes heavily depends on the physics model considered. However in a model independent approach (see for instance 21) just a few operators significantly contribute to calculating those rates. In the case of muon CLFV decays two operators have to be considered: the dipole and the four-fermion operator, whose schematic diagrams for muon decays are shown in Fig. 2.

In case of theories which prefer the dipole operator, such as the supersymmetric ones, $\mu \to e\gamma$ is favored by roughly a $1/\alpha$ factor with respect to $\mu \to e$ and $\mu \to 3e$ while in other kinds of theories such as compositness or leptoquark models, $\mu \to e\gamma$ is strongly suppressed relative to the other two processes.

It is therefore important to stress the complementarity of the different decay modes searches which would also give the possibility to distinguish among different models in case of positive detection.

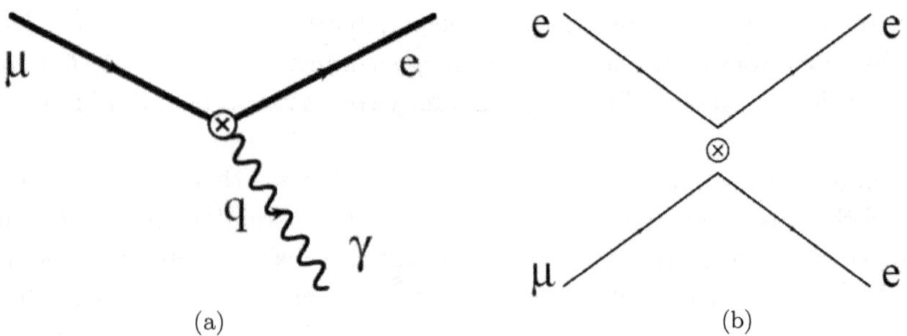

Fig. 2.   (a) Dipole operator (b) Four-fermion operator.

3. Experimental searches with muon beams

$\mu \to e\gamma$

All the results presented in Table 1 for muon decays were obtained at the PSI laboratory where the world most powerful continuous muon beam is produced by the interaction of a 590 Mev proton beam on a graphite target. 29 MeV/c muons coming from the decay at rest of pions on the surface of the target are collected by a magnetic channel and transported to the experimental areas. The intensity of tis beam is roughly $10^8\ \mu/s$.

The signature of a $\mu \to e\gamma$ event is given by a back-to-back, monoenergetic, time coincident photon-positron pair from the two body $\mu \to e\gamma$ decay. In each event, positron and photon candidates are described by four observables: the photon and positron energies (E_γ, E_e), their relative direction ($\Theta_{e\gamma}$) and emission time ($t_{e\gamma}$).

The MEG[1] detector is comprised of a positron spectrometer formed by a set of drift chambers and scintillation timing counters, located inside a superconducting solenoid with a gradient magnetic field along the beam axis, and a photon detector, located outside of the solenoid, made up of a homogeneous volume (900 liters) of liquid xenon (LXe) viewed by 846 UV-sensitive photomultiplier tubes (PMTs) submerged in the liquid.

The background has two components: one coming from the radiative muon decay (RMD) $\mu \to e\nu\bar{\nu}\gamma$ and one from the accidental superposition of energetic positrons from the standard muon Michel decay with photons from RMD, positron-electron annihilation-in-flight or bremsstrahlung. At the MEG data taking rate, 93% of events with $E_\gamma > 48$ MeV are from the accidental background.

Figure 3 shows the event distributions in the (E_γ, E_e) and ($\cos(\Theta_{e\gamma})$, $t_{e\gamma}$) planes for the full data set of the MEG experiment together with the 68%, 90% and 95% contours of the signal probability distribution function.

The final branching ratio upper limit on the $\mu \to e\gamma$ decay of 4.2×10^{-13} (90% Confidence Level) was obtained by a likekihood analysis that compared the measured distributions of the decay variables with the ones epected for the signl and backgrounds.

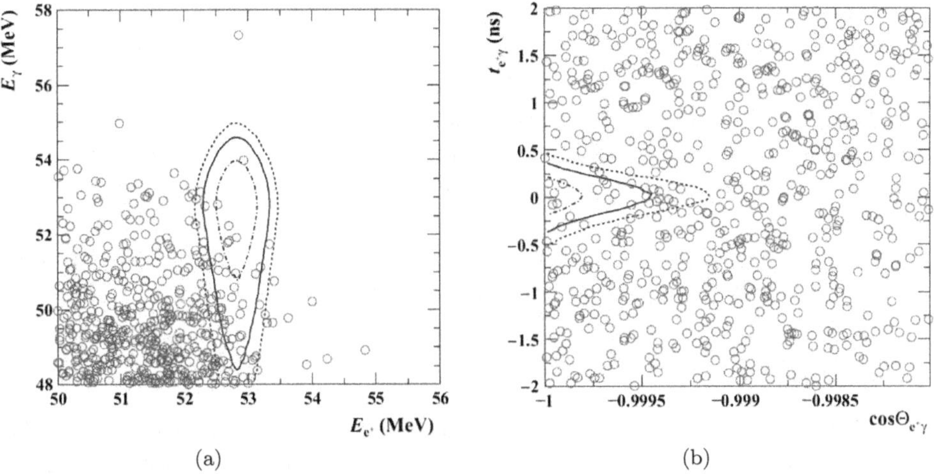

Fig. 3. MEG final results: (a) E_γ vs E_e (b) $\cos(\Theta_{e\gamma})$ vs $t_{e\gamma}$. 68%, 90% and 95% C.L. signal contour lines are shown.

While the signal region in Fig. 3 does not show any substantial excess of events it may be noted that it contains background events. This in short was the reason for deciding to stop MEG data taking and perform an upgrade of the experiment.

The MEG experiment upgrade[4], schematically shown in Fig. 4, is currently being implemented.

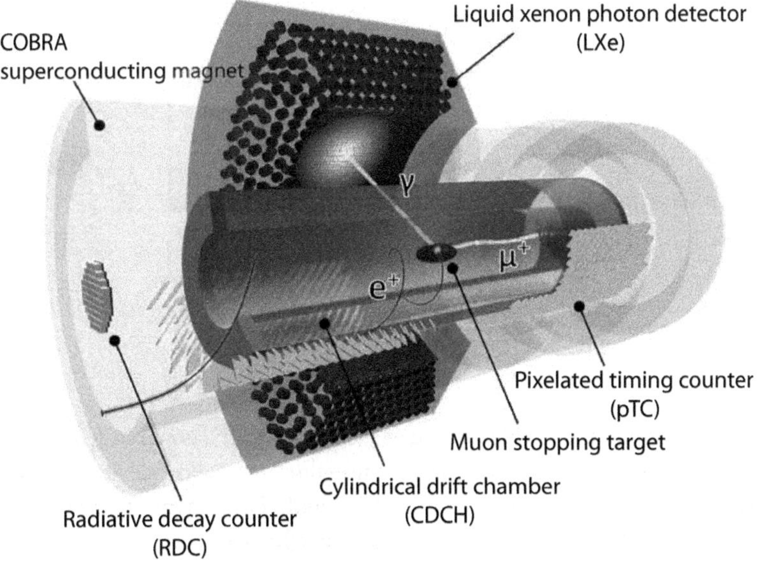

Fig. 4. Sketch of the MEG II experiment.

A new cylindrical drift chamber (CDCH) will substitute the old set of planar drift chambers; a new pixelated timing counter replaced the old scintillator bars and is currently under test; part of the photomultipliers of the LXe γ detector were substituted with smaller silicon photoultipliers. All the upgraded detectors resolutions should roughly improve by a factor two. A further detector (RDC) able to reduce the RMD background by measuring low energy positrons in coincidence with photons in the LXe detector was built and is currently under test. An improvement of one order of magnitude in sensitivity with respect to the final MEG result is foreseen in about three years of data taking which is foreseen to start in 2018.

$\mu \to 3e$

The same beam used by MEG will be used by a $\mu \to 3e$ search at the PSI laboratory[5]. In this case the signal consists of the decay into two positrons plus one electron of a positive muon at rest in a target; the total momentum of the three final state particles is therefore required to be zero and the total energy to be equal to the muon mass. The background has again two main components: a "physics" background due to the normal muon decay with an internally converted pair ($\mu \to 3e\nu\bar{\nu}$) and an accidental one in which one positron from normal muon decay is accidentally coincident in time and space with a Bhabha scattering of another Michel positron in the target. As in the $\mu \to e\gamma$ case a continuous muon beam is necessary in order to keep the accidental background under control.

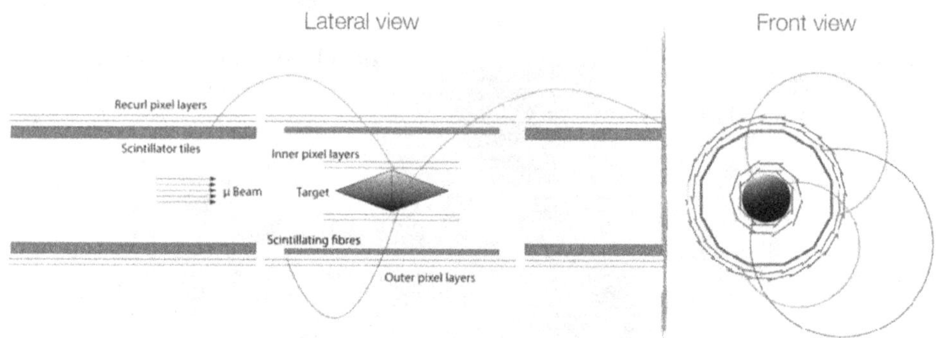

Fig. 5.   Scheme of the $\mu \to 3e$ experiment at PSI.

The planned experimental apparatus, schematically shown in Fig. 5, is composed of a muon stopping target target with the shape of a double hollow cone, placed at the center of a solenoidal 1 T magnetic field. Four layers of HVMAPS, 50 μm thick, constitute the precision tracker device which is complemented with a scintillating fiber hodoscope (composed of 250 μm scintillating fibers) and two lateral scintillating tiles (5 mm thick) detectors, all read out by silicon photomultipliers

(SiPM). The hodoscope and scintillating tiles detectors will have timing resolutions of roughly 500 and 70 ps respectively; precise timing is crucial for effectively reducing the accidental background.

The first phase of the experiment with the current PSI beam intensity shoud start around 2020, for a foreseen final sensitivity of 10^{-15} (90% CL). A second phase, for which a muon beam intensity increase of one order of magnitude is necessary, is foreseen to start later on, for a final sensitivity of 10^{-16}.

$\mu \rightarrow e$

The signature of the conversion of a muon to an electron, following the formation of a muonic atom, is the production of a single monoenergetic electron with an energy equal to the difference between the mass of the muon and the binding energy of the muon in the atom. Contrary to the previous two muon decays described above, there is no accidental background in this case. The main backgrounds are the ineliminable muon decay in orbit (DIO), producing an electron with an energy distribution extending to the signal region and beam related backgrounds, among which pion capture:

$$\pi^- + (A, Z) \rightarrow (A, Z - 1)^* \rightarrow \gamma + (A, Z - 1)$$
$$\gamma \rightarrow e^+ e^-$$

The electron energy distribution measured by the SINDRUM II experiment at PSI is shown in Fig. 6(a) together with a simulation of the searched signal (the small peak on the right). The energy distribution of electrons from beam pion capture, when not suppressed, is superimposed in red. SINDRUM II coud strongly reduce this background by using a moderator to stop pions since the pion range in matter is about half as large as the corresponding muon range. This method is not possible in future experiments whose sensitivity goal is three orders of magnitude below that obtained by SINDRUM II which in turn imposes the use of muon beams with correspondingly higher intensities, of the order of of $10^{11}\mu/s$. Beam related background is instead kept under control by using a pulsed proton beam (instead of a continuous one) to produce muons, with a time delay between two successive proton pulses of about $1\mu s$. Search for conversion electrons can then be performed, as shown in Fig. 6(b), at times far from proton pulses, thanks to the long muon decay time, thus getting rid of beam related backgrounds. The factor:

$$R_{ext} = \frac{\text{number of protons between pulses}}{\text{number of protons in a pulse}}$$

is called "extinction factor". Simulations show that in order to reach a sensitivity below 10^{-16} it is necessary to keep this factor below 10^{-9}.

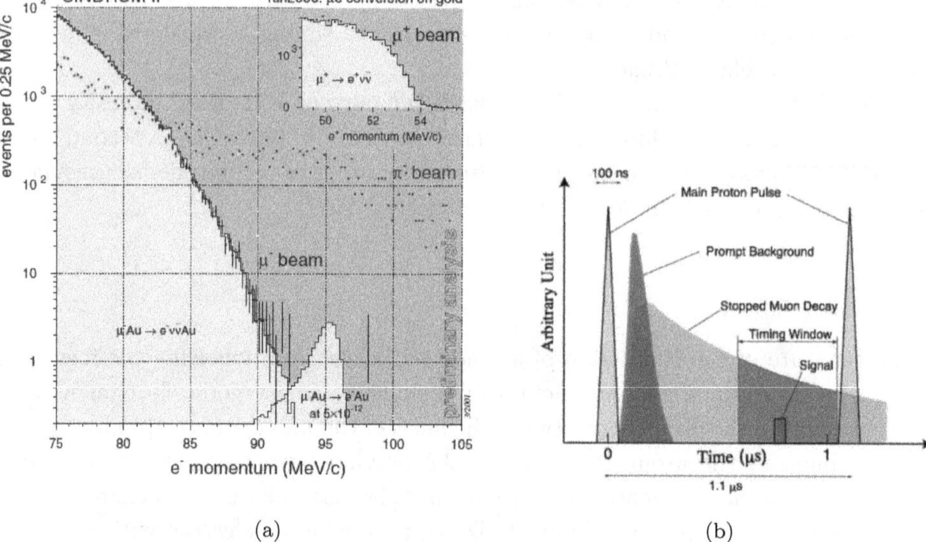

(a)   (b)

Fig. 6. (a) Measured electron energy distribution in SINDRUM II together with a simulation of the expected conversion signal. The electron distribution from beam pion capture is also shown in red. The inlet shows the positron Michel distribution from positive muons stopping in the target. (b) Timing scheme used to suppress beam related background in future $\mu \to$ e experiments (color online).

The muon beam line elements of the Mu2e experiment[7] at Fermilab are shown in Fig. 7. Eight GeV protons interact with a tungsten target to produce muons from pion decay in a production solenoid. Another solenoid is used to capture pions and subsequently muons, selecting their momentum and sign. An Aluminum stopping target and the experimental apparatus for electron detection, composed of a tracker and a calorimeter, are contained in a final detector solenoid.

A cosmic-ray veto made of four layers of scintillator must protect the detector solenoid since in its absence a backround rate of roughly one event per day would be induced by cosmic-rays in the detector. The cosmic-ray veto must further be

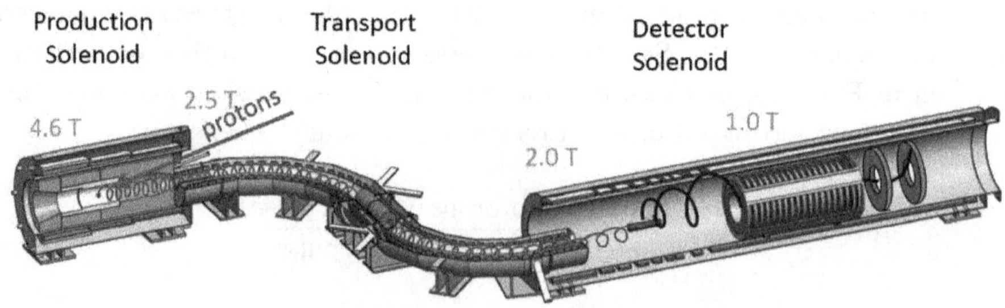

Fig. 7. Mu2e beam line elements.

protected from neutrons produced in the production solenoid which would otherwise prevent its correct operation.

The Mu2e experiment should start data taking in 2021, with the goal of reaching a sensitivity of 7×10^{-17} to $\mu \to$ e branching ratios, in few years of operation.

In Japan, the COMET[6] experiment at JPARC uses the same beam concept as Mu2e but it follows a staging approach. In a first phase, a simplified version of the transport solenoid and less performing tracking and calorimetric detectors will be used. COMET phase I should start in 2017 with the commissioning of the detectors and with the measurement of the extinction factor. This initial phase will be used for studying the backgrounds for phase II and will reach a sensitivity of about 7×10^{-15} to $\mu \to$ e branching rations . The construction of COMET phase II will start in 2020 and measurements will start in 2022 for a foreseen final sensitivity of 7×10^{-17} .

DeeMe[8], still at JPARC, is a third $\mu \to$ e experiment which should be starting in 2017. The muon beam concept is in this case very different from the previous two projects: here, muon production, muonic atom formation and conversion to an electron all take place in a pion production carbon target. An external magnetic channel and spectrometer are used to detect the possible monoenergetic conversion electrons. In four years of data taking SES should be 2.5×10^{-14}. This value could be improved to 5×10^{-15} by using different kinds of targets.

3.1. *Comparison with SUSY searches at LHC*

A nice way to understand the sensitivity of CLFV to BSM physics is presented in 21 where CLFV experimental results are used to obtain bounds within the same simplified models (defined by a subset of the relevant particles and couplings) employed by the LHC collaborations to interpret the searches for EW production of SUSY particles. Fig. 8 for instance shows the exclusion plots for the case where the only light SUSY particles are the Bino, the RH and LH sleptons, while the rest of the spectrum is decoupled. These plots make assumptions (written on the scales) on the masses of the particles involved and on the non diagonal elements of the slepton mass matrix ($(\Delta)_{LL,RR})_{ij}$) normalized to the diagonal terms:

$$(\delta_{LL,RR})_{ij} = \frac{(\Delta_{LL,RR})_{ij}}{\sqrt{(m^2_{L,R})_{ii} \, (m^2_{L,R})_{jj}}}$$

CLFV experiments set limits on $\delta_{LL,RR}$ which are shown in Fig. 8(b) as a function of the masses of the light SUSY particles of the simplified model. These limits are converted in Fig. 8(a) to an exclusion region in the parameters space of the bino mass (M_1) vs the sleptons masses (m_L, m_R). Fig. 8(a) also shows the parameter space region excluded by the LHC experiments (in violet) and the region which would explain the g-2 anomaly (in green). We may note how stringent is the constraint on the non diagonal part of the slepton mass matrix given by CLFV experiments in all the regions of this plot.

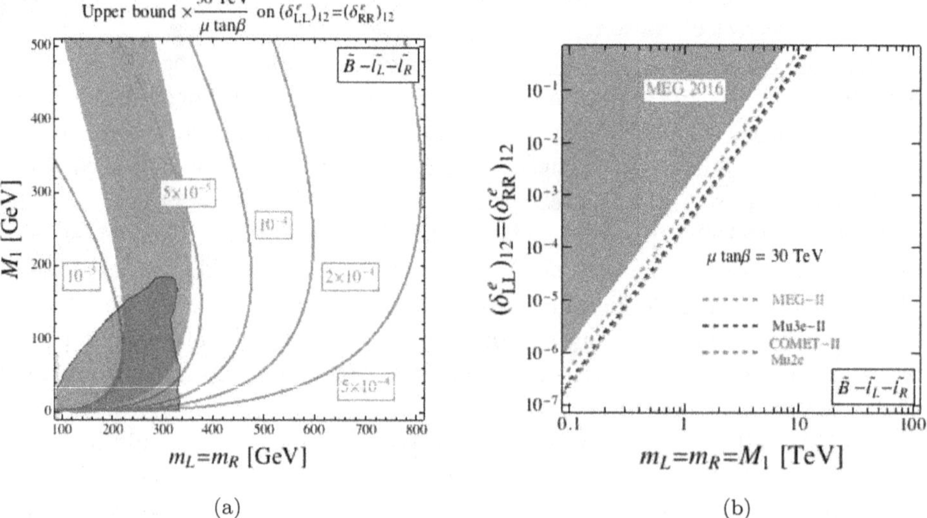

(a)   (b)

Fig. 8. Bounds and prospects from CLFV experiments for a SUSY model with Bino and both LH and RH sleptons (see text) (color online).

4. Summary

CLFV experiments are an excellent probe for BSM physics since any possible SM extension definitely gives rise to CLFV processes. There is a strong complementarity among the different possible channels since they would enable, in case of positive observation, to distinguish among the different models offered by the vast theoretical panorama. The experimental program of the next years includes a big improvement in sensitivity of all possible leptonic channels and of some meson decay channels. CLFV physics intends to explore the same kind of processes, namely extensions of the SM, which are actively searched at high energy colliders. There is optimal complementarity between these two different methods since it is true that CLFV requires no energy threshold for the production of new particles but it demands large enough lepton flavor mixing angles in order to detect a possible segnal of new physics.

References

1. A. M. Baldini et al.; Eur. Phys. J. **C 76** (2016) 434.
2. U. Bellgardt et al.; NP **B 299** (1988) 1.
3. W. H. Bertl et al.; EPJ **C 47** (2006) 337.
4. A. M. Baldini et al.; "The design of the MEG II experiment", submitted to EPJ **C**.
5. A. Blondel et al.; https://arxiv.org/abs/1301.6113.
6. D. Bryman et al.; http://comet.kek.jp/Documents_files/main-proposal.pdf (2007).
7. L. Bartoszek et al.; https://arxiv.org/abs/1501.05241.

8. DeeMe Collaboration, DeeMe KEK J-PARC Proposal `http://deeme.hep.sci.osaka-u.ac.jp/documents/deeme-proposal-r28.pdf` (2010).

9. B. Aubert et al.; PRL **104** (2010) 021802.

10. K. Hayasaka et al.; PL **B 687** (2010) 139.

11. Y. Ushiroda's talk; these proceedings.

12. D. Ambrose et al.; PRL **81** (1998) 5734.

13. A. Sher et al.; PR **D 72** (2005) 012005.

14. R. Aaij et al.; PRL **111** (2013) 141801.

15. B. Aubert et al.; PR **D 73** (2006) 092001.

16. P. Petrov, Journal of Physics; Conf. Series **800** (2017) 012039.

17. A. Belyaev et al.; `http://arxiv.org/abs/hep-ph/0008276v2`.

18. M. Pepe Altarelli's talk; these proceedings.

19. A. Crivellin et al.; `http://arxiv.org/abs/1706.08511v1`.

20. The CMS collaboration; Phys. Lett. **B 763**(2016) 472 and CMS-PAS-HIG-17-001.

21. L. Calibbi and G. Signorelli; `http://arxiv.org/abs/hep-ph/1709.00294v2`.

22. S. Antush et al.; JHEP **0611** (2006) 090.

23. R. Barbieri and L. J. Hall; Phys. Lett. **B 338** (1994) 212.

24. S. M. Bilenky, S. T. Petcov and B. Pontecorvo; Phys. Lett. **B 67** (1977) 309.

25. J. D. Bjorken, K. Lane and S. Weinberg; Phys. Rev. **D 16** (1977) 1474.

26. S. T. Petcov; Sov. J. Nucl. Phys. **25** (1977) 340.

CP Violation in Heavy-Flavour Hadrons

Greig A. Cowan

on behalf of the LHCb Collaboration*

University of Edinburgh, UK
E-mail: g.cowan@ed.ac.uk

Measurements of *CP*-violating observables in B meson decays can be used to determine the angles of the Unitarity Triangle and hence probe for manifestations of New Physics beyond the Cabibbo-Kobayashi-Maskawa Standard Model paradigm. Of particular interest are precise measurements of the angles γ and β. Also of great importance are studies of *CP*-violation involving B_s^0 mesons, in particular the phase ϕ_s, which is a golden observable in flavour physics at the LHC. Complementary to these studies is the continuing search for direct and indirect *CP*-violation in the charm system, where the experimental precision is now at the 10^{-3} level. I will present new and recent results in these topics, and in *CP*-violation searches in baryon decays, with specific emphasis on the measurement programme at the LHC.

Keywords: Flavour physics; *CP* violation; LHC.

1. Introduction

The violation of *CP* symmetry, the combination of the discrete symmetries of charge-conjugation (conjugation of all internal quantum numbers) and parity (reversing spatial coordinates), is a necessary condition to generate the baryon asymmetry of the Universe[1]. However, the level of *CP* violation allowed within the quark sector of the SM is many orders of magnitude too small[2–4] to explain astronomical observations, demanding experimental searches for new sources of violation. Heavy-quark hadrons provide an excellent laboratory to perform such searches as they allow the exploration of high energy scales well beyond the direct reach of the LHC. This approach has been successfully applied in the past with, for example, the observation of B^0 meson mixing[5,6] leading to the first estimates of the top quark mass before it was directly discovered. The large heavy-quark production cross-sections at the LHC[7,8] lead to large samples of exclusively reconstructed b and c hadron decays that have been used to make precision measurements of *CP* violating observables. These proceedings summarise the latest of these measurements, focussing on those using $3\,\mathrm{fb}^{-1}$ of data collected by the LHCb experiment[9] in pp collisions at the LHC during 2011 and 2012, unless otherwise stated. Recent comprehensive reviews can be found in Refs.[10,11] and references therein.

*28th International Symposium on Lepton Photon Interactions at High Energies, 7–12 Aug 2017
Sun Yat-sen University, Guangzhou, China

2. *CP* violation in the Standard Model

The only source of *CP* violation within the SM is due to the non-zero value of the phase in the Cabibbo-Kobayashi-Maskawa (CKM) matrix, to which all *CP*-violating observables are related. The unitarity of the matrix leads to relations between elements (e.g., $V_{ud}V_{ub}^* + V_{cd}V_{cb}^* + V_{td}V_{tb}^* = 0$), which are convenient to visualise as a triangle in the complex plane. For the triangle (*the* Unitarity Triangle) where all sides are of a similar magnitude, the angles are defined as

$$\alpha \equiv \arg\left[-\frac{V_{td}V_{tb}^*}{V_{ud}V_{ub}^*}\right], \quad \beta \equiv \arg\left[-\frac{V_{cd}V_{cb}^*}{V_{td}V_{tb}^*}\right], \quad \gamma \equiv \arg\left[-\frac{V_{ud}V_{ub}^*}{V_{cd}V_{cb}^*}\right].$$

These angles can be measured using a variety of different *CP* violating observables covering both tree-level quark transitions, where the impact of New Physics (NP) contributions is expected to be small[12], and loop-level transitions, which are sensitive to new higher-mass particles. One of the goals of studying the heavy-quark sector is to compare measurements of these quantities to check for the overall consistency of the CKM mechanism. Figure 1 shows the latest global fit of the CKM matrix parameters to experimental measurements and Lattice QCD calculations[13,14] showing that the SM is working well. However, there is still room for NP contributions at the level of $\sim 10\%$[15,16], implying that a new set of precision measurements is required.

In the quark sector, the neutral mesons (P^0) can oscillate into their antiparticles ($\overline{P}^0$), resulting in the physical states ($P_{\mathrm{H,L}}^0$) being admixtures of the flavour eigenstates: $P_{\mathrm{H}}^0 = pP^0 + q\overline{P}^0$ and $P_{\mathrm{L}}^0 = pP^0 - q\overline{P}^0$, where p and q are complex coefficients ($|p|^2 + |q|^2 = 1$). The physical states have well defined masses and lifetimes, and

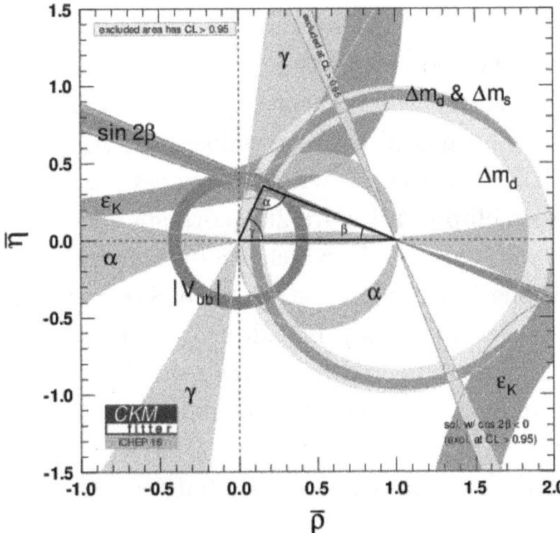

Fig. 1.   Global fit[15] to the CKM matrix parameters, showing consistency between measurements when interpreted in terms of SM quark transitions.

the parameters $\Delta m = m_H - m_L$, $\Delta\Gamma = \Gamma_L - \Gamma_H$ and $\Gamma = (\Gamma_L + \Gamma_H)/2$ control the decay-time-dependent decay rates of the mesons. In particular, Δm controls the oscillation frequency.

CP violation depends on the quantity $\lambda_f = \frac{q}{p}\frac{\overline{A}_f}{A_f}$. Here, f is the final state that the P hadron decays to with amplitude A_f and the $\overline{P}$ hadron decays to with amplitude $\overline{A}_f$. Three types of CP violation are allowed: in neutral meson mixing ($|q/p| \neq 1$); in the interference between neutral meson mixing and decay ($\arg(\lambda_f) \neq 0$) and in hadron decay ($|\overline{A}_f/A_f| \neq 1$). Only CP-violation in decay is allowed for charged mesons and baryons.

3. CP violation in B meson mixing

Semileptonic $B^0_{(s)}$ decays are dominated by tree-level quark transitions, implying that there should be no CP violation in decay, and therefore provide a clean system to search for CP in mixing. The so-called semileptonic (or flavour-specific) asymmetry is defined as $A_{sl} = \frac{\Gamma(\overline{B}^0 \to B^0 \to f) - \Gamma(B^0 \to \overline{B}^0 \to \overline{f})}{\Gamma(\overline{B}^0 \to B^0 \to f) + \Gamma(B^0 \to \overline{B}^0 \to \overline{f})} \approx \frac{\Delta\Gamma}{\Delta m}\tan\phi_M$, where ϕ_M is the mixing phase from the $B^0_{(s)}$ mixing matrix. These asymmetries are predicted to be very small in the SM, at the level of 10^{-4} or less[10], and therefore any measurement of a significantly non-zero effect would be a clear sign of beyond-the-SM physics.

Experimentally, the quantity measured is the untagged decay-time, t, dependent charge asymmetry between semileptonic $B^0_{(s)}$ decays with a positive or negatively charged muon, defined as $A_{\text{meas}}(t) = \frac{N(D^-\mu^+\nu,t) - N(D^+\mu^-\nu,t)}{N(D^-\mu^+\nu,t) + N(D^+\mu^-\nu,t)} \approx A_D + \frac{A_{sl}}{2} + \left(A_P - \frac{A_{sl}}{2}\right)\cos(\Delta mt)$. This is sensitive to A_{sl} along with other production, A_P, and particle detection, A_D, asymmetries. In the case of the measurements from the LHCb collaboration[17,18] these asymmetries can be controlled to high precision using data calibration samples and by reversing the LHCb dipole magnet, thereby allowing a precision measurement of the CP asymmetries in both the B^0 and B^0_s systems.[a] Figure 2(a) shows the current experimental situation for CP violation in B meson mixing. The global average values are $A^d_{sl} = (-0.21 \pm 0.17)\%$ and $A^s_{sl} = (-0.06 \pm 0.28)\%$[19], consistent with SM expectations.

The left-most green ellipse in Figure 2(a) corresponds to the dimuon asymmetry measured by the D0 collaboration[20], which is a measurement of a linear combination of A^d_{sl} and A^s_{sl}. Although it is inconsistent with SM expectations at $\sim 3\sigma$ it has been proposed[21,22] that there may be additional contributions from a non-zero value of $\Delta\Gamma_d/\Gamma_d$, which is expected to be very small in the SM. The most precise measurement of this quantity has recently been made by the ATLAS collaboration[23] (Figure 2(b)), obtaining $\Delta\Gamma_d/\Gamma_d = (-0.1 \pm 1.1 \pm 0.9) \times 10^{-2}$. An update of the 1 fb^{-1} measurement[24] of $\Delta\Gamma_d/\Gamma_d$ from the LHCb collaboration is eagerly anticipated.

---

[a]For the B^0_s system, the time-integrated rate can be measured as the fast B^0_s oscillations wash out the production asymmetry.

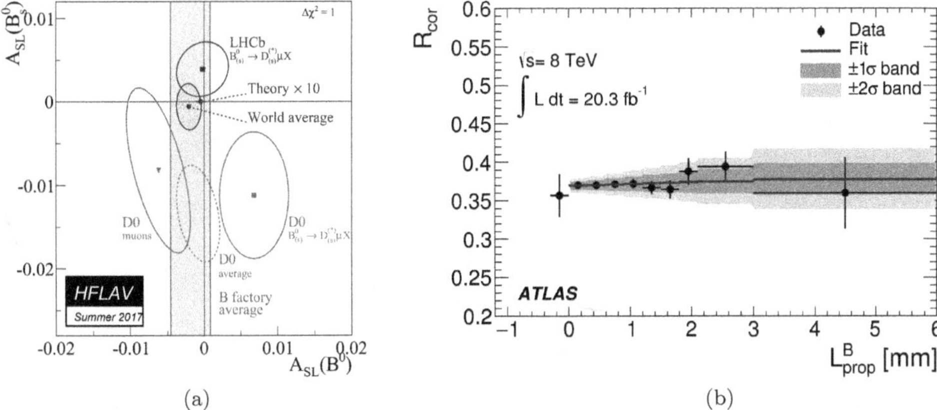

Fig. 2.   (a) HFLAV combination[19] of A^d_{sl} and A^s_{sl} from several experiments compared to the theory ($\times 10$) predictions[10]. (b) Efficiency-corrected ratio of the observed decay-length distributions for $B^0 \to J/\psi K^*$ and $B^0 \to J/\psi K^0_S$ decays[23].

4. *CP* violation in the interference of *B* meson mixing/decay

In the case where the $B^0_{(s)}$ or $\overline{B}^0_{(s)}$ mesons decay to the same final state, f, the decay-time-dependent *CP* asymmetry is given by

$$\frac{\Gamma_{\overline{B}^0 \to f}(t) - \Gamma_{B^0 \to f}(t)}{\Gamma_{\overline{B}^0 \to f}(t) + \Gamma_{B^0 \to f}(t)} = \frac{S_f \sin(\Delta mt) - C_f \cos(\Delta mt)}{\cosh(\Delta\Gamma t/2) + A^{\Delta\Gamma}_f \sinh(\Delta\Gamma t/2)},\tag{1}$$

where $|S_f|^2 + |C_f|^2 + |A^{\Delta\Gamma}_f|^2 = 1$ by definition.

4.1. *The B^0 system*

In the B^0 system, $\Delta\Gamma_d \approx 0$ and only the numerator of Eq. (1) needs to be considered. The canonical decay mode used by the B-factories to measure this asymmetry is $B^0 \to J/\psi K^0_S$, which proceeds predominately via a tree-level $b \to c\bar{c}s$ transition. In the case where the sub-dominant penguin diagrams can be neglected[25–28], $S_{J/\psi K^0_S} \approx \sin 2\beta$.

The LHCb collaboration has recently used its Run 1 data to measure S_f and C_f in $B^0 \to J/\psi(\mu^+\mu^-)K^0_S$[29], $B^0 \to J/\psi(e^+e^-)K^0_S$ and $B^0 \to \psi(2S)(\mu^+\mu^-)K^0_S$ decays[30] using a flavour-tagged[31] decay-time-dependent analysis. The asymmetry for $B^0 \to J/\psi(\mu^+\mu^-)K^0_S$ decays can be seen in Figure 3(a). The individual measurements and their combination are shown in Figure 3(b), where the systematic uncertainty is dominated by background tagging asymmetry. The LHCb-averaged values are $S_{[c\bar{c}]K^0_S} = 0.760 \pm 0.034$ and $C_{[c\bar{c}]K^0_S} = -0.017 \pm 0.029$. Together, these measurements reduce the tension between the world average value for $\sin 2\beta$ and the indirect determination from global fits[15,16]. The consistency between the results using the electron and muon channels for charmonium reconstruction also help to

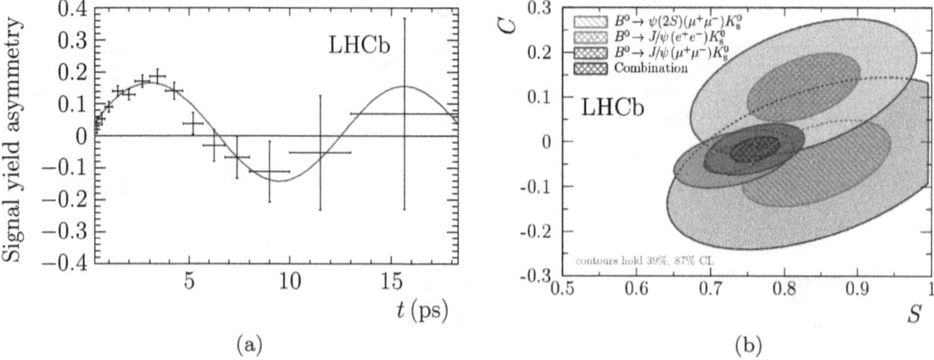

Fig. 3. (a) CP asymmetry as a function of decay time for $B^0 \to J/\psi K^0_S$ decays[29], showing a clear oscillation. (b) LHCb combination of CP violation parameters as measured using $B^0 \to J/\psi(\mu^+\mu^-)K^0_S$[29], $B^0 \to J/\psi(e^+e^-)K^0_S$ and $B^0 \to \psi(2S)(\mu^+\mu^-)K^0_S$[30] decays.

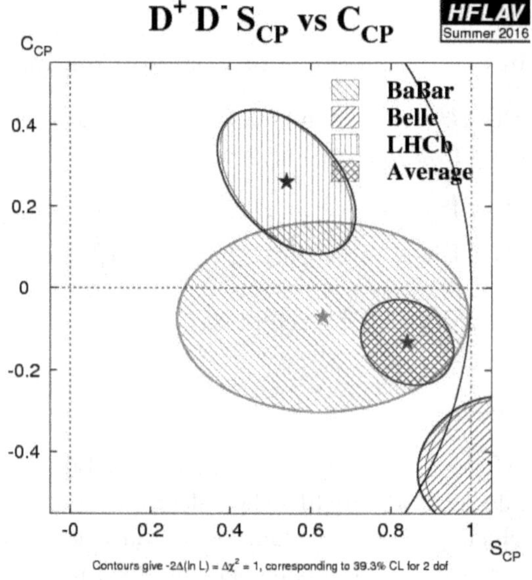

Fig. 4. HFLAV average of CP parameters in $B^0 \to D^+D^-$ decays[19].

build confidence in the electron reconstruction performance of LHCb, which is particularly relevant when viewed through the prism of recent anomalies in $b \to s\ell^+\ell^-$ transitions[32].

Decays such as $B^0 \to D^+D^-$ are governed by $b \to c\bar{c}d$ quark transitions and therefore measurements of the decay-time-dependent asymmetry gives complimentary information about $\sin 2\beta$ that can be used to constrain the size of potential penguin contributions to the decay[25–28]. Figure 4 summarises the current situation with these measurements from the BaBar, Belle and LHCb[33] collaborations. The

Belle result is outside of the physical region ($S_{DD}^2 + C_{DD}^2 < 1$), which may have been an indication of large hadronic effects. However, the latest LHCb measurement shows that these terms are small and consistent with zero, with the phase shift induced by the penguin diagrams measured to be $\Delta\phi = -0.16^{+0.19}_{-0.21}$ rad.

4.2. *The B_s^0 system*

Decay-time-dependent CP asymmetries in the B_s^0 system using $b \to c\bar{c}s$ transitions are sensitive to the CKM phase $\beta_s \equiv \arg\left[-\frac{V_{ts}V_{tb}^*}{V_{cs}V_{cb}^*}\right]$. Typically measurements are made of the experimentally observable phase ϕ_s, which is equal to $-2\beta_s$ if the penguin contributions to the decay can be neglected. Global fits give a precise Standard Model prediction for ϕ_s of -36.5 ± 1.3 mrad[15]. Deviations from this value would be a clear sign for NP, strongly motivating the need for more precise experimental measurements.

The golden mode for measuring ϕ_s is using a flavour-tagged decay-time-dependent angular analysis of the $B_s^0 \to J/\psi\,(\mu^+\mu^-)\phi(K^+K^-)$ decay. This channel has a high branching fraction and the presence of two muons in the final state leads to a high trigger efficiency at hadron colliders. An angular analysis is necessary to disentangle the interfering CP-odd and CP-even components in the final state, which arise due to the relative angular momentum between the two vector resonances. In addition, there is a small ($\sim 2\%$) CP-odd K^+K^- S-wave contribution that must be accounted for.

The CDF[34], D0[35], ATLAS[36], CMS[37] and LHCb[38] collaborations have all measured ϕ_s (in addition to other mixing-related parameters of the B_s^0 system) using the $B_s^0 \to J/\psi\phi$ decay. Additional information can be obtained by utilising the region of the K^+K^- invariant mass spectrum above the $\phi(1020)$ meson, where higher spin K^+K^- resonances are expected to contribute. Such a flavour-tagged decay-time-dependent amplitude analysis has just been performed by the LHCb collaboration[39], which finds the dominant component of the high-mass spectrum comes from the $f_2'(1525)$ meson (Figure 5) and measures $\phi_s = 119 \pm 107 \pm 34$ mrad. The LHCb detector has excellent time resolution (~ 45 fs) and tagging power ($\sim 4\%$), both of which are crucial to the measurement. Combining the LHCb results from $B_s^0 \to J/\psi\phi$ (low mass), $B_s^0 \to J/\psi K^+K^-$ (high mass) and $B_s^0 \to J/\psi\pi^+\pi^-$ decays[40] gives $\phi_s = 1 \pm 37$ mrad.

The global combination of ϕ_s and $\Delta\Gamma_s$ using the measurements referenced above in addition to $B_s^0 \to \psi(2S)\phi$[41] and $B_s^0 \to D_s^+ D_s^+$[42] decays gives average values of $\Delta\Gamma_s = 0.090 \pm 0.005\,\mathrm{ps}^{-1}$ and $\phi_s = -21 \pm 31$ mrad. The combination is dominated by the statistical uncertainty from the LHCb $B_s^0 \to J/\psi\phi$ result and are consistent with the SM predictions[10,15]. However, there remains space for new physics contributions at $O(10\%)$ and as the experimental precision improves it is essential that there is good control over hadronic effects[43,44] that could mimic the signature of beyond-the-SM physics.

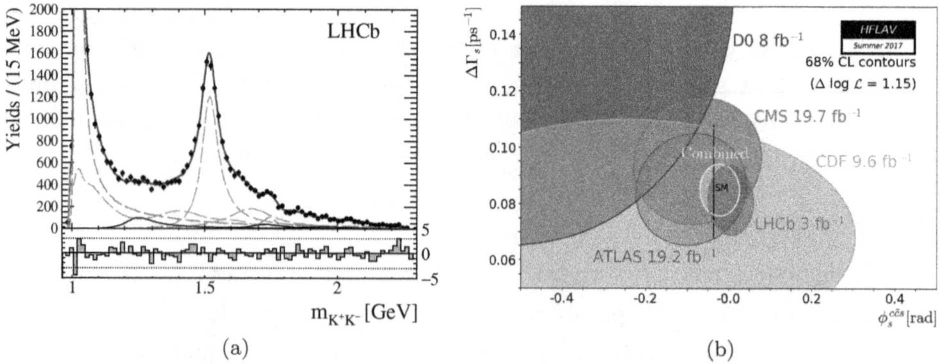

Fig. 5. (a) Distribution of $m_{K^+K^-}$ from $B_s^0 \to J/\psi\, K^+K^-$ decays. The total fit function is overlaid in blue, while the $\phi(1020)$, $f_2'(1525)$ and K^+K^- S-wave contributions are shown by the long-dashed pink, brown and green lines, respectively. (b) HFLAV combination[19] (white contour) of ϕ_s and $\Delta\Gamma_s$ from several experiments (coloured contours) as discussed in the text (color online).

A related CP-violating phase, $\phi_s^{s\bar{s}s}$, can be measured by applying similar analysis methods to B_s^0 meson decays that occur via $b \to s\bar{s}s$ transitions. The LHCb collaboration has performed such an analysis using $B_s^0 \to \phi\phi$[45], measuring $\phi_s = -0.17 \pm 0.15 \pm 0.03$ rad, which is consistent with the SM predictions, all of which are very close to zero[46–48]. Updated measurements of CP-violating parameters in charmless $B_s^0 \to K^+K^-$ decays have also recently been reported by the LHCb collaboration[49], including a first measurement of $A_{K^+K^-}^{\Delta\Gamma}$.

5. CP violation in b hadron decay

5.1. The CKM angle γ

The CKM angle γ is the only CP violating parameter that can be measured from tree-level decays[b] and the uncertainty on its theoretical prediction is constrained to be $< O(10^{-7})$[12]. Together these make measurements of γ a "standard candle" within the SM with which other loop-level determinations can be compared in order to look for the effects of new CP violating contributions. The canonical technique to measure γ is to exploit the interference between the different decay paths in $B \to DK$ decays. Depending on the final state of the D^0 meson there are different analysis methods (e.g., GLW[50,51], ADS[52,53] and GGSZ[54]) that vary in their sensitivity to γ and the other hadronic parameters that describe the strong dynamics of the B and D meson decays. Unlike the measurement of $\beta_{(s)}$, there is no dominant channel in which to measure γ and a combination of several modes is required to achieve the maximal sensitivity.

A new result[55] from the LHCb collaboration is an update, using Run 2 data, of the measurements of the CP-violating observables in $B^\pm \to D^{(*)0}K^\pm$ and

---

[b]There are some caveats.

$B^{\pm} \to D^{(*)0}\pi^{\pm}$ decays using the GLW method. Figure 6 shows the invariant mass distribution of the $D^0 h$ system. In the case of $B^{\pm} \to D^0 K^{\pm}$ decays a clear asymmetry is visible between the oppositely-charged modes. After controlling for small detector and production asymmetries using the Cabibbo-favoured $B^{\pm} \to [K^{\pm}\pi^{\mp}]_D \pi^{\pm}$ mode, the CP asymmetry is measured to be $A_K^{KK} = +0.126 \pm 0.014 \pm 0.002$, where the first uncertainty is statistical and the second systematic.

For the first time the collaboration has also used a partial reconstruction technique to measure the CP observables using the modes with excited $D^{*0} \to D^0 \gamma$ and $D^{*0} \to D^0 \pi^0$ decays where the photon or π^0 is not reconstructed. This approach avoids the efficiency penalty that would need to be paid to fully reconstruct these channels. The partially reconstructed decays correspond to the structures at lower mass in Figure 6, where the different shapes for the $D^{*0} \to D^0 \gamma$ and $D^{*0} \to D^0 \pi^0$ contributions allow the decay rates and CP asymmetries to be measure separately for each. The CP asymmetry in the $B^{\pm} \to (D^{*0} \to D^0\pi^0)K^{\pm}$ channel is measured as $A_K^{CP,\pi^0} = -0.151 \pm 0.033 \pm 0.011$, which is different from zero at 4.3σ. The corresponding asymmetry for the γ mode is $A_K^{CP,\gamma} = +0.276 \pm 0.094 \pm 0.047$.

Another new result is the measurement of the CP-violating observables in $B^{\pm} \to DK^{*\pm}$ decays, with $K^{*\pm} \to K_s^0\pi^{\pm}$. This updates Ref.[56], using two- and four-body D meson final states[57] in addition to Run 2 data. The branching ratio of the $B^{\pm} \to DK^{*\pm}$ decay is of similar magnitude to $B^{\pm} \to DK^{\pm}$ (described above), but the overall event yield in LHCb is lower due to the efficiency for reconstructing K_s^0 mesons. Figure 7 shows the invariant mass distributions of the DK^* systems for the two-body Cabibbo-favoured and ADS D^0 decay modes. The decays are isolated

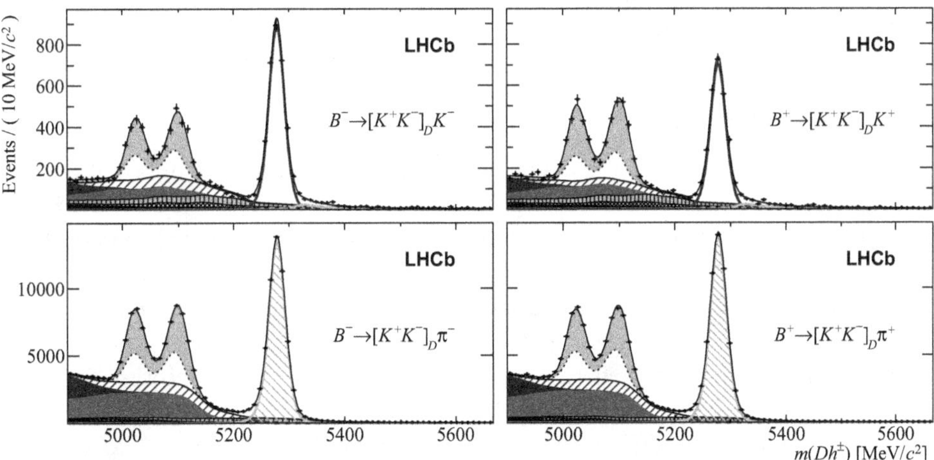

Fig. 6. Distribution of invariant mass of the $Dh^{\pm}$ system for (left) B^- and (right) B^+ decays. The bachelor hadron is a (top) kaon and (bottom) pion. In the case of the $B^{\pm} \to [K^+K^-]_D K^{\pm}$ decays a clear asymmetry is visible between the distributions. The structures at lower masses is due to partially reconstructed $D^{*0} \to D^0 \gamma$ and $D^{*0} \to D^0 \pi^0$ decays where the γ or π^0 particle is missed.

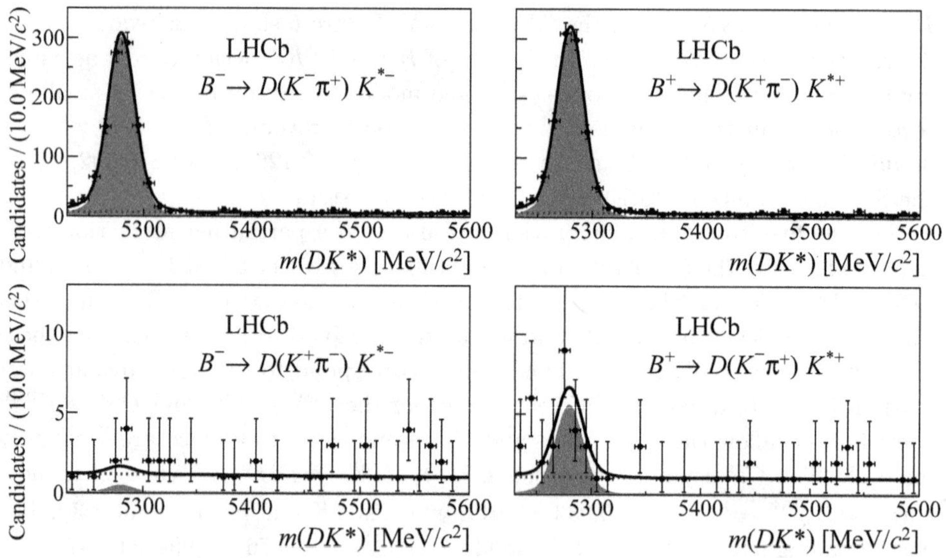

Fig. 7. Distribution of invariant mass of the DK^* system for (left) B^- and (right) B^+ decays. Both the (top) Cabibbo-favoured and (bottom) ADS $D^0 \to K\pi$ decay modes are visible.

with high signal purity and this result corresponds to the first 4.2σ evidence of the ADS mode, with the rate measured to be $R^+_{K\pi} = 0.020 \pm 0.006 \pm 0.001$. The measurements for the decay rates and CP asymmetries are consistent with and more precise than the corresponding analysis from the BaBar collaboration[58]. In the future they will help to further constrain γ and the related hadronic parameters for this system.

5.2. γ combination

As stated above the best precision on γ is obtained by combining information from many $B \to DK$ decay modes. An updated combination from LHCb has recently been performed[59] that utilises 85 observables and contains 37 parameters. It includes the $B^\pm \to D^{(*)0}K^\pm$ and $B^\pm \to DK^{*\pm}$ results described above[c] and an updated decay-time-dependent measurement of the CP asymmetry in $B^0_s \to D^+_s K^+$ decays[60]. The final measurement is $\gamma = 76.8^{+5.1°}_{-5.7}$. The HFLAV collaboration have combined this with existing measurements from the B-factories to obtain $\gamma = 76.2^{+4.7°}_{-5.0}$, where the precision is dominated by the LHCb measurement. Many more updates of $B \to DK$ channels can be expected with Run 2. The aim is to have sub-degree-level precision at the end of the LHCb phase 1 upgrade in 2024 by which point LHCb will have collected $50\,\mathrm{fb}^{-1}$.

---

[c]In fact, only the $B^\pm \to DK^{*\pm}$ results from Ref.[56] are used.

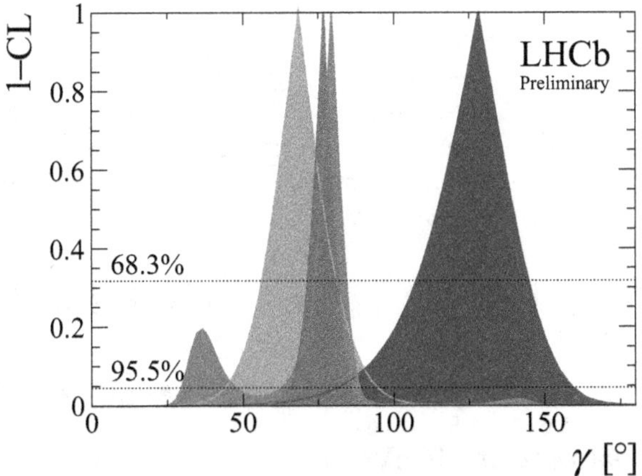

Fig. 8. 1-CL plot, using the profile likelihood method, for the γ combinations split by analysis method. (yellow) GGSZ methods, (orange) GLW/ADS methods, (blue) other methods and (green) the full combination (color online).

5.3. *First evidence of CP violation in the baryon system*

So far there has been no observation of CP violation in b baryon decays. However, since they are governed by the same quark-level transitions as meson decays there is potential for non-zero effects in the SM[61–63]. For example, charmless decays of b baryons have contributions of similar magnitude from both tree and penguin decays, which give rise to sensitivity to the CKM angle α. No sign of CP violation has been found in such charmless two[64] or three-body[65,66] Λ_b^0 or Ξ_b baryon decays, however, a recent study[67] of four-body $\Lambda_b^0 \to ph^-h^+h^-$ decays has revealed the first evidence for CP-violation in the baryon sector.

The measurement is performed by using the four-body decay topology to compute triple products, which are odd under the motion reversal operator, $\hat{T}$.[d] The triple products are defined as $C_{\hat{T}} = \vec{p}_p \cdot (\vec{p}_{h_1^-} \times \vec{p}_{h_2^+})$ and $\overline{C}_{\hat{T}} = \vec{p}_{\overline{p}} \cdot (\vec{p}_{h_1^+} \times \vec{p}_{h_2^-})$ for Λ_b^0 and $\overline{\Lambda}_b^0$ decays, respectively. The corresponding $\hat{T}$-odd asymmetries are $A_{\hat{T}}(C_{\hat{T}}) = \frac{N(C_{\hat{T}}>0)-N(C_{\hat{T}}<0)}{N(C_{\hat{T}}>0)+N(C_{\hat{T}}<0)}$ and $\overline{A}_{\hat{T}}(\overline{C}_{\hat{T}}) = \frac{N(-\overline{C}_{\hat{T}}>0)-N(-\overline{C}_{\hat{T}}<0)}{N(-\overline{C}_{\hat{T}}>0)+N(-\overline{C}_{\hat{T}}<0)}$, where N refers to the number of observed $\Lambda_b^0 \to ph^-h^+h^-$ candidates. From these $\hat{T}$-odd asymmetries is possible to build P-odd and CP-odd observables, defined as $a_P^{\hat{T}-\text{odd}} = \frac{1}{2}(A_{\hat{T}} + \overline{A}_{\hat{T}})$ and $a_{CP}^{\hat{T}-\text{odd}} = \frac{1}{2}(A_{\hat{T}} - \overline{A}_{\hat{T}})$, respectively. The CP-odd observable is insensitive to production and detection asymmetries that affect standard CP-asymmetries (e.g., those based on event yields) and is also formed from a different combination of strong and weak phases.

Figure 9(a) shows the $\Lambda_b^0 \to p\pi\pi\pi$ invariant mass from Ref.[67] that has been used to measure $a_{CP}^{\hat{T}-\text{odd}}$. The global measurement is consistent with CP symmetry

---

[d]The $\hat{T}$ operator is equivalent to the parity operation for spinless particles.

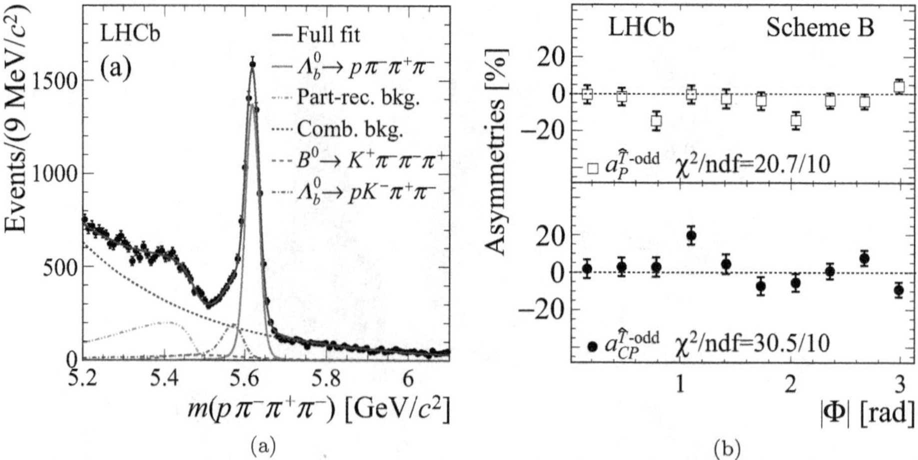

Fig. 9.   (a) The invariant mass distributions for $\Lambda_b^0 \to p\pi\pi$ used to extract the asymmetries. (b) The P-odd and CP-odd asymmetries as a function of $|\Phi|$, the magnitude of angle between the $p\pi_{\text{fast}}^-$ and $\pi_{\text{slow}}^-\pi^+$ decay planes in the Λ_b^0 rest frame.

but it has been noted[68] that there is increased sensitivity to CP violating effects by looking at differential distributions. Figure 9(b) shows $a_{CP}^{\hat{T}\text{-odd}}$ as a function of the phase space of the $\Lambda_b^0 \to p\pi\pi$ decay. Using this and an alternative binning scheme, the p-value for the CP-symmetry hypothesis is evaluated as 9.8×10^{-4}, which is equivalent to a 3.3σ deviation of $a_{CP}^{\hat{T}\text{-odd}}$ from zero. This constitutes the first evidence of CP violation in baryon decays. With the larger data samples to be collected in Run 2 and beyond it will be possible to perform a full amplitude analysis of the $\Lambda_b^0 \to p\pi\pi$ channel to understand where the CP asymmetry arises in the phase space of the decay.

Similar methods using $\hat{T}$-odd observables have been used to search for CP violation in rare Λ_b^0 decays[69,70] and the charm system[71]. In each case the results are consistent with the hypothesis of CP conservation.

6. Study of b-baryon oscillations

As noted in Section 1 the origin of the baryon asymmetry in the Universe is unclear. Baryon number violation (BNV) has never been seen experimentally, with strong constraints imposed by the measured proton and bound-neutron lifetimes. However, beyond-the-SM models containing flavour-diagonal six-fermion vertices[72–75] could permit BNV without violating existing constraints. Unambiguous experimental observation of such BNV would be the observation of baryon-antibaryon oscillations of hadrons containing quarks of all three generations (*i.e.*, usb), such as the Ξ_b^0 baryon.

A new result[76] from the LHCb collaboration measures the decay-time-dependent ratio between the rates of same-sign (SS) and opposite-sign (OS) decays of the Ξ_b^0 baryon. The ratio is defined as $R(t) = \frac{\Gamma(\Xi_b^0 \to \overline{\Xi}_c^- \pi^+)}{\Gamma(\Xi_b^0 \to \Xi_c^+ \pi^-)} \approx (\omega t)^2$, where ω is the mixing

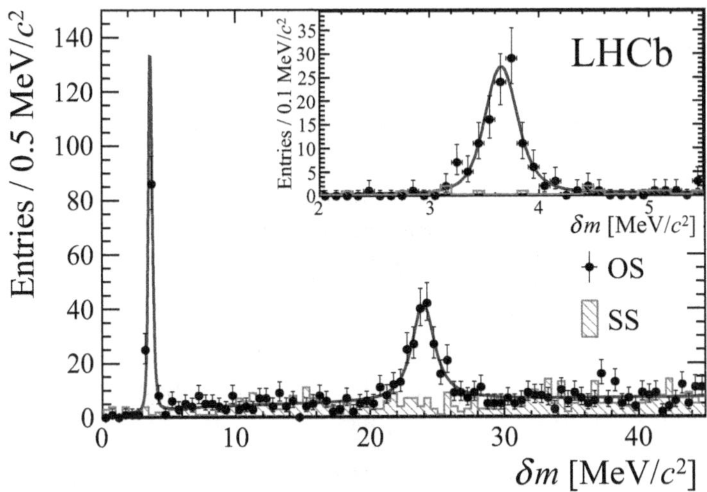

Fig. 10. Spectra of the mass difference $\delta m = m(\Xi_b^0 \pi) - m(\Xi_b^0) - m_\pi$ for the OS-tagged (black points with error bars) and SS-tagged (red, hatched histogram) decays. Inset: detail of the region $2.0 < \delta m < 5.5$ MeV/c^2 (color online).

frequency. Here, SS (OS) means that the charge of the proton from the charged $\Xi_c \to pK\pi$ decay is the same (opposite) to the charge of the pion from the strong decay of the excited $\Xi_b'^{,*}$ baryons that tags the initial flavour of the Ξ_b^0 candidate. Figure 10 shows the mass difference distribution for the LHCb data. The two narrow peaks visible near threshold in the OS-tag sample are due to the excited $\Xi_b'^{,*}$ baryons. The red histogram shows the corresponding distribution for the SS tags, which is consistent with the background-only hypothesis. In seven bins of decay time the ratio $R(t)$ is evaluated using a likelihood fit to the mass distribution, which allows an upper limit to be set on the mixing frequency. The limit is $\omega < 0.08$ ps^{-1} at 95% CL, determined using a likelihood ratio test and the CLs method.

7. Charm

Studies of the charm system provide the only way to investigate CP violation using up-type quarks. As in the neutral B meson system, neutral D mesons undergo oscillations, but in this case they are dominated by long-distance effects and the mixing parameters ($x = \Delta m/\Gamma$, $y = \Delta\Gamma/(2\Gamma)$) are expected to be small. Likewise, very small CP-violating effects are expected[77]. As yet there is no evidence for CP violation in the charm system, but the very large data samples being collected by the LHCb experiment provide an ideal place to precisely test theoretical predictions and search for new sources of CP violation.

At hadron colliders it is possible to tag the initial flavour of D^0 mesons using two different methods, depending on how they were produced. The flavour is determined either by the charge of the pion from promptly-produced $D^{*+} \to D^0\pi^+$ decays or the charge of the muon from semileptonic B decays ($B \to D^0 \mu^- X$). These independent samples have different properties in terms of background and reconstruction efficiencies, leading to different systematic uncertainties in the final measurements.

The large yields available in the charm system can also be used to search for CP violation in strong interactions. The LHCb collaboration has recently used $D^+_{(s)} \to \pi^+ \pi^+ \pi^-$ decays to search for CP violating $\eta^{(\prime)} \to \pi^+ \pi^-$ decays [78], finding no signal in the $\pi^+ \pi^-$ mass spectrum. This constrains the branching fractions to be less than $\sim 10^{-5}$ at 90% CL, which are comparable with or better than existing limits [79].

7.1. *Direct CP violation in the charm system*

Direct CP violation in the charm system can be explored by measuring the asymmetries in the decay of D^0 mesons to CP eigenstates (i.e., $D^0 \to h^+ h^-$). Experimentally the raw yield asymmetry is measured, $A_{\mathrm{raw}} \equiv \frac{N(D^0 \to h^+ h^-) - N(\overline{D}^0 \to h^+ h^-)}{N(D^0 \to h^+ h^-) + N(\overline{D}^0 \to h^+ h^-)}$, and subsequently corrected for small production and detection asymmetries using data control modes in order to determine the CP asymmetry, $A_{CP}(D^0 \to h^+ h^-) = A_{\mathrm{raw}}(D^0 \to h^+ h^-) - A_{\mathrm{P}}(D^{*+}) - A_{\mathrm{D}}(\pi_s^+)$.

The LHCb collaboration has recently measured $A_{CP}(D^0 \to K^+ K^-)$ using a prompt-tagged sample (Figure 11(a)) [80]. While still statistically limited, the dominant systematic uncertainty in the result relates to control of the nuisance asymmetries. Combining $A_{CP}(D^0 \to K^+ K^-)$ with an earlier measurement of the difference in CP asymmetries between the kaon and pion modes, $\Delta A_{CP} = A_{CP}(D^0 \to K^+ K^-) - A_{CP}(D^0 \to \pi^+ \pi^-)$, allows $A_{CP}(D^0 \to \pi^+ \pi^-)$ to

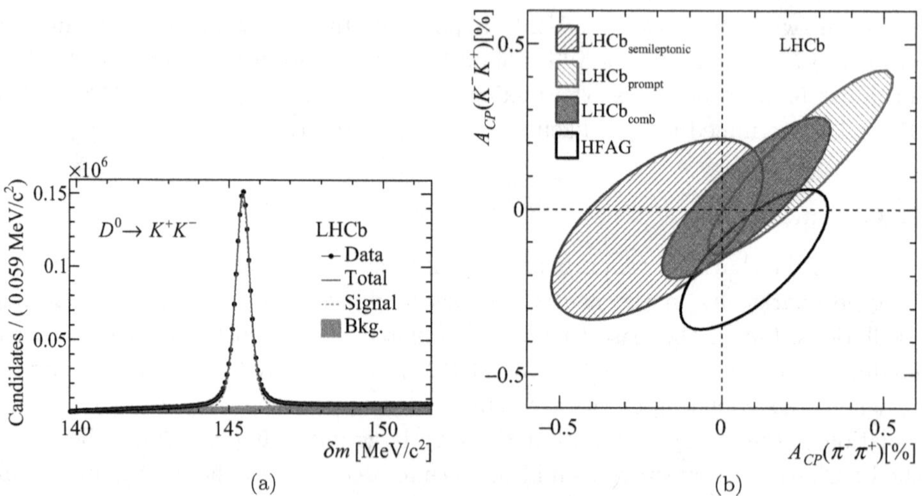

Fig. 11. (a) Distribution of the mass difference between the D^* and D^0 mesons in the prompt sample. (b) CP violating asymmetries for $D^0 \to K^+ K^-$ and $D^0 \to \pi^+ \pi^-$ decays using the prompt and semileptonic-tagged samples. The 68% confidence level contours are displayed where the statistical and systematic uncertainties are added in quadrature.

be extracted. Figure 11(b) shows the result for the two CP asymmetries, which are consistent with CP symmetry. These results have been combined with independent measurements from the semileptonic-tagged sample[81] to provide an overall LHCb result of $A_{CP}(D^0 \to K^+K^-) = (0.04 \pm 0.12 \pm 0.10)\%$ and $A_{CP}(D^0 \to K^+K^-) = (0.07 \pm 0.14 \pm 0.11)\%$, where the first uncertainty is statistical and the second systematic. These results are now approaching the per-mille level of uncertainty but still do not show any signs of CP violation. Likewise, there are no indications of CP violation in other modes such as $D^\pm \to \eta'\pi^\pm$ and $D_s^\pm \to \eta'\pi^{\pm}$[82]. A search for CP violation in $D^0 \to \pi^+\pi^-\pi^+\pi^-$ decays using an energy test[83] reports a mild tension with CP-symmetry that requires further exploration with Run 2 data.

7.2. *Indirect CP violation*

As discussed in above, the mixing parameters and decay-time-integrated CP asymmetries in the charm system are small. Based upon this, a measurement of the decay-time-dependent CP asymmetry of D^0 decays to CP eigenstates is sensitive to the indirect CP violating parameter A_Γ. Here, $A_\Gamma \equiv \dfrac{\hat{\Gamma}_{D^0 \to f} - \hat{\Gamma}_{\bar{D}^0 \to f}}{\hat{\Gamma}_{D^0 \to f} + \hat{\Gamma}_{\bar{D}^0 \to f}}$, is the asymmetry between the inverse effective lifetimes of the D^0 meson and its antiparticle. Within the SM, its magnitude is expected to be $\lesssim 5 \times 10^{-3}$ and independent of the final state, f, owing to the small CP violation in decay.

The LHCb collaboration has recently measured A_Γ using a prompt-tagged sample of $D^0 \to K^+K^-$ and $D^0 \to \pi^+\pi^-$ decays[84]. Two different methods are used to measure the asymmetry, an approach that is binned in the D^0 decay time and an unbinned method. Both approaches use the large sample of $D \to K\pi$ decays to control production and detection asymmetries. They each give consistent measurements, with the binned result being slightly more precise and chosen as the nominal result. Figure 12 shows the binned asymmetries as a function of decay time for the kaon and pion decay modes. The value of A_Γ is given by the gradient of the linear fit to these asymmetries. These values are combined to give $A_\Gamma = (-0.13 \pm 0.28 \pm 0.10) \times 10^{-3}$. They have subsequently been combined with

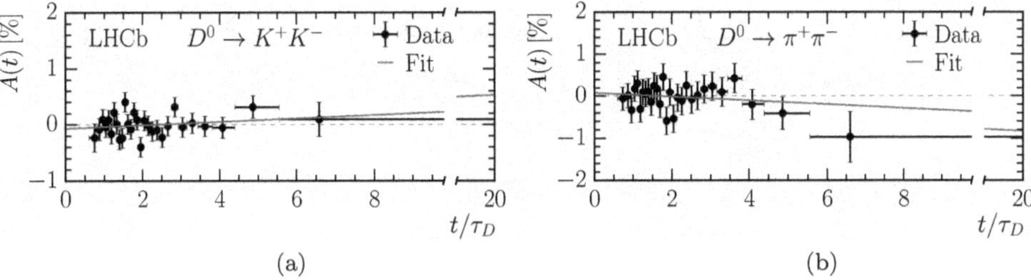

Fig. 12.   Measured asymmetry $A(t)$ in bins of t/τ_D, where τ_D is the world average value for the D^0 lifetime[19], for (a) $D^0 \to K^+K^-$ and (b) $D^0 \to \pi^+\pi^-$ decays. The solid line shows a linear fit to the data, with a slope equal to the best estimate of $-A_\Gamma$.

168

the semileptonic-tagged sample[85] to obtain $A_\Gamma = (-0.29 \pm 0.28) \times 10^{-3}$. This is the most precise measurement of a CP-violating observable in the charm system ever made.

7.3. *Charm mixing and indirect CP violation*

The measurement of charm mixing parameters and a search for CP violation can be performed by studying the decay-time-dependent ratio of yields of Cabibbo-suppressed to Cabibbo-favoured $D^0 \to K\pi$ decays. This ratio is defined separately for D^0 and $\overline{D}^0$ decays and is related to the mixing parameters via a second-order expansion of the mixing equations (assuming small mixing) given by $R(t)^\pm = R_D^\pm + \sqrt{R_D^\pm}\, y' \frac{t}{\tau} + \frac{(x'^\pm)^2 + (y'^\pm)^2}{4} \left(\frac{t}{\tau}\right)^2$, where τ is the average D^0 lifetime and the x', y' have been rotated from the nominal mixing parameters by the strong phase in the $D \to K\pi$ decay. A recent result[86] from the LHCb collaboration uses a double-tagged (DT) technique (using the charge of the pion and the muon from the semileptonic B decays $(B^0 \to (D^{*+} \to D^0\pi^+)\mu^- X)$) to provide a very pure sample of events (Figure 13(a) and (b)) that cover a complementary region of the decay time spectrum than has been previously studied using a prompt-sample[87]. Figure 13(c) shows the ratios for D^0 and $\overline{D}^0$ decays and their difference for both the DT and prompt samples. The ratios increase as a function of time, consistent with charm mixing, allowing x', y' to be measured for the DT sample alone and both samples combined. The complementary coverage of the DT and prompt sample gives an improved precision, by 10–20%, for the charm mixing parameters from the combined fit even though the DT analysis is based on almost 40 times fewer

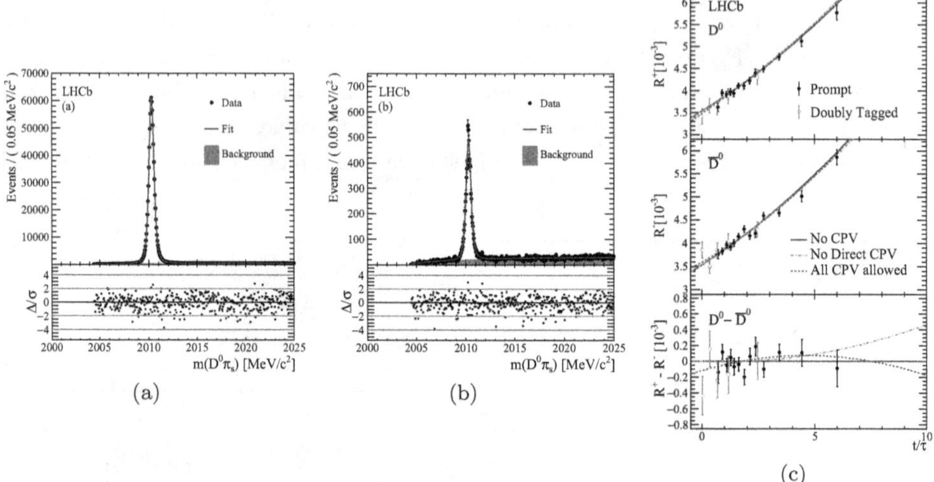

Fig. 13.   Distribution of the D^{*+} invariant mass in the double-tagged (DT) sample for (a) RS and (b) WS decays. (c) Efficiency corrected ratios of WS/RS decays and fit projections for the DT (red open circles) and prompt (black filled circles) samples (color online).

candidates than the prompt analysis. The data are consistent with the hypothesis of CP conservation (both for decay and interference of mixing and decay).

8. Looking to the future

The majority of beauty and charm sector measurements are statistically limited, which strongly motivates the case for a new set of precision measurements of CP-violating observables to constrain the size of (or perhaps find) new physics contributions. The LHCb collaboration recently submitted an expression of interest[88] regarding future phase 1b and 2 upgrades that would operate during LHC runs 4 and 5, respectively. It is expected that using the data collected with the improved detector it will be possible to improve the precision on γ to 0.4°, ϕ_s to 9 mrad (see Figure 14) and achieve precision on charm mixing and CP violation parameters at the level of 10^{-4}. These will allow unprecedented sensitivity to the small effects of new physics. Crucially, as the precision improves it is essential to control hadronic effects that can hide small non-Standard Model effects. Other studies can be found in Refs.[89,90].

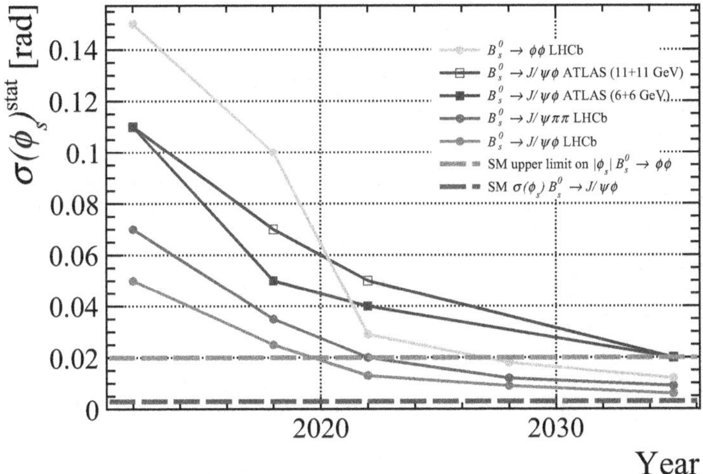

Fig. 14. Projection of how precision on ϕ_s from LHCb measurements will scale as a function of time for different decay modes. Information taken from Ref.[89,90].

9. Summary

This year is the 40th anniversary of the discovery of the b quark[91]. In that time there has been huge progress in using the physics of heavy-flavour hadrons to make detailed studies of CP-violation, both experimentally and theoretically. So far all measurements are consistent with Standard Model predictions, confirming validity of the CKM mechanism. The LHCb experiment is now leading the way in terms of precision measurements in the b and c sectors, with new explorations of CP violation in baryons just beginning. It will be fascinating to observe how the SM is able to withstand the next round of precision measurements that will be made over the next decade.

Acknowledgements

The author thanks the organisers of the Lepton Photon conference and acknowledges the support of the Science and Technology Facilities Council (UK) grant ST/K004646/1.

References

1. A. D. Sakharov, Violation of CP invariance, C asymmetry, and baryon asymmetry of the universe, *Soviet Physics Uspekhi* **34**, p. 392 (1991).
2. M. B. Gavela, P. Hernandez, J. Orloff and O. Pene, Standard model CP violation and baryon asymmetry, *Mod. Phys. Lett.* **A9**, 795 (1994).
3. P. Huet and E. Sather, Electroweak baryogenesis and standard model CP violation, *Phys. Rev.* **D51**, 379 (1995).
4. M. B. Gavela, P. Hernandez, J. Orloff, O. Pene and C. Quimbay, Standard model CP violation and baryon asymmetry. Part 2: Finite temperature, *Nucl. Phys.* **B430**, 382 (1994).
5. C. Albajar *et al.*, Search for $B^0 - \overline{B}^0$ oscillations at the CERN proton - anti-proton collider. 2., *Phys. Lett.* **B186**, 247 (1987), [Erratum: Phys. Lett. B197, 565(1987)].
6. H. Albrecht *et al.*, Observation of $B^0 - \overline{B}^0$ mixing, *Phys. Lett.* **B192**, 245 (1987).
7. R. Aaij *et al.*, Measurement of the b-quark production cross-section in 7 and 13 TeV pp collisions, *Phys. Rev. Lett.* **118**, p. 052002 (2017), Erratum in preparation.
8. R. Aaij *et al.*, Measurements of prompt charm production cross-sections in pp collisions at $\sqrt{s} = 13$ TeV, *JHEP* **03**, p. 159 (2016).
9. A. A. Alves Jr. *et al.*, The LHCb detector at the LHC, *JINST* **3**, p. S08005 (2008).
10. M. Artuso, G. Borissov and A. Lenz, CP violation in the B_s^0 system, *Rev. Mod. Phys.* **88**, p. 045002 (2016).
11. T. Gershon and V. V. Gligorov, CP violation in the B system, *Rept. Prog. Phys.* **80**, p. 046201 (2017).
12. J. Brod and J. Zupan, The ultimate theoretical error on γ from $B \to DK$ decays, *JHEP* **01**, p. 051 (2014).
13. A. Bazavov *et al.*, $B_{(s)}^0$-mixing matrix elements from lattice QCD for the Standard Model and beyond, *Phys. Rev.* **D93**, p. 113016 (2016).
14. S. Aoki *et al.*, Review of lattice results concerning low-energy particle physics, *Eur. Phys. J.* **C77**, p. 112 (2017).
15. J. Charles *et al.*, Current status of the Standard Model CKM fit and constraints on $\Delta F = 2$ New Physics, *Phys. Rev.* **D91**, p. 073007 (2015).
16. M. Bona *et al.*, The unitarity triangle fit in the Standard Model and hadronic parameters from lattice QCD: a reappraisal after the measurements of $\Delta m_{(s)}$ and $BR(B \to \tau\nu_\tau)$, *JHEP* **10**, p. 081 (2006).
17. R. Aaij *et al.*, Measurement of the CP asymmetry in B_s^0–$\overline{B}_s^0$ mixing, *Phys. Rev. Lett.* **117**, p. 061803 (2016).
18. R. Aaij *et al.*, Measurement of the semileptonic CP asymmetry in B^0–$\overline{B}^0$ mixing, *Phys. Rev. Lett.* **114**, p. 041601 (2015).

19. Y. Amhis *et al.*, Averages of b-hadron, c-hadron, and τ-lepton properties as of summer 2016 (2016), arXiv:1612.07233.

20. V. M. Abazov *et al.*, Study of CP-violating charge asymmetries of single muons and like-sign dimuons in $p\bar{p}$ collisions, *Phys. Rev.* **D89**, p. 012002 (2014).

21. G. Borissov and B. Hoeneisen, Understanding the like-sign dimuon charge asymmetry in $p\bar{p}$ collisions, *Phys. Rev.* **D87**, p. 074020 (2013).

22. Y. S. Amhis, T. Aushev and M. Jung, Mixing and mixing-related CP violation in the B system, in *8th International Workshop on the CKM Unitarity Triangle (CKM 2014) Vienna, Austria, September 8-12, 2014*, 2015. arXiv:1510.07321.

23. M. Aaboud *et al.*, Measurement of the relative width difference of the B^0-$\bar{B}^0$ system with the ATLAS detector, *JHEP* **06**, p. 081 (2016).

24. R. Aaij *et al.*, Measurements of the B^+, B^0, B_s^0 meson and Λ_b^0 baryon lifetimes, *JHEP* **04**, p. 114 (2014).

25. S. Faller, M. Jung, R. Fleischer and T. Mannel, The Golden Modes $B^0 \to J/\psi K_{(S,L)}$ in the Era of Precision Flavour Physics, *Phys. Rev.* **D79**, p. 014030 (2009).

26. M. Jung, Determining weak phases from $B \to J/\psi P$ decays, *Phys. Rev.* **D86**, p. 053008 (2012).

27. K. De Bruyn and R. Fleischer, A Roadmap to Control Penguin Effects in $B_d^0 \to J/\psi K_S^0$ and $B_s^0 \to J/\psi\phi$, *JHEP* **03**, p. 145 (2015).

28. P. Frings, U. Nierste and M. Wiebusch, Penguin contributions to CP phases in $B_{d,s}$ decays to charmonium, *Phys. Rev. Lett.* **115**, p. 061802 (2015).

29. R. Aaij *et al.*, Measurement of CP violation in $B^0 \to J/\psi K_S^0$ decays, *Phys. Rev. Lett.* **115**, p. 031601 (2015).

30. R. Aaij *et al.*, Measurement of CP violation in $B^0 \to J/\psi(e^+e^-)K_S^0$ and $B^0 \to \psi(2S)(\mu^+\mu^-)K_S^0$ decays (2017), LHCb-PAPER-2017-029, in preparation.

31. R. Aaij *et al.*, New algorithms for identifying the flavour of B^0 mesons using pions and protons, *Eur. Phys. J.* **C77**, p. 238 (2017).

32. M. Pepe Altarelli, Rare decays of heavy hadrons, *these proceedings* (2017).

33. R. Aaij *et al.*, Measurement of CP violation in $B \to D^+D^-$ decays, *Phys. Rev. Lett.* **117**, p. 261801 (2016).

34. T. Aaltonen *et al.*, Measurement of the bottom-strange meson mixing phase in the full CDF data set, *Phys. Rev. Lett.* **109**, p. 171802 (2012).

35. V. M. Abazov *et al.*, Measurement of the CP-violating phase $\phi_s^{J/\psi\phi}$ using the flavor-tagged decay $B_s^0 \to J/\psi\phi$ in 8 fb^{-1} of $p\bar{p}$ collisions, *Phys. Rev.* **D85**, p. 032006 (2012).

36. G. Aad *et al.*, Measurement of the CP-violating phase ϕ_s and the B_s^0 meson decay width difference with $B_s^0 \to J/\psi\phi$ decays in ATLAS, *JHEP* **08**, p. 147 (2016).

37. V. Khachatryan *et al.*, Measurement of the CP-violating weak phase ϕ_s and the decay width difference $\Delta\Gamma_s$ using the $B_s^0 \to J/\psi\phi(1020)$ decay channel in pp collisions at $\sqrt{s} = 8$ TeV, *Phys. Lett.* **B757**, 97 (2016).

38. R. Aaij *et al.*, Precision measurement of CP violation in $B_s^0 \to J/\psi K^+K^-$ decays, *Phys. Rev. Lett.* **114**, p. 041801 (2015).

39. R. Aaij *et al.*, Resonances and CP-violation in $\bar{B}_s^0$ and $B_s^0 \to J/\psi K^+K^-$ decays in the mass region above the $\phi(1020)$ (2017), submitted to JHEP.

40. R. Aaij *et al.*, Measurement of the CP-violating phase ϕ_s in $\bar{B}_s^0 \to J/\psi\pi^-\pi^-$ decays, *Phys. Lett.* **B736**, p. 186 (2014).

41. R. Aaij *et al.*, First study of the CP violating phase and decay-width difference in $B_s^0 \to \psi(2S)\phi$ decays, *Phys. Lett.* **B762**, p. 253 (2016).

42. R. Aaij *et al.*, Measurement of the CP-violating phase ϕ_s in $\bar{B}_s^0 \to D_s^+D_s^-$ decays, *Phys. Rev. Lett.* **113**, p. 211801 (2014).

43. S. Faller, R. Fleischer and T. Mannel, Precision physics with $B_s^0 \to J/\psi\phi$ at the LHC: the quest for new physics, *Phys. Rev.* **D79**, p. 014005 (2009).

44. B. Bhattacharya, A. Datta and D. London, Reducing penguin pollution, *Int. J. Mod. Phys.* **A28**, p. 1350063 (2013).

45. R. Aaij *et al.*, Measurement of CP violation in $B_s^0 \to \phi\phi$ decays, *Phys. Rev.* **D90**, p. 052011 (2014).

46. M. Beneke, J. Rohrer and D. Yang, Branching fractions, polarisation and asymmetries of $B \to VV$ decays, *Nucl. Phys.* **B774**, 64 (2007).

47. M. Bartsch, G. Buchalla and C. Kraus, $B \to V(L)V(L)$ decays at next-to-leading order in QCD (2008).

48. H.-Y. Cheng and C.-K. Chua, QCD factorization for charmless hadronic B_s decays revisited, *Phys. Rev.* **D80**, p. 114026 (2009).

49. Measurement of time-dependent CP violating asymmetries in $B^0 \to \pi^+\pi^-$ and $B_s^0 \to K^+K^-$ decays at LHCb (Jan 2017), LHCb-CONF-2016-018.

50. M. Gronau and D. London, How to determine all the angles of the unitarity triangle from $B^0 \to DK_S^0$ and $B_{(s)}^0 \to D\phi$, *Phys. Lett.* **B253**, 483 (1991).

51. M. Gronau and D. Wyler, On determining a weak phase from CP asymmetries in charged B decays, *Phys. Lett.* **B265**, 172 (1991).

52. D. Atwood, I. Dunietz and A. Soni, Enhanced CP violation with $B \to KD^0(\overline{D}^0)$ modes and extraction of the CKM angle γ, *Phys. Rev. Lett.* **78**, 3257 (1997).

53. D. Atwood, I. Dunietz and A. Soni, Improved methods for observing CP violation in $B^\pm \to KD$ and measuring the CKM phase γ, *Phys. Rev.* **D63**, p. 036005 (2001).

54. A. Giri, Y. Grossman, A. Soffer and J. Zupan, Determining gamma using $B^\pm \to DK^\pm$ with multibody D decays, *Phys. Rev.* **D68**, p. 054018 (2003).

55. R. Aaij *et al.*, Measurement of CP observables in $B^\pm \to D^{(*)0}K^\pm$ and $B^\pm \to D^{(*)0}\pi^\pm$ with $D^0 \to K\pi, KK, \pi\pi$ decays (2017), arXiv:1708.06370, submitted to Phys. Lett. B.

56. R. Aaij *et al.*, Study of the decay $B^\pm \to DK^{*\pm}$ with two-body D decays (Dec 2016), LHCb-CONF-2016-014.

57. R. Aaij *et al.*, Measurement of CP observables in $B^\pm \to DK^{*\pm}$ decays using two- and four-body D-meson final states (2017), LHCb-PAPER-2017-030, in preparation.

58. B. Aubert *et al.*, Measurement of CP violation observables and parameters for the decays $B^\pm \to DK^{*\pm}$, *Phys. Rev.* **D80**, p. 092001 (2009).

59. R. Aaij *et al.*, Measurement of the CKM angle γ from a combination of $B \to DK$ analyses (Jul 2017), LHCb-CONF-2017-004.

60. R. Aaij *et al.*, Measurement of CP asymmetry in $B_s^0 \to D_s^\mp K^\pm$ decays (Jan 2017), LHCb-CONF-2016-015.

61. W. Bensalem, A. Datta and D. London, T violating triple product correlations in charmless $\Lambda_{(b)}$ decays, *Phys. Lett.* **B538**, 309 (2002).

62. M. Gronau and J. L. Rosner, Triple product asymmmetries in Λ_b and Ξ_b decays, *Phys. Lett.* **B749**, 104 (2015).

63. I. I. Bigi, Bridge between hadrodynamics & HEP: regional CP violation in beauty & charm decays, *PoS* **ICHEP2016**, p. 531 (2016).

64. T. A. Aaltonen *et al.*, Measurements of Direct CP-violating asymmetries in charmless decays of bottom baryons, *Phys. Rev. Lett.* **113**, p. 242001 (2014).

65. R. Aaij *et al.*, Searches for Λ_b^0 and Ξ_b^0 decays to $K_S^0 p\pi^-$ and $K_S^0 pK^-$ final states with first observation of the $\Lambda_b^0 \to K_S^0 p\pi^-$ decay, *JHEP* **04**, p. 087 (2014).

66. R. Aaij *et al.*, Observations of $\Lambda_b^0 \to \Lambda K^+\pi^-$ and $\Lambda_b^0 \to \Lambda K^+K^-$ decays and searches for other Λ_b^0 and Ξ_b^0 decays to Λh^+h^- final states, *JHEP* **05**, p. 081 (2016).

67. R. Aaij *et al.*, Measurement of matter-antimatter differences in beauty baryon decays, *Nature Physics* **13**, p. 391 (2017).

68. G. Durieux and Y. Grossman, Probing CP violation systematically in differential distributions, *Phys. Rev.* **D92**, p. 076013 (2015).

69. R. Aaij *et al.*, Observation of the $\Lambda_b \to \Lambda\phi$ decay, *Phys. Lett.* **B759**, p. 282 (2016).

70. R. Aaij *et al.*, Observation of the decay $\Lambda_b^0 \to pK^-\mu^+\mu^-$ a search for CP violation, *JHEP* **06**, p. 1 (2017).

71. R. Aaij *et al.*, Search for CP violation using T-odd correlations in $D^0 \to K^+K^-\pi^+\pi^-$ decays, *JHEP* **10**, p. 005 (2014).

72. C. Smith, Proton stability from a fourth family, *Phys. Rev.* **D85**, p. 036005 (2012).

73. G. Durieux, J.-M. Gerard, F. Maltoni and C. Smith, Three-generation baryon and lepton number violation at the LHC, *Phys. Lett.* **B721**, 82 (2013).

74. D. McKeen and A. E. Nelson, CP violating baryon oscillations, *Phys. Rev.* **D94**, p. 076002 (2016).

75. K. Aitken, D. McKeen, A. E. Nelson and T. Neder, Baryogenesis from oscillations of charmed or beautiful baryons (2017), arXiv:1708.01259.

76. R. Aaij *et al.*, Search for baryon-number-violating Ξ_b^0 oscillations (2017), arXiv: 1708.05808, submitted to Phys. Rev. Lett.

77. B. Bhattacharya, M. Gronau and J. L. Rosner, CP asymmetries in singly-Cabibbo-suppressed D decays to two pseudoscalar mesons, *Phys. Rev.* **D85**, p. 054014 (2012), [Phys. Rev.D85,no.7,079901(2012)].

78. R. Aaij *et al.*, Search for the CP-violating strong decays $\eta \to \pi^+\pi^-$ and $\eta'(958) \to \pi^+\pi^-$, *Phys. Lett.* **B764**, p. 233 (2016).

79. C. Patrignani *et al.*, Review of Particle Physics, *Chin. Phys.* **C40**, p. 100001 (2016).

80. R. Aaij *et al.*, Measurement of CP asymmetry in $D^0 \to K^+K^-$ decays, *Phys. Lett.* **B767**, p. 177 (2017).

81. R. Aaij *et al.*, Measurement of CP asymmetry in $D^0 \to K^-K^+$ and $D^0 \to \pi^-\pi^+$ decays, *JHEP* **07**, p. 041 (2014).

82. R. Aaij *et al.*, Measurement of CP asymmetries in $D^\pm \to \eta'\pi^\pm$ and $D_s^\pm \to \eta'\pi^\pm$ decays, *Phys. Lett.* **B771**, p. 21 (2017).

83. R. Aaij *et al.*, Search for CP violation in the phase space of $D^0 \to \pi^+\pi^-\pi^+\pi^-$ decays, *Phys. Lett.* **B769**, p. 345 (2017).

84. R. Aaij *et al.*, Measurement of the CP violation parameter A_Γ in $D^0 \to K^+K^-$ and $D^0 \to \pi^+\pi^-$ decays, *Phys. Rev. Lett.* **118**, p. 261803 (2017).

85. R. Aaij *et al.*, Measurement of the time-integrated CP asymmetry in $D^0 \to K_S^0 K_S^0$ decays, *JHEP* **10**, p. 055 (2015).

86. R. Aaij *et al.*, Measurements of charm mixing and CP violation using $D^0 \to K^\pm\pi^\mp$ decays, *Phys. Rev.* **D95**, p. 052004 (2017).

87. R. Aaij *et al.*, Measurement of D^0–$\overline{D}^0$ mixing parameters and search for CP violation using $D^0 \to K^+\pi^-$ decays, *Phys. Rev. Lett.* **111**, p. 251801 (2013).

88. R. Aaij *et al.*, *Expression of Interest for a Phase-II LHCb Upgrade: Opportunities in flavour physics, and beyond, in the HL-LHC era*, Tech. Rep. CERN-LHCC-2017-003, CERN (Geneva, 2017).

89. *Impact of the LHCb upgrade detector design choices on physics and trigger performance*, Tech. Rep. LHCb-PUB-2014-040, CERN-LHCb-PUB-2014-040, CERN (Geneva, 2014).

90. *ATLAS B-physics studies at increased LHC luminosity, potential for CP-violation measurement in the $B_s^0 \to J/\psi\,\phi$ decay*, Tech. Rep. ATL-PHYS-PUB-2013-010, CERN (Geneva, 2013).

91. S. W. Herb *et al.*, Observation of a Dimuon Resonance at 9.5-GeV in 400-GeV Proton-Nucleus Collisions, *Phys. Rev. Lett.* **39**, 252 (1977).

KAGRA, Underground Cryogenic Gravitational Wave Telescope

S. Haino

on behalf of the KAGRA Collaboration

Institute of Physics, Academia Sinica,
Nankang, Taipei 11529, Taiwan
E-mail: Sadakazu.Haino@cern.ch

KAGRA is the first km-scale interferometric gravitational wave telescope constructed underground to reduce seismic noise, and the first telescope to use cryogenic cooling of test masses to reduce thermal noise. The construction of the infrastructure to house the interferometer in the tunnel, and the initial phase operation of the interferometer with a simple 3-km Michelson configuration have been completed in 2016. The first cryogenic operation is expected in 2018, and the observing runs with a full interferometer are expected in 2020s. In this article, we will discuss about the introduction of Gravitational Wave and special features, current status and future prospects of KAGRA.

Keywords: KAGRA; gravitational wave.

1. Introduction

We are witnessing the dawn of Gravitational Wave (GW) Astronomy. We now have a new way to "hear" the Universe. In this article, we would like to discuss about the experimental aspect of GW Astronomy.

It is known that the GW strain amplitude is extremely small. Usually, it is mentioned that as a ratio of the size of Hydrogen atom over the distance between Earth and the Sun. Another way to express the tininess of the GW amplitude is as follows:

$$h \sim \frac{2G}{c^4 R} M v^2 \tag{1}$$

where h is the GW strain amplitude, G is the Gravitational Constant, M and v are the mass and speed of the system, respectively, c is the speed of light, and R is the distance to the GW source. For example, let's consider a Neutron Star (NS) binary system each with the mass of Sun colliding each other in almost same as the speed of light ($v \sim c$) and located 100 Mpc away. The above formula becomes:

$$h \sim \frac{2G}{c^2 R} = \frac{r_g}{R} \tag{2}$$

where r_g is Schwarzschild radius of mass M. Schwarzschild radius of the Sun is about 3km so the above-mentioned ratio is 3km/100Mpc = 10^{-21}, which is the typical mentioned number of GW amplitude. Another point of this GW amplitude formula is that it is proportional to $1/R$. This is a key feature of the GW measurement. In case of Electro-Magnetic (EM) waves, which are the basis of the astronomy ranging from radio waves to high energy gamma-rays, we measure the EM energy (or photon) density, so it is proportional to $1/R^2$, while in case of GW observation we can measure the amplitude so it is proportional to $1/R$. However, the disadvantage of GW detection is that the detector is like an ear. We can only "hear" the GW with a single detector, because we detect the GW amplitude but not the energy density. This is the first special feature of the GW observation.

The second feature is that with EM waves we can see the Universe only after ~380,000 years since the Big Bang, which is called Cosmic Microwave Background (CMB) radiation, while GWs are expected to be coming from the much earlier Universe, even though it gets more and more difficult to detect such primordial GW Background signal after subtracting variety of foreground GW signals. In summary, GWs have a great potential to provide a new way to explore the deep into the Universe, much deeper than we could make with EM waves.

2. World-wide GW detector network

Figure 1 shows the schematic view of the current status of the ground-based GW detectors over the world. In the U.S., advanced LIGO has twin detectors (Hanford and Livingston) which have been operating since 2015 and made a historical achievement of the first GW detection [1]. In Europe, advanced Virgo locates near Pisa in Italy which joined the detector network in August/2017 and made another historical achievement of the first GW detection of the NS-NS merger [2] as well as demonstrating three-detection of BH-BH merger [3]. These are called 2nd generation GW detectors. The 1st generation GW detector (Initial LIGO and Initial Virgo) have been developed in 1990's and 2000's which couldn't succeed to catch any GW signal and later they had been up-graded to advanced 2nd generation detectors (the detailed scheme will be discussed later). There is another detector called GEO600 near Hannover in Germany. We would call it as a 1.5 generation detector. The sensitivity is not competitive with advanced LIGO and Virgo detectors but it has been operating for over 10 years in a so-called Astro-Watch mode to try to catch the possible violent GW signal (like Supernova) in near-by distance (like in our Milky Way Galaxy). The advanced technologies used in LIGO have been developed and tested in GEO600 detector. That is the reason why we would call GEO600 as a 1.5 Generation GW detector.

In Asia, there is another advanced GW detector called KAGRA. It has been developed as LCGT (Large Cryogenic Gravitational-wave Telescope) since 2010. As its name says, it is employing further advanced technologies of cryogenics. I would call KAGRA as a 2.5 Generation GW detector. We will discuss about the main advanced features in the following later. LIGO collaboration is trying to deploy yet another detector in India, (LIGO-India or Indigo) recently, after the historical discovery of GW signal by LIGO, Indian government approved the project. However, it has just started and we expect that LIGO-India will be online only after KAGRA.

There are at least couple of advantages to establish a world-wide GW detector network. One is about the sky coverage. Figure 2 shows the comparison of so-called antenna amplitude pattern of the GW signal [4]. Among several techniques of GW detectors, a Michelson-type laser interferometer is considered to be a broad-frequency-band, general-purpose detector, which are employed in above mentioned 2nd generation GW detectors (LIGO, Virgo and KAGRA). The Michelson laser interferometer is suitable to detect the differential change of the two orthogonal arms and so it has the best sensitivity for the GW coming from top or bottom of the detector, while it has less or no sensitivities for GW coming from the side. For example, if we consider a GW coming from the direction on the same plane as the two arms with 45 degree from the arms, the GW signal will change the distance of the arms simultaneously and the interferometer cannot sense it. We could call it as "blind" (or "deaf") spots. Therefore, as shown in the left of Figure 2, the antenna amplitude pattern has its maximum on top and bottom and dips in the sides. The two LIGO detectors were designed to be oriented in a similar direction in order to maximize the confidence of the GW detection itself by taking the coincidence of two detector outputs (just like they demonstrated with GW150914 [1]) instead of trying to compensate the "deaf" spots each other. Adding Virgo should help to increase the sky coverage but we will still have some remaining "deaf" spots. By combining at least four detectors like two LIGO, Virgo and KAGRA, we can compensate the sensitivity each other and enable almost full sky coverage as shown in the right of Figure 2.

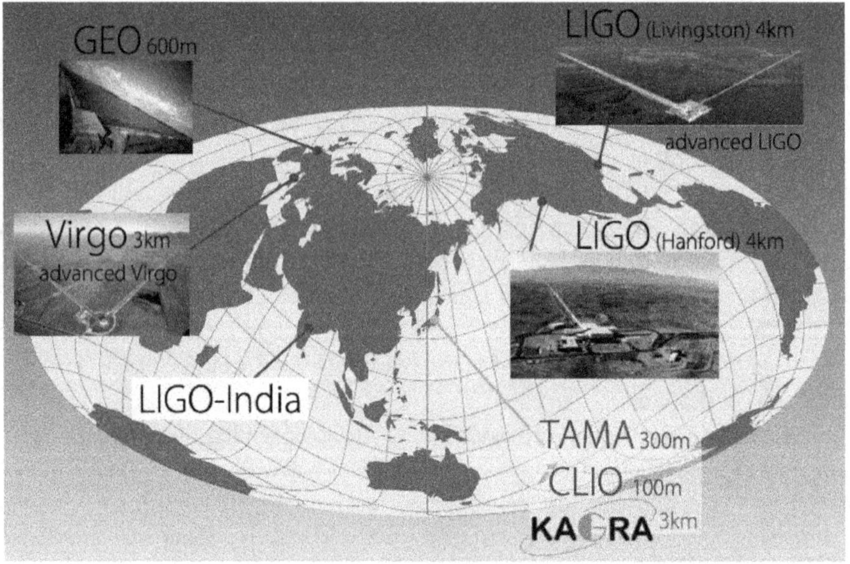

Fig. 1.　Schematic view of world-wide GW detector network.

The other advantage is about the sky localization. As recently, demonstrate by LIGO-Virgo and EM follow-up observations on NS-NS merger coinciding with Gamma-Ray Burst (GRB) [2], it is very important to locate where the GW signal come from so that EM telescopes can locate the host Galaxy. The principle to locate the GW source direction is called triangulation. Assuming that GW travels in a speed of light, by taking the arrival time differences between different GW detectors, we can locate the arrival direction. With only two detectors, we can localize as a ring on the sky, with three detectors or more we can pin-down the source direction more accurately. Figure 3 shows the schematic view to explain the sky localization.

With 4 or more detectors, we can also measure the GW polarization, with which we could test Einstein's theory of General Relativity (GR). According to GR, the GW has two polarization components, plus and cross which are 45 degree rotated each other. As shown in Figure 2, a Michelson-type laser interferometer have preferred sensitivity mostly on one polarization. As mentioned above, two LIGO detectors are located with similar orientation so both of them have the sensitivity to only one same polarization. By adding third detector, we can significantly improve the source localization, however, there still remain significant uncertainties on the measurement of each polarization component. By adding the forth detector will improve the polarization measurement, too.

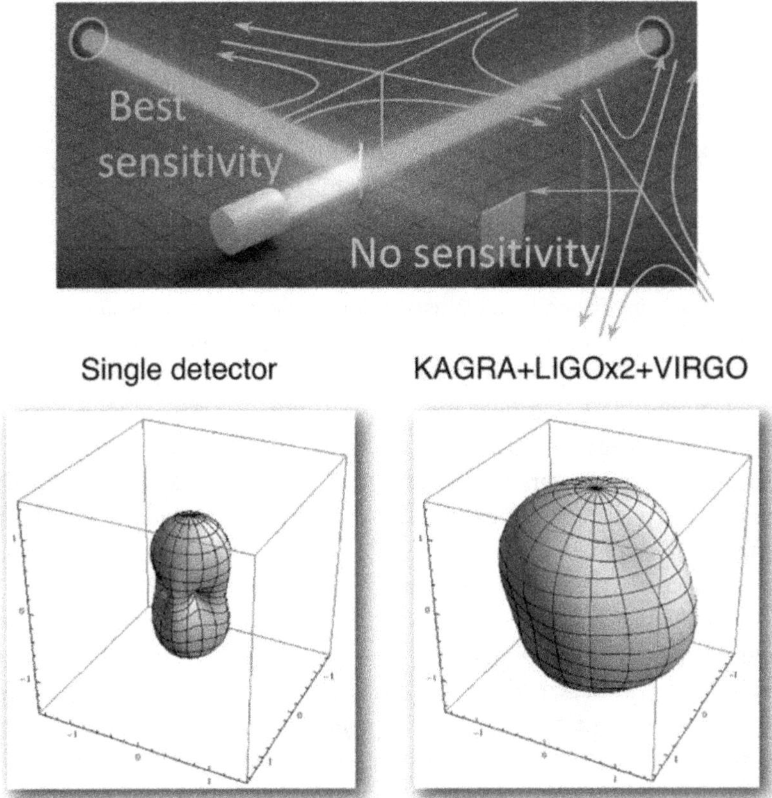

Fig. 2. (Top) Schematic view of Michelson-type laser interferometer and GW signal coming from top (blue) with the best sensitivity and side (orange) with lower sensitivity. (Bottom) Comparison of antenna amplitude pattern between two cases; (left) only single detector, (right) for detector network with KAGRA, 2 LIGO and VIRGO (color online).

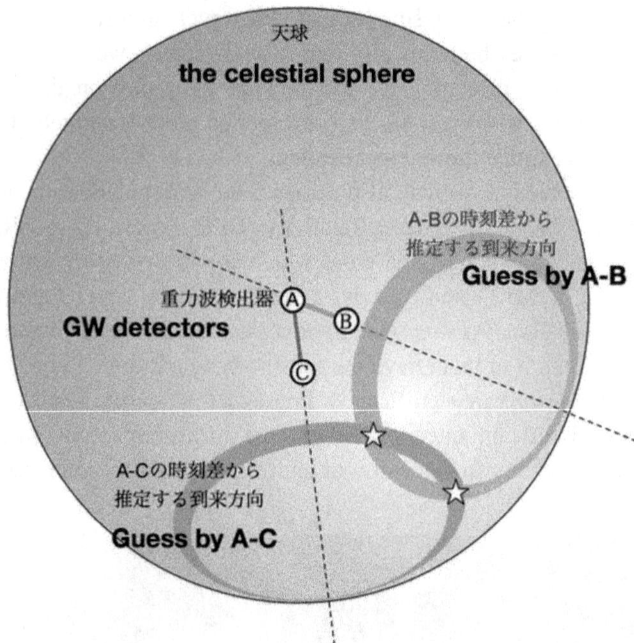

Fig. 3.   Schematic view of how we can make sky localization by taking the time difference of three GW detectors.

3.   The principle of detecting GW signal with laser interferometer

Interferometers are investigative tools used in many fields of science and engineering by merging two or more sources of light to create an interference pattern. They are often used to make very small measurements that are not achievable any other way. Interferometers were invented in the late 19th century by Albert Michelson to prove (or disprove) the existence of "Luminiferous Aether". Remarkably, the basic structure of the current GW interferometers differs little from the one that Michelson designed over 125 years ago, but with some added features, while the size and added complexity of the GW interferometers are far beyond anything Michelson could have envisioned or that his original interferometer could have achieved.

It is particularly important to have long arms to have laser light farther the laser travels. But there are obvious limitations to how long one can build an interferometer. It was fixed by adding so-called "Fabry Perot cavities". Each arm has a laser cavity created by adding mirrors near the beam splitter that continually reflect parts of each laser beam back and forth within the arms about several $100 \sim 1000$ times before they are merged together again. With Fabry Perot cavities, the interferometer arms can be effectively long by the same factor. The two mirrors facing each other to form the Fabry Perot cavity of each arm are called Test Masses (TMs).

There are several noise sources to be surpassed to have enough sensitivity to detect the GW. The first one is vibration noise due to the seismic motion of the ground. The TMs as well as most of other optics are suspended by multi-layer pendulum for the vibration isolation. Each pendulum has its own resonant frequency which depends on the lengths of the suspension wire but well above this frequency the TMs behave like a free mass. The second one is thermal noise due to thermally driven fluctuations of the mirrors. LIGO and Virgo detectors employ high quality quartz mirrors with sophisticated coatings and monolithic quartz fiber suspension to minimize the thermal noise, which KAGRA employs the high quality artificial Sapphire mirrors cooled down to cryogenic temperature (20K). The third one is the quantum noise. It refers to a broad class of noise sources that arise from the quantum nature of the light source and photo-detection process used in GW interferometers. There are two types of quantum noise; one is Radiation-pressure noise, which arises from uncertainties in the mirror positions due to quantum fluctuations exerting fluctuating radiation pressure on the mirrors. It was shown that this radiation pressure force can be attributed to vacuum fluctuations, another is Shot noise, which arises from uncertainty due to quantum mechanical fluctuations in the number of photons at the interferometer output photo-detector. The shot noise can be reduced by increasing the laser power but at the same time it increases the radiation pressure noise, while the radiation pressure noise can be reduced by using the heavier TM mirrors. Recently, a new technology to break this quantum limit is being developed using so-called the squeezed light. The detailed about its technology can be found for example in [5]. It will be employed in the next few years for the upgrades of current GW detectors.

Most of GW interferometers makes it natural to decompose the optical fields and the corresponding motion of the arm-cavity mirrors into symmetric and antisymmetric modes and they are operated on the dark fringe, ideally only optical signals induced by the antisymmetric motion of the arm-cavity mirrors exit the output port of the beam splitter. Hence most of the laser light is reflected back to the input of the interferometer. In order to increase the stored laser power in the arm cavities and to reduce the quantum (shot) noise, a power recycling mirror is put in front of the input of the interferometer to reflect back the laser light and increase the effective laser power by factor of several 10s. In addition to that there is a signal recycling mirror at the output port of the interferometer to further enhance the GW signal. This is a typical configuration of the 2nd generation GW detectors and called Dural Recycling Fabry Perot Michelson Interferometer (DRFPMI).

4. Special feature of KAGRA

KAGRA is considered to be further advanced GW detector (or 2.5 Generation detector) [6]. The basic interferometer configuration (DRFPMI) is very similar to that of LIGO and Virgo as shown in Figure 4. In addition to that, KAGRA employs two distinct features; one is the underground location, and the other is cryogenic TMs. As mentioned above, one of the major noise source is the seismic vibration. We found that the vibration noise can be about two order of magnitude less at the underground condition compared with the

180

surface. To take this advantage, KAGRA is built inside the 3km-by-3km L-shape tunnels inside the Kamioka mine in Japan. Kamioka mine is famous for neutrino detectors, called Kamiokande and Super-Kamiokande (Super-K) which contributed two Nobel Prize in Physics for the discoveries of cosmic neutrino and neutrino oscillation. Kamioka area is located around the center of Japanese main-island. Figure 5 shows the cut view of the Kamioka mountain area which hosts various neutrino detectors (Super-K, XMASS, and KamLAND), and a prototype detector of KAGRA (CLIO) in addition to KAGRA. Recently a new high speed train (Shinkansen) started the operation so one can reach the experimental site in 2 hours by the train from Tokyo to Toyama and 1 hour of drive from Toyama.

Another feature, cryogenic mirror is considered to be the key technology for the next (3rd) generation GW detectors. Thermal noise is considered as a mirror reflective surface deformation by thermal elastic vibration of mirror. It is known that there are three dominant kinds of mirror thermal noise; Substrate Brownian noise, Substrate thermo-elastic noise and the Coating Brownian noise. According to Fluctuation Dissipation Theorem, thermal noise is proportional to the product of temperature and mechanical loss of the material. We found that the mechanical loss of Sapphire gets much smaller at cryogenic temperature around 20 K. Therefore, by cooling down the Sapphire mirror we can suppress both the temperature term and mechanical loss term at the same time and resulting to suppress the mirror thermal noise. A similar technology using the Silicon mirror is being considered to be one of the best candidate for the 3rd generation GW detectors.

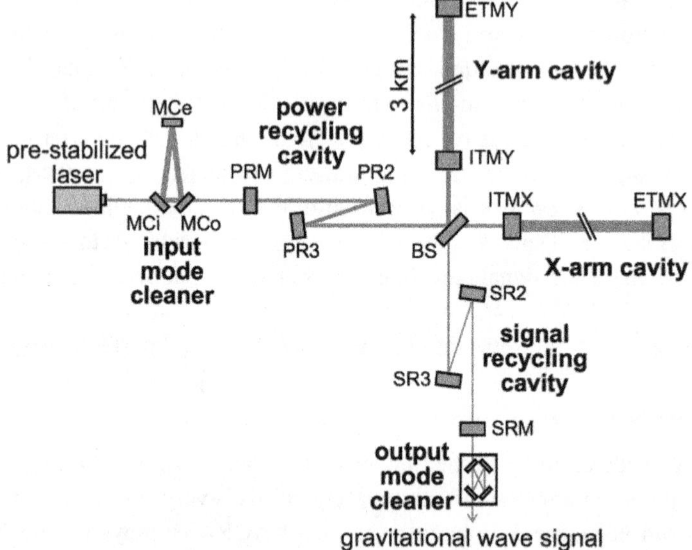

Fig. 4. Schematic view of KAGRA optical configuration (Dural Recycling Fabry Perot Michelson Interferometer).

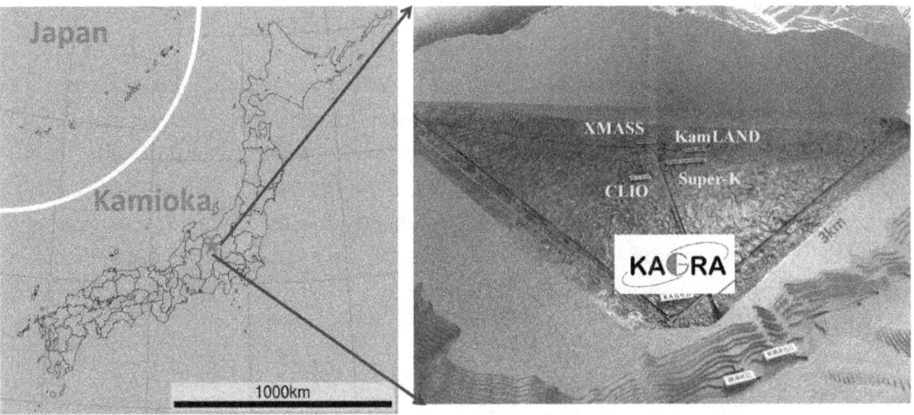

Fig. 5. (Left) Location of Kamioka area in Japan. (Right) Cut view of Kamioka mountain area which hosts various detector projects (Super-K, XMASS, KamLAND, CLIO) in addition to KAGRA.

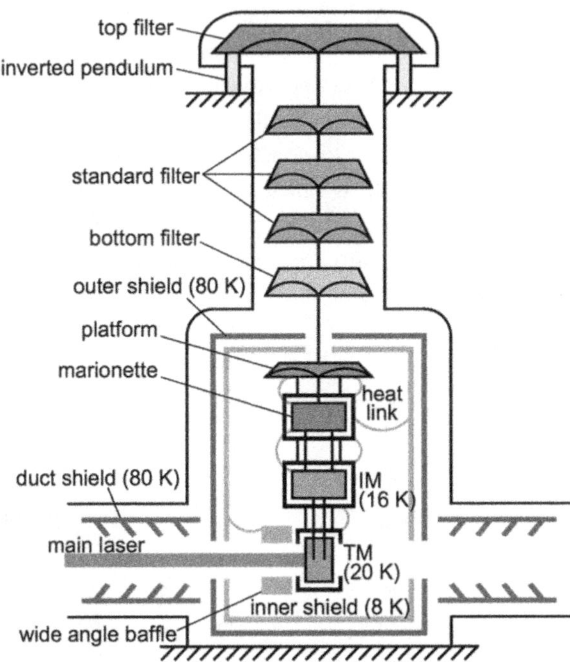

Fig. 6. Schematic view of KAGRA vibration isolation system for test masses (TMs) inside two-floor tunnel. The system consists of room-temperature part (orange) and cryogenic part (blue and purple) (color online).

In order to enjoy the advantages of such new features, we need to fight some practical difficulties. KAGRA cryogenic system used to cool down cryogenic TM mirrors. Figure 6 shows the schematic view of vibration isolation system and cryogenic suspension. The whole system is installed inside the tunnel with two floors. TMs are suspended by eight-stage pendulum suspended from a top geometric anti-spring filter on an inverted pendulum table for low frequency vertical and horizontal vibration isolation. The last four stages of the pendulum consist of platform, marionette, Intermediate Mass (IM), and TM, and cooled down to cryogenic temperatures. The heat generated by the main intra-cavity beam is extracted via sapphire fibers which suspend the TM from the intermediate mass, and via high purity (6N) aluminum heat links. The cryogenic payload is surrounded by inner and outer shields which are cooled down to 8 K and 80 K, respectively, by low vibration pulse-tube cryocoolers. The arm cavity also has different kinds of baffles and duct shields for absorbing stray light and thermal radiation from ducts at room temperature.

5. Current status of KAGRA and near future prospects

Because of the challenge to employ advanced technologies, the deployment of KAGRA is a bit behind other detectors, LIGO and Virgo, in the time scale. LIGO started the scientific observation in Sep./2015 (and discovered GW signal for the first time), and Virgo joined the GW network in Aug./2017 (and contributed the discovery of NS-NS merger signal). KAGRA is still an installation phase now and expecting to be online in 2019~2020. The funding of KAGRA was approved by Japanese government in 2010 and most of the initial time was used to excavate the 3km-3km arm tunnels under Kamioka area and to install the vacuum tubes to construct the arms. The installation of arms and cryostats finished in 2015. Then so-called initial-KAGRA (iKAGRA), which is the world-first km-scale underground room-temperature Michelson interferometer, was operated in March-April of 2016. Currently, the cryogenic suspension system is being installed to realize the world-first km-scale underground and cryogenic Michelson

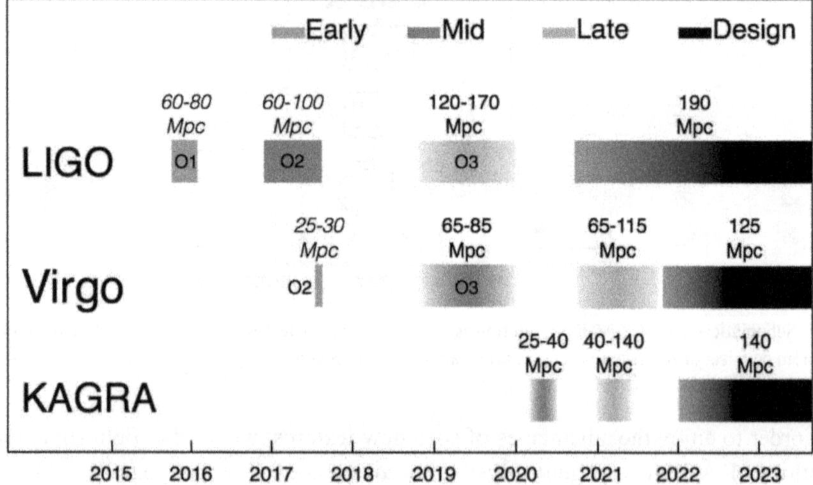

Fig. 7. Possible observation scenario of KAGRA considering also the observation plans of LIGO and Virgo. The Binary Neutron Star (BNS) horizontal distance is shown for each period.

interferometer within the first quarter of 2018. Then all the other optical components to form the Fabry Perot and dual recycling optical cavities by the end of 2018 ~ early 2019. Then, after several months of commissioning runs and noise hunting, we hope to start the possible scientific observations. The expected milestones and observation scenario together with LIGO and Virgo is being published in [7]. Figure 7 shows the schematic diagram of the observation scenario of LIGO/Virgo/KAGRA and Figure 8 shows the comparison of their strain sensitivities.

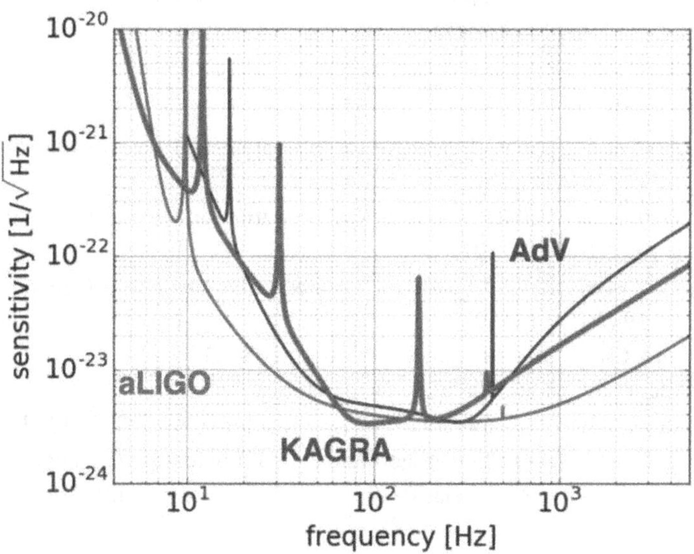

Fig. 8.   Comparison of strain sensitivities between KAGRA, LIGO (aLIGO) and Virgo (AdV).

6. Summary and conclusion

This year (2017) the Nobel Prize in Physics was given to the founders and key person of LIGO. Regrettably enough, another LIGO founder, Prof. Ronald Drever passed away last year and couldn't be awarded this year, while Prof. Barry Barish, who is not exactly the founder of LIGO won the prize. Most of the people, particularly in LIGO are happy that Prof. Barish won the prize, because without him they couldn't realize such are large and complex system. As mentioned in the beginning, the GW astronomy has just begun. The Nobel prize is not given only once to one research field. A typical example is about the neutrino physics which is another difficult signal to be detected. In 2002, Profs. Raymond Davis Jr., Masatoshi Koshiba, Riccardo Giacconi won the Nobel Prize "for pioneering contributions to astrophysics, in particular for the detection of cosmic neutrinos", then in 2015 Profs. Takaaki Kajita, and Arthur B. McDonald won another Nobel Prize "for the discovery of neutrino oscillations, which shows that neutrinos have mass". We expect young researchers who are reading this article to be ambitious to won the next Nobel Prize to discover new Physics by using GW as an advanced tool.

Acknowledgments

The KAGRA project is supported by MEXT, JSPS Leading-edge Research Infrastructure Program, JSPS Grant-in-Aid for Specially Promoted Research 26000005, MEXT Grant-in-Aid for Scientific Research on Innovative Areas 24103005, JSPS Core-to-Core Program, A. Advanced Research Networks, the joint research program of the Institute for Cosmic Ray Research, University of Tokyo, National Research Foundation (NRF) and Computing Infrastructure Project of KISTI-GSDC in Korea, the LIGO project, and the Virgo project. The author is supported by Academia Sinica and the Ministry of Science and Technology (MOST) in Taiwan.

References

1. B. P. Abbott et al., [LIGO Scientific Collaboration and Virgo Collaboration], Phys. Rev. Lett. 116 (2016) 061102.
2. B. P. Abbott et al., [LIGO Scientific Collaboration and Virgo Collaboration], Phys. Rev. Lett. 119 (2017) 161101.
3. B. P. Abbott et al., [LIGO Scientific Collaboration and Virgo Collaboration], Phys. Rev. Lett. 119 (2017) 141101.
4. B. F. Schutz, Class. Quantum Grav. 28 (2011) 125023.
5. C. B. Møller et al., Nature 547 (2017) 191–195.
6. K. Somiya [KAGRA Collaboration], Class. Quantum Grav. 29 (2012) 124007.
7. B. P. Abbott et al., [KAGRA Collaboration, LIGO Scientific Collaboration and Virgo Collaboration] Living Rev. Relativ. 19 (2016) 1; Update submitted arXiv:1304.0670.

LISA Pathfinder

M. Armano[a], H. Audley[b], J. Baird[c], P. Binetruy[d,*], M. Born[b], D. Bortoluzzi[e], E. Castelli[f],
A. Cavalleri[f], A. Cesarini[g], A. M. Cruise[h], K. Danzmann[b], M. de Deus Silva[a], I. Diepholz[b],
G. Dixon[h], R. Dolesi[f], L. Ferraioli[i], V. Ferroni[f], E. D. Fitzsimons[j], M. Freschi[a], L. Gesa[k],
F. Gibert[f], D. Giardini[i], R. Giusteri[f], C. Grimani[g], J. Grzymisch[l], I. Harrison[m], G. Heinzel[b],
M. Hewitson[b], D. Hollington[c], D. Hoyland[h], M. Hueller[f], H. Inchauspé[d], O. Jennrich[l],
P. Jetzer[n], N. Karnesis[b], B. Kaune[b], N. Korsakova[o], C. J. Killow[o], J. A. Lobo[k,†], I. Lloro[k],
L. Liu[f], J. P. López-Zaragoza[k], R. Maarschalkerweerd[m], D. Mance[i], N. Meshskar[i], V. Martín[k],
L. Martin-Polo[a], J. Martino[d], F. Martin-Porqueras[a], I. Mateos[k], P. W. McNamara[l],
J. Mendes[m], L. Mendes[a], M. Nofrarias[k], S. Paczkowski[b,‡], M. Perreur-Lloyd[o], A. Petiteau[d],
P. Pivato[f], E. Plagnol[d], J. Ramos-Castro[p], J. Reiche[b], D. I. Robertson[o], F. Rivas[k], G. Russano[f],
J. Slutsky[q], C. F. Sopuerta[k], T. Sumner[c], D. Texier[a], J. I. Thorpe[q], D. Vetrugno[f], S. Vitale[f],
G. Wanner[b], H. Ward[o], P. Wass[c,r], W. J. Weber[f], L. Wissel[b], A. Wittchen[b], and P. Zweifel[i]

[a] *European Space Astronomy Centre, European Space Agency, Villanueva de la Cañada,
28692 Madrid, Spain*

[b] *Albert-Einstein-Institut, Max-Planck-Institut für Gravitationsphysik und Leibniz Universität
Hannover, Callinstraße 38, 30167 Hannover, Germany*
‡*E-mail: sarah.paczkowski@aei.mpg.de*

[c] *High Energy Physics Group, Physics Department, Imperial College London,
Blackett Laboratory, Prince Consort Road, London, SW7 2BW, UK*

[d] *APC, Univ Paris Diderot, CNRS/IN2P3, CEA/lrfu, Obs de Paris, Sorbonne Paris Cité,
France *Deceased 30 March 2017*

[e] *Department of Industrial Engineering, University of Trento, via Sommarive 9, 38123 Trento,
and Trento Institute for Fundamental Physics and Application / INFN*

[f] *Dipartimento di Fisica, Università di Trento and Trento Institute for Fundamental Physics
and Application / INFN, 38123 Povo, Trento, Italy*

[g] *DISPEA, Università di Urbino "Carlo Bo", Via S. Chiara, 27 61029 Urbino/INFN, Italy*

[h] *The School of Physics and Astronomy, University of Birmingham, Birmingham, UK*

[i] *Institut für Geophysik, ETH Zürich, Sonneggstrasse 5, CH-8092, Zürich, Switzerland*

[j] *The UK Astronomy Technology Centre, Royal Observatory, Edinburgh, Blackford Hill,
Edinburgh, EH9 3HJ, UK*

[k] *Institut de Ciències de l'Espai (CSIC-IEEC), Campus UAB, Carrer de Can Magrans s/n,
08193 Cerdanyola del Vallès, Spain *Deceased 30 September 2012*

[l] *European Space Technology Centre, European Space Agency, Keplerlaan 1,
2200 AG Noordwijk, The Netherlands*

[m] *European Space Operations Centre, European Space Agency, 64293 Darmstadt, Germany*

[n] *Physik Institut, Universität Zürich, Winterthurerstrasse 190, CH-8057 Zürich, Switzerland*

[o] *SUPA, Institute for Gravitational Research, School of Physics and Astronomy,
University of Glasgow, Glasgow, G12 8QQ, UK*

[p] *Department d'Enginyeria Electrònica, Universitat Politècnica de Catalunya,
08034 Barcelona, Spain*

[q] *Gravitational Astrophysics Lab, NASA Goddard Space Flight Center,
8800 Greenbelt Road, Greenbelt, MD 20771, USA*

[r] *Department of Physics, University of Florida, 2001 Museum Rd,
Gainesville, FL 32603, USA*

Since the 2017 Nobel Prize in Physics was awarded for the observation of gravitational waves, it is fair to say that the epoch of gravitational wave astronomy (GWs) has begun. However, a number of interesting sources of GWs can only be observed from space. To demonstrate the feasibility of the Laser Interferometer Space Antenna (LISA), a future gravitational wave observatory in space, the LISA Pathfinder satellite was launched on December 3rd, 2015. Measurements of the spurious forces accelerating an otherwise free-falling test mass, and detailed investigations of the individual subsystems needed to achieve the free-fall, have been conducted throughout the mission. This overview article starts with the purpose and aim of the mission, explains satellite hardware and mission operations and ends with a summary of selected important results and an outlook towards LISA. From the LISA Pathfinder experience, we can conclude that the proposed LISA mission is feasible.

Keywords: Gravitational waves; interferometers; space research instruments; laser metrological applications.

1. Introduction to the LISA Pathfinder project

LISA Pathfinder is a technology demonstrator mission for the Laser Interferometer Space Antenna, LISA. To understand the necessity and the main goal of LISA Pathfinder, let us review some important aspects of LISA.

1.1. *The Laser Interferometer Space Antenna LISA*

1.1.1. *Short summary of gravitational wave sources in the LISA band*

LISA is a mission concept for a future gravitational wave observatory in space. It will not be competing with the ground-breaking discoveries of the LIGO-Virgo collaboration but opens the opportunity to observe gravitational waves at lower frequencies. To be more precise, with LISA we aim to be able to measure gravitational waves in the frequency regime from $20\,\mu\text{Hz}$ to $1\,\text{Hz}$. In this measurement band, we aim to observe gravitational waves from several very interesting sources, which are for example supermassive black hole binaries and extreme mass ratio inspirals (EMRIs).

As explained in 1, LISA is a great instrument to observe supermassive black hole binaries, especially for those with large redshifts of $z \approx 10$ and beyond with relatively small masses between $10^4 M_\odot$ and $10^7 M_\odot$. Black holes at these redshifts are difficult to observe using electromagnetic radiation in the optical regime because it is suppressed for certain frequencies and for redshifts larger than approximately 6. Also X-Ray observations may suffer from a deterioration of the signals of these sources due to crowded sources and unresolved background light, difficulties that do not apply for LISA. These supermassive black hole binaries are very important to discover the formation of seed black holes around the cosmic dawn. LISA is expected to enable us to study the black hole binaries with masses between $10^4 M_\odot$ and $10^7 M_\odot$ out to redshifts of 20, if they exist. In general, LISA can also be seen as a large black hole binary search over a vast range of masses and redshifts in nearly all directions which is, even with future observatories, only in part accessible in the electromagnetic regime. In total, a coalescence rate between 10 and 100 per year is expected to be observable with LISA. The observation of coalescences are especially interesting because they will allow us to better understand the accretion mechanisms of black holes.

EMRIs consist of a compact object, which, in the case of LISA signals, is more likely a stellar mass black hole than a neutron star, which is in a highly relativistic orbit around a massive black hole. As such, EMRIs are an excellent opportunity to study gravity in the strong-field and non-linear regime. In addition, they provide another opportunity to study the stellar dynamics around a massive black hole. This information, as well as further details, can be found in 1.

Furthermore, LISA will be sensitive to a stochastic background of gravitational waves with an energy density of $\Omega_{\mathrm{GW}} \simeq 10^{-13}$ or higher between between $1\,\mathrm{mHz}$ and $10\,\mathrm{mHz}$[2]. Conservative estimates for the previous eLISA concept required an energy density of $\Omega_{\mathrm{GW}} \simeq 10^{-10}$ in the present universe[3]. Interestingly, signals with such a feature can be originated by both astrophysical and cosmological sources[2]. For instance, from the detection of GW150914, the stochastic background from binary black holes is predicted to have an energy density $\Omega_{\mathrm{GW}} = 1.1^{+2.7}_{-0.9} \cdot 10^{-9}$ at $25\,\mathrm{Hz}$[4]. Concerning the cosmological sources, the most studied are those related to the inflationary processes, topological defects and first-order phase transition[5,6]. To create a visible signal in LISA, at the time of production in the radiation dominated era in the early universe, a fraction of $\Omega_{\mathrm{GW}} > 10^{-5}$ of the energy density of the universe must have been converted into gravitational radiation[1]. All these processes involve physics beyond the standard model of particle physics with a new physics scale from the electroweak up to the Planck mass scale. Due to the sensitivity to these scales, LISA is complementary to particle physics experiments as e.g. the LHC[2].

Moreover, with LISA, we expect to be able to predict the time to coalescence of black-hole binaries which pass from the LISA measurement band to the advanced LIGO frequency range within the LISA lifetime, with an accuracy of approximately $10\,\mathrm{s}$ and their sky location with an accuracy of one square degree[7].

Finally, it should be mentioned that the list of possible sources is by far not complete. Especially, the unpredicted sources which might be measured could open a whole new discovery space.

1.1.2. *LISA is only possible in Space*

The main reason why LISA is only possible in space is that towards lower frequencies, in a ground-based gravitational wave observatory such as LIGO, the seismic and gravity gradient noise becomes limiting. Gravity gradient noise is also known as Newtonian gravity noise and describes the disturbances caused by changes in local Newtonian gravity due to moving objects close to the extremely sensitive detector[8]. In addition, it is difficult to achieve sufficient thermal and mechanical stability for the corresponding long measurement durations. To some extent, achieving the required thermal stability is also non-trivial in space. Moreover, in space, it is much easier to have even longer arms to increase the sensitivity of the instrument.

The gravitational wave observatory LISA, as described in 9, is characterised by the triangular constellation which is formed by three satellites. This constellation is trailing approximately 20° behind the Earth on a heliocentric orbit. These orbits also introduce a 'cartwheel' rotation of the whole constellation but at the same time they keep the relative separation of the satellites as well as the angles of the triangle as constant as possible. Laser light is exchanged between the satellites, which are planned to be 2.5 million km away from each other. The sides of the triangle correspond to the arms of a ground-based gravitational wave observatory and hence, they are also called arms. Each satellite hosts two test masses which define the beginning and the end of each of the arms. Gravitational waves now cause tiny variations in the relative distance of the two such test masses. These are measured using heterodyne laser interferometry. Therefore, there is an optical bench for each test mass, so two on each satellite. The measurement of the relative distance in between two test masses on two different satellites is split into three measurements: a local test mass to satellite measurement, a satellite to satellite measurement and another local test mass to satellite measurement on the far satellite. Accordingly, there will be local heterodyne interferometry and inter-satellite heterodyne interferometry required. In addition, the laser light of the two different optical benches on the same satellite needs to be compared to enable the laser frequency noise suppression via the Time-Delay Interferometry (TDI) algorithm[10]. Therefore, in the current design, an optical fibre is linking the two optical benches on a LISA satellite.

1.1.3. *LISA requires quiet test masses*

Most importantly, to measure gravitational waves, LISA requires free-falling test masses. This means they are subject to no other force than gravity. To put it simply, we have to ensure that a measured signal is due to the effect of gravitational waves and not caused by interaction between any of the two test masses and the respective satellite or its components. The residual acceleration of an otherwise free-falling test mass, originating from a number of possibly unknown, spurious forces caused by the satellite and the satellite environment is what we call Δg. It is a noisy signal and the smaller the noise level, the less residual acceleration is present and the quieter and closer to free-fall is the test mass.

Just how very close to a perfect free-fall, each pair of test masses has to be can be seen from the requirement[9]

$$S^{\frac{1}{2}}_{g,\text{LISA}} \leq 3 \cdot 10^{-15} \frac{\text{m}}{\text{s}^2\sqrt{\text{Hz}}} \sqrt{1 + \left(\frac{0.4\,\text{mHz}}{f}\right)^2} \sqrt{1 + \left(\frac{f}{8\,\text{mHz}}\right)^4} \qquad (1)$$

that applies to the square root of the power spectral density of the residual acceleration of a single test mass g. Ideally, these fluctuations would vanish but this is not possible in reality. If the fluctuations are below the requirement, the sensitivity of LISA is good enough to answer the science questions. However, it is impossible to test the technology for LISA at these tiny acceleration levels in a standard laboratory on Earth. A torsion pendulum facility, however, gets close to the required accuracy but is then limited by mechanical thermal noise at lower frequencies and readout noise towards higher frequencies[11]. Let us note furthermore that to measure g at mHz frequencies, it is necessary to measure $\approx 17\,\text{min}$. That is why a drop-tower would not be sufficient. In addition, this requirement is more stringent than on previous drag-free missions. For example on GOCE, the residual acceleration requirement is relaxed by two orders of magnitude in comparison to LISA Pathfinder[12]. That is why a test in space, as it is done with the LISA Pathfinder Mission, is required.

1.2. *LISA Pathfinder Mission Goal*

Accordingly, the LISA Pathfinder mission goal is to demonstrate the technology for LISA, which means foremost to show that a nearly perfect free-fall is feasible. In comparison to LISA, the residual differential acceleration requirement is relaxed by one order of magnitude to[13]

$$S^{\frac{1}{2}}_{\Delta g,\text{LPF}} \leq 30 \cdot 10^{-15} \frac{\text{m}}{\text{s}^2\sqrt{\text{Hz}}} \sqrt{1 + \left(\frac{f}{3\,\text{mHz}}\right)^4} \qquad (2)$$

and restricted to the frequency range from $1\,\text{mHz}$ to $30\,\text{mHz}$. In contrast to LISA, LISA Pathfinder is not designed to observe gravitational waves because the distance in between the two test masses, as explained below, is too short.

2. The LISA Pathfinder satellite

2.1. *LISA Pathfinder Hosts Two Free-Floating Test Masses*

A first and idealised idea to prove Equation (2) could be to put a single test mass carefully into a quiet environment in space. However, to measure an acceleration, a reference point and a measurement system is needed. Thus, we have a satellite that hosts the measurement system and shields the test mass from, for example, solar radiation pressure. In theory, already the satellite could be used as a reference point for the acceleration measurement. However, this reference point itself is too noisy for the acceleration levels we want to measure on LISA Pathfinder. Therefore, we also have a quiet, free-floating reference mass.

These two free-falling test masses are at the core of LISA Pathfinder. Free-falling means, they are not physically connected or touched by anything. The test masses are quasi cubes made of a gold-platinum alloy and their edges are (46.0 ± 0.5) mm long[13]. The material has been chosen to have a low magnetic susceptibility of $\chi_M \approx 10^{-5}$ combined with a high density[14]. Each test mass weighs (1.920 ± 0.001) kg. A comparatively large mass minimises undesired gravitational interaction with the surroundings. They are (376.00 ± 0.05) mm apart from each other, as can be seen in Figure 2. The reason they are not perfect cubes is the caging and release mechanism. It has to fix the test masses during launch because a loose test mass during the rocket launch would cause severe damage. In addition, it has to release the test masses smoothly in the final orbit. The main measurement on LISA Pathfinder is now the change in relative distance in between the two free-falling test masses. The relative residual acceleration is then obtained as the second derivative of this measurement. The details will be given in Section 3.1.

2.2. *The LISA Pathfinder Drag-Free and Attitude Control System*

On LISA Pathfinder, a Drag-Free and Attitude Control System (DFACS) is being used to minimise the undesired influence of the satellite on the free-floating test masses. Our main science mode, which is used for our measurements of the residual acceleration, works like this: Along the sensitive axis, which is defined by the line that connects the two test masses, one of the test masses, usually called TM1, is not subject to any control force. The satellite is then controlled by the DFACS to follow the motion of TM1. The DFACS system obtains the position of the TM1 with respect to the satellite, as measured by the interferometer X1. The output of this interferometer is called o1, as can be seen in Figure 1. From this, it determines

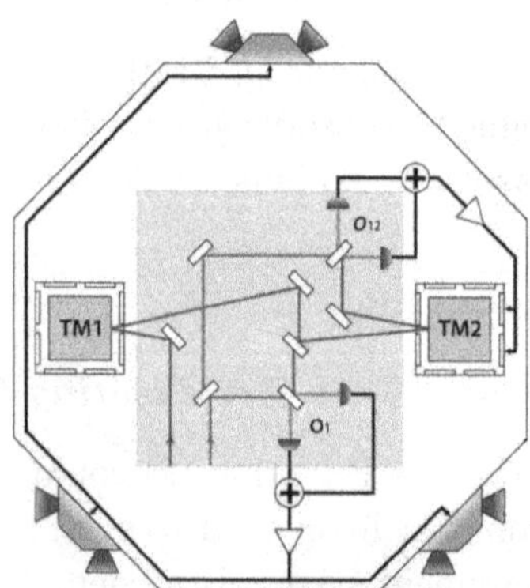

Fig. 1. A schematic representation of the LISA Pathfinder components used by the Drag-Free and Attitude Control System (DFACS). o1 is the measurement of the interferometer X1 and o12 is the measurement of the interferometer X12. Reprint from 15.

the necessary thrust that has to be applied by the thrusters to have the satellite follow TM1. This control loop is called the drag-free loop. Also the other test mass is made to follow the motion of TM1 in this mode. The measurement of the relative position of the two test masses, as measured with the interferometer X12, is called o12. Using this o12 measurement, the DFACS system also estimates the force that has to be applied to the other test mass such that it follows TM1. This loop is called the suspension loop. The necessity for the suspension can be understood by thinking of the two test masses moving in opposite directions along the sensitive axis. Then the satellite would not be able to follow them both. So one of them, usually TM2, is made to follow the other, usually TM1. However, the commanded force is known to the data analysis. It is used to estimate the true applied forces onto TM2 such that they can be subtracted in post-processing. It is also important to mention that the unity gain frequency of this so-called suspension loop is near the end of our measurement band. This minimises the impact of the control forces on the measurement of the residual acceleration noise. More details can be found in Section 3.1. In addition, a dedicated control mode is implemented on LISA Pathfinder which replaces the continuous control of TM2 along x with an intermittent control scheme. This experiment is called the drift mode or free-flight experiment (see Section 3.2).

2.3. *Key Subsystems to Achieve Free-Fall on LISA Pathfinder*

To achieve the necessary levels of free-fall, not only the DFACS as explained above is essential but it can work only together with a number of key subsystems, as shown in Figure 2. That means also that we test all these systems for LISA. A key subsystem is the Gravity Reference Sensor (GRS). This system includes the two test masses as well as the electrodes mounted on the electrode housing and the corresponding front-end electronics (FEE). Each of the two test masses is in a vacuum tank. They have an optical window such that the light can reach the test masses for the measurement of the relative position in between the two. Another key component is the discharge mechanism. In contrast to other space missions, for example MICROSCOPE[16], where the discharging is done via a wire, the test masses on LISA Pathfinder are not in contact with anything and thus have no discharge wire. Instead, a UV lamp is used for discharging via the photoelectric effect. In addition, thrusters that allow the satellite to follow the test mass, while producing minimal undesired noise, are needed. These are the cold-gas µN thrusters, which are also used on the MICROSCOPE and GAIA mission[17]. It is also important to mention that the temperature on LISA Pathfinder has to be stable and that the magnetic fields on board have to be minimised to achieve the required level of free-fall. Another key component is the Optical Metrology System (OMS). In Figure 2, the OMS is easily discernible via the optical bench that is located in the centre of the picture. The optical bench is made out of a material with ultra-low thermal expansion, which is called Zerodur©. The lines in red and blue mark the paths of the laser light through the mirrors and beam splitters.

Fig. 2. The core of LISA Pathfinder. The two golden cubes are the test masses inside their respective electrode housings inside a vacuum tank. The optical bench is discernible in the centre. Image: ESA/ATG medialab (color online).

2.4. *The LISA Pathfinder Optical Metrology System*

The main purpose of the optical metrology system is to measure the distance of the free-falling test mass with respect to the quiet reference test mass with a required precision of [18]

$$S^{\frac{1}{2}}_{\delta x} \leq 9 \, \frac{\text{pm}}{\sqrt{\text{Hz}}} \sqrt{1 + \left(\frac{3\,\text{mHz}}{f}\right)^4}. \tag{3}$$

Here, $S^{\frac{1}{2}}_{\delta x}$ is the square root of the power spectral density of the measured fluctuations. To achieve this precision, a heterodyne laser interferometry set-up has been chosen. A Nd:YAG laser with a wavelength of 1064 nm and an output power of a few tens of mW is located in the so called reference laser unit (not discernible in Figure 2). From there, the light travels to the laser modulator unit where the light is split into two beams. Each of them is frequency shifted. Then the two beams travel via optical fibres to the optical bench. They leave the fibres via the fiber injector optical sub assembly which can be recognized by the green ends in the centre of Figure 2. On the optical bench, the two beams are brought to interference again. The resulting beat note of the light is recorded by the photodiodes. These are the metallic pieces on the optical bench in Figure 2 whose cables are held by blue cable ties. The phase of this signal is the measured quantity which contains, for example, the information on the relative distance in between the two test masses. However, on the optical bench, the two beams are brought to interference not only once but in four different ways. Hence, we have four different interferometers, as shown in Figure 3.

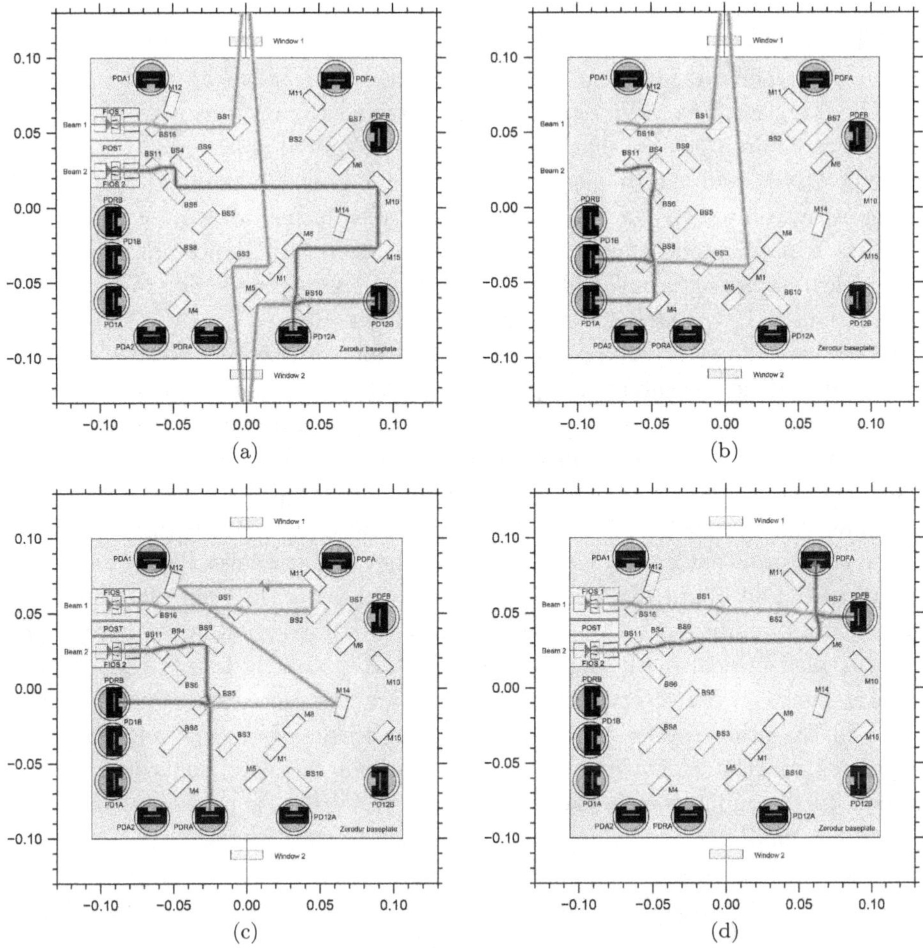

Fig. 3. The four interferometers on the LISA Pathfinder optical bench. (a) X12 interferometer. (b) X1 interferometer. (c) Reference interferometer XR. (d) Frequency interferometer XF.

The main interferometer is the X12 interferometer which measures the relative distance of the two test masses. In this interferometer, the so-called measurement beam is steered with mirrors in such a way that it hits both test masses and gets reflected on the respective surfaces. The other beam, the reference beam, stays on the optical bench. If now the relative distance in between the two test masses changes, the phase of the measurement beam will reflect this and so will the beatnote signal. In each of the four interferometers, the same beatnote is recorded by two photodiodes whose signals are combined. This scheme is called balanced detection[19]. Having two photodiodes for each interferometer is not only useful in terms of redundancy but also because with this balanced detection scheme, some phase errors that arise due to relative intensity noise in the laser cancel[19].

Similarly, the X1 interferometer measures the relative distance between the TM1 and the satellite. Here, the measurement beam hits only test mass 1 and not both test masses. The measurement beam stays on the optical bench.

There are also two auxiliary interferometers on the optical bench that are required to obtain a position measurement at the required noise level. These are the reference (XR) and the frequency interferometer (XF). In both of these, the two beams stay completely on the optical bench and do not hit any of the two test masses. To understand the reference interferometer, it is important to note that the optical path length in the fibres may fluctuate, for example due to temperature fluctuations. These fibres are not as stable as the optical bench. This noise will show up as a common mode phase noise in all four interferometers. Moreover, the two fibres for the measurement beam and the reference beam differ in optical path length by approximately $38.2\,\mathrm{cm}$[20]. The path of the light in the reference interferometer is chosen such the beam with the shorter fibre has the longer path on the optical bench to match the path length to minimise the coupling of laser frequency noise, as explained below. Thus, the reference interferometer measures exactly the phase noise resulting from the common mode path length noise. This measurement is subtracted on board LISA Pathfinder from the other three interferometer measurements. In addition to the noise mitigation by subtraction in software, there is the so-called optical path length difference (OPD) control loop[21]. Based on the measurement in the reference interferometer, two piezo-actuators adjust the optical path length in the laser modulator unit. For further details we also refer to 22. This suppression mechanism was found to be necessary in test campaigns on ground due to non-linear couplings from the radio signals that drive the acousto-optic modulators which modulate the light to have two different frequencies[23].

The fourth interferometer is the frequency interferometer (XF). The path of the two beams of the frequency interferometer on the optical bench is equally long, such that there is an intentional path length difference arising from different fibre lengths. This amplifies the laser frequency noise δf following[24]

$$\delta\phi = 2\pi\frac{\Delta s}{c}\delta f\,,\tag{4}$$

with $\delta\phi$ being the resulting phase error which cannot be distinguished from the measured signal. The coupling is determined by the path length difference, Δs, and c denotes the speed of light. This measurement of the laser frequency noise is fed into a controller that commands a piezo and a heater to stabilise the laser frequency. For details, we refer to 25.

After many years of development and testing on ground, not only for the OMS but also for all the other key systems, they were ready to be finally integrated into the satellite.

2.5. *Launch and Mission Operations*

LISA Pathfinder was launched by a ESA-VEGA rocket by Arianespace on December 3rd, 2015 at 04:04 UTC from Kourou, French Guiana. After six apogee raising manoeuvres, there was the final burn that got LISA Pathfinder out of the elliptical orbit around the Earth and onto its way to the Lagrange point, L1. This Lagrange point is located in between Earth and the Sun, at a distance of approximately 1.5 Gm. Simply speaking, objects at this point orbit the Sun simultaneously with Earth which leads to a very small local gravity field. This makes it an ideal location for a high-precision low acceleration measurement which is what we want to do with LISA Pathfinder. However, the precise orbit of LISA Pathfinder is a 500 000 km x 800 000 km Lissajous orbit around the L1 point [15]. Already while the satellite was still on its way to L1, one unit after the next was switched on and checked for its principal functionality. This process is known as the in-orbit commissioning and it started on January 11th, 2016. After the successful in-orbit commissioning of LISA Pathfinder, the nominal mission operations phase of the LISA Technology Package (LTP) began on the first of March 2016. The nominal mission duration of three month continued until June 26th, 2016. This period was longer than June 1st, since the days where station keeping manoeuvres took place were not counted for the three months of science operations. These station-keeping manoeuvres were necessary to keep the satellite on its orbit but no residual acceleration noise measurements could be performed during these days. The nominal LTP operations phase was followed by the Disturbance Reduction System (DRS) operations. The DRS is a NASA payload on board of LISA Pathfinder. It tests a slightly different drag-free and attitude control system and a different set of µN thrusters. This phase was followed by an LTP mission extension which lasted until the final shut down of the LISA Pathfinder satellite on July 18th, 2017.

The technology demonstrator mission LISA Pathfinder can be seen as our laboratory in space. From an operating point of view however, the daily routine is different. Due to the limited duration of the operations and the numerous measurements needed, the time had to be well organised. Once scientists have decided for the next experiments to take place, they are inserted into a human readable schedule which takes into account the state that the satellite has to be in and the duration of the experiment. This is transformed by the Science and Technology Operations Centre (STOC) into a series of pre-defined command-blocks. The Mission Operations Centre (MOC) then controls the expansion of the command-blocks into telecommands. These telecommands are then uploaded to the satellite for execution. The resulting data is transferred to Earth with a limited data rate. This is why, in contrast to many other experiments, the sampling frequency of many LISA Pathfinder science data channels was 1 Hz or 10 Hz. The data was transferred to the servers and stored as so-called analysis objects in LTPDA [26], a dedicated MATLAB© toolbox developed for LISA Pathfinder. A major advantage of this software is that it tracks the history of the data used through all the stages of the analysis that is done to it. The data analysis was performed during the nominal operations phase in close to real time. That means, a team of engineers and scientists was located at the European Space Operations Centre in Darmstadt, Germany to analyse the arriving data. On the basis of these results, the next experiments were chosen.

In the following section, we will describe selected measurements and results obtained with the procedure and software explained in this section.

3. The physics of LISA Pathfinder

3.1. *The Residual Acceleration Δg Explained*

As mentioned already, the main measurement on LISA Pathfinder was the measurement of the residual acceleration noise. For this measurement, we take the second derivative of the measured relative position of the two test masses, denoted as x_{12}, and subtract the estimated applied forces on TM2, $g_c(t)$:

$$\Delta g = \ddot{x}_{12} - g_c(t) \,. \tag{5}$$

All quantities in this equation are always given per unit mass in this document and thus forces have the unit of acceleration, too. Equation (5), however, is only the starting point for the estimation of the residual acceleration and is not complete yet. In addition, we have to take known cross-couplings into account. Let us note that Equation (5) is simplified in the sense that it assumes the effect of the control force to be immediate. In reality, the DFACS control loops have a non-zero delay which is neglected here for simplicity.

Two long-known terms arise from the motion of the test masses in force gradients. This coupling is called a stiffness ω^2 and it acts like a spring with a negative spring constant. That means if a test mass gets accelerated into one direction, the resulting motion will be enhanced. In contrast, the usual springs have a positive spring constant and pull back the deflected object. The stiffness of each test mass is the sum of two stiffness contributions: a gravitational contribution and an electrostatic contribution. The gravitational component is due to the fact that the satellite is balanced in such a way that the test masses are subject to a minimal gravitational force at their optimal positions. Even a slight deviation from this position causes the test masses to be pulled to the centre of mass of the satellite which leads even further away from the optimal position. The stiffness also has an electrostatic component because there is a change in the electrostatic force acting on the test mass due to a position change of the test mass. The amplitude of this coupling depends on the so-called actuation authority. This is the maximum amount of actuation forces or torques that is permitted in the current setting on the satellite. From this number follows the amplitude of the 100 kHz AC voltage that is used instead of a noisy DC position readout. And the electrostatic component of the stiffness depends on exactly these voltages. Preliminary results indicate the gravitational contribution is larger than the electrostatic[27].

Including now the stiffness, the acceleration of each of the test masses reads

$$\ddot{x}_1 = g_1 - \omega_1^2 \left(x_1 - x_{\mathrm{SC}}\right) \tag{6}$$

$$\ddot{x}_2 = g_2 - \omega_2^2 \left(x_2 - x_{\mathrm{SC}}\right) + g_{\mathrm{c}}(t) \tag{7}$$

Combining these two equations to obtain the residual differential acceleration yields:

$$\Delta g = g_2 - g_1\,, \tag{8}$$

$$= -\ddot{x}_1 - \omega_1^2 \left(x_1 - x_{\mathrm{SC}}\right) + \ddot{x}_2 + \omega_2^2 \left(x_2 - x_{\mathrm{SC}}\right) - g_{\mathrm{c}}(t)\,. \tag{9}$$

Next, we take into account that the interferometer measures $o_{12} = x_2 - x_1$ and replace x_2 with $x_2 = o_{12} + x_1$ accordingly to obtain:

$$\Delta g = \ddot{o}_{12} - \omega_1^2 \left(x_1 - x_{\mathrm{SC}}\right) + \omega_2^2 \left(o_{12} + x_1 - x_{\mathrm{SC}}\right) - g_{\mathrm{c}}(t)\,. \tag{10}$$

This equation is also simplified for in reality, the interferometer is a system with readout noise and possible cross-sensing from other degrees of freedom or measurement channels. The coupling from other degrees of freedom will be partially included later in the derivation of Δg in the cross-coupling term. Leakage from other OMS measurement channels is not considered in this summary and we refer the reader to 28. Finally, the X1 interferometer measures in fact the position of TM1 with respect to the satellite such that we can redefine $o_1 := x_1 - x_{\mathrm{SC}}$. Introducing also the differential stiffness $\Delta\omega = \omega_2^2 - \omega_1^2$ allows us to write:

$$\Delta g = \ddot{o}_{12} + \omega_2^2 o_{12} + \Delta\omega o_1 - g_c\,. \tag{11}$$

From this equation, it can be seen that as the two test masses react slightly differently to the motion of the satellite (as measured by the X1 interferometer) due to the two different stiffnesses; this acceleration that originates in the satellite looks like a differential acceleration. The precise values of these stiffnesses have been estimated via dedicated experiments. These experiments consist of signal injection into the drag-free and the suspension loop. For details, we refer to 28.

However, during the mission, three more effects that need to be taken into account have been identified. One of them is the interferometer pick-up of satellite motion $g_{\mathrm{crosstalk}}(t)$, as explained in 29. In addition, the centrifugal forces $g_\Omega(t)$ [13] and other spacecraft angular acceleration effects $g_{\mathrm{decorr}}(t)$, which will be explained in a future publication, have to be subtracted. The equation for the residual differential acceleration currently reads, therefore,

$$\Delta g = \ddot{o}_{12} + \omega_2^2 o_{12} + \Delta\omega o_1 - g_c - g_\Omega(t) - g_{\mathrm{crosstalk}}(t) - g_{\mathrm{decorr}}(t)\,. \tag{12}$$

Neglecting the spacecraft angular acceleration effects, which do not provide a significant contribution in April 2016, when the data was taken, the measured residual acceleration was found to be $(5.2 \pm 0.1)\,\mathrm{fm\,s^{-2}}\,\sqrt{\mathrm{Hz}}^{-1}$ in the frequency range from $0.7\,\mathrm{mHz}$ to $20\,\mathrm{mHz}$ [13]. Towards lower frequencies, the noise increases but remains below $12\,\mathrm{fm\,s^{-2}}\,\sqrt{\mathrm{Hz}}^{-1}$ down to $0.1\,\mathrm{mHz}$. The reasons for this increase are not identified yet but laser radiation pressure fluctuations as well as thermal gradient and magnetics force effects are not limiting the performance in this frequency range. At frequencies above 60 mHz, the interferometer readout noise of $(34.8 \pm 0.3)\,\mathrm{fm}\,\sqrt{\mathrm{Hz}}^{-1}$ is dominating[13].

These results have exceeded even the most optimistic expectations. These measurements show that LISA Pathfinder was the quietest place in the universe ever measured. The residual acceleration level is more than five times better than originally required. In this frequency range, the LISA Pathfinder results are even close, that is within a factor of 1.25, to the required LISA performance. In addition, the interferometer noise performance is more than one hundred times better than expected from ground test campaigns. This is very encouraging for the further development and construction of LISA for it shows that indeed, the technology to achieve the required levels of free-fall is available and working. This is underlined from an operational point of view by the fact that during the first 55 days of operations, already more than 650 hours of residual acceleration noise measurements could be taken[13].

In view of LISA, it is important to not only measure the residual acceleration noise level on LISA Pathfinder but to make the best use of this unique opportunity of such a laboratory in space to understand the system and the different subsystems in as much detail as possible.

3.2. *The GRS Measurements*

One example for detailed analysis is the electrostatic sensing and actuation system which is part of the GRS, as explained above. This system consists of a set of 18 electrodes as shown in Figure 4. The electrodes in green are used for sensing the position of the test masses and, at the same time, to apply the commanded forces as determined by the DFACS. The electrodes in red are used to apply the AC 100 kHz voltages to the test masses. They polarise the test masses and in fact, the position measurements are the result of the demodulation of the corresponding difference in current in a capacitive-inductive bridge, as depicted for example in 30. With this system of electrodes and the corresponding circuitry, it is possible to measure all six degrees of freedom for each of the two test masses. These measurements are, however, noisier than the measurements of the OMS, for the degrees of freedom for which the later provides measurements. For example, with the GRS, the noise in the position sensing is below $2.4\,\mathrm{nm}\,\sqrt{\mathrm{Hz}}^{-1}$ in the frequency range from 1 mHz to $0.5\,\mathrm{Hz}$[32]. With the OMS, the noise in the position sensing along the sensitive x-axis is at the level of $(34.8 \pm 0.3)\,\mathrm{fm}\,\sqrt{\mathrm{Hz}}^{-1}$ only and therefore was found to dominate Δg for frequencies above 60 mHz. When comparing the precision of the two sensing systems, we find that the OMS is roughly five orders of magnitude less noisy. A very important characteristic of the set-up shown in Figure 4 is the size of the gaps between each of the test masses and the respective electrodes. Along the sensitive x-axis, the gap is 4 mm, along the y-axis 2.9 mm have been chosen and 3.5 mm for the z-axis[31]. These are larger than all other sensing gaps implemented

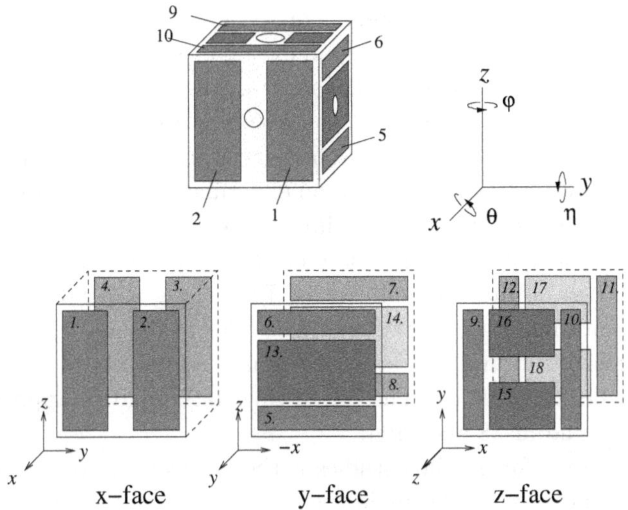

Fig. 4.   The GRS layout with the sensing and actuation electrodes in green and the injection electrodes, which apply the 100 kHz voltage, in red. Reprint from 31 (color online).

on drag-free space missions so far[32]. In nominal science operations, actuation is necessary to keep the two test masses at their nominal position, with the exception of the x-degree of freedom of the free-falling TM1, and to avoid tilts of the test masses along all rotational degrees of freedom. The maximum actuation force that could be applied during nominal science operations is the limit of the so-called high resolution mode of the actuation system, which is always switched on during nominal science operation, and whose maximum is $\approx 2.2\,\mathrm{nN}$[33]. The size of these actuation forces and torques strongly depends on the size of the local gravity field on board the satellite. That means, the better the balancing of the masses of all satellite components, the smaller the remaining static gravitational force, Δg_{DC}, that has to be compensated. Prior to launch, the satellite had to be designed such that a local gravitational acceleration of $\Delta g_{\mathrm{DC}} = 650\,\mathrm{pm\,s^{-2}}$ would not be exceeded along the sensitive axis[34]. In flight, the same quantity was found to be below $50\,\mathrm{pm\,s^{-2}}$[13]. The maximum force or torque that can be applied by the actuation system in the current setting, the so called actuation authority, is a very important quantity because it determines the level of stray force noise that is caused by the actuation. This can be understood by noting that the actuation authority determines the amplitude of the carrier voltages that are applied to the electrodes. It is by modulating these carrier voltages that the electrostatic forces are applied to the test masses. However, a certain level of fluctuations cannot be avoided. On ground, the relative amplitude stability of the actuation voltages of the flight amplifiers was found to be between 3 and $8\,\mathrm{ppm/\sqrt{Hz}}$[34]. In summary, this means the higher the necessary maximum voltages, the noisier the actuation system and the more undesired force noise is experienced by the test mass. Consequently, the very low static local gravity field allowed us to use the actuation system in an extremely low noise configuration. This enabled us to make the very quiet free-fall measurements as shown in 13. If a higher actuation authority were necessary, the noise in the frequency range from 0.1 mHz to approximately 3 mHz would have been higher. To understand if there are other effects in the electrostatic actuation system resulting from actuation noise which is independent of the actuation authority, an experiment called the drift mode or free-flight experiment is used. In this experiment, the continuous control of TM2 along x is replaced with a control scheme based on short force impulses which are followed by several minutes of uncontrolled drift of TM2 along the x direction. For details of the implementation of this experiment, we refer to 35 and 36. The results of the test of this experiment on Earth at the torsion pendulum facility at the University of Trento can be found in 37. The interesting data analysis approaches and first results can be found in 38 and 39.

Another example for detailed studies is the Optical Metrology System. Preliminary results indicate that the main measurement of the relative position of the two test masses is more than a hundred times less noisy than the tests performed on ground. This applies to frequencies above approximately 0.4 Hz. Below this frequency, the interferometer signal is dominated by the pick-up of test mass and satellite motion. Towards even lower frequencies, the motion of the test mass itself is much larger than the noise. The minimal level of the white OMS noise is determined by the noise in the phasemeter[13]. In addition, possible contributions from laser frequency noise[25] and relative intensity noise[40] are under investigation.

3.3. *Charge-Related Measurements and Results*

In addition, another set of detailed investigations has been performed to understand the influence of charge on LISA Pathfinder. There are two ways in which charge can produce an undesired force noise. One of them is the mixing of a fluctuating charge with stray DC potentials around the test masses. A certain level of charge fluctuations cannot be avoided for they result from high-energy cosmic rays and solar energetic particles which hit the TMs and charge them. Consequently, the remedy is to compensate the stray potentials as much as possible. This is done by adding small mV DC voltages to the electrodes. The second mechanism is the mixing of a static charge with noisy stray voltages. These stray voltages can be caused by surface patch potentials and noise in the GRS electronics. The influence of these stray voltages is minimised by the comparatively large gap size, compare Section 3.2. In addition, both test masses are discharged via the photoelectric effect using UV-light. Accordingly, two kinds of charge-related experiments have been performed. The charge itself was estimated by applying quasi DC modulations of the voltages to the surrounding electrodes and measuring the resulting force. Vice versa, the stray potentials have been estimated based on the resulting force from a deliberate change in charge. In summary, it can be concluded that charge-related effects are far from limiting the LISA Pathfinder performance[41].

These are only selected examples of detailed system studies and more results will be published soon.

4. The future of LISA Pathfinder: LISA

While the first proposal of LISA to ESA and NASA, at that time involving four satellites, dates back to the 1993, we will summarise only the recent development of LISA here. In March 2013, the European Space Agency called for science themes which could be addressed by future missions of their large scale class. This call was answered by many proposals, one of them being 'The Gravitational Universe'[1]. In November 2013, this science theme was selected for the L3 mission slot with a planned launch in 2034. In October 2016, ESA called for mission concepts which implement the science theme 'The Gravitational Universe'. LISA was proposed as such a mission concept in January 2017[9] and selected in June 2017. At the time of writing, the development of LISA is continuing at a high speed and with great enthusiasm.

5. Conclusions

LISA Pathfinder is the technology demonstrator mission for the future gravitational wave observatory, LISA. The satellite was launched on December 3rd, 2015 and operated successfully until the final shut down on July 18th, 2017. It was not only shown that it is possible to have a free-falling test mass in space whose residual acceleration due to spurious forces is below the required level of $30\,\mathrm{fm\,s^{-2}}\,\sqrt{\mathrm{Hz}}^{-1}$ for mHz frequencies but also that this requirement is fulfilled with a large margin. This means, we are already approaching the necessary free-fall levels for LISA. In addition, from an organisational point of view, it was shown that such a satellite, in which the payload is the science instrument, can be operated in close cooperation between the European Space Agency, industry partners and scientists. Important lessons learned during all of the mission phases from hardware development to mission operations are currently being summarised and will provide additional input for the development of LISA. We leave the details to future publications.

The performance of the Optical Metrology System with a readout noise of only $(34.8 \pm 0.3)\,\mathrm{fm}\,\sqrt{\mathrm{Hz}}^{-1}$[13] is more than one hundred times better than the latest measurement of the test campaigns on Earth. In other words: we were not only able to operate the first public laser interferometer system in space but also to measure fm instead of the required pm accuracy. This is a very successful test of the local interferometry system on board of each LISA satellite. For the inter-satellite interferometry, the noise level is expected to be above the fm level. In general, the satellite and all key subsystems to achieve free-fall, as for example the OMS, the GRS and the charge management, could be characterised in detail with dedicated experiments. A deep understanding of each subsystem is very useful for the remaining development of LISA.

LISA has been selected as a Mission Concept in June 2017 and currently work is well under way to clarify the remaining science questions and to start the industrial engineering. Even though the launch date in 2034 seems in the far future and many key components have shown excellent performance on LISA Pathfinder, some development concerning for example the inter-satellite interferometry and the required telescopes as well as the link in between the two optical benches on one satellite is still required. Even though some aspects of inter-satellite interferometry will be tested on the GRACE Follow-On mission[42], the future interferometry development for LISA will include more dedicated hardware testing. It necessarily takes a certain time as, for example, complex optical benches need to be built. That is the reason why the development has already started at the time of writing.

To conclude, LISA Pathfinder has exceeded all expectations and shown that LISA is feasible.

Acknowledgements

This work has been made possible by the LISA Pathfinder mission, which is part of the space-science program of the European Space Agency. The French contribution has been supported by CNES (Accord Specific de Projet No. CNES 1316634/CNRS 103747), the CNRS, the Observatoire de Paris and the University Paris-Diderot. E.P. and H.I. would also like to acknowledge the financial support of the UnivEarthS Labex program at Sorbonne Paris Cité (Grants No. ANR-10-LABX-0023 and No. ANR-11-IDEX-0005-02). The Albert-Einstein-Institut acknowledges the support of the German Space Agency, DLR. The work is supported by the Federal Ministry for Economic Affairs and Energy based on a resolution of the German Bundestag (Grants No. FKZ 50OQ0501 and No. FKZ 50OQ1601). The Italian contribution has been supported by Agenzia Spaziale Italiana and Instituto Nazionale di Fisica Nucleare. The Spanish contribution has been supported by Contracts No. AYA2010-15709 (MICINN), No. ESP2013-47637-P, and No. ESP2015-67234-P (MINECO). M.N. acknowledges support from Fundacion General CSIC Programa ComFuturo). F.R. acknowledges support from a Formacin de Personal Investigador (MINECO) contract. The Swiss contribution acknowledges the support of the Swiss Space Office (SSO) via the PRODEX Programme of ESA. L.F. acknowledges the support of the Swiss National Science Foundation. The UK groups wish to acknowledge support from the United Kingdom Space Agency (UKSA), the University of Glasgow, the University of Birmingham, Imperial College London, and the Scottish Universities Physics Alliance (SUPA). J.I.T. and J.S. acknowledge the support of the U.S. National Aeronautics and Space Administration (NASA).

References

1. The eLISA Consortium et al., The Gravitational Universe (2013).
2. G. Nardini, private communication.
3. C. Caprini, private communication.

4. The LIGO Scientific Collaboration and the Virgo Collaboration, GW150914: Implications for the stochastic gravitational wave background from binary black holes (2016).

5. N. Bartolo et al., Science with the space-based interferometer LISA. IV: Probing inflation with gravitational waves (2016).

6. C. Caprini et al., Science with the space-based interferometer eLISA. II: Gravitational waves from cosmological phase transitions (2015).

7. A. Sesana, Multi-band gravitational wave astronomy: Science with joint space- and ground-based observations of black hole binaries (2017).

8. P. R. Saulson, *Fundamentals of Interferometric Gravitational Wave Detectors* (World Scientific, 1994).

9. Pau Amaro-Seoane et al., Laser Interferometer Space Antenna (2017).

10. M. Otto, G. Heinzel and K. Danzmann, TDI and clock noise removal for the split interferometry configuration of LISA, *Classical and Quantum Gravity* **29**, p. 205003 (October 2012).

11. M. Hueller, A. Cavalleri, R. Dolesi, S. Vitale and W. J. Weber, Torsion pendulum facility for ground testing of gravitational sensors for LISA, *Classical and Quantum Gravity* **19**, p. 1757 (2002).

12. M. R. Drinkwater, R. Floberghagen, R. Haagmans, D. Muzi and A. Popescu, GOCE: ESA's First Earth Explorer Core Mission, *Space Sciences Series of ISSI Earth Gravity Field from Space – From Sensors to Earth Sciences*, p. 419-432 (2003).

13. Armano et al., Sub-Femto-g Free Fall for Space-Based Gravitational Wave Observatories: LISA Pathfinder Results, *Physical Review Letters* **116**, p. 231101 (June 2016).

14. F Antonucci et al., LISA Pathfinder: Mission and status, *Classical and Quantum Gravity* **28**, p. 094001 (2011).

15. Armano et al., The LISA Pathfinder Mission, *Journal of Physics: Conference Series* **610**, p. 012005 (2015).

16. J. Bergé, P. Touboul and M. Rodrigues, Status of MICROSCOPE, a mission to test the Equivalence Principle in space (2015).

17. Armano et al., A Strategy to Characterize the LISA-Pathfinder Cold Gas Thruster System, *Journal of Physics: Conference Series* **610**, p. 012026 (2015).

18. D. Bortoluzzi, C. D. Hoyle, S. Vitale, G. Heinzel and K. Danzmann, *Science Requirements and Top-level Architecture Definition for the LISA Test-flight Package (LTP) on Board LISA Path Finder (SMART-2)*, tech. rep., University of Trento and AEI Hannover (2004).

19. H. R. Carleton and W. T. Maloney, A balanced optical heterodyne detector, *Appl. Opt.* **7**, 1241 (Jun 1968).

20. D. Robertson, *3OB As Built OptoCAD Model - S2-UGL-TN-3045*, tech. rep., University of Glasgow (2013).

21. G. Hechenblaikner et al., Digital Laser Frequency Control and Phase-Stabilization Loops in a High Precision Space-Borne Metrology System, *IEEE Journal of Quantum Electronics* **47**, 651 (May 2011).

22. Michael Born for the LPF collaboration, LISA Pathfinder: OPD loop characterisation, *Journal of Physics: Conference Series* **840**, p. 012036 (2017).

23. G. Heinzel, V. Wand, A. Garcia, F. Guzman, F. Steier, C. Killow, D. Robertson, H. Ward and C. Braxmeier, *Investigation of noise sources in the LTP interferometer - Technical report S2-AEI-TN-3028*, tech. rep., AEI Hannover (2005).

24. G. Heinzel, *SMART-2 interferometer - S2-AEI-TN-3010*, tech. rep., AEI Hannover (2002).

25. Sarah Paczkowski on behalf of the LPF collaboration, Laser Frequency Noise Stabilisation and Interferometer Path Length Differences on LISA Pathfinder, *Journal of Physics: Conference Series* **840**, p. 012004 (2017).

26. Hewitson et al., Data analysis for the LISA Technology Package, *Classical and Quantum Gravity* **26**, p. 094003 (2009).

27. N. Karnesis, *System Identification experiments of LISA Pathfinder: Fitting an underlying model of the stiffnesses in all available data - S2-AEI-TN-3078*, tech. rep., AEI Hannover (2017).

28. Daniele Vetrugno and Nikolaos Karnesis on behalf of the LPF collaboration, Calibrating LISA Pathfinder raw data into femto-g differential accelerometry, *Journal of Physics: Conference Series* **840**, p. 012002 (2017).

29. Gudrun Wanner and Nikolaos Karnesis for the LISA Pathfinder collaboration, Preliminary results on the suppression of sensing cross-talk in LISA Pathfinder, *Journal of Physics: Conference Series* **840**, p. 012043 (2017).

30. Meshksar et al. for the LISA Pathfinder collaboration, Gravitational Reference Sensor Front-End Electronics Simulator for LISA, *Journal of Physics: Conference Series* **840**, p. 012041 (2017).

31. J. Smit, *Inertial Sensor Actuation model for LISA Pathfinder End-to-end Simulator*, tech. rep., Netherlands Institute for Space Research (2006).

32. Armano et al., Capacitive sensing of test mass motion with nanometer precision over millimeter-wide sensing gaps for space-borne gravitational reference sensors, *Phys. Rev. D* **96**, p. 062004 (Sep 2017).

33. N. Brandt, *Test Mass Actuation Algorithm for DFACS - S2-ASD-TN-2011*, tech. rep., EADS Astrium (2009).

34. V. Ferroni, Constraints on LISA Pathfinder's self-gravity: design requirements, estimates and testing procedures (2016).

35. A. Grynagier, W. Fichter and S. Vitale, Parabolic drag-free flight, actuation with kicks, spectral analysis with gaps, *Space Science Reviews* **151**, 183 (Mar 2010).

36. A. Grynagier, W. Fichter and S. Vitale, The LISA Pathfinder drift mode: implementation solutions for a robust algorithm, *Classical and Quantum Gravity* **26**, p. 094007 (2009).

37. G. Russano, A torsion pendulum ground test of the LISA Pathfinder Free-fall mode (2016).

38. Armano et al., Free-flight experiments in LISA Pathfinder, *Journal of Physics: Conference Series* **610**, p. 012006 (2015).

39. Roberta Giusteri on behalf of the LPF collaboration, The free-fall mode experiment on LISA Pathfinder: first results, *Journal of Physics: Conference Series* **840**, p. 012005 (2017).

40. Andreas Wittchen for the LPF Collaboration, Coupling of relative intensity noise and pathlength noise to the length measurement in the optical metrology system of LISA Pathfinder, *Journal of Physics: Conference Series* **840**, p. 012003 (2017).

41. Armano et al., Charge-Induced Force Noise on Free-Falling Test Masses: Results from LISA Pathfinder, *Phys. Rev. Lett.* **118**, p. 171101 (Apr 2017).

42. Daniel Schütze and LRI team, Measuring Earth: Current status of the GRACE Follow-On Laser Ranging Interferometer, *Journal of Physics: Conference Series* **716**, p. 012005 (2016).

The Latest Results from AMS on the Searches for Dark Matter

W. Xu

on behalf of the AMS Collaboration

Massachusetts Institute of Technology (MIT),
Cambridge, Massachusetts 02139, USA
E-mail: Weiwei.Xu@cern.ch

The Alpha Magnetic Spectrometer, AMS, is successfully operating on the International Space Station for more than 6 years and has collected over 100 billion cosmic rays. One of the main objectives of AMS is to search for Dark Matter through the precision measurements of charged elementary particles in the cosmos. The positron flux is measured in the energy range of 0.5 to 700 GeV and electron flux of 0.5 to 1000 GeV. Both positron and electron fluxes require additional sources of high energy positrons and electrons, like the Dark Matter. The antiproton flux is measured in the absolute rigidity(momentum/charge) of 1 to 450 GV. The antiproton to proton flux ratio is rigidity independent above 60 GV. In the absolute rigidity range of ~ 60 to ~ 500 GV, the positron, antiproton and proton fluxes have nearly identical rigidity dependence. These are new observations of the properties of charged elementary particles in the cosmos.

Keywords: Dark matter; positron; antiproton; elementary particles.

1. Introduction

There are four charged elementary particles traveling in cosmos, protons (p), antiprotons $(\bar{p})$, electrons (e^-) and positrons (e^+). Protons are the most abundant cosmic rays and electrons are the most abundant negatively charged cosmic rays. Positrons and antiprotons are rare components in cosmic rays, therefore, they are sensitive to probe new cosmic phenomena, like the Dark Matter.

The precision proton flux measured by AMS shows that the proton flux progressively hardens above 200 GV[1]. Out of the four elementary particles, the experimental data on the $\bar{p}$ are limited because there are 10^4 protons for each antiproton. To measure the $\bar{p}$ flux to 1% accuracy requires a proton separation power of $\sim 10^6$. AMS has published the $\bar{p}$ measurements up to 450 GV based on the first 4 years data.

Over the last decades, there has been a strong interest in the cosmic ray positrons in both particle physics and astrophysics. The first AMS result on positron fraction[2] generated widespread interest. Our previous measurements of positron fraction[3], positron flux (up to 500 GeV) and electron flux (up to 700 GeV) are based on the data collected during the first 2.5 years[4]. In this report, the latest results based on 5.5 years data will be presented. With more than double statistics, the energy reach is extended to 700 GeV for positrons and 1 TeV for electrons.

The simultaneous measurements of p, $\bar{p}$, e^- and e^+ fluxes performed with the same detector, provide precise information over an extended energy range in the study of elementary particles traveling through the cosmos.

2. Detector

The description of the AMS detector is presented in Ref.[2] and the references therein. All detector elements are used in the present analyses: the silicon tracker, the permanent magnet, the time of flight counters TOF, the anti-coincidence counters ACC, the transition radiation detector TRD, the ring imaging Čerenkov detector RICH, and the electromagnetic calorimeter ECAL.

Together with the magnet, the 9-layer tracker measures the rigidity(R) of the comsic rays and differentiate the positive and negative particles. The first layer (L1) is on top of the TRD, the second layer (L2) just above the magnet, six layers (L3 to L8) in the magnet, and the last layer (L9) is between RICH and ECAL. L2 to L8 constitute the inner tracker. For charge $|Z| = 1$ particles, the maximum detectable rigidity (MDR) is 2 TV and charge resolution is $\Delta Z = 0.05$.

The ACC rejects cosmic rays entering from the side of the inner tracker with an efficiency of 0.99999. The TOF measures $|Z|$ and velocity with a resolution of $\Delta\beta/\beta = 4\%$ for $|Z| = 1$ particles. The RICH has a velocity resolution $\Delta\beta/\beta = 0.1\%$ for $|Z| = 1$ particles.

Three main detectors provide clean and redundant separation between $e^\pm$ and $p(\bar{p})$. These are the TRD (above the magnet), the ECAL (below the magnet), and the tracker. The TRD and the ECAL are separated by the magnet and the tracker. This ensures that most of the secondary particles produced in the TRD and in the upper TOF planes are swept away and do not enter into the ECAL. Events with large angle scattering are also rejected by a quality cut on the measurement of the trajectory using the tracker. The matching of the ECAL energy, E, and the rigidity, R, measured with the tracker, greatly improves the proton rejection.

The TRD separates $p(\bar{p})$ from $e^\pm$ using the Λ_{TRD} estimator by combining the signals from the 20 layers of proportional tubes[2]. $p(\bar{p})$ which have $\Lambda_{\mathrm{TRD}} \sim 1$ are efficiently separated from $e^\pm$ which have $\Lambda_{\mathrm{TRD}} \sim 0.5$. The proton rejection power of the TRD estimator at 90% $e^\pm$ efficiency measured on orbit is 10^3 to 10^4.

To cleanly identify electrons and positrons in the ECAL, an estimator Λ_{ECAL} is constructed using the 3D shower shape in the ECAL[6]. The proton rejection power of the ECAL reaches 10^4 when combined with the energy-rigidity matching requirement $E/|R| > 0.75$.

To distinguish e^+ from charge confusion e^-, that is, e^- reconstructed in the tracker with positive rigidity due to the finite tracker resolution or interactions with the detector materials, a charge confusion estimator Λ^e_{CC} is built using the Boosted Decision Trees technique[7]. The Λ^e_{CC} estimator combines information from the tracker and TOF. With this method, e^+, which has $\Lambda^e_{\mathrm{CC}} \sim 1$, are efficiently separated from charge confusion e^-, which has $\Lambda^e_{\mathrm{CC}} \sim -1$.

Similarly, to distinguish $\bar{p}$ from charge confusion p, that is, p reconstructed with negative rigidity, a charge confusion estimator Λ^p_{CC} is built. $\bar{p}$, which has $\Lambda^p_{\mathrm{CC}} \sim 1$, are efficiently separated from charge confusion p, which has $\Lambda^p_{\mathrm{CC}} \sim -1$.

The entire detector has been extensively calibrated in a test beam at CERN with e^+ and e^- from 10 to 290 GeV/c, with protons at 180 and 400 GeV/c, and with π^- from 10 to 180 GeV/c, which produce transition radiation equivalent to p up to 1.2 TeV/c. In total, measurements with 18 different energies and particles at 2000 positions were performed[2].

Optimization of all reconstruction algorithms was performed using the test beam data. Several corrections are applied to the data to ensure long term stability of the absolute scales in the constantly varying on-orbit environment. These corrections are performed using specific samples of particles, predominantly protons. They include off-line calibrations of the amplitude response of TRD, TOF, tracker, RICH, and ECAL electronic channels. These calibrations are performed every 1/4 of an orbit with the exception of the alignment of the outer tracker planes 1 and 9 which is performed every two minutes. The stability of the electronics response is ensured by on-board calibrations of all channels every half-orbit (~ 46 min). The corrections also include the alignment of all the AMS detectors and the temperature correction of the magnetic field strength.

Monte Carlo simulated events are produced using a dedicated program developed by AMS based on GEANT-4.10.1[8]. This program simulates electromagnetic and hadronic interactions of particles in the materials of AMS and generates detector responses. The digitization of the signals, including those of the trigger, is simulated according to the measured characteristics of the electronics. The digitized signals then undergo the same reconstruction as used for the data. The Monte Carlo samples used in the present analyses have sufficient statistics such that they do not contribute to the errors.

3. Positron and Electron

The isotropic fluxes of cosmic ray electrons and positrons in the energy bin E of width ΔE are given by

$$\Phi_{e^\pm} = \frac{N_{e^\pm}}{A_{\text{eff}}(E)\epsilon_{\text{trig}}(E)T(E)\Delta E} \tag{1}$$

where N_{e^+} (N_{e^-}) is the number of e^+ (e^-), A_{eff} is the effective acceptance, ϵ_{trig} is the trigger efficiency, and T is the exposure time. The effective acceptance is defined as

$$A_{\text{eff}} = A_{\text{geom}}\epsilon_{\text{sel}}\epsilon_{\text{id}}(1 + \delta) \tag{2}$$

where A_{geom} is the geometric acceptance, ϵ_{sel} is the selection efficiency, ϵ_{id} is the $e^\pm$ identification efficiency, and δ is a minor correction to the acceptance due to the difference between MC simulation and data.

The geometric acceptance for this analysis is $A_{\text{geom}} \approx 550$ cm^{2}sr. The product $A_{\text{geom}}\epsilon_{\text{sel}}\epsilon_{\text{id}}$ is determined from Monte Carlo simulation.

The trigger efficiency ϵ_{trig} is determined from data. The data acquisition system is triggered by the coincidence of all four TOF planes. AMS also records unbiased triggers that require a coincidence of any three out of the four TOF planes to measure ϵ_{trig}. It is found to be 100% above 3 GeV decreasing to 75% at 1 GeV.

The exposure time used in this analysis includes only those seconds during which the detector was in normal operating conditions and, in addition, the AMS was pointing within 40° of the local zenith, the data acquisition live time exceeded 50% (compared to its typical value of 90%), and the ISS was outside of the South Atlantic Anomaly. Because of the influence of the geomagnetic field, see below, this exposure time for primary cosmic rays increases with increasing energy and becomes constant at 1.3×10^8 s above 30 GeV.

The bin widths ΔE are chosen to be at least 2 times the energy resolution to minimize migration effects. With increasing energy the bin width becomes smoothly wider to ensure adequate statistics in each bin.

Events are selected by requiring the presence of a shower in the ECAL and a reconstructed track in the TRD and in the tracker. To identify downward-going particles of charge $|Z| = 1$, cuts are applied on the velocity measured by the TOF and on the charge reconstructed by the tracker, the upper TOF planes, and the TRD. To reject positrons and electrons produced by the interaction of primary cosmic rays with the atmosphere, the minimum energy within the bin is required to exceed 1.2 times the Størmer cutoff[9] for either a positron or an electron at the geomagnetic location where the particle was detected and at any angle within the acceptance. The selection efficiency ϵ_{sel} is determined from the Monte Carlo simulation and found to be a smooth function of energy with a value of $\sim 70\%$ at 100 GeV.

The identification of the e^- and e^+ signal requires rejection of the p background. Cuts are applied on the $E/|R|$ matching and the reconstructed depth of the shower maximum. This makes the negatively charged sample, as determined by the rigidity, a sample of pure electrons. A cut on the ECAL estimator is applied to further reduce the proton background in the positive rigidity sample after which the numbers of positrons and protons are comparable at all energies. The identification efficiency ϵ_{id} is defined using the Monte Carlo simulation as the efficiency for e^- to pass these three cuts. It is identical for both e^- and e^+.

In order to determine the correction δ, a negative rigidity sample is selected for every cut using information from the detectors unrelated to that cut. The effects of the cut are compared between data and Monte Carlo simulation. This correction is found to be a smooth, slowly falling function of energy. It is -2% at 10 GeV and -6% at 700 GeV. The selection cut values and the identification cut values are chosen to maximize the measurement accuracy of the separate fluxes.

The resulting effective acceptance is identical for e^- and e^+ . Therefore, the positron fraction can be determined by

$$\frac{\Phi_{e^+}}{\Phi_{e^+} + \Phi_{e^-}} = \frac{N_{e^+}}{N_{e^+} + N_{e^-}} \tag{3}$$

In each energy bin, the two-dimensional reference spectra for $e^\pm$ and the background are fit to data in the ($\Lambda_{\mathrm{TRD}} - \Lambda_{\mathrm{CC}}^e$) plane by varying the normalizations of the signal and the background. This method provides a data driven control of the dominant systematic uncertainties by combining the redundant and independent TRD, ECAL, and tracker information. The reference spectra are determined from high statistics electron and proton data samples selected using tracker and ECAL information including the charge sign, track-shower axis matching, and the ECAL estimator. The purity of each reference spectrum is verified using Monte Carlo simulation. The fit is performed simultaneously for the positive and negative rigidity data samples in each energy bin yielding the number of positrons N_{e+}, the number of electrons N_{e-}, the number of protons, and the amount of charge confusion.

The proton contamination in the region populated by positrons is small. It is accurately measured using the TRD estimator. The amount of proton contamination has a negligible contribution to the statistical error.

In total, 1.08×10^6 e^+ are selected from 0.5 GeV to 700 GeV, and 1.65×10^7 e^- are selected from 0.5 GeV to 1 TeV.

The systematic uncertainties in the positron and electron fluxes come from three groups. The first is the energy scale and bin-to-bin migration, the second is effective acceptance, the third is the counting of e^+ and e^-.

The absolute energy scale is verified by using minimum ionizing particles and the ratio $E/|R|$. These results are compared with the test beam values where the beam energy is known to high precision. This comparison limits the uncertainty of the absolute energy scale to 2% in the range covered by the beam test results, $10 - 290$ GeV. Below 10 GeV it increases to 5% at 0.5 GeV and above 290 GeV to 5% at 1 TeV. This is treated as an uncertainty of the bin boundaries for the e^+ and e^- fluxes. The systematic error from the absolute energy scale is mostly canceled and negligible in the positron fraction.

The bin-to-bin migration error on e^+ and e^- fluxes is $\sim 1\%$ at 1 GeV; it decreases to 0.2% above 10 GeV. This error is also mostly canceled in the positron fraction, and only noticeable bellow 5 GeV.

The systematic error on the effective acceptance is given by the uncertainties on δ. For every cut, this uncertainty is derived from the comparison between data and the Monte Carlo simulation. This includes an overall scaling uncertainty of 2%, which introduces a correlation between energy bins and between the electron and positron fluxes. The acceptance uncertainty is the leading contribution to the systematic error below 300 GeV. The systematic error on the effective acceptance is totally canceled in the positron fraction.

The systematic errors in the counting of e^+ and e^- comes from the event selection, charge confusion, and the reference spectra.

To evaluate the systematic uncertainty related to event selection, the complete analysis is repeated in every energy bin over 1000 times with different cut values, such that the selection efficiency varies up to 30%. The distribution of the flux values and positron fraction resulting from these 1000 analyses contains both statistical and systematic effects. For e^+ flux, e^- flux, and positron fraction, the difference between the width of their distribution from data and from Monte Carlo simulation quantifies their associated systematic uncertainty from event selection.

Two sources of charge confusion dominate. The first source is related to the finite resolution of the tracker and multiple scattering. It is mitigated by the $E/|R|$ matching and quality cuts of the trajectory measurement including the track $\chi^2/d.f.$, charge measured in the tracker, and charge measured in the TOF. The second source is related to the production of secondary tracks along the path of the primary e^- in the tracker. It was studied using control data samples of electron events where the ionization in the lower TOF counters corresponds to at least two traversing particles. Both sources of charge confusion are found to be well reproduced by the Monte Carlo simulation and their reference spectra are derived from the Monte Carlo simulation. The systematic uncertainties due to these two effects are obtained by varying the background normalizations within the statistical limits and comparing the results with the Monte Carlo simulation. They were examined in each energy bin.

The systematic error associated with the uncertainty of the data derived reference spectra arises from their finite statistics. It is measured by varying the shape of the reference spectra within the statistical uncertainties. Its contribution to the overall error is small compared to the statistical uncertainty of data and is included in the total systematic error.

4. Antiproton and Proton

Over 65 billion cosmic ray events have been recorded by AMS in its first 4 years of operations. The data analyses of antiproton flux and antiproton-to-proton flux ration are performed in 57 absolute rigidity bins. The same binning as in our proton flux measurement[1] is chosen below 80.5 GV. Above 80.5 GV two to four bins are combined to ensure sufficient antiproton statistics.

Events are selected by requiring a track in the TRD and in the inner tracker and a measured velocity $\beta > 0.3$ in the TOF corresponding to a downward-going particle. To maximize the number of selected events while maintaining an accurate rigidity measurement, the acceptance is increased by releasing the requirements on the external tracker layers, L1 and L9. Below 38.9 GV neither L1 nor L9 is required. From 38.9 to 147 GV either L1 or L9 is required. From 147 to 175 GV only L9 is required. Above 175 GV both L1 and L9 are required. The $\chi^2/d.f.$ of the reconstructed track fit is required to be less than 10 both in the bending and non-bending projections. The dE/dx measurements in the TRD, the TOF, and the inner tracker must be consistent with $|Z| = 1$. To select only primary cosmic rays, the measured rigidity is required to exceed the maximum IGRF cutoff[10] by a factor of 1.2 for either positive or negative particles within the AMS field of view.

Events satisfying the selection criteria are classified into two categories – positive and negative rigidity events. A total of 2.42×10^9 events with positive rigidity are selected as protons. They are 99.9% pure protons with almost no background.

The negative rigidity event category comprises both antiprotons and several background sources: electrons, light negative mesons (π^- and a negligible amount of K^-) produced in the interactions of primary cosmic rays with the detector materials, and charge confusion protons. The combination of information from the TRD, TOF, tracker, RICH, and ECAL enables the efficient separation of the antiproton signal events from these background sources using a template fitting technique.

Three overlapping rigidity regions with different types of template function are defined to maximize the accuracy of the analysis: low absolute rigidity region (1.00–4.02 GV), intermediate region (2.97–18.0 GV) and high absolute rigidity region (16.6–450 GV). In the overlapping rigidity bins, the results with the smallest error are selected.

At low rigidities, a cut on the TRD estimator Λ_{TRD} is applied to remove bulk of the e^- and π^- background. The mass distribution, calculated from the rigidity measurement in the inner tracker and the velocity measured by the TOF, is used to construct the templates and to differentiate between the antiproton signal and the residual background.

At intermediate rigidities, Λ_{TRD} and the velocity measured with the RICH β_{RICH} are used to separate the antiproton signal from light particles (e^- and π^-). To determine the number of antiproton signal events, the π^- background is removed by a rigidity dependent β_{RICH} cut and the Λ_{TRD} distribution is used to construct the templates and to differentiate between the $\bar{p}$ signal and e^- background.

In the high rigidity region, the two-dimensional ($\Lambda_{\mathrm{TRD}} - \Lambda_{\mathrm{CC}}^p$) distribution is used to determine the number of antiproton events. To fit the data three template shapes are defined. The first two are for antiprotons and electrons with correctly reconstructed charge-sign and the last one is for charge confusion protons. The background templates (i.e., electrons and charge confusion protons) are from the Monte Carlo simulation.

Overall, results for all 57 rigidity bins give a total of 3.49×10^5 antiproton events in the data.

The isotropic antiproton flux for the absolute rigidity bin R_i of width ΔR_i is given by:

$$\Phi_i^{\bar{p}} = \frac{N_i^{\bar{p}}}{A_i^{\bar{p}} \, T_i \, \Delta R_i} \, , \qquad (4)$$

where the rigidity is defined on top of the AMS, $N_i^{\bar{p}}$ is the number of antiprotons in the rigidity bin i corrected with the rigidity resolution function (see below), $A_i^{\bar{p}}$ is the corresponding effective acceptance that includes geometric acceptance as well as the trigger and selection efficiency, and T_i is the exposure time.

Detector resolution effects cause migration of events $N_i^{\bar{p}}$ from rigidity bin R_i to the measured rigidity bins $\tilde{R}_j$ resulting in the observed number of events $\tilde{N}_j^{\bar{p}}$. To

212

account for this event migration, an iterative unfolding procedure is used to correct the number of observed events. It is described in detail in our publication on the proton flux[1]. After completion of this procedure, and the measured flux can be expressed as

$$\Phi_i^{\bar{p}} = \frac{\tilde{N}_i^{\bar{p}}}{\tilde{A}_i^{\bar{p}} \, T_i \, \Delta R_i} \; .$$ (5)

The same procedure is used to unfold the observed number of 2.42×10^9 proton events in this analysis.

The $(\bar{p}/p)$ flux ratio is defined for each absolute rigidity bin by

$$\left(\frac{\bar{p}}{p}\right)_i \equiv \frac{\Phi_i^{\bar{p}}}{\Phi_i^{p}} = \frac{\tilde{N}_i^{\bar{p}}}{\tilde{N}_i^{p}} \cdot \frac{\tilde{A}_i^{p}}{\tilde{A}_i^{\bar{p}}} \; .$$ (6)

With 3.49×10^5 antiproton events, the detailed study of systematic errors of the $\bar{p}$ flux and $(\bar{p}/p)$ flux ratio is the key part of the present analysis.

There are four sources of systematic errors on the $\bar{p}$ flux and $(\bar{p}/p)$ flux ratio. The first source affects mostly $\tilde{N}_i^{\bar{p}}$ and, to a much lesser extent, $\tilde{N}_i^{p}$. It includes uncertainties in the definition of the geomagnetic cutoff factor, in the event selection, and in the shape of the templates. The second source affects $\tilde{A}_i^{\bar{p}}$ and $\tilde{A}_i^{p}$. It includes uncertainties in the inelastic cross sections of protons and antiprotons in the detector materials and in the migration matrices. The third source is the uncertainty in the absolute rigidity scale. The fourth source, relevant only for the $\bar{p}$ flux, is the uncertainty in the normalization of the effective folded acceptance $\tilde{A}_i^{\bar{p}}$. Contributions of these four sources are added quadratically to arrive at the systematic errors.

The detail of the systematic error analysis is described in Ref.[5]. The uncertainty from the charge confusion proton template becomes significant for $|R| > 30$ GV. To ensure that the shape of the charge confusion proton background templates from the Monte Carlo simulation does not introduce bias into the antiproton identification, we also performed a completely independent data driven analysis based on the linear regression method for $|R| > 30$ GV over the acceptance which includes L1, L9, and ECAL. The results of this data driven analysis agree within the systematic errors with those published in Ref.[5].

Most importantly, in addition to the linear regression analysis, several other independent analyses were performed on the same data sample by different study groups. The results of those analyses are consistent with Ref.[5].

5. Results

5.1. *The latest positron and electron fluxes and positron fraction*

The positron fraction is shown in the Fig. 1. Our previous result[3] in which we observed that, above ~ 200 GeV, the positron fraction no longer exhibits an increase with energy, is consistent with this preliminary result based on 5.5 years data. The new result extends the energy range to 700 GeV with more than double statistics.

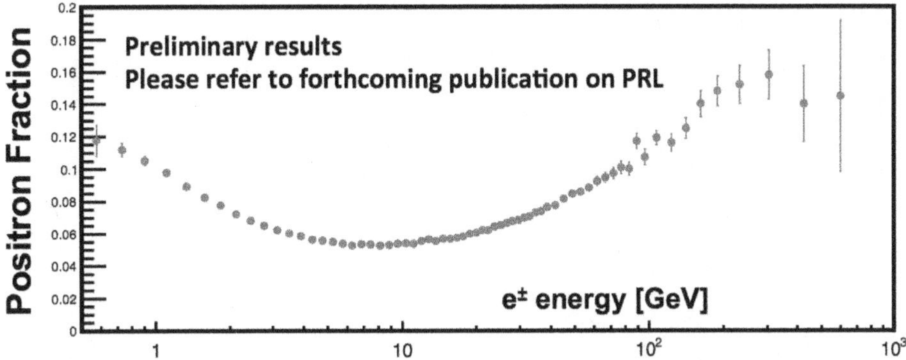

Fig. 1. Preliminary positron fraction from 0.5 to 700 GeV measured by AMS based on 5.5 years data. The error bars correspond to the quadratic sum of the statistical and systematic errors. Above ~ 200 GeV, the positron fraction no longer increases with increasing energy.

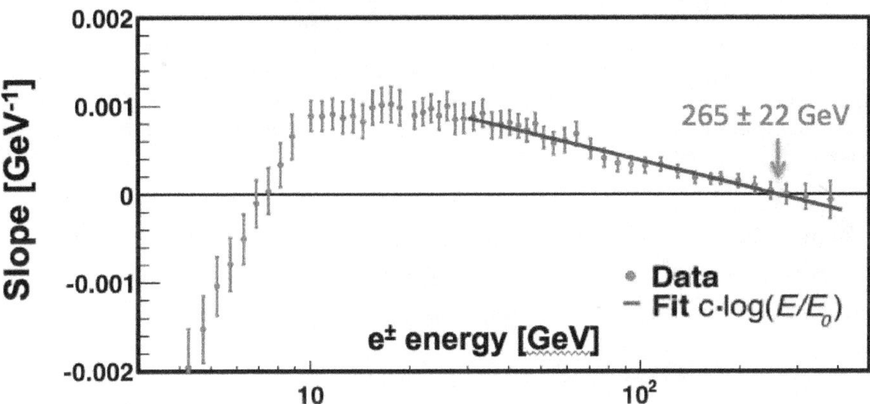

Fig. 2. The slope of the positron fraction versus energy over the entire energy range (the values of the slope below 4 GeV are off scale). The line is a logarithmic fit to the data above 30 GeV. The arrow indicates the position where the slope crosses 0, i.e., the positron fraction reaches its maximum.

As seen in Fig. 2, using the model independent method described in Ref.[3], we determine that the positron reaches its maximum at 265 ± 22 GeV.

The preliminary results of electron flux and positron flux are shown in Fig. 3. The fluxes are multiplied by $\tilde{E}^3$ for display purpose. The $\tilde{E}$ is calculated according to Ref.[11] for a flux $\propto E^3$. As seen, the magnitude and energy dependence are very different for electron and positron fluxes. Above ~ 20 GeV and up to ~ 200 GeV the electron flux decreases more rapidly with energy than the positron flux, that is, the electron flux is softer than the positron flux. This is not consistent with only the secondary production of positrons. Neither the electron flux nor the positron flux can be described by single power laws ($\Phi \propto E^\gamma$) over the entire range.

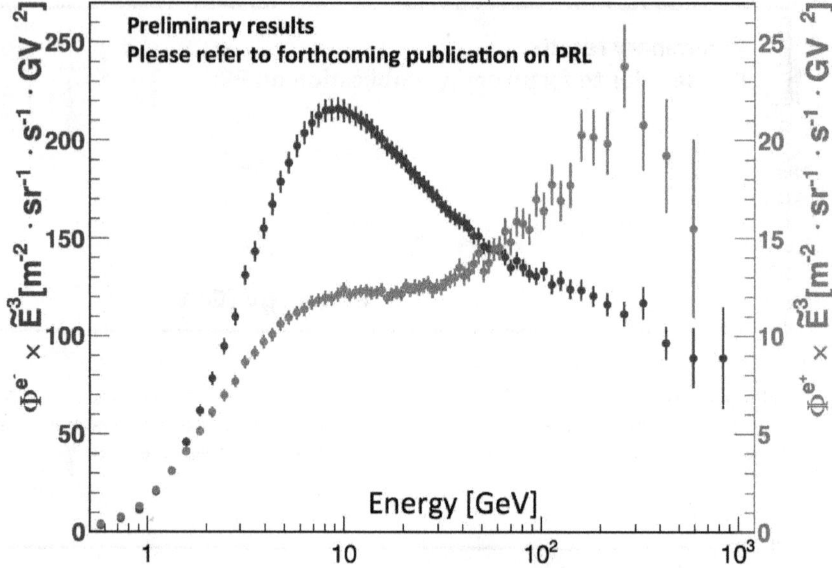

Fig. 3. Preliminary electron(left, blue scale) and positron(right, red scale) fluxes measured by AMS based on 5.5 years data. The error bars correspond to the quadratic sum of the statistical and systematic errors. Both the magnitude and energy dependence are very different for electron and positron fluxes (color online).

To quantitatively examine the energy dependence of the fluxes in a model independent way, each of them is fit with a spectral index $\gamma_{e\pm}$ as

$$\gamma_{e\pm} = d[\log[(\Phi_{e\pm})]/d[\log(E)] \tag{7}$$

(E is energy in GeV) over a sliding energy window, where the width of the window varies with energy to have sufficient sensitivity to determine the spectral index. The resulting energy dependencies of the fitted spectral indices are shown in Fig. 4, where the shading indicates the correlation between neighboring points due to the sliding energy window. The steep softening of the spectral indices below 10 GeV is due to solar modulation. Above 20 GeV, that is, above the effects of solar modulation, the spectral indices for positrons and electrons are significantly different. From 20 to 200 GeV, γ_{e+} is significantly harder than γ_{e-}. This demonstrates that the increase with energy observed in the positron fraction above 10 GeV is due to the hardening of positron spectrum and not to the softening of the electron spectrum.

Figure 4 shows the existence of structures in the spectral indices γ_{e+} and γ_{e-}. γ_{e-} progressive hardens with increasing energy above ~ 30 GeV and then levels off. γ_{e+} progressively hardens with increasing energy from ~ 20 to ~ 200 GeV and shows a tendency to soften above. This is consistent with the observation that the positron fraction no longer increase with energy above ~ 200 GeV. The differing behavior of the spectral indices versus energy indicates that high energy positrons have a different origin from that of electrons.

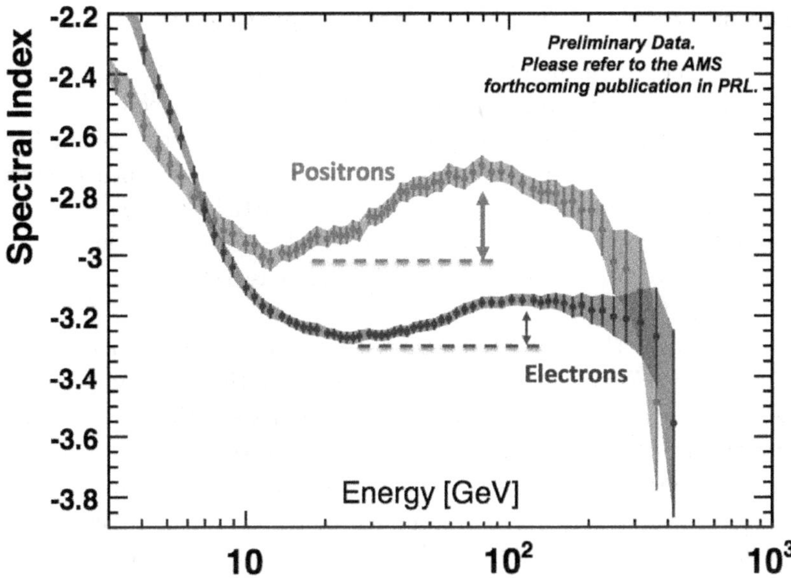

Fig. 4. Preliminary results of the electron(blue) and positron(red) spectra indices as a function of energy. The shaded regions indicate the 68% C.L. intervals including the correlation between neighboring points due to the sliding energy window. The data shows the existence of structures in the spectra indices (color online).

An analysis of the arrival directions of positrons and electrons was presented in Ref.[2]. The same analysis was performed including the additional data. The positron to electron ratio remains consistent with isotropy; the upper limit on the amplitude of the dipole anisotropy for positrons is $\delta \leq 0.012$ at the 95% C.L. for energies above 16 GeV.

The current AMS positron flux and positron fraction can be explained by collision of dark matter particles[12], astrophysical sources[13], and collision of cosmic rays[14].

The measurement of positron anisotropy will help to distinguish the dark matter models from the astrophysical models, such as pulsars. In the case of pulsar models, after flattening out with energy, a dipole anisotropy should be observed. In the dark matter models, no dipole anisotropy will be observed. Over its lifetime till at least 2024, AMS will reach a dipole anisotropy sensitivity of $\delta < 0.01$ at the 95% C.L., which will provide important information on the origin of the unexpected observation in the positron flux, electron flux, and positron fraction.

In addition, some of these models also include specific predictions on the antiproton flux and antiproton-to-proton flux ratio. The precision measurement of the protons and antiprotons fluxes will provide complementary information to understand the underlying physics of the new observation of cosmic ray positrons and electrons.

5.2. *The properties of charged elementary particles in cosmos*

The precision measurement of proton flux to 1.8 TV[1] is shown in Fig. 5. The previous important result from PAMELA[15] is also shown for comparison. As seen, the AMS proton flux deviates from a single power law and progressively hardens at high rigidities above ~ 200 GV. This is different from previous measurement[15].

The published results of the four elementary particles[5], p, $\bar{p}$, e^+, and e^- are shown together in Fig. 6. The fluxes of p, e^-, and e^+ are based on the same time period, May 19, 2011 to November 26, 2013, while $\bar{p}$ is based on the data from May 19, 2011 to May 26, 2015.

We have studied the time dependent solar effects on these fluxes during the 6 years operation[16]. Within our current accuracy, the time dependent solar effects are observable for antiprotons, electrons, and positrons with $|R| < 10$ GV and for protons with $|R| < 20$ GV. For the study of the flux ratio dependence of elementary particles (see below), we chose $|R| > 10$ GV where the time dependent solar effects for protons are small and the uncertainties are dominated by the accuracies of the measurements of e^-, e^+, and $\bar{p}$. This enables us to study the overall rigidity dependent behavior of different fluxes as shown in Fig. 6 above 10 GV. The points are placed along the abscissa at $\tilde{R}$ calculated for a flux $\propto R^{-2.7}$ according to Ref.[11].

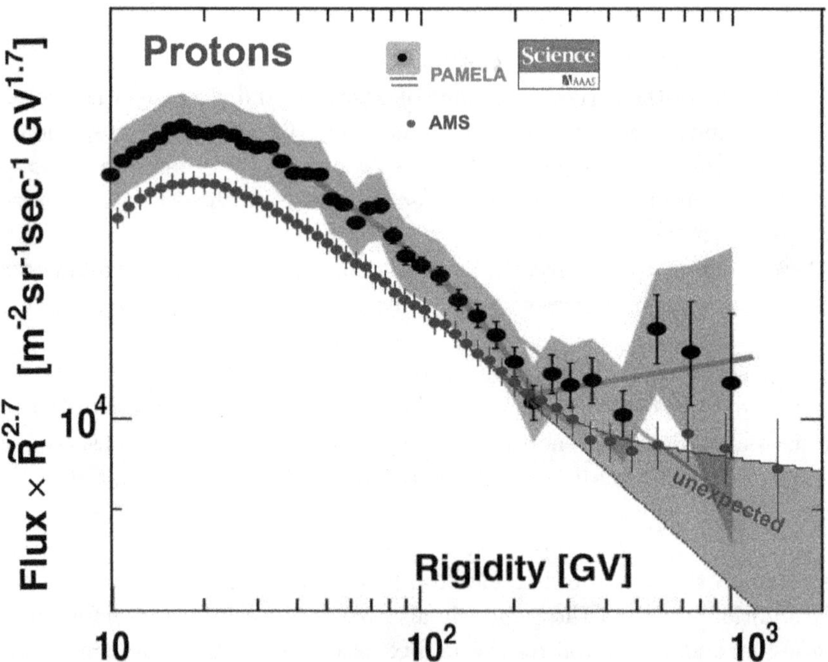

Fig. 5.   The proton flux measured by AMS up to 1.8 TV. The previous measurement by PAMELA is also shown[15]. The AMS proton flux deviates from a single power law and progressively hardens at high rigidities above ~ 200 GV. This is different from previous measurement[15].

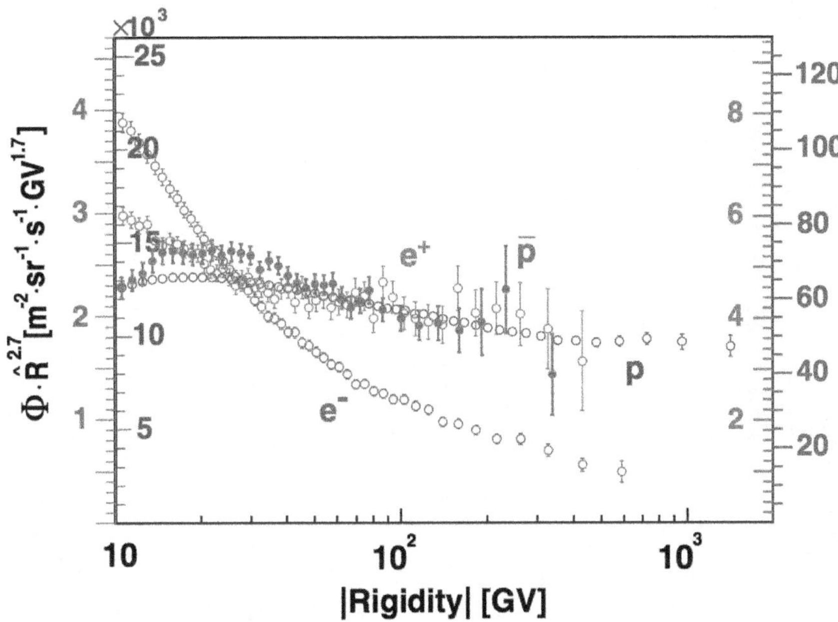

Fig. 6. The measured fluxes of four charged elementary particles, antiproton (red, left axis), proton (blue, left axis), electron (purple, right axis) and positron (green, right axis). All the fluxes are multiplied by $\tilde{R}^{2.7}$ for display purpose. The fluxes show different behavior at low rigidities while at $|R|$ from ~ 60 to ~ 500 GV the functional behavior of the antiproton, proton, and positron fluxes are nearly identical and distinctly different from the electron flux. The error bars correspond to the quadratic sum of the statistical and systematic errors (color online).

As seen from Fig. 6, the rigidity dependence of the fluxes for $\bar{p}$, e^+, and p are nearly identical above ~ 60 GV whereas the rigidity dependence of the e^- flux is different.

Figure 7(a) presents the measured $(\bar{p}/p)$ flux ratio. Compared with earlier experiment[17], the AMS results extend the rigidity range to 450 GV with increased precision. To minimize the systematic error for this flux ratio we have used the 2.42×10^9 protons selected with the same acceptance, time period, and absolute rigidity range as the antiprotons. As seen from Fig. 7(a), above ~ 60 GV the ratio appears to be rigidity independent. Further analysis[5] shows that, for $60.3 \leq |R| < 450$ GV, $\bar{p}/p$ flux ratio is rigidity independent with a mean value of $(1.81 \pm 0.04) \times 10^{-4}$. Fitting a straight line

$$\bar{p}/p = C + k(|R| - R_0) \tag{8}$$

over the rigidity range of 60.3 to 450 GV yields a slope $k(\bar{p}/p) = (-0.7 \pm 0.9) \times 10^{-7}$ GV^{-1}, which is consistent with 0. The solid red line in Fig. 7(a) shows this best straight line fit above 60.3 GV, as determined above, together with the 68% C.L. range of the fit parameters (shaded region).

Figure 7(b) presents the measured $\bar{p}/e^+$ and p/e^+ using the same data as Fig. 6. For the $\bar{p}/e^+$ ratio the rigidity independent interval is $60.3 \leq |R| < 450$ GV with

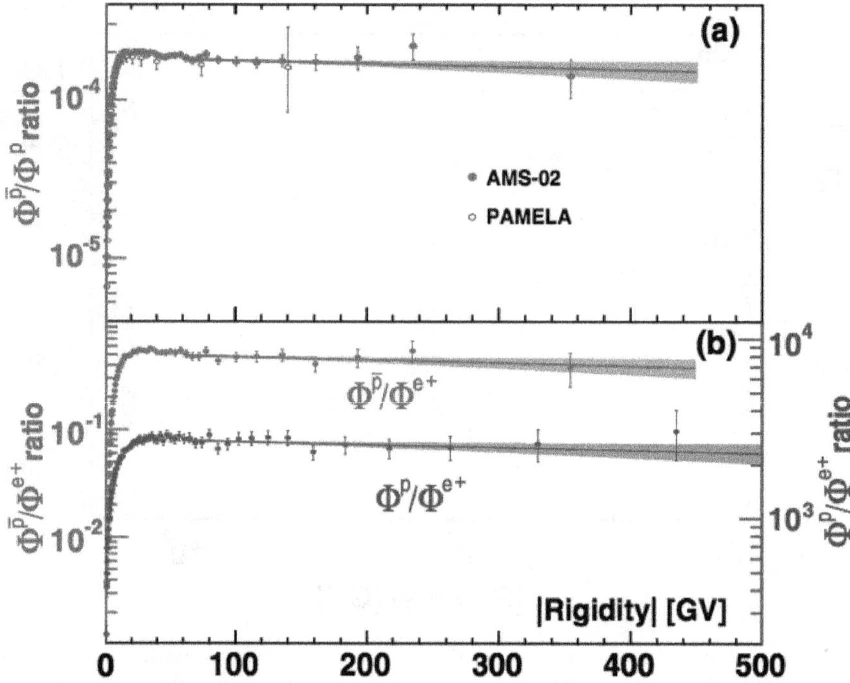

Fig. 7. (a) The measured $(\bar{p}/p)$ flux ratio as a function of the absolute value of the rigidity from 1 to 450 GV. The PAMELA[17] measurement is also shown. (b) The measured $(\bar{p}/e^+)$ (red, left axis) and (p/e^+) (blue, right axis) flux ratios. The solid lines show the best fit of Eq. (4) to the data above the lowest rigidity consistent with rigidity independence together with the 68% C.L. ranges of the fit parameters (shaded regions). For the AMS data, the error bars are the quadratic sum of statistical and systematic errors. Horizontally, the data points are placed at the center of each bin (color online).

a mean value of 0.479 ± 0.014. Fitting Eq. (8) over this interval yields $k(\bar{p}/e^+) = (-2.8 \pm 3.2) \times 10^{-4}$ GV^{-1}. For the p/e^+ ratio the rigidity independent interval is $59.13 \leq |R| < 500$ GV with a mean value of $(2.67 \pm 0.05) \times 10^3$ and $k(p/e^+) = (-0.9 \pm 1.0)$ GV^{-1}. Both results are shown in Fig. 7(b) together with the 68% C.L. range of the fit parameters (shaded regions).

In the rigidity range $\sim 60 < |R| < 500$ GV, the $\bar{p}/p$, $\bar{p}/e^+$, and p/e^+ flux ratios do not depend on rigidity, that is, the $\bar{p}$, p, and e^+ fluxes have nearly identical rigidity dependence. These are new observations of the charged elementary particles in the cosmos.

Current measurements at high rigidities are limited by the statistics of $\bar{p}$ and e^+. By collecting data in the life time of the International Space Station, AMS will extend the measurements to higher rigidity (or energy) and greatly improve the accuracy. In addition, the precision measurements of the cosmic ray nuclei by AMS is providing comprehensive information with unprecedented accuracy[18], to understand the origin and propagation of cosmic rays.

6. Conclusion

The preliminary results of the positron flux, electron flux, and positron fraction are presented. The new results are based on 1.08 million e^+ and 16.5 million e^- events collected during the first 5.5 years of operation. Comparing with our previous publication[3,4], the measurements are extended to 700 GeV for positron flux and positron fraction, and to 1 TeV for electron flux, with improved accuracy.

The positron fraction reaches its maximum at 265 ± 22 GeV. The positron and electron fluxes are different both in magnitude and in energy dependence. The spectral indices of positron and electron fluxes shows the existence of structures. Neither positron nor electron flux can be described by a single power law. The positron flux hardens starting from ~ 20 GeV to ~ 200 GeV and shows a tendency to soften above. The electron flux hardens starting from ~ 30 GeV to ~ 200 GeV and levels off.

The arrival direction of positrons are consistent with isotropic. With 5.5 years data, the 95% C.L. upper limit of dipole anisotropy is $\delta \leq 0.012$ for energies above 16 GeV.

The precision proton flux measured by AMS shows that it progressively hardens above 200 GV, which is different from previous PAMELA results[15].

The antiproton flux and antiproton-to-proton flux ratio is measured based on 349 thousands $\bar{p}$ collected during the first 4 years of operation. Together with the proton, positron, and electron fluxes measured by the same detector, an unexpected phenomenon is experimentally observed for the first time. Unexpectedly, the proton, antiproton, and positron fluxes are found to have nearly identical rigidity dependence from 60 to 500 GV, while electron flux has distinctly different rigidity dependence.

Acknowledgements

This work has been supported by acknowledged person and institutions in Ref.[5].

References

1. M. Aguilar *et al.*, Phys. Rev. Lett. **114**, 171103 (2015).
2. M. Aguilar *et al.*, Phys. Rev. Lett. **110**, 141102 (2013).
3. L. Accardo *et al.*, Phys. Rev. Lett. **113**, 121101 (2014).
4. M. Aguilar *et al.*, Phys. Rev. Lett. **113**, 121102 (2014); M. Aguilar *et al.*, Phys. Rev. Lett. **113**, 221102 (2014).
5. M. Aguilar *et al.*, Phys. Rev. Lett. **117**, 091103 (2016).
6. A. Kounine *et al.*, Nucl. Instrum. Methods Phys. Res., Sect. A 869, 110 (2017).
7. B. P. Roe *et al.*, Nucl. Instrum. Methods Phys. Res., Sect. A 543, 577 (2005).
8. J. Allison *et al.*, IEEE Trans. Nucl. Sci. 53, 270 (2006); S. Agostinelli *et al.*, Nucl. Instrum. Methods Phys. Res., Sect. A 506, 250 (2003).
9. C. Størmer, The Polar Aurora (Oxford University Press, London, 1950).
10. J. Alcaraz *et al.*, Phys. Lett. B 484, 10 (2000); C. C. Finlay *et al.*, Geophys. J. Int. 183, 1216 (2010); E. Thébault *et al.*, Earth Planets Space 67, 79 (2015).

11. G. D. Lafferty and T. R. Wyatt, Nucl. Instrum. Methods Phys. Res., Sect. A 355, 541 (1995). We have used Eq. (6) to obtain $\tilde{E}$ with $\tilde{E} = x_{lw}$.

12. J. Kopp, Phys. Rev. D 88, 076013 (2013); L. Feng, R-Z. Yang, H-N. He, T-K. Dong, Y-Z. Fan, and J. Chang, Phys. Lett. B 728, 250 (2014); L. Bergström, T. Bringmann, I. Cholis, D. Hooper, and C. Weniger, Phys. Rev. Lett. 111, 171101 (2013); C. H. Chen, C. W. Chiang, and T. Nomura, Phys. Lett. B 747, 495 (2015); H. B. Jin, Y. L. Wu, and Y.-F. Zhou, Phys. Rev. D 92, 055027 (2015).

13. I. Cholis and D. Hooper, Phys. Rev. D 88, 023013 (2013); T. Linden and S. Profumo, Astrophys. J. 772, 18 (2013); P. Mertsch and S. Sarkar, Phys. Rev. D 90, 061301 (2014); K. Kohri, K. Ioka, Y. Fujita, and R. Yamazaki, Prog. Theor. Exp. Phys. 2016, 021E01 (2016).

14. K. Blum, B. Katz, and E. Waxman, Phys. Rev. Lett. 111, 211101 (2013); R. Cowsik, B. Burch, and T. Madziwa-Nussinov, Astrophys. J. 786, 124 (2014); R. Kappl and M. W. Winkler, J. Cosmol. Astropart. Phys. 09 (2014) 051.

15. O. Adriani et al., Science 332, 69 (2011).

16. AMS Collaboration, Measurement of the Time Dependent Solar Modulation of Primary Cosmic Rays Fluxes with the Alpha Magnetic Spectrometer on the International Space Station (to be published).

17. O. Adriani et al., Phys. Rev. Lett. 102, 051101 (2009); O. Adriani et al., Phys. Rev. Lett. 105, 121101 (2010); JETP Lett. 96, 621 (2013).

18. M. Aguilar et al., Phys. Rev. Lett. 115, 211101 (2015); M. Aguilar et al., Phys. Rev. Lett. 117, 231102 (2016).

Hard QCD Measurements at the LHC

Gabriella Pásztor

on behalf of the ATLAS, CMS and LHCb Collaborations

MTA-ELTE Lendület CMS Particle and Nuclear Physics Group,
Eötvös Loránd University, Budapest, Hungary
E-mail: gabriella.pasztor@cern.ch
http://pasztor.web.cern.ch/gpasztor/

The rich proton-proton collision data of the LHC allow the study of QCD processes in a previously unexplored region with ever improving precision. This paper summarises recent results of the ATLAS, CMS and LHCb Collaborations using primarily multi-jet and vector boson plus jet data collected at $\sqrt{s} = 8$ and 13 TeV. Comparisons to higher-order theoretical calculations and sophisticated Monte Carlo predictions are presented, as well as the impact of the data on the determination of the parton distribution functions and the measurement of the strong coupling constant, α_s.

Keywords: LHC; ATLAS; CMS; LHCb; QCD; pQCD; jet; photon; Drell-Yan; parton distribution function; pdf; strong coupling constant.

1. Introduction

Processes governed by the strong interaction are dominant in the high-energy proton-proton collisions at the LHC, making the machine a jet factory. The rich jet data allows us to study perturbative QCD (pQCD) in a previously unexplored region.

Cross-sections at LHC — including the coveted New Physics processes such as supersymmetric partner, dark matter or other exotic particle production — are affected by QCD processes. The cross-sections can be written as the convolution of three terms. The initial state is described by universal parton distribution functions (PDFs) that are fitted from data using Dokshitzer–Gribov–Lipatov–Alterelli–Parisi (DGLAP) evolution. The hard scattering matrix element (ME) is typically calculated to (next-to-)next-to-leading-order ((N)NLO) in perturbative QCD α_s expansion with the resummation of large logarithms due to soft and collinear gluon corrections up to (next-to-)next-to-leading-log ((N)NLL). Final-state partons then form hadrons, the process is described by universal fragmentation functions (FFs) fitted from data using DGLAP evolution.

The extraction of the strong coupling constant, the proton PDFs and the FFs are important thus not only from a QCD perspective but for the whole physics program of the LHC. PDF uncertainty is dominant systematics for Higgs boson and top production, and it limits the precision measurements of the strong coupling constant and the top mass as well as the search for new heavy particles. The running of α_s is affected if new colored particles exist, and thus its measurement constrains New Physics.

QCD processes also give significant background to new particle searches, thus the test of their higher-order calculations, the validation of sophisticated Monte Carlo (MC) tools[a], including parton shower (PS) modelling and ME-PS matching, also contribute to the success of the New Physics program of the LHC.

The impressive agreement between LHC data and theoretical predictions over many orders of magnitude in cross-section (from inclusive jet production to the rare electroweak production of two bosons accompanied by two jets as shown in Fig. 1) is a feat. It is thanks to the advances in experimental methods (such as improved object reconstruction and calibration that are robust against the number of proton-proton collisions per bunch crossing, high-performance triggers, sophisticated data-driven background predictions, boosted techniques) and in theoretical predictions (including NNLO and N^3LO pQCD calculations, resummation of large logs, improved treatment of parton showers and their matching to matrix element calculations, electroweak corrections, versatile MC tools).

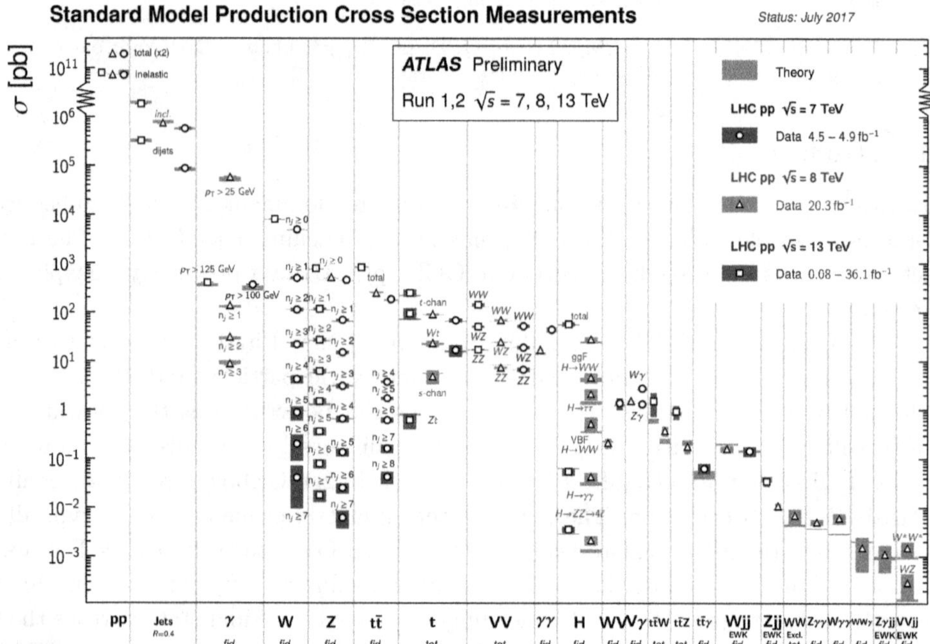

Fig. 1. Total and fiducial production cross-section measurements of Standard Model processes corrected for leptonic branching fractions and compared to the corresponding theoretical expectations calculated at NLO or higher order [1].

---

[a] For references of the Monte Carlo tools, the parton distribution functions and tunes used in the analyses, see the references in the quoted experimental papers.

The LHC delivered about 23 fb^{-1} data at a proton-proton center-of-mass energy of $\sqrt{s} = 8$ TeV in 2012, as well as about 4 fb^{-1} and 40 fb^{-1} data at $\sqrt{s} = 13$ TeV in 2015 and 2016 respectively. In the following, selected results are shown from the ATLAS, CMS and LHCb Collaborations concentrating on multi-jet and vector boson plus jet final states based on these datasets. The emphasis is on detailed data comparisons to theoretical calculations and Monte Carlo predictions validating the recent advances, as well as on the extraction of α_s and the proton PDFs.

The LHC can reach down to the low-x and up to the high-Q^2 region in parton kinematics extending the phase space previously explored by fixed target experiments and the HERA electron-proton collider, as shown in Fig. 2. We have many handles that provide complementary information. Studies of jet, top-quark pair and prompt photon production test the gluon and valence quark PDFs and are sensitive to α_s. W and Z production cross-sections and cross-section ratios help to differentiate light-flavour quark PDFs. Off-peak Drell-Yan production reaches low- and high-x regions for u- and d-quark PDFs. While Z/W+jets measurements are especially sensitive to the gluon PDF, the Z/W + heavy flavour processes constrain the strange, charm and bottom quark distributions.

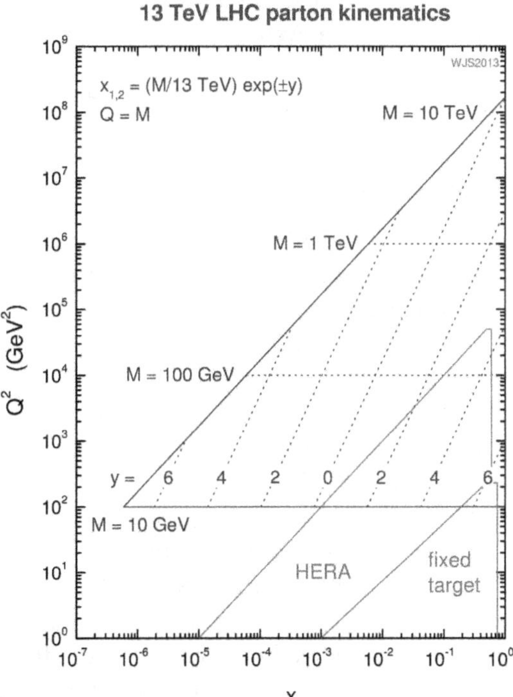

Fig. 2.   Parton kinematics at the LHC at $\sqrt{s} = 13$ TeV in the $x - Q^2$ plane[2]. The coverage of previous fixed target experiments and HERA is also shown.

2. Multi-jet results

2.1. *Jet reconstruction at LHC*

Jets are typically formed from reconstructed particles, so called particle-flow[3] objects that result from the combined information of charged particle tracks in the inner tracking detectors and muon spectrometers and of energy deposits in the electromagnetic and hadronic calorimeters. They aim to precisely reconstruct the physics objects (photons, electrons, muons, tau leptons, charged and neutral hadrons) and their interactions as they traverse the detector. To reconstruct the jets, the infrared- and collinear-safe anti-kt clustering algorithm[4] is used with size parameter typically in the $R = 0.4 - 0.8$ range.

The reconstructed hadronic jets are corrected for detector effects and are frequently unfolded to estimate the kinematic distributions of the original "true" particle jets to facilitate comparisons with theoretical predictions and with other experimental results.

In most measurements the dominant experimental systematics arise from jet energy calibration. As an example, the size of the jet energy correction is around 10% at a jet transverse momentum (p_T) of 100 GeV in the CMS measurements at $\sqrt{s} = 8$ TeV[5]. The resolution improves from about 15% at $p_T = 10$ GeV to about 4% at $p_T = 1$ TeV. These induce a systematic uncertainty on the jet cross-section of 2-4% at sub-TeV and 20% at the highest transverse momenta from the energy scale, as well as about 30% at low and 1-5% at high p_T from the energy resolution.

2.2. *Inclusive jet cross-section*

The LHC data covers more than 12 orders of magnitude in jet cross-section measured differentially in jet transverse momenta and absolute jet rapidity ($|y|$) as illustrated in Fig. 3(left). The experimental uncertainties being smaller than the theoretical ones, these measurements are powerful in constraining the proton PDFs in particular for gluons and valence quarks. Measurements of cross-section ratios at different center-of-mass energies profit from partial cancellation of systematic uncertainties. The CMS results[5] comparing jet production at 2.76 TeV and 8 TeV clearly disfavour ABM11 PDF set as shown in Fig. 3(right).

The measured data shows a reasonable agreement with next-to-leading order (NLO) theoretical predictions when studied in individual p_T bins but show some tension when fitting all bins together for all tested PDF sets as illustrated by the ATLAS measurement[6] at $\sqrt{s} = 8$ TeV on Fig. 4 for four PDF sets. When considering all jets above 70 GeV for all rapidity bins, the χ^2 ranges from 340 to 424 for 171 degrees of freedom, depending on the PDF set tested, the value of R in the anti-kt clustering (R=0.4 or 0.6) and the choice of central renormalisation and factorisation scales.

The gluon density at high Bjorken x is poorly constrained. While quark-gluon (qg) interactions dominate the dijet cross-section for almost the full

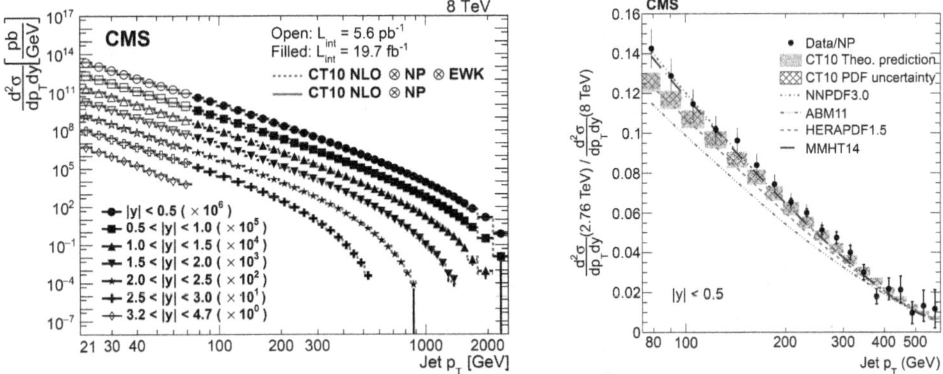

Fig. 3. (left) Double-differential inclusive jet cross-sections as function of jet p_T in bins of rapidity for anti-kt jets with R=0.7[5]. The data are compared to NLO predictions corrected for non-perturbative (NP) and electroweak (EWK) effects. (right) Ratio of the inclusive jet production cross-sections at $\sqrt{s} = 2.76$ TeV and 8 TeV as function of jet p_T for jet rapidities of $|y| < 0.5$[5].

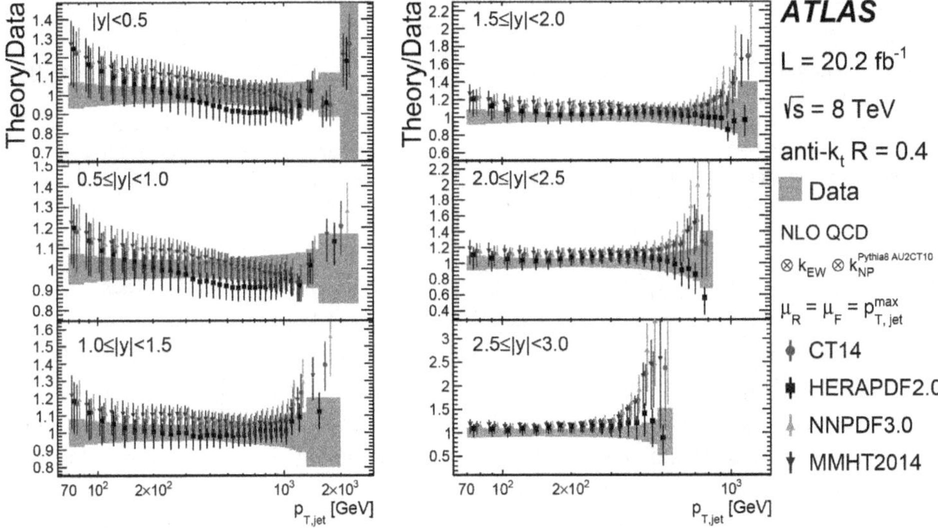

Fig. 4. Ratio of the inclusive jet cross-section predicted by NLO QCD corrected for non-perturbative and electroweak effects to the cross-section observed in data as a function of the jet p_T in each jet rapidity bin for anti-kt jets with R=0.4[6]. Various PDF sets are compared with the data. The central renormalisation and factorisation scales are set to the maximum jet p_T in the event. The points are slightly offset along the jet p_T axis for better visibility.

kinematic range, at high p_T quark-quark (qq) processes take over. The impact of the 8 TeV CMS data on the gluon and light quark PDFs is shown on Fig. 5[5]. The improvement is most notable for the gluon densities. If the strong coupling constant is left free in the fit, the derived value of $\alpha_s(M_Z) =$

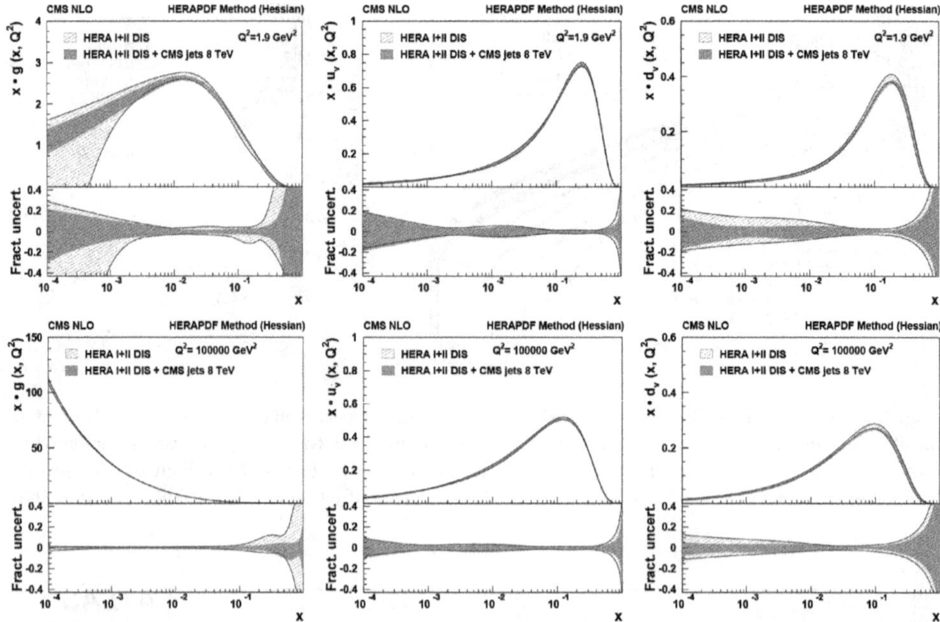

Fig. 5. Gluon (left), u valence quark (middle), and d valence quark (right) distributions as functions of Bjorken x at the starting scale $Q^2 = 1.9$ GeV^2 (top) and at high $Q^2 = 10^5$ GeV^2 (bottom)[5]. The results of the fit to the HERA data and inclusive jet measurements at 8 TeV by CMS (shaded band), and to HERA data only (hatched band) are compared with their total uncertainties as determined by using the HERAPDF method. In the bottom panels the fractional uncertainties are shown.

$0.1184^{+0.0019}_{-0.0021}$ (exp) $^{+0.0002}_{-0.0015}$ (model) $^{+0.0000}_{-0.0004}$ (param) $^{+0.0022}_{-0.0018}$ (scale) $= 0.1184^{+0.0019}_{-0.0021}$ [5] is compatible with the world average[7] of $\alpha_s(M_Z) = 0.1181 \pm 0.0011$.

The inclusive jet cross-section is also measured at the highest available energy of 13 TeV with jet transverse momentum up to 2 TeV and rapidity up to $|y| = 4.7$. Comparisons with various LO and NLO Monte Carlo generators, complemented with electroweak and non-perturbative corrections, using the anti-kt algorithm with R=0.4 (see Fig. 6) and 0.7 show that data reconstructed with a larger jet size are better described by the fixed order NLO prediction[8]. When NLO is matched to parton shower (Powheg interfaced to Pythia8) the description for reconstruction with a lower jet size value is also improved.

Recent next-to-next-to-leading order (NNLO) calculations are also compared to the LHC data. In Fig. 7 the ATLAS 13 TeV results[9] show that while scale uncertainties decrease in general for NNLO, the central scale choice (either the jet p_T or the maximum jet p_T in the event) has a large impact on the prediction. At NNLO using $\mu_R = \mu_F = p_T$ for the central scale describes the data better, especially at low jet transverse momenta.

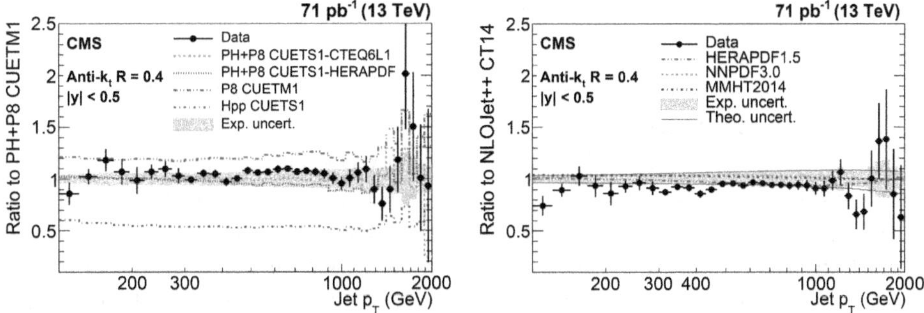

Fig. 6.   Ratio of measured values of the inclusive jet cross-section to theoretical predictions[8].

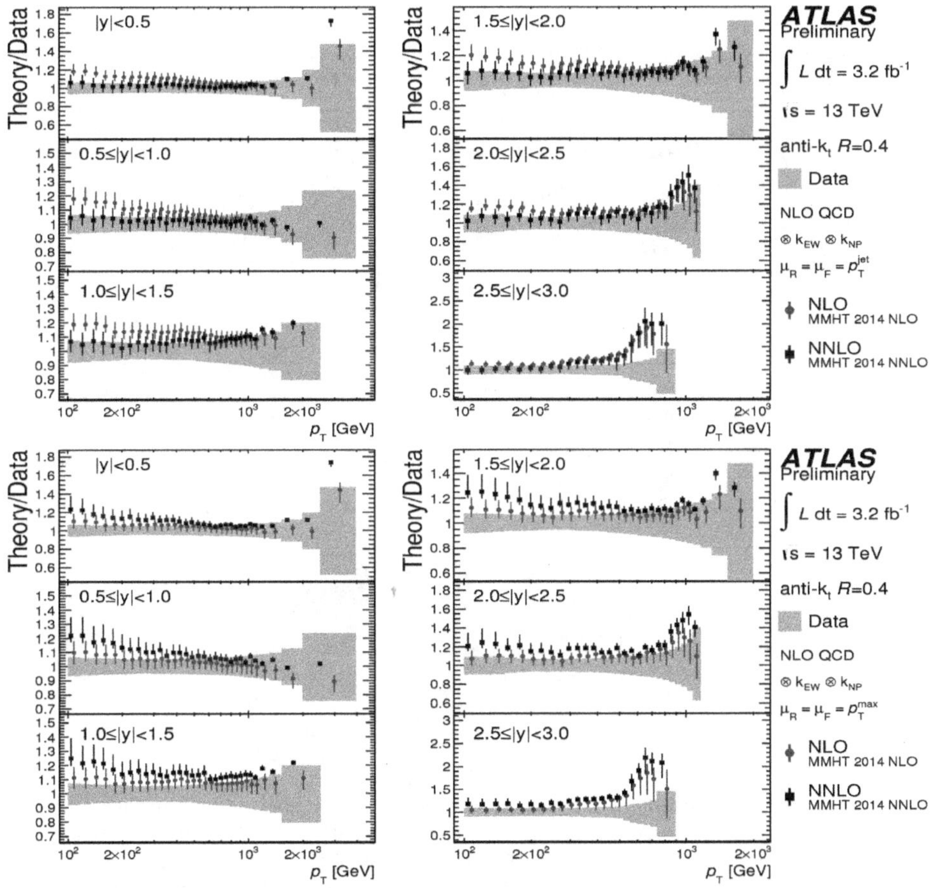

Fig. 7.   Ratios of pQCD predictions corrected for non-perturbative and electroweak effects to the measured inclusive jet cross-sections as a function of the jet p_{T} in $|y|$ bins for anti-kt jets with R=0.4. The NLO predictions are calculated using NLOJET++, while the NNLO predictions are provided by NNLOJET, both using MMHT 2014 PDF sets and $p_{\mathrm{T}}^{\mathrm{jet}}$ (upper panel) or $p_{\mathrm{T}}^{\mathrm{max}}$ (lower panel) as the central QCD scale[9].

2.3. *Dijet cross-section*

The double differential dijet cross-section[9] manifests in Fig. 8(top right) a reasonable agreement as function of the dijet mass in rapidity bins with NLO pQCD calculations, with the tested proton PDF sets showing similar behaviour. Measurements[10] are also performed triple differentially by CMS as a function of the average jet transverse momentum $p_{\mathrm{T,avg}}$, the average jet rapidity y_b and the rapidity difference $y^* = 0.5|y_1 - y_2|$, as presented in Fig. 8(top left). AMB11, due to its soft gluon PDF and low α_s, underestimates the data in most of the phase space. At high $p_{\mathrm{T,avg}}$ and high y_b (i.e. in the previously less studied high-x region) the theoretical prediction overestimates the data by up to a factor of two (see Fig. 8(bottom)). As the experimental precision is better than the theoretical one, the proton PDFs can

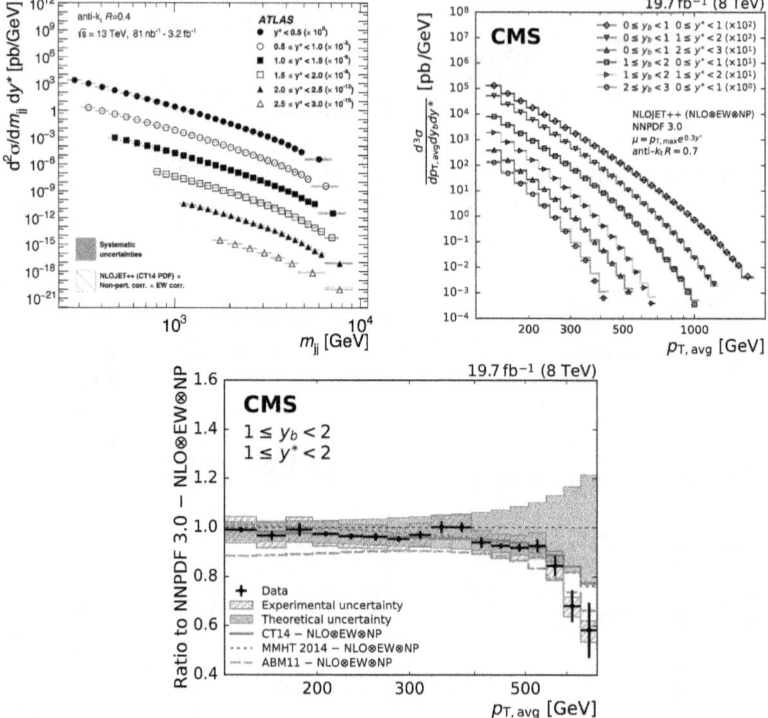

Fig. 8.   (top left) Double differential dijet cross-section as a function of the dijet mass m_{jj} and rapidity difference $y^* = |y_1 - y_2|/2$, for anti-kt jets with R=0.4[9]. The data are compared to NLO pQCD predictions calculated using NLOJET++ with $p_{\mathrm{T}}^{\max}e^{0.3y^*}$ as the QCD scale and the CT14 NLO PDF set, to which non-perturbative and electroweak corrections are applied. In most bins the experimental systematic uncertainty is smaller than the theory uncertainties and is not visible. (top right) Triple differential dijet cross-section for anti-kt jets with R=0.7 as a function of the average jet transverse momentum $p_{\mathrm{T,avg}}$, average jet rapidity y_b and rapidity difference y^* compared to the NLOJET++ prediction using the NNPDF3.0 PDF set[10]. (bottom) Ratio of the triple-differential dijet cross-section to the NLOJET++ prediction in a selected (y^*, y_b) bin[10]. The ratios calculated with the predictions for different PDF sets are also shown.

be constrained, in particular the uncertainty on the gluon PDF is decreased. The constraints are similar to those obtained from the inclusive jet data.

The jet mass is sensitive to parton shower modelling. It is used to reconstruct the mass of heavy particles in analyses of boosted topologies, such as searches for new heavy states or studies of Higgs boson or top quark production. The double differential cross-section is presented in Fig. 9 unfolded in jet p_T and jet mass[11] in dijet events, for anti-kt jets with R=0.8, with and without soft drop jet grooming[12]. With grooming the Sudakov peak is suppressed and a better agreement is seen at moderate values of jet mass and p_T as the soft radiation contribution is removed. Pythia8 agrees best with the data. Semi-analytical NNLL calculations[13] predict the data well except at high mass and p_T as illustrated in Fig. 10.

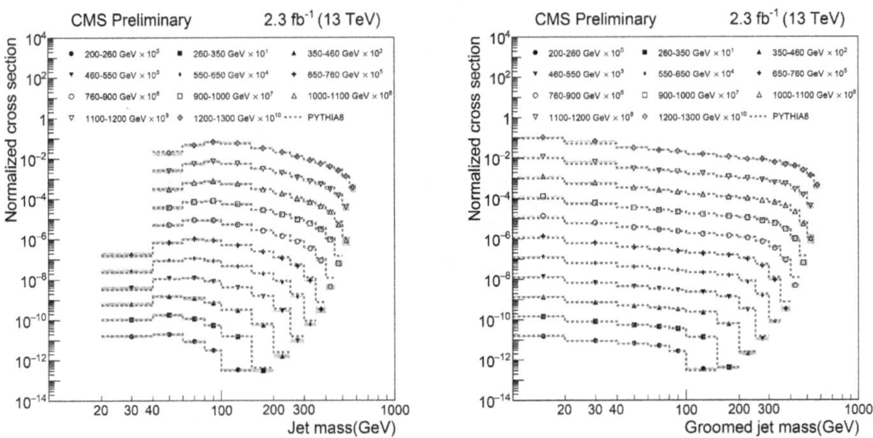

Fig. 9. Differential cross-sections in the jet mass unfolded for (left) ungroomed and (right) groomed jets in p_T bins[11]. The data with total uncertainties are compared to Pythia8 predictions.

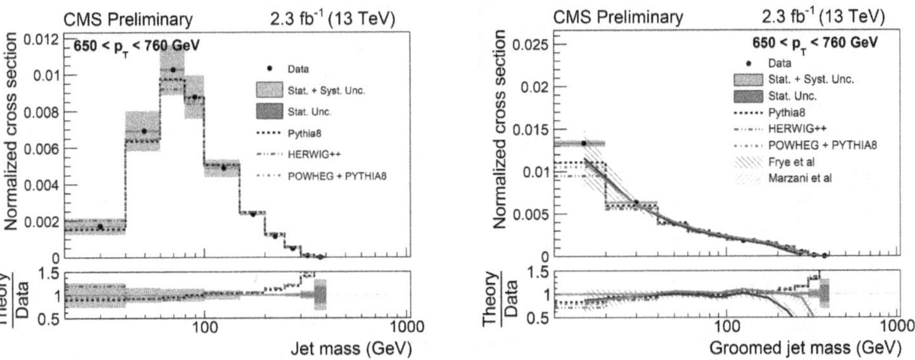

Fig. 10. Unfolded jet mass distribution for (left) ungroomed and (right) groomed jets for 650 GeV $< p_T <$ 760 GeV compared to various Monte Carlo predictions and for groomed jets also to semi-analytical calculations[11].

The observed azimuthal correlations in two-, three-, four-jet topologies[14], such as the azimuthal difference of the two leading jets $\Delta\phi_{1,2}$ shown in Fig. 11 or the minimal azimuthal difference in the event $\Delta\phi_{2j}^{\min}$, are generally not well described by MC simulations. LO Pythia8 and Herwig++ show large deviations. The tree-level multijet event generator MadGraph in combination with Pythia8 for shower-

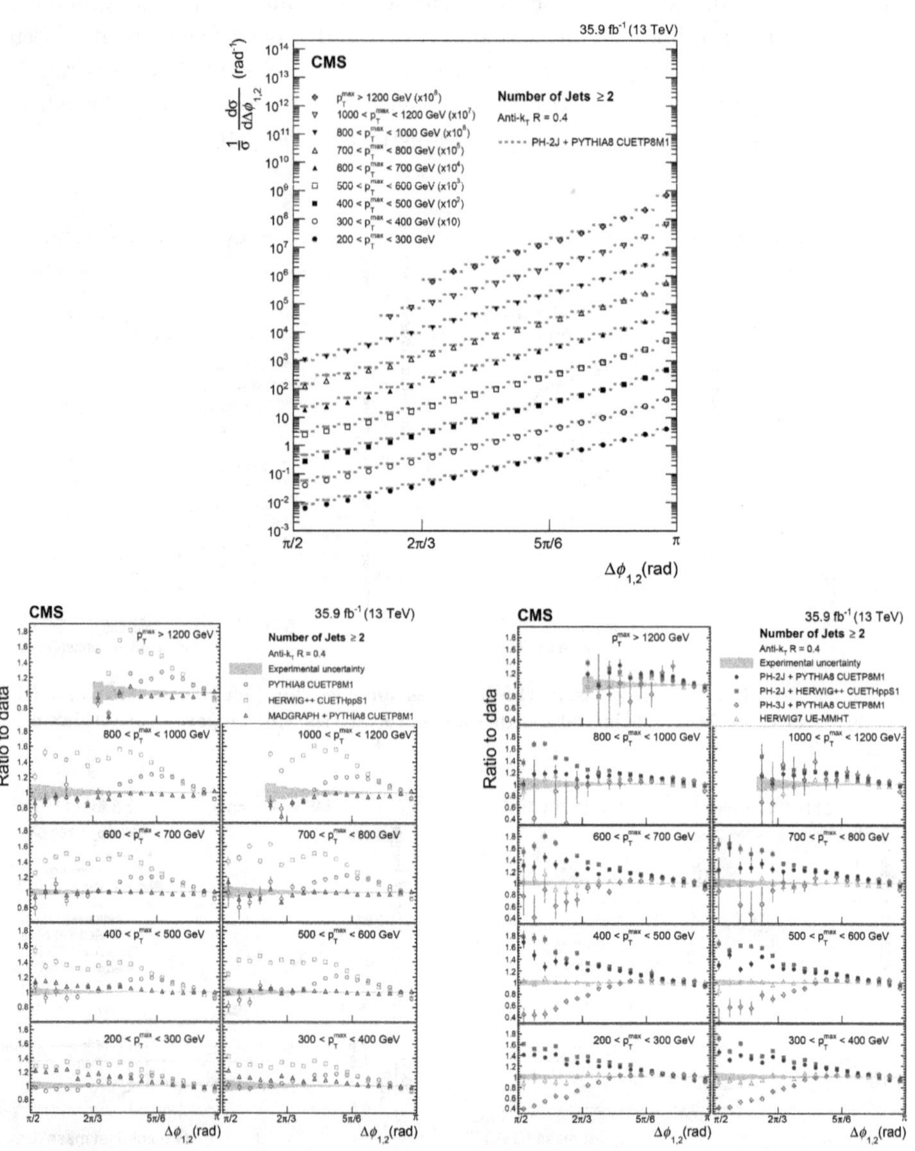

Fig. 11. (top) Normalised inclusive 2-jet cross-section differential in $\Delta\phi_{1,2}$ in different regions in the maximum jet p_T[14]. The size of the data symbol includes the total uncertainties. (bottom) Ratios of various Monte Carlo predictions to the normalised inclusive 2-jet cross-section[14].

ing, hadronisation, and multiparton interactions provides a good overall description (except for $\Delta\phi_{2j}^{\min}$ in four-jet topologies). The NLO dijet and three-jet simulations with Powheg fail to describe the measured correlations. The NLO dijet sample by Herwig7 matched to parton shower contributions with the MC@NLO method provides a very good description of $\Delta\phi_{1,2}$. The results prove that the azimuthal correlations are sensitive to details of the modelling of the non-perturbative effects and emphasize the need to improve predictions for multijet production.

All Monte Carlo generators have difficulty reproducing the $b\bar{b}$ dijet data. The ATLAS measurement [15] shown in Fig. 12 requires a jet with $p_T > 270$ GeV which enhances the sample with events of three-jet topology thus raising the contribution of processes of gluon-splitting and flavor excitation. The NLO Powheg interfaced to Pythia6 provides a better description than MC@NLO, but significant differences are visible in regions dominated by flavour creation (low transverse momentum, high mass, large angular separation of the $b\bar{b}$ system). The rapidity distribution is sensitive to the PDF and is well described by Powheg. LO Pythia6 dijet simulation (scaled by 0.61 to correct the overall normalisation) describes the shapes somewhat better than LO Sherpa including matrix elements up to $2\to3$ processes.

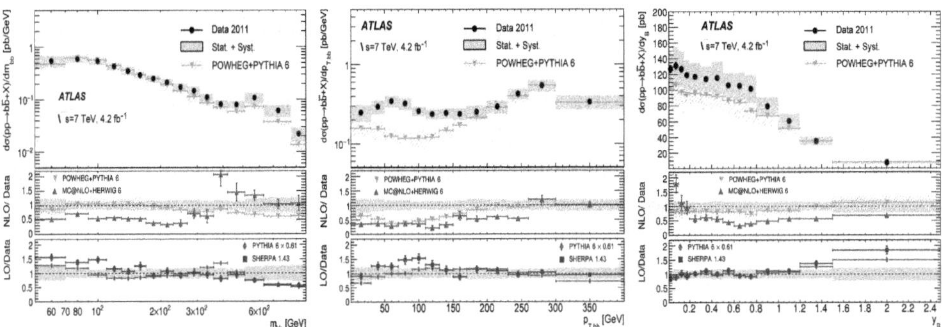

Fig. 12. Differential cross-section for $b\bar{b}$ production as a function of (left) the mass, (middle) the transverse momentum, (right) the average boost of the $b\bar{b}$ system compared to various MC predictions [15].

2.4. *Determination of the strong coupling constant*

The strong coupling constant can be determined in a variety of ways at the LHC.

The inclusive jet cross-section $\mathrm{d}^2\sigma/\mathrm{d}p_T\mathrm{d}\eta$ is proportional to α_s^2. Various experimental and theoretical errors cancel in inclusive cross-section ratios $R_{mn} = \sigma(pp \to m$ jets$)/\sigma(pp \to n$ jets$)$ that are proportional to α_s^{m-n}. CMS has measured the strong coupling constant at NLO with both methods (see Fig. 13) and arrived to

$$\alpha_s(M_Z) = 0.1162^{+0.0014}_{-0.0015} \text{ (exp) } ^{+0.0053}_{-0.0028} \text{ (scale) } ^{+0.0025}_{-0.0029} \text{ (PDF) } \pm 0.0001 \text{ (NP)}$$

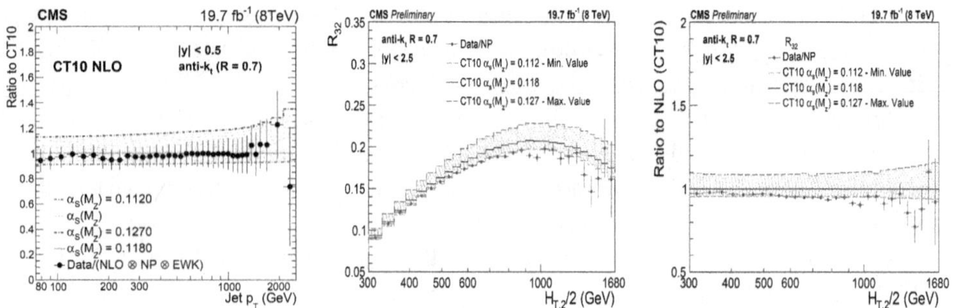

Fig. 13. (left) Ratio of the observed inclusive jet cross-section over theory prediction using the CT10 NLO PDF set as a function of the jet p_T for anti-kt jets with $R = 0.7$ with the default $\alpha_s(M_Z) = 0.118$ value[5]. Ratios of the predictions obtained with the CT10 PDF set evaluated with different values of $\alpha_s(M_Z)$ indicate the sensitivity of data to the strong coupling constant. (middle) Cross-section ratio R_{32} as a function of average jet transverse momentum[16]. (right) Ratio of measured R_{32} over NLO theory with non-perturbative corrections[16].

from the inclusive cross-section[5] and to

$$\alpha_s(M_Z) = 0.1142 \pm 0.0010 \ (\text{exp}) \ ^{+0.0049}_{-0.0006} \ (\text{scale}) \pm 0.0013 \ (\text{PDF}) \ \pm 0.0014 \ (\text{NP})$$

from the R_{32} ratio measurements[16]. These are compatible with the world average $\alpha_s(M_Z) = 0.1181 \pm 0.0011$[7].

The running of α_s from the R_{32} measurement is shown in Fig. 14 using the MSTW2008 NLO PDF set evolving the extracted $\alpha_s(M_Z)$ values using the 2-loop 5-flavour renormalisation group equations. The result agrees well with previous measurements and the world average.

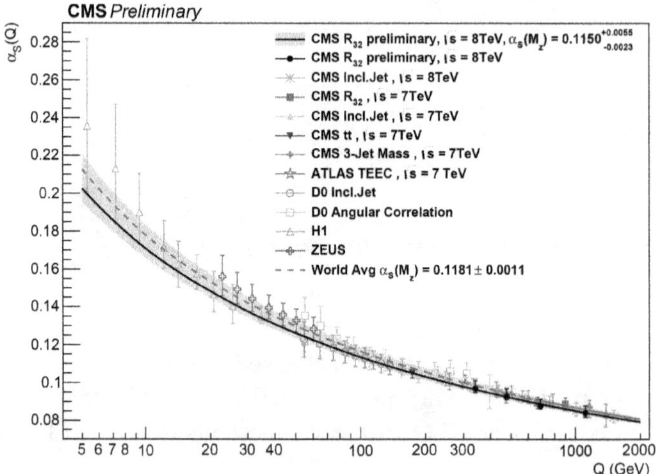

Fig. 14. The running $\alpha_s(Q)$ as a function of the scale Q as obtained by using the MSTW2008 NLO PDF set[16]. The dashed line represents the evolution of the world average.

ATLAS uses transverse energy – energy correlation (TEEC), the energy weighted angular distribution of jet pairs to measure the strong coupling constant[17]. TEEC, shown in Fig. 15(top left) gets only moderate NLO corrections and is also robust with respect to experimental uncertainties, thus well-suited for precision tests of pQCD. One can also use the asymmetry (ATTEC) between the forward and backward contributions of TEEC (see Fig. 15(top right) for data - MC comparisons). Both TEEC and ATTEC are proportional to α_s^2 at NLO. Both distributions are well described by NLO pQCD predictions within the uncertainties. The extracted α_s values from the TEEC and ATTEC distribution respectively are

$$\alpha_s(M_Z) = 0.1162 \pm 0.0011 \text{ (exp) } ^{+0.0076}_{-0.0061} \text{ (scale) } \pm 0.0018 \text{ (PDF)} \pm 0.0003 \text{ (NP)}$$

$$\alpha_s(M_Z) = 0.1196 \pm 0.0013 \text{ (exp) } ^{+0.0061}_{-0.0013} \text{ (scale) } \pm 0.0017 \text{ (PDF)} \pm 0.0004 \text{ (NP)}$$

The running of α_s is shown in Fig. 15 for the slightly more precise ATEEC measurement. Both results agree with the world average[7].

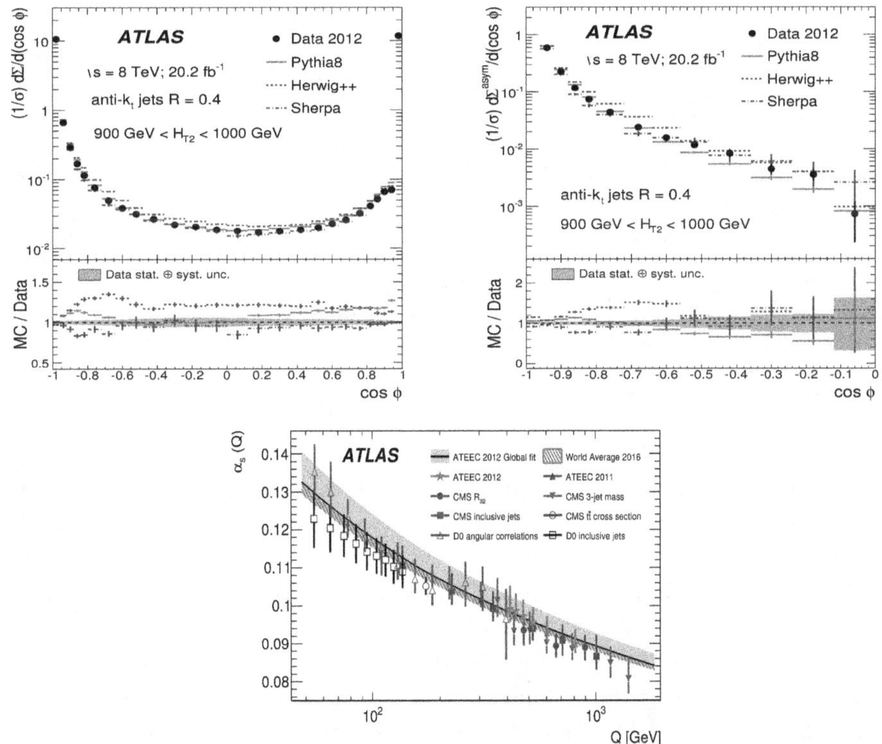

Fig. 15.   Particle-level distributions for (top left) the TEEC and (top right) the ATEEC functions for the 900 GeV to 1000 GeV bin in the average jet transverse momentum of the two leading jets $H_{T,2}/2$, compared to various MC predictions[17]. (bottom) Comparison of the values of $\alpha_s(Q)$ obtained from fits to the TEEC functions at the energy scales given by $< H_{T,2} > /2$ with the uncertainty band from the global fit and the world average[17].

234

3. Probing QCD with photons

Photonic final states allow the test of pQCD with a hard, colorless probe.

Prompt photon production ($qg \rightarrow q\gamma, q\bar{q} \rightarrow g\gamma$) is sensitive to the gluon PDF at leading order. NLO predictions by JETPHOX provide a reasonable description of the inclusive photon data[18] of the ATLAS experiment as illustrated in Fig. 16(left, middle). It, however, can not describe the angular difference between the photon and the closest jet ($\Delta\phi^{\gamma-\text{jet}}$)[19] due to limitations in the number of final state partons. Sherpa with up to $2 \rightarrow 5$ matrix elements calculated at NLO agrees well with data for $\Delta\phi^{\gamma-\text{jet}}$, but has difficulty describing the leading jet transverse momentum distribution which is well reproduced by JETPHOX as shown in Fig. 16(right)[19]. The experimental uncertainties are lower than the theoretical ones, thus NNLO calculations are desirable to decrease the scale uncertainties.

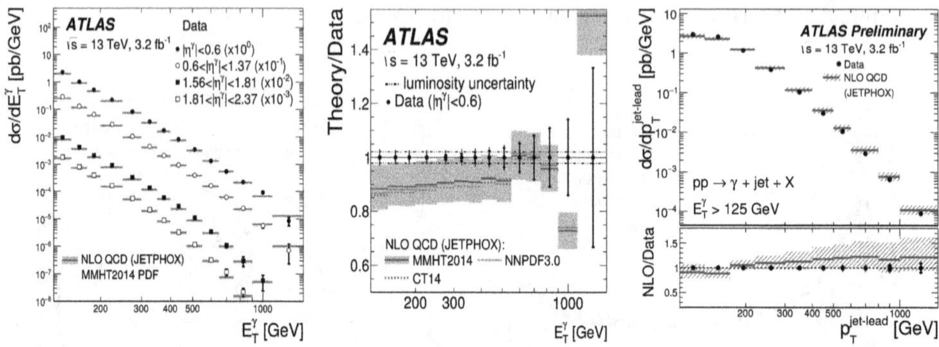

Fig. 16.   (left) Measured cross-sections for isolated-photon production as functions of the photon transverse energy in different photon pseudorapidity ranges[18]. (middle) Ratio of the NLO pQCD predictions from JETPHOX based on the MMHT2014 PDFs to the measured cross-sections[18]. (right) Measured cross-sections for isolated-photon plus one-jet production as a function of the leading jet transverse momentum[19].

The ATLAS Collaboration in its $\sqrt{s} = 8$ TeV measurement[20] observed for the first time the difference in QCD radiation pattern around the leading jet and the photon, in agreement with theoretical predictions.

The diphoton production cross-section is sensitive to pQCD corrections. Fixed order calculations are lower than the observed data[21], though NNLO brings significant improvement as shown in Fig. 17. However, especially in regions sensitive to infrared emission, soft-gluon resummation is needed at NNLL. The best description of the data is provided by SHERPA with NLO matrix element calculation up to $2\rightarrow5$ processes and parton shower simulation.

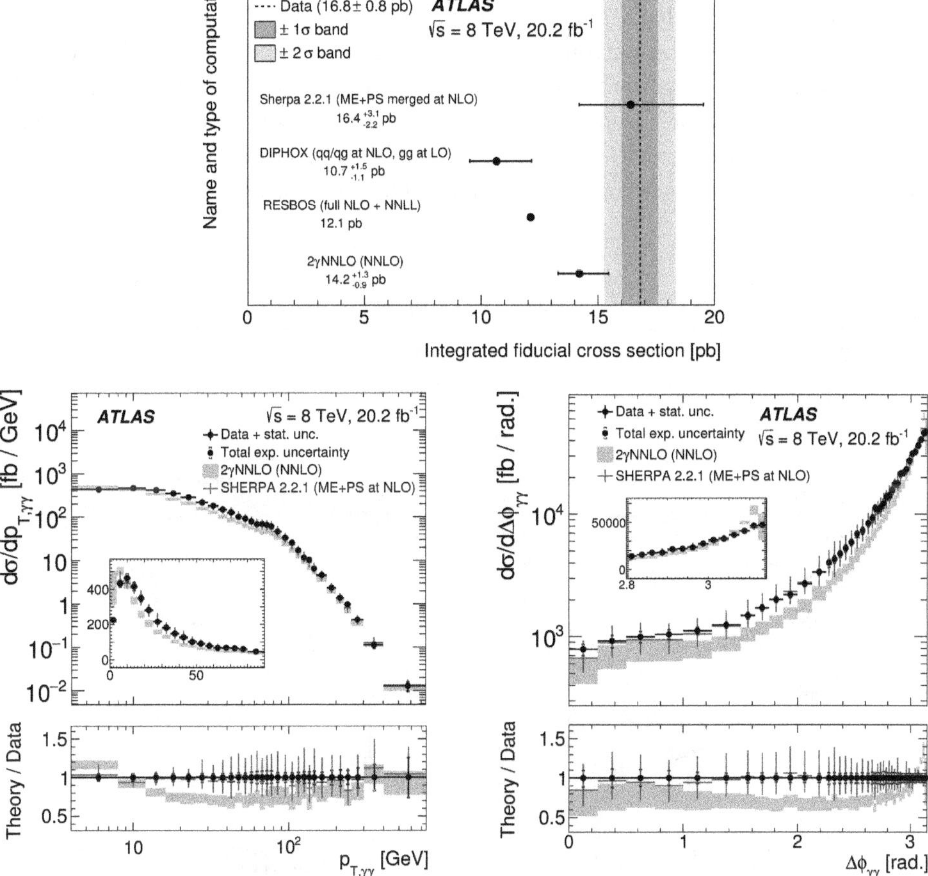

Fig. 17. (top) Measured fiducial cross-section of diphoton production compared to theoretical predictions [21]. (bottom) Differential cross-sections as functions of the diphoton transverse momentum and the angle between the two photons compared to NLO and NNLO predictions [21].

4. Drell-Yan production

A large number of studies are carried out for Drell-Yan lepton-pair production. The LHCb data [22,23] reaching large rapidities complements the ATLAS and CMS kinematic coverage.

The cross-section is measured differentially in the dilepton mass from 15 GeV to more than 2 TeV over nine orders of magnitude in cross-section at the highest energy of $\sqrt{s} = 13$ TeV by CMS and described well by theoretical predictions [24]. These final states offer clean signatures and large statistics. Together with precise detector calibrations, they allow reaching an experimental systematics of about 1%. The uncertainty on the luminosity (2-3%) as well as other contributions can be cancelled out in cross-section ratio measurements.

236

The Drell-Yan data samples thus help understanding the proton structure, probe higher-order pQCD calculations, allow the study of non-perturbative effects like soft gluon resummation and parton shower modelling, test the available Monte Carlo tools. All these are essential both for precision physics and searches at the LHC.

Electroweak vector boson production was reviewed by Q. Lee on the conference. It is, however, important to comment on the huge impact of the precision W and Z data on our knowledge of the proton PDFs. This is illustrated in Fig. 18.

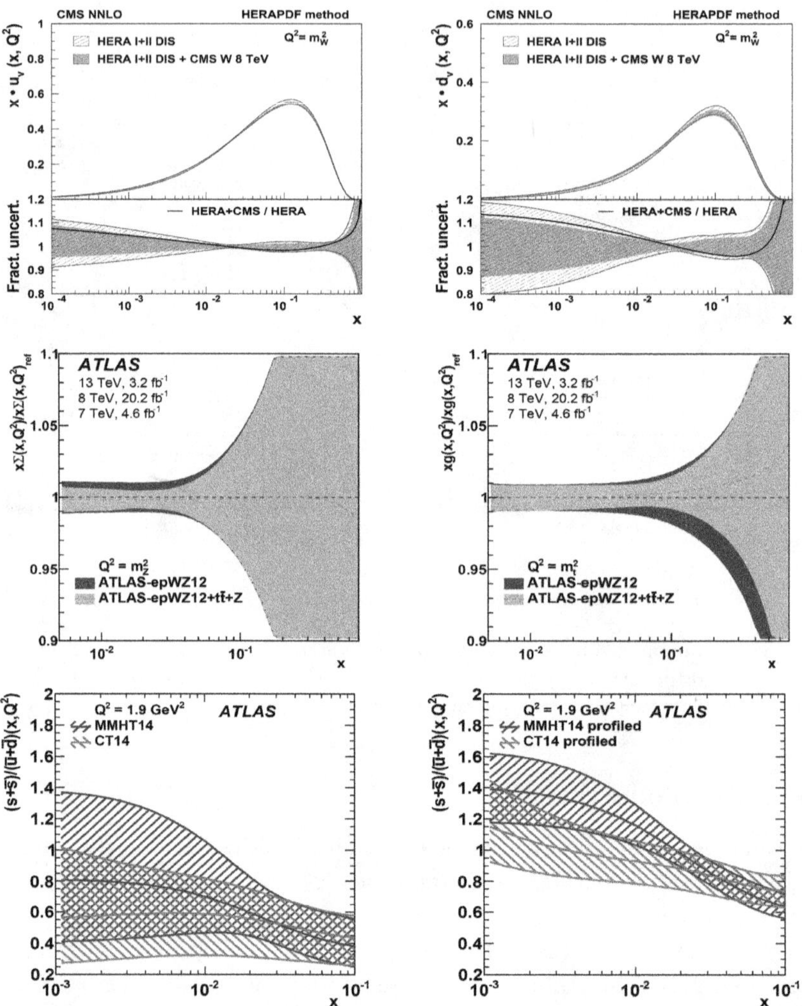

Fig. 18.   (top) Constraints from the CMS $\sqrt{s} = 8$ TeV W boson charge asymmetry measurement on the valence quark PDFs [25]. (middle) Constraints from the ATLAS $t\bar{t}$ and Z measurements on the sea quark and gluon PDFs [26]. (bottom) Constraints from the ATLAS $\sqrt{s} = 7$ TeV Z and W boson precision cross-section measurements on the strange quark distribution [29]. The left plot shows the results before, while the right one after the inclusion of the precision W and Z data.

The W charge asymmetry is measured as a function of the lepton pseudorapidity. It is sensitive to the $u(x)/d(x)$ ratio as the main production mechanisms are $u\bar{d} \to W^+$ and $d\bar{u} \to W^-$. The additional constraints on the valence quark PDFs from the $\sqrt{s} = 8$ TeV CMS measurement[25], when added to the HERA deep inelastic scattering data, are shown in the top plots of Fig. 18.

The measurement of the $t\bar{t}$ to Z production cross-section ratio allows the cancellation of the luminosity uncertainty. The largest challenge is to take correlations between the $t\bar{t}$ and Z measurements precisely into account. Double ratios like $[\sigma_{t\bar{t}(\sqrt{s}_1)}/\sigma_{Z(\sqrt{s}_1)}]/[\sigma_{t\bar{t}(\sqrt{s}_2)}/\sigma_{Z(\sqrt{s}_2)}]$ allow most uncertainties to cancel out. The $\sqrt{s} = 7$ TeV $t\bar{t}$ ATLAS data causes a tension in these measurements[26]. The constraints on the sea quark and gluon distributions are presented in the middle row plots of Fig. 18.

The strange quark distributions can be measured studying W+charm production. While the $\sqrt{s} = 7$ TeV Wc results[27,28] do not show enhanced s-quark content, the global fit to the ATLAS precision W and Z data[29] does, as illustrated on the bottom plots of Fig. 18.

Z boson production in association with jets (Z+jets) is used to study kt-splitting for anti-kt jet reconstruction algorithm with distance parameters R=0.4 and R=1[30]. The distributions of $\sqrt{d_0}$ quantifying the p_T of the leading jet and the distances $\sqrt{d_N}$ ($N > 0$), where the N-jet event resolves to N+1 jets, are studied. At high values of $\sqrt{d}$ the perturbative model is tested, while in the low $\sqrt{d}$ region non-perturbative effects like hadronisation and parton shower modelling, multi-parton interactions also play a role. Fig. 19 shows that at high $\sqrt{d}$, the SHERPA generator (NLO for up to 2 jets and LO for up to 4 jets) performs better using the improved CKKW method for ME-PS matching. At low values, DYNNLO interfaced to Powheg and Pythia8 underestimates the data less, though neither Monte Carlo tool gives a good description. These data can be used to tune the non-perturbative stages of the Monte Carlo models.

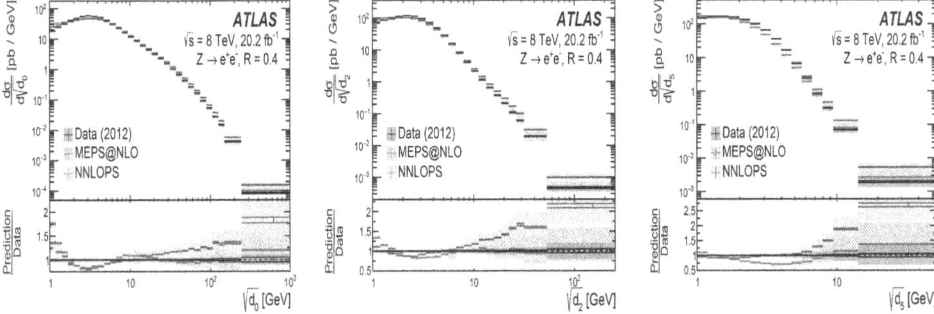

Fig. 19. Charged-only distributions for the splitting scales (left) $\sqrt{d_0}$, (middle) $\sqrt{d_2}$ and (right) $\sqrt{d_5}$ in the $(Z \to e^+e^-)$+jets analysis using the anti-kt jet-radius parameter R=0.4 compared to theoretical predictions from Sherpa (MEPS@NLO) and DYNNLO (NNLOPS)[30].

5. Jet properties

Jet properties, such as particle multiplicities and charge, depend on the type of the fragmenting parton.

Gluon jets are expected to have higher particle multiplicities due to the gluons' higher colour charge. Moreover the particle multiplicity also grows faster for gluon jets with the jet transverse momentum. These properties can be measured on dijet events, as more forward jets typically belong to the partons with higher longitudinal momentum and thus less likely to come from a gluon. ATLAS has measured the charged multiplicity for central and forward jets and extracted the gluon and quark jets' average charged multiplicities as a function of the jet transverse momentum[31] shown in Fig. 20(top). The N^3LO pQCD calculation can not give absolute predictions but describes well the p_T dependence.

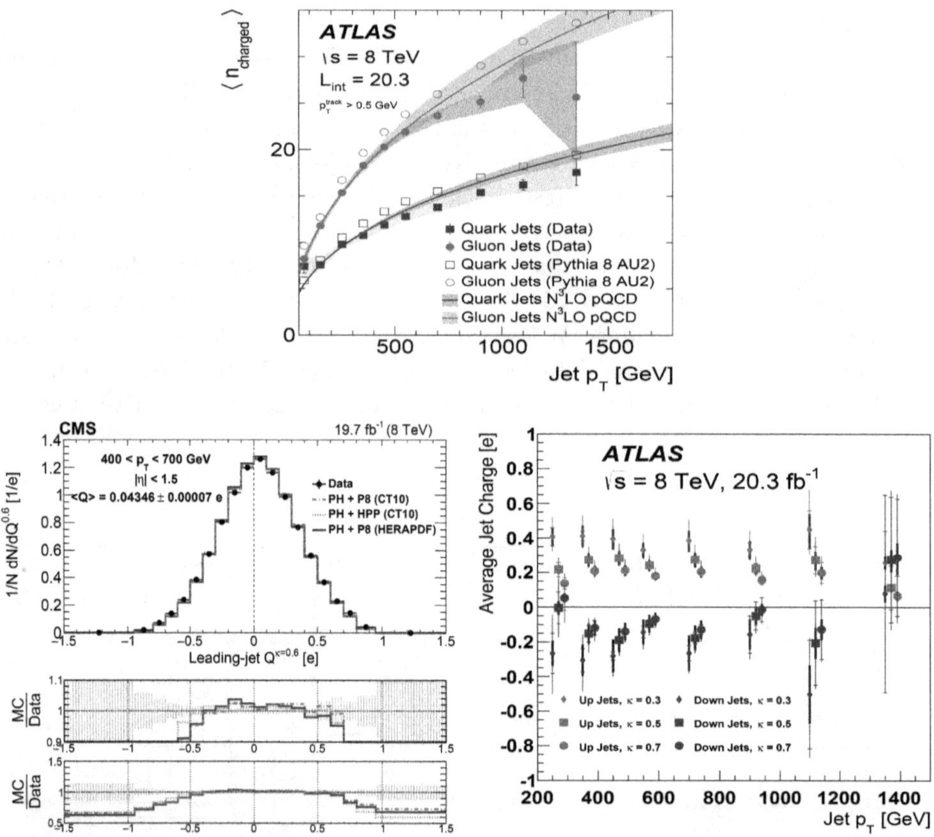

Fig. 20. Jet properties in dijet events. (top) Charged multiplicity of quark and gluon jets as a function of the jet transverse momentum[31]. (bottom left) Jet charge distribution of the leading jet[32]. (bottom right) Average jet charge for up and down quarks as a function of the jet transverse momentum[33].

The jet charge is defined as the momentum-weighted sum of particle charges above a transverse momentum threshold: $Q_J = (p_{T,J})^{-\kappa} \sum_{i:\text{tracks}} q_i \cdot (p_{T,i})^{\kappa}$, where κ is a free regularisation parameter (typically chosen between 0.2 and 1.0). The jet charge is expected to be correlated with the charge of the parton thus could be used to identify the origin of a jet. The average jet charge in inclusive dijet events should increase with the parton center-of-mass energy due to the increasing u valence quark fraction. The jet charge is sensitive to the proton PDFs and the fragmentation modelling, both of these contributing to its p_T dependence.

CMS extracted the jet charge distribution of the leading jet in dijet events (an example is shown in Fig. 20(bottom left)) and compared it to theoretical models[32] using various definitions of the jet charge. ATLAS studied the average jet charge and the jet charge distribution's standard deviation for the more central and more forward jets[33]. Using PDF information, the average jet charge for up and down quarks are extracted as a function of p_T, shown in Fig. 20(bottom right). The data shows that the jet charge slowly decreases with p_T and this decrease increases with κ, as predicted.

6. Summary

As demonstrated in this paper, the ATLAS, CMS and LHCb proton-proton collision data provide a wealth of precise information about the strong interaction, accessing new regions in the phase space. The LHC entered its precision era. In general, good agreement is observed with pQCD theoretical predictions but deviations are present in certain kinematic regions. (N)NLO predictions provide an improved description of the measurements but non-perturbative effects play also a major role and their careful modelling is necessary to understand the observations.

The experimental precision for certain processes (jet, photon+jet, diphoton production) exceeds the theoretical one. Improved (NNLO+NNLL, N^{3}LO) theoretical calculations are desirable for these processes. Certain calculations (like that of the dijet cross-section) show a strong dependence on the choice of the central renormalisation and factorisation scales at NNLO, exceeding the size of the uncertainty bands.

The jet and Drell-Yan data allow placing significant constraints on the proton PDFs. New measurements of the strong coupling constant at high Q are available and give good agreement with the world average.

Acknowledgements

The author wishes to thank for their support the National Research, Development and Innovation Fund, NKFIA of Hungary (research grant no. 124845) and the Lendület (Momentum) Programme of the Hungarian Academy of Sciences (contract no. LP 2015-7/2015).

References

1. ATLAS Collaboration, `https://twiki.cern.ch/twiki/bin/view/AtlasPublic/StandardModelPublicResults` (July 2017 version).
2. W. J. Stirling, `http://www.hep.ph.ic.ac.uk/~wstirlin/plots/plots.html`.
3. CMS Collaboration, JINST 12 (2017) P10003.
4. M. Cacciari, G. P. Salam, and G. Soyez, JHEP 04 (2008) 063
5. CMS Collaboration, *JHEP* **03** (2017) 156.
6. ATLAS Collaboration, *JHEP* **09** (2017) 020.
7. C. Patrignani and others (Particle Data Group), Review of Particle Physics, *Chin. Phys. C* **40** (2016) 100001.
8. CMS Collaboration, *Eur. Phys. J. C* **76** (2016) 451
9. ATLAS Collaboration, ATLAS-CONF-2017-048 (July 2017), `https://cds.cern.ch/record/2273864/`.
10. CMS Collaboration, *Eur. Phys. J. C* **77** (2017) 746.
11. CMS Collaboration, CMS PAS SMP-16-010 (July 2017), `https://cds.cern.ch/record/2273393`.
12. A. J. Larkoski, S. Marzani, G. Soyez, and J. Thale, *JHEP* 05 (2014) 146.
13. C. Frye, A. J. Larkoski, M. D. Schwartz, and K. Yan, *JHEP* 07 (2016) 064; S. Marzani, L. Schunk, and G. Soyez, *JHEP* 07 (2017) 132.
14. CMS Collaboration, CMS-PAS-SMP-16-014 (March 2017), `http://cds.cern.ch/record/2257685`.
15. ATLAS Collaboration, *Eur. Phys. J. C* **76** (2016) 670.
16. CMS Collaboration, CMS-PAS-SMP-16-008 (February 2017), `https://cds.cern.ch/record/2253091`.
17. ATLAS Collaboration, *Eur. Phys. J. C* **77** (2017) 872.
18. ATLAS Collaboration, *Phys. Lett. B* **770** (2017) 473.
19. ATLAS Collaboration, ATLAS-CONF-2017-059 (July 2017), `https://cds.cern.ch/record/2273875`.
20. ATLAS Collaboration, *Nucl. Phys. B* **918** (2017) 257.
21. ATLAS Collaboration, *Phys. Rev. D* **95** (2017) 112005.
22. LHCb Collaboration, *JHEP* **5** (2016) 001.
23. LHCb Collaboration, *JHEP* **01** (2016) 155, *JHEP* **10** (2016) 030.
24. CMS Collaboration, CMS-PAS-SMP-16-009 (August 2016), `https://cds.cern.ch/record/2205152`.
25. CMS Collaboration, *Eur. Phys. J. C* **76** (2016) 469.
26. ATLAS Collaboration, *JHEP* **02** (2017) 117.
27. ATLAS Collaboration, *JHEP* **02** (2014) 013.
28. CMS Collaboration, *JHEP* **05** (2014) 068.
29. ATLAS Collaboration, *Eur. Phys. J. C* **77** (2017) 367.
30. ATLAS Collaboration, *JHEP* **08** (2017) 026.
31. ATLAS Collaboration, *Eur. Phys. J. C* **76** (2016) 1.
32. CMS Collaboration, *JHEP* **10** (2017) 131.
33. ATLAS Collaboration, *Phys. Rev. D* **93** (2016) 052003.

Soft QCD Measurements at LHC

M. Taševský

on behalf of the ALICE, ATLAS, CMS,
LHCb, LHCf and TOTEM Collaborations

*Institute of Physics of the Czech Academy of Sciences,
Na Slovance 2, Prague, 18221, Czech Republic
E-mail: Marek.Tasevsky@cern.ch*

Results from recent soft QCD measurements by LHC experiments ALICE, ATLAS, CMS, LHCb, LHCf and TOTEM are reported. The measurements include total, elastic and inelastic cross sections, inclusive and identified particle spectra, underlying event and hadronic chains. Results from particle correlations in all three collision systems, namely pp, pPb and $PbPb$, exhibit unexpected similarities.

Keywords: Soft QCD; total cross section; inclusive particle spectra; identified particle spectra; underlying event; particle correlations; hadronization.

1. Introduction

Soft Quantum Chromodynamics (QCD) physics is a domain of particle physics which is characterized by a low momentum transfer, typically a low transverse momentum, p_T. It is usually used to describe that part of the scattering which dominates at soft scales and where perturbative QCD cannot be applied. One example of a process which is entirely governed by soft QCD physics is the process of hadronization. Since there is no uniform description of the phenomena that occur at low p_T, there is a variety of models trying to explain them through comparisons with extracted data. There is a wealth of LHC measurements that probe the soft QCD region — basically all LHC experiments measure soft QCD phenomena. This text reports on results from experiments ALICE[1], ATLAS[2], CMS[3], LHCb[4], LHCf[5] and TOTEM[6] and tries to select those which are recent and illustrative at the same time. We will discuss measurements of inclusive total cross sections, inclusive and identified particle spectra, underlying event, particle correlations and it will also be shown that there are surprising similarities between results from all three collision system: pp, pPb and $PbPb$ collisions. The models which will be occasionally mentioned are based on multi-parton interactions (MPI), color reconnections (CR), hadronization and hydrodynamical laws or gluon saturation in the proton. We will also see that there are very interesting links between three big domains of particle physics, namely particle collisions, heavy-ion collisions and cosmic rays. They used to be studied separately in the last decades but it turns out that a wise synergy pays off.

242

2. Total inclusive cross sections

The total cross section is an important ingredient to estimate the number of pile-up events at the LHC but also to model interactions in cosmic rays. The amount of pile-up events, occurring usually at a soft scale, increases with increasing instantaneous luminosity, and may amount to several tens for values around the nominal luminosity of 10^{34}cm^{-2}s^{-1}. Total, elastic as well as inelastic cross sections can be measured using special forward proton detectors which are placed very far from the interaction point and very close to the beam since their aim is to detect a forward-going proton which scatters under a very small polar angle. The experiment dedicated for such measurements is called TOTEM and the special forward proton detector at the ATLAS side is called ALFA. They are placed more than 200 m from the interaction point and for the purpose of measuring total and elastic cross sections (which are large and dominate at low t values, where t is a four-momentum transfer squared) we use special LHC optics (characterized mainly by the betatron function, β^*) and inject only a few bunches with low proton intensity. This leads to a small number of proton-proton interactions per bunch crossing (often termed "low pile-up"). The larger β^*, the lower t values can be reached. Figure 1 left shows

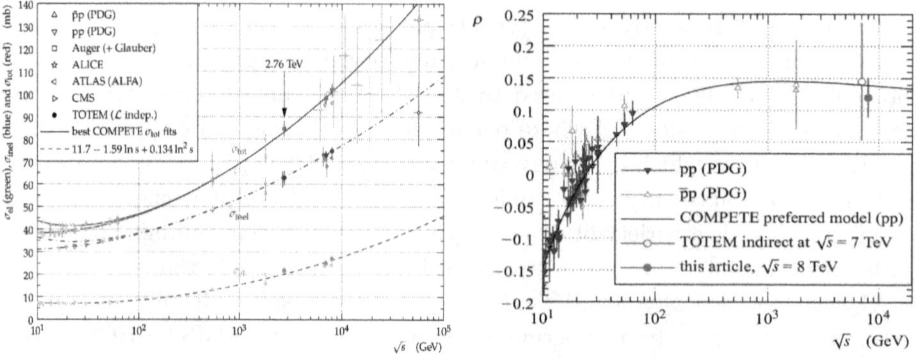

Fig. 1. (Left) Compilation of the total, inelastic and elastic cross-section measurements. The continuous black lines (lower for pp, upper for $\bar{p}$p) represent the best fits of the total cross-section data by the COMPETE collaboration[7]. The dashed line results from a fit of the elastic scattering data. The dash-dotted lines refer to the inelastic cross section and are obtained as the difference between the continuous and dashed fits. (Right) The energy dependence of the ρ parameter. The hollow red circle stands for the earlier indirect determination by TOTEM[8]. The filled red circle represents the result from TOTEM[9]. The black curve gives the preferred pp model by COMPETE[7], obtained without using LHC data (color online).

a compilation of all total, inelastic and elastic cross section measurements so far together with a preliminary point from TOTEM[10] for the collision energy, $\sqrt{s}$, of 2.76 TeV. The new ATLAS results for the 8 TeV point[11] are also included. The inelastic cross section was recently measured by ATLAS at 13 TeV with the central detector only[12] and a good consistency was reached with the results based on the special forward proton detectors as well as with predictions of PYTHIA 8[13], EPOS LHC[14] and QGSJET-II[15]. An interesting feature of the elastic cross section measurement was documented by TOTEM in the 8 TeV measurement[16]: in the region where Coulomb interactions can be safely neglected, the exponential form of the t-slope was excluded at the level of 7.2 σ. Similar non-exponential t-slopes were also observed at 7 and 13 TeV measurements. The explanation of this non-exponentiality is still lively discussed among theorists. If t values of the order of 10^{-4} can be reached, one can then study the so called Coulomb-nuclear interference region which then enables us to measure the ρ parameter which is the ratio of real to imaginary part of the forward amplitude. Figure 1 right shows the $\sqrt{s}$ dependence of the ρ parameter. A direct measurement of the ρ parameter by TOTEM at 8 TeV ($\beta^\star = 1$ km and $t \sim 10^{-4}$)[9] agrees within experimental uncertainties with the COMPETE fit, and another direct measurement, namely at 13 TeV and $\beta^\star = 2.5$ km, will appear soon and thus will provide a valuable level-arm for the energy dependence.

3. Inclusive charged particle spectra in pp

Measurements of inclusive and identified particle spectra belong to basic items in the physics programs of high-energy experiments. They are usually measured regularly at each collision energy. The multiplicity of charged particles is one of the key characteristics of high-energy hadron collisions and has been the subject of many experimental and theoretical studies because although quite simple to measure, it is quite difficult to describe it in the full measured range. Measurements of charged particle distributions probe the non-perturbative region of QCD where QCD-inspired models implemented in MC event generators are used to describe the data. Measurements are used to constrain free parameters of these models. Accurate description of low-energy strong interaction processes is essential for simulating single pp as well as multiple pp interactions in the same bunch crossing at higher instantaneous luminosities. Such pp measurements are also used as input in many models trying to describe heavy-ion results. The ALICE analysis[17] presents a comprehensive set of measurements of pseudorapidity density and multiplicity distributions in pp collisions over the LHC energy range from 0.9 up to 8 TeV, in 5 energy points. Three event selections are used, namely INEL which means all inelastic events, then INEL > 0 which means events with at least 1 charged particle in the $|\eta| < 1$ range, and Non-Single-Diffractive events. Figure 2 shows the multiplicity distributions at 0.9 TeV (left) and 7 TeV (middle). All models (EPOS LHC, PHOJET, PYTHIA 6 and PYTHIA 8) show big troubles in describing the whole spectrum in data but the best agreement is achieved with EPOS. Similar observations are made for the remaining $\sqrt{s}$ points. It is also very useful to measure the energy dependence of the charged particle multiplicity at mid-rapidity, i.e. for $|\eta| < 0.5$, since it is related to the average energy density in the interaction of protons and it gives a reference for heavy-ion collisions. Alternatively, we can calculate

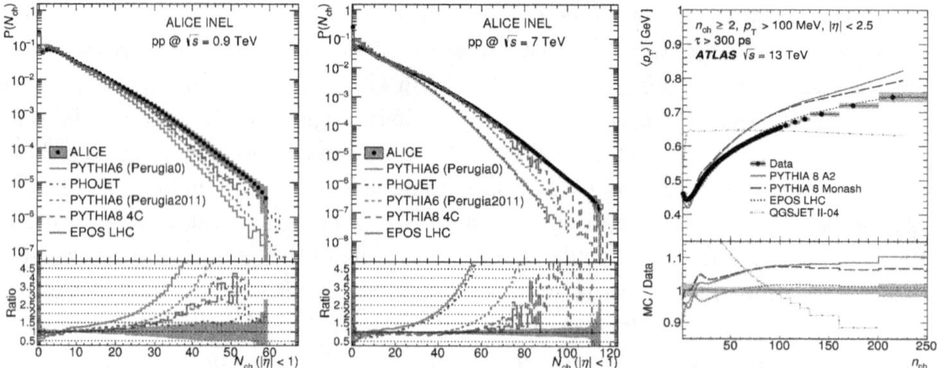

Fig. 2. Primary charged-particle multiplicities measured by: ALICE at 0.9 TeV [17] (left), 7 TeV [17] (middle) for the INEL event class in the pseudorapidity range $|\eta| < 1.0$ and (right) by ATLAS at 13 TeV [18] as a function of the mean p_T for events with at least two primary charged particles with $p_T > 100$ MeV and $|\eta| < 2.5$, each with a lifetime $\tau > 300$ ps. The black dots represent the data and the coloured curves the different MC model predictions. The vertical bars represent the statistical uncertainties, while the shaded areas show statistical and systematic uncertainties added in quadrature. The lower panel in each figure shows the ratio of the data to MC simulation (left and middle) or vice versa (right) (color online).

normalized q-moments (C_q) and measure their energy dependence. KNO (Koba, Nielsen and Olesen) scaling then states that C_q stays energy-independent. Since it was reported in the past as dependent on the event selection and η range, ALICE came up with a comprehensive analysis covering three event classes and three η ranges. The conclusion of the measurement is that the KNO scaling violation increases with increasing $\sqrt{s}$ and at a given $\sqrt{s}$, with increasing η interval.

A similar study, now at 13 TeV energy, has recently been performed and published by ATLAS [18]. Here only charged particles with lifetime smaller than 30 ps and larger than 300 ps are used since those with lifetime larger than 30 ps are usually strange baryons which have a low reconstruction efficiency. For comparison to previous measurements which used also these strange baryons, an extrapolation is used to include these particles. The data are compared to PYTHIA 8, EPOS LHC and QGSJET-II. PYTHIA 8 and EPOS include the effects of color coherence which is important in dense parton environments and effectively reduces the number of particles produced in MPI. PYTHIA 8 splits the generation into diffractive and non-diffractive processes, the latter dominated by t-channel gluon exchange, the former is described by Pomeron-based approach. EPOS implements a parton-based Gribov-Regge theory, effective field theory describing both hard and soft scattering at the same time. QGSJET-II is based on Reggeon-field theory framework. EPOS and QGSJET-II do not rely on parton density function (PDF). Averaged p_T as a function of multiplicity increases as modeled by a colour reconnection mechanism in PYTHIA 8 and by the hydrodynamical evolution model in EPOS. QGSJET-II model which has no model for colour coherence effects describes the data poorly.

This analysis also reported that multiplicity distribution is not described perfectly by any of the models, there are large discrepancies especially at large multiplicities. Having observed similar discrepancies at all measured energies, we conclude that for every collision energy, model parameters usually need to be re-tuned in every MC generator. Unlike the multiplicity distributions, the mean particle multiplicity at mid-rapidity ($|\eta| < 0.2$) measured at several $\sqrt{s}$ points was found to be well described by PYTHIA 8 Monash and EPOS models for three event selections.

3.1. *Forward energy flow of charged particles in pp*

Measurements of the forward energy flow of charged and neutral particles serve to tune two types of MC models: i) those used at hadron colliders where the measurements are used to tune MPI and other soft characteristics, ii) the measurements have also an impact on the total number of muons in the extensive air showers at the ground whose measurements are still not well-described by models for cosmic rays. The CMS measurement described in [19] is based on using a forward calorime-

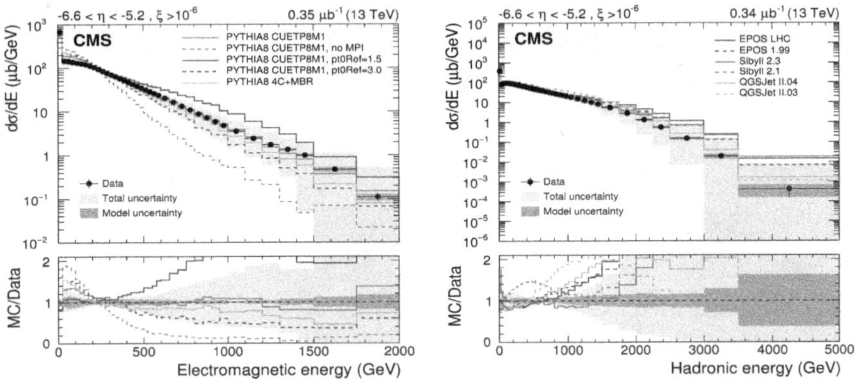

Fig. 3.   (Left) Differential cross section as a function of the electromagnetic energy in the region $-6.6 < \eta < -5.2$ for events with $\xi > 10^{-6}$ (xi is fractional momentum loss of the incident proton). The panel shows the data compared to different PYTHIA 8 tunes. (Right) Differential cross section as a function of the hadronic energy in the region $-6.6 < \eta < -5.2$ for events with $\xi > 10^{-6}$. The panel shows the data compared to MC event generators mostly developed for cosmic ray induced air showers. Plots taken from Ref. 19.

ter at negative rapidities only, called CASTOR which has an electromagnetic and hadronic part, so is able to distinguish electromagnetic particles (which are mostly electrons and photons from π^0 decays) and hadrons (mostly $\pi^{\pm}$). The data were taken at a very low instantaneous luminosity to suppress the pile-up. In Fig. 3 left, the energy spectra of electromagnetic particles are compared to predictions of models used to describe the multihadron production at hadron colliders and from the comparisons to various MPI tunings, we conclude that the data are very sensitive to the MPI modeling. The right side of Fig. 3 documents huge deficiencies in describing the spectrum of hadrons by all models used to model cosmic rays (EPOS, SYBILL and QGSJET II).

3.2. *Forward energy flow of neutral particles in pp*

The very forward energy flow of neutral particles is measured by a set of hodoscopes, in a dedicated experiment LHCf located at 140 m downstream of the ATLAS detector. The aim of such measurements is to improve hadronic interaction models, especially used in generators for cosmic rays. For example, modeling of X_{max} (the position of shower maximum) needs the total cross section for the collision of proton with air, but also the spectra of identified particles going very forward. And modeling of the hadronic interactions in those generators is based on correlations between spectra of particles going central versus those going forward. This is perfectly possible since LHCf uses the same interaction point as the ATLAS detector which measures very precisely the central production. A first common analysis of the same event sample analyzed by ATLAS and LHCf is in preparation. Figure 4

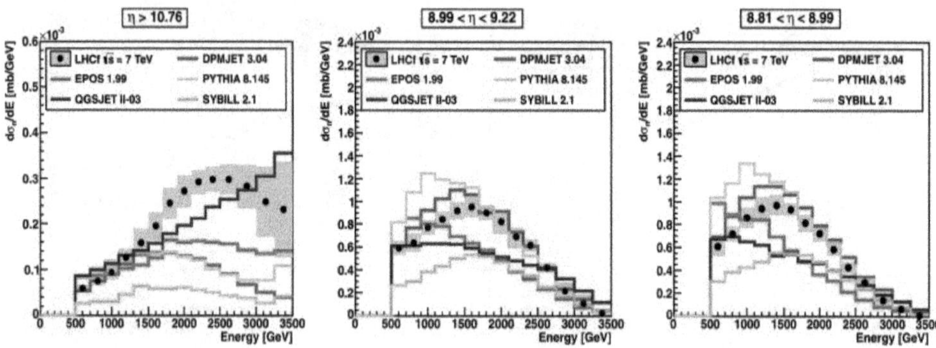

Fig. 4. A comparison of the LHCf neutron energy spectra measured in 7 TeV pp collisions[20] with model predictions. The black markers and gray shaded areas show the corrected data and the systematic uncertainties, respectively.

shows the energy spectra of neutrons from the 7 TeV pp collisions[20] where large deficiencies in all models used for modeling cosmic rays (EPOS, QGSJET, DPMJET[21] and SIBYLL[22] and also PYTHIA 8) are observed. The p_T spectra of π^0 produced in 7 TeV pp collisions[23] in a fine η binning are quite well described by EPOS, less well described by DPMJET. Recently, photon energy spectra have been measured in 13 TeV pp collisions[24] and again all models are observed to have difficulties in describing the whole measured spectrum.

4. Identified particle spectra in *pp*, *pPb* and *PbPb*

In order to identify particle species, each experiment has sophisticated identification procedures usually based on the ionization energy loss, dE/dx, or other techniques.

In the ALICE analysis[25], strange hadrons with different strangeness content are identified and a ratio of multiplicity of strange hadrons to pions is plotted in Fig. 5 left as a function of multiplicity density in the center of the detector for all three collision systems: going from relatively low multiplicity densities occurring in pp collisions over middle densities seen in pPb collisions up to large densities corresponding to $PbPb$ collisions. We observe that the integrated yields of strange and multi-strange particles relative to pions, increase significantly with the charged particle multiplicity and that the yield ratios measured in pp collisions agree with pPb measurements at similar multiplicities. This is a first observation of a strangeness

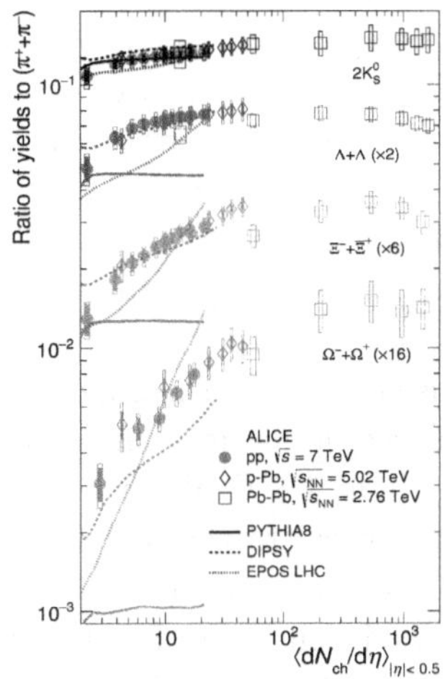
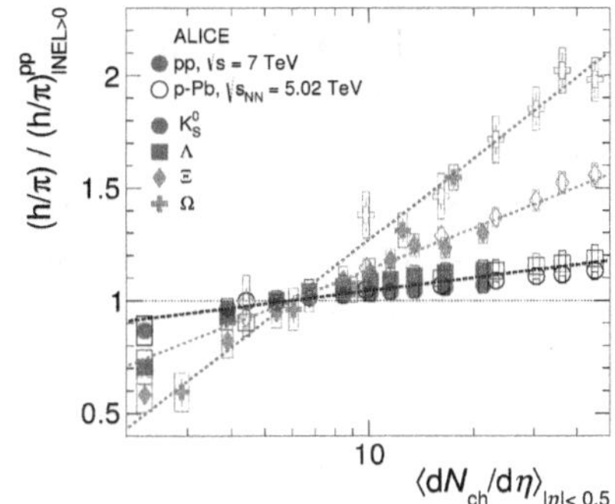

Fig. 5. (Left) The p_T integrated yield ratios to pions as a function of $\langle dn_{ch}/d\eta \rangle$ measured in $|\eta| < 0.5$. The error bars show the statistical uncertainty, whereas the empty and dark-shaded boxes show the total systematic uncertainty and the contribution uncorrelated across multiplicity bins, respectively. The values are compared to calculations from MC models[13,14,27] and to previous results obtained by ALICE in pPb and $PbPb$ collisions. (Right) Particle yield ratios to pions normalized to the values measured in the inclusive pp sample. The error bars show the statistical uncertainty. The common systematic uncertainties cancel in the double-ratio. The empty boxes represent the remaining uncorrelated uncertainties. The lines represent a simultaneous fit of the results with an empirical scaling formula. Plots taken from Ref. 25.

enhancement in high-multiplicity pp collisions. In very high-multiplicity pPb collisions, the strangeness production reaches values similar to those observed in $PbPb$ collisions, where a quark-gluon plasma (QGP) is formed. QGP is matter in a phase of deconfined quarks and gluons which is created at sufficiently high temperatures and energy densities which is usually reached in collisions of heavy ions with high energies. Strangeness enhancement has been proposed as one of the main signatures of formation of QGP[26]. We see that none of the three models is able to describe the data but DIPSY[27], a special model using color ropes, describes data best. In the analysis[25], the ratio of baryon to meson yields is also studied and it is observed that none of the models is able to describe the data. In Fig. 5 right, the strangeness enhancement with respect to inclusive samples is plotted: here we see a clear strangeness hierarchy, namely the slope increasing with increasing strangeness content, and in addition, the same hierarchy as measured already for pPb data is observed. We conclude that mass and multiplicity dependencies of the strangeness enhancement as well as of the spectral shapes (not shown here, see[25]) remind the patterns seen for pPb and $PbPb$ collisions which can be understood assuming a collective expansion of the system in the final state.

Absolute yields and p_T spectra of identified particles in hadron-hadron collision are usually used to improve the modeling of various key ingredients of MC event generators, such as MPI, parton hadronization, and final state effects such as parton correlations in color, p_T, spin, baryon and strangeness number, and collective flow. Parton hadronization and final state effects are mostly constrained from e^+e^- data whose final states are dominated by simple $q\bar{q}$ states, whereas low p_T hadrons at LHC come from fragmentation of multiple gluons, so called minijets. This is also the reason why the production of baryons and strange hadrons in pp collisions is not well reproduced by current generators, and hence makes a good motivation for this study. In addition, identified particle spectra serve as an important reference for high-energy HI studies. This CMS study[28] used data with negligible pile-up. Pions, kaons and protons are identified using dE/dx, a specific ionization which works very well for momenta lower than 1.2, 1.1 and 1.7 GeV for pions, kaons and protons, respectively. All particle species are limited to the region $|\eta| < 1.0$. And because transverse momenta as low as 0.1 GeV are measured, special tracking algorithms were used with high reconstruction efficiency and low background. These algorithms feature special track seeding and cleaning, hit cluster shape

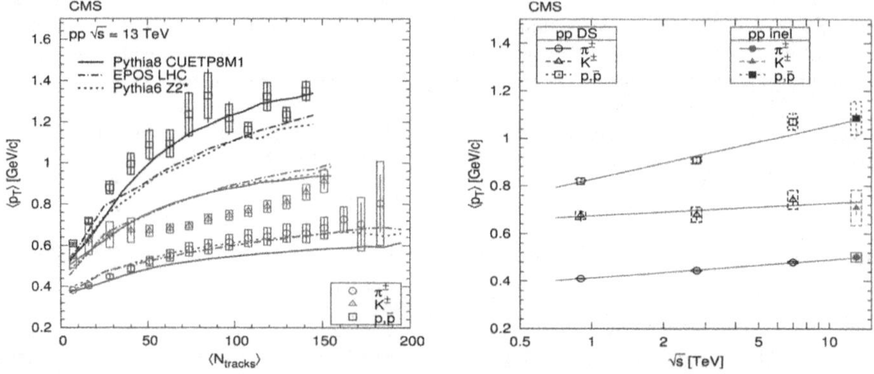

Fig. 6.   Average p_T of charged-averaged pions, kaons and protons in the range $|\eta| < 1$ as functions: (left) of the corrected track multiplicity for $|\eta| < 2.4$, computed assuming a Tsallis-Pareto distribution in the unmeasured range. Curves indicate predictions from PYTHIA 8, EPOS and PYTHIA 6; (right) of $\sqrt{s}$. The curves show linear fit in $\ln s$. Error bars indicate the uncorrelated combined uncertainties, while boxes show the uncorrelated systematic uncertainties. Plots taken from Ref. 28.

filtering, modified trajectory propagation and track quality requirements. Figure 6 left shows the average p_T as a function of multiplicity in the event. We can notice that for kaons, all models overshoot the data. While the low multiplicity region is well modeled by the generators, the high multiplicity region needs some tuning of baryon and/or strangeness production. Figure 6 right shows a rising evolution of the average p_T with $\sqrt{s}$. This collision energy evolution of the average p_T provides useful information on the so called saturation scale of the gluons in proton.

5. Underlying event in pp

The hard scattering is accompanied by interactions of a soft nature, namely those coming from the rest of the proton-proton collision. They can come from the initial state radiation (ISR), final state radiation (FSR), MPI and from color reconnections (CR, namely from the QCD evolution of colour reconnections between the hard scatter and beam remnants). A combination of contributions from all these processes is called underlying event (UE) which is important to consider and measure (or estimate) since:

- These processes can not be completely described by perturbative QCD, and require phenomenological models, whose parameters are tuned by means of fits to data.
- Final state (or its part) coming from UE can mimic a signal final state, for example the same-sign WW production from MPI can mimic final state of the same-sign dilepton SUSY searches.
- It can affect isolation criteria applied to photons and charged leptons.
- It can affect the vertex reconstruction efficiency. For example the primary vertex in the process $H \to \gamma\gamma$ can be partly determined from the charged particles originating from UE.

An usual procedure of estimating the amount of UE is spatially dividing tracks in each event according to their azimuthal angle to the Towards region (where the highest p_T jet points), the Away region (where the second highest p_T jet points) and then we have two Transverse regions where one of them contains imprints of ISR and FSR and the other one has the least activity (so called Transverse-min) — this one is believed to be the most sensitive to UE. The usual observables are average track multiplicity per unit area and average scalar sum of track p_T per unit area.

Figures 7 left and middle concentrate on the UE-dominated region studied in ATLAS events containing at least one charged particle with $p_T > 1$ GeV[29]. The left plot shows the multiplicity evolution of the average p_T, it can also be seen as a correlation between two "soft" properties. It shows a balance between p_T sum and multiplicity. This balance is affected in some models by CR which typically increases the p_T per particle. The description by all models is within 5%. EPOS gives in general best description at low p_T of the leading particle, but is the worst

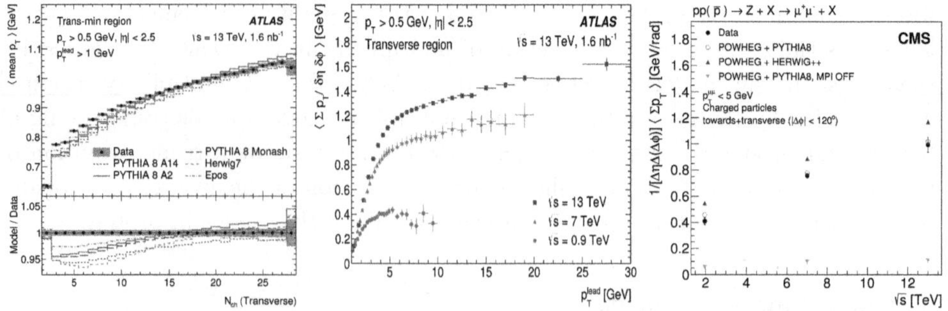

Fig. 7. (Left) Mean charged-particle average p_T as a function of charged-particle multiplicity in the transverse-min azimuthal region[29]. The error bars on data points represent statistical uncertainty and the blue band the total combined statistical and systematic uncertainty. (Middle) Mean sum p_T densities as a function of transverse momentum of the leading charged particle measured for $\sqrt{s} = 0.9$, 7 TeV[30] and 13 TeV[29]. (Right) Sum p_T density, with $p_{\mu\mu} < 5$ GeV as a function of $\sqrt{s}$ for data[31] and predictions from simulations by POWHEG+PYTHIA 8 and POWHEG+HERWIG++. The predictions of POWHEG+PYTHIA 8 without MPI are also shown. Error bars represent the statistical and systematic uncertainties added in quadrature (color online).

when the p_T of the leading particle is above 10 GeV (not shown here). The middle plot shows the evolution of the p_T sum density with p_T of the leading particle, for three collision energies. The initial rapid rise up to $p_T^{lead} \approx 5$ GeV after which the density stabilizes around one charged particle or 1 GeV per unit $\phi - \eta$ area is known as the "pedestal effect". It reflects a reduction of the pp impact parameter with increasing p_T^{lead} and hence the transition between the minimum bias and hard scattering regimes. The right plot shows a similar trend, namely the evolution of p_T average per event with increasing collision energy. It is a CMS[31] analysis based on Drell-Yan events selected using a detection of a dimuon pair. From a comparison to various models and mainly to PYTHIA with and without MPI, we can clearly see the importance of MPI in modeling the UE activity.

6. 2-particle azimuthal correlations in pp, pPb and $PbPb$

Soft processes are also in the heart of heavy-ion collisions[32]. In the HI collisions at LHC, it is believed that quark-gluon plasma is created. The conditions for the phase transition are created, the system gets close to the thermal equilibrium and expands collectively. The expansion means that the matter cools down and hadrons are formed. The aim is to measure macroscopic properties of the QGP and study its microscopic laws. Usually $PbPb$ collisions serve to create and study QGP, the pPb collisions serve as a sort of control experiment (cold nuclear matter effects (e.g. modifications to PDF)) and the pp collisions serve as reference data. But recently, striking similarities were observed between these three collision systems. Phenomena considered fundamental for QGP are now seen also in pPb and even *pp*. These were discovered in high-multiplicity events but they may be relevant also for minimum bias events which would, in turn, have important consequences for all hadronic collisions. We speak especially about the ridge observation. The ridge was first observed in central HI collisions as 2-particle long-range correlations ($\Delta\eta > 2$) on a near side ($\Delta\phi \approx 0$) and it is believed to be a result of collective hydrodynamic expansion of hot and dense nuclear matter created in the overlap region. The 2-particle correlations exhibit also the so-called away-side peak coming from the other lower-energy jet. The ridge is usually described by Fourier decomposition using a term $\cos(n\Delta\phi)v_n$ where v_n is a single-particle anisotropy harmonics. Unexpectedly ridge structures were also measured in pPb and even in pp at high multiplicities. Currently the origin of the ridges in these small collision systems is lively debated. Is it hydrodynamics like in QGP which would mean a final state effect? Or is it due to initial state fluctuations (as embedded in Color Glass Condensate (CGC) model[33] using gluon saturation)? Or perhaps these long range correlations come from hadronization described by ropes[27]? Or from collisions of thin flux tubes[34]? Definitely the ridge is a testing ground to study complementarity between dynamical and hydrodynamical models.

Particle correlations are a very powerful tool to study properties of multihadron production in general. Several production mechanisms can be studied simultaneously. The baseline mechanism underlying all correlations is a global conservation of momentum and energy as well strangeness, baryon number and electric charge. Other phenomena, including mini-jets, elliptic flow, Bose-Einstein correlations (BEC) and resonance decays are source of additional correlations and all those sum up.

By studying the $\Delta\phi$ projections of 2-particle correlations around the near-side and away-side peak, it was observed e.g. by LHCb[35] using 5.02 TeV pPb data and by ALICE[36] using 2.76 TeV $PbPb$ data that both peaks increase with multiplicity and that the size of the near-side peak is maximal for particle with $1 < p_T < 2$ GeV. To study the long-range correlations, however, we have to subtract 2-particle correlations coming from the so called "non-flow" which includes resonance decays and dijets. There are several methods to do that, one class of methods tries to subtract the non-flow using low-multiplicity events. It also turns out that the extraction of collective flow in pp collisions strongly depends on the event selection and also on the purity of the non-flow extraction as documented e.g. in the analysis[37].

Figure 8 left shows an almost linear increase of 2-particle correlations with event multiplicity for all three collision systems measured by CMS[38]. The size of these correlations is biggest for HI collisions, while it is smallest for pp collisions. Figure 8 right then shows a measurement by ATLAS[39] of the v_2 quantity, the elliptic flow harmonics, again for all three collision systems now as a function of multiplicity. The measurement shows that v_2 from pPb is smaller than that extracted from the $PbPb$ collisions but originally much smaller values from pPb collisions were expected reflecting the big difference in sizes of the $PbPb$ and pPb systems.

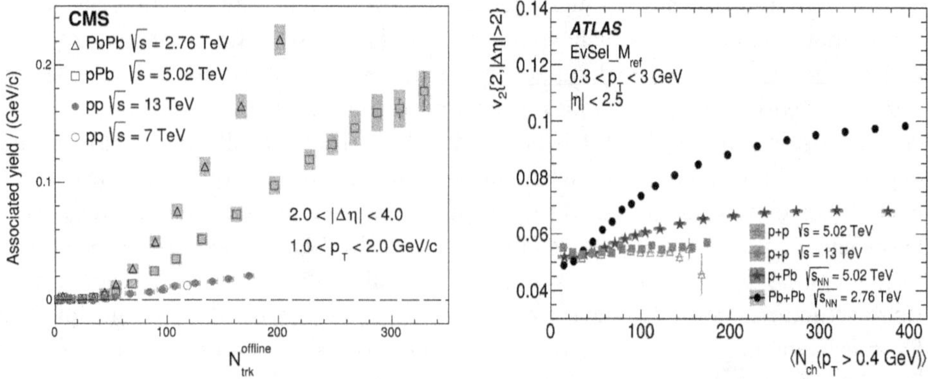

Fig. 8. (Left) Associated yield of long-range near-side two-particle correlations for $1 < p_T < 2$ GeV in pp collisions at $\sqrt{s} = 13$ and 7 TeV, pPb collisions at $\sqrt{s} = 5.02$ TeV, and $PbPb$ collisions at $\sqrt{s} = 2.76$ TeV. The error bars correspond to the statistical uncertainties, while the shaded areas denote the systematic uncertainties. Plot taken from Ref. 38. (Right) Comparison of $v_2\{2, |\Delta\eta| > 2\}$ as a function of $\langle N_{ch}\rangle(p_T > 0.4$ GeV$)$ for pp collisions at $\sqrt{s} = 5.02$ and 13 TeV, p+Pb collisions at $\sqrt{s} = 5.02$ TeV and low-multiplicity Pb+Pb collisions at $\sqrt{s} = 2.76$ TeV, and for reference particles with $0.3 < p_T < 3$ GeV. The error bars and shaded boxes denote statistical and systematic uncertainties, respectively. Plot taken from Ref. 39.

7. Multi-particle azimuthal correlations in pp, pPb and $PbPb$

In the light of difficulties with the residual non-flow which the 2-particle correlations suffer from, methods based on multi-particle correlations started to be more widely used. It was proved that multi-particle correlations are more robust with respect to the non-flow but it is clear that they are also more statistically demanding. The method often used is to build cumulants $c_n\{2k\}$ (of the n-th order based on $2k$-particle correlations) and calculate flow harmonics $v_n\{2k\}$ from them. A novel method, the three-subevent method, has recently been proposed by ATLAS[37] which turns out to be quite effective in reducing the non-flow and moreover it provides negative values of cumulants $c_2\{4\}$, c_2 calculated from 4-particle correlations, which is required to get positive $v_2\{4\}$. Figure 9 from the multi-particle correlation study by CMS[40] shows v_2 coefficients calculated using 2-, 3-, 4-, 6- and even 8-particle correlations as functions of multiplicity for all three collision systems. First the results say that v_2 from 4-particle correlations are smaller than v_2 from 2-particle correlations in pPb and $PbPb$ collisions — that was expected for long-range correlations. We however observe smaller $v_2\{4\}$ than $v_2\{2\}$ values for the pp system (even larger differences are reported using the three-subevent method in the analysis[37]) and a similarity between $v_2\{4\}$ and $v_2\{6\}$ values in all three collision systems. All these findings suggest again that some collective effects are occurring even in pp collisions.

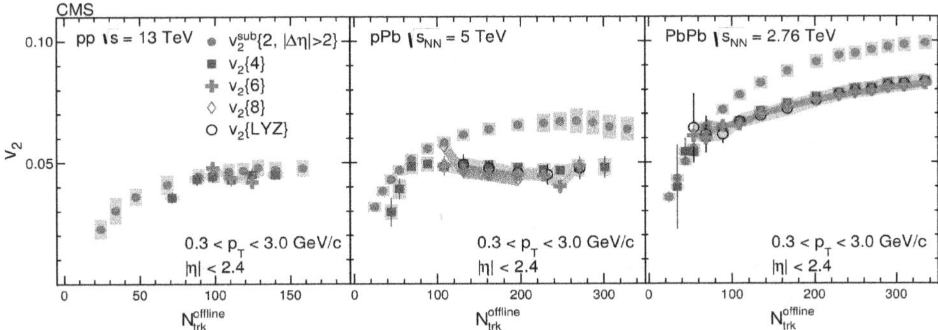

Fig. 9. (Left) The v_2 values calculated from 2-, 4-, 6- and 8-particle correlations as functions of multiplicity of charged particles, averaged over $0.3 < p_T < 3.0$ GeV and $|\eta| < 2.4$, in pp collisions at $\sqrt{s} = 13$ TeV (left), in pPb collisions at $\sqrt{s} = 5.02$ TeV[41] (middle) and in $PbPb$ collisions at $\sqrt{s} = 2.76$ TeV[41] (right). The error bars correspond to the statistical uncertainties, while the shaded areas denote the systematic uncertainties. Plot taken from Ref. 40.

8. Angular correlations of identified particles in pp

The 2-particle correlations can also be studied with identified particles. By choosing specific particle types, we select a specific combination of quantum numbers (strangeness, baryon number) that may manifest in the measured correlations. The correlations should also be sensitive to the details of particle production, including the parton fragmentation. In this ALICE study[42] the near-side peak structure is studied. It is a combination of at least three effects: i) fragmentation of hard-scattered partons, ii) resonance decays and iii) femtoscopic correlations (BEC for identical bosons, Fermi-Dirac anticorrelations for identical fermions, Coulomb and strong final-state interactions). For pairs of same mesons, the near-side peak comes from the minijet mechanism and BEC. The correlations of particle-antiparticle pairs also include a minijet like structure on the near-side as well as on the away-side. For pairs of non-identical particles Bose-Einstein and Fermi-Dirac effects are not present, however, resonances play a significant role. In contrast to same-sign meson correlations, the baryon-baryon (or antibaryon-antibaryon) distributions for identical proton and lambda baryon pairs show a qualitatively different effect, namely a near-side depression instead of the peak. The correlations for same-sign pions and kaons as well as same-sign protons and Λ baryons are shown in Fig. 10, projected onto the $\Delta\phi$ axis and compared with several MC generators. While for mesons (a-b), the description of the apparent near-side and small away side peaks is reasonable, all event generators fail by giving positive correlations for protons and Λ baryons (c-d), where data show a significant depression. It should be noted that these generators conserve the local baryon number and do not include quantum statistical effects such as Fermi-Dirac anticorrelations. Several sources of this depression were studied in this analysis and it was shown that neither Fermi-Dirac anticorrelations nor strong final state effects nor local baryon number conservation

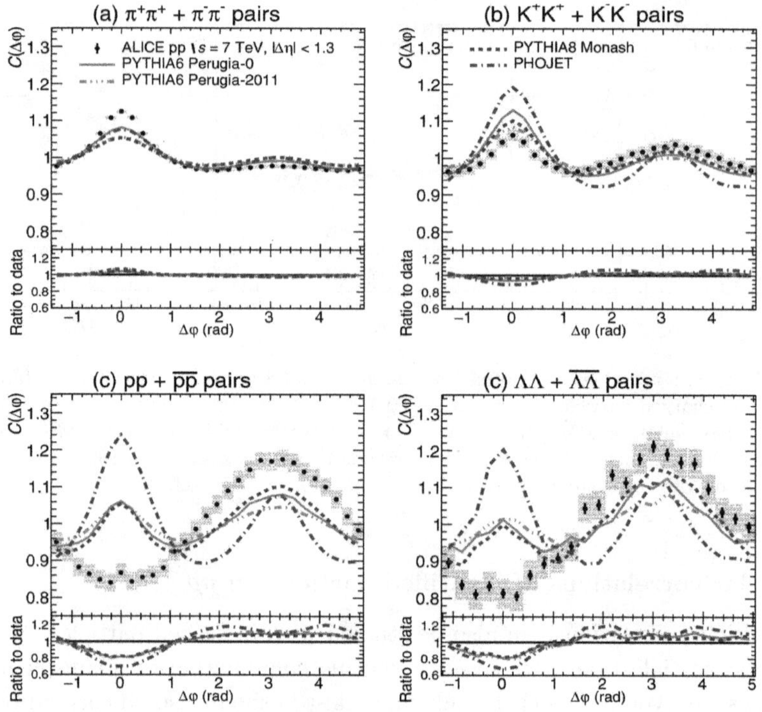

Fig. 10. The $\Delta\eta$ integrated projections of correlation functions for combined pairs of (a) $\pi^+\pi^+ + \pi^-\pi^-$, (b) $K^+K^+ + K^-K^-$, (c) $pp + \bar{p}\bar{p}$ and (d) $\Lambda\Lambda + \bar{\Lambda}\bar{\Lambda}$, obtained from ALICE data and four Monte Carlo models (PYTHIA 6 Perugia-0, PYTHIA 6 Perugia-2011, PYTHIA 8 Monash, PHOJET). Bottom panels show ratios of MC models to ALICE data. Statistical (bars) and systematic (boxes) uncertainties are plotted. Plot taken from Ref. 42.

can explain this depression. So we conclude that something essential is missing in the string fragmentation.

9. Bose-Einstein correlations in pp, pPb and $PbPb$

ATLAS has recently published a BEC study with 7 TeV data based on particles with p_T as low as 100 MeV[43]. BE correlations are defined using c_2, the 2-particle cumulants for identical particles, and are usually plotted as a ratio of same-sign to opposite-sign c_2 cumulants as a function of Q which is the momentum difference of the particles in a pair. This $c_2(Q)$ dependence is then fitted using the function $C_2 = [1 + \Omega(\lambda, R)](1 + \epsilon Q)$ where λ represents a correlation strength and R represents a size of the correlation source. The size of the source has been measured as a function of event multiplicity and particle p_T. The multiplicity dependence represented in Fig. 11 left shows an interesting feature, namely a rise and then a sudden stop and a saturation from multiplicities of about 50. This saturation at high multiplicities is seen for the first time at all. The p_T dependence of R was measured to decrease

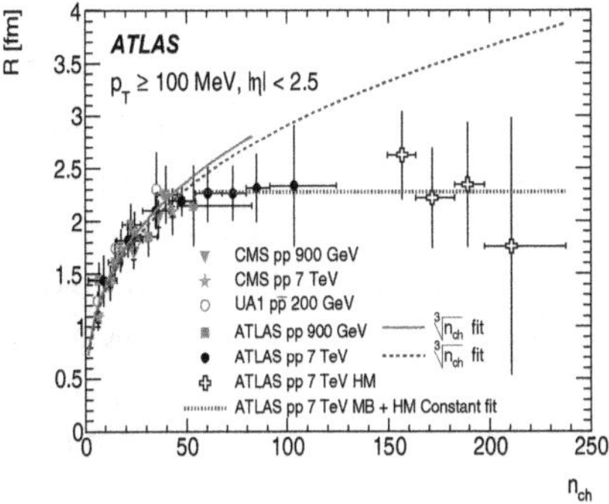
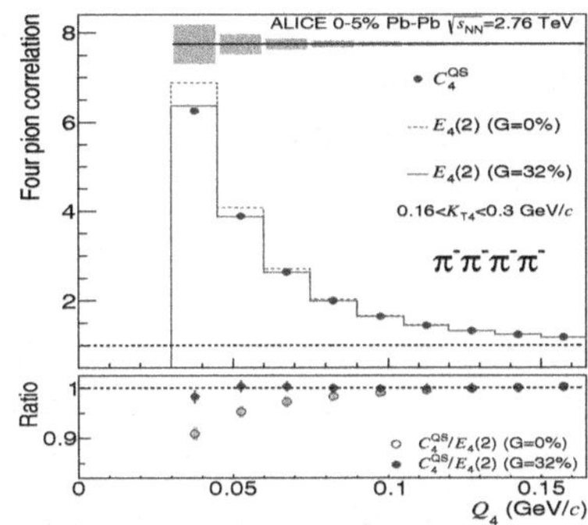

Fig. 11. (Left) Multiplicity n_{ch} dependence of the parameter R obtained from the exponential fit to the two-particle double-ratio correlation functions $R_2(Q)$ at $\sqrt{s} = 0.9$ and 7 TeV, compared to the equivalent measurements of the CMS[44] and UA1[45] experiments. The solid and dashed curves are the results of the $\sqrt[3]{n_{ch}}$ for $n_{ch} < 55$ fits. The dotted line is a result of a constant fit to minimum-bias and high-multiplicity events data at 7 TeV for $n_{ch} \geq 55$. The error bars represent the quadratic sum of the statistical and systematic uncertainties. Plot taken from Ref. 43. (Right) Same-charge four-pion full correlations versus Q_4 (see the text). Measured (points) and expected (histograms) correlations of the first type are shown. Dashed and solid block histograms show the expected correlations based on G=0 and G=32% fraction of coherent correlations, respectively. Systematic uncertainties are shown at the top. The bottom panel shows the ratio of measured to the expected correlations. The systematic uncertainties on the ratio are shown with a shaded blue band (G=0) and with a thick blue line (G=32%). Plot taken from Ref. 48 (color online).

for all multiplicity classes. A similar tendency was measured in another ATLAS analysis of pPb data[46]. Also in the LHCb study[47], R was measured to increase with multiplicity, while λ to decrease with multiplicity, so in accordance with observations made by the other LHC experiments, we can conclude that larger sources are more coherent.

Multi-pion BE correlations have also been measured in all three collision systems by ALICE[48]. In Fig. 11 right we can see the 4-pion correlations as a function of Q between these pions and the ratio of these measured 4-particle correlations to the expected ones from 2-pion correlations. While for pp and pPb collisions, no suppression is seen, in $PbPb$, a clear suppression of both the 4-pion and 3-pion (not shown here) correlations is observed. Interestingly, if a 32%-fraction of coherent correlations is assumed, the suppression is explained for the 4-pion correlation, while it does not help to the 3-pion correlations.

10. Hadronic chains

The origin of the enhanced production of pairs of identical particles is usually studied using BEC, believed to come from an incoherent particle production (see the previous section). An alternative approach to BEC is a causality-respecting model of quantized fragmentation of a 3D QCD string as a consequence of coherent hadron emission. This approach is based on studying hadronic chains[49] using helical strings[50] and utilized in a recent ATLAS analysis[51]. It studies hadronization effects using the same ATLAS data as in the BEC study[43], namely again same-sign and opposite-sign identical particle pairs. In the Lund hadronization, used in PYTHIA, there is a randomly broken 1D string and no cross-talk between break-up vertices. In the model of quantized helical (3D) string, one tries to make use of causality (in other words the cross-talk between break-up vertices) which gives two parameters: κR and $\Delta\phi$. This model predicts that the hadron spectra follow a simple quantized pattern: $m_T = n\kappa R\Delta\phi$, so that κR and $\Delta\phi$ can be fixed using known masses of pseudoscalar mesons. Then one can predict momentum difference Q for pairs of ground-state hadrons for various pair rank differences, r. Adjacent (opposite-sign) pions are then predicted to be produced with p_T difference of 266 MeV, while the like-sign pion pairs with rank difference 2 should have 91 MeV. Figure 12 left rep-

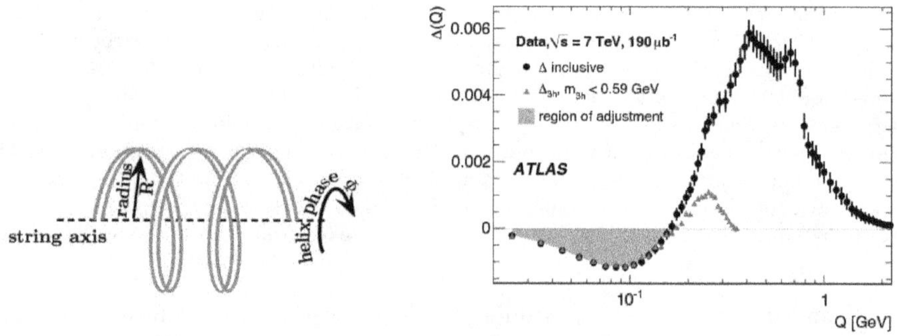

Fig. 12.   (Left) Parameterization of the helical shape of the QCD string. (Right) The corrected $\Delta(Q)$ is compared with the corrected contribution from low-mass three-hadron chains $\Delta_{3h}(Q)$. The chain mass limit is set to a value of $m_{3h}^{cut} = 0.59$ GeV, which reproduces the excess in the inclusive like-sign pair production at low Q (shaded area). Bin errors indicate the combined statistical and reconstruction uncertainty. Plots taken from Ref. 51.

resents a parameterization of the helical shape of the QCD string, while the plot on right showing a ratio of differences between number of opposite-sign and same-sign charged particles to the number of charged particles, [N(OS)-N(SS)]/N$_{ch}$, plotted as a function of Q, documents that the low-Q region is described by this new model of fragmentation using helical strings. The very low-Q region is predicted to be populated exclusively by the same-sign pairs (which means the rank $r = 2$). The further study then showed that the source of these correlations are 3-hadron chains.

Summary

This text summarizes recent soft QCD measurements from six LHC experiments and attempts to document the importance of soft QCD physics in a number of aspects. For example, the soft QCD results serve as input for modeling pile-up at LHC and interactions in cosmic rays. They also greatly help in a better understanding of the hadronization mechanism, typically a mixture of processes dominating at low transverse momenta. Work on improving hadronization models is currently ongoing and involves a better treatment of color-confining objects such as strings and ropes. It is also evident that the effect of underlying event is non-negligible at LHC energies. Finally it was indicated that by studying particle correlations, we gain not only a powerful tool to learn more about multihadron production, but to investigate striking similarities between three collision systems at LHC, namely *PbPb*, *pPb* and *pp*. With a limited set of analyses we demonstrated that all collision systems are useful for soft QCD studies, nicely complementing each other. Let us also note that the performant LHC machine and experiments unprecedently provide high-statistics and at the same time high-precision data samples which enable us to estimate reliably many sources of systematic uncertainties. Sophisticated techniques are used to measure very low p_T particles, to efficiently subtract many sources of background and to make use of unfolding techniques in several dimensions. This way, the precision data help to understand unexplained phenomena and to develop or to reject various theoretical models. We reported about similar phenomena observed in *PbPb*, *pPb* and *pp* (high multiplicity) collisions, namely about strangeness enhancement and collectivity effects. The primordial question currently discussed in the HI community is why these effects are observed in small systems such as *pPb* and *pp*. We also conclude that the near-side ridge may be a good testing ground to study complementarity between hydrodynamics/QGP and dynamics models (CGC/saturation/ropes).

Acknowledgement

Supported by the project LG15052 of the Ministry of Education, Youth and Sports of the Czech Republic. Author wishes to thank Edward Sarkisyan-Grinbaum for helpful discussions.

References

1. ALICE Collab., *The ALICE Experiment at the CERN LHC, JINST* **3**, S08002 (2008).
2. ATLAS Collab., *The ATLAS Experiment at the CERN Large Hadron Collider, JINST* **3**, S08003 (2008).
3. CMS Collab., *The CMS Experiment at the CERN LHC, JINST* **3**, S08004 (2008).
4. LHCb Collab., *The LHCb Detector at the LHC, JINST* **3**, S08005 (2008).
5. LHCf Collab., *The LHCf Detector at the CERN Large Hadron Collider, JINST* **3**, S08006 (2008).
6. TOTEM Collab., *The TOTEM Experiment at the CERN Large Hadron Collider, JINST* **3**, S08007 (2008).
7. COMPETE Collab., *Benchmarks for the forward observables at RHIC, the Tevatron Run II and the LHC, Phys. Rev. Lett.* **89**, 201801 (2002).
8. TOTEM Collab., *Luminosity-independent measurements of total, elastic and inelastic cross-sections at $\sqrt{s} = 7$ TeV, Europhys. Lett.* **101**, 21004 (2013).
9. TOTEM Collab., *Measurement of Elastic pp Scattering at $\sqrt{s} = 8$ TeV in the Coulomb-Nuclear Interference Region — Determination of the ρ Parameter and the Total Cross-Section, Eur. Phys. J.* **C 76**, 661 (2016).

10. M. Deile for the TOTEM Collab., talk at the conference EDS Blois 2017, Prague, Czech Rep.

11. ATLAS Collab., *Measurement of the total cross section from elastic scattering in pp collisions at $\sqrt{s}$ = 8 TeV with the ATLAS detector, Phys. Lett.* **B 761**, 158 (2016).

12. ATLAS Collab., *Measurement of the Inelastic Proton-Proton Cross Section at $\sqrt{s}$ = 13 TeV with the ATLAS Detector at the LHC, Phys. Rev. Lett.* **117**, no. 18, 182002 (2016).

13. T. Sjöstrand, S. Mrenna, and P. Z. Skands, *A Brief Introduction to PYTHIA 8.1, Comput. Phys. Commun.* **178**, 852 (2008).

14. T. Pierog *et al.*, *EPOS LHC: Test of collective hadronization with data measured at the CERN Large Hadron Collider, Phys. Rev.* **C 92** no. 3, 034906 (2015).

15. S. Ostapchenko, *Monte Carlo treatment of hadronic interactions in enhanced Pomeron scheme: I. QGSJET-II model, Phys. Rev.* **D 83**, 014018 (2011).

16. TOTEM Collab., *Evidence for Non-Exponential Elastic Proton-Proton Differential Cross-Section at Low $|t|$ and $\sqrt{s}$ = 8 TeV by TOTEM, Nucl. Phys.* **B 899**, 527 (2015).

17. ALICE Collab., *Charged-particle multiplicities in proton-proton collisions at $\sqrt{s}$ = 0.9 to 8 TeV, Eur. Phys. J.* **C 77**, 33 (2017).

18. ATLAS Collab., *Charged-particle distributions at low transverse momentum in $\sqrt{s}$ = 13 TeV pp interactions measured with the ATLAS detector at the LHC Eur. Phys. J.* **C 76**, 502 (2016).

19. CMS Collab., *Measurement of the inclusive energy spectrum in the very forward direction in proton-proton collisions at $\sqrt{s}$ = 13 TeV, J. High En. Phys.* **08**, 046 (2017).

20. LHCf Collab., *Measurement of very forward neutron energy spectra for 7 TeV proton-proton collisions at the Large Hadron Collider, Phys. Lett.* **B 750**, 360 (2015).

21. F. W. Bopp *et al.*, *Antiparticle to Particle Production Ratios in Hadron-Hadron and d-Au Collisions in the DPMJET-III Monte Carlo, Phys. Rev.* **C 77**, 014904 (2008).

22. E.-J. Ahn *et al.*, *Cosmic ray interaction event generator SIBYLL 2.1, Phys. Rev.* **D 80**, 094003 (2009).

23. LHCf Collab., *Measurements of longitudinal and transverse momentum distributions for neutral pions in the forward-rapidity region with the LHCf detector, Phys. Rev.* **D 94**, 032007 (2016).

24. LHCf Collab., *Measurement of forward photon-energy spectra for $\sqrt{s}$ = 13 TeV proton-proton collisions with the LHCf detector, CERN-EP-2017-051, arXiv:1703.07678 [hep-ex].*

25. ALICE Collab., *Enhanced production of multi-strange hadrons in high-multiplicity proton-proton collisions, Nat. Phys.* **13**, 535 (2017).

26. P. Koch, B. Muller and J. Rafelski, *Strangeness in Relativistic Heavy Ion Collisions, Phys. Rept.* **142**, 167 (1986).

27. C. Bierlich and J. R. Christiansen, *Effects of Colour Reconnection on Hadron Flavour Observables, Phys. Rev.* **D 92**, 094010 (2015).

28. CMS Collab., *Measurement of charged pion, kaon, and proton production in proton-proton collisions at $\sqrt{s}$ = 13 TeV, CERN-EP-2017-091, arXiv:1706.10194 [hep-ex].*

29. ATLAS Collab., *Measurement of charged-particle distributions sensitive to the underlying event in $\sqrt{s}$ =13 TeV proton–proton collisions with the ATLAS detector at the LHC, J. HEP* **03**, 157 (2017).

30. ATLAS Collab., *Measurement of underlying event characteristics using charged particles in pp collisions at $\sqrt{s}$ = 900 GeV and 7 TeV with the ATLAS detector, Phys. Rev.* **D 83**, 112001 (2011).

31. CMS Collab., *Measurement of the underlying event using the Drell-Yan process in proton-proton collisions at $\sqrt{s}$ = 13 TeV*, CERN-EP-2017-249, arXiv:1711.04299 [hep-ex].

32. M. Sumbera and R. Pasechnik, *Phenomenological Review on Quark-Gluon Plasma: Concepts vs. Observations*, Universe **3**, no. 1, 7 (2017).

33. L.D. McLerran and R. Venugopalan, *Computing quark and gluon distribution functions for very large nuclei*, Phys. Rev. **D 49**, 2233 (1994).

34. J. D. Bjorken *et al.*, *Possible multiparticle ridge-like correlations in very high multiplicity proton-proton collisions*, Phys. Lett. **B 726**, 344 (2013).

35. LHCb Collab., *Measurements of long-range near-side angular correlations in $\sqrt{s}$ = 5.02 TeV proton-lead collisions in the forward region*, Phys. Lett. **B 762**, 473 (2016).

36. ALICE Collab., *Evolution of the longitudinal and azimuthal structure of the near-side jet peak in Pb-Pb collisions at $\sqrt{s}$ = 2.76 TeV*, Phys. Rev. **C 96**, no. 3, 034904 (2017).

37. ATLAS Collab., *Measurement of multi-particle azimuthal correlations with the subevent cumulant method in pp and p+Pb collisions with the ATLAS detector at the LHC*, CERN-EP-2017-160, arXiv:1708.03559 [hep-ex].

38. CMS Collab., *Measurement of Long-Range Near-Side Two-Particle Angular Correlations in pp Collisions at $\sqrt{s}$ = 13 TeV*, Phys. Rev. Lett. **116**, 172302 (2016).

39. ATLAS Collab., *Measurement of multi-particle azimuthal correlations in pp, p+Pb and low-multiplicity Pb+Pb collisions with the ATLAS detector*, Eur. Phys. J. **C 77**, 428 (2017).

40. CMS Collab., *Evidence for collectivity in pp collisions at the LHC*, Phys. Lett. **B 765**, 193 (2017).

41. CMS Collab., *Multiplicity and transverse momentum dependence of two- and four-particle correlations in pPb and PbPb collisions*, Phys. Lett. **B 724**, 213 (2013).

42. ALICE Collab., *Insight into particle production mechanisms via angular correlations of identified particles in pp collisions at $\sqrt{s}$ = 7 TeV*, Eur. Phys. J. **C 77**, no. 8, 569 (2017).

43. ATLAS Collab., *Two-particle Bose–Einstein correlations in pp collisions at $\sqrt{s}$ = 0.9 and 7 TeV measured with the ATLAS detector*, Eur. Phys. J. **C 75**, 466 (2015).

44. CMS Collab., *Measurement of Bose-Einstein Correlations in pp Collisions at $\sqrt{s}$ = 0.9 and 7 TeV*, J. High Energy Phys. **05**, 029 (2011).

45. UA1 Collab., *Bose-Einstein Correlations in $\bar{p}p$ Interactions at $\sqrt{s}$ = 0.2 to 0.9 TeV*, Phys. Lett. **B 226**, 410 (1989).

46. ATLAS Collab., *Femtoscopy with identified charged pions in proton-lead collisions at $\sqrt{s}$ = 5.02 TeV with ATLAS*, CERN-EP-2017-004, arXiv:1704.01621 [hep-ex].

47. LHCb Collab., *Bose-Einstein correlations of same-sign charged pions in the forward region in pp collisions at $\sqrt{s}$ = 7 TeV*, J. High Enery Phys. **1712** 025 (2017).

48. ALICE Collab., *Multipion Bose-Einstein correlations in pp, p-Pb, and Pb-Pb collisions at energies available at the CERN Large Hadron Collider*, Phys. Rev. **C 93**, 054908 (2016).

49. S. Todorova-Nova, *Quantization of the QCD string with a helical structure*, Phys. Rev. **D 89**, no. 1, 015002 (2014).

50. B. Andersson *et al.*, *Is there screwiness at the end of the QCD cascades?*, J. High Energy Phys. **9809**, 014 (1998).

51. ATLAS Collab., *Study of ordered hadron chains with the ATLAS detector*, Phys. Rev. **D 96**, no. 9, 092008 (2017).

Recent Experimental Highlights in Heavy-Ion Physics

A. Kalweit

CERN, EP Department, Geneva, Switzerland
E-mail: Alexander.Philipp.Kalweit@cern.ch

This article summarises a personal selection of recent experimental highlight results in the field of ultra-relativistic heavy-ion physics which have advanced our understanding of hadronic matter under extreme conditions. After a general introduction, the production of light flavour hadrons at Large Hadron Collider energies is addressed going from large systems as created in Pb-Pb collisions to small systems in pp and p-Pb collisions. This discussion of particle production as function of system size is complemented by a short discussion of the search for the QCD critical point and the onset of de-confinement which is addressed by studying particle production as a function of beam energy at the Relativistic Heavy-Ion Collider. Furthermore, the physics of heavy-flavour hadrons and recent findings concerning jets in heavy-ion collisions are very briefly discussed.

1. Introduction to heavy-ion physics

At the high energy frontier, experimental research in heavy-ion physics is nowadays mainly taking place at the Large Hadron Collider (LHC) at CERN in Geneva since its first Pb-Pb run in November 2010. While ALICE is a dedicated heavy-ion experiment, also the other major LHC experiments (ATLAS, CMS, LHCb) pursue a dedicated heavy-ion program. LHCb joined for the first time in heavy-ion data taking in November 2015. At the moment, the analysis of LHC Run 2 (2015-2019) data is in full swing. The currently already available approximately four times larger integrated luminosity with respect to Run 1 (2009-2013) allows for a more precise investigation of rare probes. In addition, the various collisions systems (pp, pPb, Xe-Xe, Pb-Pb) which were taken at several centre-of-mass energies are ideally suited for systematic studies of particle production. At lower beam energies, several existing (RHIC, GSI, SPS) and planned (FAIR, NICA) facilities contribute to the search for the QCD critical point and the onset of de-confinement.

Generally speaking, the physics of ultra-relativistic heavy-ion collisions is the physics of the strong interaction at high energy density. We have today strong reasons to believe that quantum chromodynamics (QCD) is the appropriate theory of the strong interaction. QCD is one of the main pillars of the Standard Model. The dynamics of the theory are best described by its Lagrangian,

$$\mathcal{L}_{\text{QCD}} = \bar{q}(i\gamma^{\mu}D_{\mu} - m)q - \frac{1}{4}F^{a}_{\mu\nu}F^{\mu\nu}_{a} \ , \tag{1}$$

where q corresponds to the quark field, γ^{μ} are the Dirac matrices, m is the quark mass (or masses), and $F^{a}_{\mu\nu}$ is the gluon field strength tensor. We note that it is formulated only in terms of quarks and gluons and no equivalent formulation can be derived for hadrons which are the particles that we observe in nature. The first years of data taking at the Large Hadron Collider (LHC) at CERN have dramatically confirmed the validity of the standard model. With the discovery of the Higgs Boson and the investigation of its properties the last major missing piece of the Standard Model has been completed. However, no strong indications for physics beyond the standard model have been observed so far within the current experimental precision of high energy physics experiments. Future experimental work for the observation of new phenomena thus goes in two directions:

(1) Continuation of precision tests of the standard model and searches for dark matter candidates as well as super-symmetric particles with increased statistics and precision.
(2) Exploring the multi-body interactions of the fundamental matter constituents of the Standard Model (quarks and gluons) in order to investigate their collective behaviour. Such studies find their historic analogy in condensed matter physics: While the fundamental interactions of an electron with the atomic nucleus are understood within the framework of Quantum Electrodynamics (QED) to an almost unlimited precision, completely new phenomena arise once one looks at a multi-body system like a crystal consisting of millions of atoms.

The properties of QCD give rise to a rich and unique phase structure in multi-body systems such as the system which is created in heavy-ion collisions. The main features of QCD with relevance for heavy-ion physics are *confinement* (quarks and gluons are bound in color neutral mesons $q\bar{q}$ or baryons qqq), *asymptotic freedom*[1,2] (the coupling constant α_S and thus the interaction strength decreases with increasing momentum transfer Q^2, i.e. $\alpha_S \longrightarrow 0$ for $Q^2 \longrightarrow \infty$), and *chiral symmetry* (the interaction between left and right handed quarks disappears for massless quarks).

If one considers extremely high temperatures T beyond the TeV regime where each significant momentum transfer occurs in the asymptotically free regime of the strong interaction, quarks and gluons must behave free and cannot be localised to individual hadrons anymore[3,4]. Such a state of matter, which corresponds to the equilibrium state of QCD at high temperature, is called the quark-gluon-plasma (QGP). Since at low temperatures like room temperature ($T_0 \approx 1/40$ eV), only bound quarks and gluons are observed, a phase transition between the two regimes must occur somewhere in between. Nowadays, the transition temperature can be calculated ab-initio based on Lattice QCD and values of around $T_C = 156$ MeV with a few MeV uncertainty are found by recent calculations[5,6] for the critical temperature T_C. While there is strong evidence that such extreme temperatures are reached in accelerator based heavy-ion experiments, it is still an open topic in current research if the phase-transition temperature is the same for all quark flavours.

The temperatures prevailing in heavy-ion collisions are experimentally accessible via the measurement of direct photons, which are partially emitted as the black body radiation from the QGP, but also in the later (hadronic) stages of the system created in the collision. The main challenge of the measurement is the subtraction of the background induced by electromagnetic decays such as $\pi^0 \to \gamma\gamma$ which is dominating over the signal. Based on model comparisons with the measured photon spectrum, an effective temperature of $T_{eff} = 304 \pm 11 \pm 40\,\mathrm{MeV}$ is observed at LHC energies as a result of a high initial temperature and the blue-shift caused by the radial expansion of the system[7].

The geometry of a heavy-ion collision, i.e. the impact parameter of the impinging nuclei, is a key feature for their classification and understanding. In the experiment, the nuclei collide with random impact parameters, but it can be determined a posteriori based on the Glauber model[8]. The events are then typically divided into centrality percentiles (see for instance[9]), in which for example the 0-5% class corresponds to the most central collisions (small impact parameter, many particles are produced at midrapidity) and the 80-90% class corresponds to the most peripheral collisions (large impact parameter, only few particles are produced at midrapidity). For each of these classes, the average number of binary nucleon-nucleon collisions N_{coll} and the average number of participating nucleons N_{part} is determined based on the Glauber model.

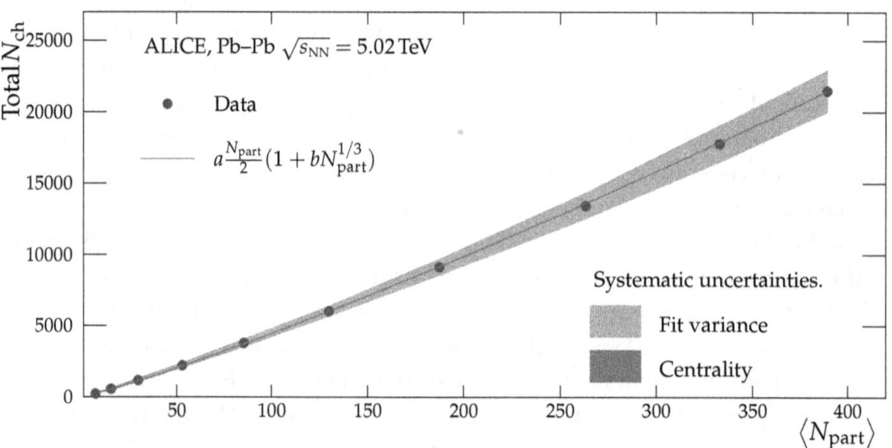

Fig. 1. Total number of charged particles produced in Pb-Pb collisions at 5.02 TeV as a function of the number of nucleons participating in the collision. Figure taken from[10].

The most basic measurement in high energy nuclear and particle physics is always the measurement of the average charged particle multiplicity, i.e. the number of particles produced in the collision. While in a proton-proton (pp) collision at LHC energies only about 5-6 particles are produced per unit pseudo-rapidity η, a value of $\mathrm{d}N_{ch}/\mathrm{d}\eta = 1943 \pm 54$ is found for central Pb-Pb collisions[10]. It is remarkable, that — even at LHC energies — about 95% of all these particles are produced with $p_T < 2$ GeV/c in both pp and Pb–Pb collisions. Bulk particle production, which is naturally most relevant for the study of many-body (collective) phenomena, is therefore associated with so-called "soft physics" in the non-perturbative regime of QCD. As shown in Fig. 1, the integration over the entire rapidity range yields a total number of around $N_{ch} = 21000$ charged particles which is produced in a central heavy-ion collision at the LHC. While thermodynamic approaches are typically applied to systems with about 1mol of particles, we also observe that the system created in heavy-ion collisions with $1 \ll N_{ch} \ll$ 1mol appears in apparent local thermodynamic equilibrium. This concept is strongly supported by the success of thermal-statistical models[11,12] describing the yields of hadrons containing up, down, and strange (so-called *light flavour*) hadrons with an approach based on chemical equilibrium and by the success of hydrodynamic models in explaining spectral shapes and azimuthal anisotropies (kinetic equilibrium). One of the interesting current research topics at the moment is to clarify how small a system can be for this picture to hold, i.e. if and why it also describes particle production in systems like pp and p–Pb.

2. Soft probes in heavy-ion physics

2.1. *Soft particle production in Pb-Pb collisions at LHC energies*

Figure 2 shows recent LHC Run 2 measurements of transverse momentum spectra as well as of the elliptic flow. Elliptic flow is a result of the initial spatial anisotropy of the collisions which is converted by scatterings into an anisotropy in momentum space. It is quantified by a Fourier decomposition

$$E\frac{\mathrm{d}^3N}{\mathrm{d}p^3} = \frac{\mathrm{d}^2N}{2\pi p_T \mathrm{d}p_T \mathrm{d}y}\left(1 + 2\sum_{n=1}^{\infty} v_n(p_T)\cos[n(\varphi - \Psi_n)]\right) \tag{2}$$

where v_2 is the elliptic flow coefficient and Ψ_n is n-th order symmetry plane of the event. In both the elliptic flow measurement as well as in the transverse momentum spectra, one observes a clear mass ordering. In heavy-ion collisions, this mass ordering and the hardening of the spectra for more central collisions is commonly interpreted as a signature of the collective radial expansion of the system with a common expansion velocity profile $\beta(r)$ which can be described by hydrodynamics. The ordering in momentum p according to the particle mass m is then understood as a larger blue-shift for heavier particles following $p = \beta\gamma \cdot m$.

In two-particle correlation studies in heavy-ion collisions, elliptic flow causes a symmetric long-range azimuthal correlation commonly known as the double ridge. The near-side (close to the trigger particle) component was first observed by the CMS collaboration also in high multiplicity pp collisions while its symmetric nature was first seen by the ALICE collaboration in high multiplicity p-Pb collisions. Extending these studies as shown in Fig. 3 to multi-particle correlations shows clear

264

signs of collectivity in pp and p-Pb collisions. Similar observations hold true for many other typical *kinetic* heavy-ion observables measured in high multiplicity pp and pPb collisions (for a recent review see for instance [15]).

While the spectral shapes and elliptic flow of light flavour hadrons are well described by hydrodynamics (kinetic equilibrium), the p_T-integrated particle yields dN/dy are well described by statistical-thermal models (chemical equilibrium) based on a hadron resonance gas description. In this approach, the production

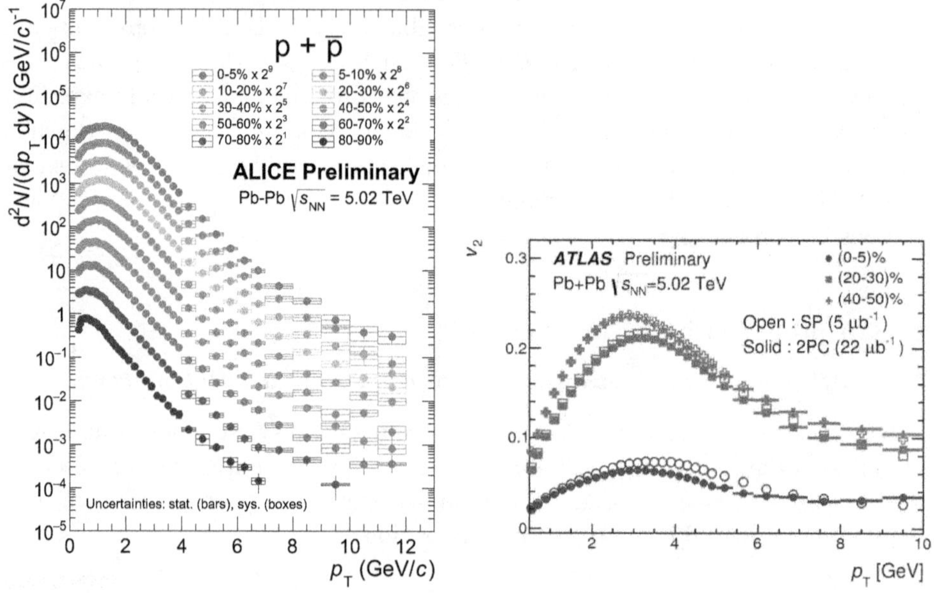

Fig. 2. (left) Transverse momentum spectra of protons in Pb-Pb collisions measured by the ALICE collaboration for several centrality classes. (right) Elliptic flow in heavy-ion collisions measured by the ATLAS collaboration. Figures taken from [13] (left) and [14] (right).

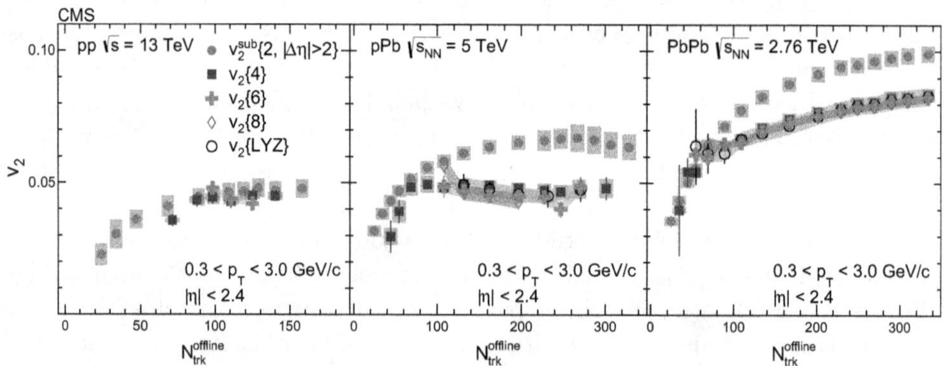

Fig. 3. Elliptic flow coefficients from higher order particle correlations. Figure taken from [16].

yields of light flavour hadrons roughly scale as $dN/dy \propto \exp(-m/T_{\rm ch})$ where $T_{\rm ch}$ corresponds to the chemical freeze-out temperature. As shown in Fig. 4, the particle yields in central Pb-Pb collisions are well described over seven orders of magnitude with a common chemical freeze-out temperature of about 156 MeV coinciding with the previously mentioned Lattice QCD estimates. Finding a T_{ch} which is not higher than the phase transition temperature is an important consistency check for the field. Finding a T_{ch} that is close to the phase transition temperature might indicate a fast equilibration process caused by multi-body collisions close to the phase transition[17].

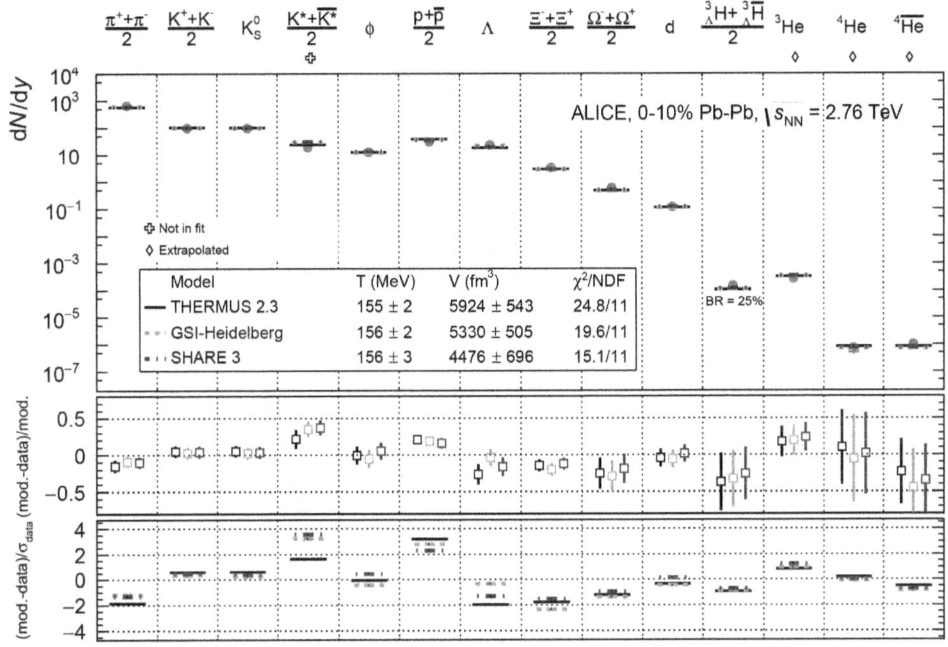

Fig. 4.   Comparison of measured particle yields with thermal-statistical models. Figure taken from[18].

2.2. *Soft particle production in small systems at LHC energies*

In Pb-Pb collisions, the thermal-statistical hadron resonance gas description of particle production based on the assumption of a fireball in chemical equilibrium also holds for the hadrons containing strangeness which are slightly rarer produced than those containing only u and d quarks: approximately every fourth to fifth valence quark is a strange quark in Pb-Pb collisions while it is only approximately every tenth in minimum bias pp collisions.

A recent measurement of the ALICE collaboration now shows that there is a smooth evolution of hadrochemistry from pp to p-Pb to Pb-Pb collisions as a function of charged particle multiplicity. As shown in Fig. 5, a significant enhancement

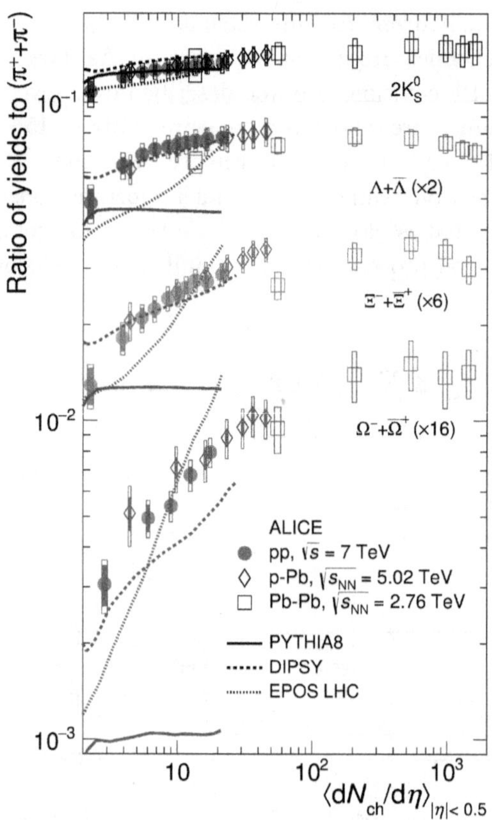

Fig. 5.   Comparison of abundances of strange hadrons relative to pions with models. Figure taken from [19].

of strange to non-strange particle production is observed in pp collisions. Since at the same time, the proton-to-pion and lambda-to-kaon ratio does not show such an increase, the effect appears to be strangeness related and not mass or baryon number related. With respect to similar effects previously observed in p-Pb collisions, the pp collision data allows to compare to a large set of QCD inspired event generators. While PYTHIA8 completely misses the behaviour of the data (independent of switching on and off color reconnection), the DIPSY model (color ropes) describes the increase in strangeness production qualitatively, but failed to predict the proton yield correctly in its original version. The EPOS-LHC model only qualitatively describes the data based on a core-corona approach.

2.3. *Measurements of (anti-)(hyper-)nuclei*

Some of the most intriguing objects which can be studied at LHC energies are anti- and hyper-nuclei which are produced in a large amount in collisions at the LHC [18]. Due to the very small baryon transport from beam- to mid-rapidity at LHC energies, matter and anti-matter is produced in equal abundance at mid-rapidity at LHC energies. Surprisingly, the production yields of these exotic objects are in agreement with statistical-thermal model predictions as shown in Fig. 4 even though they should be dissolved in such a hot medium with temperatures around 156 MeV which is much larger than the binding energy (≈ 2.2 MeV in the anti-deuteron case) of these objects.

These measurements also impact physics beyond the community of heavy-ion physics. Firstly, the heavy-ion measurements may help in constraining the not well known lifetime of the hyper-triton which is sensitive to the hyperon-nucleon interaction in nuclear physics[20]. Secondly, collider measurements are used for background estimations in searches for anti-nuclei of galactic and dark matter origin such as in AMS[21].

2.4. *Search for the QCD critical point and onset of de-confinement*

The thermodynamics of QCD is commonly summarised in a phase diagram which is schematically represented in Fig. 6. Different regions of the phase diagram are probed by running experiments at different $\sqrt{s_{NN}}$. This is investigated done systematically within the Beam Energy Scan (BES) at RHIC[22]. In this context, critical fluctuations within the vicinity of the critical point are of particular importance. In general, phase transitions are often connected to critical phenomena. A classical

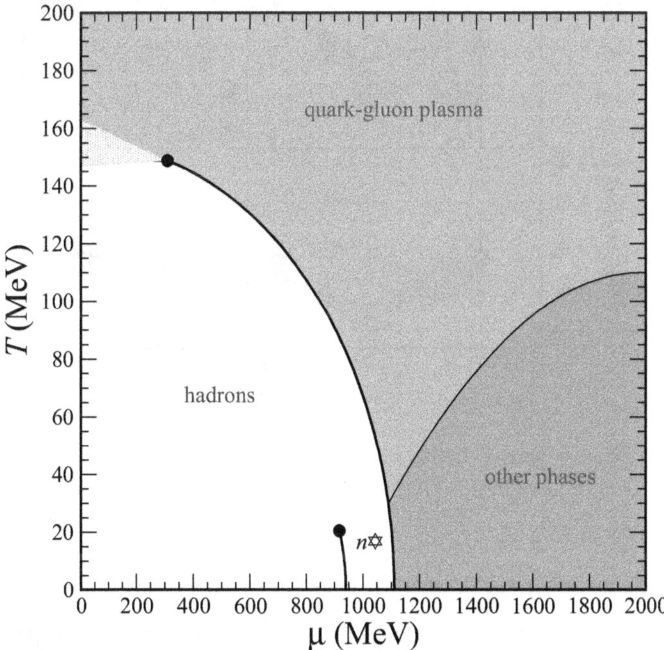

Fig. 6.   The QCD phase diagram. Figure taken from[23].

example is given by the opalescence of many substances due to the divergence of the correlation length.

In the QCD case, event-by-event fluctuations in the conserved charges of QCD (baryon number, strangeness, electric charge) are investigated. In this context, the fluctuations in the baryon number represent the key observable[24]. In practice, the higher moments (skewness, kurtosis) of the net-proton distribution are investigated. As shown in Fig. 7, a hint for a possible deviation from Poisson behaviour around $\sqrt{s_{NN}}$ 20 GeV. Future studies with increased statistics across all beam energies, including those at the LHC, which carefully address also the rapidity window dependence of the observable are needed in order to clarify the significance of the signal.

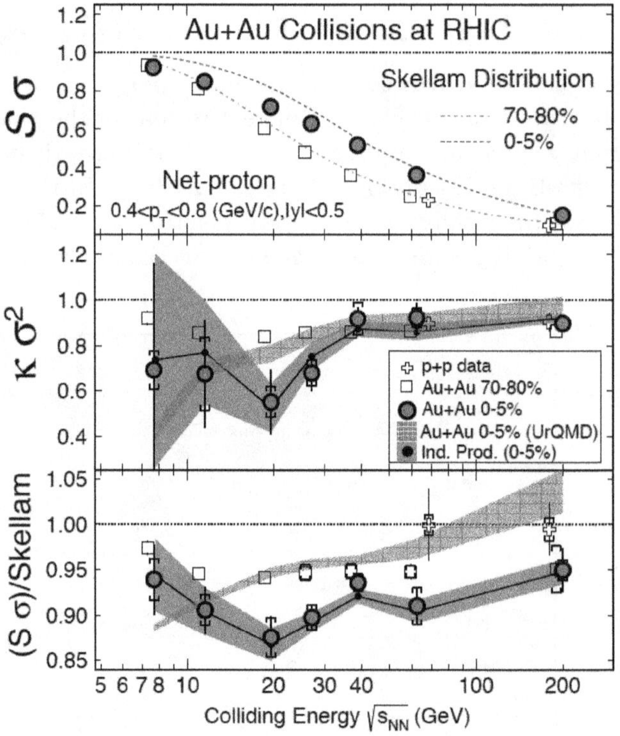

Fig. 7. Collision energy and centrality dependence of the net-proton skewness and kurtosis as measured by the STAR experiment. Figure taken from [24].

2.5. *Heavy-flavour production in heavy-ion collisions*

In heavy-ion collisions, the heavy c and b-quarks are dominantly produced in the initial hard binary nucleon-nucleon scatterings. This production cross-sections are in principle calculable ab-initio from perturbative QCD. Afterwards, these quarks

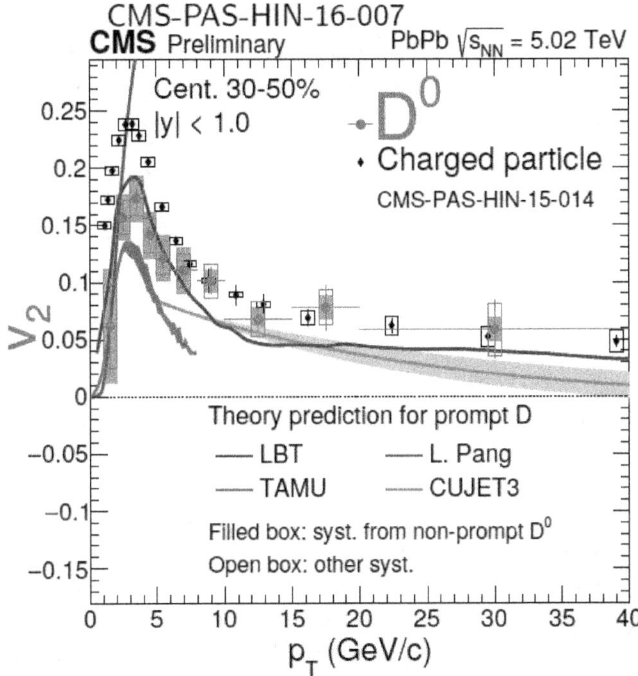

Fig. 8. Elliptic flow of D^0-mesons as a function of transverse momentum. Figure taken from [25].

interact with the medium which mainly consists of light quarks. Given their theoretically accessible primordial production and their negligible production during the evolution of the medium makes them unique probes of the quark-gluon-plasma.

By now, there is strong evidence that charm quarks thermalise in the medium, i.e. by re-scatterings with particles in the medium, they also follow the common hydrodynamic expansion. This behaviour is experimentally for instance observed in the elliptic flow of D-Mesons as shown in Fig. 8 or the baryon-to-meson enhancement seen in the Λ_c to D-meson ratio. In this context, electroweak probes serve as a crucial control experiment. As expected, they do not show any interaction with the QCD medium. Thus, as shown in Fig. 9, the Z^0-boson does not exhibit any elliptic flow.

One of the most interesting observables in heavy-ion physics is the production of the J/ψ-meson which is sensitive to the de-confinement$\leftrightarrow$confinement phase transition. As a $c\bar{c}$ bound state, the J/ψ-meson is expected not to be bound in the QGP phase[27], but it can re-generate at the QCD phase boundary[28]. For a quantitative study, the nuclear modification factor R_{AA} is investigated:

$$R_{AA} = \frac{\mathrm{d}N_{AA}/\mathrm{d}p_T}{\langle N_{coll}\rangle \cdot \mathrm{d}N_{pp}/\mathrm{d}p_T}, \tag{3}$$

270

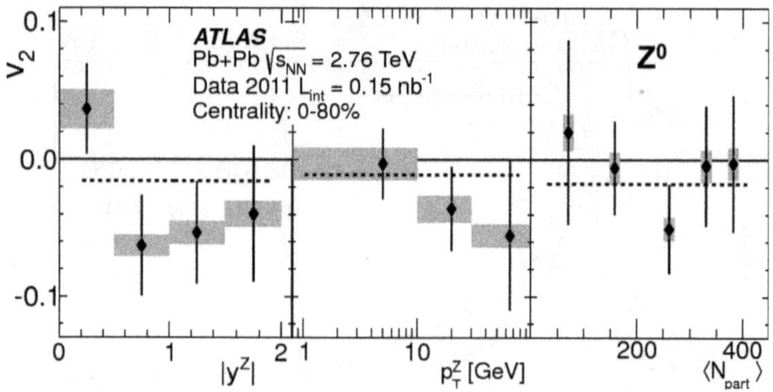

Fig. 9.   Elliptic flow of Z^0-bosons as a function of transverse momentum. Figure taken from [26].

where $\langle N_{coll} \rangle$ corresponds to the average number of binary nucleon-nucleon collisions and $\mathrm{d}N_{AA}/\mathrm{d}p_\mathrm{T}$ and $\mathrm{d}N_{pp}/\mathrm{d}p_\mathrm{T}$ correspond the spectra measured in AA and pp collisions, respectively. An R_{AA} value < 1 signals a suppression with respect to pp collisions while a value > 1 indicates an enhancement with respect to pp collisions. As shown in Fig. 10, the R_{AA} exhibits the tendency to rise above unity at midrapidty for low momentum particles. A detailed analysis shows that this implies not only less suppression with respect to the R_{AA} values found at RHIC, but also that the R_{AA} at midrapidity is larger than at forward rapidities where the recombination probability is lower. Both observations strongly confirm the J/ψ recombination picture and can be seen interpreted as a signature of de-confinement.

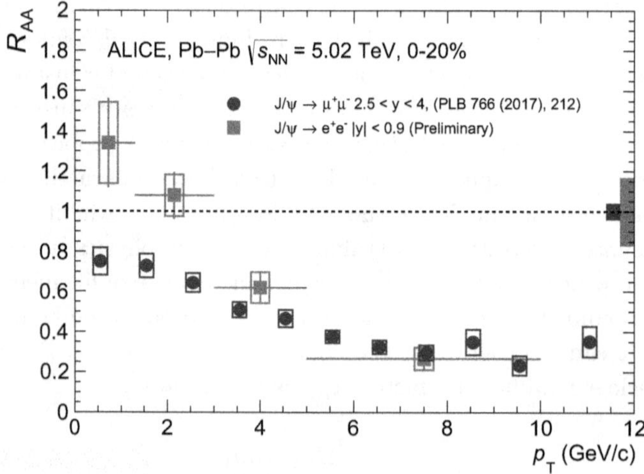

Fig. 10.   Nuclear modification factor for J/ψ as a function of transverse momentum as measured by the ALICE collaboration. Figure taken from [29].

2.6. *Jet substructure measurements in heavy-ions*

One of the most striking observations made immediately after the first LHC heavy-ion run was the suppression of high energetic jets[30]. Following the advances in the proton-proton jet physics community, several state-of-the-art jet substructure measurements are as well carried out in the heavy-ion community. Interestingly, jets loose energy in the medium, but their structure seems — from a qualitative point of view — unmodified. From a quantitative point of view, cancellation effects need to be taken properly into account in future studies in order to answer the question of sub-structure modification. Figure 11 shows two examples, the N-subjettiness (sensitive to two-prongness) and the jet mass (sensitive to broadening), in which the heavy-ion data is in agreement with the PYTHIA expectations for pp collisions.

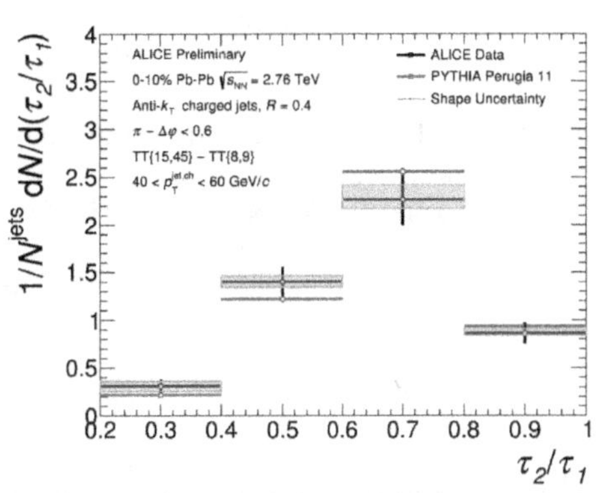
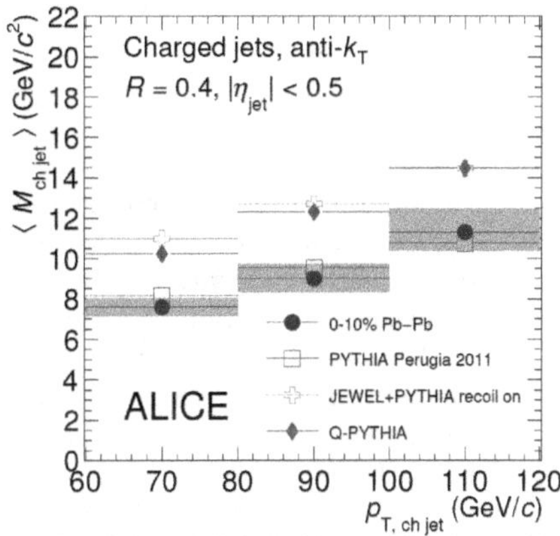

Fig. 11. (left) N-subjettiness distribution in Pb-Pb collisions compared to PYTHIA. (right) Average jet mass as a function of transverse momentum compared to PYTHIA. Figure taken from [31] (left) and [32] (right).

2.7. *Summary and outlook*

The field of (ultra-relativistic) heavy-ion physics has seen a wealth of beautiful new results in the last years from LHC and the beam energy scan at RHIC. In addition, New facilities are coming up at the high net-baryon density frontier at lower center-of-mass energies: NICA and FAIR. At the high energy frontier, the LHC Run2 data analysis is in full swing. Maybe one of the most important recent results in the field is the fact, that pp and p-Pb data samples are not only reference samples anymore, but show at high multiplicities similar features to Pb-Pb collisions.

References

1. D. J. Gross and F. Wilczek, Ultraviolet Behavior of Nonabelian Gauge Theories, *Phys. Rev. Lett.* **30**, 1343 (1973).

2. H. D. Politzer, Reliable Perturbative Results for Strong Interactions?, *Phys. Rev. Lett.* **30**, 1346 (1973).

3. N. Cabibbo and G. Parisi, Exponential Hadronic Spectrum and Quark Liberation, *Phys. Lett.* **59B**, 67 (1975).

4. J. C. Collins and M. J. Perry, Superdense Matter: Neutrons Or Asymptotically Free Quarks?, *Phys. Rev. Lett.* **34**, p. 1353 (1975).

5. T. Bhattacharya *et al.*, QCD Phase Transition with Chiral Quarks and Physical Quark Masses, *Phys. Rev. Lett.* **113**, p. 082001 (2014).

6. S. Borsanyi, Z. Fodor, C. Hoelbling, S. D. Katz, S. Krieg and K. K. Szabo, Full result for the QCD equation of state with 2+1 flavors, *Phys. Lett.* **B730**, 99 (2014).

7. J. Adam *et al.*, Direct photon production in Pb-Pb collisions at $\sqrt{s_{NN}} = 2.76$ TeV, *Phys. Lett.* **B754**, 235 (2016).

8. A. Bialas, M. Bleszynski and W. Czyz, Multiplicity Distributions in Nucleus-Nucleus Collisions at High-Energies, *Nucl. Phys.* **B111**, 461 (1976).

9. B. Abelev *et al.*, Centrality determination of Pb-Pb collisions at $\sqrt{s_{NN}} = 2.76$ TeV with ALICE, *Phys. Rev.* **C88**, p. 044909 (2013).

10. J. Adam *et al.*, Centrality dependence of the pseudorapidity density distribution for charged particles in Pb-Pb collisions at $\sqrt{s_{NN}} = 5.02$ TeV, *Phys. Lett.* **B772**, 567 (2017).

11. A. Andronic, P. Braun-Munzinger, K. Redlich and J. Stachel, Decoding the phase structure of QCD via particle production at high energy (2017).

12. S. Wheaton and J. Cleymans, THERMUS: A Thermal model package for ROOT, *Comput. Phys. Commun.* **180**, 84 (2009).

13. N. Jacazio, Production of identified charged hadrons in PbPb collisions at $\sqrt{s_{NN}} = 5.02$ TeV, *Nucl. Phys.* **A967**, 421 (2017).

14. *Measurement of the azimuthal anisotropy of charged particles produced in 5.02 TeV Pb+Pb collisions with the ATLAS detector*, Tech. Rep. ATLAS-CONF-2016-105, CERN (Geneva, 2016).

15. C. Loizides, Experimental overview on small collision systems at the LHC, in *25th International Conference on Ultra-Relativistic Nucleus-Nucleus Collisions (Quark Matter 2015) Kobe, Japan, September 27-October 3, 2015*, 2016.

16. V. Khachatryan *et al.*, Evidence for collectivity in pp collisions at the LHC, *Phys. Lett.* **B765**, 193 (2017).

17. P. Braun-Munzinger, J. Stachel and C. Wetterich, Chemical freezeout and the QCD phase transition temperature, *Phys. Lett.* **B596**, 61 (2004).

18. S. Acharya *et al.*, Production of ^{4}He and $^4\overline{\text{He}}$ in Pb-Pb collisions at $\sqrt{s_{NN}} = 2.76$ TeV at the LHC, *Nucl. Phys.* **A971**, 1 (2018).

19. J. Adam *et al.*, Enhanced production of multi-strange hadrons in high-multiplicity proton-proton collisions, *Nature Phys.* **13**, 535 (2017).

20. S. Trogolo and ALICE Collaboration, Recent results on (anti-)(hyper-)nuclei production in pp, p-Pb and Pb-Pb collisions with ALICE, *PoS* **EPS-HEP2017**, p. 200 (2017).

21. K. Blum, K. C. Y. Ng, R. Sato and M. Takimoto, Cosmic rays, antihelium, and an old navy spotlight, *Phys. Rev.* **D96**, p. 103021 (2017).

22. L. Adamczyk *et al.*, Bulk Properties of the Medium Produced in Relativistic Heavy-Ion Collisions from the Beam Energy Scan Program, *Phys. Rev.* **C96**, p. 044904 (2017).

23. A. S. Kronfeld, Twenty-first Century Lattice Gauge Theory: Results from the QCD Lagrangian, *Ann. Rev. Nucl. Part. Sci.* **62**, 265 (2012).

24. L. Adamczyk *et al.*, Energy Dependence of Moments of Net-proton Multiplicity Distributions at RHIC, *Phys. Rev. Lett.* **112**, p. 032302 (2014).

25. A. M. Sirunyan *et al.*, Measurement of prompt D^0 meson azimuthal anisotropy in PbPb collisions at $\sqrt{s_{NN}} = 5.02$ TeV (2017).

26. G. Aad *et al.*, Measurement of Z boson Production in Pb+Pb Collisions at $\sqrt{s_{NN}} = 2.76$ TeV with the ATLAS Detector, *Phys. Rev. Lett.* **110**, p. 022301 (2013).

27. T. Matsui and H. Satz, J/ψ Suppression by Quark-Gluon Plasma Formation, *Phys. Lett.* **B178**, 416 (1986).

28. P. Braun-Munzinger and J. Stachel, (Non)thermal aspects of charmonium production and a new look at J/psi suppression, *Phys. Lett.* **B490**, 196 (2000).

29. D. Weiser, Centrality and transverse momentum dependence of J/ψ production in Pb-Pb collisions at $\sqrt{s_{NN}} = 5.02$ TeV at mid-rapidity with ALICE, in *17th International Conference on Strangeness in Quark Matter (SQM 2017) Utrecht, the Netherlands, July 10-15, 2017*, 2017.

30. G. Aad *et al.*, Observation of a Centrality-Dependent Dijet Asymmetry in Lead-Lead Collisions at $\sqrt{s_{NN}} = 2.77$ TeV with the ATLAS Detector at the LHC, *Phys. Rev. Lett.* **105**, p. 252303 (2010).

31. N. Zardoshti, Investigating the Role of Coherence Effects on Jet Quenching in Pb-Pb Collisions at $\sqrt{s_{NN}} = 2.76$ TeV using Jet Substructure, *Nucl. Phys.* **A967**, 560 (2017).

32. S. Acharya *et al.*, First measurement of jet mass in PbPb and pPb collisions at the LHC, *Phys. Lett.* **B776**, 249 (2018).

Theoretical Implications of Precision Measurements

Jens Erler

Instituto de Física, Universidad Nacional Autónoma de México,
Apartado Postal 20–364, CDMX 01000, México
E-mail: erler@fisica.unam.mx
www.unam.mx

Results of an updated global electroweak fit to the Standard Model (SM) are presented, where special attention is paid to some key observables, such as the weak mixing angle, the W boson mass, and the anomalous magnetic moment of the muon. Implications for new physics beyond the SM are also discussed.

Keywords: Electroweak global fit; weak mixing angle; oblique parameters; high-precision measurements.

1. The Electroweak Fit

As of this writing the Standard Model (SM) is almost exactly 50 years old. In the fall of 1967 Steven Weinberg proposed a model of leptons[1] which was both inconsistent and incomplete due to the lack of the quark degrees of freedom, rendering it gauge anomalous but this was unknown at the time. It was then modified several times, and evolved into what we call the SM today, which despite of its senior age, appears almost immortal.

And, of course, on the 4th of July was the 5th birthday of the Higgs boson. To celebrate it let us time travel back to the time of ICHEP 2012 where the discovery was announced. If one took all of the available data at that time except for the LHC reconstruction data of the Higgs boson itself, namely electroweak precision measurements and the exclusion regions mapped out at other facilities, one found[2] the probability distribution of the Higgs boson mass, M_H, shown in Fig. 1(a). Including then the direct measurements of M_H at the time, one obtained the distribution in Fig. 1(b), showing a remarkable consistency and demonstrating the important role that the precision data played in terms of predicting M_H before its discovery and contributing to its rapid acceptance.

This brings us to old physics implications of precision measurements, also known as the global electroweak fit. Some of the results presented here are from the 2016 edition of the *Review of Particle Physics*[3], but I also include updates due to new results of this year. Besides the Yukawa sector, there are basically five parameters needed to fix the bosonic sector of the SM, namely the $SU(3)_C \times SU(2)_L \times U(1)_Y$ gauge couplings and two parameters from the Higgs potential. One combination of these is the fine structure constant, α, which one can take, for example, from the Rydberg constant, and doing so would leave the even more precise determination from the anomalous magnetic moment of the electron as a derived quantity and an

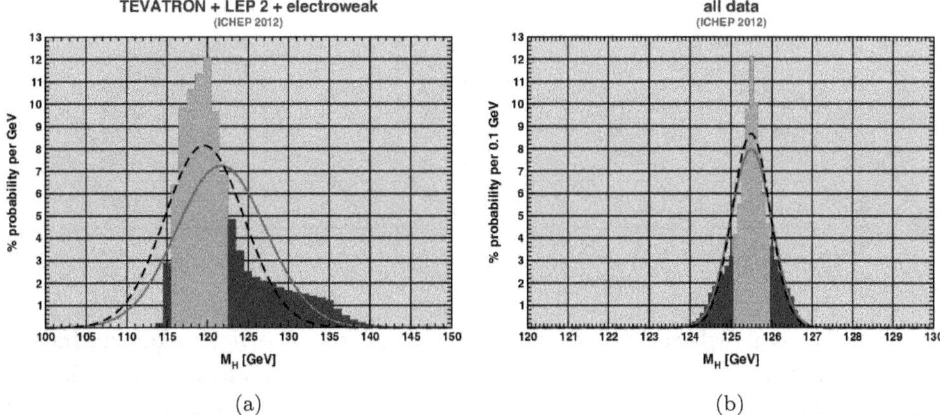

Fig. 1. Normalized probability distributions of M_H where the 68% CL highest probability density regions are marked in green. Also shown are two reference Gaussian: the dashed ones (in black) are centered around the median and have the same width as the 68% CL *central* probability intervals while the solid ones (in red) are based on mean and variance. (a) Electroweak precision data supplemented with exclusion constraints from the Tevatron and LEP 2 as of ICHEP 2012. (b) All data at that time (color online).

extra SM test. Another combination is the Fermi constant, G_F, which is extracted from the precise muon lifetime measurement[4] at the PSI in Switzerland. One obtains the Z boson mass, M_Z, from the Z line shape fit[5] at LEP, and nowadays M_H from the LHC[6] (see Sec. 1.3). On the other hand, the strong coupling constant, α_s, is left as an unconstrained fit parameter and is thus an output.

1.1. *Weak probes of the strong coupling*

There are basically two opportunities to obtain α_s from electroweak observables. One is again from the Z line shape, namely from the total decay width and the hadronic peak cross section, in addition to various branching ratios. The resulting constraint,

$$\alpha_s(M_Z) = 0.1203 \pm 0.0028 \,, \tag{1}$$

is the only one not limited by QCD theory. Incidentally, another way to use the Z line shape is to fit to the number of neutrinos, N_ν. In the past the result has been somewhat low compared to the SM prediction of $N_\nu = 3$, but now we find[7],

$$N_\nu = 2.992 \pm 0.007 \,. \tag{2}$$

The W boson width is also strongly dependent on α_s, but due to its lower precision it rather serves as a first plus second row CKM unitarity test.

The other possibility to obtain a determination of α_s from a weak process is through the τ lepton lifetime and its branching ratios. The τ mass is at a much

Table 1. Combinations of measurements of the top quark mass[a].

	Central value	Statistical error	Systematic error	Total error
ATLAS	172.84	0.34	0.61	0.70
Tevatron	174.30	0.35	0.54	0.64
CMS	172.43	0.13	0.46	0.48
grand average	172.97	0.13	0.38	0.41

Note: [a]All values in GeV.

lower energy scale where QCD is at the verge of a perturbative breakdown. Indeed, one can express the perturbative expansion in two different ways, either strictly in fixed-order perturbation theory, or by re-summing certain terms to all orders which is referred to as contour-improved perturbation theory[8]. Both of these expansions work quite well, but while they are seemingly converging they do not appear to converge to the same value. The reason for this is not at all understood and introduces an additional uncertainty. Including higher order terms in the operator product expansion and quark-hadron duality violating corrections[9,10], as well as the associated uncertainties, we find,

$$\alpha_s(m_\tau) = 0.314^{+0.016}_{-0.013} \,, \qquad\qquad \alpha_s(M_Z) = 0.1174^{+0.0019}_{-0.0017} \,. \qquad (3)$$

In total, the global fit which also contains many other α_s-dependent quantities returns the result[7],

$$\alpha_s(M_Z) = 0.1182 \pm 0.0016 \,. \qquad (4)$$

1.2. *Top quark mass*

Let us now turn to measurements[11] of the top quark pole mass, m_t. ATLAS[12], CMS[13], as well as the Tevatron Electroweak Working Group[14] each produced a combination (shown in Table 1) of their various determinations of m_t, based on different decay modes and analysis details. Assuming a common theoretical uncertainty for all three combinations equal to the smallest one, which is the quoted 0.29 GeV uncertainty from ATLAS, it is not difficult to perform the grand average,

$$m_t = 172.97 \pm 0.28_{\text{uncorr.}} \pm 0.29_{\text{corr.}} \pm 0.50_{\text{QCD}} \text{ GeV} \,, \qquad (5)$$

where the first and second error are the uncorrelated and correlated components, respectively. However, to split the total error of this average up into statistical and systematic components as done in Table 1 is less straightforward, and for the issues involved I refer the interested reader to Ref. 15. The last error in Eq. (5) from QCD includes the uncertainty from the ambiguity which top quark mass definition is actually measured at a hadron collider, and given that it is presumably close to the pole mass one also needs to include an extra error from the conversion[16] from the pole mass to the short-distance $\overline{\text{MS}}$-mass, where the latter actually enters

Table 2.   Determinations of M_H.

Method	Result
event kinematics	125.09 ± 0.24 GeV
Higgs branching ratios	126.1 ± 1.9 GeV
electroweak fit	90^{+18}_{-16} GeV

the radiative corrections to precision observables. There is an on-going debate how much this theory error can be reduced in the future. Improving the top quark mass determination still matters considering that the change from the previous value, $m_t = 173.34 \pm 0.81$ GeV, which was only slightly higher, lowers the extracted value of M_H by about 3 GeV.

If one performs a fit to the precision data only, ignoring the direct mass measurements at the hadron colliders, one finds a somewhat higher value[7],

$$m_t = 176.7 \pm 2.1 \text{ GeV} \qquad \text{(indirect)}. \tag{6}$$

We will encounter the reason for this in Sec. 2.2.

1.3. *Higgs boson mass*

M_H can be measured[6] precisely and cleanly at the LHC by the reconstruction of Higgs decays into two photons and *via* intermediate states involving a real Z boson and a virtual Z^* into four charged leptons. But there are two other strategies to determine M_H. One invokes the Higgs decay branching ratios which for a Higgs boson of $\mathcal{O}(125$ GeV$)$ strongly vary with M_H. Using ratios of the decay rates[17] into two gauge bosons, $\gamma\gamma$, WW^* and ZZ^*, in which some of the Higgs production uncertainties cancel, we find the result[7] on the second line in Table 2, in perfect agreement with the kinematically reconstructed Higgs boson mass and with an error of less than 2 GeV. And then there is the electroweak fit itself which returns a much lower central value of M_H, about 1.9 σ below those of the other two methods.

Figure 2 shows a breakdown of the constraints of the global fit. Indicated are the direct measurements of m_t and M_H, the Z pole asymmetries which are basically measurements of the weak mixing angle, the W boson mass, M_W, and a set of further Z pole observables which by themselves map out a bounded part of the parameter space. Overall, very good agreement is observed, except that M_W is seen to favor lower values of M_H.

2. Key Observables

2.1. *Weak mixing angle*

This section provides some details on the most important observables entering the global electroweak analysis. The first one to discuss is the weak mixing angle, which

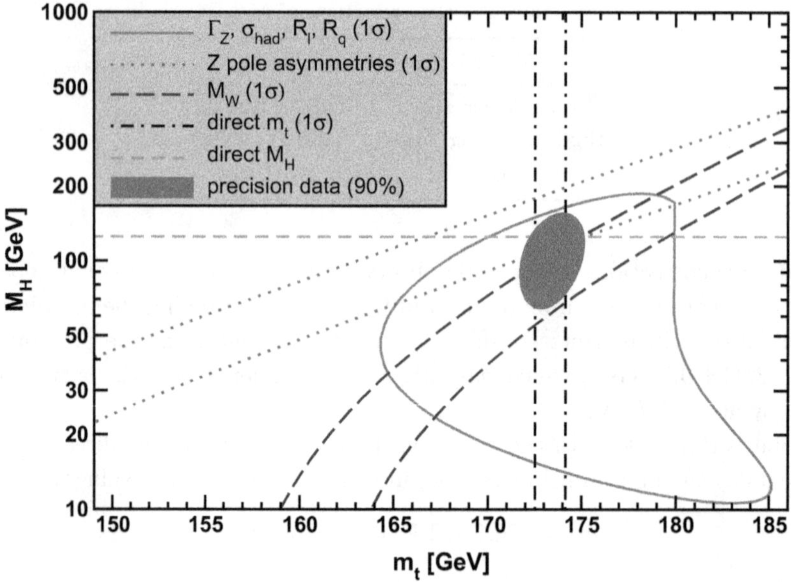

Fig. 2. One standard deviation constraints[7] on M_H as functions of m_t from various sets of precision data and the combined 90% CL region.

besides M_W is one of the most precise *derived* quantities in the electroweak sector. At the SM tree level it is given by

$$\sin^2 \theta_W = \frac{g'^2}{g^2 + g'^2} = 1 - \frac{M_W^2}{M_Z^2} \ . \tag{7}$$

Experimental results are often reported as measurements of an effective weak mixing angle, defined in terms of the vector and axial-vector Z boson couplings to leptons, v_ℓ and a_ℓ, as

$$\sin^2 \theta_{\text{eff}}^\ell \equiv \frac{1}{4} \left(1 - \frac{v_\ell}{a_\ell} \right). \tag{8}$$

Figure 3 shows all measurements of $\sin^2 \theta_{\text{eff}}^\ell$ that achieved a precision of better than 1%. Strictly parity-violating observables are marked by diamonds, others by circles. The first group is from the LEP Collaborations[5], featuring the forward-backward asymmetries into bottom and charm quark pairs which are measured on the low side of the SM predictions, hence favoring values of $\sin^2 \theta_W$ on the high side. The next group is from the SLD Collaboration[5] at the SLC, where the left-right polarization asymmetries into hadronic and leptonic final states both favor lower values of $\sin^2 \theta_W$.

The forward-backward asymmetries for e^+e^- and $\mu^+\mu^-$ final states have been measured by CDF and DØ at the Tevatron[18], and by ATLAS and CMS at the LHC[19]. The asymmetry with muon final states was also measured by the LHCb

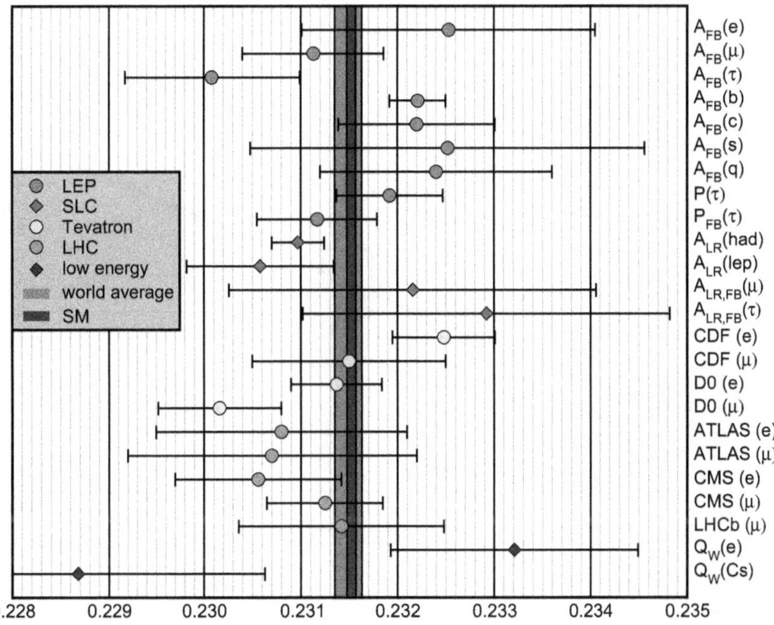

Fig. 3. Sub-% measurements of the effective leptonic weak mixing angle. The wider one of the two vertical bands is the world average of all measurements, while the narrower one represents the SM prediction based on the input values for m_t, M_H, etc.

Collaboration[19], providing valuable complementarity due to the different kinematics focused on by their detector.

Finally, there are various weak charges, Q_W. The weak charge of the electron, $Q_W(e)$, has been measured by the SLAC–E–158 Collaboration in polarized Møller scattering using the SLC electron beam[20]. Qweak was the analogous experiment measuring the weak charge of the proton[21], $Q_W(p)$, in elastic e^-p scattering using the polarized electron beam at Jefferson Lab, and has been completed recently[22]. Atomic parity violation (APV) is sensitive to the weak charges of heavy nuclei. The most precise result was achieved in Cs in an experiment at Boulder[23]. The very complicated atomic theory[24] is also best understood in Cs.

Measurements of $\sin^2\theta_W$ are important for a variety of reasons even in the absence of physics beyond the SM. They represent a unique type of test of the electroweak symmetry breaking sector, because as summarized in Eq. (7) it is a parameter that can be written either in terms of gauge couplings, or in terms of vector boson masses. They also serve as a test of the still poorly studied Higgs sector, since values of $\sin^2\theta_W$ can be translated into values of M_H and then confronted with the corresponding LHC results. And finally there is a $3\,\sigma$ conflict between the most precise results from $A_{LR}(\text{had})$ and $A_{FB}(b)$ as illustrated in Fig. 4.

Measurements of $\sin^2\theta_W$ are even more important in the context of new physics beyond the SM, which can enter in very different ways as sketched in Fig. 5.

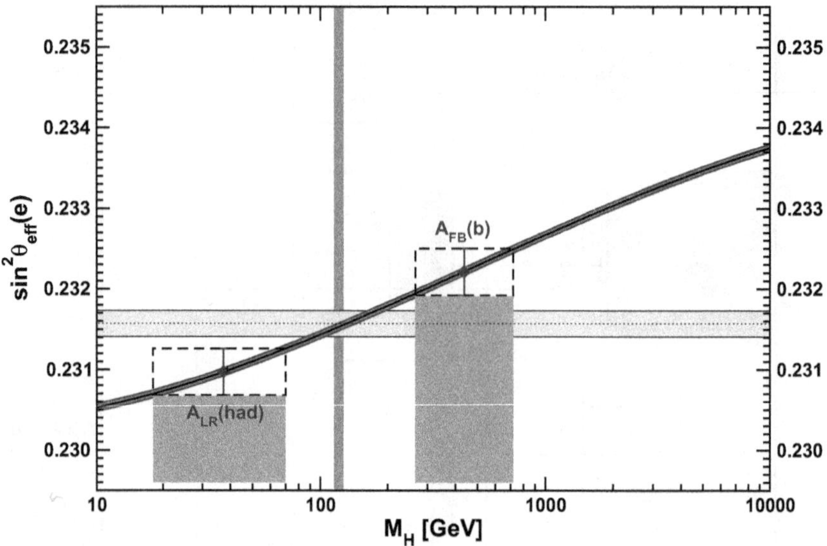

Fig. 4. Higgs mass predictions derived from A_{LR}(had) (SLC) and $A_{FB}(b)$ (LEP). The former predicts Higgs boson masses of the order of tens of GeV, while the latter prefers M_H values of order hundreds of GeV. Only the average of those and other measurements of $\sin^2 \theta_W$ is truly consistent with the SM.

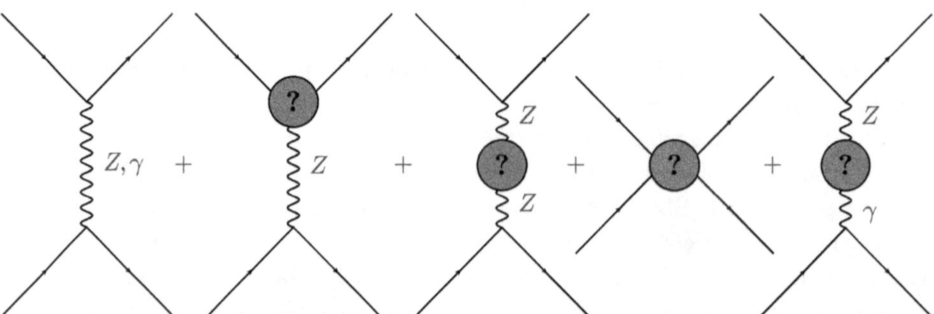

Fig. 5. Sketch of how new physics may affect the extracted values of $\sin^2 \theta_W$. The first diagram represents the SM tree level, while the remaining ones represent, respectively, vertex corrections, oblique corrections, a non-standard four-fermion contact interaction from heavy new physics, and a change in the renormalization group running of $\sin^2 \theta_W$ from light new physics. One needs to vary measurement types and energy scales to disentangle these possibilities experimentally [25].

(i) One way is through Z-Z' mixing. If there is an extra neutral gauge boson [26,27], Z', exhibiting mass mixing with the ordinary Z, there may be very significant modifications of its vector couplings. These would manifest themselves in determinations of $\sin^2 \theta_W$ seemingly disagreeing with the SM. This is also the reason why the extracted limits on Z-Z' mixing angles are very strong, and typically at the few per-mille level.

(ii) Another important way is through the interpretation of the so-called oblique parameters, which are discussed in Sec. 3.1.

(iii) New amplitudes may also be present, *e.g.*, from an additional Z' boson. Some new four-Fermi operator could produce a measurable effect by means of interference with the photon at low momentum transfer, but would go unnoticed in the context of measurements around the Z resonance under which it would be buried. If one then compares on with off Z pole measurements of $\sin^2\theta_W$ one may be able to isolate this kind of new contact interaction.

(iv) Finally, there is the possibility of a change in the renormalization group evolution of $\sin^2\theta_W$. If there was a new light particle with a mass somewhere between zero and M_Z, this could have an effect on the β function[28] of $\sin^2\theta_W$.

The renormalization group running of the weak mixing angle within the SM is illustrated in Fig. 6. The calculation faces similar issues and problems as the calculation of the electromagnetic coupling at the Z scale in terms of α in the Thomson limit. In the case of $\sin^2\theta_W$ one starts at the Z pole from where the most precise measurements derive, and moves to lower scales to compare with the extractions from Qweak[22] or other processes involving parity-violation.

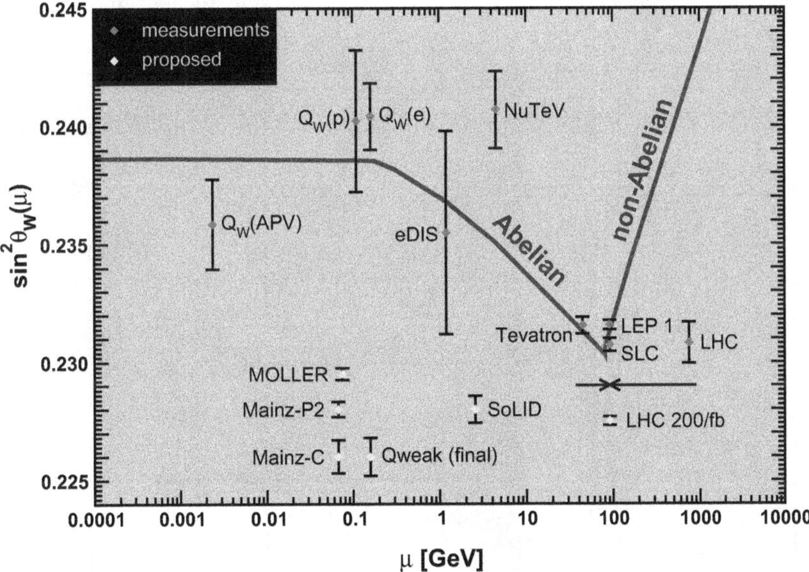

Fig. 6.   Renormalization group evolution[28] of $\sin^2\theta_W$ in the modified minimal subtraction scheme, $\overline{\text{MS}}$. At the scale of $\mu = M_W$ the β-function changes sign, signaling the change from an effectively Abelian theory to a non-Abelian one. Indicated are also various existing and upcoming measurements. For more details on some of the lower energy measurements, see Ref. 22. The data points around the Z pole (for lack of space, the Tevatron and LHC points have been shifted horizontally) are the averages of the individual determinations displayed in Fig. 3, taking into account correlated systematic errors.

One employs perturbative QCD wherever possible, *i.e.*, down to $\mu \approx 2$ GeV, for which one needs precise input values of the charm and bottom quark masses. In the region where one cannot rely on perturbation theory one can try to relate the hadronic contribution that is not calculable from first principles to the corresponding result of α. While there is a part which contributes in the same way to both, $\sin^2 \theta_W$ and α, there is a complication because the ratio of Z vector couplings to up-type and down-type quarks differs from the ratio of their electric charges, and a flavor separation is in order. This can be achieved to sufficient precision by constructing upper and lower bounds on the strange quark contribution[28]. Another separation is needed for the singlet piece, *i.e.*, the OZI rule violating part where one has a quark current connecting to a set of gluons and then connecting further to a another quark-anti-quark pair. This piece is small but in principle introduces some additional uncertainty. Fortunately, there is a lattice gauge theory calculation of the singlet contribution to the anomalous magnetic moment of the muon[29] that can be adapted to this case[30].

2.2. *W boson mass*

Another key observable is M_W where the status is almost the opposite from $\sin^2 \theta_W$. As can be seen in Fig. 7, the most precise measurements are in perfect agreement with each other, but the central values of all available measurements except for

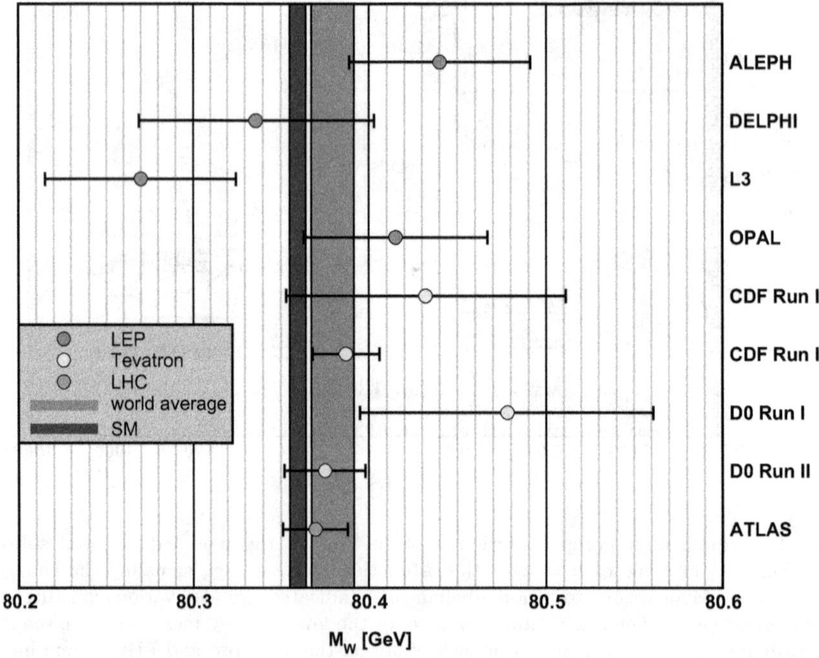

Fig. 7. Measurements of the W boson mass from LEP[31], the Tevatron[32,33] and ATLAS[19].

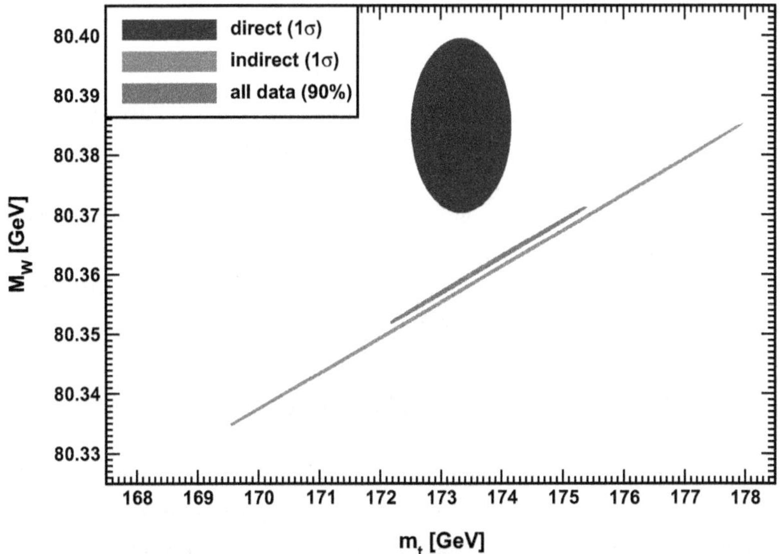

Fig. 8. Constraints[7] on M_W as functions of m_t. The blue ellipse represents the direct measurements, while the long green line is the 1 σ contour from all other data. The shorter red line is the combination of all data at the 90% CL. The approximately 2 σ SM deviation in M_W is apparent (color online).

DELPHI and L3 are higher than the SM prediction. As a result, the world average is off by about two standard deviations. This is also transparent from Fig. 8.

There is a very interesting interpretation of an enhanced M_W within the Minimal Supersymmetric Standard Model (MSSM). While the size of a possible shift in M_W is not clearly predicted, the overall sign of the MSSM contributions[34] is expected to increase M_W relative to the SM prediction, in agreement with what is currently seen. This is regardless of whether the boson that the LHC has discovered was the lighter or the heavier of the two CP-even Higgs eigenstates that are present in the MSSM, but the latter case is much more constrained[35].

2.3. *Anomalous magnetic moment of the muon*

The final key observable is the anomalous magnetic moment of the muon,

$$a_\mu \equiv \frac{g_\mu - 2}{2} = \frac{\alpha}{2\pi} + \mathcal{O}(\alpha^2) \ . \tag{9}$$

It has been measured by the Muon $g - 2$ (BNL–E–821) Collaboration[36],

$$a_\mu = (1165920.91 \pm 0.63) \times 10^{-9}, \tag{10}$$

and there will be a follow-up experiment at Fermilab[37], as well as a conceptually different experiment at J-PARC[38], each attempting to reduce the experimental uncertainty by a factor of about four. The SM prediction,

$$a_\mu = (1165917.63 \pm 0.46) \times 10^{-9}, \tag{11}$$

deviates by 4.2 σ from Eq. (10), but there are some issues regarding the SM value.

One is the hadronic vacuum polarization contribution which enters first at the two-loop level and is depicted in Fig. 9(a). Again, one can use perturbative QCD for part of the effect[39], and just as in Sec. 2.1 one needs values of the bottom and charm quark masses as inputs. Here most of the contribution is from the lower energy hadronic regime, where one has to resort to data from experiments or lattice gauge theory simulations. Currently, experimental data come from three different types of sources:

(i) energy scans measuring e^+e^- annihilation cross sections into hadronic final states;

(ii) radiative returns[40] from resonances such as the ϕ or the $\Upsilon(4S)$ (also produced in e^+e^- annihilation), *i.e.*, decays where the energy is shared between the hadronic system and an additional photon (this method is dominated by systematic and theoretical uncertainties);

(iii) spectral functions of $\tau^\pm \to \nu_\tau \pi^\pm \pi^0$ decays, which are related by an isospin rotation to $e^+e^- \to \pi^+\pi^-$. Final states with four pions can also be used. The method assumes isospin symmetry and isospin violating effects have to be corrected for introducing an extra uncertainty.

The experimental results from e^+e^- annihilation and τ decays are discussed in detail in Ref. 41. In each of the two classes, the results are in very good agreement with each other, but the analysis based on τ decays showed a smaller deviation from the SM which prefers higher values of the hadronic contribution to a_μ. However, a few years ago it was pointed out[42] that another isospin-breaking effect originating from γ-ρ mixing needs to be included. Ref. 42 obtained a parameter-free prediction for the size this correction and applying it moves the τ based analysis into very good agreement with the e^+e^- data. Averaging them together implies a larger deviation compared to considering the e^+e^- data alone.

There are interesting proposals to reduce the hadronic vacuum polarization uncertainty in a_μ in the future. They involve to measure the vacuum polarization in the space-like region from Bhabha scattering[43] or from μe scattering[44]. Such an experiment would measure the running of α, and the corresponding contribution to a_μ would be achieved by means of the convolution[45],

$$a_\mu = \frac{\alpha}{\pi} \int_0^1 dx (1-x) \Delta\alpha \left(\frac{x^2 m_\mu^2}{x-1} \right). \tag{12}$$

Another difficult issue is the hadronic light-by-light contribution shown in Fig. 9(b). Entering first at the three-loop level, it is much smaller than that of the hadronic vacuum polarization. However, its uncertainty is nevertheless numerically comparable and crudely estimated to be roughly 30% of itself. The reason is

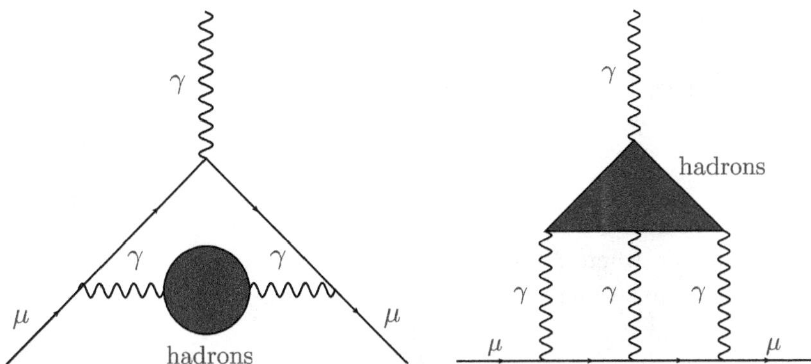

Fig. 9. Hadronic contributions to the anomalous magnetic moment of the muon. (a) Two-loop vacuum polarization contribution. (b) Three-loop light-by-light scattering contribution.

that there are not have enough data to constrain the associated dispersion integral which in this case is two-dimensional. Fortunately, there are promising efforts from lattice gauge theory[29]. A recent calculation of the quark-connected and leading disconnected contributions[46] achieved a 25% statistical precision while the systematics is currently under investigation. There is a valuable and encouraging cross-check in which the muonic light-by-light contribution has been simulated on the lattice, and the result agrees within about 2% with perturbative QED[47].

There are also recent lattice results[48] on the vacuum polarization contribution. Here the uncertainty could be reduced to 6%, and sub-% precision may be achievable in the not too distant future. Note that about half a percent accuracy would be needed to be competitive with the dispersion result.

3. New Physics Implications

3.1. *Oblique physics beyond the Standard Model*

The oblique parameters describe radiative corrections to the vector boson two-point correlation functions, and in one specific formalism[49] are called the S, T and U parameters. They may be defined to describe new physics only, such that $S = T = U = 0$ in the SM. To extract them from the data one needs measurements of $\sin^2 \theta_W$, M_W, as well as at least one other derived observable, such as the Z boson width, Γ_Z, or a low-energy neutral current observable.

The oblique parameters are easily affected by most models of physics beyond the SM, especially those addressing naturalness issues of the Higgs potential (the hierarchy problem). The oblique approximation neglects possible additional new physics effects, such as direct contributions to fermion couplings. Even where this is not a reasonable approximation, it still provides a valuable reference case.

The T parameter breaks the accidental (custodial) $SO(4)$ symmetry present in the Higgs potential. Its effects are indistinguishable from tree-level new physics corrections to the ρ_0 parameter, defined as the ratio of the neutral-current to charged-current interaction strengths.

A new multiplet of heavy *degenerate* chiral fermions would produce a constant contribution,

$$\Delta S = \sum_i \frac{N_C^i}{3\pi}(t_{3L}^i - t_{3R}^i)^2, \qquad (13)$$

to the S parameter, where N_C^i is the color factor of multiplet i. E.g., an additional complete and degenerate fermion generation or mirror generation would contribute

$$\Delta S = \frac{2}{3\pi} \approx 0.21 \ .$$

Another way to think about S and T is that they correspond to dimension six operators in the SM effective field theory, in which one supplements the SM with additional gauge-invariant, but non-renormalizable operators. The U parameter corresponds to a combination of dimension eight operators, so that U is expected to be more suppressed than S and T. This expectation is indeed borne out in concrete models.

3.2. *Non-degenerate doublets*

A simple example is given by a set of extra non-degenerate doublets which would contribute to $\rho_0 = 1 + \alpha T$ as[50],

$$\Delta\rho_0 = \frac{G_F}{8\sqrt{2}\pi^2} \sum_i N_C^i \Delta m_i^2 \ , \qquad (14)$$

where Δm_i^2 is not simply the difference of the masses squared of the two members of each doublet, but rather a more complicated function with the property that $\Delta m_i^2 \geq (m_1 - m_2)^2$. Thus, it is a positive definite function, and despite appearances, *there is* decoupling in this formula. At first sight one could have the impression that a very heavy doublet with, say, Planck scale masses and a mass difference of electroweak size, could give a measurable effect even at present day colliders. However, this is not the case, because in actual models in turns out that Δm_i^2 itself will experience a see-saw type of suppression and will tend to zero, consistent with the interpretation of T as a dimension six operator.

With the latest results on $\sin^2\theta_W$, *etc.*, I now find,

$$\rho_0 = 1.00039 \pm 0.00019 \ , \qquad (15)$$

which differs from the SM prediction, $\rho_0 = 1$, by 2.0 standard deviations. It is amusing that this implies a *non-vanishing new physics contribution* at the 90% CL, for which I find the range,

$$(15 \text{ GeV})^2 \leq \sum_i \frac{N_C^i}{3} \Delta m_i^2 \leq (47 \text{ GeV})^2. \qquad (16)$$

Note, that the future Circular Electron Positron Collider (CEPC) under consideration in China, could measure ρ_0 with a precision of 8×10^{-5}, and assuming that the central value would not change from today, the quantity in Eq. (16) would be discovered to be non-zero at the $5\ \sigma$ level.

3.3. *The S and T parameters*

The result of a simultaneous fit to S and T is shown in Fig. 10. Notice that with the exception of the constraint from atomic parity violation (which is from Cs and Tl experiments) all constraints have a similar positive slope as they are all related to $\sin^2 \theta_W$. The various classes of constraints and the combined fit are in reasonable agreement with the SM prediction, $S = T = 0$, but at the best fit values,

$$S = 0.06 \pm 0.08 \ , \tag{17}$$

$$T = 0.09 \pm 0.06 \ , \tag{18}$$

the fit is moderately better, showing a decrease of $\Delta\chi^2 = -4.0$ compared to the SM global fit. It would be interesting to improve the precision in such a fit. This could be done, for example, at the CEPC which could achieve uncertainties of ± 0.014 and ± 0.017 in S and T, respectively.

The S parameter by itself rules out QCD-like technicolor models, and only much more complicated models may still be viable. S also incontrovertibly rules out a degenerate fourth fermion generation, and even a non-degenerate one is highly disfavored[51].

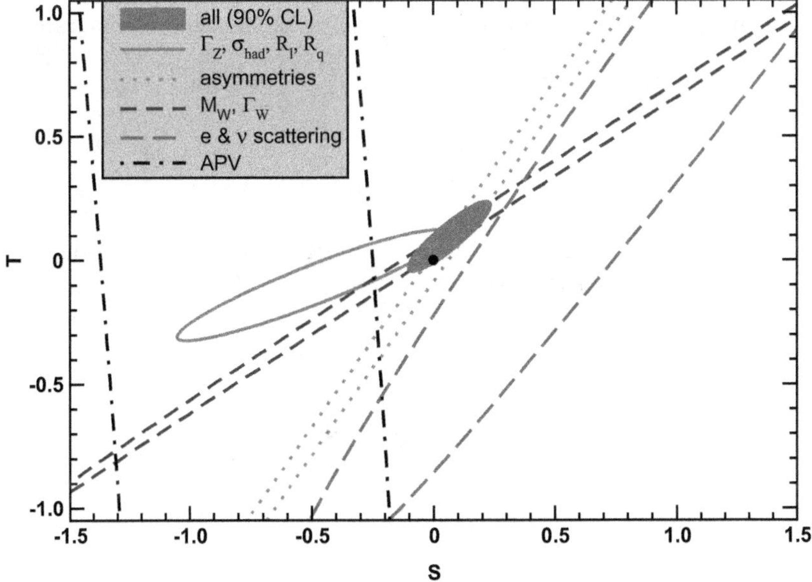

Fig. 10.   Constraints[7] on T as functions of S.

3.4. *Non-oblique parameters*

As already mentioned in Sec. 2.1, there is a long-standing SM deviation in the forward-backward asymmetry, $A_{FB}(b)$, as measured at LEP. Interpreting $A_{FB}(b)$ instead of a measurement of $\sin^2 \theta_W$ as a measurement of the flavor-dependent form factor $\Delta\kappa_b$ multiplying the weak mixing angle relevant for b quarks, and performing a fit simultaneously with another form factor $\Delta\rho_b$ which can be thought of as the ρ_0 parameter for b quarks, we find[7],

$$\Delta\rho_b = 0.056 \pm 0.020 , \tag{19}$$

$$\Delta\kappa_b = 0.182 \pm 0.068 . \tag{20}$$

This represents a 2.7 σ deviation from the SM prediction $\Delta\rho_b = \Delta\kappa_b = 0$, and is driven by the deviation in $A_{FB}(b)$. However, it is difficult to explain this deviation in terms of new physics without also shifting the Z boson branching ratio into b quarks, R_b, which is in reasonable agreement with the SM.

The CEPC may be able to achieve precisions of ± 0.005 and ± 0.007 in $\Delta\rho_b$ and $\Delta\kappa_b$, respectively. Note that the results in Eq. (20) are essentially independent of S, T, and U, meaning that whether you allow them to also vary in the fits or fix them to the SM, makes little difference in these extractions.

3.5. *Compositeness scales from low energies*

From polarized electron-proton[22], Møller, and deep inelastic scattering (DIS), as well as APV, one can obtain constraints in a three-dimensional coupling space of parity-violating four-fermion operators, appearing in the effective Lagrangian[52],

$$\mathcal{L}_{\rm NC}^{eq} = - \frac{2}{v^2} \frac{\bar{e}\gamma^5\gamma^\mu e}{2} \left[g_{AV}^{eu} \frac{\bar{u}\gamma_\mu u}{2} + g_{AV}^{ed} \frac{\bar{d}\gamma_\mu d}{2} \right]$$
$$- \frac{2}{v^2} \frac{\bar{e}\gamma^\mu e}{2} \left[g_{VA}^{eu} \frac{\bar{u}\gamma^5\gamma_\mu u}{2} + g_{VA}^{ed} \frac{\bar{d}\gamma^5\gamma_\mu d}{2} \right] , \tag{21}$$

where $v = (\sqrt{2}G_F)^{-1/2} = 246.22$ GeV is the Higgs vacuum expectation value. The SM tree-level relations for the real-valued coefficients g_{AV}^{eq} and g_{VA}^{eq} are given by

$$g_{AV}^{eu} \equiv \cos^2 \theta_W g_A^e g_V^u = -\frac{1}{2} + \frac{4}{3}\sin^2 \theta_W, \tag{22}$$

$$g_{AV}^{ed} \equiv \cos^2 \theta_W g_A^e g_V^d = \frac{1}{2} - \frac{2}{3}\sin^2 \theta_W, \tag{23}$$

$$g_{VA}^{eu} \equiv \cos^2 \theta_W g_V^e g_A^u = -\frac{1}{2} + 2\sin^2 \theta_W, \tag{24}$$

$$g_{VA}^{ed} \equiv \cos^2 \theta_W g_V^e g_A^d = \frac{1}{2} - 2\sin^2 \theta_W. \tag{25}$$

The results and their combinations are shown in Fig. 11. For Fig. 12 these have been translated into sensitivities to new physics scale. In the future, such scales may reach up to about 50 TeV provided one assumes the strong coupling case, as *e.g.*, in compositeness models.

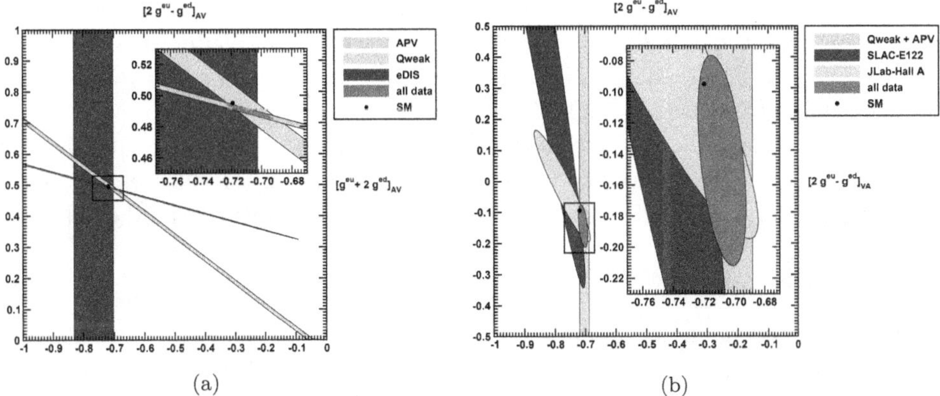

Fig. 11.   Constraints on parity-violating four-fermion operators[53]. Notice that the abscissa is identical in the two figures. (a) Operators containing the axial-vector electron bilinear $\bar{e}\gamma^5\gamma^\mu e$. (b) Operators containing the charge-weighted quark combinations.

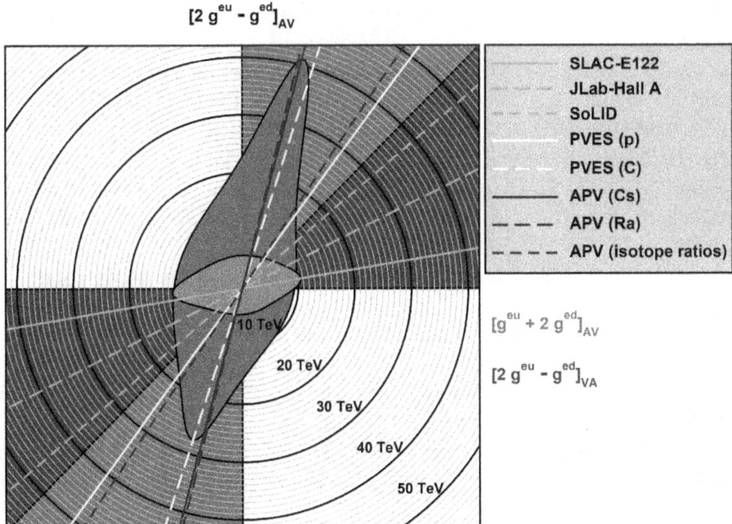

Fig. 12.   Reach to new physics scales[53] from the four-fermion operator constraints in Fig. 11. The blue segment is accessible to DIS experiments (yellow lines) and defines a plane (containing the brown 95% CL exclusion contour) perpendicular to the plane containing the red segment, the green contour and the white and maroon lines. Thus, the two planes are subspaces of a three-dimensional parameter space which intersect along the horizontal direction. The lines indicate the coupling combinations of the various experiments relative to the common horizontal direction (color online).

4. Conclusions

To summarize, the SM is almost 50 years old, but in remarkable health. It is over-constrained, where derived quantities like $\sin^2\theta_W$, M_W, $g_\mu - 2$, and weak charges have been both computed and measured. The precision in future measurements of $\sin^2\theta_W$, M_W, $Q_W(e)$ and $Q_W(p)$ will challenge theory, and a major global effort is needed to keep the theory uncertainties well below the projected experimental ones. Contact interactions derived from comparing future measurements of $\sin^2\theta_W$ at low energies[54] with determinations near the Z pole will be able to test new physics scales up to around 50 TeV.

Currently, the indirectly determined Higgs boson mass from the global fit is 1.9 σ below the directly reconstructed mass, while the ρ_0 parameter is 2.0 σ high. Both observations can be traced to the measured M_W which is above the SM prediction.

Acknowledgements

I would like to thank the organizers for the kind invitation to a very enjoyable symposium. This work is supported by CONACyT (México) project 252167–F.

References

1. S. Weinberg, A Model of Leptons, *Phys. Rev. Lett.* **19**, 1264 (1967).
2. J. Erler, Weighing in on the Higgs, arXiv:1201.0695 [hep-ph].
3. Particle Data Group: C. Patrignani *et al.*, Review of Particle Physics, *Chin. Phys. C* **40**, 100001 (2016).
4. MuLan Collaboration: D. M. Webber *et al.*, Measurement of the Positive Muon Lifetime and Determination of the Fermi Constant to Part-per-Million Precision, *Phys. Rev. Lett.* **106**, 041803 (2011), arXiv:1010.0991 [hep-ex].
5. ALEPH, DELPHI, L3, OPAL, and SLD Collaborations, LEP Electroweak Working Group, and SLD Electroweak and Heavy Flavour Groups: S. Schael *et al.*, Precision electroweak measurements on the Z resonance, *Phys. Rept.* **427**, 257 (2006), hep-ex/0509008.
6. Elisabetta Pianori, Higgs boson measurements in diboson decay modes, these proceedings.
7. J. Erler and A. Freitas, Electroweak Model and Constraints on New Physics, pp. 151–171 in Ref. 3.
8. F. Le Diberder and A. Pich, The perturbative QCD prediction to R_τ revisited, *Phys. Lett. B* **286**, 147 (1992).
9. D. Boito, M. Golterman, M. Jamin, A. Mahdavi, K. Maltman, J. Osborne and S. Peris, An Updated determination of α_s from τ decays, *Phys. Rev. D* **85**, 093015 (2012), arXiv:1203.3146 [hep-ph].
10. D. Boito, M. Golterman, K. Maltman, J. Osborne and S. Peris, Strong coupling from the revised ALEPH data for hadronic τ decays, *Phys. Rev. D* **91**, 034003 (2015), arXiv:1410.3528 [hep-ph].
11. Yuji Yamazaki, Top-Quark Measurements, these proceedings.
12. R. Nisius, Measurements of the top quark mass with the ATLAS detector, arXiv:1709.09845 [hep-ex].
13. S. Spannagel, Top quark mass measurements with the CMS experiment at the LHC, *PoS DIS* **2016**, 150 (2016), arXiv:1607.04972 [hep-ex].

14. CDF and DØ Collaborations, and Tevatron Electroweak Working Group: T. Aaltonen *et al.*, Combination of CDF and D0 results on the mass of the top quark using up 9.7 fb^{-1} at the Tevatron, arXiv:1608.01881 [hep-ex].

15. J. Erler, On the Combination Procedure of Correlated Errors, *Eur. Phys. J. C* **75**, 453 (2015), arXiv:1507.08210 [physics.data-an].

16. P. Marquard, A. V. Smirnov, V. A. Smirnov, M. Steinhauser and D. Wellmann, $\overline{\text{MS}}$-on-shell quark mass relation up to four loops in QCD and a general $SU(N)$ gauge group, *Phys. Rev. D* **94**, 074025 (2016), arXiv:1606.06754 [hep-ph].

17. ATLAS and CMS Collaborations: G. Aad *et al.*, Combined Measurement of the Higgs Boson Mass in pp Collisions at $\sqrt{s} = 7$ and 8 TeV with the ATLAS and CMS Experiments, *Phys. Rev. Lett.* **114**, 191803 (2015), arXiv:1503.07589 [hep-ex], and `cds.cern.ch/record/2052552/files/ATLAS-CONF-2015-044.pdf`.

18. CDF and DØ Collaborations and the Tevatron Electroweak Working Group, Tevatron combination of the effective leptonic electroweak mixing angles (2017), `https://tevewwg.fnal.gov/wz/sw2eff17/drafts/Fermilab_Conf_17_201_E.pdf`.

19. Qiang Li, EW Measurements from LHC and Past Experiments, these proceedings.

20. SLAC–E–158 Collaboration: P. L. Anthony *et al.*, Precision measurement of the weak mixing angle in Møller scattering, *Phys. Rev. Lett.* **95**, 081601 (2005), hep-ex/0504049.

21. J. Erler, A. Kurylov and M. J. Ramsey-Musolf, The Weak charge of the proton and new physics, *Phys. Rev. D* **68**, 016006 (2003), hep-ph/0302149.

22. Wouter Deconinck, Precision Measurements with Leptons and Kaons and Nuclei, these proceedings.

23. S. C. Bennett and C. E. Wieman, Measurement of the $6S \to 7S$ transition polarizability in atomic cesium and an improved test of the Standard Model, *Phys. Rev. Lett.* **82**, 2484 (1999), hep-ex/9903022.

24. J. S. M. Ginges and V. V. Flambaum, Violations of fundamental symmetries in atoms and tests of unification theories of elementary particles, *Phys. Rept.* **397**, 63 (2004), physics/0309054.

25. J. Erler and M. J. Ramsey-Musolf, Low energy tests of the weak interaction, *Prog. Part. Nucl. Phys.* **54**, 351 (2005), hep-ph/0404291.

26. P. Langacker, The Physics of Heavy Z' Gauge Bosons, *Rev. Mod. Phys.* **81**, 1199 (2009), arXiv:0801.1345 [hep-ph].

27. J. Erler, P. Langacker, S. Munir and E. Rojas, Improved Constraints on Z-prime Bosons from Electroweak Precision Data, *JHEP* **0908**, 017 (2009), arXiv:0906.2435 [hep-ph].

28. J. Erler and M. J. Ramsey-Musolf, The Weak mixing angle at low energies, *Phys. Rev. D* **72**, 073003 (2005), hep-ph/0409169.

29. Takashi Kaneko, LQCD: Flavor Physics and Spectroscopy, these proceedings.

30. J. Erler and R. Ferro-Hernández, Reduced Uncertainties in the Weak Mixing Angle at Low Energies, in preparation.

31. ALEPH, DELPHI, L3, and OPAL Collaborations and LEP Electroweak Working Group: S. Schael *et al.*, Electroweak Measurements in Electron-Positron Collisions at W-Boson-Pair Energies at LEP, *Phys. Rept.* **532**, 119 (2013), arXiv:1302.3415 [hep-ex].

32. DØ Collaboration: V. M. Abazov *et al.*, Measurement of the W Boson Mass with the D0 Detector, *Phys. Rev. Lett.* **108**, 151804 (2012), arXiv:1203.0293 [hep-ex].

33. CDF and DØ Collaborations and the Tevatron Electroweak Working Group, 2012 Update of the Combination of CDF and D0 Results for the Mass of the W Boson, arXiv:1204.0042 [hep-ex].

34. S. Heinemeyer, W. Hollik, G. Weiglein and L. Zeune, Implications of LHC search results on the W boson mass prediction in the MSSM, *JHEP* **1312**, 084 (2013), arXiv:1311.1663 [hep-ph].

35. P. Bechtle, H. E. Haber, S. Heinemeyer, O. Stål, T. Stefaniak, G. Weiglein, L. Zeune, The Light and Heavy Higgs Interpretation of the MSSM, *Eur. Phys. J. C* **77**, 67 (2017), arXiv:1608.00638 [hep-ph].

36. Muon g-2 Collaboration: G. W. Bennett *et al.*, Final Report of the Muon E821 Anomalous Magnetic Moment Measurement at BNL, *Phys. Rev. D* **73**, 072003 (2006), hep-ex/0602035.

37. Muon g-2 Collaboration: J. Grange *et al.*, Muon $g-2$ Technical Design Report, arXiv:1501.06858 [physics.ins-det].

38. M. Aoki *et al.*, Conceptual Design Report for The Measurement of the Muon Anomalous Magnetic Moment $g-2$ and Electric Dipole Moment at J-PARC, `https://g2sakura.kek.jp/public/doc/MCDR-submit.pdf`.

39. J. Erler and M. Luo, Hadronic loop corrections to the muon anomalous magnetic moment, *Phys. Rev. Lett.* **87**, 071804 (2001), hep-ph/0101010.

40. G. Rodrigo, H. Czyz, J. H. Kühn and M. Szopa, Radiative return at NLO and the measurement of the hadronic cross-section in electron positron annihilation, *Eur. Phys. J. C* **24**, 71 (2002), hep-ph/0112184.

41. M. Davier *et al.*, The Discrepancy Between τ and e^+e^- Spectral Functions Revisited and the Consequences for the Muon Magnetic Anomaly, *Eur. Phys. J. C* **66**, 127 (2010), arXiv:0906.5443 [hep-ph].

42. F. Jegerlehner and R. Szafron, ρ^0-γ mixing in the neutral channel pion form factor F_π^e and its role in comparing e^+e^- with τ spectral functions, *Eur. Phys. J. C* **71**, 1632 (2011), arXiv:1101.2872 [hep-ph].

43. C. M. Carloni Calame, M. Passera, L. Trentadue and G. Venanzoni, A new approach to evaluate the leading hadronic corrections to the muon $g-2$, *Phys. Lett. B* **746**, 325 (2015), arXiv:1504.02228 [hep-ph].

44. G. Abbiendi *et al.*, Measuring the leading hadronic contribution to the muon $g-2$ via μe scattering, *Eur. Phys. J. C* **77**, 139 (2017), arXiv:1609.08987 [hep-ex].

45. B. E. Lautrup, A. Peterman and E. de Rafael, Recent developments in the comparison between theory and experiments in quantum electrodynamics, *Phys. Rept.* **3**, 193 (1972).

46. T. Blum *et al.*, Connected and Leading Disconnected Hadronic Light-by-Light Contribution to the Muon Anomalous Magnetic Moment with a Physical Pion Mass, *Phys. Rev. Lett.* **118**, 022005 (2017), arXiv:1610.04603 [hep-lat].

47. T. Blum *et al.*, Using infinite volume, continuum QED and lattice QCD for the hadronic light-by-light contribution to the muon anomalous magnetic moment, *Phys. Rev. D* **96**, no. 3, 034515 (2017), arXiv:1705.01067 [hep-lat].

48. M. Della Morte *et al.*, The hadronic vacuum polarization contribution to the muon $g-2$ from lattice QCD, *JHEP* **1710**, 020 (2017), arXiv:1705.01775 [hep-lat].

49. M. E. Peskin and T. Takeuchi, Estimation of oblique electroweak corrections, *Phys. Rev. D* **46**, 381 (1992).

50. M. J. G. Veltman, Limit on Mass Differences in the Weinberg Model, *Nucl. Phys. B* **123**, 89 (1977).

51. J. Erler and P. Langacker, Precision Constraints on Extra Fermion Generations, *Phys. Rev. Lett.* **105**, 031801 (2010), arXiv:1003.3211 [hep-ph].

52. J. Erler and S. Su, The Weak Neutral Current, *Prog. Part. Nucl. Phys.* **71**, 119 (2013), arXiv:1303.5522 [hep-ph].

53. J. Erler, C. J. Horowitz, S. Mantry and P. A. Souder, *Weak Polarized Electron Scattering*, *Ann. Rev. Nucl. Part. Sci.* **64**, 269 (2014), arXiv:1401.6199 [hep-ph].
54. K. S. Kumar, S. Mantry, W. J. Marciano and P. A. Souder, *Low Energy Measurements of the Weak Mixing Angle*, *Ann. Rev. Nucl. Part. Sci.* **63**, 237 (2013), arXiv:1302.6263 [hep-ex].

Lattice QCD: Hadron Spectroscopy and Flavor Physics

Takashi Kaneko

High Energy Accelerator Research Organization (KEK), Ibaraki 305-0801, Japan

School of High Energy Accelerator Science, SOKENDAI (The Graduate University for Advanced Studies), Ibaraki 305-0801, Japan
E-mail: takashi.kaneko@kek.jp

We review recent progress on hadron spectroscopy and flavor physics from lattice QCD. Recent rapid progress on the muon anomalous magnetic moment is also discussed.

1. Introduction

Lattice QCD plays a key role in the intensity frontier for the search of new physics. Interpretation of experimental measurements within and beyond the Standard Model (SM) requires precise knowledge on the relevant hadronic matrix elements describing nonperturbative QCD effects in the underlying processes. Moreover, high statistics data produced at flavor factories brought about rich outcome for the hadron spectrum, such as the discoveries of exotic hadrons. Nonperturbative dynamics of QCD characterizes their properties, which do not fit into the simple quark model prediction. Lattice QCD is the only known method for ab initio studies of these nonperturbative aspects with systematically improvable uncertainties.

Lattice QCD is a regularization of QCD on a discrete Euclidean space-time lattice. On a finite volume lattice, the QCD path integral is reduced into a finite-dimensional integral and can be numerically evaluated by a Monte Carlo sampling of field configurations on a computer. While pioneering simulations had been limited to small and coarse lattices, such limitations have been gradually lifted by continuous development of lattice QCD formulations, simulation algorithm, and advances in computing power. We note that the lattice formulation is not unique: the action has a degree of freedom to add irrelevant terms, which vanish in the limit of zero lattice spacing $a \to 0$ (the continuum limit), in order to improve its properties. We can exploit this freedom to firmly establish lattice predictions and postdictions by cross-checking using different lattice actions.

Lattice simulations can straightforwardly study the spectrum and transition amplitude of hadrons which stable under QCD. The light hadron spectrum, for instance, has been reproduced with good precision by including dynamical up, down and strange quarks with their masses close to their physical values[1-4]. While implementation of QED is not straightforward on a finite periodic space-time volume[5], even the permille-level neutron–proton mass splitting is now reproduced by taking account of electromagnetic (EM) effects and strong isospin breaking due to the mass difference of the up and down quarks $m_u - m_d$[6,7]. The accuracy of simple kaon matrix elements, namely the decay constant, form factors and bag parameters, also attains the percent level or even better. A main target in precision lattice simulations for flavor physics is decay properties of heavy-flavored hadrons, which offer rich probes of new physics.

The study of hadronic decays, however, meets technical difficulties: a central problem is that there is no simple relation between the strong decay amplitudes and correlation functions on the Euclidean lattice as suggested by Maiani–Testa theorem[8]. Theoretical framework is under active development, and is being applied to the $K \to \pi\pi$ decay as well as the heavy quarkonia and exotic states, which generally lie above thresholds.

In this article, we review highlights of recent lattice studies on the hadron spectrum and flavor physics. We also briefly discuss recent rapid progress on the muon anomalous magnetic moment, which is a key quantity in the search of new physics. More detailed reviews on these topics can be found in Refs. 9–13.

2. Hadron spectroscopy

When a hadron H is stable under QCD, its energy can be straightforwardly calculated on the lattice. We prepare an interpolating field $\mathcal{O}_H$ with quantum numbers of the hadron, and extract the energy E_H from the asymptotic behavior of the two-point function towards the large temporal separation $t \to \infty$

$$\langle \mathcal{O}_H(t)\mathcal{O}_H^\dagger(0) \rangle \to \frac{|Z_H|^2}{2E_H} e^{-E_H t}, \tag{1}$$

where $Z_H = \langle 0|\mathcal{O}_H|H \rangle$ represents the overlap of the lattice interpolating field and the physical state $|H\rangle$. This simple method forms a basis for the recent postdictions for the low-lying hadron spectrum mentioned in Sec. 1.

This can also provide illuminating insight into the nature of yet-unestablished states which are stable under the strong interaction. The doubly charmed baryons Ξ_{cc}^+ and Ξ_{cc}^{++}, for instance, are expected to have large branching fraction to flavor changing decay modes[23]. The first observation was reported by the SELEX experiment at $M_{\Xi_{cc}^+} = 3519(2)$ MeV[24,25] and $M_{\Xi_{cc}^{++}} \sim 3460$ MeV[26]. This however poses puzzles of the large isospin splitting $M_{\Xi_{cc}^{++}} - M_{\Xi_{cc}^+} \sim -60$ MeV and short lifetimes in contrast to phenomenological analyses[23,27,28]. The left-panel of Fig. 1 is a compilation of recent lattice QCD predictions for the doubly-charmed baryon spectra in the isospin limit[14–21]. These studies employing different lattice actions have led to reasonable agreement around $M_{\Xi_{cc}} \sim 3600$ MeV, which is systematically higher than the SELEX results but favors recent observation $M_{\Xi_{cc}^{++}} = 3621.40(0.78)$ MeV by LHCb (right panel of Fig. 1). A lattice estimate of the isospin splitting $M_{\Xi_{cc}^{++}} - M_{\Xi_{cc}^+} = 2.2(0.2)$ MeV also contradicts the old measurement.

While studying unstable particles is more involved, there has been considerable progress in recent years. Let us consider a resonance strongly decaying into two

296

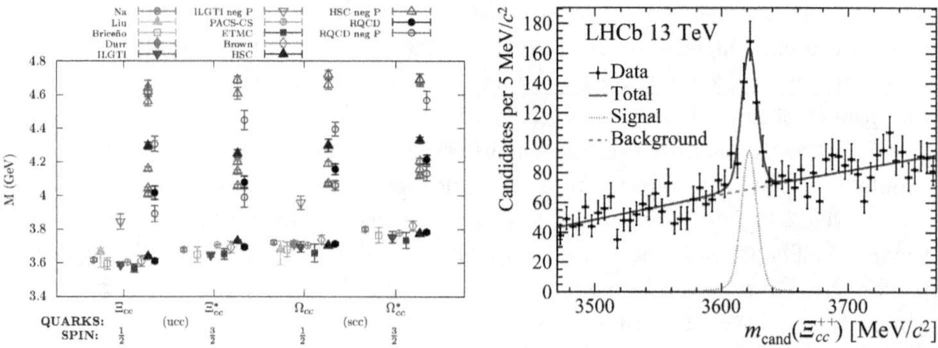

Fig. 1.   Left panel: compilation of recent lattice predictions for doubly charmed baryon spectra (figure from Ref. 14). The lowest positive-parity states have been studied by many groups[14–21], whereas less results are available for the excited and/or negative-parity states[14,17,21]. Right panel: invariant mass distribution of $(\Xi_{cc}^{++} \to)\Lambda_c^+ K^- \pi^+ \pi^+$ decay candidates from LHCb[22]. pp data sample corrected at a center-of-mass energy 13 TeV with an integrated luminosity of 1.7 fb^{-1} is analyzed. The dotted, dashed and solid lines are fit curves for the signal, background ant their total.

particles A and B. Its mass and width can be determined from the scattering amplitude of A and B, but it is not directly given by the amplitudes of the Euclidean correlation function[8]

$$\langle \mathcal{O}'_{AB}(t)\mathcal{O}^{\dagger}_{AB}(0)\rangle = \sum_n \frac{|Z_{AB,n}|^2}{2E_{AB,n}} e^{-E_{AB,n}t}, \tag{2}$$

where $\mathcal{O}_{AB}$ and $\mathcal{O}'_{AB}$ are interpolating fields for the two-particle state AB, and n is the index of the energy levels. However, the energy spectrum $\{E_{AB,1}, E_{AB,2}, ...\}$ deviates from the non-interacting energy levels due to the AB scattering on the lattice, and hence encodes the scattering amplitude. Lüscher derived a formula to extract the phase shift from the discrete spectrum on the finite volume[29–31]. Later the HALQCD Collaboration developed another method to study a multi-particle system[32–34]. By using an interpolating field $O'_{AB} = O_A(\mathbf{x},t)O_B(\mathbf{x} + \mathbf{r},t)$ with distance $|\mathbf{r}|$, the HALQCD method extracts the Nambu-Bethe-Salpeter wave function, from which the potential and scattering amplitude of the two-particle system AB can be deduced.

A good application of these methods is rigorous understanding of the resonant $\pi\pi$ and $K\pi$ scatterings directly from QCD, which is of fundamental importance in hadron physics. While the original Lüscher formula was limited to a single channel problem for two identical scalar particles in the center-of-mass frame, efforts over the past few decades have generalized it to arbitrary two-particle systems also in moving frames[37–42]. This theoretical development has made rapid stride in the lattice study of the ρ and K^* resonances in the isovector channels[10,43]. The left panel of Fig. 2 shows the scattering phase shift from a most realistic calculation by the RQCD Collaboration at a pion mass $M_\pi \sim 150$ MeV close to its physical value

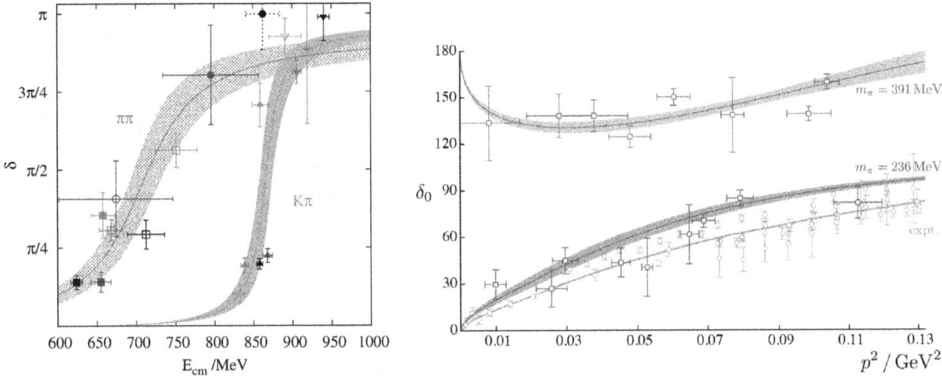

Fig. 2. Left panel: phase shifts of isovector $\pi\pi$ (circles and squares) and $K\pi$ (triangles) scatterings plotted against the center-of-mass frame energy (figure from Ref. 35). The black shaded and orange bands show Breit-Wigner type parametrization for $\pi\pi$ and $K\pi$ data. Right panel: isoscalar $\pi\pi$ scattering phase shift as a function of scattering momentum $p^2 = (E_{\rm cm}/2)^2 - M_\pi^2$. (Figure from Ref. 36) (color online).

$M_{\pi,\rm phys}$[35]. A Breit-Wigner parametrization leads to resonance masses, $M_{\{\rho,K^*\}}$, and width Γ_ρ consistent with experiment within 2σ. A slightly smaller value for Γ_{K^*} may be attributed to the 10% larger M_π than the real world. We also note that a study of ρ using the HALQCD method is in progress[44].

Lattice QCD has been also applied to the isoscalar channel, which is much more challenging than the iso-nonsinglet ones. The relevant two-point function involves the so-called quark-disconnected diagrams (see Fig. 1 of Ref. 36), which are computationally very expensive. The Hadron Spectrum Collaboration recently published the first full calculation of the isoscalar scattering phase shift[36]. The σ resonance is a bound state at $M_\pi \sim 390$ MeV, but turns into a broad resonance at ~ 240 MeV as seen in the right panel of Fig. 2. The mass and width of σ approach their experimental values[45], as M_π decreases. While a simulation at the physical pion mass $M_{\pi,\rm phys}$ is needed to make a direct comparison with experiment, those at unphysical M_π's deepen our understanding of the existence form of the hadrons.

Extension to heavy hadrons, particularly exotic hadrons discovered at flavor factories, is intriguing but still challenging task of lattice QCD. These hadrons in general have significant branching fraction to states containing three or more particles. Generalization of Lüchser's framework capable of these high multiplicity states is an active area of lattice QCD[46–49]. It is not unreasonable to hope that such a general framework will become available in five years.

There are however good examples that the current methodology can gain insight into the nature of exotic hadrons. One example is a recent study of $X(5568)$[50]. The D0 Collaboration recently reported a narrow peak ($\Gamma \sim 22$ MeV) in the $B_s\pi^+$ invariant mass of their $p\bar{p}$ collision data[53], while the later LHCb measurements did not confirm the peak structure[54]. This state, if exists, is an interesting exotic content

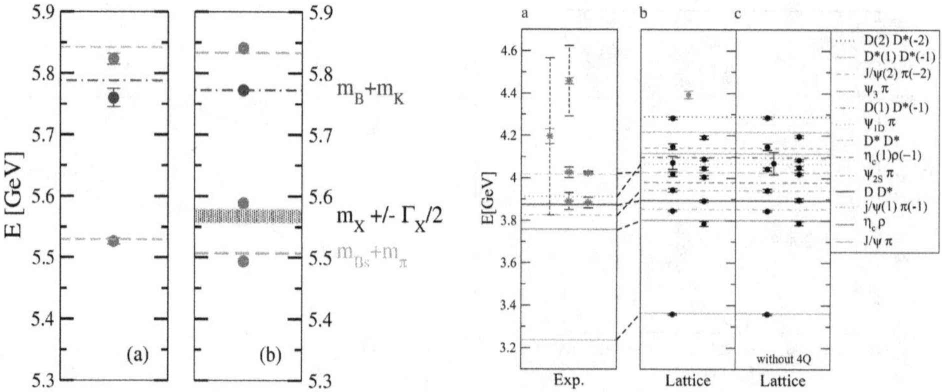

Fig. 3.   Left panel: finite volume spectrum for $X(5568)$ (figure from Ref. 50). The sub-panel (a) shows simulation data at $M_\pi \sim 156$ MeV, whereas an analytic prediction at $M_{\pi,\mathrm{phys}}$ is plotted in the sub-panel (b) by assuming the existence of $X(5568)$. The finite volume spectrum ($L = 2.9$ fm) is plotted by circles. The horizontal lines show non-interacting levels of $B_s(0)\pi^+(0)$, $B^+(0)\bar{K}^0(0)$ and $B_s(1)\pi^+(1)$, where the arguments are lattice momenta in units of $2\pi/L$. Note that the finite volume level (red circle) near the experimental mass (shaded band in sub-panel (b)) is missing in the simulation data (sub-panel (a)). Right panel: finite volume spectrum for $Z_c^+(3900)$ (figure from Ref. 51). The experimental masses of the Z_c candidates discussed in Ref. 52 ($Z_c(3885, 3900, 4020, 4025, 4430)$) are plotted in the sub-panel "a". The vertical dashed lines represent twice the widths. The sub-panels "b" and "c" show finite volume spectrum at $M_\pi = 266$ MeV obtained with different choices of the lattice interpolating fields. We note that black symbols are identified as scattering states shown by horizontal lines, and there is not extra energy level corresponding to $Z_c(3900)$ (color online).

with four different quark flavors $\bar{b}s\bar{d}u$. It strongly decays into one two-particle state $B_s\pi^+$ and lies significantly below other thresholds. Therefore lattice investigation is rather straightforward than for other exotics. Through the Lüscher formula, Lang *et al.* calculated the finite volume spectrum in Eq. (2) by numerical simulation near $M_{\pi,\mathrm{phys}}$. The left panel of Fig. 3 compares it with an analytical estimate at $M_{\pi,\mathrm{phys}}$ assuming the existence of $X(5568)$. The energy level corresponding to $X(5568)$ is not confirmed in their simulation data. This disfavors the existence of $X(5568)$ in agreement with the recent LHCb result.

There has also been recent interesting progress for $Z_c^+(3900)$, another candidate for four quark exotic states. Peak structure slightly above $D\bar{D}^*$ threshold has been found in $J/\psi\pi^+$ and $D\bar{D}^*$ invariant mass distributions of BESIII, Belle and CLEO-c data[55–57]. The finite volume spectrum has already been studied[51,58,59]. As shown in the right panel of Fig. 3, any extra energy level corresponding to Z_c has not been confirmed suggesting the possibility of the Z_c peak of kinematical origin. Recently, the scattering matrix among three states, J/ψ, $\bar{D}D^*$ and $\rho\eta_c$, has been determined at unphysical $M_\pi \geq 410$ MeV by the HALQCD method[60]. As shown in the left panel of Fig. 4, it reproduces the Z_c peak structure, which however disappears by turning off the strong coupling between $J/\psi\pi$ and $\bar{D}D^*$ shown in the right panel. This suggests that the Z_c peak is threshold cusp effects due to

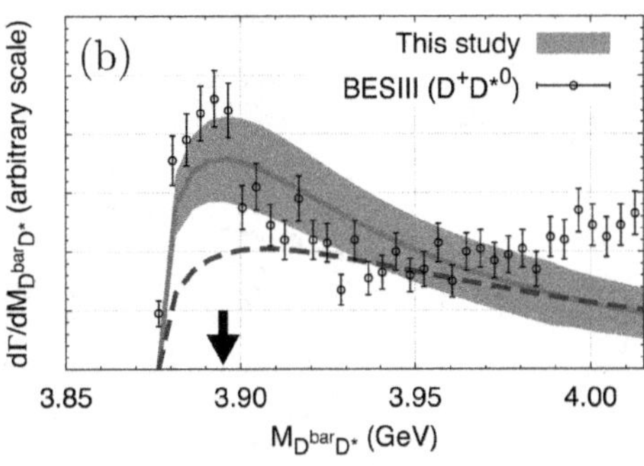

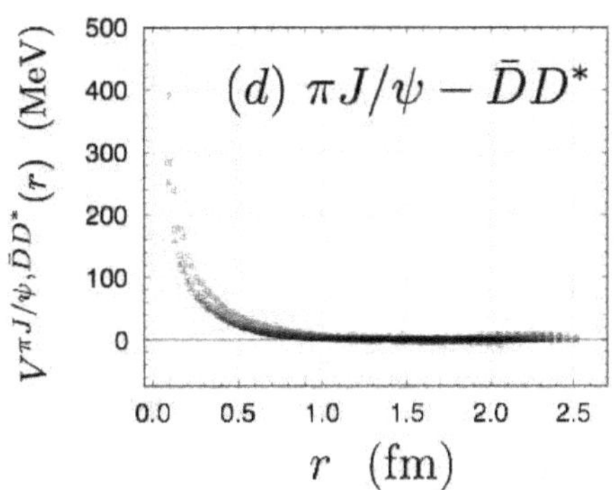

Fig. 4. Right panel: $\bar{D}D^*$ invariant mass distribution. Open circles are BESIII data[55]. The red band shows result obtained by using the HALQCD method, The peak structure disappears at the blue dashed line by turning off the off-diagonal couplings. Left panel: coupled-channel potential between $J/\psi\pi$ and $\bar{D}D^*$ states (color online).

the strong $J/\psi\pi - \bar{D}D^*$ coupling. This interesting observation should be confirmed at $M_{\pi,\mathrm{phys}}$, since unphysically heavy M_π significantly disorders possibly relevant thresholds $\Psi_{\{2S,1D,3\}}\pi$. A similar coupled-channel analysis using the Lüscher formula would provide an important cross-check to firmly establish the origin of the Z_c peak structure.

3. Flavor physics

There has been steady progress in precision lattice study of the (semi)leptonic decays and neutral meson mixings. These processes provide determination of relevant Cabibbo-Kobayashi-Maskawa (CKM) matrix elements and a wealth of probes of new physics, such as the differential distribution, angular observables and decay asymmetries. When we can safely ignore the final state interaction, the relevant hadronic matrix elements can be straightforwardly calculated from correlation functions on the lattice. For the kaon leptonic decays, for instance, the matrix element $\langle 0|A_4|K(p)\rangle = M_K f_K$ parametrized by the kaon decay constant f_K, is given by Z_H of two-point function (1) of the weak axial current $\mathcal{O}_H = A_4$. The form factors for the semileptonic decays and the bag parameters for the mixings can be similarly extracted from amplitudes of tree-point functions. This is rather straightforward procedure, and hence good simulation setup, namely the choice of the lattice formulation and simulation parameters, is a crucial issue to reach a target accuracy.

There have been independent simulations with sufficiently large and fine lattices near $M_{\pi,\mathrm{phys}}$ for the kaon (semi)leptonic decay and mixing. For instance, the current accuracy reaches the sub % level for the $K \to \pi$ form factor and the ratio of the kaon and pion decay constants f_K/f_π[61]. These are important hadronic inputs required to precise determination of $|V_{us}|$ and $|V_{us}/V_{ud}|$, which now confirm CKM unitarity in the first row $\sum_{q=d,s,b} |V_{uq}|^2 = 1$ at 0.1% accuracy as shown in the left panel of

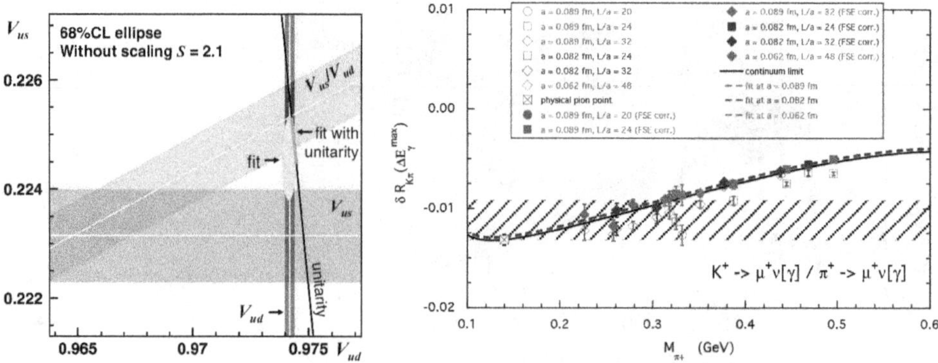

Fig. 5. Left panel: test of CKM unitarity in $(|V_{ud}|, |V_{us}|)$ plane (figure from Ref. 64). The horizontal and oblique bands indicate $|V_{us}|$ determined from $K \to \pi\ell\nu$ decay and $|V_{us}/V_{ud}|$ from the kaon and pion leptonic decays, respectively. The vertical band is $|V_{ud}|$ from the super-allowed nuclear β decays. A fit to these data yields $(|V_{ud}|, |V_{us}|)$ shown in the yellow region. The black solid line satisfies CKM unitarity in the first row, where $|V_{ub}|$ has small effects. Right panel: isospin correction $\delta R_{K\pi}$ to $\Gamma(K \to \ell\nu)/\Gamma(\pi \to \ell\nu)$ (figure from Ref. 65). Lattice simulation results are plotted by symbols, and lines are their fit curves at finite lattice spacings and in the continuum limit. The shaded band represents the ChPT estimate[66,67] (color online).

Fig. 5. This is one of the most precise unitarity test, and can probe new physics at a scale $\lesssim 10$ TeV[62,63].

At the impressive accuracy of the hadronic inputs, the uncertainty of the EM and strong isospin breaking corrections to the decay rate is no longer negligible. These corrections have been conventionally estimated within chiral perturbation theory (ChPT), and their uncertainty is typically $0.1 - 0.4\%$[68]. It is however difficult to improve the ChPT calculation by extending to higher orders where many additional unknown low energy constants appear. Lattice QCD calculation of these corrections becomes increasingly important and is actively pursued[5]. Recently, a new strategy was proposed to calculate the EM corrections to hadronic processes where the presence of the infrared divergences makes the calculation procedure much more complicated than for the hadron spectrum[69]. This has been successfully applied to the leptonic decay rate ratio $\Gamma(K \to \ell\nu)/\Gamma(\pi \to \ell\nu)$, which provides the determination of $|V_{us}|/|V_{ud}|$. As shown in the right panel of Fig. 5, their preliminary estimate $\delta R_{K\pi} = -0.0137(13)$[65] is in good agreement with the conventional ChPT estimate $-0.0112(21)$[66,67], and their accuracy is systematically improvable by more realistic simulations near $M_{\pi,\mathrm{phys}}$ with higher statistics. We also note that this strategy is also applicable to semileptonic decays.

Another thrust of recent lattice efforts in kaon physics is application to more involved processes, namely the $K \to \pi\pi$ hadronic decay and rare decays. Similar to the scattering amplitudes discussed in Sec. 2, the $K \to \pi\pi$ decay amplitudes are not directly given by the correlation functions on the lattice. Lellouch and Lüscher derived a formula to relate the finite volume matrix elements to the physical amplitudes[70].

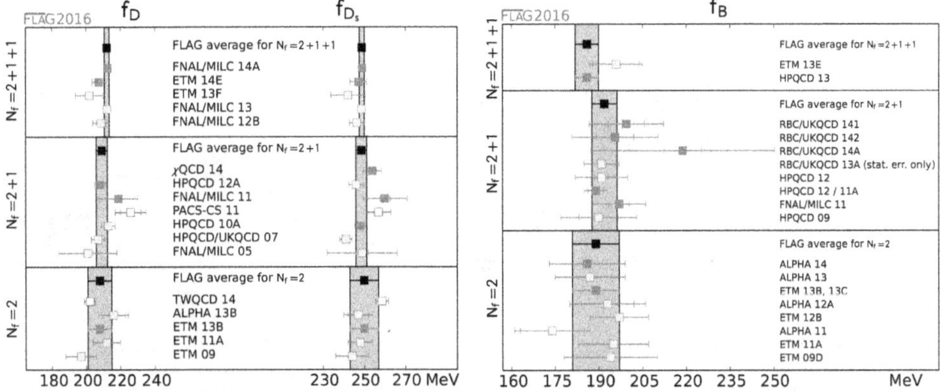

Fig. 6. Compilation of recent lattice results for heavy-light decay constants f_D and f_{D_s} (left panel) and f_B (right panel) (figures from the review of FLAG[61]). Green symbols are obtained from simulations with sufficiently good setup to control systematics, whereas reds are not. Black squares are average of the filled green squares at each N_f, the number of flavors of dynamical quarks (color online).

This has been successfully implemented in recent works by the RBC/UKQCD Collaborations for the $\Delta I = 3/2$ [71] and $3/2$ [72] channels. They obtain the direct CP violation parameter re $[\epsilon'/\epsilon] = 1.4(5.2)_{\text{stat}}(4.4)_{\text{sys}} \times 10^{-4}$. A similar but less precise value $8(25)_{\text{stat}} \times 10^{-4}$ was obtained by an independent calculation with a different lattice formulation [73]. A slight tension with the experimental value [74–76] $16.6(2.3) \times 10^{-4}$ is of great phenomenological interest [77,78]. The RBC/UKQCD Collaboration is making continuing efforts towards a more precise determination with increased statistics [79], better implementation of renormalization [80], and new method to calculate the Wilson coefficients [81]. We may therefore expect an improved estimate of re $[\epsilon'/\epsilon]$ in the near future. Note also that a first exploratory study is available both for $K \to \pi \ell \ell$ [82] and $K \to \pi \nu \bar{\nu}$ [83] decays.

The accuracy of the heavy-light meson decay constants has been steadily improved over the years. Figure 6 shows that there are independent calculations with sufficiently good setup to control systematics. The world average quoted by the Flavor Lattice Averaging Group (FLAG)[61] has uncertainty of $\approx 0.6\%$, for $f_{D_{(s)}}$ and $\lesssim 2\%$ for f_B. These accuracies are better than current experiments $(f_{D_{(s)}}, \text{BESIII}[84])$, or comparable to experimental precision expected in the future $(f_B, \text{Belle II}[85])$. We note that isospin corrections become increasingly important at this level of accuracy.

There have been a long standing problem and possible hint of new physics related to the $B \to \pi \ell \nu$ and $B \to D^{(*)} \ell \nu$ [84] semileptonic decays as shown in Fig. 7. These are the conventional decay modes to determine the CKM matrix elements $|V_{ub}|$ and $|V_{cb}|$, which however contradict with those from the inclusive decays, $B \to X_u \ell \nu$ and $X_c \ell \nu$. Note also that more than 3σ deviation between the SM and experiment has been reported for the decay rate ratios $R(D^{(*)}) = \Gamma(B \to D^{(*)} \tau \nu)/\Gamma(B \to D^{(*)} \ell \nu)$

302

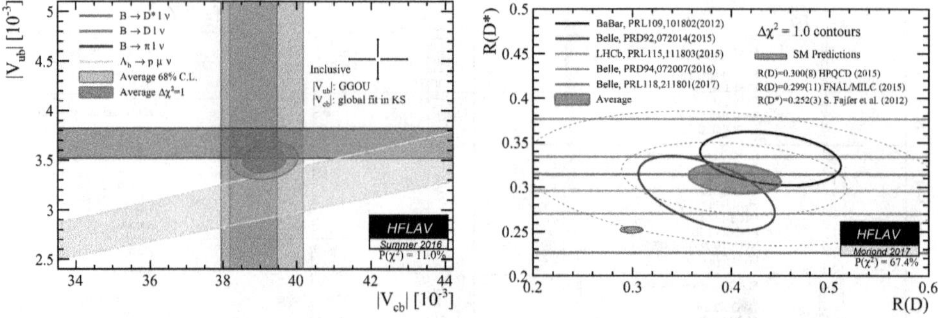

Fig. 7. Left panel: $|V_{ub}|$ versus $|V_{cb}|$. Horizontal and vertical bands represent $|V_{ub}|$ from $B \to \pi \ell \nu$ and $|V_{cb}|$ from $B \to D^{(*)} \ell \nu$, respectively. The oblique band is $|V_{ub}|/|V_{cb}|$ from $\Lambda_b \to p\mu\nu$ and $\Lambda_c \ell \nu$. The average of these estimates is plotted by red region. These should be compared with the point with the error bars obtained from the inclusive decays. Right panel: $R(D^*)$ versus $R(D)$. Filled red region is the average of BaBar, Bell and LHCb data, whereas the SM prediction is shown by the grey region. Both figures are taken from Ref. 84 (color online).

($\ell = e, \mu$). More precise and reliable calculation of the relevant hadronic matrix elements in lattice QCD is an important subject for precision search of new physics.

The $B \to \pi \ell \nu$ and $B \to D\ell \nu$ decays proceed only through the weak vector current V_μ due to parity symmetry, and its matrix element is parametrized by two form factors as

$$\langle \pi(p')|V_\mu|B(p)\rangle = \left\{ p + p' - \frac{M_B^2 - M_\pi^2}{q^2} q \right\}_\mu f_+(q^2) + \frac{M_B^2 - M_\pi^2}{q^2} q_\mu f_0(q^2), \quad (3)$$

where $q^2 = (p - p')^2$ is the momentum transfer to the lepton pair $\ell \nu$. Figure 8 shows FLAG's fit to recent lattice and experimental data for $B \to \pi$ [86–88] and $B \to D$ [89,90]. Their q^2 dependence is fitted to a model independent parametrization [91] in terms of a small parameter $z(q^2, t_{\rm opt}) = (\sqrt{t_+ - q^2} - \sqrt{t_+ - t_{\rm opt}})/(\sqrt{t_+ - q^2} + \sqrt{t_+ - t_{\rm opt}})$, where t_+ is the threshold $t_+ = (M_B + M_{\pi(D)})^2$ and $t_{\rm opt} = (M_B +$

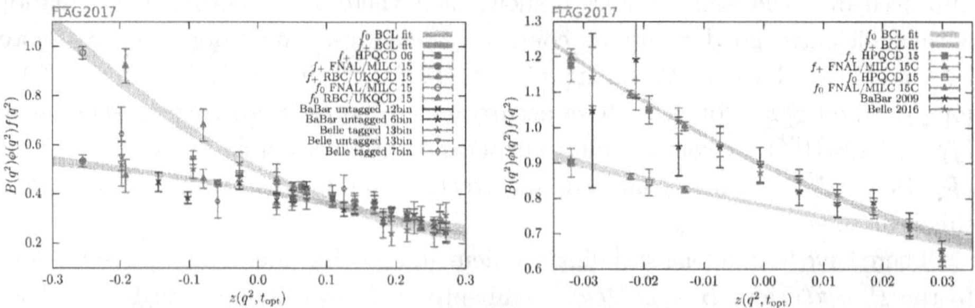

Fig. 8. Recent lattice estimates of vector and scalar form factors, $f_+(q^2)$ and $f_0(q^2)$, as a function of z-parameter. The left and right panels show data for for $B \to \pi \ell \nu$ and $B \to D\ell \nu$ decays, respectively. The bands represent fit to lattice and experimental data to determine the CKM matrix elements $|V_{ub}|$ and $|V_{us}|$.

$M_{\pi(D)})(\sqrt{M_B} - \sqrt{M_{\pi(D)}})^2$. Then the relevant CKM element can be determined as a relative normalization factor. While the state-of-the-art calculations start to achieve good accuracy competitive to experiments, the number of such calculations are rather limited and independent calculations are highly welcome[92,93].

The $B \to D^*\ell\nu$ decay also receives the contribution of the weak axial current, and have four form factors. At zero recoil, the differential decay rate is described by the single form factor in

$$\langle D^*(\epsilon, p')|A_\mu|B(p)\rangle = i\sqrt{M_B M_{D^*}}\epsilon_\mu(1 + w)h_{A_1}(w) \quad (w = 1), \qquad (4)$$

where ϵ is the polarization vector of D^*, and the recoil parameter is defined as $w = vv'$ with the velocities $v = p/M_B$ and $v' = p'/M_{D^*}$. The previous lattice calculation focused only on h_{A_1} at zero recoil $w = 1$[94]. However, it is recently argued that uncertainty due to the phenomenological parametrization of the momentum transfer dependence of the form factors is not fully understood and this leads to the discrepancy in $|V_{cb}|$ from the inclusive decays[95–97]. In order to resolve this issue, the lattice calculation of the four form factors at non-zero recoil is crucial, and a first preliminary analysis was recently reported by the Fermilab/MILC Collaboration[98].

Lattice simulation is being actively applied to other decay modes, which potentially provide alternative determinations of the CKM matrix elements to resolve the discrepancy from the inclusive decays. Modern calculations are available for $B_s \to K\ell\nu$[87,100,101] by HPQCD, RBC/UKQCD and ALPHA Collaborations as well as for $B_s \to D_s\ell\nu$[99,102,103] by the Fermilab/MILC, ETM and HPQCD Collaborations. Simulations techniques for $B \to \pi\ell\nu$ and $D^{(*)}\ell\nu$ can be straightforwardly applied to control systematics. In addition, $B_s \to K\ell\nu$ involves less valence light quarks, which lead to larger statistical fluctuation and longer chiral extrapolation than with valence strange quarks. Therefore, these modes can achieve comparable or even better accuracy compared to the conventional modes $B \to \pi\ell\nu$ and $B \to D\ell\nu$ as

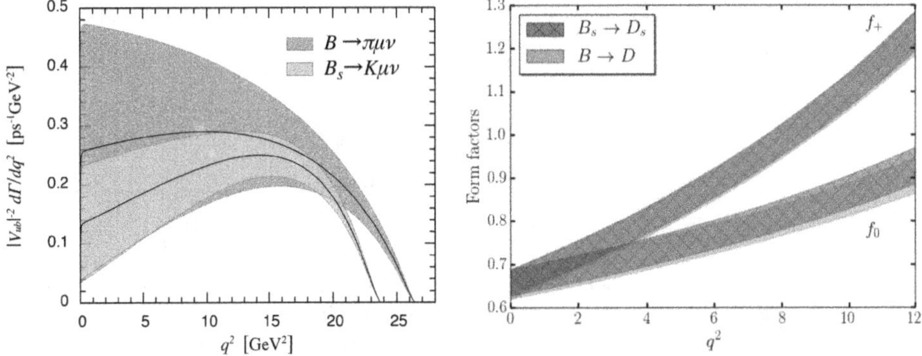

Fig. 9. Left panel: $B_s \to K\mu\nu$ (blue band) and $B_s \to K\mu\nu$ (red band) differential decay rates predicted by using recent lattice estimate of the relevant form factors (figure from Ref. 87). Right panel: vector (f_+) and scalar (f_0) form factors for $B_s \to D_s\ell\nu$ (shaded dark blue band) and $B \to D\ell\nu$ (pale blue band) decays (figure from Ref. 99) (color online).

304

seen in recent results in Fig. 9. Therefore, the most crucial issue for the alternative modes is feasibility of precise measurements at on-going and future experiments.

Baryon decays also enable determinations of the CKM matrix elements with systematics different from the meson decays. The first lattice calculation of $\Lambda_b \to p\ell\nu$ and $\Lambda_b \to \Lambda_c\ell\nu$ form factors at the physical b quark mass obtained $|V_{ub}|/|V_{cb}|$ shown in Fig. 7[104]. An example of the relevant form factors is given in the left panels of Fig. 10. Recently, the first study of $\Lambda_c \to \Lambda\ell\nu$ was also reported[105] and yields $|V_{cs}|$ consistent with those from D meson decays. However baryons are known to be more challenging in controlling systematics, in particular finite volume effects and chiral extrapolation. Since the target accuracy is high, more thorough study of systematics is highly desirable to firmly establish the CKM matrix elements from the baryon decays.

For rare decays, however, such theoretical uncertainty could be well below the experimental one. Reference 106 recently presented the first calculation of all ten form factors for $\Lambda_b \to \Lambda\ell\ell$. This decay may shed new light on the so-called $B \to K^*\ell\ell$ anomaly, namely more than 3σ tension in its angular distribution[108], since these two decays share the $b \to s\ell\ell$ effective weak Hamiltonian. As shown in the right panel of Fig. 10, the SM prediction from the form factors and the LHCb result[107] show reasonable consistency for the differential branching fraction. A simple new physics scenario for $B \to K^*\ell\ell$, namely negative new physics coupling for an effective interaction operator $(\bar{s}_L\gamma_\mu b_L)(\bar{\ell}\gamma^\mu\ell)$, worsens this consistency. Therefore the authors suggest that the $B \to K^*\ell\ell$ anomaly is due to incomplete treatment of the charmonium resonance contribution to the decay rate.

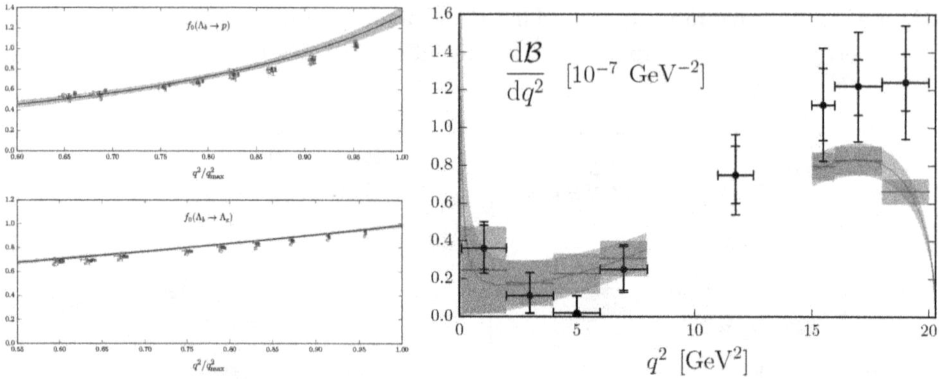

Fig. 10.   Left panels: an example of $\Lambda_b \to p$ (top left panel) and $\Lambda_b \to \Lambda_c$ (bottom left panel) form factors appearing in the matrix element of the vector current $\langle p(\Lambda_c)|V_\mu|\Lambda_b\rangle$ (figure from Ref. 104). The horizontal axis represents q^2 normalized by $q^2_{\max} = (M_{\Lambda_b} - M_{p(\Lambda_c)})^2$. Right panel: differential branching fraction of $\Lambda_b \to \Lambda\ell\ell$ as a function of q^2 (figure from Ref. 106). The black circle shows the LHCb result[107] with two error bars including and excluding the uncertainty of the normalization mode $\Lambda_b \to J/\Psi\Lambda$. The continuous blue band is the SM prediction as a function of q^2, whereas magenta band show the average at each q^2 bin (color online).

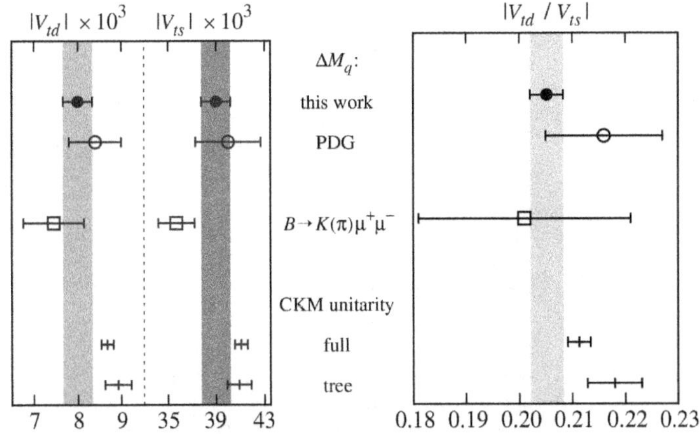

Fig. 11. Comparison of $|V_{td}|$ and $|V_{ts}|$ (left panel) and their ratio (right panel) (figure from Ref. 109). The filled circles and vertical bands show results from the Fermilab/MILC Collaboration. The open circles are obtained by using previous lattice estimates of the matrix elements, where as the square from the rare decays $B \to \pi\ell\ell$ and $B \to K\ell\ell$. The plus symbols are obtained by a global fit to all available input and that limited to inputs from tree-level processes.

For the neutral B meson mixing, the Fermilab/MILC Collaboration recently published a precise calculation of all matrix elements in and beyond the SM[109]. High statistics, realistic simulation parameters (a and M_π), and a better implementation of the renormalization of the lattice operators led to at most threefold reduction in the total uncertainty. The B (B_s) meson mixing provides determination of V_{td} (V_{tb}) through the mass difference ΔM_d (ΔM_s). Figure 11 compares the Fermilab/MILC's estimate with others, and suggests 2σ tension of $|V_{td}/V_{ts}|$ with CKM unitarity. Note also that there is a tension in a matrix element beyond the SM between the Fermilab/MILC and ETM Collaborations' results[110], while the latter ignores effects of dynamical strange quarks. More independent calculations with similar accuracies are needed to resolve and understand these tensions.

While generalization of the Lellouch-Lušcher framework is actively pursued, it is not yet sufficient to allow lattice simulations of hadronic B or D decays which have various multi-particle final states. However, lattice simulation is now being applied to inclusive decays[111–114], for which all possible final hadronic states are summed up. For the hadronic τ decays $\tau \to X_s\nu$, for instance, the optical and Cauchy theorems relate a decay rate ratio to a contour integral of the weak current correlator over the complex energy variable s

$$R_s = \frac{\Gamma(\tau \to X_s\nu_\tau)}{\Gamma(\tau \to e\bar{\nu}_e\nu_\tau)} = -\frac{1}{2\pi i}|V_{us}|^2 \oint_{|s|=s_0} ds\, w(s)\langle 0|J_\mu J_\mu|0\rangle, \tag{5}$$

where s_0 and $w(s)$ are appropriately chosen parameter and weight function, respectively, and J_μ represents the relevant weak current. By employing a pole type weight function $w(s) = 1/\Pi_{n=1}^N(s + Q_n^2)$ ($Q_n^2 > 0$), the integral can be evaluated by the current correlator at space like points $s = Q_1^2, \ldots, Q_N^2$, which can be precisely

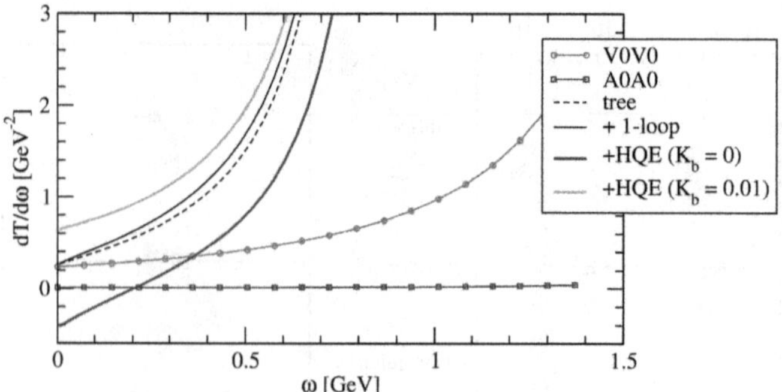

Fig. 12.   First derivative of the matrix element $T_{\mu\nu}$ as a function of kinematical variable $\omega = M_B - q_0$. Red circles and Blue squares are weak vector $(J_\mu = V_\mu)$ and axial vector $(J_\mu = A_\mu)$ current contributions, respectively. Here we plot $dT_{\mu\nu}/d\omega$ instead of $T_{\mu\nu}$ itself to avoid contamination from contact terms. Black dashed line shows the estimate from the heavy quark expansion at the leading order in α_s and $1/m_b$. The next-to-leading order contributions are added to the solid lines (color online).

determined on the lattice. This is RBC/UKQCD's implementation to determine $|V_{us}|$ from the hadronic τ decays[111]. They obtain $|V_{us}| = 0.223 - 0.225$[115,116], which is in good agreement with those from the kaon decays shown in the left panel of Fig. 5.

Inclusive semileptonic B decays are more involved but important application to resolve the discrepancy in $|V_{ub}|$ and $|V_{cb}|$ from the exclusive decays[112]. Through the optical theorem, the hadronic part of the inclusive decay amplitude can be rewritten using the forward scattering matrix element

$$T_{\mu\nu} = i \int d^4x e^{-iqx} \frac{1}{2M_B} \langle B|T\left[J_\mu^\dagger(x) J_\nu(0)\right]|B\rangle. \tag{6}$$

While the calculation of the relevant four-point function is challenging, a preliminary analysis for $B \to X_c \ell\nu$ at zero recoil has been reported by Hashimoto *et al.* in Ref. 117. Figure 12 compares the matrix element $T_{\mu\nu}$ between lattice QCD and the heavy quark expansion used in the conventional inclusive determination of $|V_{cb}|$. They are reasonably consistent in the perturbative region of the kinematical variable $\omega = M_B - q_0 \sim 0$, but deviate from each other towards large ω due to the non-perturbative resonance contribution. This method provides a detailed comparison between the inclusive and exclusive determinations in the same simulations, and may give us a hint to resolve their discrepancy.

4. Muon anomalous magnetic moment

The past several years have witnessed rapid progress in calculating the muon anomalous magnetic moment $a_\mu = (g - 2)/2$ on the lattice. This quantity is known to great precision ≈ 0.5 ppm both in the experimental measurement[118] and SM prediction[119–121]. Their $3 - 4\sigma$ deviation may be a signal of new physics. New experiments,

E989 at Fermilab[122] and E34 at J-PARC[123], aim to improve the experimental accuracy by a factor of four. On the theory side, the accuracy is currently limited by hadronic uncertainties coming from the hadronic vacuum polarization (HVP) and hadronic light-by-light (HLbL) contributions shown in Fig. 13. The former has been most precisely calculated from the dispersive analysis[119–121] of experimental data of the ratio $R(s) = \sigma(s, e^+e^- \to \text{hadrons})/\sigma_{\text{tree}}(s, e^+e^- \to \mu^+\mu^-)$, where $\sqrt{s}$ is the center-of-mass energy. A similar dispersive approach is under development[124–126], and model estimates are currently quoted for the HLbL[127–129]. An ultimate goal of recent lattice efforts is to provide first-principle prediction for these hadronic contributions with reduced uncertainty.

The leading order HVP contribution a_μ^{HVP} is expressed as an integral over the Euclidean momentum Q^2 [130,131]

$$a_\mu^{\text{HVP}} = 4\alpha^2 \int_0^\infty dQ^2 K(Q^2, m_\mu^2)\hat{\Pi}(Q^2), \quad \hat{\Pi}(Q^2) = \Pi(Q^2) - \Pi(0), \tag{7}$$

where K is a known weight. In pioneering works[131–134], the vacuum polarization function $\Pi(Q^2)$ was calculated from the EM current correlator on the lattice

$$\left(Q_\mu Q_\nu - \delta_{\mu\nu}Q^2\right) \Pi(Q^2) = \int d^4x e^{iQx} \langle J_\mu^{\text{EM}}(x) J_\nu^{\text{EM}}(0)\rangle. \tag{8}$$

It was pointed out[135–137] that the subtracted function $\hat{\Pi}(Q^2)$ can be obtained from the correlator $G(t) = \sum_{\mathbf{x}}\langle J_\mu^{\text{EM}}(\mathbf{x}, t) J_\mu^{\text{EM}}(0)\rangle$ with zero spatial momentum, which is less noisy than the right-handed side of (8). Then, a_μ^{HVP} is written as an integral over the Euclidean time

$$a_\mu^{\text{HVP}} = 4\alpha^2 \int_0^\infty dt \tilde{K}(t, m_\mu^2)G(t), \tag{9}$$

which is often employed in recent lattice studies[138–140]. We note that the integral (7) receives large contribution from the infrared regime $Q^2 \approx O(m_\mu^2)$. References

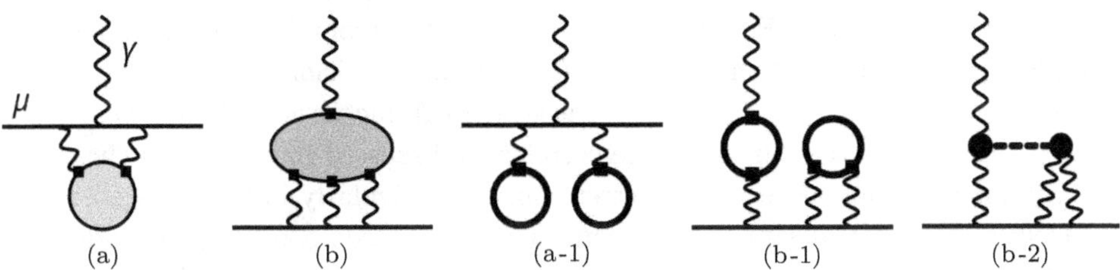

Fig. 13. The leading hadronic vacuum polarization (HVP) (a) and hadronic light-by-light (HLbL) (b) contributions to a_μ. Wavy and solid lines represent the photon and muon, respectively. The shaded circle for the HVP (HLbL) is the two- (four-)point function of the quark EM current J_μ^{EM} (squares) in QCD. (a-1) An example of quark-disconnected diagrams for the HVP. Two quark loops (thick lines) are connected by gluons. (b-1) An example of leading disconnected diagrams for the HLbL. We note that a quark loop with a single J_μ^{EM} vertex vanishes in the SU(3) limit, since the up, down and strange quark charges sum up to zero. (b-2) The π^0 contribution to the HLbL. The thick dashed line and the solid circles represent the π^0 propagator and $\pi^0 \to \gamma^*\gamma^*$ transition form factor, respectively.

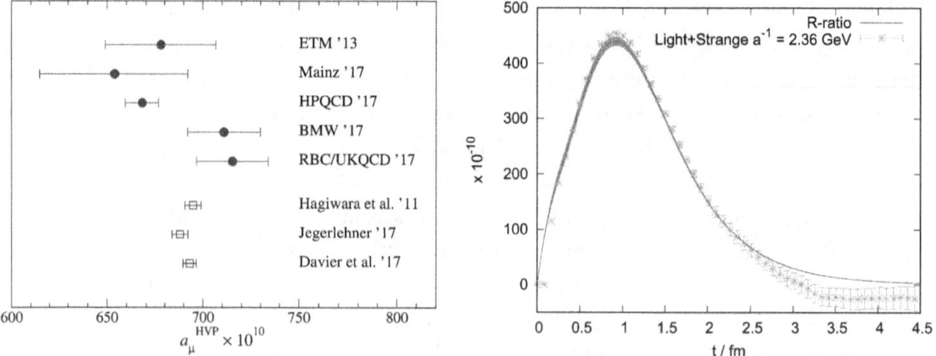

Fig. 14. Left panel: recent lattice estimates of a_μ^{HVP} (blue solid circles)[138–140,144,145]. We also plot a_μ^{HVP} from the dispersive method[119–121]. Right panel: comparison of integrand $\hat{K}(t, m_\mu^2)G(t)$ in Eq. (9) (figure from Ref. 12). The crosses are lattice data at a finite lattice spacing $a \sim 0.08$ fm, whereas the band is obtained from the experimental data of the R-ratio (color online).

141–143 employ another strategy to reconstruct $\hat{\Pi}(Q^2)$ in the infrared region by using its derivatives at $Q^2 = 0$ calculated from time moments $G_{2n} = \sum_t t^{2n} G(t) = (-1)^n \left(\partial^{2n} Q^2 \hat{\Pi}(Q^2)/\partial Q^{2n} \right)_{Q^2=0}$.

The challenge in the lattice calculation of a_μ^{HVP} is to control all uncertainties below 1% to compete with the current best estimate using the R-ratio. The dominant contribution comes from the connected diagram with the light quark current $(2/3)\bar{u}\gamma_\mu u - (1/2)\bar{d}\gamma_\mu d$. It has been calculated in the isospin limit with controlled continuum and chiral extrapolations[134,138–140,143,144], whereas finite volume effects are corrected by employing effective field theories or directly examined by simulating multiple volume sizes. Strange and charm quark contributions have been also calculated[138,139,141,146,147]. The current lattice accuracy for a_μ^{HVP} has reached a few % level as summarized in Fig. 14[a]. In has been confirmed that bottom quarks[148], the disconnected diagrams (Fig. 13 (a-1))[138,139,142,149], strong isospin breaking[140,145] and EM interaction[140,147] have small effects at this precision.

Currently, a largest uncertainty is the statistical fluctuation, though $G(t)$ is less noisy compared to the multi-particle correlators discussed in Sec. 2 and three- and four-point functions in Sec. 3. A better control particularly in the infrared region, namely small Q^2 and large t, is a crucial issue towards a more precise determination and is being actively studied[138,143,150–152]. Another interesting possibly is to combine $G(t)$ on the lattice and experimental R-ratio data. The time integral (7) is decomposed into $a_\mu^{\mathrm{HVP}} = a_{\mu,\mathrm{SD}}^{\mathrm{HVP}} + a_{\mu,\mathrm{ID}}^{\mathrm{HVP}} + a_{\mu,\mathrm{LD}}^{\mathrm{HVP}}$. The R-ratio is then used to evaluate the short ($a_{\mu,\mathrm{SD}}^{\mathrm{HVP}}$) and long ($a_{\mu,\mathrm{LD}}^{\mathrm{HVP}}$) distance contributions to avoid possibly

---

[a]Here we quote results from Refs. 139,140,145 published after this symposium. Their preliminary results had been available beforehand at *the 35th International Symposium on Lattice Field Theory (Lattice 2017)*.

large discretization effects and statistical fluctuation, respectively. In the intermediate region, $a_{\mu,\mathrm{ID}}^{HVP}$ can be calculated from $G(t)$. This combined analysis in Ref. 140 demonstrates good consistency in the integrand of Eq. (9) between lattice and experimental data (Fig. 14), and led to a most precise estimate $a_\mu^{\mathrm{HVP}} = 692.5(2.7) \times 10^{-10}$. Simulating finer lattices and better control of the statistical accuracy can expand the intermediate t window, and will eventually lead to a pure theoretical estimate of a_μ^{HVP} with reduced uncertainty.

There has also been remarkable progress in the lattice calculation of a_μ^{HLbL}. Theoretical calculation thereof is challenging, because the relevant four-point function of J_μ^{EM} involves many connected and disconnected diagrams. Therefore two approaches have been currently employed to calculate presumably dominant contributions. Ref. 153 focuses on the connected and leading disconnected diagrams which survives in the SU(3) limit (Fig. 13 (b-1)). With algorithmic improvements [154], they obtain statistically significant estimate $a_\mu^{\mathrm{HLbL}} = 5.4(1.4) \times 10^{-10}$ at the physical point but at a single lattice spacing. Another approach in Ref. 155 estimates presumably dominant π^0 contribution (Fig. 13 (b-2)) through the lattice calculation of the $\pi^0 \to \gamma^* \gamma^*$ transition form factor, and yields $a_\mu^{\mathrm{HLbL}} = 6.5(0.8) \times 10^{-10}$. While the quote errors are statistical only, the reasonable agreement between the different approaches is encouraging and motivates more realistic simulations and a survey of systematics, particularly finite volume effects due to massless photons on the lattice [156].

5. Conclusions

In this review, we have presented highlights of recent progress on hadron spectrum and flavor physics from lattice QCD. Masses and transition amplitudes can be straightforwardly calculated from lattice correlation functions for hadrons stable under QCD. Recent realistic simulations can yield deep insight into the nature of yet-unestablished states as in the case of Ξ_{cc}. Decay constants, kaon form factors and bag parameters are now calculated with fully controlled uncertainties, and represent the precision frontier of lattice QCD, where isospin corrections have to be taken into account. The number of the precision calculation is currently rather limited for heavy hadron form factors and bag parameters. However, we can expect more independent calculations in the next few years.

Lattice QCD is now ready to study coupled-channel two-body scatterings. This leads to recent interesting progress on light resonances, ρ, K^*, σ and κ, as well as heavy exotics such as $X(5568)$ and $Z_c(3900)$. These studies are however often limited to unphysically heavy pion masses, which may significantly raise thresholds including pions and turn resonances into bound-states. In order to make a direct comparison with experiment, studies at the physical pion mass are highly recommended.

There has been continuous progress on $K \to \pi\pi$ leading to slight tension with experiment, which is of great phenomenological interest. General framework to deal with three-particle states is necessary for hadronic B and D decays and under active development. However, it was proposed that inclusive decays can be straightforwardly studied without such framework. This may offer useful hints to resolve long-standing tension in $|V_{ub}|$ and $|V_{cb}|$ between the exclusive and inclusive determinations.

References

1. S. Durr *et al.*, Ab-Initio Determination of Light Hadron Masses, *Science* **322**, 1224 (2008).
2. S. Aoki *et al.*, 2+1 Flavor Lattice QCD toward the Physical Point, *Phys. Rev.* **D79**, p. 034503 (2009).
3. A. Bazavov *et al.*, Nonperturbative QCD Simulations with 2+1 Flavors of Improved Staggered Quarks, *Rev. Mod. Phys.* **82**, 1349 (2010).
4. W. Bietenholz *et al.*, Flavour blindness and patterns of flavour symmetry breaking in lattice simulations of up, down and strange quarks, *Phys. Rev.* **D84**, p. 054509 (2011).
5. A. Patella, QED Corrections to Hadronic Observables, *PoS* **LATTICE2016**, p. 020 (2017).
6. S. Borsanyi *et al.*, Ab initio calculation of the neutron-proton mass difference, *Science* **347**, 1452 (2015).
7. R. Horsley *et al.*, Isospin splittings of meson and baryon masses from three-flavor lattice QCD + QED, *J. Phys.* **G43**, p. 10LT02 (2016).
8. L. Maiani and M. Testa, Final state interactions from Euclidean correlation functions, *Phys. Lett.* **B245**, 585 (1990).
9. C. Liu, Review on Hadron Spectroscopy, *PoS* **LATTICE2016**, p. 006 (2017).
10. D. J. Wilson, Resonances in Coupled-Channel Scattering, *PoS* **LATTICE2016**, p. 016 (2016).
11. X. Feng, Recent progress in applying lattice QCD to kaon physics, in *35th International Symposium on Lattice Field Theory (Lattice 2017) Granada, Spain, June 18-24, 2017*, (2017).
12. C. Lehner, A precise determination of the HVP contribution to the muon anomalous magnetic moment from lattice QCD, in *35th International Symposium on Lattice Field Theory (Lattice 2017) Granada, Spain, June 18-24, 2017*, (2017).
13. G. Colangelo, M. Hoferichter, M. Procura and P. Stoffer, Hadronic light-by-light contribution to $(g-2)_\mu$: a dispersive approach, in *35th International Symposium on Lattice Field Theory (Lattice 2017) Granada, Spain, June 18-24, 2017*, (2017).
14. P. Perez-Rubio, S. Collins and G. S. Bali, Charmed baryon spectroscopy and light flavor symmetry from lattice QCD, *Phys. Rev.* **D92**, p. 034504 (2015).
15. H. Na and S. A. Gottlieb, Charm and bottom heavy baryon mass spectrum from lattice QCD with 2+1 flavors, *PoS* **LATTICE2007**, p. 124 (2007).
16. L. Liu, H.-W. Lin, K. Orginos and A. Walker-Loud, Singly and Doubly Charmed J=1/2 Baryon Spectrum from Lattice QCD, *Phys. Rev.* **D81**, p. 094505 (2010).
17. S. Basak, S. Datta, M. Padmanath, P. Majumdar and N. Mathur, Charm and strange hadron spectra from overlap fermions on HISQ gauge configurations, *PoS* **LATTICE2012**, p. 141 (2012).
18. S. Durr, G. Koutsou and T. Lippert, Meson and Baryon dispersion relations with Brillouin fermions, *Phys. Rev.* **D86**, p. 114514 (2012).
19. Y. Namekawa *et al.*, Charmed baryons at the physical point in 2+1 flavor lattice QCD, *Phys. Rev.* **D87**, p. 094512 (2013).

20. Z. S. Brown, W. Detmold, S. Meinel and K. Orginos, Charmed bottom baryon spectroscopy from lattice QCD, *Phys. Rev.* **D90**, p. 094507 (2014).
21. M. Padmanath, R. G. Edwards, N. Mathur and M. Peardon, Spectroscopy of doubly-charmed baryons from lattice QCD, *Phys. Rev.* **D91**, p. 094502 (2015).
22. R. Aaij *et al.*, Observation of the doubly charmed baryon Ξ_{cc}^{++}, *Phys. Rev. Lett.* **119**, p. 112001 (2017).
23. V. V. Kiselev, A. K. Likhoded and A. I. Onishchenko, Lifetimes of doubly charmed baryons: Xi(cc)+ and Xi(cc)++, *Phys. Rev.* **D60**, p. 014007 (1999).
24. M. Mattson *et al.*, First observation of the doubly charmed baryon Xi+(cc), *Phys. Rev. Lett.* **89**, p. 112001 (2002).
25. A. Ocherashvili *et al.*, Confirmation of the double charm baryon Xi+(cc)(3520) via its decay to p D+ K-, *Phys. Lett.* **B628**, 18 (2005).
26. J. S. Russ, First observation of a family of double charm baryons, in *Proceedings, 1st International Workshop on Frontier Science: Charm, beauty and CP: Frascati, Italy, October 6-11, 2002*, 2002. [25(2002)].
27. C.-W. Hwang and C.-H. Chung, Isospin mass splittings of heavy baryons in HQS, *Phys. Rev.* **D78**, p. 073013 (2008).
28. B. Guberina, B. Melic and H. Stefancic, Inclusive decays and lifetimes of doubly charmed baryons, *Eur. Phys. J.* **C9**, 213 (1999), [Eur. Phys. J. C13, 551(2000)].
29. M. Luscher, Volume Dependence of the Energy Spectrum in Massive Quantum Field Theories. 2. Scattering States, *Commun. Math. Phys.* **105**, 153 (1986).
30. M. Luscher, Two particle states on a torus and their relation to the scattering matrix, *Nucl. Phys.* **B354**, 531 (1991).
31. M. Luscher, Signatures of unstable particles in finite volume, *Nucl. Phys.* **B364**, 237 (1991).
32. N. Ishii, S. Aoki and T. Hatsuda, The Nuclear Force from Lattice QCD, *Phys. Rev. Lett.* **99**, p. 022001 (2007).
33. N. Ishii, S. Aoki, T. Doi, T. Hatsuda, Y. Ikeda, T. Inoue, K. Murano, H. Nemura and K. Sasaki, Hadron-hadron interactions from imaginary-time Nambu-Bethe-Salpeter wave function on the lattice, *Phys. Lett.* **B712**, 437 (2012).
34. S. Aoki, B. Charron, T. Doi, T. Hatsuda, T. Inoue and N. Ishii, Construction of energy-independent potentials above inelastic thresholds in quantum field theories, *Phys. Rev.* **D87**, p. 034512 (2013).
35. G. S. Bali, S. Collins, A. Cox, G. Donald, M. Gockeler, C. B. Lang and A. Schafer, ρ and K^* resonances on the lattice at nearly physical quark masses and $N_f = 2$, *Phys. Rev.* **D93**, p. 054509 (2016).
36. R. A. Briceno, J. J. Dudek, R. G. Edwards and D. J. Wilson, Isoscalar $\pi\pi$ scattering and the σ meson resonance from QCD, *Phys. Rev. Lett.* **118**, p. 022002 (2017).
37. K. Rummukainen and S. A. Gottlieb, Resonance scattering phase shifts on a nonrest frame lattice, *Nucl. Phys.* **B450**, 397 (1995).
38. C. h. Kim, C. T. Sachrajda and S. R. Sharpe, Finite-volume effects for two-hadron states in moving frames, *Nucl. Phys.* **B727**, 218 (2005).
39. N. H. Christ, C. Kim and T. Yamazaki, Finite volume corrections to the two-particle decay of states with non-zero momentum, *Phys. Rev.* **D72**, p. 114506 (2005).
40. M. Lage, U.-G. Meissner and A. Rusetsky, A Method to measure the antikaon-nucleon scattering length in lattice QCD, *Phys. Lett.* **B681**, 439 (2009).
41. S. He, X. Feng and C. Liu, Two particle states and the S-matrix elements in multi-channel scattering, *JHEP* **07**, p. 011 (2005).
42. R. A. Briceno, Two-particle multichannel systems in a finite volume with arbitrary spin, *Phys. Rev.* **D89**, p. 074507 (2014).

43. D. Mohler, Review of lattice studies of resonances, *PoS* **LATTICE2012**, p. 003 (2012).

44. D. Kawai, ρ resonance from the $I = 1$ $\pi\pi$ potential in lattice QCD, presented at the 35th International Symposium on Lattice Field Theory (Lattice 2017) `http://wpd.ugr.es/~lattice2017/`.

45. J. R. Pelaez, From controversy to precision on the sigma meson: A review on the status of the non-ordinary $f_0(500)$ resonance, *Phys. Rept.* **658**, p. 1 (2016).

46. K. Polejaeva and A. Rusetsky, Three particles in a finite volume, *Eur. Phys. J.* **A48**, p. 67 (2012).

47. M. T. Hansen and S. R. Sharpe, Multiple-channel generalization of Lellouch-Luscher formula, *Phys. Rev.* **D86**, p. 016007 (2012).

48. R. A. Briceno and Z. Davoudi, Moving multichannel systems in a finite volume with application to proton-proton fusion, *Phys. Rev.* **D88**, p. 094507 (2013).

49. R. A. Briceno, M. T. Hansen and S. R. Sharpe, Relating the finite-volume spectrum and the two-and-three-particle S matrix for relativistic systems of identical scalar particles, *Phys. Rev.* **D95**, p. 074510 (2017).

50. C. B. Lang, D. Mohler and S. Prelovsek, $B_s\pi^+$ scattering and search for X(5568) with lattice QCD, *Phys. Rev.* **D94**, p. 074509 (2016).

51. S. Prelovsek, C. B. Lang, L. Leskovec and D. Mohler, Study of the Z_c^+ channel using lattice QCD, *Phys. Rev.* **D91**, p. 014504 (2015).

52. N. Brambilla *et al.*, QCD and Strongly Coupled Gauge Theories: Challenges and Perspectives, *Eur. Phys. J.* **C74**, p. 2981 (2014).

53. V. M. Abazov *et al.*, Evidence for a $B_s^0\pi^\pm$ state, *Phys. Rev. Lett.* **117**, p. 022003 (2016).

54. R. Aaij *et al.*, Search for Structure in the $B_s^0\pi^\pm$ Invariant Mass Spectrum, *Phys. Rev. Lett.* **117**, p. 152003 (2016), [Addendum: Phys. Rev. Lett. 118, no. 10, 109904(2017)].

55. M. Ablikim *et al.*, Observation of a Charged Charmoniumlike Structure in $e^+e^- \to \pi^+\pi^- J/\psi$ at $\sqrt{s} = 4.26$ GeV, *Phys. Rev. Lett.* **110**, p. 252001 (2013).

56. Z. Q. Liu *et al.*, Study of $e^+e^- \to \pi^+\pi^- J/\psi$ and Observation of a Charged Charmoniumlike State at Belle, *Phys. Rev. Lett.* **110**, p. 252002 (2013).

57. T. Xiao, S. Dobbs, A. Tomaradze and K. K. Seth, Observation of the Charged Hadron $Z_c^\pm(3900)$ and Evidence for the Neutral $Z_c^0(3900)$ in $e^+e^- \to \pi\pi J/\psi$ at $\sqrt{s} = 4170$ MeV, *Phys. Lett.* **B727**, 366 (2013).

58. Y. Chen *et al.*, Low-energy scattering of the $(D\bar{D}^*)^\pm$ system and the resonance-like structure $Z_c(3900)$, *Phys. Rev.* **D89**, p. 094506 (2014).

59. S.-h. Lee, C. DeTar, H. Na and D. Mohler, Searching for the $X(3872)$ and $Z_c^+(3900)$ on HISQ Lattices (2014).

60. Y. Ikeda, S. Aoki, T. Doi, S. Gongyo, T. Hatsuda, T. Inoue, T. Iritani, N. Ishii, K. Murano and K. Sasaki, Fate of the Tetraquark Candidate $Z_c(3900)$ from Lattice QCD, *Phys. Rev. Lett.* **117**, p. 242001 (2016).

61. S. Aoki *et al.*, Review of lattice results concerning low-energy particle physics, *Eur. Phys. J.* **C77**, p. 112 (2017).

62. V. Cirigliano, J. Jenkins and M. Gonzalez-Alonso, Semileptonic decays of light quarks beyond the Standard Model, *Nucl. Phys.* **B830**, 95 (2010).

63. M. Gonzalez-Alonso and J. Martin Camalich, Global Effective-Field-Theory analysis of New-Physics effects in (semi)leptonic kaon decays, *JHEP* **12**, p. 052 (2016).

64. M. Moulson, Experimental determination of V_{us} from kaon decays, *PoS* **CKM2016**, p. 033 (2017).

65. V. Lubicz, G. Martinelli, C. T. Sachrajda, F. Sanfilippo, S. Simula, N. Tantalo and C. Tarantino, Electromagnetic corrections to the leptonic decay rates of charged pseudoscalar mesons: Lattice results, *PoS* **LATTICE2016**, p. 290 (2016).

66. V. Cirigliano and H. Neufeld, A note on isospin violation in Pl2(gamma) decays, *Phys. Lett.* **B700**, 7 (2011).

67. J. L. Rosner, S. Stone and R. S. Van de Water, Leptonic Decays of Charged Pseudoscalar Mesons - 2015, *Submitted to: Particle Data Book* (2015).

68. M. Antonelli *et al.*, An Evaluation of $|V_{us}|$ and precise tests of the Standard Model from world data on leptonic and semileptonic kaon decays, *Eur. Phys. J.* **C69**, 399 (2010).

69. N. Carrasco, V. Lubicz, G. Martinelli, C. T. Sachrajda, N. Tantalo, C. Tarantino and M. Testa, QED Corrections to Hadronic Processes in Lattice QCD, *Phys. Rev.* **D91**, p. 074506 (2015).

70. L. Lellouch and M. Luscher, Weak transition matrix elements from finite volume correlation functions, *Commun. Math. Phys.* **219**, 31 (2001).

71. T. Blum *et al.*, $K \to \pi\pi \, \Delta I = 3/2$ decay amplitude in the continuum limit, *Phys. Rev.* **D91**, p. 074502 (2015).

72. Z. Bai *et al.*, Standard Model Prediction for Direct CP Violation in $K \to \pi\pi$ Decay, *Phys. Rev. Lett.* **115**, p. 212001 (2015).

73. N. Ishizuka, K. I. Ishikawa, A. Ukawa and T. Yoshie, Calculation of $K \to \pi\pi$ decay amplitudes with improved Wilson fermion action in lattice QCD, *Phys. Rev.* **D92**, p. 074503 (2015).

74. A. Alavi-Harati *et al.*, Measurements of direct CP violation, CPT symmetry, and other parameters in the neutral kaon system, *Phys. Rev.* **D67**, p. 012005 (2003), [Erratum: Phys. Rev. D70, 079904(2004)].

75. J. R. Batley *et al.*, A Precision measurement of direct CP violation in the decay of neutral kaons into two pions, *Phys. Lett.* **B544**, 97 (2002).

76. E. Abouzaid *et al.*, Precise Measurements of Direct CP Violation, CPT Symmetry, and Other Parameters in the Neutral Kaon System, *Phys. Rev.* **D83**, p. 092001 (2011).

77. A. J. Buras, M. Gorbahn, S. Jager and M. Jamin, Improved anatomy of ϵ'/ϵ in the Standard Model, *JHEP* **11**, p. 202 (2015).

78. T. Kitahara, U. Nierste and P. Tremper, Singularity-free next-to-leading order $\Delta S = 1$ renormalization group evolution and ϵ'_K/ϵ_K in the Standard Model and beyond, *JHEP* **12**, p. 078 (2016).

79. C. Kelly, Progress in the improved lattice calculation of direct CP-violation in the Standard Model, presented at the 35th International Symposium on Lattice Field Theory (Lattice 2017) `http://wpd.ugr.es/~lattice2017/`.

80. C. Kelly, Progress in the calculation of ϵ' on the lattice, *PoS* **LATTICE2016**, p. 308 (2016).

81. M. Bruno, Weak Hamiltonian Wilson Coefficients from Lattice QCD, *EPJ Web Conf.* **175**, p. 13014 (2018).

82. N. H. Christ, X. Feng, A. Juttner, A. Lawson, A. Portelli and C. T. Sachrajda, First exploratory calculation of the long-distance contributions to the rare kaon decays $K \to \pi\ell^+\ell^-$, *Phys. Rev.* **D94**, p. 114516 (2016).

83. Z. Bai, N. H. Christ, X. Feng, A. Lawson, A. Portelli and C. T. Sachrajda, Exploratory Lattice QCD Study of the Rare Kaon Decay $K^+ \to \pi^+\nu\bar{\nu}$, *Phys. Rev. Lett.* **118**, p. 252001 (2017).

84. Y. Amhis *et al.*, Averages of b-hadron, c-hadron, and τ-lepton properties as of summer 2016, *Eur. Phys. J.* **C77**, p. 895 (2017).

85. Belle II Theory Interface Platform, `https://confluence.desy.de/display/BI/B2TiP+WebHome`.

86. E. Dalgic, A. Gray, M. Wingate, C. T. H. Davies, G. P. Lepage and J. Shigemitsu, B meson semileptonic form-factors from unquenched lattice QCD, *Phys. Rev.* **D73**, p. 074502 (2006), [Erratum: Phys. Rev. D75, 119906(2007)].

87. J. M. Flynn, T. Izubuchi, T. Kawanai, C. Lehner, A. Soni, R. S. Van de Water and O. Witzel, $B \to \pi\ell\nu$ and $B_s \to K\ell\nu$ form factors and $|V_{ub}|$ from 2+1-flavor lattice QCD with domain-wall light quarks and relativistic heavy quarks, *Phys. Rev.* **D91**, p. 074510 (2015).

88. J. A. Bailey *et al.*, $|V_{ub}|$ from $B \to \pi\ell\nu$ decays and (2+1)-flavor lattice QCD, *Phys. Rev.* **D92**, p. 014024 (2015).

89. J. A. Bailey *et al.*, $B \to D\ell\nu$ form factors at nonzero recoil and $-V_{cb}-$ from 2+1-flavor lattice QCD, *Phys. Rev.* **D92**, p. 034506 (2015).

90. H. Na, C. M. Bouchard, G. P. Lepage, C. Monahan and J. Shigemitsu, $B \to D l\nu$ form factors at nonzero recoil and extraction of $|V_{cb}|$, *Phys. Rev.* **D92**, p. 054510 (2015), [Erratum: Phys. Rev. D93, no. 11, 119906(2016)].

91. C. Bourrely, B. Machet and E. de Rafael, Semileptonic Decays of Pseudoscalar Particles ($M \to M'\ell\nu_\ell$) and Short Distance Behavior of Quantum Chromodynamics, *Nucl. Phys.* **B189**, 157 (1981).

92. Z. Gelzer *et al.*, Semileptonic B-Meson Decays to Light Pseudoscalar Mesons on the HISQ Ensembles, *EPJ Web Conf.* **175**, p. 13024 (2018).

93. B. Colquhoun, S. Hashimoto and T. Kaneko, $B \to \pi\ell\nu$ with Möbius Domain Wall Fermions, *EPJ Web Conf.* **175**, p. 13004 (2018).

94. J. A. Bailey *et al.*, Update of $|V_{cb}|$ from the $\bar{B} \to D^*\ell\bar{\nu}$ form factor at zero recoil with three-flavor lattice QCD, *Phys. Rev.* **D89**, p. 114504 (2014).

95. D. Bigi, P. Gambino and S. Schacht, A fresh look at the determination of $|V_{cb}|$ from $B \to D^*\ell\nu$, *Phys. Lett.* **B769**, 441 (2017).

96. B. Grinstein and A. Kobach, Model-Independent Extraction of $|V_{cb}|$ from $\bar{B} \to D^*\ell\bar{\nu}$, *Phys. Lett.* **B771**, 359 (2017).

97. F. U. Bernlochner, Z. Ligeti, M. Papucci and D. J. Robinson, Tensions and correlations in $|V_{cb}|$ determinations, *Phys. Rev.* **D96**, p. 091503 (2017).

98. A. Vaquero Avilés-Casco, C. DeTar, D. Du, A. El-Khadra, A. S. Kronfeld, J. Laiho and R. S. Van de Water, $\overline{B} \to D^*\ell\bar{\nu}$ at Non-Zero Recoil, in *35th International Symposium on Lattice Field Theory (Lattice 2017) Granada, Spain, June 18-24, 2017*, (2017).

99. C. J. Monahan, H. Na, C. M. Bouchard, G. P. Lepage and J. Shigemitsu, $B_s \to D_s\ell\nu$ Form Factors and the Fragmentation Fraction Ratio f_s/f_d, *Phys. Rev.* **D95**, p. 114506 (2017).

100. C. M. Bouchard, G. P. Lepage, C. Monahan, H. Na and J. Shigemitsu, $B_s \to K\ell\nu$ form factors from lattice QCD, *Phys. Rev.* **D90**, p. 054506 (2014).

101. F. Bahr, D. Banerjee, F. Bernardoni, A. Joseph, M. Koren, H. Simma and R. Sommer, Continuum limit of the leading-order HQET form factor in $B_s \to K\ell\nu$ decays, *Phys. Lett.* **B757**, 473 (2016).

102. J. A. Bailey *et al.*, $B_s \to D_s/B \to D$ Semileptonic Form-Factor Ratios and Their Application to BR($B_s^0 \to \mu^+\mu^-$), *Phys. Rev.* **D85**, p. 114502 (2012), [Erratum: Phys. Rev. D86, 039904(2012)].

103. M. Atoui, V. Morenas, D. Becirevic and F. Sanfilippo, $B_s \to D_s\ell\nu_\ell$ near zero recoil in and beyond the Standard Model, *Eur. Phys. J.* **C74**, p. 2861 (2014).

104. W. Detmold, C. Lehner and S. Meinel, $\Lambda_b \to p\ell^-\bar{\nu}_\ell$ and $\Lambda_b \to \Lambda_c\ell^-\bar{\nu}_\ell$ form factors from lattice QCD with relativistic heavy quarks, *Phys. Rev.* **D92**, p. 034503 (2015).

105. S. Meinel, $\Lambda_c \to \Lambda l^+\nu_l$ form factors and decay rates from lattice QCD with physical quark masses, *Phys. Rev. Lett.* **118**, p. 082001 (2017).

106. W. Detmold and S. Meinel, $\Lambda_b \to \Lambda \ell^+ \ell^-$ form factors, differential branching fraction, and angular observables from lattice QCD with relativistic b quarks, *Phys. Rev.* **D93**, p. 074501 (2016).
107. R. Aaij *et al.*, Differential branching fraction and angular analysis of $\Lambda_b^0 \to \Lambda \mu^+ \mu^-$ decays, *JHEP* **06**, p. 115 (2015).
108. S. Descotes-Genon, J. Matias and J. Virto, Understanding the $B \to K^* \mu^+ \mu^-$ Anomaly, *Phys. Rev.* **D88**, p. 074002 (2013).
109. A. Bazavov *et al.*, $B_{(s)}^0$-mixing matrix elements from lattice QCD for the Standard Model and beyond, *Phys. Rev.* **D93**, p. 113016 (2016).
110. N. Carrasco *et al.*, B-physics from $N_f = 2$ tmQCD: The Standard Model and beyond, *JHEP* **03**, p. 016 (2014).
111. K. Maltman, R. Hudspith, R. Lewis, T. Izubuchi, H. Ohki and J. Zanotti, $|V_{us}|$ from τ decays in theory, *PoS* **CKM2016**, p. 030 (2017).
112. S. Hashimoto, Inclusive semi-leptonic B meson decay structure functions from lattice QCD, *PTEP* **2017**, p. 053B03 (2017).
113. K.-F. Liu, Evolution equations for connected and disconnected sea parton distributions, *Phys. Rev.* **D96**, p. 033001 (2017).
114. M. T. Hansen, H. B. Meyer and D. Robaina, From deep inelastic scattering to heavy-flavor semileptonic decays: Total rates into multihadron final states from lattice QCD, *Phys. Rev.* **D96**, p. 094513 (2017).
115. P. Boyle, R. J. Hudspith, T. Izubuchi, A. Jüttner, C. Lehner, R. Lewis, K. Maltman, H. Ohki, A. Portelli and M. Spraggs, —Vus— determination from inclusive strange tau decay and lattice HVP, *EPJ Web Conf.* **175**, p. 13011 (2018).
116. P. Boyle, R. J. Hudspith, T. Izubuchi, A. Jüttner, C. Lehner, R. Lewis, K. Maltman, H. Ohki, A. Portelli and M. Spraggs, Novel $|V_{us}|$ Determination Using Inclusive Strange τ Decay and Lattice HVPs (2018).
117. S. Hashimoto, B. Colquhoun, T. Izubuchi, T. Kaneko and H. Ohki, Inclusive B decay calculations with analytic continuation, *EPJ Web Conf.* **175**, p. 13006 (2018).
118. G. W. Bennett *et al.*, Final Report of the Muon E821 Anomalous Magnetic Moment Measurement at BNL, *Phys. Rev.* **D73**, p. 072003 (2006).
119. K. Hagiwara, R. Liao, A. D. Martin, D. Nomura and T. Teubner, $(g-2)_\mu$ and $\alpha(M_Z^2)$ re-evaluated using new precise data, *J. Phys.* **G38**, p. 085003 (2011).
120. F. Jegerlehner, Muon $g-2$ theory: The hadronic part, *EPJ Web Conf.* **166**, p. 00022 (2018).
121. M. Davier, A. Hoecker, B. Malaescu and Z. Zhang, Reevaluation of the hadronic vacuum polarisation contributions to the Standard Model predictions of the muon $g-2$ and $\alpha(m_Z^2)$ using newest hadronic cross-section data, *Eur. Phys. J.* **C77**, p. 827 (2017).
122. J. Grange *et al.*, Muon $(g-2)$ Technical Design Report (2015).
123. N. Saito, Muon $(g-2)$ Technical Design Report, *AIP Conf. Proc.* **1467**, p. 45 (2012).
124. G. Colangelo, M. Hoferichter, M. Procura and P. Stoffer, Dispersive approach to hadronic light-by-light scattering, *JHEP* **09**, p. 091 (2014).
125. V. Pauk and M. Vanderhaeghen, Two-loop massive scalar three-point function in a dispersive approach (2014).
126. G. Colangelo, M. Hoferichter, M. Procura and P. Stoffer, Rescattering effects in the hadronic-light-by-light contribution to the anomalous magnetic moment of the muon, *Phys. Rev. Lett.* **118**, p. 232001 (2017).
127. J. Bijnens, E. Pallante and J. Prades, Analysis of the hadronic light by light contributions to the muon $g-2$, *Nucl. Phys.* **B474**, 379 (1996).

128. M. Hayakawa, T. Kinoshita and A. I. Sanda, Hadronic light by light scattering effect on muon $g - 2$, *Phys. Rev. Lett.* **75**, 790 (1995).

129. K. Melnikov and A. Vainshtein, Hadronic light-by-light scattering contribution to the muon anomalous magnetic moment revisited, *Phys. Rev.* **D70**, p. 113006 (2004).

130. B. E. Lautrup and E. De Rafael, Calculation of the sixth-order contribution from the fourth-order vacuum polarization to the difference of the anomalous magnetic moments of muon and electron, *Phys. Rev.* **174**, 1835 (1968).

131. T. Blum, Lattice calculation of the lowest order hadronic contribution to the muon anomalous magnetic moment, *Phys. Rev. Lett.* **91**, p. 052001 (2003).

132. M. Gockeler, R. Horsley, W. Kurzinger, D. Pleiter, P. E. L. Rakow and G. Schierholz, Vacuum polarization and hadronic contribution to muon $g - 2$ from lattice QCD, *Nucl. Phys.* **B688**, 135 (2004).

133. C. Aubin and T. Blum, Calculating the hadronic vacuum polarization and leading hadronic contribution to the muon anomalous magnetic moment with improved staggered quarks, *Phys. Rev.* **D75**, p. 114502 (2007).

134. X. Feng, K. Jansen, M. Petschlies and D. B. Renner, Two-flavor QCD correction to lepton magnetic moments at leading-order in the electromagnetic coupling, *Phys. Rev. Lett.* **107**, p. 081802 (2011).

135. D. Bernecker and H. B. Meyer, Vector Correlators in Lattice QCD: Methods and applications, *Eur. Phys. J.* **A47**, p. 148 (2011).

136. X. Feng, S. Hashimoto, G. Hotzel, K. Jansen, M. Petschlies and D. B. Renner, Computing the hadronic vacuum polarization function by analytic continuation, *Phys. Rev.* **D88**, p. 034505 (2013).

137. A. Francis, B. Jaeger, H. B. Meyer and H. Wittig, A new representation of the Adler function for lattice QCD, *Phys. Rev.* **D88**, p. 054502 (2013).

138. M. Della Morte, A. Francis, V. Gulpers, G. Herdoiza, G. von Hippel, H. Horch, B. Jager, H. B. Meyer, A. Nyffeler and H. Wittig, The hadronic vacuum polarization contribution to the muon $g - 2$ from lattice QCD, *JHEP* **10**, p. 020 (2017).

139. S. Borsanyi *et al.*, Hadronic vacuum polarization contribution to the anomalous magnetic moments of leptons from first principles (2017).

140. T. Blum, P. A. Boyle, V. Gulpers, T. Izubuchi, L. Jin, C. Jung, A. Juttner, C. Lehner, A. Portelli and J. T. Tsang, Calculation of the hadronic vacuum polarization contribution to the muon anomalous magnetic moment (2018).

141. B. Chakraborty, C. T. H. Davies, G. C. Donald, R. J. Dowdall, J. Koponen, G. P. Lepage and T. Teubner, Strange and charm quark contributions to the anomalous magnetic moment of the muon, *Phys. Rev.* **D89**, p. 114501 (2014).

142. B. Chakraborty, C. T. H. Davies, J. Koponen, G. P. Lepage, M. J. Peardon and S. M. Ryan, Estimate of the hadronic vacuum polarization disconnected contribution to the anomalous magnetic moment of the muon from lattice QCD, *Phys. Rev.* **D93**, p. 074509 (2016).

143. B. Chakraborty, C. T. H. Davies, P. G. de Oliviera, J. Koponen, G. P. Lepage and R. S. Van de Water, The hadronic vacuum polarization contribution to a_μ from full lattice QCD, *Phys. Rev.* **D96**, p. 034516 (2017).

144. F. Burger, X. Feng, G. Hotzel, K. Jansen, M. Petschlies and D. B. Renner, Four-Flavour Leading-Order Hadronic Contribution To The Muon Anomalous Magnetic Moment, *JHEP* **02**, p. 099 (2014).

145. B. Chakraborty *et al.*, Strong-isospin-breaking correction to the muon anomalous magnetic moment from lattice QCD at the physical point (2017).

146. T. Blum *et al.*, Lattice calculation of the leading strange quark-connected contribution to the muon $g - 2$, *JHEP* **04**, p. 063 (2016), [Erratum: JHEP05, 034(2017)].

147. D. Giusti, V. Lubicz, G. Martinelli, F. Sanfilippo and S. Simula, Strange and charm HVP contributions to the muon $(g-2)$ including QED corrections with twisted-mass fermions, *JHEP* **10**, p. 157 (2017).

148. B. Colquhoun, R. J. Dowdall, C. T. H. Davies, K. Hornbostel and G. P. Lepage, Υ and Υ' Leptonic Widths, a_μ^b and m_b from full lattice QCD, *Phys. Rev.* **D91**, p. 074514 (2015).

149. T. Blum, P. A. Boyle, T. Izubuchi, L. Jin, A. Juttner, C. Lehner, K. Maltman, M. Marinkovic, A. Portelli and M. Spraggs, Calculation of the hadronic vacuum polarization disconnected contribution to the muon anomalous magnetic moment, *Phys. Rev. Lett.* **116**, p. 232002 (2016).

150. C. Aubin, T. Blum, M. Golterman and S. Peris, Model-independent parametrization of the hadronic vacuum polarization and $g-2$ for the muon on the lattice, *Phys. Rev.* **D86**, p. 054509 (2012).

151. M. Golterman, K. Maltman and S. Peris, A Hybrid Strategy for the Lattice Evaluation of the Leading Order Hadronic Contribution to $(g-2)_\mu$, *Nucl. Part. Phys. Proc.* **273-275**, 1650 (2016).

152. S. Borsanyi, Z. Fodor, T. Kawanai, S. Krieg, L. Lellouch, R. Malak, K. Miura, K. K. Szabo, C. Torrero and B. Toth, Slope and curvature of the hadronic vacuum polarization at vanishing virtuality from lattice QCD, *Phys. Rev.* **D96**, p. 074507 (2017).

153. T. Blum, N. Christ, M. Hayakawa, T. Izubuchi, L. Jin, C. Jung and C. Lehner, Connected and Leading Disconnected Hadronic Light-by-Light Contribution to the Muon Anomalous Magnetic Moment with a Physical Pion Mass, *Phys. Rev. Lett.* **118**, p. 022005 (2017).

154. T. Blum, N. Christ, M. Hayakawa, T. Izubuchi, L. Jin and C. Lehner, Lattice Calculation of Hadronic Light-by-Light Contribution to the Muon Anomalous Magnetic Moment, *Phys. Rev.* **D93**, p. 014503 (2016).

155. A. Gerardin, H. B. Meyer and A. Nyffeler, Lattice calculation of the pion transition form factor $\pi^0 \to \gamma^*\gamma^*$, *Phys. Rev.* **D94**, p. 074507 (2016).

156. T. Blum, N. Christ, M. Hayakawa, T. Izubuchi, L. Jin, C. Jung and C. Lehner, Using infinite volume, continuum QED and lattice QCD for the hadronic light-by-light contribution to the muon anomalous magnetic moment, *Phys. Rev.* **D96**, p. 034515 (2017).

PUBLIC SHOW

Exploring Dark Matter: A Phantom of the Universe[*]

R. Michael Barnett[†]

Mailstop 50R-6008,
Lawrence Berkeley National Laboratory,
1 Cyclotron Road, Berkeley, CA 94720, USA
†E-mail: rmbarnett@lbl.gov
www.lbl.gov

Kaushik De

Physics Dept., University of Texas at Arlington,
701 S. Nedderman Drive, Arlington, TX 76019, USA
E-mail: kaushik@uta.edu

Reinhard Schwienhorst

Physics Dept., Michigan State University,
603 Wilson Road, East Lansing, MI 48824, USA
E-mail: schwier@pa.msu.edu

A multi-year project with funding from many sources was led by physicists who recruited world-class experts in producing a state-of-the-art planetarium show about dark matter. It has found worldwide success and even won an award. It has been translated into many languages and is showing on all five continents.

Keywords: Outreach; education; planetarium; film.

1. Planetarium Show on Dark Matter

1.1. *Introduction*

An international team of physicists and planetarium show producers created a state-of-the-art planetarium show about dark matter called "Phantom of the Universe: The Hunt for Dark Matter". The website for it is at http://PhantomOfTheUniverse.org. The show is distributed for free to planetariums. The show is in 16 languages and has been or is being presented in about 40 countries and in about 200 planetariums. 110 million people watch planetarium shows each year so there is a potentially large audience.

[*]This work is supported by NSF, DOE, ATLAS, LBNL, Michigan State U., U. of Texas Arlington, CERN, STFC (UK), IFIC (Valencia), Institute of High Energy Physics in Vienna, Natural History Museum in Vienna, DESY, CEA, IN2P3, and more.
[†]Work supported by the Director, Office of Science, Office of High Energy Physics of the U.S. Department of Energy under Contract No. DE-AC02-05CH11231.

2. A Worldwide Project

Production of Phantom of the Universe was a worldwide project, with the director in Switzerland, the animators in Spain, narrator in Britain, many participants in the US, and with executive producers as follows:

- Michael Barnett – Lawrence Berkeley National Lab
- Kaushik De – University of Texas, Arlington
- Reinhard Schwienhorst – Michigan State University
- Carmen Garcia – University of Valencia
- Markus Nordberg – CERN
- George Smoot – University of California, Berkeley
- And many others for other languages

3. Why a Planetarium Show

3.1. *Our original ideas*

When we began this project our motivations were primarily the following:

- More dramatic than IMAX (it surrounds you).
- Great way to show galaxies, and a scene with a spiral galaxy blowing apart (without dark matter).
- Great way to present the LHC ring.
- Great way to present collisions (like fireworks raining down the planetarium walls).
- New avenue of public outreach

3.2. *What we learned*

Through the multi-year process of producing the show, we learned other good reasons to have produced the show:

- There are hundreds of planetariums with an interest in a dark matter show (broader than ATLAS or LHC).
- They will present our show for months at a time.
- Planetariums have the perfect science — interested audience for us — general public and K-12.
- 110 million people a year watch planetarium shows.

Much of this would not have been possible with an ordinary film.

4. Content of the Show

We had to pick a few highlights for our story; we could not cover everything. A typical planetarium show is 25–30 minutes, and this one is 25 minutes plus credits. These were the topics we focused on.

- Dark matter created in the Big Bang.
- Evidence for dark matter in Galaxies, etc.
- Underground search for dark matter.
- Search for dark matter at the LHC.

5. Professional Team

Creating the planetarium show involved many international professional teams:

- Academy Award — winning actress as narrator
- Academy Award — winning sound effects by Skywalker Sound (Star Wars, Avatar…)
- Hollywood scriptwriter and producer
- Outstanding Director and team of Animators
- Nobel Prize — winning cosmologist
- Experts from seven planetariums
- Sound editors with substantial experience
- And many more…

Fig. 1.   The team of animators in Valencia, Spain, plus the Director Joao Pequenao and Michael Barnett.

324

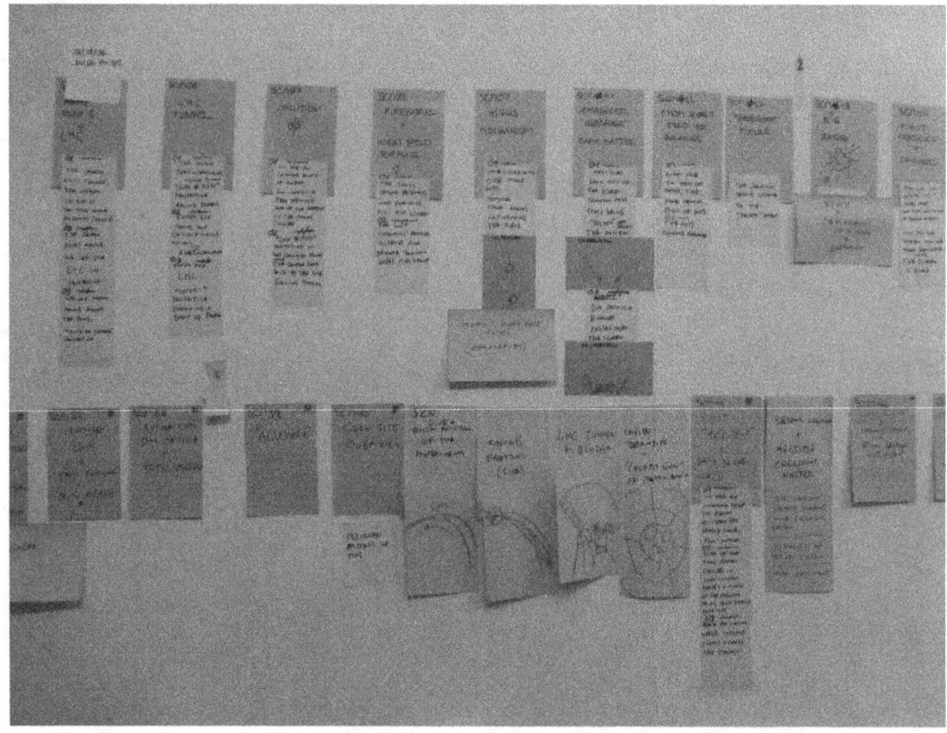

Fig. 2. The post-it layout of the full show.

5.1. *Narrator*

The narrator is Tilda Swinton who lives in Nairn, Scotland. Barnett traveled to Nairn, and then 8 kilometers out of town, down a one-lane road to a recording studio. The Academy Award-winning actress is known for roles such as:

Doctor Strange	Julia
Hail, Caesar!	Moonrise Kingdom
A Bigger Splash	We Need to Talk About Kevin
Trainwreck	Chronicles of Narnia
The Grand Budapest Hotel	Michael Clayton
Snowpiercer	Galápagos

And much more.

Fig. 3.   After recording the narration, actress Tilda Swinton posed with Executive Producer Michael Barnett.

5.2. *Sound effects and sound mixing*

The wonderful sound effects and sound mixing were done by Skywalker Sound, a company whose experts have won many Academy Awards for films such as:

Star Wars	Pirates of the Caribbean
Bridge of Spies	Titanic
Lincoln	Forrest Gump
Toy Story	Jurassic Park
Avatar	etc.

6. Vienna Premiere

The Vienna premiere of the show was coupled with a new exhibit at the Natural History Museum. The premiere was attended by Peter Higgs, George Smoot, the Austrian minister of culture, and other dignitaries. It occurred on 18 October 2016.

The show received many compliments there (as it has in many other locations), including:

"Visually stunning and scientifically fascinating".

--Director of Natural History Museum, Vienna

"People love the Show! Everybody who has seen it tells me that it is the best show they have seen in the planetarium yet."

--Director of planetarium at Natural History Museum, Vienna

7. Award

At the 2017 annual International Full Dome Festival in Germany, Phantom of the Universe won:

"The Full Dome Festival Honorable Mention for Outstanding and Innovative Production"

8. Other Language Versions

The show has been translated and recorded in 16 languages: Bulgarian, Chinese, Czech, English, Finnish, French, German, Greek, Italian, Japanese, Korean, Polish, Portuguese, Russian, Spanish (Mexico), Spanish (Spain), Swedish, and Turkish.

There are also subtitles in Romanian, Vietnamese and other languages.

We welcome those who wish to translate the show into further languages.

9. Dark Matter Day

A number major laboratories worldwide organized Dark Matter Day centered on 31 October 2017 (and also on earlier days in October). These labs also prepared various materials and resources. As part of this, many planetariums presented our dark matter show, Phantom of the Universe. After the show, they typically had a local expert to answer questions or give a presentation.

10. Distribution

The show is currently or soon to be shown in about 200 planetariums in about 40 countries on 5 continents. In the US it is in about 30 states. Thirty planetariums in France are presenting it.

11. Summary

A state-of-the-art planetarium show on dark matter is available now for free to planetariums worldwide. We invite everyone to bring it to the attention of your local planetarium. Full information is at our website at: http://PhantomOfTheUniverse.org

Acknowledgments

We are extremely grateful to our prime collaborators in this project: Joao Pequenao (Director, CERN), Carey Ann Strelecki (Producer), Jesus Nuevo (Lead Animator), and Levent Gurdemir (Associate Producer, University of Texas Arlington).

POSTER SESSIONS

String Self-Interactions Near the Deconfinement Point in SU(3) Yang-Mills Theory

A. Bakry[*], X. Chen[†], M. Deliyergiyev[‡], S. Xu[§], P.M. Zhang[¶]

Department of High Energy Nuclear Physics, Institute of Modern Physics, Chinese Academy of Sciences, Lanzhou, 730000, China
**E-mail: ahmed.bakry@physics.net*
†E-mail: xchen@impcas.ac.cn
‡E-mail: deliyergiyev@impcas.ac.cn
§E-mail: xsq234@163.com
¶E-mail: zhpm@impcas.ac.cn
http://english.imp.cas.cn

A. Galal[‖] and A. Khalaf

Physics Department, Faculty of Science, Al-Azhar University, Cairo 11651, Egypt
‖E-mail: aagalal@yahoo.com

We compare the predictions of Nambu-Goto effective string model at two loop-order for the Casimir energy and the width of the quantum delocalization of the string to the corresponding quark-antiquark potential and width profile of the color tube in pure $SU(3)$ Yang-Mills LGT at 4-dimensions near the deconfinement point. Minor corrections are returned, when considering NLO terms for both the $Q\bar{Q}$ potential and broadening of the color tube, at the temperature $T/T_c = 0.8$. At a closer temperature to the critical point $T/T_c = 0.9$, we found that the NLO contributions, from the expansion of the Nambu-Goto string, have a significant effect in improving the match to lattice data in the intermediate distances scales–before the string breaks in full QCD. In addition to decrease in the value of the string tension, the string's self-interactions account for the suppressed broadening of the string's width and the non-curved/squeezed profile along the string in the intermediate distance.

Keywords: QCD phenomenology; finite-temperature; color flux-tube; bosonic strings; Monte-Carlo methods; Nambu-Goto; Lüscher–Weiss action; Lattice Gauge Theory.

1. Introduction

Precise lattice measurements of the $Q\bar{Q}$ potential in $SU(3)$ gauge model are in consistency with the Lüscher subleading term for color source separation commencing from distance $R = 0.5$ fm [1] – a typical distance scale where the effects of the intrinsic thickness of the flux tube at finite temperature, $1/T_c$ [2], diminishes and the effective description is expected to hold. The Lüscher correction to the $Q\bar{Q}$ potential have unambiguously identified with unprecedented accuracy in many models [3–8]. In addition, the string model predicts a logarithmic broadening [9] for the width profile of the string delocalization. The observations of this effect have been reported in several lattice simulations corresponding to the different gauge groups [3, 10–20].

However, in the intermediate distances and high temperatures this simplified picture of the free bosonic string, derived on the basis of the LO formulation of the Nambu-Goto(NG) action, does not provide a good description. For instance, substantial deviations from the string behavior have been found for the lattice data corresponding to temperatures near the deconfinement point [21–24]. The region extends behind source separation distances at which LO string model predictions are valid at zero temperature regime. A comparison with the lattice Monte-Carlo data showed the validity of the LO approximation at separations larger than $R = 0.9$ fm [21, 22, 24] for both the $Q\bar{Q}$ potential and flux-tube width profile. The authors of Ref. [25] come to a similar conclusion by considering the length of the Y-string between any two quarks in the baryon [26–28].

The fact that free string theory poorly explains the lattice data in the intermediate distance and high temperatures has induced interest in the numerical experiments to verify the validity of higher order corrections to the NG action [29, 30]. However, these higher order terms in the NG action are non-universal, that is depend on the gauge model under consideration [29, 31]. Even in the NG framework not all the orders of the power expansion are believed to give a good description of the correct behavior of the strings in the intermediate region.

This calls for a discussion concerning the distance and temperature scales for which the effective NG string description in the two loop order approximation is valid and which is the target of this report.

In this proceeding, we discuss the lattice data corresponding to the $Q\bar{Q}$ potential from $SU(3)$ Yang-Mills theory in four dimensions, and corresponding predictions of the LO and NLO Nambu-Goto potential [30] in Sec. 2. In Sec. 3, the width profile of the density distribution is compare to the mean-square width of the string fluctuations in both approximations. The main conclusions are drawn in the last section.

2. Quark-Antiquark Potential at LO and NLO

The linearly rising property of the confining potential induced the conjecture [32] that the Yang-Mills (YM) vacuum admits the existance of an idealized one dimensional string object transmitting the strongly interacting forces between the color sources.

The partition function of the NG model in the physical gauge[a] is a functional integral over all the worldsheet configurations swept by the string,

$$Z(R, T) = \int_C [D\,X] \exp(-S_{NG}(X)), \tag{1}$$

---

[a]The physical gauge is required for the path integral Eq. (1) to be well defined with respect to Weyl and reparameterization invariance.

where the NG action after gauge fixing reads

$$S_{NG}[X] = \sigma \frac{R}{T} + \frac{\sigma}{2} \int_0^{L_T} d\zeta_1 \int_0^R d\zeta_2 \, (\nabla X)^2 + \ldots \tag{2}$$

With the string collective variables $X^\mu = (\sigma^\alpha, X^i(\sigma))$, which allows one to embed the string worldsheet into the target space, where $\sigma^\alpha (\alpha = 1, 2)$ is the worldsheet coordinates, X^i is the two-dimensional Goldstone bosons. For the periodic boundary condition along the time direction with extend equals to the inverse temperature, $L_T = 1/T$, the string collective variables can be defined as $X(\zeta_1 - 0, \zeta_2) = X(\zeta_1 = L_T, \zeta_2)$. The Dirichlet boundary condition at the source position $X(\zeta_1, \zeta_2 = 0) = X(\zeta_1, \zeta_2 = R) = 0$.

The Casimir potential can be extracted from the string partition function Eq. (1) as

$$V(R, T) = -\frac{1}{T} \log \left(Z(R, T) \right) \tag{3}$$

Solving the path integral of Eq. (2) and using Eq. (3) with ζ-regularization scheme [12] yields the model-independent static potential at the LO approximation

$$V_{\ell o}(R, T) = \sigma R + (D - 2)T \ln \eta \, (i\tau) + \mu(T), \tag{4}$$

where $\mu(T)$ is a renormalization parameter, $\tau = \frac{L_T}{2R}$ is the modular parameter of the cylinder, η is the Dedekind eta function defined as $\eta(\tau) = q^{1/24} \prod_{n=1}^\infty (1 - q^n)$ with $q = e^{-\frac{\pi L_T}{R}}$.

The second model-independent correction to the Casimir effect has been derived in Ref. [12] from the explicit calculation of the two-loop approximation as

$$V_{n\ell o}(R, T) = V_{\ell o}(R, T) - T \ln \left(1 - \frac{(D - 2)\pi^2 T}{1152\sigma_o R^3} \left[2E_4(\tau) + (D - 4)E_2^2(\tau) \right] \right), \tag{5}$$

where E is the Eisenstein series given by

$$E_{2k}(\tau) = 1 + (-1)^k \frac{4k}{B_k} \sum_{n=1}^\infty \frac{n^{2k-1}q^n}{1 - q^n}. \tag{6}$$

From Eq. (5) one can derive the string tension as a function of the temperature [31] up to the NLO as

$$\sigma(T) = \sigma_0 - \frac{\pi(D - 2)}{6}T^2 - \frac{\pi^2(D - 2)^2}{72\sigma_0}T^4 - \frac{\pi^3(D - 2)^3}{432\sigma_0}T^6 + \mathcal{O}(T^7). \tag{7}$$

The Monte-Carlo evaluation of the temperature-dependent $Q\bar{Q}$ potential

$$V_{Q\bar{Q}}(R, T) = -\frac{1}{T} \log\langle P(x)P(x + R)\rangle \tag{8}$$

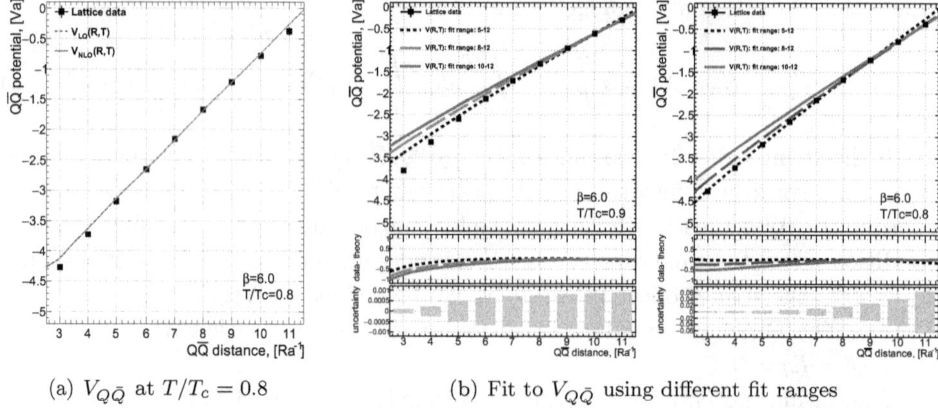

(a) $V_{Q\bar{Q}}$ at $T/T_c = 0.8$

(b) Fit to $V_{Q\bar{Q}}$ using different fit ranges

Fig. 1. (a) The quark-antiquark potential, $V_{Q\bar{Q}}$, measured at temperature $T/T_c = 0.8$, the solid and dashed lines correspond to fits to the LO and NLO string potential of Eq. (4) and Eq. (5), respectively. (b) $V_{Q\bar{Q}}$ measured at temperature $T/T_c = 0.9$ (left), and at temperature $T/T_c = 0.8$ (right), the lines correspond to the different fit ranges with a fixed end point at $R = 12$ fm – the different string tensions. Here we consider the string model solution only at NLO, see Eq. (5).

at each R is calculated through the Polyakov loop correlators

$$\mathcal{P}_{2Q} = \int d[U] P(0) P^\dagger(R) e^{-S_w} = e^{-\frac{V(R,T)}{T}}, \qquad (9)$$

with the single Polyakov loop given by $P(\vec{r}_i) = \frac{1}{3}\text{Tr}\left[\prod_{n_t=1}^{N_t} U_{\mu=4}(\vec{r}_i, n_t)\right]$.

The lattice data of the $V_{Q\bar{Q}}$, normalized to its value at $R = 1.2$, at the temperature $T/T_c = 0.8$, is shown in Fig. 1. To set comparison with the string model predictions, we fit the lattice results from Eq. (8) with the theoretical predictions of the NG string model at the LO and NLO separately, see Eqs. (4) and (5). The string tension σa^{-2} and the renormalization constant $\mu(T)$ was used as free parameters.

A large value of χ^2 is returned for fits of color sources separations from $R = 0.4$ to $R = 1.1$ fm. For the separation distance $R \leq 0.4$ fm the NG string description is expected to show increasingly significant deviations from the lattice data due to the short distance physics and the idealized one dimensional NG string. Excluding the point $R = 0.4$ fm from the fits significantly improved the data description for both the LO and the NLO approximations [33]. We observed a manifesting stability in the returned values of the string tension parameter even by the exclusion of further points at short distances $R = 0.5$ fm and $R = 0.6$ fm from the fit range [33]. The string tension settles a stable value of $\sigma a^{-2} = 0.0445$ measured in lattice units at $T/T_c = 0.8$, see Ref. [33].

The fit returns acceptable values of χ^2 for the different fit ranges with a fixed end point at $R = 12$ fm, as shown in Fig. 1(b). The values of the string tension depict a subtle corrections to the string tension compared to fits of the LO approximation of Eq. (4). Within the standard deviations of the measurement, the zero temperature string tension stabilizes around the same value of $\sigma a^{-2} = 0.046460$ for all data sets, for more details see Ref. [33]. These results points out the minor role of the higher order modes due to the string's excited spectrum in the vicinity of the QCD plateau, $T/T_c = 0.8$.

Thermal effects become more noticeable in the YM model [34, 35] considering the lattice with $N_t = 8$ slices in the temporal direction, this scales the temperature to $T/T_c = 0.9$ – sufficiently close to the critical point. The lattice data corresponding to the measured $V_{Q\bar{Q}}$ is depicted in Fig. 1(b-left). As a consistency check, we reproduced the same value of the string tension taken as a fit parameter as in Ref. [34, 36] with two different parameterizations of Ref. [1, 37] and fit domain. We follow different connive to scrutinize the fit behavior of the lattice data at this temperature scale with respect to the LO and NLO approximation, we schematically inspect the returned values of χ^2 for an interval of selected values of the string tension $\sigma a^{-2} \in [0.036, 0.045]$. The residuals and normalization constant $\mu(T)$ for the corresponding σa^{-2} are listed in Ref. [33]. The fits with the exclusion of points at short color distance separations reduces the value of χ^2 for most of the selected values of the string tension.

In spite of the fit result stretching out the role for the string's self interactions beyond the Gaussian approximation. The inclusion of the NLO terms, Eq. (5), provides a better χ^2, limited to a color separation distance $R = 0.5$ fm where indications for the validity of the string picture has been reported at zero temperature [1]. The consideration of the two-loop approximation at $T/T_c = 0.9$ finely corrects the free-parameter σa^{-2} interpreted as the zero temperature limit of the string tension for the value measured at $T/T_c = 0.8$.

3. The String Width Profile

In the following, we measure the mean-square width of the action density in $SU(3)$ gluonic configurations. The action density is related to the chromo-electromagnetic fields via $\frac{1}{2}(E^2 - B^2)$ and is measured through a three-loop improved lattice field-strength tensor [38]. A color-averaged infinitely-heavy static $Q\bar{Q}$ state has been constructed by means of two Polyakov lines $\mathcal{P}_{2Q}(\vec{r}_1, \vec{r}_2) = P(\vec{r}_1)P^\dagger(\vec{r}_2)$. A scalar field characterizing the action density distribution in the Polyakov vacuum or in the presence of color sources [39] is defined as

$$C(\vec{\rho}; \vec{r}_1, \vec{r}_2) = \frac{\langle \mathcal{P}_{2Q}(\vec{r}_1, \vec{r}_2) \, S(\vec{\rho}) \rangle}{\langle \mathcal{P}_{2Q}(\vec{r}_1, \vec{r}_2) \rangle \, \langle S(\vec{\rho}) \rangle}, \tag{10}$$

with the vector $\vec{\rho}$ referring to the spatial position of the energy probe with respect to some origin, and the brackets $\langle ... \rangle$ stands for averaging over gauge configurations and lattice symmetries. We make use of the symmetry of the four dimensional torus, that is, the measurements taken at a fixed color source's separations R are repeated at each point of the three-dimensional torus and time slice then averaged. The lattice size is sufficiently large to avoid mirror effects or correlations from the other side of the lattice due to the periodicity of the discretized space-time mesh. Cluster decomposition of the operators leads to $C \to 1$ away from the quarks.

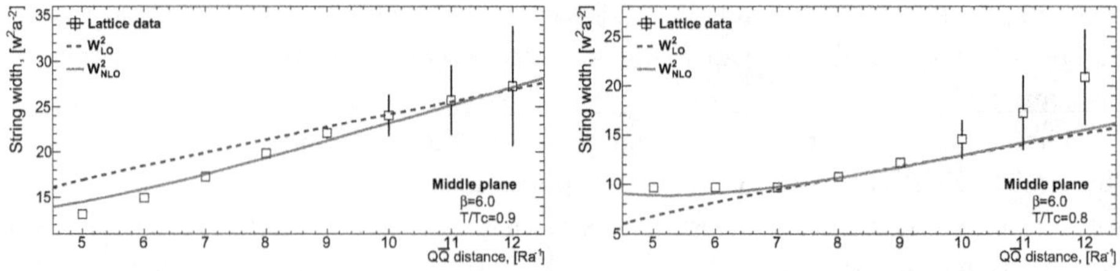

Fig. 2. Plot of the mean-square width $W^2 a^{-2}$ of the density distribution at the center of the tube, $z = R/2$, at temperature $T/T_c = 0.9$ (left) and $T/T_c = 0.8$ (right). The solid and dashed lines correspond to the free and self-interacting NG string Eq. (12) and Eq. (13).

A measurement of the width of the string's action density may be taken by fitting the density distribution $\mathcal{C}(\vec{\rho}(r, \theta; z))$ from Eq. (10) to the following form

$$\mathcal{C}(\vec{\rho}(r, \theta; z)) = 1 - G(r, \theta; z) = 1 - \left[A(e^{-r^2/\sigma_1^2} + e^{-r^2/\sigma_2^2}) + \kappa\right], \qquad (11)$$

taking into consideration the axial cylindrical symmetry of the tube, $r^2 = x^2 + y^2$ in each selected transverse plane $\vec{\rho}(r, \theta; z)$ to quark axis z.

The second moment of the action density distribution with respect to the cylinder's axis z joining the two quarks is then given by $W^2(z) = \frac{\int dr r^3 G(r,\theta;z)}{\int dr r G(r,\theta;z)}$, which defines the mean-square width of the tube. The localization of the color sources corresponds to $z = 0$ or $z = R$, respectively.

The width is estimated in accord to Eq. (11) and $W^2(z)$ at each selected plane z_i fixed with respect to one color source. The obtained values in middle plane $z = R/2$ are shown in Fig. 2. Notice, that the mean-square width of the string at all middle planes exhibit a broadening as the color sources are pulled apart. The width at consecutive transverse planes z_i, see Fig. 3, more clearly depict an increasing slop in the pattern of growth as one considers farther planes from the quark sources up to the middle plane.

The highest values of χ^2 are retrieved if the whole source separations $R = 4a$ to $R = 12a$ are included for both LO and NLO approximations [20, 40, 41]:

$$W_{\ell o}^2(\xi, \tau) = \frac{D-2}{2\pi\sigma} \log\left(\frac{R}{R_0(\xi)}\right) + \frac{D-2}{2\pi\sigma} \log\left|\frac{\theta_2(\pi\,\xi/R; \tau)}{\theta_1'(0; \tau)}\right|, \qquad (12)$$

$$W_{n\ell o}^2(\xi, \tau) = \frac{\pi}{12\sigma R^2}\left[E_2(i\tau) - 4E_2(2i\tau)\right]\left(W_{lo}^2(\xi) - \frac{D-2}{4\pi\sigma}\right)$$

$$+\frac{(D-2)\pi}{12\sigma^2 R^2}\left\{\tau\left(q_2\frac{d}{dq_2} - \frac{D-2}{12}E_2(i\tau)\right)\left[E_2(2i\tau) - E_2(i\tau)\right] - \frac{D-2}{8\pi}E_2(i\tau)\right\}. \qquad (13)$$

With the data points at short distances excluded from the fit the values of χ^2 decrease gradually, indicating that only the data points at large source separation are parameterized by the string model formula.

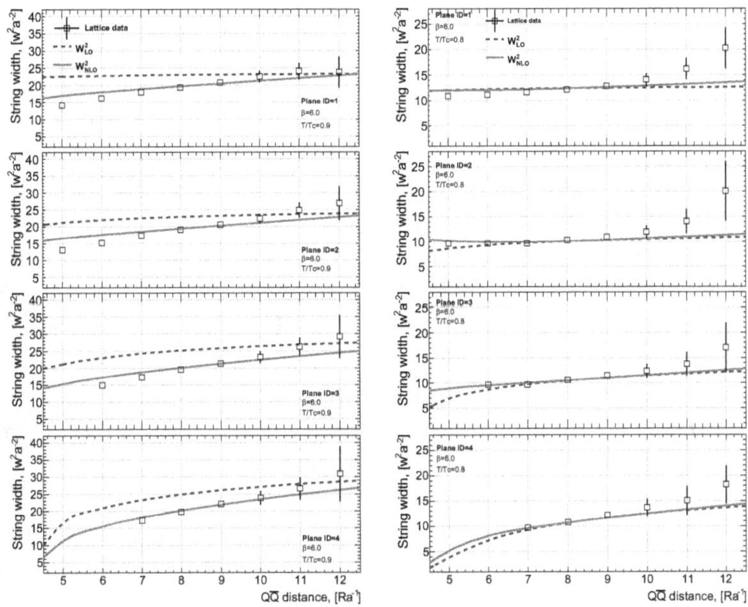

Fig. 3. The mean-square width of the string $W^2(z)$ versus quark-antiquark separations measured in the transverse planes at $z = 1$, $z = 2$, $z = 3$, and $z = 4$, from top to bottom. Measurements taken at temperature $T/T_c = 0.9$ (left), $T/T_c = 0.8$ (right). The dashed and solid lines denote the LO (Eq. (4)) and NLO (Eq. (5)) one parameter string model, respectively.

Most of the considerations on our discussion concerning the validity of both approximations to the $Q\bar{Q}$ potential at this temperature seem to hold for the string profile. The fits at $T/T_c = 0.9$, however, entails that the LO approximation Eqs. (12) are fitted on large distances to approach the NLO approximation in the asymptotic region. The increase of the uncertainties in the string's width for color source separation $R > 1.0$ fm does not loose our argument that the string model in both approximation scheme provides a good description for the string profile at this temperature.

In spite of the improvements in the parameterization behavior at high temperatures $T/T_c = 0.9$ in the intermediate distance region, the yet large values of χ^2 raises the question whether it is sufficient to consider the NG action up to NLO or a more general string action encompassing the leading terms of NG action as a limiting approximation ought be considered. Nevertheless, the resolution of our lattice data is enough to disclose the roughness of the NG action expanded up to the fourth derivative terms in the precise description of the stringy color tube profile.

4. Summary and Conclusion

We consider the predictions of the NG string for the mean-square width profile near the end of the QCD plateau region, $T/T_c = 0.8$. The obtained results indicate that the form of the string-vibration-like shapes of action density contours are independent from the temperature driven geometrical width effects suggesting that it can manifest even at lower temperatures. In addition, one may conclude that the potential extracted from the $Z(R,T)$ up to NLO are consistent with the lattice data for color source separations commencing from $R = 0.5$ fm.

At the close enough temperature to the deconfinement point, $T/T_c = 0.9$, the fits of the string model for either approximation schemes to the $Q\bar{Q}$ potential data return large values of χ^2 considering a fit region spanning the whole intermediate region $R = 0.5$ fm to $R = 1.0$ fm. Fits considering both intermediate and asymptotic color source separation distances show significant improvement with respect to that obtained on the basis of the free string approximation. Nevertheless, we found that the NLO approximation does not provide an accurate match with the numerical data. This result is manifested in the returned values of χ^2 which is still significantly large when considering distances less than $R<0.8$ fm.

The NG effective description does not accurately describe the $V_{Q\bar{Q}}$ potential which manifest as a small deviation from the standard value of the string tension and the width profile of the quantum fluctuations of the string. This drives motives to scrutinize the fine structure of the strings up to the NNLO corrections or even higher order terms in the action with coefficients respecting the open-closed string duality [19], or the general effective string actions of Lüscher-Weiss action with boundary terms.

Acknowledgments

This work has been funded by the Chinese Academy of Sciences President's International Fellowship Initiative grants No. 2015PM062 and No. 2016PM043, the Recruitment Program of Foreign Experts, NSFC grants (Nos. 11035006, 11175215, 11175220) and the Hundred Talent Program of the Chinese Academy of Sciences (Y101020BR0). We thank Thomas Filk for helpful comments.

References

[1] M. Luscher and P. Weisz, Quark confinement and the bosonic string, *JHEP* **07**, p. 049 (2002), [arXiv:hep-lat/0207003].

[2] M. Caselle and P. Grinza, On the intrinsic width of the chromoelectric flux tube in finite temperature LGTs, *JHEP* **1211**, p. 174 (2012), [arXiv:hep-th/1207.6523].

[3] K. J. Juge, J. Kuti and C. Morningstar, Fine structure of the QCD string spectrum, *Phys. Rev. Lett.* **90**, p. 161601 (2003), [arXiv:hep-lat/0207004].

[4] N. D. Hari Dass and P. Majumdar, Continuum limit of string formation in 3d SU(2) LGT, *Phys. Lett. B* **658**, 273 (2008), [arXiv:hep-lat/0702019].

[5] M. Caselle, M. Panero and D. Vadacchino, Width of the flux tube in compact U(1) gauge theory in three dimensions, *JHEP* **02**, p. 180 (2016), [arXiv:hep-lat/1601.07455].

[6] M. Caselle, M. Panero, P. Provero and M. Hasenbusch, String effects in Polyakov loop correlators, *Nucl. Phys. Proc. Suppl.* **119**, 499 (2003), [arXiv:hep-lat/0210023].

[7] P. Pennanen, A. M. Green and C. Michael, Flux-tube structure and beta-functions in SU(2), *Phys. Rev.* **D56**, 3903 (1997).

[8] B. B. Brandt and M. Meineri, Effective string description of confining flux tubes, *Int. J. Mod. Phys.* **A31**, p. 1643001 (2016), [arXiv:hep-th/1603.06969].

[9] M. Luscher, G. Munster and P. Weisz, How Thick Are Chromoelectric Flux Tubes?, *Nucl. Phys.* **B180**, 1 (1981).

[10] M. Caselle, F. Gliozzi, U. Magnea and S. Vinti, Width of long colour flux tubes in lattice gauge systems, *Nucl. Phys.* **B460**, 397 (1996), [arXiv:hep-lat/9510019].

[11] C. Bonati, Finite temperature effective string corrections in (3+1)D SU(2) lattice gauge theory, *Phys. Lett. B* **703**, 376 (2011), [arXiv:hep-lat/1106.5920].

[12] K. Dietz and T. Filk, Renormalization of string functionals, *Phys. Rev.* **D27**, 2944 (Jun 1983).

[13] M. Hasenbusch, M. Marcu and K. Pinn, High precision renormalization group study of the roughening transition, *Physica A: Statistical Mechanics and its Applications* **208**, 124 (1994).

[14] M. Caselle, M. Hasenbusch and M. Panero, High precision Monte-Carlo simulations of interfaces in the three-dimensional Ising model: A Comparison with the Nambu-Goto effective string model, *JHEP* **03**, p. 084 (2006), [arXiv:hep-lat/0601023].

[15] B. Bringoltz and M. Teper, Closed k-strings in SU(N) gauge theories : 2+1 dimensions, *Phys. Lett. B* **663**, 429 (2008), [arXiv:hep-lat/0802.1490].

[16] A. Athenodorou, B. Bringoltz and M. Teper, On the spectrum of closed k = 2 flux tubes in D=2+1 SU(N) gauge theories, *JHEP* **05**, p. 019 (2009), [arXiv:hep-lat/0812.0334].

[17] N. D. Hari Dass and P. Majumdar, String-like behaviour of 4-D SU(3) Yang-Mills flux tubes, *JHEP* **10**, p. 020 (2006), [arXiv:hep-lat/0608024].

[18] P. Giudice, F. Gliozzi and S. Lottini, Quantum broadening of k-strings in gauge theories, *JHEP* **01**, p. 084 (2007), [arXiv:hep-th/0612131].

[19] M. Luscher and P. Weisz, String excitation energies in SU(N) gauge theories beyond the free-string approximation, *JHEP* **07**, p. 014 (2004), [arXiv:hep-th/0406205].

[20] M. Pepe, String effects in Yang-Mills theory, *PoS* **LATTICE2010**, p. 017 (2010).

[21] A. Bakry *et al.*, String effects and the distribution of the glue in static mesons at finite temperature, *Phys. Rev.* **D82**, p. 094503 (Nov 2010), [arXiv:hep-lat/1004.0782].

[22] A. S. Bakry, D. B. Leinweber and A. G. Williams, Bosonic stringlike behavior and the Ultraviolet filtering of QCD, *Phys. Rev.* **D85**, p. 034504 (2012).

[23] A. S. Bakry, D. B. Leinweber and A. G. Williams, The thermal delocalization of the flux tubes in mesons and baryons, *AIP Conf. Proc.* **1354**, 178 (2011).

[24] A. S. Bakry, D. B. Leinweber and A. G. Williams, Gluonic fields as unraveled with Polyakov lines and predicted by bosonic strings, *PoS* **LATTICE2012**, p. 271 (2012).

[25] Bakry, Ahmed S., Chen, Xurong and Zhang, Peng-Ming, Confining potential of Y-string on the lattice at finite T, *EPJ Web Conf.* **126**, p. 05001 (2016).

[26] A. S. Bakry, X. Chen and P.-M. Zhang, Y-stringlike behavior of a static baryon at finite temperature, *Phys. Rev.* **D91**, p. 114506 (2015), [arXiv:hep-lat/1412.3568].

[27] A. S. Bakry, X. Chen and P.-M. Zhang, The confining baryonic Y-strings on the lattice, *AIP Conf. Proc.* **1701**, p. 030001 (2016).

[28] A. S. Bakry, D. B. Leinweber and A. G. Williams, Gluonic profile of static baryon at finite temperature and the Y baryonic string, *PoS* **LATTICE2011**, p. 256 (2011).

[29] M. Caselle, M. Hasenbusch and M. Panero, Short distance behavior of the effective string, *JHEP* **05**, p. 032 (2004).

[30] M. Caselle, M. Pepe and A. Rago, Static quark potential and effective string corrections in the (2+1)-d SU(2) Yang-Mills theory, *JHEP* **10**, p. 005 (Oct 2004).

[31] P. Giudice, F. Gliozzi and S. Lottini, The Confining string beyond the free-string approximation in the gauge dual of percolation, *JHEP* **03**, p. 104 (2009), [arXiv:hep-lat/0901.0748].

[32] M. Luscher, K. Symanzik and P. Weisz, Anomalies of the free loop wave equation in the WKB approximation, *Nucl. Phys.* **B173**, p. 365 (1980).

[33] A. Bakry, X. Chen, M. Deliyergiyev, A. Galal, S. Xu and P. M. Zhang, Stiff self-interacting string near QCD deconfinement point (2017), [arXiv:hep-lat/1707.02962].

[34] N. Cardoso and P. Bicudo, Lattice QCD computation of the SU(3) string tension critical curve, *Phys. Rev.* **D85**, p. 077501 (Apr 2012), [arXiv:hep-lat/1111.1317].

[35] T. Doi, N. Ishii, M. Oka and H. Suganuma, The lattice QCD simulation of the quark-gluon mixed condensate at finite temperature and the phase transition of QCD, *Nucl. Phys. B - Proc. Suppl.* **140**, 559 (2005), LATTICE 2004 - Proceedings of the XXIInd International Symposium on Lattice Field Theory.

[36] O. Kaczmarek, F. Karsch, E. Laermann and M. Lutgemeier, Heavy quark potentials in quenched QCD at high temperature, *Phys. Rev.* **D62**, p. 034021 (Jul 2000).

[37] M. Gao, Heavy quark potential at finite temperature from a string picture, *Phys. Rev.* **D40**, p. 2708 (1989).

[38] S. O. Bilson-Thompson, D. B. Leinweber and A. G. Williams, Highly-improved lattice field-strength tensor, *Ann. Phys.* **304**, 1 (2003), [arXiv:hep-lat/0203008].

[39] F. Bissey *et al.*, Gluon flux-tube distribution and linear confinement in baryons, *Phys. Rev.* **D76**, p. 114512 (2007), [arXiv:hep-lat/0606016].

[40] A. Allais and M. Caselle, On the linear increase of the flux tube thickness near the deconfinement transition, *JHEP* **01**, p. 073 (2009), [arXiv:hep-lat/0812.0284].

[41] F. Gliozzi, M. Pepe and U.-J. Wiese, The Width of the Confining String in Yang-Mills Theory, *Phys. Rev. Lett.* **104**, p. 232001 (2010), [arXiv:hep-lat/1002.4888].

Charm Physics Prospects at the Belle II Experiment[*]

Y. Q. Chen,[*,†,‡] L. K. Li[§], W. B. Yan[‡] and Z. P. Zhang[‡]

[‡]*State Key Laboratory of Particle Detection and Electronics,*
University of Science and Technology of China(USTC),
Hefei, Baohe District, China
[*]*E-mail: chenyq15@mail.ustc.edu.cn*
www.ustc.edu.cn

[§]*Experimental Physics Division,*
Institution of High Energy of Physics(IHEP), CAS,
Beijing, Shijingshan District, China
www.ihep.cas.cn

With a total integrated luminosity of 50 ab^{-1} and the improved vertex resolution and particle identification performances at Belle II, it will allow to increase the precision of time-dependent measurements significantly in the charm sector. The expected sensitivity of Belle II for D^0-$\bar{D}^0$ mixing, CP violation and CP asymmetries measurements are discussed. A new flavour-tagging techniques ROE method of prompt D^0 is presented.

Keywords: Belle II; D^0-$\bar{D}^0$ mixing; CP violation; ROE method.

1. Introduction

Belle II experiment is a major upgrade of Belle experiment and will perform at the B-factory SuperKEKB, located at the KEK laboratory in Tsukuba, Japan [1]. Although Belle II has been designed to perform precise measurements in the b-quark sector, it will also be an ideal laboratory to study the properties of the c-quark.

Given the environment of the e^+e^- SuperKEKB collider and the hermiticity of the detector, Belle II is expected to have great performances in the reconstruction of final states with neutral particles (e.g. γ, π^0, η) and missing energy. The data taking will start in 2018 and Belle II is expected to collect a total integrated luminosity of 50 ab^{-1} within the next decade, which will provide abundant $c\bar{c}$ events data sample. With more $c\bar{c}$ datasets, it will provide a rich charm physics program to improve the precision of D^0-$\bar{D}^0$ mixing, time-dependent CP violation, time-integrated CP asymmetries and leptonic decay measurements.

2. D^0 Proper Time Resolution

Belle II will have a six-layer silicon vertex detector (2 layers of pixel detectors + 4 layers of double-sided silicon strip detectors) whose innermost layer will be 2 times closer to the interaction point with respect to Belle's [1]. And Belle II vertex

---

[*]Supported by National Natural Science Foundation of China (11475164, 11475169, 11675166).
[†]On behalf of the Belle II Collaboration.

detector allows to reconstruct the D^0 decay vertex with a great improvement of efficiency with respect to B-Factories, because of the reduced distance between the first layer and the interaction point. Thanks to the vertex detector system, the vertex resolution will be improved significantly. Time-dependent measurements will benefit from this improved reconstruction, in particular the precise of the determination of D^0 proper time, which is essential in time-dependent measurements.

According to Monte Carlo (MC) simulations, Belle II will have a D^0 proper time resolution $t = 140$ fs [2], which is a factor of two improvement over Belle and BaBar ($t = 270$ fs). The distribution of D^0 proper time resolution t and proper time error σ_t are shown in Fig. 1.

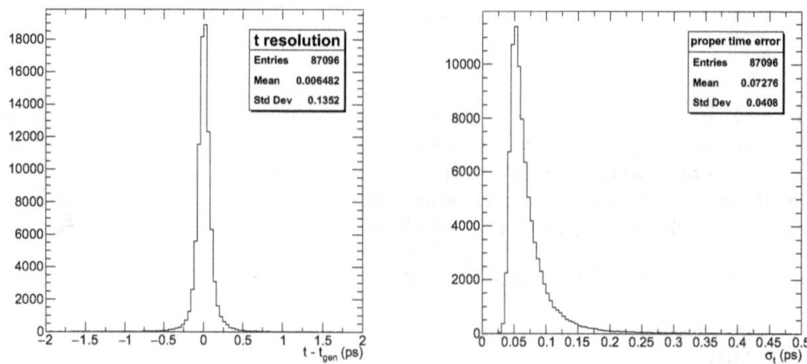

Fig. 1. D^0 proper time resolution(left) and its error(right) for D^*-tagged $D^0 \to K^+ K^-$.

3. D^0-$\bar{D}^0$ Mixing and CP Violation

One emphasis of the Belle II charm physics program is to make high precision measurements of D^0-$\bar{D}^0$ mixing and search for CP violation. These measurements typically depend on measuring the decay time of D^0 mesons.

In order to study the sensitivity on Belle II of D^0-$\bar{D}^0$ mixing parameters $x = \Delta M/\Gamma$ and $y = \Delta\Gamma/2\Gamma$, and of CP violating parameters $|q/p|$ and $Arg(q/p) = \phi$, two studies have been performed with Toy-MC samples.

3.1. $D^0 \to K^+\pi^-$ decays

Given the much larger samples of flavor-tagged $D^0 \to K^+\pi^-$ decays, Belle II will collect over those collected by Belle and BaBar. For this study, an ensemble of 1000 Toy-MC experiments are generated with separate samples of $D^{*+} \to D^0\pi_s^+$, $D^0 \to K^+\pi^-$ and $D^{*-} \to \bar{D}^0\pi_s^-$, $\bar{D}^0 \to K^-\pi^+$ decays corresponding to 5 ab^{-1}, 20 ab^{-1}, and 50 ab^{-1} of data, whose schematic diagram of wrong-sign (WS) and right-sign (RS) are presented in Fig. 2. Their decay times smeared by the expected decay time resolution of Belle II (140 fs), and the resulting decays times fitted for mixing parameters x', y' and CP-violating parameters $|q/p|$, ϕ. The probability density

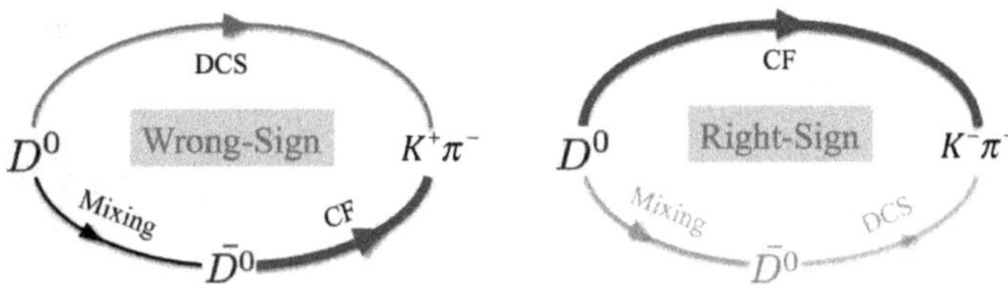

Fig. 2.   The schematic diagrams of $D^0 \to K^+\pi^-$ WS decay (left), and $D^0 \to K^-\pi^+$ RS decay (right).

Table 1.   Belle measurements and Belle II estimations of expected sensitivities of mixing parameters x', y' and CP-violating parameters $|q/p|$, ϕ of flavor tagged $D^0 \to K^+\pi^- + c.c.$ decays, under situation of CPV and CPV-allowed.

	Parameter	Belle 976 fb^{-1}	Belle II 5 ab^{-1}	20 ab^{-1}	50 ab^{-1}
no CPV	$\delta(x'^2)$ (10^{-5})	22	7.5	3.7	2.3
	$\delta(y')$ (%)	0.34	0.11	0.056	0.035
CPV-allowed	$\delta(x')$ (%)		0.37	0.23	0.15
	$\delta(y')$ (%)		0.26	0.17	0.10
	$\delta(|q/p|)$		0.197	0.089	0.051
	$\delta(\phi)$ (°)		15.5	9.2	5.7

functions of WS decay rates with D^0-$\bar{D}^0$ mixing or CPV-allowed are calculated by Eqn (1) and Eqn (2)

$$\frac{N(D^0 \to f)}{dt} = e^{-\Gamma t}\left[R_D + \left|\frac{q}{p}\right|\sqrt{R_D}(y'\cos\phi - x'\sin\phi)(\Gamma t) + \left|\frac{q}{p}\right|^2 \frac{(x'^2+y'^2)}{4}(\Gamma t)^2\right], \quad (1)$$

$$\frac{N(D^0 \to f)}{dt} = e^{-\Gamma t}\left[R_D + \sqrt{R_D}y'(\Gamma t) + \frac{(x'^2+y'^2)}{4}(\Gamma t)^2\right] (no\ CPV), \quad (2)$$

where effective mixing parameters $x' = x\cos\delta + y\sin\delta$, $y' = y\cos\delta - x\sin\delta$ with strong phase difference δ between $D^0 \to K^+\pi^-$ and $\bar{D}^0 \to K^-\pi^+$ amplitudes. We subsequently smear these decay times with using a Gaussian resolution function (with a width $\sigma = 140$ fs) for 1000 experiments. The resulting D^0 and $\bar{D}^0$ time distributions are simultaneously fitted, with fitted parameters x', y' in the case of no CPV and allowing for CPV, together with $|q/p|$, and ϕ allowing for CPV. The previous Belle measurements and the preliminary estimations of expected sensitivity of mixing and CP violating parameters with several luminosity are listed in Table 1.

3.2. $D^0 \to K^+\pi^-\pi^0$ decays

We also estimate the Belle II sensitivity to x'' and y'' in WS $D^0 \to K^+\pi^-\pi^0$ decay by performing an MC simulation study, generating 10 independent data sets of 225,000 WS events each [3], by assuming similar $D^0 \to K^+\pi^-\pi^0$ efficiency with

344

BaBar. Under the assumptions of $|x|, |y| << 1$ and CP conservation ($q/p = 1$), the decay rate of WS decays for $D^0 \to \bar{f}$ is defined as:

$$|\mathcal{M}(\bar{f}, t)|^2 = e^{-\bar{\Gamma}t} \left\{ r_0^2 \left| \mathcal{A}_{\bar{f}}^{DCS} \right|^2 - r_0(Ay'' + Bx'')(\bar{\Gamma}t) + \frac{x''^2 + y''^2}{4} \left| \bar{\mathcal{A}}_{\bar{f}}^{CF} \right|^2 (\bar{\Gamma}t)^2 \right\} \quad (3)$$

where $A = \Re(\mathcal{A}_{\bar{f}}^{DCS} \bar{\mathcal{A}}_{\bar{f}}^{*CF})$, $B = \Im(\mathcal{A}_{\bar{f}}^{DCS} \bar{\mathcal{A}}_{\bar{f}}^{*CF})$, r_0 is the modulus of the relative complex number between the CF and DCS amplitudes, and $\mathcal{A}_{\bar{f}}^{DCS}$ and $\bar{\mathcal{A}}_{\bar{f}}^{CF}$ are normalized shapes on the Dalitz plot region. The effective mixing parameters x'' and y'' could be extracted by performing a time-dependent Dalitz plot fitting of the WS sample.

The mixing parameters $(x'', y'', \delta, 1/r_0) = (2.58\%, 0.39\%, 10°, 13.8)$ in Eqn (3) are fixed as WS generator input parameters. We also generate corresponding samples of RS events, used to determine the ratio of the magnitudes of the WS and RS decay amplitudes. For both samples the decay times are smeared by $t = 140$ fs.

We fit the samples for parameters x'' and y'' as well as the magnitudes and phases of the intermediate states. A typical time-dependent fit to Dalitz plot is illustrated in Fig. 3, and resulting fit residuals for the 10 experiments are plotted in Fig. 4. From the fitting results, the precision are obtained: $\sigma_{x''} = 0.057\%$ and $\sigma_{y''} = 0.049\%$ without including systematic uncertainties and the effect of backgrounds.

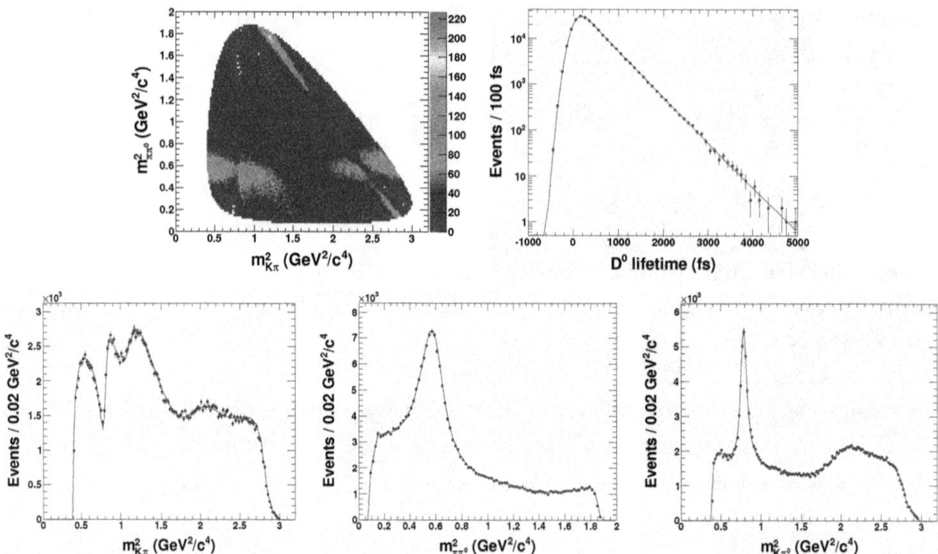

Fig. 3. Time-dependent fit to the Dalitz plot of WS $D^0 \to K^+\pi^-\pi^0$ decays. The decay times are smeared by the expected Belle II decay-time resolution of 140 fs. The second row shows projections of the fitted Dalitz variables $m^2_{K^+\pi^-}$ (left), $m^2_{\pi^-\pi^0}$ (middle), and $m^2_{K^+\pi^0}$ (right).

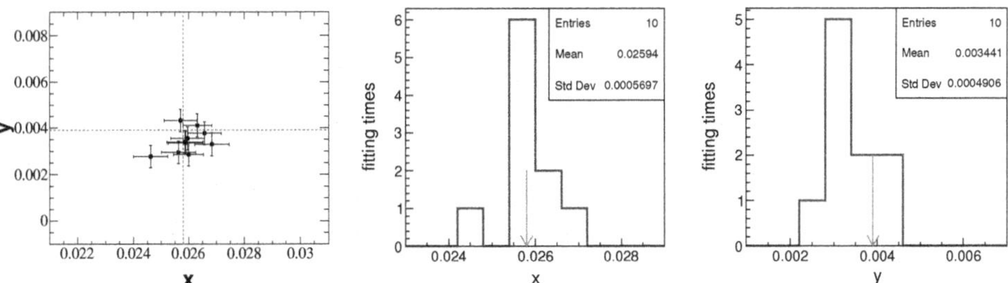

Fig. 4. Residuals resulting from fitting the time-dependent Dalitz plot of WS $D^0 \to K^+\pi^-\pi^0$ decays, for 10 experiments corresponding to 50 ab^{-1} of data (each set with 225,000 events). The projections of x and y are shown with the standard deviation value as our estimation of x and y errors (the red lines and arrows show the input (true) values of x'' and y'') (color online).

By scaling the presence of backgrounds of approximately $< 40\%$ on x'' and y''[a] by estimation on Belle data, we get the uncertaity of $\sigma_{x''} = 0.080\%$ and $\sigma_{y''} = 0.070\%$, which are almost an order of magnitude smaller than those obtained by BaBar.

4. Time-integrated CP Asymmetries

As discussed above, Belle II will have excellent efficiency for reconstructing multi-body final states with very small detector-based asymmetries. Thus the experiment is ideal for searching for time-integrated CP violation in a variety of final states. A listing of D^0, D^+, and D_s^+ decay modes that Belle has investigated are given in Table 2. The table lists the CP asymmetry $A_{CP}^f = [\Gamma(D^0 \to f) - \Gamma(\bar{D}^0 \to \bar{f})]/[\Gamma(D^0 \to f) + \Gamma(\bar{D}^0 \to \bar{f})]$ measured by Belle, and the precision expected for Belle II [2]. The latter is estimated by scaling the Belle statistical error(σ_{stat}) and the systematic error(σ_{syst}) by the ratio of integrated luminosities and those irreducible systematic uncertainties. The overall error estimate is calculated as

$$\sigma_{Belle\,II} = \sqrt{(\sigma_{stat}^2 + \sigma_{syst}^2) \cdot (\mathcal{L}_{Belle}/50\ ab^{-1}) + \sigma_{irred}^2}. \tag{4}$$

In particular, $D^0 \to K_S^0 K_S^0$, a promising golden channel amongst the SCS decays for CP violation, as the CP asymmetry may be enhanced to an observable level within the SM [4], which is expected to be 0.20% at Belle II. For the decay $D^+ \to \pi^0\pi^+$, there is no CPV by theory prediction [5], so there will be possible enhancement from NP and Belle II expects a precision of 0.40%.

5. Leptonic Decays

The low backgrounds of an e^+e^- experiment and the knowledge of the center-of-mass (CM) energy will allow Belle II to measure the branching fractions of the

---

[a]The backgrounds effect should be smaller at Belle II than at Belle due to improved vertex resolution and improved particle identification.

Table 2. Time-integrated CP asymmetries measurements from Belle, and the precision expected for Belle II in 50 ab^{-1} of data.

Channel	$\mathcal{L}$(fb^{-1})	Current measurement value(%)	References	Scaled 50 ab^{-1}
$D^0 \to K^+ K^-$	976	$-0.32 \pm 0.21 \pm 0.09$	PoS ICHEP2012 (2013) 353	± 0.03
$D^0 \to \pi^+ \pi^-$	976	$+0.55 \pm 0.36 \pm 0.09$	PoS ICHEP2012 (2013) 353	± 0.05
$D^0 \to \pi^0 \pi^0$	966	$-0.03 \pm 0.64 \pm 0.10$	PRL **112**, 211601 (2014)	± 0.09
$D^0 \to K^0_S \pi^0$	966	$-0.21 \pm 0.16 \pm 0.07$	PRL **112**, 211601 (2014)	± 0.03
$D^0 \to K^0_S \eta$	791	$+0.54 \pm 0.51 \pm 0.16$	PRL **106**, 211801 (2011)	± 0.07
$D^0 \to K^0_S \eta'$	791	$+0.98 \pm 0.67 \pm 0.14$	PRL **106**, 211801 (2011)	± 0.09
$D^0 \to K^0_S K^0_S$	921	$-0.02 \pm 1.53 \pm 0.17$	PRL **119**, 171801 (2017)	± 0.20
$D^0 \to \pi^+ \pi^- \pi^0$	532	$+0.43 \pm 1.30$	PLB **662**, 102 (2008)	± 0.13
$D^0 \to K^+ \pi^- \pi^0$	281	-0.60 ± 5.30	PRL **95**, 231801 (2005)	± 0.40
$D^0 \to K^+ \pi^- \pi^+ \pi^-$	281	-1.80 ± 4.40	PRL **95**, 231801 (2005)	± 0.33
$D^+ \to \pi^0 \pi^+$	921	$+0.89 \pm 1.98 \pm 0.22$	Belle Preliminary	± 0.40
$D^+ \to \phi \pi^+$	955	$+0.51 \pm 0.28 \pm 0.05$	PRL **108**, 071801 (2012)	± 0.04
$D^+ \to \eta \pi^+$	791	$+1.74 \pm 1.13 \pm 0.19$	PRL **107**, 221801 (2011)	± 0.14
$D^+ \to \eta' \pi^+$	791	$-0.12 \pm 1.12 \pm 0.17$	PRL **107**, 221801 (2011)	± 0.14
$D^+ \to K^0_S \pi^+$	977	$-0.36 \pm 0.09 \pm 0.07$	PRL **109**, 021601 (2012)	± 0.03
$D^+ \to K^0_S K^+$	977	$-0.25 \pm 0.28 \pm 0.14$	JHEP **02** (2013) 098	± 0.05
$D_s^+ \to K^0_S \pi^+$	673	$+5.45 \pm 2.50 \pm 0.33$	PRL **104**, 181602 (2010)	± 0.29
$D_s^+ \to K^0_S K^+$	673	$+0.12 \pm 0.36 \pm 0.22$	PRL **104**, 181602 (2010)	± 0.05

leptonic decays $D^-_{(q)} \to \ell^- \bar{\nu}$ (with $q = d, s$) precisely and to extract the quantities $|V_{cq}| f_{D_q}$, where f_D and f_{D_s} are decay constants, $|V_{cd}|$ and $|V_{cs}|$ are CKM matrix elements. Belle has measured the branching fraction for $D_s^- \to \ell^- \bar{\nu}$ [6]. By using the f_{D_s} values computed with lattice QCD techniques [7], Belle II will significantly improve the Belle measurement of $|V_{cs}|$ and determine $|V_{cd}|$ with less than 2% uncertainty.

The method used by Belle to reconstruct $D^-_{(q)} \to \ell^- \bar{\nu}$ will surely be applied to Belle II. To study leptonic decays with missing energy from neutrino in $c\bar{c} \to D_{tag}^{0+} D^-_{(q)} X^{-0}_{frag}$, $D^-_{(q)} \to \ell^- \bar{\nu}$ ($\ell = e/\mu$). Firstly, one need to reconstruct the "tag side" D_{tag}, like D^0, D^+ or Λ_c^+, nominally recoiling against the "signal side" $D^-_{(q)} \to \ell^- \bar{\nu}$. In addition, remaining pions, kaons and protons are grouped together into what is referred to "fragmentation side" X_{frag}. The missing momentum of neutrino $P_{miss} = P_{CM} - P_{tag} - P_{frag} - P_{\ell^-}$ is then constructed, and for signal processes $D^-_{(q)} \to \ell^- \bar{\nu}$, the signal yield can be obtained by fitting the variable of neutrino $U_{miss} = E_{miss} - |\vec{P}_{miss}|$, which should peak at zero. This method can also be used in Belle II to improve search for the invisible final state decays of D^0 mesons, like $D^0 \to \nu \bar{\nu}$.

6. Flavour Tagging: ROE Method

Different from the widespread flavour-tagging method, in which the flavour of D^0 or $\bar{D}^0$ is tagged by the charge of π_s in decay $D^{*+} \to D^0\pi_s^+$ or $D^{*-} \to \bar{D}^0\pi_s^-$, a new flavour tagging method aims at increasing size of D^0-tagged candidates sample, and adding D^0 produced in $c\bar{c}$ events that are not coming from D* decays.

This new method consists in looking at so-called Rest Of the Event (ROE) [8] with respect to the neutral D meson. The principle of the ROE method is illustrated in Fig. 5. Since Cabibbo-favored transition for a charm quark is $\bar{c} \to \bar{s}$, we expect to find at least one strange meson in the ROE, namely a $K^+(u\bar{s})$ or a $K^0(d\bar{s})$. The flavour tagging is performed by selecting events with only one $K^\pm$ in the ROE, and using the charge of the kaon to determine the flavour of D^0 at the time of its production.

The amount of D^0 mesons tagged with the ROE method are expected to be roughly the same with respect to the one tagged with the D^{*+} method, $\frac{N_{tag}^{ROE}}{N_{tag}^{*}} \sim 1$ [8], which means larger statistics with an additional D^0 sample is available for D^0-$\bar{D}^0$ mixing and CP violation measurements. It has been estimated that, by combining the results obtained using D^* tag and ROE method, a 15% reduction of the statistical uncertainty on $A_{CP}(D^0 \to K^-\pi^+)$ can be achieved.

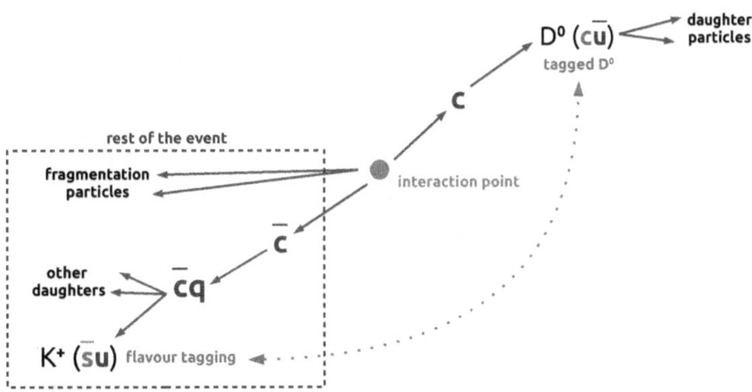

Fig. 5. The sketches of the ROE method. The events with only one $K^\pm$ in the ROE are selected; the flavour of the neutral D meson is determined by the charge of the kaon.

7. Conclusions

Belle II at SuperKEKB will start to collect data in 2018 and has a rich charm physics program. Considering impact of the improved vertex resolution and PID performances, dataset of 50 ab^{-1} collected at Belle II will allow improved about one magnitude precision of D^0-$\bar{D}^0$ mixing and CPV parameters, more precise CP asymmetries, improved leptonic decay measurements. ROE method will gain as much D^0 sample as D^* tag method, and it will increase our data sample for more precise CP violation and D^0-$\bar{D}^0$ mixing analyses.

References

[1] T. Abe *et al.*, "Belle II Technical Design Report", KEK-REPORT-2010-1 (2010) [arXiv:1011.0352].

[2] Belle II Collaboration & B2TiP Theory Community, "The Belle II Physics Book" (to be published on PTEP).

[3] L. K. Li *et al.*, Chin. Phys. C **41**, 023001 (2017).

[4] U. Nierste and A. Schacht, Phys. Rev. D **92**, 054036 (2015).

[5] F. Buccella *et al.*, Phys. Lett. B **302**, 319-325 (1993).

[6] A. Zupanc *et al.* (Belle Collaboration), J. High Energy Phys. **09** (2013) 139.

[7] S. Aoki *et al.* (FLAG Working Group), arXiv:1607.00299 (2016).

[8] G. De Pietro and G. Casarosa, BELLE2-NOTE-PH-2017-001 (2017).

Neutron Production from Cosmic-Ray Muons at Daya Bay

Jie Cheng

on behalf of Daya Bay Collaboration

Shandong University
E-mail: chengjie@ihep.ac.cn

Neutrons induced by cosmic-ray muons are a significant background for underground experiments studying neutrino oscillations, neutrino-less double beta decay, dark matter and other rare-event signals. The Daya Bay Reactor Antineutrino experiment consists of 8 antineutrino detectors (AD) placed in three experimental halls at different baselines from six nuclear reactors. Each AD contains 20 tons of Gd-doped liquid scintillator, serving as the main target for antineutrinos interacting via the inverse beta-decay (IBD) reaction. The data from Daya Bay allows to make a competitive measurement of neutron production by cosmogenic muons at depths of 250, 265 and 860 meters-water-equivalent.

Keywords: neutron; cosmic-ray muons; Daya Bay.

1. Introduction

Neutrons and isotopes produced by interactions between muons and nuclei in a detector are significant backgrounds for low background experiments. Therefore, the rate estimation for these muon-induced particles is very crucial for Daya Bay as well as other experiments. Daya Bay is designed to detect inverse beta decay (IBD) to study neutrino oscillations, obtaining the world's most precise measurement of θ_{13}[1-5] and $|\Delta m^2_{ee}|$[3-5]. The Daya Bay experiment is located near the the city of Shenzhen in the Guangdong province in China. Fig. 1 shows a diagram of the Daya Bay experimental site. The Daya Bay Nuclear Power Plant complex consists of six reactors, including Daya Bay [1-2] and Ling Ao [1-4]. The experiment has three experimental halls (EHs), including near halls (EH1 and EH2) and the far hall (EH3). The experiment employs eight functionally-identical antineutrino detectors (AD) to decrease detector-related uncertainties, placed in three experimental halls at different baselines from six nuclear reactors. Muon detectors, including resistive plate chambers (RPCs) and two individual water Cherenkov detectors, inner water shield (IWS) and outer water shield (OWS), are also employed in all EHs to detect muons. These detectors provide good detection for IBD neutrons which capture on gadolinium (nGd), as well as spallation neutrons induced by muons, allowing a precise neutron yield measurement at Daya Bay.

A sample of 10^6 simulated muons for each EH is generated with the modified Gaisser formula[6] according to the muon flux at the sea level and transported with MUSIC code[7] through rocks to the underground experimental hall. Table 1 shows the underground muon flux simulation results which have been published[8]. The error in the simulated flux is about 10%. The simulated muon trajectories and energy spectra are shown in Fig. 2 and Fig. 3 respectively. Zenith is defined as the angle from vertical, and azimuth is the horizontal compass angle from true North.

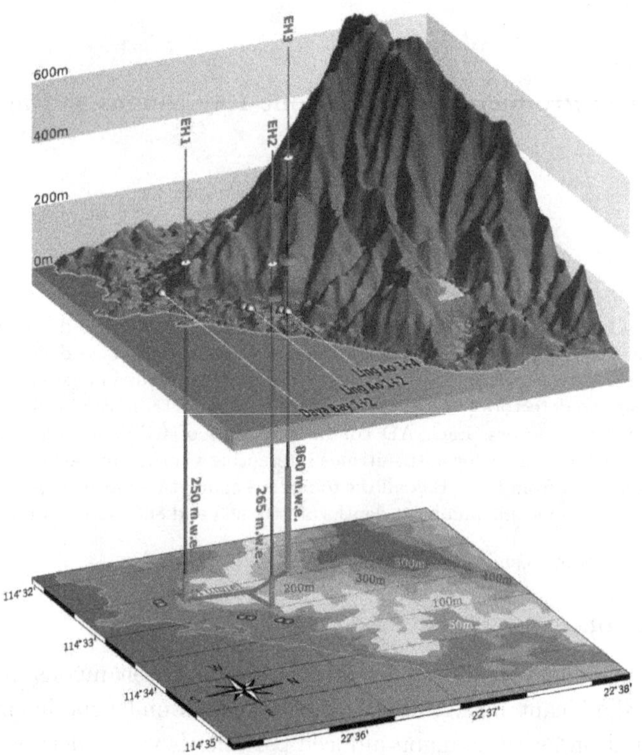

Fig. 1. A 3D representation (above) and 2D projection (below) of the layout of the Daya Bay Reactor Neutrino Experiment, including six reactor cores (Daya Bay 1-2, Ling Ao 1-4) and three experimental halls (EH1-3). The antineutrino detectors (ADs) are located in the underground experimental halls, with two ADs in EH1, two in EH2, and four in EH3.

Table 1. Underground muon flux simulation results. All values have been transformed into a detector-independent spherical geometry.

Hall	Overburden m	Overburden mwe	Muon flux (Hz/m^2)	Average enegy (GeV)
EH1	93	250	1.27	57
EH2	100	265	0.95	58
EH3	324	860	0.056	137

2. Neutron production from muons

In our measurement, the target region is Gd-doped liquid scintillator (GdLS). The neutron yield is defined as the number of neutrons produced per muon, per path length of the muon through the GdLS, per density of the GdLS. Neutrons captured on Gd following an identified AD muon are selected in the data. Thanks to the design of Daya Bay's detectors, muons and muon-induced neutrons are detected with a high efficiency and the selected sample is almost background-free.

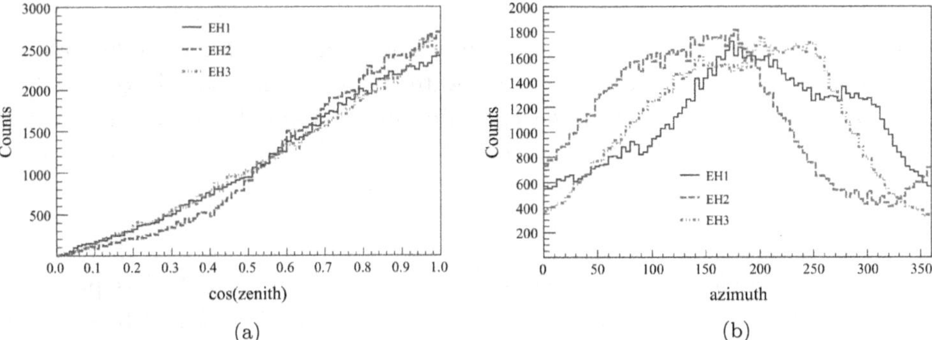

Fig. 2. Simulated muon trajectories (a) Zenith is defined as the angle from vertical (b) Azimuth is the horizontal compass angle from true North.

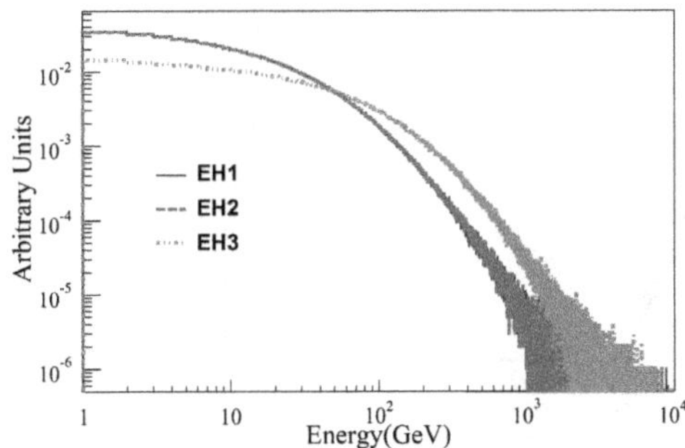

Fig. 3. Simulated muon energy spectra.

The expression of neutron yield Y_n can be written as shown in the formula below,

$$Y_n = \frac{N_n}{N_\mu L_{avg} \rho},$$ (1)

where N_μ is the number of muons traversing the GdLS target, N_n is the number of neutrons produced in association with N_μ, L_{avg} is the average path length of muons in the GdLS from the simulation, and ρ is the measured density of Daya Bay's GdLS.

The Daya Bay detector MC simulation is based on GEANT4[9,10]. In order to do a cross-check of the default GEANT4 simulation, FLUKA[11] is also used. The main aims of both simulation tools are background subtraction, efficiency correction and uncertainty estimation.

3. Event selection

3.1. *Muon event selection*

A cosmic muon going through Daya Bay's detectors is detected by the OWS, IWS and AD. In the water pool, muons are tagged by the PMT multiplicity. A water

352

pool tagged muon is defined by more than 12 combined PMT hits in the IWS or OWS. An AD-tagged muon is defined by visible energy greater than 20 MeV and within a $[-2\,\mu s, 2\,\mu s]$ time window of a water pool tagged muon. In our analysis, AD-tagged muons ($N_{\mu,Obs}$) are selected. Because this sample includes muons that don't traverse the GdLS, the parameter P_μ calculated from simulation is used to obtain N_μ.

$$N_\mu = N_{\mu,Obs}P_\mu \tag{2}$$

The selected muon deposited energy distribution in data and MC is plotted in Fig. 4(a) and shows a good agreement between data and MC. Fig. 4(b) shows the distribution of muon path length through GdLS from simulation.

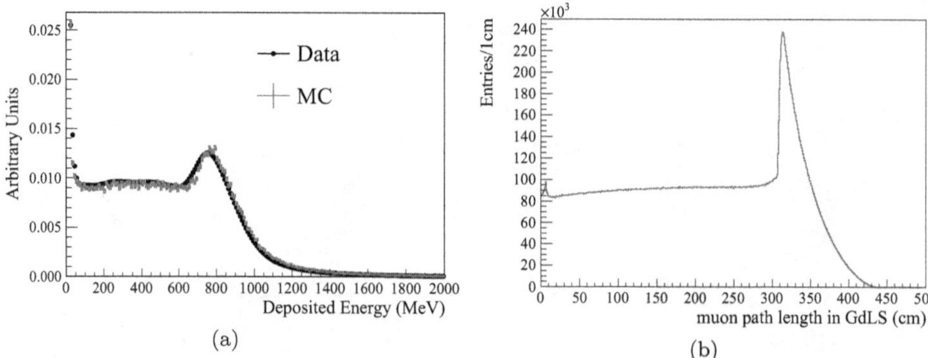

Fig. 4.   (a) Selected muon deposited energy in data and MC(EH1). (b) Distribution of muon path length through the GdLS from simulation (EH1).

3.2. *Neutron event selection*

In our analysis, spallation neutrons captured on Gd are selected with the following criteria: AD triggers are chosen with energy between 6 and 12 MeV occurring in a signal time window, at least $10\,\mu s$ and no more than $200\,\mu s$, after an AD-tagged muon. The time window $[10, 200\,\mu s]$ is defined as the signal window. A background window with the same time width $[1010, 1200\,\mu s]$ is also defined to reduce random background in the signal window. The number of selected neutron captures (N_{cap}) is calculated by Eq. (3).

$$N_{cap} = N_{10-200\mu s} - \alpha N_{1010-1200\mu s} \tag{3}$$

Where $N_{10-200\mu s}$ is the number of selected events in the signal window, $N_{1010-1200\mu s}$ is the number of selected events in the background window, and α is the correction factor. The correction factor is necessary because the background distribution is an exponential with a time constant equal to the muon rate. Fig. 5(a) shows energy of the delayed events which is neutron capture energy and Fig. 5(b) shows time between the muon and delayed events which is the neutron capture time.

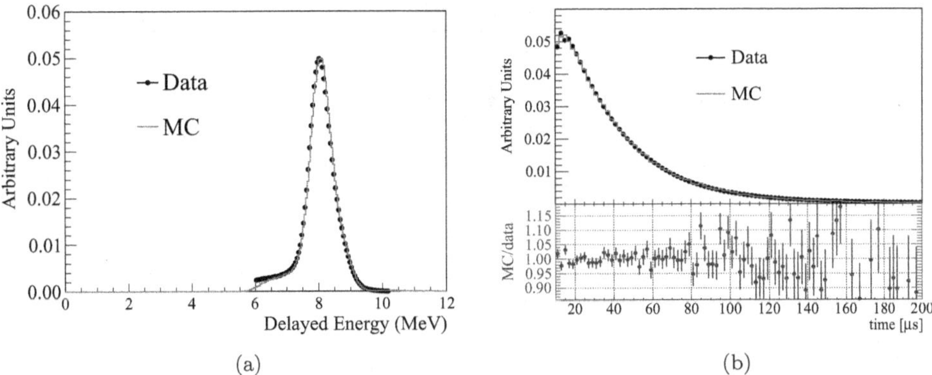

Fig. 5. (a) Energy of the delayed events (neutron capture energy) for data and MC in EH1. (b) Time between the muon and delayed events (neutron capture time) for data and MC(EH1).

The select candidates should be corrected for the efficiency of the neutron selection cuts: time cut efficiency (ε_t), energy cut efficiency (ε_e), Gd capture ratio (ε_{Gd}), and electronics readout time window efficiency (ε_{ro}). ε_{ro} is due to the fact that within a $1.2\,\mu s$ window, only the first trigger will be read out by the electronics. For high multiplicity events, any subsequent triggers within that $1.2\,\mu s$ buffer time will be lost and not counted. For the selected neutron sample, there are two kinds of backgrounds. f_{nSSV} corrects for neutrons that are produced by muons but captured on the stainless steel vessel (SSV) instead of Gd, and $f_{fake\mu}$ corrects for neutrons captured on Gd but induced by fake AD-tagged muons, particles other than muons that mimic the AD muon criteria. Then, two parameters (R_{spill} and R_{det}) are introduced to determine the number of neutron produced due to a muon's trajectory through the GdLS. R_{spill} represents the spill-in/spill-out effect. The corrects for the fact that we measure the number of neutrons captured in GdLS, but we want to know the number of neutrons generated in GdLS. R_{det} accounts for the finite detector size. Therefore, N_n is given by Eq. (4). Note that in Eq. (4), all parameters are from simulation.

$$N_n = N_{cap} \times \frac{(1 - f_{nSSV})(1 - f_{fake\mu})}{\varepsilon_e \varepsilon_t \varepsilon_{Gd} \varepsilon_{ro}} \times \frac{R_{spill}}{R_{det}} \tag{4}$$

4. Results

This proceeding presents the simulated neutron yield results at the three different experimental sites of the Daya Bay experiment, shown in Table 2. The predicted results at Daya Bay from GEANT4 and FLUKA are also shown in Fig. 6 as function of average muon energy.

Table 2. Simulated results.

	EH1	EH2	EH3
MC Predictions ($\times 10^{-5}\mu^{-1}g^{-1}cm^2$)			
Y_n(GEANT4)	7.53±0.01	7.47±0.05	13.35±0.03
Y_n(FLUKA)	8.3±0.02	8.7±0.03	17.2±0.04

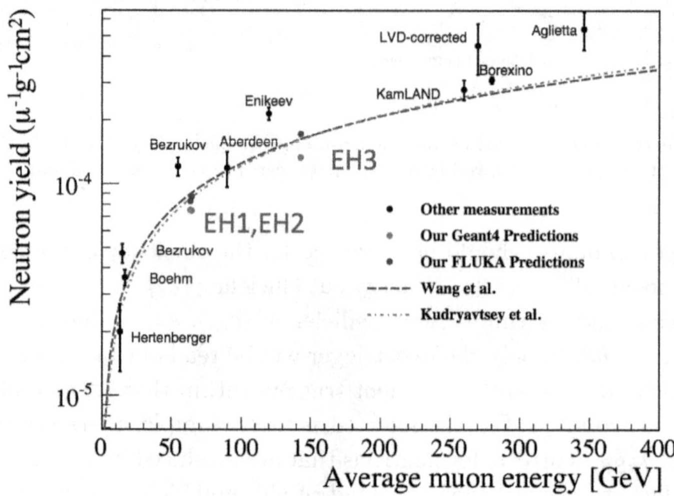

Fig. 6. Predicted neutron yield vs. average muon energy from the three Daya Bay experimental halls compare to the other experiments. Experimental data is shown from Hertenberger[12], Boehm[13], Bezrukov[14], Aberdeen Tunnel[15], Enikeev[16], KamLAND[17], LVD-corrected[18], Borexino[19] and Aglietta[20]. The dashed line and dash-dotted lines show Fluka-based predictions for the dependence of the neutron yield on muon energy from Wang et al.[21] and Kudryavtsey et al.[22]

Acknowledgement

Part of the work in this paper is supported by the National Natural Science Foundation of China (11175107) and the Joint Large-Scale Scientific Facility Funds of the NSFC and CAS (U1532258).

References

1. F. An *et al.* (DAYA-BAY Collaboration), Phys. Rev. Lett. 108, 171803 (2012), arXiv:1203.1669 [hep-ex].
2. F. An *et al.* (Daya Bay), Chin. Phys. C37, 011001 (2013), arXiv:1210.6327 [hep-ex].
3. F. An *et al.* (Daya Bay), Phys. Rev. Lett. 112, 061801 (2014), arXiv:1310.6732 [hep-ex].
4. F. P. An *et al.* (Daya Bay), Phys. Rev. Lett. 115, 111802 (2015), arXiv:1505.03456 [hep-ex].
5. F. P. An *et al.* (Daya Bay), Phys. Rev. D95, 072006 (2017), arXiv:1610.04802 [hep-ex].

6. M. Guan, *et al.*, Muon Simulation at the Daya Bay Site, Lawrence Berkeley National Laboratory Paper LBNL-4262E.

7. F. An *et al.* (Daya Bay Collaboration), Physics Research A 773 (2015) 820.

8. F. An, *et al.* (Daya Bay Collaboration), Nucl. Instrum. Meth. A773 (2015) 8-20.

9. S. Agostinelli *et al.* (GEANT4), Nucl. Instrum. Meth. A506, 250 (2003).

10. J. Allison *et al.*, IEEE Trans. Nucl. Sci. 53, 270 (2006).

11. T. Boehlen *et al.*, Nuclear Data Sheets 120, 211 (2014).

12. R. Hertenberger, M. Chen, and B. L. Dougherty, Phys. Rev. C52, 3449 (1995).

13. F. Boehm *et al.*, Phys. Rev. D 62, 092005 (2000).

14. L. B. Bezrukov *et al.*, Sov. J. Nucl. Phys. 17, 51 (1973).

15. S. C. Blyth *et al.* (Aberdeen Tunnel Experiment), Phys. Rev. D93, 072005 (2016), arXiv:1509.09038 [physics.insdet].

16. R. I. Enikeev *et al.*, Sov. J. Nucl. Phys. 46, 1492 (1987).

17. S. Abe *et al.* (KamLAND Collaboration), Phys. Rev. C81, 025807 (2010), arXiv:0907.0066 [hep-ex].

18. D. Mei and A. Hime, Phys. Rev. D73, 053004 (2006), arXiv:astro-ph/0512125 [astro-ph].

19. G. Bellini *et al.* (Borexino), JCAP 1308, 049 (2013), arXiv:1304.7381 [physics.ins-det].

20. M. Aglietta *et al.*, Nuovo Cimento Soc. Ital. Fis. C 12, 467 (1989).

21. Y. F. Wang, V. Balic, G. Gratta, A. Fasso, S. Roesler, and A. Ferrari, Phys. Rev. D64, 013012 (2001), arXiv:hep-ex/0101049 [hep-ex].

22. V. A. Kudryavtsev, N. J. C. Spooner, and J. E. McMillan, Nucl. Instrum. Meth. A505, 688 (2003), arXiv:hep-ex/0303007 [hep-ex].

Estimation of Sensitivity to Potential Electron Anti-Neutrinos Associated with Gravitational Waves in Some Neutrino Experiments

Zhaokan Cheng* and Jingbo Zhang

*School of Science, Harbin Insititute of Technology,
No. 92 Xidazhi Street, Harbin, Heilongjiang Province, P. R. China
*E-mail: 14B911018@hit.edu.cn
www.hit.edu.cn*

Wei Wang, Jiajie Ling

*School of Physics, Sun Yat-sen University,
No. 135 Xingangxi Road, Guangzhou, Guangdong Province, P. R. China*

Five gravitational wave events labelled GW150914, GW151226, GW170104, GW170814 and GW170817 were observed by the Advanced LIGO detectors at the past two years, successively. A sensitive search for coincident neutrino in some neutrino experiments can provide an exciting opportunity of additional identification for GW sources. Without a observed signals, the sensitivity on 90% confidence level (C.L.) can be estimated by using the observed neutrino backgrounds. At Daya Bay, Double-CHOOZ, RENO, KamLand and Borexino experiment, the sensitivities on 90% C.L. are roughly located in a level of 2-3 events within a time window of 1000 s. With an assumption of a monochromatic spectrum of coincident neutrinos, the corresponding fluence upper limits on 90% C.L., ranging from $\sim 10^{14}$ to $\sim 10^{8}$ cm^{-2}, are obtained and found to mainly depend on the target mass and are inversely propotional to cross section $\sigma(E_\nu)$. Assume other energy spectrum, such as the Fermi-Dirac distribution, it is more clear the upper limits have a negative relation with the target mass, which implys a larger detector will be more sensitive to find coincedent neutrinos associated with GW sources.

Keywords: Neutrino; sensitivity; fluence.

1. Introduction

Since the first gravitational wave event (GW150914) was detected by LIGO on September 14, 2015, four other events (GW151226, GW170104, GW170814 and GW170817) were successively observed at the past two years, referred to the Ref.[1-5] The first four GW events were astronomically derived from the coalesence or merge of a binary black-hole system, while the lastest one, GW170817, was observed as a production from a binary neutron-star inspiral.

At recent, there is a lack of the mechanism to describe a production of electromagnetic and weak interactions in such coalesence or merge systems. However, two coincedent $\gamma-$ray bursts found by Fermi-GBM, having a time latency of 0.4 s and 1.7 s after GW150914 and GW170817, filled into the blank of experimental observation and revealed a powerful evidence to the electromagnetic-weak production from the binary black-hole / neutron-star system. Thus, a multi-messenger search, inolved in neutrinos, may provide more opportunities to glimpsing the dynamics of these astrophysical sources. Experimentally, it is extremely difficult to measure coincedent neutrinos due to a great statistic of backgrounds. Nevertheless, a sensitivity estimation from experimental observation will be feasible to constrain the associated neutrino flux from these sources.

With the observation of GW150914, ANTARRES and IceCube searched for the high-energy-neutrino candidates (> 100 GeV) at first, see the Ref.[6] Whereafter, KamLAND, Super-Kamiokande and Borexino also performed a search in a energy range of [0.9, 100] MeV, [3.5 MeV, 100 PeV] and > 250 keV, successively. The three experiments reported one more result from GW151226 not only GW150914, referred to the Ref.[7-9] All searches provided the same result that there wasn't neutrino candidates found in both temporal and spatial coincidence within a time window of $\pm$ 500 s corresponding to GW150914 and GW151226. Maybe, it is more significant to calculate the sensitivity on 90% C.L. by using the observed neutrino backgrounds. In this paper, we simply calculate the signal sensitivity and translate it into the fluence on 90% C.L. at Super-Kamiokande, KamLAND, Borexino, Daya Bay, Double-CHOOZ and RENO experiment.

2. Expected Backgrounds Inner 1000 s

Super-Kamiokande experiment uses a water Cherenkov detector within $\sim$50 kilotons of ultra-pure water to detect the neutrinos. It is uniquely sensitive to these neutrinos in the energy level of MeV. KamLAND, Borexino, Daya Bay, Double-CHOOZ, and RENO are similar to each other within one or several liquid scintillator detectors containing different volumes. Basicly, Borexino is designed to measure solar neutrinos, while the other four are established to detect reactor neutrinos within a configuration of single-detector or multi-detectors. All these experiments can observe electron antineutrinos via the inverse $\beta-$decay interaction (IBD, $\bar{\nu}_e + p \to e^+ + n$), which has a largest cross section and is used in the paper.

For a massive neutrino, the time-of-flight delay associated with astrophysical observation is given by, see the Ref.[10]:

$$\Delta t = \frac{D}{v} - D \simeq \frac{m^2}{2E^2}D = 5.15\text{ms} \left(\frac{m}{\text{eV}}\right)^2 \left(\frac{E}{10\text{MeV}}\right)^{-2} \frac{D}{10\text{kpc}} \tag{1}$$

where D is the neutrino travelled distance, E is the neutrino energy, and m is neutrino mass. Assume a mass of $\sim$1 eV and a energy of $\sim$10 MeV, the delay is $\sim$250 s within a distance of 500 Mpc. Thereby, a time range of $\pm$ 500 s around detection time of GW events is usually selected in these coincedent searches.

Within a real-time of 1000 s, Super-Kamiokande experiment can measure $\sim$2.90 $\bar{\nu}_e$ in an energy range of [3.5, 75] MeV, referred to the Ref.,[8] while KamLAND and Borexino can detect $\sim$0.03 and $\sim$0.0005 $\bar{\nu}_e$ corresponding to an energy range of [0.9, 100] MeV and [0.25, 15] MeV, respectively, referred to the Ref.[7,9] The observed $\bar{\nu}_e$ events are taken as the background events in these three experiments. For Daya Bay, Double-CHOOZ and RENO experiments, the observed neutrinos are dominated by the reactor neutrino backgrounds in an energy range of [1.8, 10] MeV and by the fast neutron backgrounds in an energy range of [10, 100] MeV. To substract the reactor neutrinos, we use an extrapolation from the fast neutron spectrum of [10, 100] MeV to the energy range of [1.8, 10] MeV, referred to the Ref.[11–13]. As a result, such extrapolation will provide the background events from muons and astrophysical source in the energy range of [1.8, 100] MeV. Within a real-time of 1000 s, the corresponding backgrounds within an energy range from 1.8 to 100 MeV were summarized in the Table 1.

Table 1. Within a real-time of 1000 s, the backgrounds in an energy range of [1.8, 100] MeV at Daya Bay, Double-CHOOZ and RENO.

	Daya Bay	Double-CHOOZ	RENO
Backgrounds	0.3	0.47	0.04

3. Sensitivity Estimation

In a small background case, a Poisson process is generally adopted to calculate confidence belts, which is given by:

$$P(n|\mu) = (\mu + b)^n \exp(-(\mu + b))/n! \tag{2}$$

where n is the total number of observed events, μ is the mean value of test signal events and b is the background events. Testing each value of signal n, a μ_{best} that is a μ value maximized $P(n|\mu)$ will be physically obtained by $\mu_{best} = \max(0, n - b)$. Then the ratio of $P(n|\mu)$ to $P(n|\mu_{best})$ is given by, see the Ref.[14]:

$$R = P(n|\mu)/P(n|\mu_{best}) \tag{3}$$

In physics, a 90% confidence belt corresponds to the sum of $P(n|\mu)$ by a decreasing order of R that is higher than 0.9. In the Fig. 1, the left panel shows this confidence belts obtained on 90% C.L., i.e. the two same color ladders, which imply a signal upper limit and lower limit on 90% C.L. Using the background events b to replace the total observed events n, a similar upper limit and lower limit will be equivalently obtained on 90% C.L. Usually, we name the upper limit as the sensitivity on 90% C.L., summarized into the Table 2.

The sensitivity was translated into a fluence upper limit without oscillation, which is given in neutrinos per cm^{-2} by:

$$F_{\mathrm{UL}}(E_\nu) = \frac{\text{sensitivity}}{N_p \cdot \int \sigma(E_\nu)\lambda(E_\nu)dE_\nu} \tag{4}$$

where N_p is the essential number of target mass, $\lambda(E_\nu)$ is the normalized neutrino energy spectrum, referred to Ref.[15]. $\sigma(E_\nu)$ is the total neutrino cross section, which is obtained from the Ref.[16] for IBD.

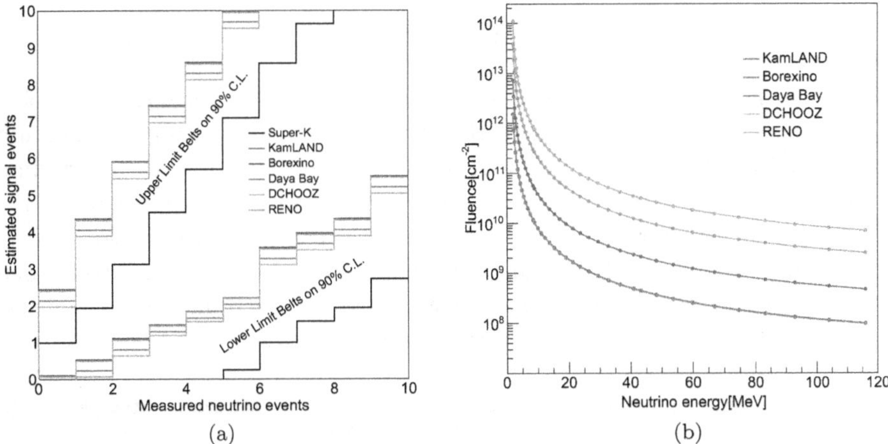

Fig. 1. (a) 90% confidence belts, two same color ladders corresponds to the upper limit and lower limit on 90% C.L. Assume measured neutrino events are totally the backgrounds, the estimated signal upper limit is usually named as the sensitivity. (b) Fluence upper limits on 90% C.L. based on a monochromatic spectrum of neutrino energy. The fluence upper limits on 90% C.L. are translated from the corresponding sensitivities on 90% C.L. by the Eq. (5). The fluence upper limits are inversely proportional to cross section $\sigma(E_\nu)$.

Table 2. Within a real-time of 1000 s, the sensitivity on 90% C.L. obtained by using the background events at Super-Kamiokande, KamLAND, Borexino, Daya Bay, Double-CHOOZ and RENO.

	Super-Kamiokande	KamLAND	Borexino	Daya Bay	Double-CHOOZ	RENO
Sensitivity (90% C.L.)	3.12	2.41	2.44	2.14	2.39	1.97

Assume $\lambda(E_\nu)$ is a monochromatic spectrum by using δ-function, the fluence upper limit of the Eq. (4) is simplified into the below formula:

$$F_{\mathrm{UL}}(E_\nu) = \frac{\text{sensitivity}}{N_p \cdot \sigma(E_\nu)} \tag{5}$$

The corresponding fluence upper limits depend on the target mass and efficiency, decreasing by the neutrino energy for these neutrino experiments, see the right panel in the Fig. 1.

Another hypothesis $\lambda(E_\nu)$ is the normalized pinched Fermi-Dirac distribution is also used. For zero chemical potential and pinching factor $\eta = 0$, this Fermi-Dirac distribution is given by:

$$\lambda_{\mathrm{FD}}(E_\nu) = \frac{1}{T^3 F_2(\eta)} \frac{E_\nu^2}{e^{E_\nu/T} + 1} \tag{6}$$

where, the complete Fermi-Dirac integral $F_n(\eta)$ is given by:

$$F_n(\eta) = \int_0^\infty \frac{x^n}{e^{x-\eta}+1} dx \tag{7}$$

The temperature is obtain by $T = <E>/3.15$. And the average energy is chosen as $<E> = 12.7$ MeV, referred to the Ref.[17]. The corresponding fluence upper limits on 90% C.L. are estimated and filled into the Table 3 by each experiment. It reveals a negative relation between target mass and fluence upper limit, which implys a larger detector will improve the measurement of neutrinos associated with GW events. Thus, KamLAND and Borexino are more easy and sensitive to detect the coincident neutrino signals. The fluence wasn't estimated in Super-Kamiokande experiment due to the unknown detection efficiency.

Table 3. The target mass and fluence upper limits based on the sensitivities 90% C.L. at KamLAND, Borexino, Daya Bay, Double-CHOOZ and RENO.

	KamLAND	Borexino	Daya Bay	Double-CHOOZ	RENO
Target mass ($\times 10^{31}$)	5.98	5.50	0.197	0.0679	0.238
Fluence upper limit ($\times 10^{12}$cm^{-2})	0.00365	0.00358	0.119	1.79	0.636

4. Conclusion

With the observation of five gravitational wave events and two associated $\gamma-$ray bursts, a multi-messenger search has begun in earnest. In the previous searches, ANTARRES, IceCube, KamLAND, Super-Kamiokande and Borexino didn't find any coincedent neutrino signals within a large energy range from $\sim$MeV to $\sim$PeV. Without a observed signals, it makes sense to estimate the sensitivity on 90% C.L. by using the background events in each detector. For Daya Bay, Double-CHOOZ, RENO, KamLand and Borexino experiment, the obtained sensitivities on 90% C.L. are roughly located in a level of 2-3 events within a time window of 1000 s. With an assumption of a monochromatic spectrum of coincident neutrinos, the fluence upper limits on 90% C.L., ranging from $\sim 10^{14}$ to $\sim 10^8$ cm^{-2}, mainly depend on the target mass and are inversely proportional to cross section $\sigma(E_\nu)$. Assume other energy spectrum, such as the Fermi-Dirac distribution, it is more clear the upper limits have a negative relation with the target mass, which implys a larger detector will be more easy and sensitive to find coincedent neutrinos associated with GW sources.

References

1. B. P. Abbott *et al.* *GW150914: The Advanced LIGO Detectors in the Era of First Discoveries. Phys. Rev. Lett.* 116, 131103(2016).
2. B. P. Abbott *et al.* *GW151226: Observation of Gravitational Waves from a 22-Solar-Mass Binary Black Hole Coalescence. Phys. Rev. Lett.* 116, 241103(2016).
3. B. P. Abbott *et al.* *GW170104: Observation of a 50-Solar-Mass Binary Black Hole Coalescence at Redshift 0.2. Phys. Rev. Lett.* 118, 221101(2017).

4. B. P. Abbott *et al.* *GW170814: A Three-detector Observation of Gravitational Waves from a Binary Black Hole Coalescenc. Phys. Rev. Lett.* 119, 141101(2017).

5. B. P. Abbott *et al.* *GW170817: Observation of Gravitational Waves from a Binary Neutron Star Inspiral. Phys. Rev. Lett.* 119, 161101(2017).

6. M. S. Adrian *et al.* *High-energy Neutrino Follow-up Search of Gravitational Wave Event GW150914 with ANTARES and IceCube. Phys. Rev. D* 93, 122010(2016).

7. A. Gando *et al.* *A Search for Electron Antineutrinos Associated with Gravitational-wave Events GW150914 and GW151226 Using KamLand. Ap. J. L* 829, L34(2016).

8. K. Abe *et al.* *Search for Neutrinos in Super-kamiokande Associated with Gravitational-wave Events GW150914 and GW151226. Ap. J. L* 830, L11(2016).

9. M. Agostini *et al.* *A Search for Low-energy Neutrinos Correlated with Gravitational Wave Events GW150914, GW151226 and GW170104 with the Borexino Detector.* arXiv: 1706.10176.

10. C. Giunti and C. Kim, Fundamentals of Neutrino Physics and Astrophysics. Oxford University Press, 2007.

11. F. P. An *et al.* *Measurement of electron antineutrino oscillation based on 1230 days of operation of the Daya Bay experiment. Phys. Rev. D* 95 (2016), 072006.

12. Y. Abe *et al.* *Improved measurements of the neutrino mixing angle θ_{13} with the Double Chooz detector. JHEP* 10 (2014), 086.

13. S. B. Kim *et al.* *Measurement of neutrino mixing angle θ_{13} and mass difference Δm_{ee}^2 from reactor antineutrino disappearance in the RENO experiment. Nucl. Phys. B* 908 (2016), 94-115.

14. G. J. Feldman and R. D. Cousins. *A Unified approach to the classical statistical analysis of small signals. Phys. Rev. D* 57 (1998), pp. 3873-3889.

15. S. Fukuda *et al.* *Search for Neutrinos from Gamma-Ray Bursts Using Super-Kamiokande. Ap. J.* 578, 1(2002).

16. A. Strumia and F. Vissani, *Precise Quasielastic Neutrino/Nucleon Cross-section. Phys. Lett. B* 564, 1-2(2003).

17. O. L. Caballero *et al.* *Black Hole Spin Influence on Accretion Disk Neutrino Detection. Phys. Rev. D* 93, 123015(2016).

362

Experimental Study of the Wave Packet Treatment of Neutrino Oscillation at Daya Bay

Zhaokan Cheng

on behalf of the Daya Bay Collaboration

School of Science, Harbin Insititute of Technology,
No. 92 Xidazhi Street, Harbin, Heilongjiang Province, P. R. China
E-mail: 14B911018@hit.edu.cn
www.hit.edu.cn

Steven Wong and Wei Wang

School of Physics, Sun Yat-sen University,
No. 135 Xingangxi Road, Guangzhou, Guangdong Province, P. R. China

The conventional formula for neutrino oscillation probability is based on the neutrino plane-wave assumption, which is a good approximation but not theoretically self-consistent. Wave-packet treatment is necessary for a self-consistent description of neutrino oscillation phenomenon. The oscillation of reactor neutrino observed by the Daya Bay experiment is examined in the framework of a model in which the neutrino is described by a wave packet with a relative intrinsic momentum dispersion σ_{rel}. The modified survival probability formula based on neutrino wave-packet (WP) treatment was used to fit the Daya Bay data, providing the first quantitative limits on 95% confidence level: $2.38 \cdot 10^{-17} < \sigma_{\mathrm{rel}} < 0.232$. Furthermore, the effect due to the wave packet nature of neutrino oscillation is found to be insignificant in the Daya Bay experiment, which ensures an unbiased measurement of the oscillation parameters $\sin^2 2\theta_{13}$ and Δm_{32}^2 within the plane wave model.

Keywords: Wave-packet; decoherence; Daya Bay.

1. Introduction

In the plane-wave assumption, neutrino flavor states are coherent superpositions of mass eigenstates $|\nu_k(p)\rangle$ with the same momentum p: $|\nu_\alpha(p)\rangle = \sum_{k=1}^{3} V_{\alpha k}^* |\nu_k(p)\rangle$. In this case, the neutrino oscillation probability of $(\nu_\alpha \to \nu_\beta)$ can be described by

$$P_{\alpha\beta}(L) = \sum_{k,l=1}^{3} V_{\alpha k}^* V_{\beta l}^* V_{\beta k} V_{\alpha l} \mathrm{e}^{-i2\pi L/L_{kl}^{\mathrm{osc}}}, \tag{1}$$

where $L_{kl}^{\mathrm{osc}} = 4\pi p/\Delta m_{kl}^2$ is the oscillation length corresponding to the mass squared difference $\Delta m_{kl}^2 = m_k^2 - m_l^2$. p is the momentum of neutrino and L is the distance between the detector and neutrino source.

The conventional quantum-mechanical approach within a set of hypotheses has been successful in explaining a wide range of neutrino experiments, nevertheless, this approximation is not self-constisting [1,2]:

- A hypothesis of a coherent production/detection of neutrino mass eigenstates must be justified.

- A hypothesis that neutrino mass eigenstates have definite three-momentum $\mathbf{p_i}$ implies that one can not define a distance neutrino travelled.
- A hypothesis that $\mathbf{p_i}$ are all equal to each other contradict to Lorentz-invariance.
- A hypothesis that time t in the evolution of neutrino states could be replaced by a travelled distance L is not justified. Small correction taking into account the neutrino state speed $v_i \neq 1$ leads to a spurious factor two in the oscillation formula.

A wave packet model is necessary for a self-consistent consideration[3–5].

2. Wave-packet approach to neutrino oscillations

In wave-packet models, the intrinsic momentum dispersion σ_p of the neutrino wave packet is an effective quantity comprising the microscopic momenta dispersions of all particles involved in the production and detection of the neutrino. A non-zero value of σ_p leads to the decoherence with time evolution for a quantum superposition of massive neutrinos. Actually, a relative intrinsic momentum dispersion $\sigma_{\rm rel} \equiv \sigma_p/p$ is adopted to describe the decoherence effect in general.

Currently, there are no solid estimations of $\sigma_{\rm rel}$, which could vary by orders of magnitude, depending on the assumptions of spatial wave packet dispersion σ_x; for example, uranium nucleus diameter ($\sigma_x \simeq 10^{-12}$ cm, $\sigma_{\rm rel} \simeq 1$), atomic or inter-atomic distances ($\sigma_x \simeq 10^{-8}$ cm, $\sigma_{\rm rel} \simeq 10^{-3}$), pressure broadening ($\sigma_x \simeq 10^{-4}$ cm, $\sigma_{\rm rel} \simeq 10^{-3}$), etc. Furthermore, there is no quantitative experimental support for these estimations. Hence, an experimental study of the WP impact is necessary.

In the neutrino WP description, the plane wave eigenstate $|\nu_k(p)\rangle$ is replaced by a wave packet with a mean momentum p':

$$|\widetilde{\nu}_k(p')\rangle = \int \frac{dp}{2\pi} f(p,p') |\nu_k(p)\rangle \tag{2}$$

where, $f(p,p')$ is the wave function of the neutrino in momentum space and is assumed to be Gaussian. Projection of $|\widetilde{\nu}_\alpha\rangle$ onto a detection state $|\widetilde{\nu}_\beta\rangle$ defined similarly is used to calculate the transition apmplitude and the probability which depends on the effective momentum dispersion σ_p given by $\frac{1}{\sigma_p^2} = \frac{1}{\sigma_{p,prod}^2} + \frac{1}{\sigma_{p,det}^2}$. $\sigma_{p,prod}$ and $\sigma_{p,det}$ represent the momentum dispersion at the production and detection of neutrinos respectively.

Finally, the neutrino flavor transition probability can be described as:

$$P_{\alpha\beta}(L) = \sum_{k,l=1}^{3} \frac{V_{k\beta} V_{\alpha k}^* V_{l\alpha} V_{\beta l}^*}{\sqrt[4]{1 + \left(L/L_{kl}^{\rm d}\right)^2}} e^{-\frac{\left(L/L_{kl}^{\rm coh}\right)^2}{1+\left(L/L_{kl}^{\rm d}\right)^2} - {\rm D}_{kl}^2} e^{-i\widetilde{\varphi}_{kl}}, \tag{3}$$

where, $\widetilde{\varphi}_{kl} = \varphi_{kl} + \varphi_{kl}^d$, $\varphi_{kl} = 2\pi L/L_{kl}^{\rm osc}$ represents the plane wave phase and φ_{kl}^d represents the correction caused by the dispersion of the wave packet that is

364

expressed as:

$$\varphi_{kl}^{\mathrm{d}} = -\frac{L/L_{kl}^{\mathrm{d}}}{1 + \left(L/L_{kl}^{\mathrm{d}}\right)^2}\left(\frac{L}{L_{kl}^{\mathrm{coh}}}\right)^2 + \frac{1}{2}\arctan\frac{L}{L_{kl}^{\mathrm{d}}}, \tag{4}$$

In Eq. (3), L_{kl}^{osc}, represents the usual oscillation length of a pair of neutrino states $|\nu_k\rangle$ and $|\nu_l\rangle$; L_{kl}^{coh} is the neutrino coherence length, i.e. a distance at which the interference of neutrino mass eigenstates vanishes; L_{kl}^{d} is the dispersion length, which describes the distance at which the wave packet is doubled in its spatial dimension due to the dispersion of waves moving with different velocities, respectively:

$$L_{kl}^{\mathrm{osc}} = \frac{4\pi p}{\Delta m_{kl}^2}, \qquad L_{kl}^{\mathrm{coh}} = \frac{L_{kl}^{\mathrm{osc}}}{\sqrt{2}\pi\sigma_{\mathrm{rel}}}, \qquad L_{kl}^{\mathrm{d}} = \frac{L_{kl}^{\mathrm{coh}}}{2\sqrt{2}\sigma_{\mathrm{rel}}}, \tag{5}$$

Finally the term D_{kl}^2 corresponds to the delocalization and suppresses the oscillation if the relative momentum dispersion σ_{rel} is so small that $\sigma_x = 1/2\sigma_p \gg L_{kl}^{\mathrm{osc}}$, which is given by:

$$\mathrm{D}_{kl}^2 = \frac{1}{2}\left(\frac{\Delta m_{kl}^2}{4p^2\sigma_{\mathrm{rel}}}\right)^2 = \frac{1}{4}\left(\frac{\sqrt{2}\pi\sigma_x}{L_{kl}^{\mathrm{osc}}}\right)^2. \tag{6}$$

Moreover, the terms in Eq. (3) which correspond to the interference of ν_k and ν_l states also get suppressed by the denominator $\sqrt[4]{1 + \left(L/L_{kl}^{\mathrm{d}}\right)^2}$ and vanish for both limits $\sigma_p \to 0$ and $\sigma_p \to \infty$, so that the oscillation will be destroyed. Details of Eq. (3) can be found in Refs. [6–8]

3. Analysis and results

The Daya Bay data were analyzed with the wave-packet treatment taken into account. The Daya Bay Experiment was designed to measure the neutrino mixing angle θ_{13} by detecting the disappearance of $\bar{\nu}_e$ stemmed from six nuclear reactors, two located in the Daya Bay Nuclear Power Plant and the other four in the Ling Ao Nuclear Power Plant. It consists of eight antineutrino detectors (ADs), two couples respectively clustered in two near-sites, and the remaining four detectors instrumented into one far-site away from the two plants. There are three experimental halls (EHs) totally in the two near-sites and one far-site, see the left panel of Fig. 1. The distances between detectors and reactors are in a range from 0.5 to 1.9 km. The analysis used the data with a statistics of $\sim 1.2 \times 10^6$ $\bar{\nu}_e$ under the operation of 621 days [9]. The ratios of the observed to the predicted $\bar{\nu}_e$ events without oscillation are displayed in the right panel of Fig. 1. It reveals the plane wave approch fits the data very well.

As a goodness-of-fit measure, a function $\chi^2(\eta) = (\mathbf{d} - \mathbf{t}(\eta))^T V^{-1}(\mathbf{d} - \mathbf{t}(\eta))$ was constructed and used in the analysis, where $\mathbf{d}$ is a data vector containing detected numbers of $\bar{\nu}_e$ in energy bins and in different detectors, while $\mathbf{t}(\eta)$ is the corresponding theoretical model vector which depends on constrained and free parameters η. All statistical and systematic uncertainties were quantified into

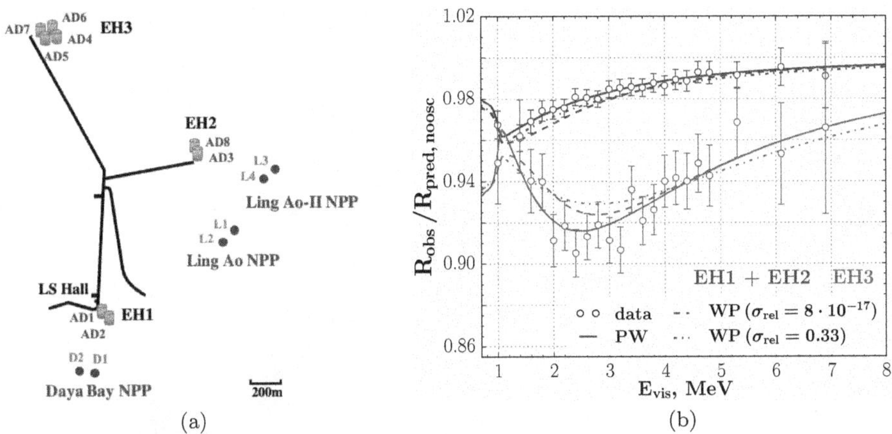

Fig. 1. (a) Layout of the Daya Bay experiment. The dots represent reactors, labeled as D1, D2, L1, L2, L3, and L4. Eight ADs, AD1−AD8, are installed in three EHs. (b) Predicted and observed ν_e spectra in EH1, EH2 and EH3, as ratios to spectra without oscillation. The solid, dashed and dot-dashed curves correspond to the cases of PW approximation, extremely small value (8×10^{-17}) and large value (0.33) of σ_{rel} respectively.

the convariance matrix V. The analysis was done with four free parameters σ_{rel}, Δm_{32}^2, $\sin^2 2\theta_{13}$ and reactor flux normalization N. The confidence regions are obtained by means of two statistical methods: the conventional fixed-level $\Delta\chi^2$ analysis and the Feldman-Cousins method[10]. The marginalized $\Delta\chi^2$ statistic is: $\Delta\chi^2(\eta') = \min_{\eta \backslash \eta'} \chi^2(\eta) - \min_{\eta} \chi^2(\eta)$, where $\eta = (\sigma_{\mathrm{rel}}, \Delta m_{32}^2, \sin^2 2\theta_{13}, N)$ and η' is its subspace with parameters of interest ($\eta' = \sigma_{\mathrm{rel}}$ for one dimensional interval, and $\eta' = (\sigma_{\mathrm{rel}}, \Delta m_{32}^2)$ or $\eta' = (\sigma_{\mathrm{rel}}, \sin^2 2\theta_{13})$ for two dimensional regions), and both are used to determine the p-value of the observed dataset and the model.

The allowed regions of $(\sigma_{\mathrm{rel}}, \Delta m_{32}^2)$ and $(\sigma_{\mathrm{rel}}, \sin^2 2\theta_{13})$ are shown in the top and middle panel of Fig. 2. The dashed and solid contours correspond to fixed-level $\Delta\chi^2$ and to the Feldman-Cousins methods respectively. The constraints from these methods are in good agreement. The $\Delta\chi^2$ variation of σ_{rel} is also displayed in the bottom panel of Fig. 2. The allowed region for the parameter σ_{rel} at 95% C.L. is found to be:

$$2.38 \cdot 10^{-17} < \sigma_{\mathrm{rel}} < 0.232. \tag{7}$$

In the region of $10^{-16} \ll \sigma_{\mathrm{rel}} \ll 0.1$, the wave-packet impact is negligible for the Daya Bay experiment. In the region of $\sigma_{\mathrm{rel}} \lesssim 10^{-16}$, σ_{rel} is strongly correlated to Δm_{32}^2 and $\sin^2 2\theta_{13}$, yielding smaller values of Δm_{32}^2 and $\sin^2 2\theta_{13}$. These correlations arise from D_{kl}^2 term in Eq. (3). For $\sigma_{\mathrm{rel}} \gtrsim O(0.1)$, the absolute values of the corresponding correlation coefficients are smaller than 10^{-5}.

Taking the average momentum $p = 4$ MeV, the upper bound of Eq. (7) corresponds to $L_{32}^{\mathrm{coh}} > 0.98 L_{32}^{\mathrm{osc}}$. The lower bound can be interpreted in terms of σ_x which corresponds to the spatial width of neutrino wave packet. Taking as an estimate the

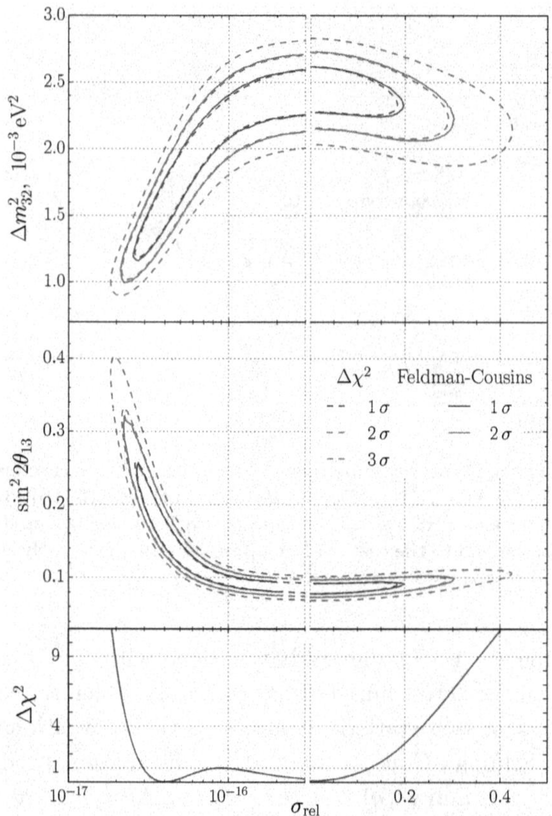

Fig. 2.   Allowed regions of $(\sigma_{\rm rel}, \Delta m^2_{32})$ (top) and of $(\sigma_{\rm rel}, \sin^2 2\theta_{13})$ (middle) parameters obtained with fixed-level $\Delta\chi^2$ (contours corresponding to 1σ, 2σ, 3σ C.L., dashed lines) and within the Feldman-Cousins (contours corresponding to 1σ, 2σ C.L., solid lines) methods. Bottom panel shows the marginalized $\Delta\chi^2(\sigma_{\rm rel})$ profile vs $\sigma_{\rm rel}$. Note the break in the abscissa and the change from a logarithmic to linear scale.

neutrino momentum $p = 4$ MeV, one can find that our lower bound corresponds to $\sigma_x \sim 1$ km. Such value of σ_x fits the data, however an obvious constrain that σ_x should not exceed the dimensions of reactor cores and an AD detector, excludes this value. Using this constrain we provide an upper limit at 95% C.L. ignoring D^2_{kl} term in Eq. (3):

$$\sigma_{\rm rel} < 0.2 \tag{8}$$

4. Conclusion

The first experimental and quantitative constraints on the neutrino wave packet momentum dispersion are obtained. The wave-packet impact is found to be insignificant in the Daya Bay experiment. This ensures an unbiased measurement of the parameters $\sin^2 2\theta_{13}$ and Δm^2_{32} within the conventional plane-wave approximation.

Acknowledgements

This work is developed in Daya Bay Reactor Neutrino Experiment and supported by the National Natural Science Foundation of China (NSFC11675273).

References

1. E. K. Akhmedov and A. Y. Smirnov. *Paradoxes of neutrino oscillations. Phys. Atom. Nucl.* 72 (2009), pp. 1363-1381. arXiv: 0905.1903.
2. C. Giunti. *Coherence and wave packets in neutrino oscillations. Found. Phys. Lett.* 17 (2004), 81 pp. 103-124. arXiv: hep-ph/0302026.
3. B. Kayser. *On the Quantum Mechanics of Neutrino Oscillation. Phys. Rev. D.* 24 (1981), p. 110. DOI: 10.1103/PhysRevD.24.110.
4. K. Kiers, S. Nussinov, and N. Weiss. *Coherence effects in neutrino oscillations. Phys. Rev. D.* 53 (1996), pp. 537-547. DOI: 10.1103/PhysRevD.53.537. arXiv: hep-ph/9506271 [hep-ph].
5. E. Akhmedov, D. Hernandez, and A. Smirnov. *Neutrino production coherence and oscillation experiments. JHEP* 1204 (2012), pp. 052. DOI: 10.1007/JHEP04(2012)052. arXiv: 1201.4128 [hep-ph].
6. Y. L. Chan, M. C. Chu, K. M. Tsui, C. F. Wong and J. Xu. *Eur. Phys. J. C.* 76 (2016), 310. arXiv: 1507.06421.
7. F. P. An *et al. Study of the wave packet treatment of neutrino oscillation at Daya Bay. Eur. Phys. J. C.* 77 (2017), no.9, 606. arXiv: 1608.01661.
8. D. V. Naumov and V. A. Naumov. *A diagrammatic treatment of neutrino oscillations. J. Phys. G.* 37 (2010), 105014.
9. F. P. An *et al. New Measurement of Antineutrino Oscillation with the Full Detector Configuration at Daya Bay. Phys. Rev. Lett.* 115. 11 (2015), pp. 111802. arXiv: 1505.03456.
10. G. J. Feldman and R. D. Cousins. *A Unified approach to the classical statistical analysis of small signals. Phys. Rev. D.* 57 (1998), pp. 3873-3889. DOI: 511 10.1103/PhysRevD.57.3873. arXiv: physics/9711021 [physics.data-an].

Physics Potential with Borexino Phase-II Data

X. F. Ding[1,2]

on behalf of Borexino Collaboration

[1]*Gran Sasso Science Institute, 67100 L'Aquila, Italy*
[2]*INFN Laboratori Nazionali del Gran Sasso, 67010 Assergi (AQ), Italy*
E-mail: xuefeng.ding@gssi.infn.it   https://dingxf.cn

Borexino collaboration: M. Agostini, K. Altenmüller, S. Appel, V. Atroshchenko, Z. Bagdasarian, D. Basilico, G. Bellini, J. Benziger, D. Bick, G. Bonfini, D. Bravo, B. Caccianiga, F. Calaprice, A. Caminata, S. Caprioli, M. Carlini, P. Cavalcante, A. Chepurnov, K. Choi, L. Collica, D. D'Angelo, S. Davini, A. Derbin, X.F. Ding, A. Di Ludovico, L. Di Noto, I. Drachnev, K. Fomenko, A. Formozov, D. Franco, F. Gabriele, C. Galbiati, C. Ghiano, M. Giammarchi, A. Goretti, M. Gromov, D. Guffanti, C. Hagner, T. Houdy, E. Hungerford, Aldo Ianni, Andrea Ianni, A. Jany, D. Jeschke, V. Kobychev, D. Korablev, G. Korga, D. Kryn, M. Laubenstein, E. Litvinovich, F. Lombardi, P. Lombardi, L. Ludhova, G. Lukyanchenko, L. Lukyanchenko, I. Machulin, G. Manuzio, S. Marcocci, J. Martyn, E. Meroni, M. Meyer, L. Miramonti, M. Misiaszek, V. Muratova, B. Neumair, L. Oberauer, B. Opitz, V. Orekhov, F. Ortica, M. Pallavicini, L. Papp, Ö. Penek, N. Pilipenko, A. Pocar, A. Porcelli, G. Ranucci, A. Razeto, A. Re, M. Redchuk, A. Romani, R. Roncin, N. Rossi, S. Schönert, D. Semenov, M. Skorokhvatov, O. Smirnov, A. Sotnikov, L.F.F. Stokes, Y. Suvorov, R. Tartaglia, G. Testera, J. Thurn, M. Toropova, E. Unzhakov, A. Vishneva, R.B. Vogelaar, F. von Feilitzsch, H. Wang, S. Weinz, M. Wojcik, M. Wurm, Z. Yokley, O. Zaimidoroga, S. Zavatarelli, K. Zuber, G. Zuzel

Borexino is an experiment located at Laboratori Nazionali del Gran Sasso (LNGS) in Italy and its primary goal is to detect solar neutrinos with unprecedented high sensitivity. After Phase-I data-taking, purifications were conducted and backgrounds were significantly reduced. After that Borexino has started a new round of data-taking, Phase-II, since 2011 December. Besides Monte Carlo based detector model, analytical detector models, which consists of nonlinearity models, resolution models, and mono-energetic line shapes, are also used. The energy range in the analysis extends up to 2 MeV, and this allows Borexino to measure the fluxes of all low energy solar neutrino components: pp, ^{7}Be, *pep*, and CNO. The multivariate likelihood techniques based on the radial distribution and pulse-shape discriminator distribution are adopted to suppress external gammas and cosmogenic ^{11}C, the major backgrounds of *pep*/CNO neutrinos. Graphic Processing Unit(GPU) based fitting tool has been developed and significantly shortens the converging time. In this proceeding the techniques for suppressing backgrounds, the analytical approach, the method to estimate the detector response model related systematic uncertainties and expected uncertainties are presented.

Keywords: Borexino Phase-II; solar neutrino; analytical response function.

1. Introduction

Solar neutrinos released with fusion reactions in the core, unlike the high energy gamma rays, can pass through the solar plasma immediately and make it possible to study the core of the Sun, and Borexino has demonstrated that it is possible to measure the individual branches of the *pp*-chain above 190 keV[1-4]. Moreover, abundances of heavy elements in the core region is the key to solve the solar abundance problem[5]. Using new and old abundances leads to different predictions for the neutrino flux[5]. This is especially true for the neutrinos from the CNO cycle, that are to date not yet measured.

Solar neutrinos also provide a unique probe to study the matter effect of neutrino oscillations thanks to the high electron density in the Sun[6]. Wide energy range of solar neutrino spectra have been observed covering both the adiabatic regime (the pp, ^{7}Be and pep) and vacuum regime (high energy part of ^{8}B) of MSW effect. Borexino can even uniquely reach the transition zone by the detection of ^{8}B neutrino with the lowest threshold as low as 3 MeV[4,7].

Detection of solar neutrinos and measurement of their fluxes are not easy tasks. Solar neutrinos are detected via elastic scattering channel. To achieve expected sensitivity and precision, it is of prime importance to purify the liquid scintillator and remove the natural radio-active elements in the liquid scintillator. Borexino as the pioneer of low background control has achieved unprecedented radioactive-purity level. In Phase-I from 2008 to 2010, the concentration of ^{238}U and ^{232}Th in the liquid scintillator are $(5.3 \pm 0.5) \times 10^{-18}$ g/g and $(3.8 \pm 0.8) \times 10^{-18}$ g/g, respectively[8]. In data reported by this proceeding, their concentrations are even reduced to $< 9.4 \times 10^{-20}$ g/g (95% C.L.) and $< 5.7 \times 10^{-19}$ g/g (95% C.L.). Such radio-purity levels were achieved by the extensive purification campaign in 2011.

After Phase-I, more data has been collected between December 14th, 2011 and May 21st, 2016, which corresponds to an exposure of 1291.51 days $\times$ 71.3t. With larger exposure and better radio-purity levels, as well as improved detector response function and improved Monte Carlo[9], we are expected to improve the precision of fluxes of pp, ^{7}Be and pep solar neutrinos fluxes and to reach the same level of upper limit for CNO neutrinos with weaker assumptions. In this proceeding, we report the analysis techniques to suppress backgrounds and the response functions, the methods for estimating the systematic uncertainties related to detector response and expected uncertainties on neutrino interaction rate measurements.

2. Borexino experiment

Borexino is a solar neutrino experiment located at hall C of Laboratori Nazionali del Gran Sasso (LNGS). LNGS provides 3800 meters water equivalent passive shield against cosmic muons and is schematically shown in Figure 1(a). Borexino detector consists of the outer detector, a water Cherenkov detector for the muon tagging, and the inner detector, a liquid scintillator based unsegmented calorimetor. It is shown schematically in Figure 1(b)[10]. Liquid scintillator used in the sensitive region consists of the solvent PC (pseudocumene, 1,2,4-trimethylbenzene $C_6H_3(CH_3)_3$) and the solute PPO (2,5-diphenyloxazole, $C_{15}H_{11}NO$) at a concentration of 1.5 g/L (0.17% by weight).

When neutrinos undertake elastic scattering interaction, the recoiled electrons can be detected with scintillation light. Around 500 p.e./MeV/2000 PMTs is col-

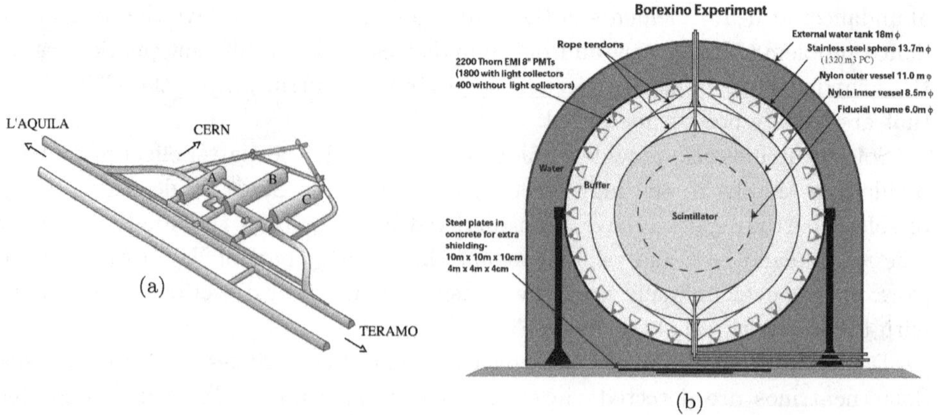

Fig. 1.   Schematic drawing of (a) Laboratori Nazionali del Gran Sasso and (b) Borexino detector.

lected. The charge and arrival time of each hit are recorded by front-end electronics and digitizers[11], then they are used to reconstruct the vertex and energy of the event in the offline software. Multi-variate fits are performed with the a package developed based on `m-stats`[12] and a package developed based on `GooStats`[13].

3. The analytical fitting approach

The detector response is the critical information for extracting the event rates through spectrum fitting. To avoid risk of spoiling the extremely good radio-purity, we did not undertake frequent calibrations. Internal contaminations, such as the ^{210}Po α-decay and cosmogenic ^{11}C positron decay, have been used to monitor the detector stability and the instability is treated as systematics. If we extract the detector response parameters, such as the light yield, from the spectrum itself, the uncertainty depends on the amount of information inside rather than the understanding of calibrations and the precision of Monte Carlo tuning, and thus is an independent way of treating detector modelling related uncertainties.

This method is carefully validated. First, we fit random spectra sampled according to analytical response functions. The median fit values are unbiased for not only the rates, but also the detector response parameters, so our spectra contains enough information to extract the chosen parameters. Then, we fit spectra sampled according to the Monte Carlo detector emulation. In this way we can estimate the discrepancies induced by the simplification of the high order terms of the detector response model, and they are treated as the model related systematics.

The response function is used through convolution to map the spectrum of the kinetic energy of the recoiled electrons or that of the decay products, hereafter as the kinetic energy, to the spectrum of the energy estimators, hereafter as the observed energy. The response function is determined by three components: the non-linearity model $\mu(E)$, which maps the kinetic energy to the expected observed energy, the variance/resolution model $\mathrm{var}(\mu)$, which maps the expectation to the variance, and the line shape, which is the probability density function given the expectation and variance.

3.1. *The non-linearity model*

The non-linearity model is a composite function. The core function maps the kinetic energy to the expected number of photoelectrons produced on the photocathode, and the shell functions map it further to the energy estimators.

The core function is shared among different energy estimators and includes only the quenching effect and the Cherenkov photon contribution. For β-like events[8],

$$\mathrm{npe}(E_k) = \mathrm{LY} \cdot [\mathrm{Quenching}(E) + \mathrm{fCher} \cdot \mathrm{Cherenkov}(E)] \tag{1}$$

$$\mathrm{Cherenkov}(E) = (A_0 + A_1 \cdot x + A_2 \cdot x^2 + A_3 \cdot x^3)(1 + A_4 \cdot E) \qquad (E > E_0) \tag{2}$$

$$\mathrm{Quenching}(E) = \frac{q_1 + q_2 \ln(E) + q_3 \ln(E)^2}{1 + q_4 \ln(E) + q_5 \ln(E)^2} \tag{3}$$

where E is the energy in MeV, $x = \ln(E)$, E_0 is the threshold for having Cherenkov contribution, LY is the light yield and a free parameter, fCher is the relative contribution of Cherenkov photons compared with sctillation photons and also a free parameter. Table of A_0 to A_4 is built by by integrating the scintillation lights over energy loss trajectories considering the quenching effect according to Birk's law given certain quenching coefficients k_B, and the best combinations is chosen by fitting the events simulated in detector center, selecting only photoelectron originated from scintillations. q_1 to q_4 are determined by fitting directly the number of photoelectrons originated from Cherenkov radiation, and they are very close to the theoretically calculated original-number-of-photons–energy dependences. Parameters As and qs are fixed because of not enough information to extract them. For positron events and $^{210}\mathrm{Po}$ α-decays, additional free parameters are used to describe the quenching effect.

The shell functions are different for different energy estimators. For charge it is a second order polynomials:

$$\frac{N_Q}{\mathrm{npe}} = \frac{p_{\mathrm{dn}} + (1 + p_{\mathrm{miscalib}}) \cdot \mathrm{npe} + p_{\mathrm{quadr}} \cdot \mathrm{npe}^2}{\mathrm{npe}} \tag{4}$$

where p_{dn} is the effective dark noise contribution, p_{miscalib} is the effective calibration factor for the increase of average charge due rejection of hits with charge under threshold, p_{quadr} is the high order term from the non-linearity of the PMT, front-end, digitizer and offline reconstruction algorithms. They are all effective parameters. For example, one version of charge reconstruction algorithms applied compensations and p_{dn} is reduced to almost zero, while in other versions it is of the order of magnitude of 1. For the number of fired PMT in fixed windows, it is a modified binomial expectation formula:

$$N_p^{dt_{1(2)}} = N_0 \left[1 - \exp\left(-\frac{\text{npe}}{N_0}\right) \cdot (1 + p_t \cdot \frac{\text{npe}}{N_0}) \right] \cdot \left(1 - g_C \cdot \frac{\text{npe}}{N_0}\right) \tag{5}$$

where N_0 is the total number of channels used for counting the number of fired channels. p_t is the probability for a hit to be rejected by the discriminator due to low amplitude. g_C is the correction for the effect that for events further from center, more photons are consumed firing the same PMTs that are close to them, and thus overall number of fired channels decrease.

3.2. *The variance/resolution model*

For both N_Q and $N_p^{dt_{1(2)}}$ variables, given the kinetic energy, their distributions can be regarded as the compound probability distributions of npe, N_x, the event vertex position $\vec{x}$ and the total number of channels N_{live}. The variance can be calculated with the sum rule of variance.

Consider npe follows Poisson distribution, charge is the sum of charge of each p.e., which follows approximately the Gaussian distribution with width as σ_1, and the expected npe is approximately a linear function of z, we have

$$\text{var}(N_{pe}) = f_{\text{eq}} \cdot (1 + \sigma_1^2) \cdot \mu_N + v_T \cdot \mu_N^2 \tag{6}$$

where $f_{\text{eq}} = \frac{2000}{N_{\text{live}}}$ is the normalization factor and N_{live} is the total number of channels used for summing charge. v_T is $(1/21)k^2 R^6$ when the fiducial volume is a sphere of radius R and the npe non-uniformity shape is $1 + kz$. Here σ_1 is free and mainly extracted from the ^{210}Po peak, and v_T is fixed to 0 after the non-uniformity correction. The feq is different between ^{210}Po and other species, because it's temporally-weighted.

For $N_p^{dt_{1(2)}}$ variable, it approximatedly follows the binomial distribution and

$$\text{var}(N_p^{dt_{1(2)}}) = f_{\text{eq}} \cdot \mu_N \cdot (1 - p_1 - v \cdot p_1) + v_N + \beta_2 \cdot v_q^2 + \beta_0 \cdot \mu_N^3 \cdot f_{\text{eq}}^{-1} \tag{7}$$

$$p_1 = \frac{\mu_N}{2000} \tag{8}$$

$$v_q = \mu_N \cdot (1 - p_1) \cdot \frac{\ln(1 - p_1)}{p_1} \tag{9}$$

$$v_N = \beta_1 \cdot \mu_N \cdot f_{\text{eq}} \qquad \text{only for } \beta\text{s} \tag{10}$$

where $\mu_N \cdot (1 - p_1)$ is the variance from the bimomial distribution, v is the correction factor for the fluctuaion of the p_t (probabily of the hit being rejected due to low magnitude) among different channels and v_N is the variance from the δ-electrons that only exist in β-like events. $\beta_2 \cdot v_q^2 + \beta_0 \cdot \mu_N^3 \cdot f_{\text{eq}}^{-1}$ accounts for the variance due to both the non-uniformity of the npe and of the npe $\mapsto N_p^{dt_{1(2)}}$ non-linearity. They are relavant for *pep* and CNO neutrinos, and it is validated against Monte Carlo that higher order terms are not needed considering the fit range we used. Based on whether or not our spectrum has sensitivity on these parameters, we decide to fix v and β_2, and leave β_0 and β_1 free. Besides, unlike charge, independent β_0 is used for ^{210}Po.

For the line shape, the modified gaussian[8] is used for charge, and the scaled Poisson[8] is used for $N_p^{dt_{1(2)}}$.

3.3. *The Graphic Processing Unit based fitting package*

Due to the heavy computation for the convolution of the energy spectrum and the response function, the fitting time of the analytical approach is rather long. Although special log-likelihood terms are used to avoid multi-dimensional fitting, the multivariate fit is still too slow and the original CPU version fitter needs more than one week to converge. We developed a dedicated fitter for Borexino analysis based on the open source project `GooStats`[13]. The package is designed to be object oriented and interface based. Dedicated implementation class parses the config file exactly the same way as the original fitter and makes it zero-difficulty for the collaboration to use it. The fitting time is reduced because of two factors. On one hand the memory management and computation logic are optimized, and on the other hand the program benefits well from the power of parallel computing and the modern nVidia General Purpose Graphic Processing Unit (GPGPU) architecture. The overall speedup is more than 2000, and the fitting time is reduced from more than one week to around four minutes for the multivariate fit, and the fitting time without multivariate terms is reduced from around one hour to around seven seconds.

4. Detector response model related systematic uncertainties

Simplifications made in models and fixed parameters will introduce systematic biases. For the detector response model, we have estimated the corresponding systematic uncertainties with Monte Carlo methods. Random spectra are sampled according to the detector response model used, then corresponding distortions are applied to account for the uncertainty on the non-linearity, detector response non-uniformity, the pulse shape parameter modelling and theoretical ^{210}Bi shape. Magnitudes of distortions are determined by independent study with help of calibration data and literatures. Undistorted models are used to fit them and the expansions of the 1–σ range is considered as the systematic uncertainty.

5. Results and outlook

For illustration a random spectrum is fitted and shown in Figure 2. The expected uncertainties of neutrino interaction rate measurements based on fit of tens of thousands of randomly generated spectra are listed in Table 1, and great improvements are expected for all species. For CNO neutrino flux, we constrain *pep* neutrino flux through the well known theoretical *pp/pep* flux ratio. Assuming HZ metallicity

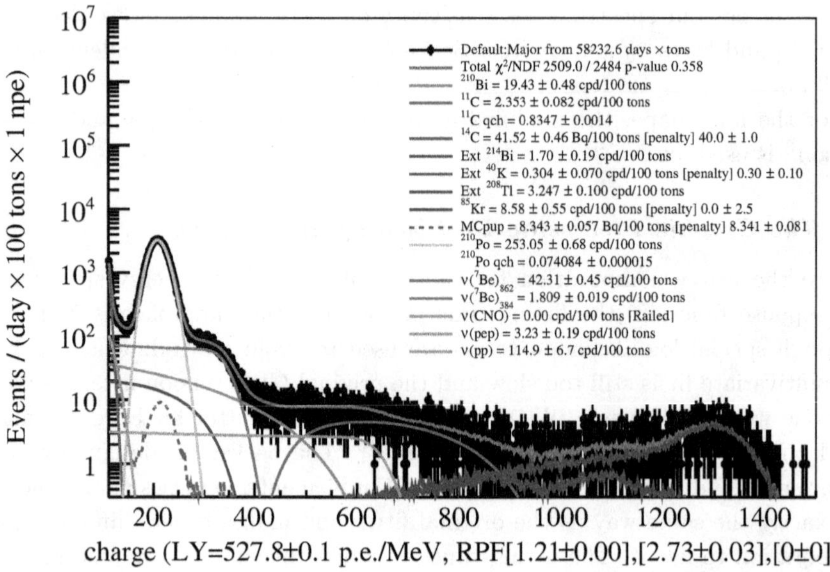

Fig. 2. A spectrum sampled according to the Monte Carlo based detector response and determined background rates and neutrino rates according to SSM and LMA-MSW model.

model, the expected median upper limit got is 8.6 cpd/100t without systematics and 10.1 cpd/100t considering detector modelling related systematic uncertainties, and it is slightly worse compared with Phase-I results where *pep* flux is fixed.

Table 1. Expected statistical uncertainty and detector response related systematics.

Source of uncertainty	*pp* −%	+%	^{7}Be −%	+%	*pep* −%	+%
Statistical (Expected)	-7.5	7.5	-2.3	2.3	-15	15
Detector response	-4.5	0.5	-1.0	0.2	-6.8	2.8

We are currently preparing publishing the results of analyses on the collected data. When analyzing data, unknown sources of systematic uncertainties is included by comparing results with different fitting strategies. Such results will be published in the near future.

Acknowledgments

The Borexino program is made possible by funding from INFN (Italy), NSF (USA), BMBF, DFG, HGF, and MPG (Germany), RFBR (Grants 16-02-01026 A, 15-02-02117 A, 16-29-13014 ofim, 17-02-00305 A) (Russia), and NCN (Grant No. UMO

2013/10/E/ST2/00180) (Poland). We acknowledge also the computing services of LNGS Computing and Network Service (Italy), of Jülich Supercomputing Centre at FZJ (Germany), and of ACK Cyfronet AGH Cracow (Poland).

X.F. Ding thanks Francesco Vissani and Le Li for polishing the manuscript, and thanks Eugenio Coccia, Francesco Vissani, and GSSI for the support which made this conference contribution possible.

References

1. G. Bellini *et al.*, Precision measurement of the Be7 solar neutrino interaction rate in Borexino, *Physical Review Letters* **107**, 1 (*Preprint* 1104.2150) (2011).
2. G. Bellini *et al.*, First evidence of pep solar neutrinos by direct detection in borexino, *Physical Review Letters* **108**, 1 (*Preprint* 1110.3230) (2012).
3. G. Bellini *et al.*, Neutrinos from the primary proton-proton fusion process in the Sun, *Nature* **512**, 383 (*Preprint* 1508.05379) (Aug 2014).
4. G. Bellini *et al.*, Measurement of the solar ^{8}B neutrino rate with a liquid scintillator target and 3 MeV energy threshold in the Borexino detector, *Physical Review D* **82**, p. 033006 (*Preprint* 0808.2868) (Aug 2010).
5. N. Vinyoles *et al.*, A new Generation of Standard Solar Models, *The Astrophysical Journal* **835**, p. 202 (*Preprint* 1611.09867) (2017).
6. F. Vissani, Joint analysis of Borexino and SNO solar neutrino data and reconstruction of the survival probability, 1 (*Preprint* 1709.05813) (Sep 2017).
7. The Borexino Collaboration *et al.*, Improved measurement of 8B solar neutrinos with 1.5 kt y of Borexino exposure, **016**, 1 (*Preprint* 1709.00756) (Sep 2017).
8. G. Bellini *et al.*, Final results of Borexino Phase-I on low-energy solar neutrino spectroscopy, *Physical Review D - Particles, Fields, Gravitation and Cosmology* **89**, 1 (*Preprint* 1308.0443) (2014).
9. M. Agostini *et al.*, The Monte Carlo simulation of the Borexino detector, *Astroparticle Physics* **888**, p. 012193 (*Preprint* 1704.02291) (Oct 2017).
10. G. Alimonti *et al.*, The Borexino detector at the Laboratori Nazionali del Gran Sasso, *Nuclear Instruments and Methods in Physics Research, Section A: Accelerators, Spectrometers, Detectors and Associated Equipment* **600**, 568 (*Preprint* 0806.2400) (2009).
11. V. Lagomarsino and G. Testera, A gateless charge integrator for Borexino energy measurement, *Nuclear Instruments and Methods in Physics Research Section A: Accelerators, Spectrometers, Detectors and Associated Equipment* **430**, 435 (Jul 1999).
12. M. Agostini and O. Schulz, M-STATS: Framework for frequentist statistical analysis (2015), https://github.com/mmatteo/m-stats.
13. X. Ding, GooStats: A Graphic Processing Unit based statistical analysis toolkit (2017), https://github.com/DingXuefeng/GooStats.

GooStats Based Analytical Multivariate Analysis in Borexino Phase-II Precision Measurement of Low Energy Solar Neutrino Flux

X. F. Ding[1,2,**], A. Vishneva[3], Ö. Penek[4,5], S. Marcocci[1,2,6,*]

on behalf of the Borexino Collaboration

[1]*Gran Sasso Science Institute, 67100 L'Aquila, Italy*
[2]*INFN Laboratori Nazionali del Gran Sasso, 67010 Assergi (AQ), Italy*
[3]*Joint Institute for Nuclear Research, 141980 Dubna, Russia*
[4]*IKP-2 Forschungzentrum Jülich, 52428 Jülich, Germany*
[5]*RWTH Aachen University, 52062 Aachen, Germany*
[6]*Dipartimento di Fisica, Università degli Studi e INFN, 16146 Genova, Italy*
**Now at: Fermi National Accelerator Laboratory, Batavia, IL 60510, USA*
***E-mail: xuefeng.ding@gssi.infn.it   https://dingxf.cn*

Borexino collaboration: M. Agostini, K. Altenmüller, S. Appel, V. Atroshchenko, Z. Bagdasarian, D. Basilico, G. Bellini, J. Benziger, D. Bick, G. Bonfini, D. Bravo, B. Caccianiga, F. Calaprice, A. Caminata, S. Caprioli, M. Carlini, P. Cavalcante, A. Chepurnov, K. Choi, L. Collica, D. D'Angelo, S. Davini, A. Derbin, X.F. Ding, A. Di Ludovico, L. Di Noto, I. Drachnev, K. Fomenko, A. Formozov, D. Franco, F. Gabriele, C. Galbiati, C. Ghiano, M. Giammarchi, A. Goretti, M. Gromov, D. Guffanti, C. Hagner, T. Houdy, E. Hungerford, Aldo Ianni, Andrea Ianni, A. Jany, D. Jeschke, V. Kobychev, D. Korablev, G. Korga, D. Kryn, M. Laubenstein, E. Litvinovich, F. Lombardi, P. Lombardi, L. Ludhova, G. Lukyanchenko, L. Lukyanchenko, I. Machulin, G. Manuzio, S. Marcocci, J. Martyn, E. Meroni, M. Meyer, L. Miramonti, M. Misiaszek, V. Muratova, B. Neumair, L. Oberauer, B. Opitz, V. Orekhov, F. Ortica, M. Pallavicini, L. Papp, Ö. Penek, N. Pilipenko, A. Pocar, A. Porcelli, G. Ranucci, A. Razeto, A. Re, M. Redchuk, A. Romani, R. Roncin, N. Rossi, S. Schönert, D. Semenov, M. Skorokhvatov, O. Smirnov, A. Sotnikov, L.F.F. Stokes, Y. Suvorov, R. Tartaglia, G. Testera, J. Thurn, M. Toropova, E. Unzhakov, A. Vishneva, R.B. Vogelaar, F. von Feilitzsch, H. Wang, S. Weinz, M. Wojcik, M. Wurm, Z. Yokley, O. Zaimidoroga, S. Zavatarelli, K. Zuber, G. Zuzel

Multivariate analysis technique is developed for Borexino Phase-II analysis to take the advantage of shape information of observables other than energies. When using the analytical detector response function, however, the fitting time is unacceptably slow. A new spectral analysis package is developed based on an open source project GooStats to overcome this challenge and it is found to be able to shorten the fitting time to a superior level compared to the original package. In this proceeding, developed algorithms in the package, its validation and benchmarking against the original package are presented.

Keywords: Borexino Phase-II; solar neutrino; analytical response function.

1. Introduction

In neutrino experiments, observables other than energies, such as reconstructed vertex positions and time to last muons, are used in cut-based analysis for suppressing backgrounds. Sometimes it is difficult to improve the signal/background ratio via this method, while easy through the shape analysis. For example, solar neutrino interaction rates are seasonally modulated[1] while rates of dissolved natural radioactive backgrounds decay according to certain half lives or are stable, and thus time stamps of events in principle can be used to improve the sensitivity of measurements of solar neutrino interaction rates. To combine the shape information of both the energy and other observables, we can build a multi-observable likelihood. However, this will lead to increase of fitting time by several orders of magnitude, and for the binned-likelihood analysis, Monte Carlo produced for building the one-dimensional probability density functions will be insufficient for building the multi-dimensional probability density functions and required production is unaffordable.

In Borexino Phase-I analysis, it is proposed to put the shape information of each extra observable into a likelihood term built on the one-dimensional distribution and model of that observable and then minimize the product of all pieces of likelihoods[2]. Correlation introduced by double-counting in both energy and other observable was removed by introducing a scaling factor. Although mathematical basis is not understood, the test statistics is observed to be unbiased and consistent when fitting spectra sampled according to the probability density function used and thus this method is reliable reliable. Although acceptably slow using the Monte Carlo approach which describes the detector response with GEANT4 based simulation codes, such method is too slow using the analytical approach[3] which describes detector response with an analytical response function calibrated in situ, and thus multivariate analysis is not performed in the analytical approach.

Six years has passed since the end of Phase-I. During these years Borexino has dedicated one year to a purification and a calibration campaign and another five years as Phase-II period to collect more data. In the meanwhile, computing architectures have taken revolutionary evolutions, and General Purpose Graphic Processing Unit (GPGPU) has become more and more popular in high energy physics community[4–6]. The heavy part of computation of fitting tasks is to evaluate the likelihood in each bin, which fulfils the single instruction multiple data (SIMD) model, thus the fitting task can benefit well from the power of parallel computing, so we have rewritten the original fitting package based on GooStats[7] which runs on nVidia GPGPUs and is accelerated from parallel computing techniques, and this new package becomes widely used in Borexino[8,9]. In this proceeding we introduce the algorithms developed based on GooStats for describing detector response analyticaly, improving sensitivity and suppressing backgrounds developed in this package and presente its performance.

2. Borexino experiment

Borexino is a multi-purpose neutrino experiment. The primary goal is to measure solar ^{7}Be neutrino flux from the pp chain precisely, which is achieved within 5%[10]. It is also demonstrated that with its unprecedented good radioactivity it can measure all the low energy branches from the pp chain, including the pp neutrinos[11], pep and CNO neutrinos[12], and ^{8}B neutrinos with lowest threshold as low as 3 MeV[13,14]. Borexino also gave the detection of geo-neutrinos[15] and provided various rare processes studies. Additionaly, it is planned to search for sterile neutrino using an artificial ^{144}Ce-^{144}Pr source via short baseline neutrino oscillation within the SOX project[16], which is schematically illustrated in Figure 1(b).

Borexino detector consists of the outer detector and the inner detector. The former is a water Cherenkov detector and it tags muons and events subsequent to the tagged muons within 300 ms are vetoed to reject backgrounds from decay of short lived cosmogenic isotopes. 3800 meters water equivalent rocks provide passive shield to further suppress muons and cosmogenic backgrounds. The later is a liquid scintillator detector. Liquid scintillator is contained in a stainless steel sphere (SSS) with photomultipliers (PMT's) installed on its inner surface. Its structure is shown schematically in Figure 1(a)[17]. Between the liquid scintillator and SSS, there is a buffer buffer region to absorb radioactivity from the PMT's and SSS. Two nylon vessels, one used to separate the buffer region and liquid scintillator and another one separating the buffer into another two shells, are installed to prevent radon from entering the liquid scintillator. Only the inner most region, the fiducial volume, is used in the analysis to further suppress γ's from the PMT's and SSS. The radius of the inner vessel, outer vessel and the stainless steel sphere are 4.25 m, 5.5 m and 6 m, respectively. The liquid scintillator used is an organic solution. Its solvent is PC (pseudocumene, 1,2,4-trimethylbenzene $C_6H_3(CH_3)_3$), and the solute is PPO (2,5-diphenyloxazole, $C_{15}H_{11}NO$) at a concentration of 1.5 g/L (0.17% by weight).

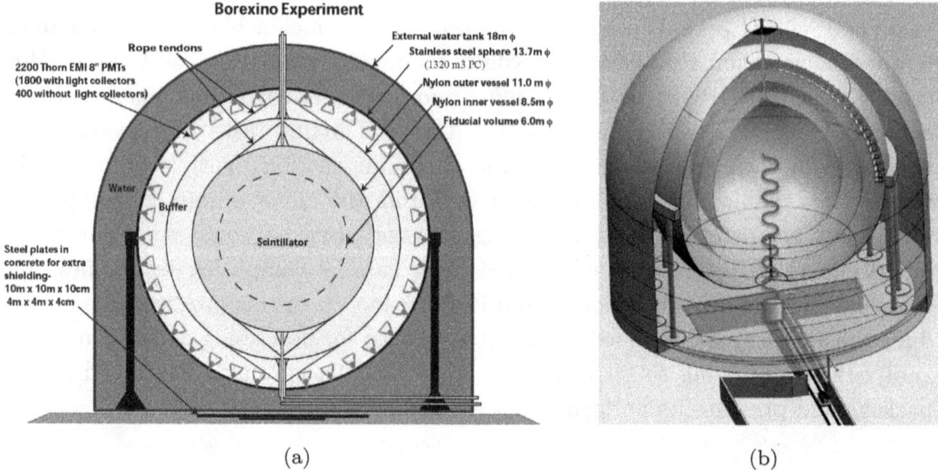

(a)   (b)

Fig. 1. Schematic drawing of (a) Borexino detector and (b) setup for CeSOX experiment.

Scintillation photons and Cherenkov photons are collected by PMT's. The number of photonelectrons collected is around 500 p.e./MeV/2000 PMTs. Charge and arrival time of each hit are produced by a specially designed gateless charge integrator and an amplifier-discrminator circuit[18], respectively, then the energy and vertex are reconstructed based on them. In Phase-II analysis, three energy estimators are used: The sum of charge of all collected hits (N_Q), the total number of detected photons, including multiple photons on the same PMT (N_h), and the total number of triggered PMTs in a fixed window interval of 230 (400) ns ($N_p^{dt_{1(2)}}$). For N_Q and $N_p^{dt_{1(2)}}$, analytical detector response functions are studied extensively and validated against Monte Carlo based detector emulation.

3. GooStats and Graphic Processing Unit based parallel computing

In our multivariate analysis, GooStats[7], an open-source package, serves as the middle layer between the minimization engine running on General Purpose Graphic Processing Unit (GPGPU) and the algorithms implemented for Borexino analysis. It also provides services such as producing plots with data histograms and best-fit spectra for inspections and producing compact files containing thousands of best-fit parameters to validate the implemented algorithms, to study biases of models, and to study statistical sensitivities and systematic uncertainties of measurements. The package loads the input files and configurations through interfaces, and thus it's easy to keep the format of configurations, so when Borexino collaborators move from using the original fitting tool to this tool, no extra learning is required.

The structure of GooStats is illustrated in Figure 2. AnalysisManager class is responsible for controlling the workflow. InputManager class collects all necessary inputs, such as the data spectra, the probability density functions of all species, the configuration files, then composes the product of all likelihood terms and provides it to the minimization engine GooFit[19], which communicates with MINUIT algorithm and GPGPU's.

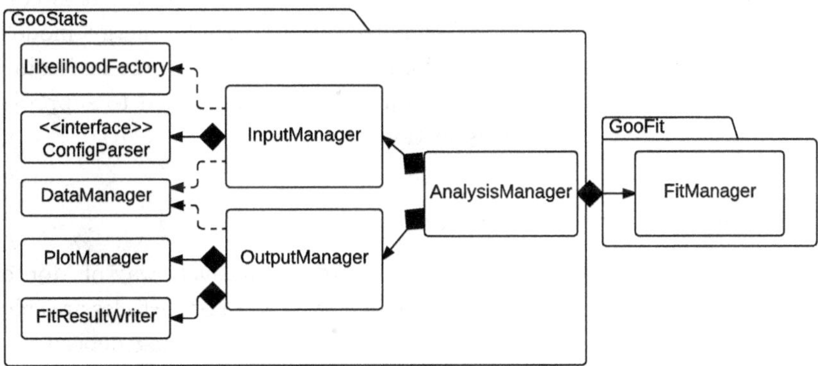

Fig. 2.   Unified Modeling Language graph for the package GooStats and GooFit.

An example of figures produced by PlotManager in GooStats is shown in Figure 3. Detailed legends make it self-explaining. Color, line style and line shapes of each species can be easily set through configuration files.

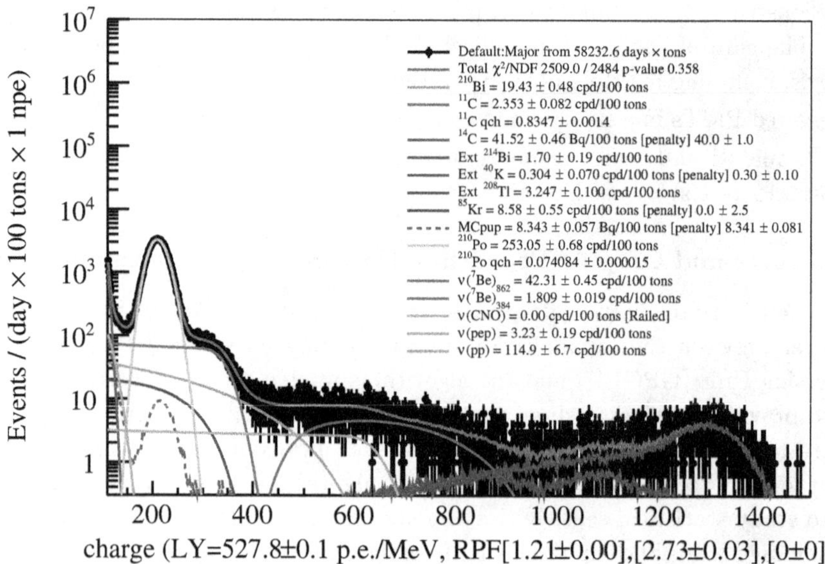

Fig. 3. An example output with Monte Carlo dataset. The dataset is plotted as the black dot with error bars. Components are drawn labeled by colours and line styles with names indicated in the legend. The best fit and error of the rates of components are in the legends. The exposure, goodness of the fit is also in the legend. On the x-axis, you can find the optimized detector response parameters, such as the light yield (color online).

4. Developed algorithms for Borexino multivariate analysis

Various algorithms are developed to support Borexino's multivaraite analysis. The analytical response function describes the detector response. It is convolved with the particle kinetic energy spectra to get the visible energy models used in the maximum-likelihood method. Simultaneous fitting, three-fold-coincidence technique, complementary fit and multivariate fit are developed to suppress backgrounds. They are described in this section.

4.1. *The analytical response function*

Analytical response functions is developed as an alternatively way of Monte Carlo based detector emulation. To use it, kinetic-energy spectra of particles are calculated priorly, then visible energy spectra are predicted by convolving deposited energy spectra with the response function:

$$f_{\mathrm{vis}}(N_x) = \int f_{\mathrm{k}}(E) \cdot R_x(N_x, E; p_0, p_1, ...) \, \mathrm{d}E \qquad (1)$$

where N_x is the chosen energy estimators, $f_{\mathrm{vis}}(N_x)$ is the spectrum of N_x, $f_{\mathrm{k}}(E)$ is the kinetic-energy spectrum, R_x is the corresponding response function, and p_0, p_1, ... are its parameters[3].

To account for pile-up effects, such as the ^{14}C–^{14}C pile-up effect that for a small fraction of events due to the high rate of ^{14}C, two ^{14}C event may frequently fall into the same DAQ window and mimic a pp solar neutrino interaction event, additional convolution is applied:

$$f_{\text{w/ pile-up}}(N_{x0}) = \int f_{\text{w/o pile-up}}(N_x) \cdot g_{\text{random trigger}}(N_{x0} - N_x) \, \mathrm{d}N_x \qquad (2)$$

where $g_{\text{random trigger}}$ is the spectrum of N_x variable from random trigger.

4.2. *Physics based background suppression techniques*

For N_Q, correction functions are built based on the total geometrical acceptance solid angle and valiadated on Monte Carlo and calibration sources, or built based on ^{210}Po α-decay events inside the liquid scintillator tagged by a pulse shape discriminator (PSD) based on the multi-layer perception (MLP) method. After applying such corrections, the resolution of N_Q is improved, especially for events with energy around or higher than the ^{7}Be neutrino shoulder, or around 0.7 MeV.

Three-fold-coincidence (TFC) technique[12] is used to suppress the cosmogenic ^{11}C positron decay backgrounds, which sit on the critical region of pep neutrinos. The idea of TFC is that events associated with a muon and a subsequent neutron are believed more likely to contain the ^{11}C decay, and thus should be vetoed. Veto criteria is optimized to maximize the exposure and to suppress as much as possible ^{11}C according to the sensitivity of pep neutrino interaction rate as the merit.

A so-called "complementary fit" method is used combined with the TFC technique to improve the sensitivity of other species. Instead of abandoning TFC-vetoed events, a likelihood term is built on them and is multiplied to the one built on events after vetos, and the product is minimized.

$$\mathcal{L} = \mathcal{L}^{\text{TFC vetoed}} \times \mathcal{L}^{\text{TFC tagged}} \qquad (3)$$

In this way we saved the exposure for measuring pp and ^{7}Be neutrinos interaction rates, which is not affected by the ^{11}C decay backgrounds.

The multivariate fitting is used to further disentangle the signals from the backgrounds. Radial distribution is used to suppress the impact of external gammas. The neutrino elastic scattering signals and cosmogenic ^{11}C position decay backgrounds are uniform distributed, while the flux of gammas penetrating the buffer into the fiducial volume originated from the radioactivities on the PMT glasses and stainless stell sphere reduces exponentially along the radial axis toward the center, thus can be distinguished through different radial distribution. For ^{210}Bi, dedicated studies are performed and no strong non-uniformity is observed. The involved energy range is carefully chosen to reduce the systematics from the light yield as well as to avoid ^{210}Po, which is strongly non-uniform. e^+/e^- pulse shape discriminator PS-$\mathcal{L}_{\text{PR}}$, which is the minimized likelihood of the position reconstruction[3], is used to further suppress the ^{11}C. Its distributions for position, electron are built based on

Monte Carlo and validated against data. Distributions for electrons are validated against the ^{214}Bi events selected through fast coincidence and those for positrons are validated against special TFC selected ^{11}C events. The observables radial and pulse-shape are included by mutiplying additional terms of likelihood[2].

$$\mathcal{L}^{\text{multivariaite}} = \mathcal{L}^{\text{TFC vetoed}} \times \mathcal{L}^{\text{TFC tagged}} \times \prod_i \mathcal{L}_i^{\text{radial}} \times \prod_j \mathcal{L}_j^{\text{pulse-shape}} \tag{4}$$

where i and j indicate different pieces of likelihoods selected based on the energy of the events.

5. Validations and benchmarking

The package is validated against the original CPU version of the fitter. Fits are performed with two fitters under same configuration on the same dataset. The difference of the minimized NLL is less than 10^{-6}, and the difference on rate is less than $10^{-3} \sim 10^{-5}$, which is due to the fact that the minimization stopped as soon as the likelihood change between consequent trials reaches the pre-set precision limit.

We conduct the benchmarking with a AMD Opteron(tm) Processor 6320 and nVidia K40. The fitting time is reduced by 99.86% without multivariate terms and reduced by 99.95% with them. On one hand the memory management and computation logic are optimized, and on the other hand the program benefits well from the power of parallel computing and the modern nVidia GPGPU architecture.

6. Outlook

The analysis package based on `GooStats` has been validated and has shown powerful computing abilities. Search for neutrino Non Standard Interactions, study on global fitting etc. are under discussions and required new features in this package might be brought up in the near future.

Acknowledgments

The Borexino program is made possible by funding from INFN (Italy), NSF (USA), BMBF, DFG, HGF, and MPG (Germany), RFBR (Grants 16-02-01026 A, 15-02-02117 A, 16-29-13014 ofim, 17-02-00305 A) (Russia), and NCN (Grant No. UMO 2013/10/E/ST2/00180) (Poland). We acknowledge also the computing services of INFN-CNAF data centre (Bologna) and LNGS Computing and Network Service (Italy), of Jülich Supercomputing Centre at FZJ (Germany), and of ACK Cyfronet AGH Cracow (Poland).

X.F. Ding thanks Le Li for polishing this proceeding, thanks Eugenio Coccia, Francesco Vissani, and the GSSI for the support which made this conference contribution possible.

References

1. M. Agostini *et al.*, Seasonal modulation of the 7 Be solar neutrino rate in Borexino, *Astroparticle Physics* **92**, 21 (*Preprint* arXiv:1701.07970v1) (Jun 2017).
2. S. Davini, Measurement of the pep and CNO solar neutrino interaction rates in Borexino-I, PhD thesis, University of Genova (Springer International Publishing, Cham, 2013).
3. G. Bellini *et al.*, Final results of Borexino Phase-I on low-energy solar neutrino spectroscopy, *Physical Review D — Particles, Fields, Gravitation and Cosmology* **89**, 1 (*Preprint* 1308.0443) (2014).
4. M. A. Clark, R. Babich, K. Barros, R. C. Brower and C. Rebbi, Solving lattice QCD systems of equations using mixed precision solvers on GPUs, *Computer Physics Communications* **181**, 1517 (*Preprint* 0911.3191) (2010).
5. D. Emeliyanov and J. Howard, GPU-based tracking algorithms for the ATLAS high-level trigger, *Journal of Physics: Conference Series* **396** (2012).
6. S. Gorbunov *et al.*, ALICE HLT high speed tracking on GPU, *IEEE Transactions on Nuclear Science* **58**, 1845 (2011).
7. X. Ding, Non-linearity in the analytical model: The source of non-linearity, 1 (2017).
8. M. Agostini *et al.*, First Simultaneous Precision Spectroscopy of pp, ^{7}Be, and pep Solar Neutrinos with Borexino Phase-II, 1 (*Preprint* 1707.09279) (Jul 2017).
9. M. Agostini *et al.*, Limiting neutrino magnetic moments with Borexino Phase-II solar neutrino data, 1 (*Preprint* 1707.09355) (Jul 2017).
10. G. Bellini *et al.*, Precision measurement of the Be7 solar neutrino interaction rate in Borexino, *Physical Review Letters* **107**, 1 (*Preprint* 1104.2150) (2011).
11. G. Bellini *et al.*, Neutrinos from the primary proton-proton fusion process in the Sun, *Nature* **512**, 383 (*Preprint* 1508.05379) (Aug 2014).
12. G. Bellini *et al.*, First evidence of pep solar neutrinos by direct detection in borexino, *Physical Review Letters* **108**, 1 (*Preprint* 1110.3230) (2012).
13. G. Bellini *et al.*, Measurement of the solar ^{8}B neutrino rate with a liquid scintillator target and 3 MeV energy threshold in the Borexino detector, *Physical Review D* **82**, p. 033006 (*Preprint* 0808.2868) (Aug 2010).
14. The Borexino Collaboration *et al.*, Improved measurement of 8B solar neutrinos with 1.5 kt y of Borexino exposure, **016**, 1 (*Preprint* 1709.00756) (Sep 2017).
15. M. Agostini *et al.*, Spectroscopy of geoneutrinos from 2056 days of Borexino data, *Physical Review D* **92**, p. 031101 (*Preprint* 1506.04610) (Aug 2015).
16. G. Bellini *et al.*, SOX: Short distance neutrino Oscillations with BoreXino, *Journal of High Energy Physics* **2013**, p. 38 (*Preprint* 1304.7721) (Aug 2013).
17. G. Alimonti *et al.*, The Borexino detector at the Laboratori Nazionali del Gran Sasso, *Nuclear Instruments and Methods in Physics Research, Section A: Accelerators, Spectrometers, Detectors and Associated Equipment* **600**, 568 (*Preprint* 0806.2400) (2009).
18. V. Lagomarsino and G. Testera, A gateless charge integrator for Borexino energy measurement, *Nuclear Instruments and Methods in Physics Research Section A: Accelerators, Spectrometers, Detectors and Associated Equipment* **430**, 435 (Jul 1999).
19. A. David *et al.*, GooFit: Version 2.0 (2017), DOI: 10.5281/zenodo.804881.

Search for New Physics via Baryon EDM at LHC

L. Henry[1], D. Marangotto[2], A. Merli[2,3], N. Neri[2,3], J. Ruiz[1], F. Martinez Vidal[1]

[1]*IFIC, Universitat de València-CSIC, Valencia, Spain*
[2]*INFN Sezione di Milano and Università di Milano, Milan, Italy*
[3]*CERN, Geneva, Switzerland*

Permanent electric dipole moments (EDMs) of fundamental particles provide powerful probes for physics beyond the Standard Model. We propose to search for the EDM of strange and charm baryons at LHC, extending the ongoing experimental program on the neutron, muon, atoms, molecules and light nuclei. The EDM of strange Λ baryons, selected from weak decays of charm baryons produced in pp collisions at LHC, can be determined by studying the spin precession in the magnetic field of the detector tracking system. A test of CPT symmetry can be performed by measuring the magnetic dipole moment of Λ and $\overline{\Lambda}$ baryons. For short-lived Λ_c^+ and Ξ_c^+ baryons, to be produced in a fixed-target experiment using the 7 TeV LHC beam and channeled in a bent crystal, the spin precession is induced by the intense electromagnetic field between crystal atomic planes. The experimental layout based on the LHCb detector and the expected sensitivities in the coming years are discussed.

Keywords: Baryons (including antiparticles); electric and magnetic moments.

1. Introduction

The magnetic dipole moment (MDM) and the electric dipole moment (EDM) are static properties of particles that determine the spin motion in an external electromagnetic field, as described by the T-BMT equation[1-3].

The EDM is the only static property of a particle that requires the violation of parity (P) and time reversal (T) symmetries and thus, relying on CPT invariance, the violation of CP symmetry. The amount of CP violation in the weak interactions of quarks is not sufficient to explain the observed imbalance between matter and antimatter in the Universe. CP-violation in strong interactions is strongly bounded by the experimental limit on the neutron EDM[4]. In the Standard Model (SM), contributions to the EDM of baryons are highly suppressed but can be largely enhanced in some of its extensions. Hence, the experimental searches for the EDM of fundamental particles provide powerful probes for physics beyond the SM.

Since EDM searches started in the fifties[5,6], there has been an intense experimental program, leading to limits on the EDM of leptons[7-9], neutron[4], heavy atoms[10], proton (indirect from ^{199}Hg)[11], and Λ baryon[12]. New experiments are ongoing and others are planned, including those based on storage rings for muon[13,14], proton and light nuclei[15-17]. Recently we proposed to improve the limit on strange baryons and extend it to charm and bottom baryons[18,19].

EDM searches of fundamental particles rely on the measurement of the spin precession angle induced by the interaction with the electromagnetic field. For unstable particles this is challenging since the precession has to take place before the decay. A solution to this problem requires large samples of high energy polarized particles traversing an intense electromagnetic field.

Here we reviewed the unique possibility to search for the EDM of the strange Λ baryon and of the charmed baryons at LHC. Using the experimental upper limit of the neutron EDM, the absolute value of the Λ EDM is predicted to be $< 4.4 \times 10^{-26}$ $e\,\mathrm{cm}$[20-23], while the indirect constraints on the charm EDM are weaker, $\lesssim 4.4 \times 10^{-17}$ $e\,\mathrm{cm}$[24]. Any experimental observation of an EDM would indicate a new source of CP violation from physics beyond the SM. The EDM of the long-lived Λ baryon was measured to be $< 1.5 \times 10^{-16}$ $e\,\mathrm{cm}$ (95% C.L.) in a fixed-target experiment at Fermilab[12]. No experimental measurements exist for short-lived charm baryons since negligibly small spin precession would be induced by magnetic fields used in current particle detectors.

2. Experimental setup

The magnetic and electric dipole moment of a spin-1/2 particle is given (in Gaussian units) by $\boldsymbol{\mu} = g\mu_B \mathbf{s}/2$ and $\boldsymbol{\delta} = d\mu_B \mathbf{s}/2$, respectively, where $\mathbf{s}$ is the spin-polarization vector[a] and $\mu_B = e\hbar/(2mc)$ is the particle magneton, with m its mass. The g and d dimensionless factors are also referred to as the gyromagnetic and gyroelectric ratios. The experimental setup to measure the change of the spin direction in an elextromagnetic field relies on three main elements:

(i) a source of polarized particles whose direction and polarization degree are known;
(ii) an intense electromagnetic field able to induce a sizable spin precession angle during the lifetime of the particle;
(iii) the detector to measure the final polarization vector by analysing the angular distribution of the particle decays.

2.1. Λ and $\overline{\Lambda}$ case

Weak decays of heavy baryons (charm and beauty), mostly produced in the forward/backward directions at LHC, can induce large longitudinal polarization due to parity violation. For example, the decay of unpolarized Λ_c^+ baryons to the $\Lambda\pi^+$ final state[25], produces Λ baryons with longitudinal polarization $\approx -90\%$. Another example is the $\Lambda_b^0 \to \Lambda J/\psi$ decay where Λ baryons are produced almost 100% longitudinally polarized[26,27].

The spin-polarization vector $\mathbf{s}$ of an ensemble of Λ baryons can be analysed through the angular distribution of the $\Lambda \to p\pi^-$ decay[28,29],

$$\frac{dN}{d\Omega'} \propto 1 + \alpha \mathbf{s} \cdot \hat{\mathbf{k}} \,, \tag{1}$$

---

[a]The spin-polarization vector is defined such as $\mathbf{s} = 2\langle \mathbf{S}\rangle/\hbar$, where $\mathbf{S}$ is the spin operator.

where $\alpha = 0.642 \pm 0.013$[30] is the decay asymmetry parameter. The *CP* invariance in the Λ decay implies $\alpha = -\overline{\alpha}$, where $\overline{\alpha}$ is the decay parameter of the charge-conjugate decay. The unit vector $\hat{\mathbf{k}} = (\sin\theta'\cos\phi', \sin\theta'\sin\phi', \cos\theta')$ indicates the momentum direction of the proton in the Λ helicity frame, with $\Omega' = (\theta', \phi')$ the corresponding solid angle.

For the particular case of Λ flying along the z axis in the laboratory frame, an initial longitudinal polarization s_0, *i.e.* $\mathbf{s}_0 = (0, 0, s_0)$, and $\mathbf{B} = (0, B_y, 0)$, the solution of the T-BMT equation is[18]

$$\mathbf{s} = \begin{cases} s_x = -s_0 \sin\Phi \\ s_y = -s_0 \dfrac{d\beta}{g}\sin\Phi \\ s_z = s_0 \cos\Phi \end{cases} \tag{2}$$

where $\Phi = \frac{D_y\mu_B}{\beta\hbar c}\sqrt{d^2\beta^2 + g^2} \approx \frac{gD_y\mu_B}{\beta\hbar c}$ with $D_y \equiv D_y(l) = \int_0^l B_y dl'$ the integrated magnetic field along the Λ flight path. The polarization vector precesses in the xz plane, normal to the magnetic field, with the precession angle Φ proportional to the gyromagnetic factor of the particle. The presence of an EDM introduces a non-zero s_y component perpendicular to the precession plane of the MDM, otherwise not present. At LHCb, with a tracking dipole magnet providing an integrated field $D_y \approx \pm 4\,\mathrm{T\,m}$[31], the maximum precession angle for particles traversing the entire magnetic field region yields $\Phi_{\max} \approx \pm\pi/4$, and allows to achieve about 70% of the maximum s_y component. Moreover, a test of *CPT* symmetry can be performed by comparing the g and $-\overline{g}$ factors for Λ and $\overline{\Lambda}$ baryons, respectively, which precess in opposite directions as g and d change sign from particle to antiparticle.

2.2. *Charm baryon case*

The Λ_c^+ baryon EDM can be extracted by measuring the precession of the polarization vector of channeled particles in a bent crystal. There, a positively-charged particle channeled between atomic planes moves along a curved path under the action of the intense electric field between crystal planes. In the instantaneous rest frame of the particle the electromagnetic field causes the spin rotation. The signature of the EDM is a polarization component perpendicular to the initial baryon momentum and polarization vector, otherwise not present, similarly to the case of the Λ baryon.

The phenomenon of spin precession of positively-charged particles channeled in a bent crystal was firstly observed by the E761 collaboration that measured the MDM of the strange Σ^+ baryon[32]. The feasibility of the measurement at LHC energies offers clear advantages with respect to lower beam energies since the estimated number of channeled charm baryons is proportional to $\gamma^{3/2}$, where γ is the Lorentz factor of the particles[33].

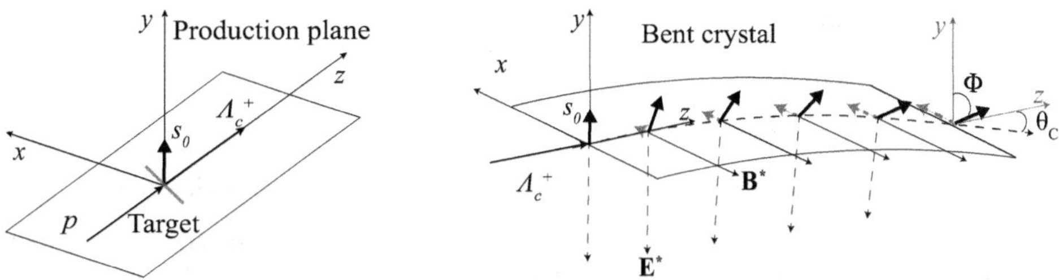

Fig. 1. (Left) Production plane of the Λ_c^+ baryon defined by the proton and the Λ_c^+ momenta. The initial polarization vector s_0 is perpendicular to the production plane, along the y axis, due to parity conservation in strong interactions[34]. (Right) Deflection of the baryon trajectory and spin precession in the yz and xy plane induced by the MDM and the EDM, respectively. The red (dashed) arrows indicate the (magnified) s_x spin component proportional to the particle EDM. Φ is the MDM precession angle and θ_C is the crystal bending angle. $\mathbf{E}^*$ and $\mathbf{B}^*$ are the intense electromagnetic field in the particle rest frame[35,36] which induce spin precession (color online).

In the limit of large boost with Lorentz factor $\gamma \gg 1$, the precession angle Φ, shown in Fig. 1, induced by the MDM is[37]

$$\Phi \approx \frac{g-2}{2}\gamma\theta_C, \tag{3}$$

where g is the gyromagnetic factor, $\theta_C = L/\rho_0$ is the crystal bending angle, L is the circular arc of the crystal and ρ_0 the curvature radius.

In presence of a non-zero EDM, the spin precession is no longer confined to the yz plane, originating a s_x component proportional to the particle EDM represented by the red (dashed) arrows in (Right) Fig. 1. The polarization vector, after channeling through the crystal is[18]

$$\mathbf{s} = \begin{cases} s_x \approx s_0 \dfrac{d}{g-2}(\cos\Phi - 1) \\[2mm] s_y \approx s_0 \cos\Phi \\[2mm] s_z \approx s_0 \sin\Phi \end{cases}, \tag{4}$$

where Φ is given by Eq. (3). The MDM and EDM information can be extracted from the measurement of the spin polarization of channeled baryons at the exit of the crystal, via the study of the angular distribution of final state particles. For Λ_c^+ decaying to two-body final states such as $f = \Delta^{++}K^-, pK^{*0}, \Delta(1520)\pi^+$ and $\Lambda\pi^-$, the angular distribution is described by Eq. (1). A Dalitz plot analysis would provide the ultimate sensitivity to the EDM measurement.

The initial polarization s_0 would require in principle the measurement of the angular distribution for unchanneled baryons. In practice this is not required since the measurement of the three components of the final polarization vector for channeled baryons allows a simultaneous determination of g, d and s_0, up to discrete ambiguities. These can be solved exploiting the dependence of the angular distribution with the Λ_c^+ boost γ, as discussed in Ref. [19].

3. Sensitivity studies

3.1. Λ and $\overline{\Lambda}$ case

The number of Λ particles produced can be estimated as

$$N_\Lambda = 2\mathcal{L}\sigma_{q\bar{q}}f(q \to H)\mathcal{B}(H \to \Lambda X')\mathcal{B}(\Lambda \to p\pi^-)\mathcal{B}(X' \to \text{charged}), \qquad (5)$$

where $\mathcal{L}$ is the total integrated luminosity, $\sigma_{q\bar{q}}$ ($q = c, b$) are the heavy quark production cross sections from pp collisions at $\sqrt{s} = 13\,\text{TeV}$ [38–41], and f is the fragmentation fraction into the heavy baryon H [42–45]. In Table 1 the dominant

Table 1. Dominant Λ production mechanisms from heavy baryon decays and estimated yields produced per fb^{-1} at $\sqrt{s} = 13\,\text{TeV}$, shown separately for SL and LL topologies. The Λ baryons from Ξ^- decays, produced promptly in the pp collisions, are given in terms of the unmeasured production cross section.

SL events	N_Λ/fb^{-1} ($\times 10^{10}$)	LL events, $\Xi^- \to \Lambda\pi^-$	N_Λ/fb^{-1} ($\times 10^{10}$)
$\Xi_c^0 \to \Lambda K^-\pi^+$	7.7	$\Xi_c^0 \to \Xi^-\pi^+\pi^+\pi^-$	23.6
$\Lambda_c^+ \to \Lambda\pi^+\pi^+\pi^-$	3.3	$\Xi_c^0 \to \Xi^-\pi^+$	7.1
$\Xi_c^+ \to \Lambda K^-\pi^+\pi^+$	2.0	$\Xi_c^+ \to \Xi^-\pi^+\pi^+$	6.1
$\Lambda_c^+ \to \Lambda\pi^+$	1.3	$\Lambda_c^+ \to \Xi^-K^+\pi^+$	0.6
$\Xi_c^0 \to \Lambda K^+K^-$ (no ϕ)	0.2	$\Xi_c^0 \to \Xi^-K^+$	0.2
$\Xi_c^0 \to \Lambda\phi(K^+K^-)$	0.1	Prompt Ξ^-	$0.13 \times \sigma_{pp\to\Xi^-}$ [μb]

production channels and the estimated yields are summarised. Only the decays where it is experimentally possible to determine the production and decay vertex of the Λ are considered. Overall, there are about 1.5×10^{11} Λ baryons per fb^{-1} produced directly from heavy baryon decays (referred hereafter as short-lived, or SL events), and 3.8×10^{11} from charm baryons decaying through an intermediate Ξ^- particle (long-lived, or LL events). The yield of Λ baryons experimentally available can then be evaluated as $N_\Lambda^{\text{reco}} = \epsilon_{\text{geo}}\epsilon_{\text{trigger}}\epsilon_{\text{reco}}N_\Lambda$, where ϵ_{geo}, $\epsilon_{\text{trigger}}$ and ϵ_{reco} are the geometric, trigger and reconstruction efficiencies of the detector system. The geometric efficiency for SL topology has been estimated to be about 16% using a Monte Carlo simulation of pp collisions at $\sqrt{s} = 13\,\text{TeV}$ and the decay of heavy hadrons.

To assess the EDM sensitivity, pseudo-experiments have been generated using a simplified detector geometry that includes an approximate LHCb magnetic field mapping [31,46]. Λ baryons decaying towards the end of the magnet provide most of the sensitivity to the EDM and MDM, since a sizeable spin precession could happen. The decay angular distribution and spin dynamics have been simulated using Eq. (1) and the general solution as a function of the Λ flight length [18], respectively. For this study the initial polarization vector $\mathbf{s}_0 = (0, 0, s_0)$, with s_0 varying between 20% and 100%, and factors $g = -1.458$ [30] and $d = 0$, were used. Each generated sample was fitted using an unbinned maximum likelihood method with d, g and

s_0 as free parameters. The d-factor uncertainty scales with the number of events $N_\Lambda^{\rm reco}$ and the initial longitudinal polarization s_0 as $\sigma_d \propto 1/(s_0\sqrt{N_\Lambda^{\rm reco}})$. The sensitivity saturates at large values of s_0, as shown in (Left) Fig. 2, and it partially relaxes the requirements on the initial polarizations. Similarly, (Right) Fig. 2 shows the expected sensitivity on the EDM as a function of the integrated luminosity, summing together SL and LL events, assuming global trigger and reconstruction efficiency $\epsilon_{\rm trigger}\epsilon_{\rm reco}$ of 1% (improved LHCb software-based trigger and tracking for the upgrade detector[47,48]) and 0.2% (current detector[31]), where the efficiency estimates are based on a educated guess. An equivalent sensitivity is obtained for the gyromagnetic factor. Therefore, with 8 fb^{-1} a sensitivity $\sigma_d \approx 1.5 \times 10^{-3}$ could be achieved (current detector), to be compared to the present limit, 1.7×10^{-2} [12]. With 50 fb^{-1} (upgraded detector) the sensitivity on the gyroelectric factor can reach $\approx 3 \times 10^{-4}$.

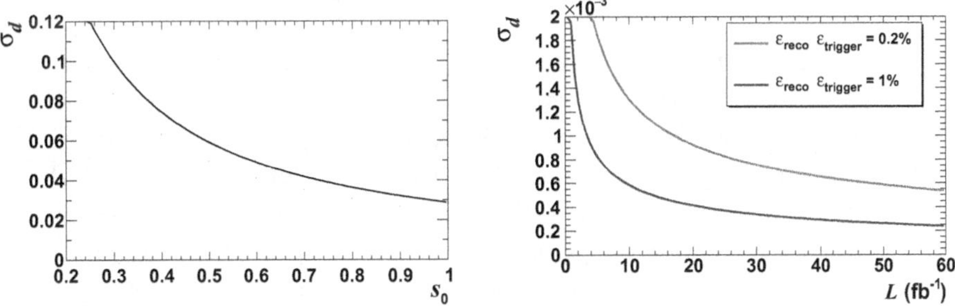

Fig. 2. (Left) Dependence of the Λ gyroelectric factor uncertainty with the initial polarization for $N_\Lambda^{\rm reco} = 10^6$ events, and (Right) as a function of the integrated luminosity assuming reconstruction efficiency of 0.2% and 1%.

3.2. *Charm baryon case*

We propose to search for charm baryon EDMs in a dedicated fixed-target experiment at the LHC to be installed in front of the LHCb detector. The target should be attached to the crystal to maximize the yield of short-lived charm baryons to be channeled. The rate of Λ_c^+ baryons produced with 7 TeV protons on a fixed target can be estimated as

$$\frac{dN_{\Lambda_c^+}}{dt} = \frac{F}{A}\sigma(pp \to \Lambda_c^+ X)N_T, \tag{6}$$

where F is the proton rate, A the beam transverse area, N_T the number of target nucleons, and $\sigma(pp \to \Lambda_c^+ X)$ is the cross-section for Λ_c^+ production in pp interactions at $\sqrt{s} = 114.6$ GeV center-of-mass energy. The number of target nucleons is $N_T = N_A\rho A T A_N/A_T$, where N_A is the Avogadro number, ρ (T) is the target density (thickness), and A_T (A_N) is the atomic mass (atomic mass number). For our estimates we consider a target of tungsten thick $T = 0.5$ cm with den-

390

sity $\rho = 19.25\,\mathrm{g/cm}$. The rate of Λ_c^+ particles channeled in the bent crystal and reconstructed in the LHCb detector is estimated as

$$\frac{dN_{\Lambda_c^+}^{\mathrm{reco}}}{dt} = \frac{dN_{\Lambda_c^+}}{dt}\mathcal{B}(\Lambda_c^+ \to f)\varepsilon_{\mathrm{CH}}\varepsilon_{\mathrm{DF}}(\Lambda_c^+)\varepsilon_{\mathrm{det}}, \tag{7}$$

where $\mathcal{B}(\Lambda_c^+ \to f)$ is the branching fraction of Λ_c^+ decaying to f, $\varepsilon_{\mathrm{CH}}$ is the efficiency of channeling Λ_c^+ inside the crystal, $\varepsilon_{\mathrm{DF}}\,(\Lambda_c^+)$ is the fraction of Λ_c^+ decaying after the crystal and $\varepsilon_{\mathrm{det}}$ is the efficiency to reconstruct the decays. A 6.5 TeV proton beam was extracted from the LHC beam halo by channeling protons in bent crystals[49]. A beam with intensity of 5×10^8 proton/s, to be directed on a fixed target, is attainable with this technique[50].

The Λ_c^+ cross section is estimated from the total charm production cross section[51], rescaled to $\sqrt{s} = 114.6\,\mathrm{GeV}$ assuming a linear dependence on $\sqrt{s}$, and Λ_c^+ fragmentation function[44] to be $\sigma_{\Lambda_c^+} \approx 18.2\,\mu\mathrm{b}$, compatible with theoretical predictions[52].

The channeling efficiency in silicon crystals, including both channeling angular acceptance and dechanneling effects, is estimated to be $\varepsilon_{\mathrm{CH}} \approx 10^{-3}$[53], while the fraction of Λ_c^+ baryons decaying after the crystal is $\varepsilon_{\mathrm{DF}}(\Lambda_c^+) \approx 19\%$, for $\gamma = 1000$ and 10 cm crystal length. The geometrical acceptance for $\Lambda_c^+ \to pK^-\pi^+$ decaying into the LHCb detector is $\varepsilon_{\mathrm{geo}} \approx 25\%$ according to simulation studies. The LHCb software-based trigger for the upgrade detector[47] is expected to have efficiency for charm hadrons comparable to the current high level trigger[31], $i.e.$ $\varepsilon_{\mathrm{trigger}} \approx 80\%$. The tracking efficiency is estimated to be 70% per track, leading to an efficiency $\varepsilon_{\mathrm{track}} \approx 34\%$ for a Λ_c^+ decay with three charged particles. The detector reconstruction efficiency, $\varepsilon_{\mathrm{det}} = \varepsilon_{\mathrm{geo}}\varepsilon_{\mathrm{trigger}}\varepsilon_{\mathrm{track}}$, is estimated to be $\varepsilon_{\mathrm{det}}(pK^-\pi^+) \approx 5.4 \times 10^{-2}$ for $\Lambda_c^+ \to pK^-\pi^+$ decays.

Few Λ_c^+ decay asymmetry parameters α_f are known. At present, they can be computed from existing $\Lambda_c^+ \to pK^-\pi^+$ amplitude analysis results[54] yielding $\alpha_{\Delta^{++}K^-} = -0.67 \pm 0.30$ for the $\Lambda_c^+ \to \Delta^{++}K^-$ decay[18].

For the sensitivity studies we assume $s_0 = 0.6$ and $(g-2)/2 = 0.3$, according to experimental results and available theoretical predictions, respectively, quoted in Ref.[55]. The d and $g-2$ values and errors can be derived from Eq. (4). The estimate assumes negligibly small uncertainties on θ_C, γ.

Given the estimated quantities we obtain $dN_{\Lambda_c^+}^{reco}/dt \approx 5.9 \times 10^{-3}\ \mathrm{s}^{-1} = 21.2\ \mathrm{h}^{-1}$ for $\Lambda_c^+ \to \Delta^{++}K^-$. A data taking of 1 month will be sufficient to reach a sensitivity of $\sigma_\delta = 1.3 \times 10^{-17}$ on the Λ_c^+ EDM. Therefore, a measurement of Λ_c^+ EDM is feasible in Λ_c^+ quasi two-body decays at LHCb.

The dependence of the sensitivity to Λ_c^+ EDM and MDM as a function of the number of incident protons on the target is shown in Fig. 3. The same technique could be applied to any other heavy charged baryon, for instance containing b quark. The production rate is lower than Λ_c^+ and the estimates have been studied and discussed in Ref.[19].

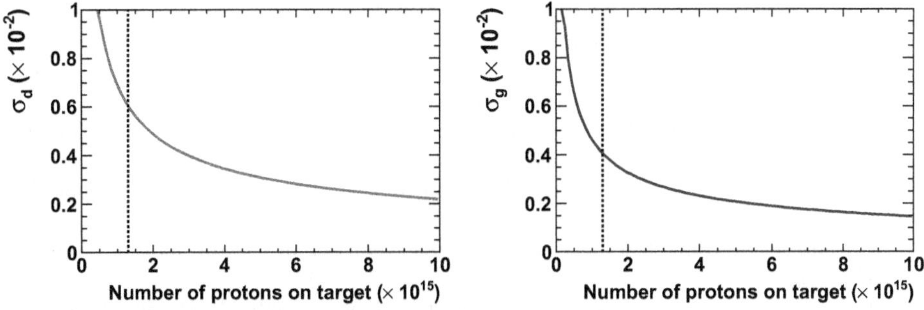

Fig. 3.   Dependence of the (Left) d and (Right) g uncertainties for the Λ_c^+ baryon, reconstructed in the $\Delta^{++}K^-$ final state, with the number of protons on target. One month of data taking corresponds to 1.3×10^{15} incident protons (dashed line), according to the estimated quantities.

4. Conclusions

The unique possibility to search for the EDM of strange and charm baryons at LHC is discussed, based on the exploitation of large statistics of baryons with large Lorentz boost and polarization. The Λ strange baryons are selected from weak charm baryon decays produced in pp collisions at ≈ 14 TeV center-of-mass energy, while Λ_c^+ charm baryons are produced in a fixed-target experiment to be installed in the LHC, in front of the LHCb detector. Signal events can be reconstructed using the LHCb detector in both cases. The sensitivity to the EDM and the MDM of the strange and charm baryons arises from the study of the spin precession in intense electromagnetic fields. The long-lived Λ precesses in the magnetic field of the detector tracking system. Short-lived charm baryons are channeled in a bent crystal attached to the target and the intense electric field between atomic planes induces the spin precession. Sensitivities for the Λ EDM at the level of 1.3×10^{-18} e cm can be achieved using a data sample corresponding to an integrated luminosity of 50 fb^{-1} to be collected during the LHC Run 3. A test of CPT symmetry can be performed by measuring the MDM of Λ and $\overline{\Lambda}$ baryons with a precision of about 4×10^{-4} on the g factor. The EDM of the Λ_c^+ can be searched for with a sensitivity of 2.1×10^{-17} e cm in 11 days of data taking. The proposed experiment would allow about two orders of magnitude improvement in the sensitivity for the Λ EDM and the first search for the charm baryon EDM, expanding the search for new physics through the EDM of fundamental particles.

References

1. L. H. Thomas, The motion of a spinning electron, *Nature* **117**, p. 514 (1926).
2. L. H. Thomas, The kinematics of an electron with an axis, *Phil. Mag.* **3**, 1 (1927).
3. V. Bargmann, L. Michel and V. L. Telegdi, Precession of the polarization of particles moving in a homogeneous electromagnetic field, *Phys. Rev. Lett.* **2**, 435 (May 1959).
4. J. M. Pendlebury *et al.*, Revised experimental upper limit on the electric dipole moment of the neutron, *Phys. Rev.* **D92**, p. 092003 (2015).

5. E. M. Purcell and N. F. Ramsey, On the Possibility of Electric Dipole Moments for Elementary Particles and Nuclei, *Phys. Rev.* **78**, 807 (1950).

6. J. H. Smith, E. M. Purcell and N. F. Ramsey, Experimental limit to the electric dipole moment of the neutron, *Phys. Rev.* **108**, 120 (1957).

7. J. Baron *et al.*, Order of Magnitude Smaller Limit on the Electric Dipole Moment of the Electron, *Science* **343**, 269 (2014).

8. G. W. Bennett *et al.*, An Improved Limit on the Muon Electric Dipole Moment, *Phys. Rev.* **D80**, p. 052008 (2009).

9. K. Inami *et al.*, Search for the electric dipole moment of the tau lepton, *Phys. Lett.* **B551**, 16 (2003).

10. W. C. Griffith, M. D. Swallows, T. H. Loftus, M. V. Romalis, B. R. Heckel and E. N. Fortson, Improved Limit on the Permanent Electric Dipole Moment of Hg-199, *Phys. Rev. Lett.* **102**, p. 101601 (2009).

11. V. F. Dmitriev and R. A. Sen'kov, Schiff moment of the mercury nucleus and the proton dipole moment, *Phys. Rev. Lett.* **91**, p. 212303 (2003).

12. L. Pondrom, R. Handler, M. Sheaff, P. T. Cox, J. Dworkin, O. E. Overseth, T. Devlin, L. Schachinger and K. J. Heller, New Limit on the Electric Dipole Moment of the Λ Hyperon, *Phys. Rev.* **D23**, 814 (1981).

13. J. Grange *et al.*, *Muon (g-2) Technical Design Report*, tech. rep. (2015).

14. N. Saito, A novel precision measurement of muon g-2 and EDM at J-PARC, *AIP Conf. Proc.* **1467**, 45 (2012).

15. V. Anastassopoulos *et al.*, A Storage Ring Experiment to Detect a Proton Electric Dipole Moment, 2015 (2015).

16. J. Pretz, Measurement of electric dipole moments at storage rings, *Physica Scripta* **2015**, p. 014035 (2015).

17. I. B. Khriplovich, Feasibility of search for nuclear electric dipole moments at ion storage rings, *Phys. Lett.* **B444**, 98 (1998).

18. F. J. Botella, L. M. Garcia Martin, D. Marangotto, F. M. Vidal, A. Merli, N. Neri, A. Oyanguren and J. R. Vidal, On the search for the electric dipole moment of strange and charm baryons at LHC, *Eur. Phys. J.* **C77**, p. 181 (2017).

19. E. Bagli *et al.*, Electromagnetic dipole moments of charged baryons with bent crystals at the LHC, *Eur. Phys. J.* **C77**, p. 828 (2017).

20. F.-K. Guo and U.-G. Meissner, Baryon electric dipole moments from strong *CP* violation, *JHEP* **12**, p. 097 (2012).

21. D. Atwood and A. Soni, Chiral perturbation theory constraint on the electric dipole moment of the Λ hyperon, *Phys. Lett.* **B291**, 293 (1992).

22. A. Pich and E. de Rafael, Strong CP violation in an effective chiral Lagrangian approach, *Nucl. Phys.* **B367**, 313 (1991).

23. B. Borasoy, The electric dipole moment of the neutron in chiral perturbation theory, *Phys. Rev.* **D61**, p. 114017 (2000).

24. F. Sala, A bound on the charm chromo-EDM and its implications, *JHEP* **03**, p. 061 (2014).

25. J. M. Link *et al.*, Study of the decay asymmetry parameter and CP violation parameter in the $\Lambda_c^+ \to \Lambda \pi^+$ decay, *Phys. Lett.* **B634**, 165 (2006).

26. R. Aaij *et al.*, Measurements of the $\Lambda_b^0 \to J/\psi \Lambda$ decay amplitudes and the Λ_b^0 polarisation in pp collisions at $\sqrt{s} = 7$ TeV, *Phys. Lett.* **B724**, 27 (2013).

27. G. Aad *et al.*, Measurement of the parity-violating asymmetry parameter α_b and the helicity amplitudes for the decay $\Lambda_b^0 \to J/\psi \Lambda^0$ with the ATLAS detector, *Phys. Rev.* **D89**, p. 092009 (2014).

28. T. D. Lee and C.-N. Yang, General Partial Wave Analysis of the Decay of a Hyperon of Spin 1/2, *Phys. Rev.* **108**, 1645 (1957).

29. J. D. Richman, *An experimenter's guide to the helicity formalism*, Tech. Rep. CALT-68-1148, Calif. Inst. Technol. (Pasadena, CA, 1984).

30. C. Patrignani, Review of Particle Physics, *Chin. Phys.* **C40**, p. 100001 (2016).

31. R. Aaij *et al.*, LHCb detector performance, *Int. J. Mod. Phys.* **A30**, p. 1530022 (2015).

32. D. Chen *et al.*, First observation of magnetic moment precession of channeled particles in bent crystals, *Phys. Rev. Lett.* **69**, 3286 (1992).

33. V. G. Baryshevsky, The possibility to measure the magnetic moments of short-lived particles (charm and beauty baryons) at LHC and FCC energies using the phenomenon of spin rotation in crystals, *Phys. Lett.* **B757**, 426 (2016).

34. M. Jacob and G. C. Wick, On the general theory of collisions for particles with spin, *Annals Phys.* **7**, 404 (1959).

35. V. Baryshevsky, Spin rotation and depolarization of high-energy particles in crystals at LHC and FCC energies. The possibility to measure the anomalous magnetic moments of short-lived particles and quadrupole moment of Ω-hyperon, *Nuclear Instruments and Methods in Physics Research Section B: Beam Interactions with Materials and Atoms* **402**, 5 (2017).

36. I. J. Kim, Magnetic moment measurement of baryons with heavy flavored quarks by planar channeling through bent crystal, *Nucl. Phys.* **B229**, 251 (1983).

37. V. L. Lyuboshits, The Spin Rotation at Deflection of Relativistic Charged Particle in Electric Field, *Sov. J. Nucl. Phys.* **31**, p. 509 (1980).

38. R. Aaij *et al.*, Measurements of prompt charm production cross-sections in pp collisions at $\sqrt{s} = 13$ TeV, *JHEP* **03**, p. 159 (2016), [Erratum: JHEP09,013(2016)].

39. M. Cacciari, FONLL Heavy Quark Production `http://www.lpthe.jussieu.fr/~cacciari/fonll/fonllform.html`, Accessed: 17.05.2016.

40. R. Aaij *et al.*, Measurement of $\sigma(pp \to b\bar{b}X)$ at $\sqrt{s} = 7$ TeV in the forward region, *Phys. Lett.* **B694**, 209 (2010).

41. R. Aaij *et al.*, Measurement of forward J/ψ production cross-sections in pp collisions at $\sqrt{s} = 13$ TeV, *JHEP* **10**, p. 172 (2015).

42. M. Lisovyi, A. Verbytskyi and O. Zenaiev, Combined analysis of charm-quark fragmentation-fraction measurements, *Eur. Phys. J.* **C76**, p. 397 (2016).

43. L. Gladilin, Fragmentation fractions of c and b quarks into charmed hadrons at LEP, *Eur. Phys. J.* **C75**, p. 19 (2015).

44. Y. Amhis *et al.*, Averages of b-hadron, c-hadron, and τ-lepton properties as of summer 2016, *Eur. Phys. J.* **C77**, p. 895 (2017).

45. M. Galanti, A. Giammanco, Y. Grossman, Y. Kats, E. Stamou and J. Zupan, Heavy baryons as polarimeters at colliders, *JHEP* **11**, p. 067 (2015).

46. A. Hicheur and G. Conti, Parameterization of the LHCb magnetic field map, *Proceedings, 2007 IEEE Nuclear Science Symposium and Medical Imaging Conference (NSS/MIC 2007): Honolulu, Hawaii, October 28-November 3, 2007*, 2439 (2007).

47. LHCb collaboration, LHCb Trigger and Online Technical Design Report (2014), LHCb-TDR-016.

48. LHCb collaboration, LHCb Tracker Upgrade Technical Design Report (2014), LHCb-TDR-015.

49. W. Scandale *et al.*, Observation of channeling for 6500 GeV/c protons in the crystal assisted collimation setup for LHC, *Phys. Lett.* **B758**, 129 (2016).

50. J. P. Lansberg *et al.*, A Fixed-Target ExpeRiment at the LHC (AFTER@LHC) : luminosities, target polarisation and a selection of physics studies, *PoS* **QNP2012**, p. 049 (2012).

51. A. Adare *et al.*, Measurement of High-p_T Single Electrons from Heavy-Flavor Decays in $p + p$ Collisions at $\sqrt{s} = 200$ GeV, *Phys. Rev. Lett.* **97**, p. 252002 (2006).

52. B. A. Kniehl and G. Kramer, D^0, D^+, D_s^+, and Λ_c^+ fragmentation functions from CERN LEP1, *Phys. Rev.* **D71**, p. 094013 (2005).

53. V. M. Biryukov *et al.*, *Crystal Channeling and Its Application at High-Energy Accelerators* (Springer-Verlag Berlin Heidelberg, 1997).

54. E. M. Aitala *et al.*, Multidimensional resonance analysis of $\Lambda_c^+ \to pK^-\pi^+$, *Phys. Lett.* **B471**, 449 (2000).

55. V. M. Samsonov, On the possibility of measuring charm baryon magnetic moments with channeling, *Nucl. Instrum. Meth.* **B119**, 271 (1996).

Search for a Light Sterile Neutrino with the Daya Bay, Bugey-3 and MINOS Experiments

Zhuojun Hu

on behalf of Daya Bay Collaboration

*Department of Physics, Sun Yat-sen University,
Guangzhou, Guangdong Province, P. R. China
E-mail: huzhj3@mail2.sysu.edu.cn*

A sterile neutrino at the eV or sub-eV scale was suggested as a possible explanation of the electron antineutrino excess observed in the LSND and MiniBooNE experiments. Searches for a light sterile neutrino have been independently performed by the MINOS and the Daya Bay experiments using the muon (anti)neutrino and electron antineutrino disappearance channels, respectively. In a recent analysis, results from both experiments are combined with those from Bugey-3 reactor neutrino experiment to constrain oscillations into light sterile neutrinos, setting stringent limits on $\sin^2 2\theta_{\mu e}$ over six orders of magnitude in the sterile mass-squared splitting Δm_{41}^2. The parameter space allowed by the LSND and MiniBooNE experiments is excluded for $\Delta m_{41}^2 < 0.8\text{eV}^2$ at 95% CL$_\text{s}$.

Keywords: Light sterile neutrino; Daya Bay; MINOS.

1. Introduction

Since the discovery of neutrino oscillations [1, 2], a large amount of neutrino oscillation data from both natural (atmospheric neutrinos, solar neutrinos) and artificial sources (accelerator neutrinos, reactor neutrinos) have been collected. Most of the neutrino oscillation data can be well described by the current 3-flavor framework, yet some of them, including those from LSND [3] and MiniBooNE [4], cannot be explained, suggesting the existence of an additional light neutrino state at the eV or sub-eV scale. However, the existence of more than three active flavor neutrinos with mass smaller than half the Z-boson mass has been excluded by precision electroweak measurements, so that neutrino states beyond the known three active states are called "sterile". Sterile neutrinos are hypothesized gauge particles that do not participate in standard weak interactions and thus cannot be directly detected with known methods. Nevertheless, mixing between sterile neutrinos and active neutrinos could have detectable effects on neutrino oscillation.

Searches for a light sterile neutrino have been independently performed by the MINOS [5] and the Daya Bay [6] experiments. In a recent analysis, the muon (anti)neutrino disappearance measurement from MINOS is combined with the measurements of electron antineutrino disappearance performed by Daya Bay and Bugey-3 [7]. The joint analysis, based on the CLs method [8, 9], is motivated by the indications of the existence of sterile neutrino from the LSND and MiniBooNE experiments. The results of the three disappearance experiments are not affected by CP-violation, thus the combined results constrain both neutrino and antineutrino appearance.

The minimal extension of the standard model where only one sterile neutrino is considered, is adopted in this analysis. In the 3+1 sterile neutrino model, the appearance probability of muon to electron neutrino can be expressed using a 4×4 unitary mixing matrix, U, by

$$P_{\nu_l \rightarrow \nu_{l'}}\left(L/E\right) = \left| \sum_i U_{li} U_{l'i}^* e^{-i\left(m_i^2/2E\right)L} \right|^2 . \tag{1}$$

A non-zero $4|U_{e4}|^2|U_{\mu4}|^2$ is a possible explanation for the MiniBooNE and LSND results. In this analysis, with the measurements of electron antineutrino disappearance by Daya Bay and Bugey-3 experiments, the matrix element $|U_{e4}|^2$ can be constrained, while $|U_{\mu4}|^2$ can be constrained with the measurement of muon (anti)neutrino disappearance by MINOS experiment. For these experiments, the survival probabilities in the 3+1 sterile neutrino model can by expressed by

$$P_{\bar{\nu}_e \rightarrow \bar{\nu}_e}\left(L/E\right) = 1 - 4\sum_{k>j} |U_{ek}|^2 |U_{ej}|^2 \sin^2\left(\frac{\Delta m_{kj}^2 L}{4E}\right), \tag{2}$$

$$P_{\overset{(-)}{\nu_\mu} \rightarrow \overset{(-)}{\nu_\mu}}\left(L/E\right) = 1 - 4\sum_{k>j} |U_{\mu k}|^2 |U_{\mu j}|^2 \sin^2\left(\frac{\Delta m_{kj}^2 L}{4E}\right). \tag{3}$$

The parameterization of the mixing matrix is chosen as [10]

$$U = R_{34} R_{24} R_{14} R_{23} R_{13} R_{12} , \tag{4}$$

where R_{ij} represents a 4×4 rotation in the (i, j) plane, yielding

$$|U_{e4}|^2 = \sin^2 \theta_{14} ,$$

$$|U_{\mu4}|^2 = \sin^2 \theta_{24} \cos^2 \theta_{14} , \tag{5}$$

$$4|U_{e4}|^2 |U_{\mu4}|^2 = \sin^2 2\theta_{14} \sin^2 \theta_{24} \equiv \sin^2 2\theta_{\mu e} .$$

The CL_s method [8, 9] is a two-hypothesis test that compares the 3ν hypothesis to the 4ν hypothesis. A test statistic $\Delta\chi^2 = \chi_{4\nu}^2 - \chi_{3\nu}^2$ is constructed to determine whether the 4ν hypothesis can be rejected. $\chi_{4\nu}^2$ is the χ^2 value from a fit to the 4ν hypothesis, and $\chi_{3\nu}^2$ is the χ^2 value from a fit to the 3ν hypothesis. The CL_s value is defined as

$$CL_b = P\left(\Delta\chi^2 \geq \Delta\chi_{obs}^2 \mid 3\nu\right),$$

$$CL_{s+b} = P\left(\Delta\chi^2 \geq \Delta\chi_{obs}^2 \mid 4\nu\right), \tag{6}$$

$$CL_s = \frac{CL_{s+b}}{CL_b},$$

where $\Delta\chi^2_{obs}$ is the $\Delta\chi^2$ observed with data. CL_b tests agreement between data and the 3ν hypothesis, and CL_{s+b} measures consistency with the 4ν hypothesis. The 4ν hypothesis is excluded at the α confidence level if $CL_s \leq 1 - \alpha$.

2. MINOS results

The MINOS experiment [10] is an accelerator neutrino experiment which measured the muon (anti)neutrino disappearance channel. Two detectors, with the 1 kton Near Detector (ND) located 1 km downstream, and the 5.4 kton Far Detector (FD) located at the Soudan Underground Laboratory [10], sample the NuMI neutrino beam. The analysis used data from an exposure of 10.56×10^{20} protons-on-target, for which the neutrino beam composition is 91.8% ν_μ, 6.9% $\bar{\nu}_\mu$, and 1.3% ($\nu_e + \bar{\nu}_e$).

To analyze the MINOS results, CL_s method is adopted as well as Feldman-Cousins method. Figure 1 shows the comparison of the MINOS 90% C.L. contours obtained using the Feldman-Cousins method [5] with those obtained using the CL_s method. The contours are in good agreement with each other. Note that the 3+1 sterile neutrino model is degenerate with the three-flavor model when $\Delta m^2_{41} = 2\Delta m^2_{31}$ or $\Delta m^2_{41} \ll \Delta m^2_{31}$ and $\sin^2\theta_{23} = \sin^2\theta_{34} = 1$, leading to regions of parameters space that cannot be excluded.

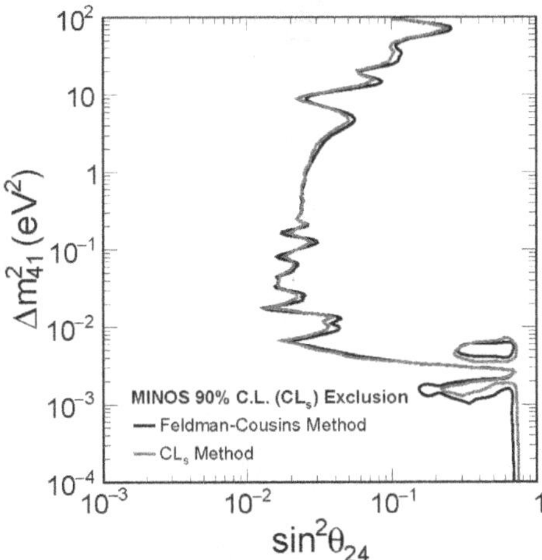

Fig. 1. Comparison of the MINOS 90% C.L. contour using the Feldman-Cousins method [5] and the CL_s method. The region to the right of the curve is excluded at the 90% C.L. (CL_s).

3. Daya Bay results

The Daya Bay experiment is a reactor neutrino experiment measuring electron antineutrinos from six reactor cores via inverse beta decay (IBD): $\bar{\nu}_e + p \rightarrow e^+ + n$. Eight identical Gadolinium-doped liquid-scintillator detectors (ADs), with two ADs situated in underground experimental hall (EH) 1, two ADs in EH2, and four ADs in EH3, weigh 20 tons each. The flux averaged baselines for EH1, EH2, and EH3 are 520,

398

570, and 1590m, respectively. Details of the IBD event selection, background estimates, and assessment of systematic uncertainties can be found in [11, 12].

The analysis is based on data collected in 217 days during the 6AD configuration and 404 days during the 8AD configuration. The sensitivity in 6+8AD period increases by a factor of nearly two compared to the sensitivity in 6AD period. The contribution of each EH to the sensitivity of the sterile neutrino search in Daya Bay is shown in Fig. 2. EH2 contributes to the $\Delta m_{41}^2 > 10^{-2} eV^2$ region, while EH3 contributes to the $\Delta m_{41}^2 < 10^{-2} eV^2$ range.

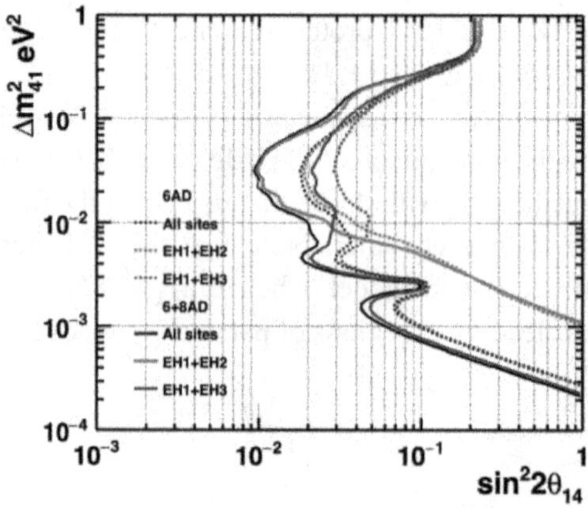

Fig. 2.   Sensitivities of different EH combinations in 6 and 6+8 AD period. A factor of 2 improvement from 6 to 6+8 AD period is observed.

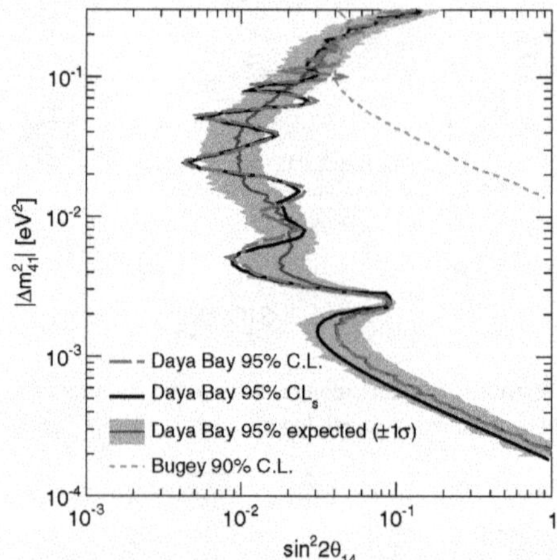

Fig. 3.   Exclusion contours in the $(\sin^2 2\theta_{14}, |\Delta m_{41}^2|)$ plane, under the assumption of $\Delta m_{32}^2 > 0$ and $\Delta m_{41}^2 > 0$. The red long-dashed curve represents the 95% C.L. exclusion contour with the Feldman-Cousins method. The black solid curve represents the 95% CL$_s$ exclusion contour with the CL$_s$ method. The expected 95% C.L. 1σ band in yellow is centered around the sensitivity curve, shown as a thin blue line. The region of parameter space to the right side of the contours is excluded. For comparison, Bugey's [7] 90% C.L. limit on $\bar{\nu}_e$ disappearance is also shown as the green dashed curve (color online).

Feldman-Cousins method and CL_s method are adopted to analyze the Daya Bay results [6]. Figure 3 shows the 95% C.L. contour from the Feldman-Cousins procedure and the 95% CL_s contour. The contours are consistent with each other, and are mostly within the $\pm 1\sigma$ band of the 95% C.L. expectation. The dip structure at $|\Delta m_{41}^2| \approx |\Delta m_{32}^2| \approx 2.4 \times 10^{-3}$ eV^2 is due to the degeneracy between $\sin^2 2\theta_{14}$ and $\sin^2 2\theta_{13}$.

4. Bugey-3 results

The Bugey-3 experiment is a reactor neutrino experiment performed in the early 1990s, aiming to search for neutrino oscillations. The experiment operated two ^{6}Li-doped liquid scintillator detectors, measuring electron antineutrino generated from two reactor cores at three different baselines (15, 40 and 95m) [7]. Bugey-3 measured electron antineutrino via IBD interactions with the recoil neutron capturing on ^{6}Li. Due to the short baselines, Bugey-3 is sensitive to regions of parameter space with larger Δm_{41}^2 values compared to Daya Bay, thus the two experiments are sensitive to complementary regions.

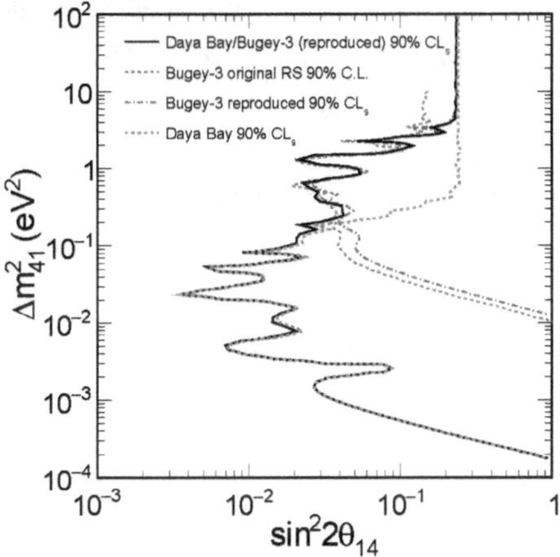

Fig. 4. Excluded regions for the original Bugey-3 raster scan (RS) result [7], for the reproduced Bugey-3 with adjusted IBD cross section and antineutrino flux, for the Daya Bay result [6], and for the combined Daya Bay and reproduced Bugey-3 results. The region to the right of the curve is excluded at the 90% CL_s.

The original Bugey-3 results are successfully reproduced first using the raster scan technique, following the procedure in [7]. The uncertainty on the overall normalization is set to 5% to be consistent with the constraint employed in Daya Bay. Two changes are made in the reproduction of Bugey-3 analysis in order to be consistent with the Daya Bay analysis [6].

(i) the change in cross section of the IBD process due to the updated neutron decay time [13] is applied.

(ii) the antineutrino flux is adjusted from ILL+Vogel [14, 15] model to that of Huber [16] and Mueller [17].

The reproduced Bugey-3 contour, together with the contour obtained by combining Bugey-3 and Daya Bay are shown in Fig. 4.

5. Combined analysis

In the combined analysis, $\Delta\chi^2_{obs}$, as well as $\Delta\chi^2_{3\nu}$ and $\Delta\chi^2_{4\nu}$ distributions are obtained for each $(\sin^2 2\theta_{14}, \Delta m^2_{41})$ grid point of the Daya Bay and Bugey-3 combination, and for each $(\sin^2 2\theta_{24}, \Delta m^2_{41})$ grid point from MINOS. Since the systematic uncertainties of accelerator and reactor experiments are uncorrelated, a combined $\Delta\chi^2_{obs}$ for each $(\sin^2 2\theta_{14}, \sin^2 2\theta_{24}, \Delta m^2_{41})$ grid point is constructed by summing over the corresponding MINOS and Daya Bay/Bugey-3 $\Delta\chi^2_{obs}$ values. Similarly, the combined $\Delta\chi^2_{3\nu}$ and $\Delta\chi^2_{4\nu}$ distributions are constructed by adding random samples drawn from the corresponding MINOS and Daya Bay/Bugey-3 distributions. Thus a CL_s value can be calculated for each $(\sin^2 2\theta_{14}, \sin^2 2\theta_{24}, \Delta m^2_{41})$ grid point. Then for each $(\sin^2 2\theta_{\mu e}, \Delta m^2_{41})$ grid point, CL_s values are calculated (cf. Eq. (6)), and the largest CL_s value is conservatively chosen. Figure 5 presents the combined 90% CL_s limit on $\sin^2 2\theta_{\mu e}$ from this analysis. The combined results of MINOS and Daya Bay/Bugey-3 set constraints on $\sin^2 2\theta_{\mu e}$, with $\sin^2 2\theta_{\mu e} < [3.0 \times 10^{-4}(90\%\,CL_s), 4.5 \times 10^{-4}(95\%\,CL_s)]$ for $\Delta m^2_{41} = 1.2 eV^2$.

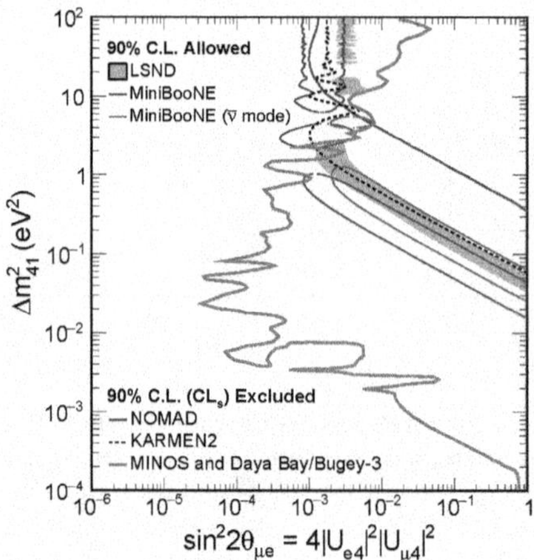

Fig. 5. MINOS and Daya Bay/Bugey-3 combined 90% CL_s limit on $\sin^2 2\theta_{\mu e}$ compared to the LSND and MiniBooNE 90% C.L. allowed regions. Regions of parameter space to the right of the red contour are excluded. The regions excluded at 90% C.L. by KARMEN2 [18] and NOMAD [19] are also shown. Note that the excursion to small mixing in the exclusion contour at around $\Delta m^2_{41} \sim 7 \times 10^{-3} eV^2$ is originated from the island in Fig. 1 (color online).

6. Conclusion and outlook

In conclusion, the analysis combined constraints on $\sin^2 2\theta_{14}$ derived from electron antineutrino disappearance measurements by the Daya Bay and Bugey-3 reactor experiments with constraints on $\sin^2 2\theta_{24}$ derived from muon (anti)neutrino disappearance measurement in the NuMI beam at the MINOS experiment. No evidence of a light sterile neutrino is found. Sterile neutrino mixing phase space allowed by the LSND and MiniBooNE is excluded for $\Delta m_{41}^2 < 0.8 \text{eV}^2$ at a 95% CL$_s$. The results set the strongest constraint on $\sin^2 2\theta_{\mu e}$ to date and agrees well with results from global fits [20, 21]. An updated result is expected to be released in 2018.

References

1. Y. Fukuda *et al.* (Super-Kamiokande), *Phys. Rev. Lett.* **81**, 1562 (1998).
2. Q. R. Ahmad *et al.* (SNO), *Phys. Rev. Lett.* **87**, 071301 (2001).
3. A. Aguilar *et al.* (LSND), *Phys. Rev. D* **64**, 112007(2001).
4. A. Aguilar-Arevalo *et al.* (MiniBooNE), *Phys. Rev. Lett.* **110**, 161801 (2013).
5. P. Adamson *et al.* (MINOS), *Phys. Rev. Lett.* **117**, 151803 (2016).
6. F. P. An *et al.* (Daya Bay), *Phys. Rev. Lett.* **117**, 151802 (2016).
7. B. Achkar *et al.*, *Nucl. Phys. B* **434**, 503 (1995).
8. A. L. Read, *J. Phys. G* **28**, 2693 (2002).
9. T. Junk, *Nucl. Instrum. Meth. A* **434**, 435 (1999).
10. H. Harari and M. Leurer, *Phys. Lett.* **B181**, 123 (1986).
11. F. P. An *et al.*, *Phys. Rev. Lett.* **115**, 111802 (2015).
12. F. An *et al.* (Daya Bay), *Phys. Rev. Lett.* **112**, 061801 (2014).
13. K. Olive *et al.* (Particle Data Group), *Chinese Phys. C* **38**, 090001 (2014).
14. K. Schreckenbach *et al.*, *Phys. Lett. B* **160**, 325 (1985).
15. P. Vogel, *Phys. Rev. D* **29**, 1918 (1984).
16. P. Huber, *Phys. Rev. C* **84**, 024617 (2011), [Erratum: *Phys. Rev. C* **85**, 029901 (2012)].
17. T. A. Mueller *et al.*, *Phys. Rev. C* **83**, 054615 (2011).
18. B. Armbruster *et al.* (KARMEN), *Phys. Rev. D* **65**, 112001 (2002).
19. P. Astier *et al.* (NOMAD), *Phys. Lett. B* **570**, 19 (2003).
20. J. Kopp, P. A. N. Machado, M. Maltoni, and T. Schwetz, *JHEP* **05**, 050 (2013).
21. S. Gariazzo, C. Giunti, M. Laveder, Y. F. Li, and E. M. Zavanin, *J. Phys.* **G43**, 033001 (2016).

Update of the Detector Geometry System and Visualization for the New TOF System in BESIII[*]

Shuhui Huang[†], Zhengyun You[‡]

on behalf of the BESIII Collaboration

*School of Physics, Sun Yat-sen University,
Guangzhou, 510275, China
E-mail: †huangshh28@mail2.sysu.edu.cn,   ‡youzhy5@mail.sysu.edu.cn*

The Beijing Spectrometer (BESIII) has operated on Beijing Electron-Positron Collider (BEPCII) at the τ-charm energy region since 2008. The recent upgrades of endcap Time-Of-Flight system (TOF) in BESIII provides better flight time measurement for particle identification. The event display tool plays an important role for data monitoring, reconstruction algorithms tuning and physics analysis in BESIII, it is necessary to update the event display with the new sub-detectors. In this paper, the updated detector geometry system and event visualization in the BESIII event display software is discussed, with focus on implementation of the new endcap TOF system.

Keywords: BESIII; event display; detector geometry; visualization.

1. Introduction

1.1. *BEPCII and BESIII*

BEPCII[1], the upgrade of the first generation of Beijing Electron-Positron Collider, is a double-ring multi-bunch collider located at the Institute of High Energy Physics (IHEP) in Beijing, China. The designed luminosity of BEPCII is 10^{33} cm^{-2} s^{-1} and the center-of-mass energy is designed to be between 2.0 and 4.6 GeV. BESIII[2] is a general-purpose experiment running on BEPCII. In the past ten years, BESII has accumulated large data samples in the τ-charm energy region.

During the past few years, BESIII has make a few surprising discoveries like Zc(3900)[3] and it provides vast and diverse opportunities to better understand the Standard Model. As shown in Fig. 1, the BESIII detector consists of 4 sub-detectors, including Main Drift Chamber (MDC), Time-Of-Flight System (TOF) based on scintillators and the new Multigap-Resistive-Plate-Chamber (MRPC) on its endcaps, Electro-Magnetic Calorimeter (EMC) made with CsI crystals and Muon Counter (MUC). A uniform 1 Tesla magnetic field is produced by Superconducting Solenoid Magnet (SSM), which is located between EMC and MUC. The details about the design and performance of the BESIII detector is available in Ref. [2].

---

[*]This work is supported by National Natural Science Foundation of China (11675275) and the Major Science and Technology Infrastructure Opening Research Program of Chinese Academy of Sciences.

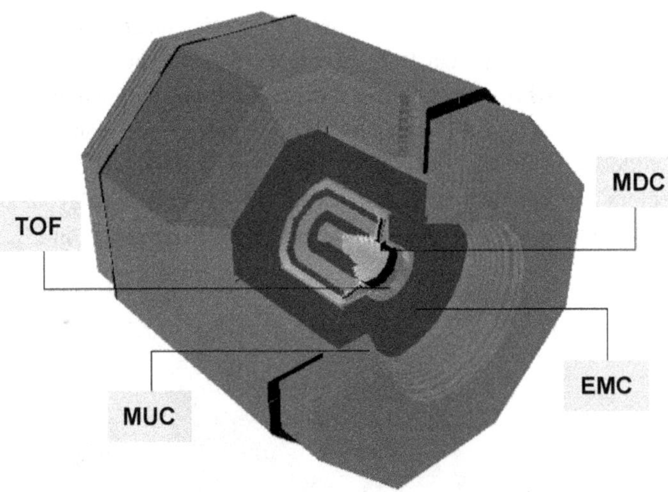

Fig. 1.   3D model of the BESIII detector.

1.2. *BESIII Offline Software System*

BESIII Offline Software System (BOSS)[4] is based on the GAUDI[5] framework, which uses C++ and object-oriented techniques to realize the functions of data processing and physics analysis in BESIII, including simulation, calibration, reconstruction and analysis tools.

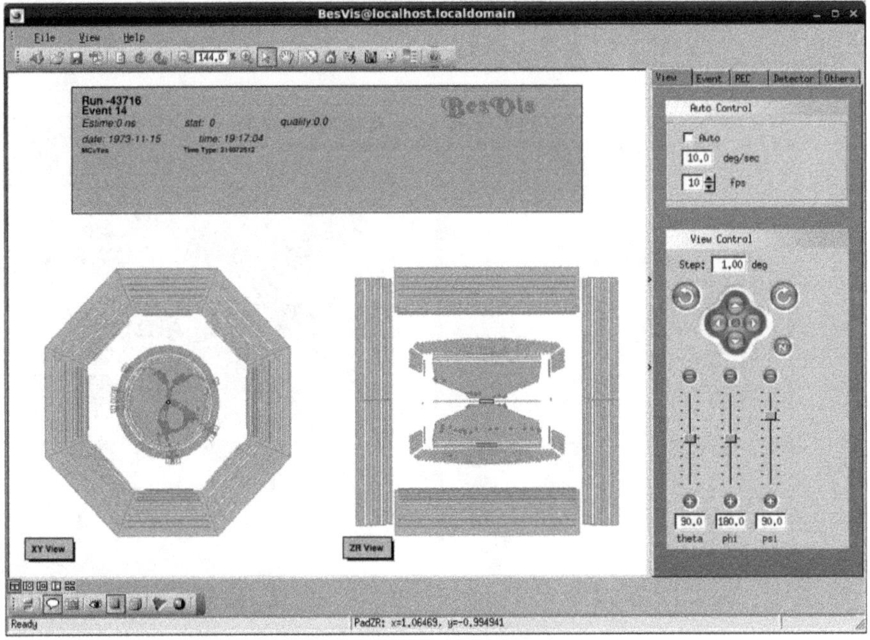

Fig. 2.   The GUI of BesVis.

Event Display is an important module in BOSS. The visualization software BesVis was first released in 2005. Its Graphical User Interface (GUI) is shown in Fig. 2. The main window of BesVis consists of a menu bar at the top, a status window in blue, XY view and ZR view window at the bottom, and a tool bar on the right column. Especially, in XY view and ZR view window, users can zoom in or zoom out, use the widgets in the toolbar to rotate the detectors and to choose which part of detectors to be displayed. Besides, a 3D display of the detector and event is also provided in BesVis, as shown in Fig. 3.

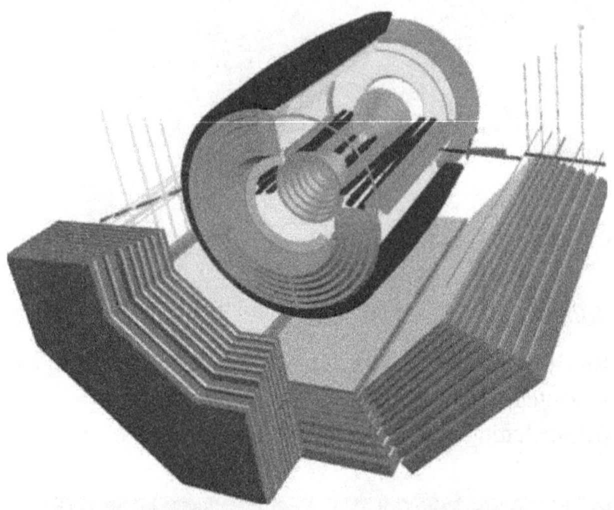

Fig. 3.   3D display of an event by OpenGL in BesVis.

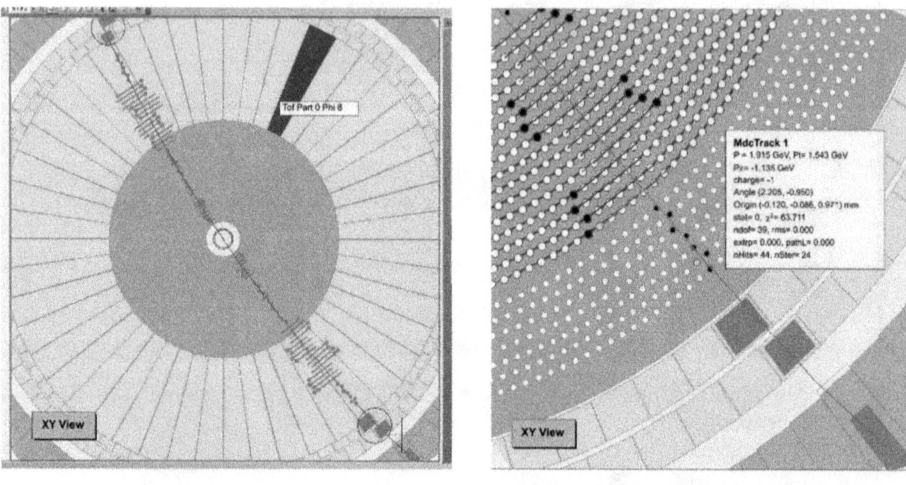

(a) Details of TOF   (b) Details of MdcTrack

Fig. 4.   Display of the detector geometry and event hits.

BesVis can intuitively display the BESIII detector geometry as well as the hits and reconstruction output from event file, as shown in Fig. 4. The geometry management in BesVis is based on GDML[6] and ROOT[7]. Both GDML files and ROOT files can be used to initialize detector geometry but the original geometry is described with GDML. In BesVis, the GDML geometry can be converted into ROOT objects and then be saved as ROOT files[8,9].

2. Endcap TOF Upgrade

The endcap TOF (ETOF) in BESIII was upgraded into Multigap-Resistive-Plate-Chamber (MRPC)[10] and the replacement was finally completed in 2015. Compared to the old endcap TOF with 48 single layer scintillators on east or west part, the structures of new endcap TOF with MRPC are more complicated, which contains 36 modules on each part with 12 strips in each module. These modules are placed in 2 layers with 18 modules in each layer. After the upgrade, the time resolution of endcap TOF can reach 60 ps for double-end MRPCs and 69 ps for single-end MRPCs[11], in comparison to the 100 ps time resolution of old endcap TOF.

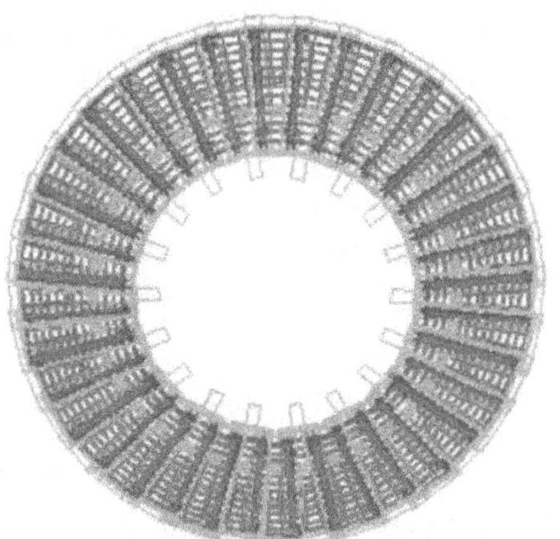
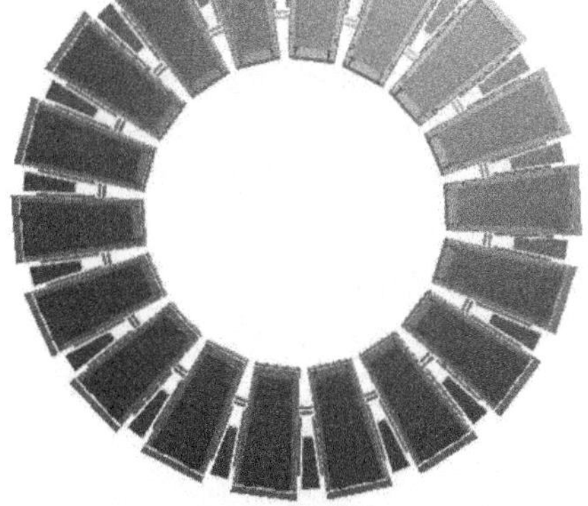

(a) Complete 2D view with double layers of 36 modules

(b) 3D view with single layer of 18 modules

Fig. 5.   The geometry of new endcap TOF with MRPC.

3. Visualization updates for Endcap TOF

The updates of ETOF in BesVis consist of two major work, display of the ETOF geometry from GDML file and display of the hits on ETOF from data event files. They are discussed below respectively.

406

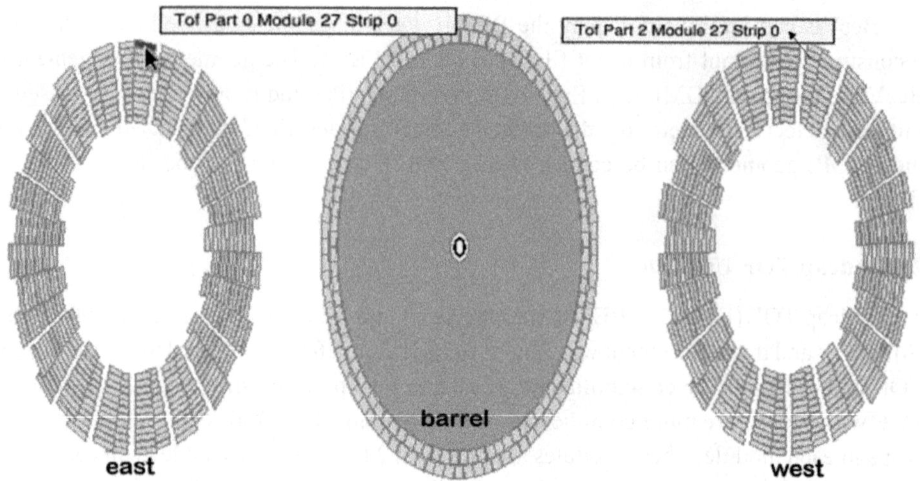

Fig. 6.   Overview of new ETOF geometry in BesVis.

3.1. *Update of Geometry*

The hierarchy of new ETOF geometry can be describe as an array with 3 dimensions part-module-strip. Part 0 and 2 represents to the east part and west part, respectively. Module 0 to 35 represent the 36 modules on each part. Strip 0 to 11 correspond to the 12 strips on each module. To allow the existence of both old and new ETOF geometry, a parameter has been implemented to switch between the two geometries conveniently. After these changes, the new ETOF geometry has been successfully displayed in BesVis, as shown in Fig. 7.

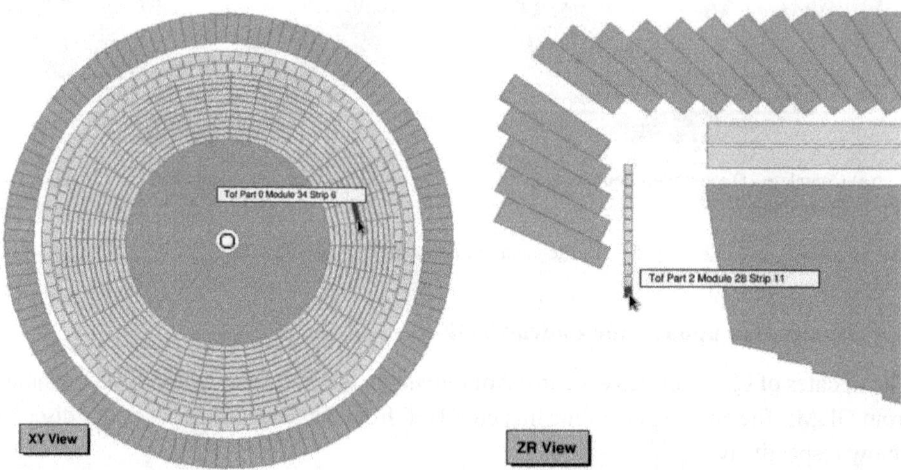

Fig. 7.   New ETOF geometry in XY View and ZR View in BesVis, with yellow MRPC strips and blue transverse shape of barrel EMC (color online).

3.2. *Update of Hits Display*

Since the new ETOF with MRPC contains different read-out system compared to the old one, we need to consider this problem when displaying hits information in one strip. The read-out system of MRPC is a new version with single end and double sided so the read-out information is leading time and trailing time, instead of time and charge information of old ETOF, as shown in Fig. 8.

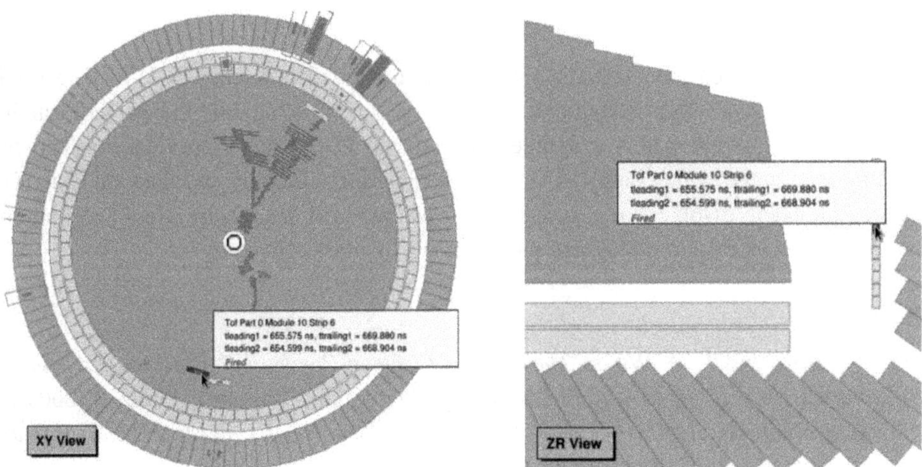

Fig. 8. Display of event hits and hits information in XY view and ZR view in BesVis, with one fired MRPC strip.

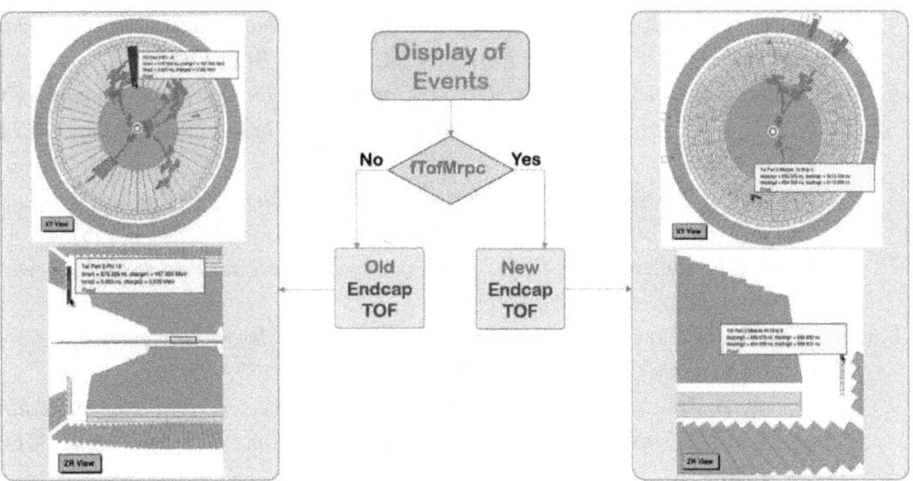

Fig. 9. Comparison between old and new ETOF.

3.3. *Comparison Between Old and New ETOF*

From Fig. 9, we can clearly see the biggest difference between old and new ETOF is geometry. The more complex hierarchy of geometry structure in new ETOF provides more precise time measurement than before. When displaying events, the read-out information of fired strips is time and charge information in old ETOF, but now it is replaced by leading time and trailing time of ETOF with MRPC, which makes better time resolution available.

4. Summary

The geometry system and visualization for the upgraded new endcap TOF detector in BESIII software has been introduced. BESIII experiment is the only experiment running at τ-charm energy region now. Upgrade of the endcap TOF detector provides more precise time measurement, which helps to improve particle identification and data analysis. The update of event display with new ETOF detector will also help users working on reconstruction algorithm tuning and physics analysis more efficiently.

References

1. C. Zhang and J. Q. Wang, *The Beijing Electron-Positron Collider and its second phase construction, Proceedings of EPAC* (Luzern, Switzerland), 230–232 (2004).
2. M. Ablikim et al., *Design and construction of the BESIII detector, Nucl. Instrum. Meth.* A **614**, 345–399 (2010).
3. M. Ablikim et al. (BESIII Collaboration), *Observation of a charged charmoniumlike structure in* $e^{+}e^{-} \rightarrow \pi^{+}\pi^{-}J/\psi$ *at* $\sqrt{s} = 4.26\,GeV$, *Phys. Rev. Lett.* **110**, 252001 (2013).
4. W. D. Li, Y. J. Mao and Y. F. Wang, *The BES-III Detector and Offline Software, International Journal of Modern Physics A* **24**, Suppl. 1, 9–21 (2009).
5. G. Barrand, I. Belyaev et al., *GAUDI — A software architecture and framework for building HEP data processing applications, Computer Physics Communications* **140**, 45–55 (2001).
6. R. Chytracek et al., *Geometry Description Markup Language for Physics Simulation and Analysis Applications, IEEE Trans. Nucl. Sci.* **53**(5), 2892–2896 (2006).
7. R. Brun and F. Rademakers, *ROOT – An object oriented data analysis framework, Nucl. Instrum. Meth. A* **389**, 81–86 (1997).
8. Z. Y. You, Y. T. Liang, Y. J. Mao, *A method for detector description exchange among ROOT GEANT4 and GEANT3, Chin. Phys. C* **32**(7), 572–575 (2008).
9. Y. T. Liang, B. Zhu, Z. Y. You, et al., *A uniform geometry description for simulation, reconstruction and visualization in the BESIII experiment, Nucl. Inst. Meth.* A **603**, 325–327 (2009).
10. Y. J. Sun et al., *A prototype MRPC beam test for the BESIII ETOF upgrade. Chin. Phys. C* **26**, 429–433 (2012).
11. Z. Wu, S. S. Sun, Y. K. Heng et al., *First results of the new endcap TOF commissioning at BESIII. JINST* **11**(07), C07005 (2016).

Understanding Backgrounds Using Deep Learning at the Daya Bay Experiment

S. Kohn

Department of Physics, University of California, Berkeley
Berkeley, CA 94720, USA
E-mail: skohn@lbl.gov

The Daya Bay experiment uses reactor antineutrino disappearance to measure the neutrino mixing parameter θ_{13}. A variety of deep neural networks are tested with a well-understood uncorrelated accidental background to the inverse beta decay signal to assess the utility of deep learning approaches for characterizing and discriminating backgrounds. Crucially, the training procedures are data-driven and do not rely on simulated events to train the neural networks. The eventual goal of this technique is to reduce the correlated β-n background, which results from the decay of ^{9}Li produced by cosmic-ray muons. This background contributes the largest systematic uncertainty in the determination of θ_{13}.

Keywords: Neutrinos; machine learning; Daya Bay.

1. Introduction

The Daya Bay Reactor Neutrino Experiment is a precision neutrino oscillation experiment that measures the disappearance of $\bar{\nu}_e$ at the far site relative to the measurements obtained at the near sites[1]. The value of the θ_{13} neutrino mixing angle can be extracted from the difference in $\bar{\nu}_e$ rates and energy spectra between the different detector sites.

The IBD reaction, $\bar{\nu}_e + p \rightarrow n + e^+$, is used to identify when antineutrinos interact in the antineutrino detectors (ADs), which are filled with Gd-doped liquid scintillator. The positron loses energy through scintillation and eventually annihilates within a nanosecond. The combination of these processes causes a prompt burst of scintillation light. The neutron thermalizes and captures on a Gd nucleus (or, less frequently, on hydrogen) after approximately $25\mu s$, releasing a delayed burst of scintillation light. The characteristic prompt-delayed double coincidence of the IBD reaction is the primary discriminator used for rejecting background.

Although the double coincidence is a useful discriminator, there are other physical processes in the ADs which also cause a double coincidence, including accidental coincidences and the β-n decay of ^{9}Li. Accidental coincidences are a result of the natural decay of radiocontaminants in and near the ADs, which causes a steady background of single flashes. Occasionally, two uncorrelated flashes accidentally occur within the coincidence time window and are accepted as an IBD event.

This background rate is straightforward to predict since the single-flash (singles) rate is well-measured in each AD and the process is Poissonian. Figure 1 shows the spectrum of singles in each AD. Approximately 1%-2% of the IBD sample is

410

accidental background, and that percentage is known to approximately 1%, for a total contribution to the rate uncertainty of $\lesssim 0.01\%$ [2]. As a consequence, accidentals do not contribute a significant uncertainty to the θ_{13} result and are an ideal trial data set for the convolutional autoencoder technique.

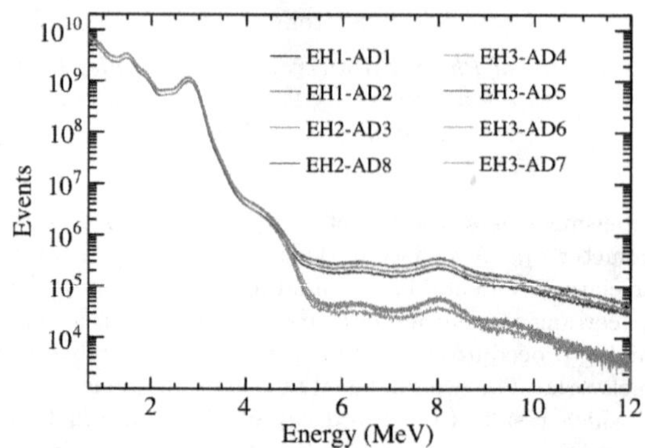

Fig. 1.   Spectrum of uncorrelated events (singles).

Because accidentals are formed from two uncorrelated events, it is simple to create an "artificial" set of accidental background out of real data, even if the true physical origin of each event is unknown. Modulo the energy selection, all that is necessary is for random singles events—displaced significantly in time—to be paired up and decreed to be accidentals. Such an artificial data set is by construction 100% pure and statistically similar to the true accidentals data set.

On the other hand, the more difficult background to identify is ^{9}Li decay, which results in an electron and neutron being released. Since the electron signal in liquid scintillator is similar to a positron signal—and the electron energy in this decay is within the expected range of positron energies for IBD—this signature is almost indistinguishable from a true IBD signal.

The rate of ^{9}Li background can be estimated since it is known that ^{9}Li is created when cosmic-ray muons traverse the AD, and decays with a lifetime of 257.2 ms. By measuring the rate of double coincidence events as a function of time since the last muon event, the number of ^{9}Li events contaminating the data set can be estimated. Figure 2 shows the "time since last muon" distribution, and the resulting measured spectrum of ^{9}Li events. The final rate is $< 1\%$ of the IBD rate, but this rate has a large uncertainty. The contribution to the final IBD rate uncertainty is 0.1%—approximately 10 times larger than the uncertainty due to any other source of background [2].

The Daya Bay analysis filters out the vast majority of these background events using the double coincidence timing (for accidentals) and the muon veto (for ^{9}Li). The remaining contamination is subtracted statistically, which, for ^{9}Li, leads to a large unertainty. An accurate event-by-event filter for these remaining events—

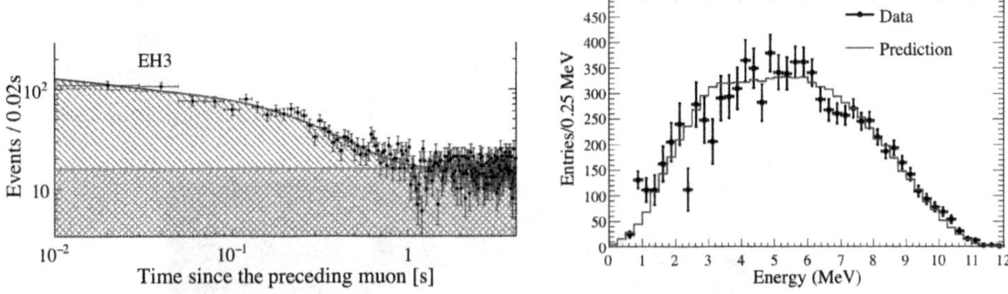

Fig. 2.   Distribution of IBD-like events as a function of time since the preceding muon (left) and the resulting ^{9}Li background spectrum (right).

particularly for ^{9}Li—would reduce the dominant systematic uncertainty and improve the precision of θ_{13}.

2. Machine learning methods

Although the underlying physical processes are different between these background interactions and IBD, existing analysis methods have not yet succeeded at reliably distinguishing these event types (after the coincidence timing and muon vetos are applied). Machine learning techniques allow further investigation of any distinct signatures of these background events. For Daya Bay, machine learning could eventually supplement the existing selection process to filter out as much of the remaining background contamination as possible while retaining a high efficiency and low bias.

To investigate the differences in geometry, energy and timing between these different classes of events, an unsupervised neural network approach is used. Such an approach does not rely on *a priori* classification information (i.e. labels) for training. Since the neural network can be trained using unlabeled data, unsupervised learning is not subject to biases from simulated training events, such as from timing or electronics noise, that may degrade the performance of the trained neural network.

The convolutional autoencoder neural network is such an unsupervised technique which can process images (or other arrays of pixels) and form groups of similar images[3,4]. Convolutional networks (whether in an autoencoder architecture or not) operate on data where spatial relationships are important, such as photographs or data from an array of physics detectors. The goal of the training is to create a compressed encoding scheme which preserves "important" features of the input data, and whose encodings can be decoded to a close "reconstructed" approximation of the original input. The importance of a feature is determined by whether knowledge of that feature is useful in creating an accurate reconstruction of the original input image.

One difficulty with this approach is that the final trained neural network is not a classifier, so it cannot be directly used to categorize events as signal or background.

412

Instead, the encodings and the input data must be examined to help determine which data features caused the autoencoder to associate different events into groups. Once these important features can be isolated, an analysis can be constructed which classifies events based on the presence or absence of those particular features.

Fig. 3.   Inside of a cylindrical Daya Bay antineutrino detector module (AD).

The data used for this analysis is collected from photomultiplier tubes (PMTs) in the Daya Bay antineutrino detectors (ADs). As shown in Figure 3, each AD contains a cylindrical array of PMTs in 8 rings with 24 columns per ring, for a total of 192 PMTs per AD. Hence each input "image" for the neural network will be an image with 8×24 "pixels" containing data (such as hit time or number of photons) from each PMT in the AD. The cylindrical geometry is unrolled into a rectangle, so that rows of pixels correspond to rings of PMTs, and columns of pixels correspond to columns of PMTs. The total antineutrino data set is approximately 2×10^6 antineutrinos in 4 years.

A preliminary study[5] indicated that an autoencoder can successfully encode and decode these images. It also suggests that autoencoders are able to distinguish IBD signal events from accidental background to a certain extent. This new study will show that the autoencoder can recognize geometric features of these PMT images, and assign similar images to similar encodings.

The training set was formed out of 20, 000 events taken from a single AD during normal operation of the Daya Bay experiment. The training set was split evenly between events tagged as IBD and the artificial accidentals described earlier. The value for each pixel was taken to be the amount of photoelectron charge collected by the corresponding PMT, scaled so that the smallest charge of the whole data set is -1 and the largest charge is $+1$ (in arbitrary units).

One way to convert the prompt and delayed event images into an input to the autoencoder is to concatenate them, as shown in Figure 4, into a single image with 16 rows and 24 columns. However, the autoencoder would not be aware that there is a correlation between the top half of the pixels and the bottom half in this image: namely, each pixel in the top half has a pixel in the bottom half which corresponds

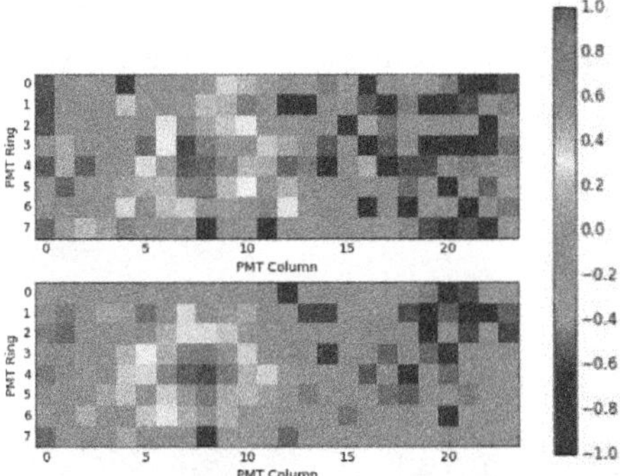

Fig. 4. Example prompt (top) and delayed (bottom) flashes for an IBD event. The color scale corresponds to the charge on the PMT (color online).

to the same physical PMT, just at a later time. To encourage the network to pay attention to these correlations, the concept of image channels is employed. An image with multiple channels can be thought of as assigning a vector value rather than a scalar value to each pixel. These channels are often utilized in the machine learning community to process red, green, and blue information in digital images. For this study, two channels are used: prompt charge and delayed charge. Thus there is an explicit connection in the autoencoder between the prompt and delayed charges for a given PMT.

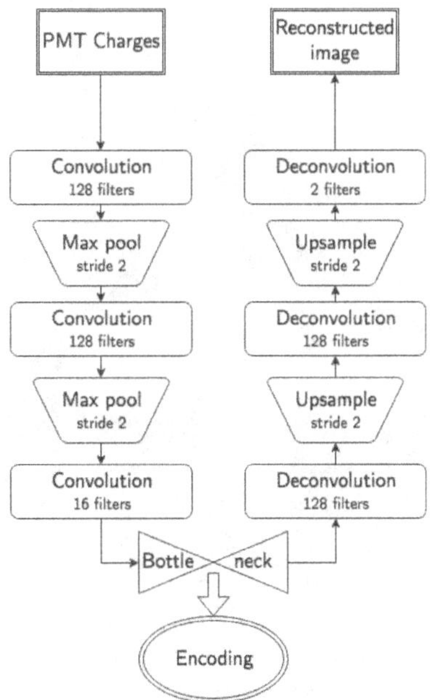

Fig. 5. Architecture of the convolutional autoencoder used in this study.

The architecture used for the neural network is detailed in Figure 5. It was chosen as an adaptation of a supervised convolutional network studied previously[6]. In the diagram, the bottleneck layer is where the encoding is extracted. The encoding size is 16 pixels, hence the information stored in the original input (384 pixels) must be summarized in 16 pixels.

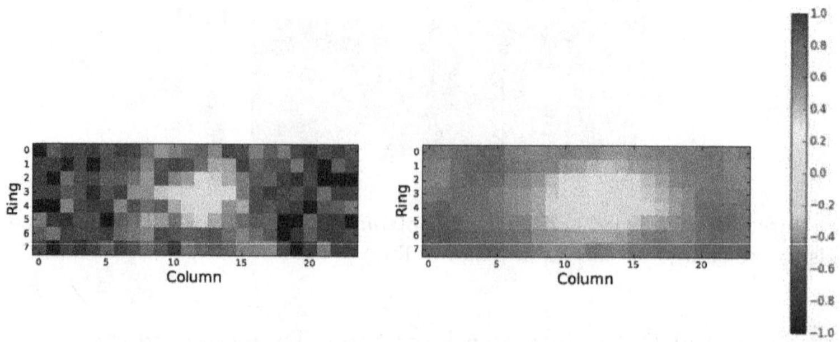

Fig. 6.   Input (left) and reconstructed (right) images for a prompt light flash.

If the convolutional autoencoder is trained well, then it should be able to reconstruct the most important parts of the images given just the encodings. An example reconstructed image is shown in Figure 6. It is clear that the location of the bright spot is preserved, indicating that it is considered an important feature by the autoencoder. By examining the 16-pixel encodings, it should be possible to

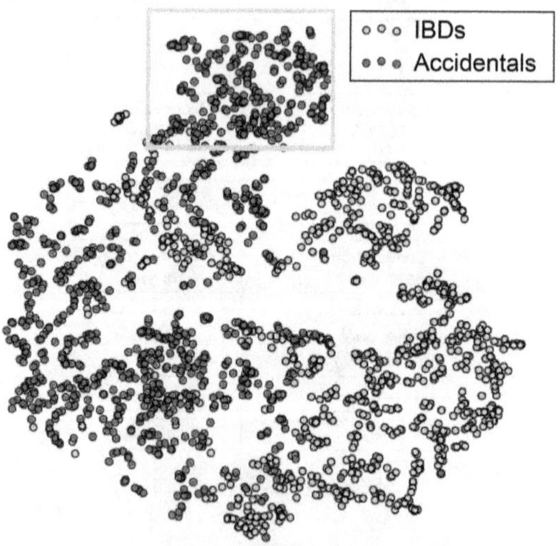

Fig. 7.   t-SNE embedding of events from the validation data set. Each point represents one IBD or accidental event. The colors represent the data set the event was drawn from—IBDs or artificial accidentals. (This classification was not provided to the autoencoder.) The green box encloses events which were investigated further (color online).

distinguish between different classes of 384-pixel images. Visualization of these encodings is a useful way to quickly determine if there are any patterns. The technique used for visualization is known as t-distributed Stochastic Neighbor Embedding, or t-SNE[7]. This technique converts a set of points in a high-dimensional space (such as 16 dimensions) into a set of points in a low-dimensional space (such as 2) in such a way that the Euclidean distances between nearby points are preserved as well as possible. This technique allows for the easy identification of clusters of points.

The t-SNE visualization for a random subset of the validation data set is shown in Figure 7. Each point in the plot represents one event pair, and the position of each point represents the t-SNE representation of the 16-pixel encoding, also known as "semantic space." Each point is colored according to whether the event it represents came from the IBD data set or the artificial accidental data set.

From the figure, it is clear that the convolutional autoencoder assigned different encodings to images representing IBD events compared to images representing accidentals. This is evident in the split between red points and blue points. Since the true identity of each event was not provided to the convolutional autoencoder at any point, it can be concluded that the autoencoder identified features which help determine whether an event is an IBD or an accidental.

Clearly, the discriminating power of the autoencoder is not exemplary, given the contamination of blue points in the red region and vice versa. However, the autoencoder learned on its own that some events had a different set of features compared to other events.

More investigation can uncover reasons why the autoencoder grouped the neutrino events the way it did. For example, Fig. 8 shows the distribution of accidental events as a function of vertical position in the AD, as reconstructed using a standard Daya Bay position reconstruction algorithm. The events are split into two data sets,

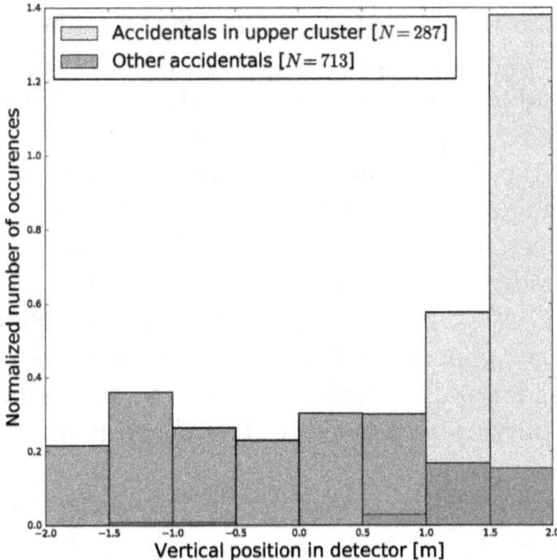

Fig. 8. Distribution of vertical position in AD of accidentals, based on their location on the t-SNE plot.

based on their position in the t-SNE plot of Fig. 7. The green bars show accidental events (red points) with a t-SNE representation inside the green box drawn on Fig. 7. The blue bars show all other accidental events (red points). That all of the points in the identified cluster represent events near the top of the AD indicates that one feature the autoencoder considers important is the vertical position of the event.

3. Discussion

The information gained from understanding the discriminating features identified by the autoencoder could be used to construct a traditional analysis which can tag accidentals and IBDs and has well-understood systematics and biases (unlike an unsupervised neural network). Such an analysis is the subject of a further study. The same analysis pipeline could be used to improve identification of ^{9}Li if convolutional autoencoders are also successful at distinguishing those background events from IBDs.

This technique could be improved by using more sophisticated neural network architectures, including adversarial techniques[8], and by incorporating time information into the input data so the autoencoder can make decisions based on the timing of the light flashes.

Acknowledgment

This work was supported by the U.S. Department of Energy, Office of Science, Office of High Energy Physics.

References

1. F. P. An *et al.*, New measurement of antineutrino oscillation with the full detector configuration at daya bay, *Phys. Rev. Lett.* **115**, p. 111802 (Sep 2015).
2. F. P. An *et al.*, Measurement of electron antineutrino oscillation based on 1230 days of operation of the daya bay experiment, *Phys. Rev. D* **95**, p. 072006 (Apr 2017).
3. Y. LeCun, L. Bottou, Y. Bengio and P. Haffner, Gradient-based learning applied to document recognition, *Proc. of the IEEE* **86** (Nov 1998).
4. R. Fergus, M. D. Zeiler, G. W. Taylor and D. Krishnan, Deconvolutional networks, *IEEE Conf. on Computer Vision and Pattern Recognition*, 2528 (2010).
5. S. Kohn, E. Racah, C. Tull, D. Dwyer, Prabhat and W. Bhimji, Deep learning with raw data from daya bay, *Journal of Physics: Conference Series* **898**, p. 072050 (2017).
6. E. Racah, S. Ko, P. Sadowski, W. Bhimji, C. Tull, S.-Y. Oh, P. Baldi and Prabhat, Revealing fundamental physics from the daya bay neutrino experiment using deep neural networks (Jan 2016).
7. L. van der Maaten and G. Hinton, Visualizing data using t-SNE, *Journal of Machine Learning Research* **9**, 2579 (Nov 2008).
8. A. Makhzani, J. Shlens, N. Jaitly and I. J. Goodfellow, Adversarial autoencoders, (2015).

The JUNO Central Detector Filling and Overflow System for the Liquid Scintillator

Shuaijie Li*, Baobiao Yue†, Jiajie Ling

Sun Yat-sen University,
Guangzhou, Guangdong Province, P. R. China
*E-mail: *lishi26@mail2.sysu.edu.cn, †yuebb@mail2.sysu.edu.cn*

Zhi Wu, Xiao Tang, Yuekun Heng

Institute of High Energy Physics, Chinese Academy of Sciences,
Beijing, P. R. China

On behalf of the JUNO Collaboration

The Jiangmen Underground Neutrino Observatory is a multi-purpose underground experiment and its liquid scintillator detector will be one of the largest detectors in the world. The liquid level control during filling and after filling is essential to the safety of the central detector. The process of filling and controlling the overflow is crucial for the safe and stable operation of the central detector. We present here details on the filling and overflow system design. It has multi-function, multi-monitor, multi-control level, multi-emergency solutions.

Keywords: JUNO; LS filling.

1. Introduction

The Jiangmen Underground Neutrino Observatory (JUNO) is a multi-purpose neutrino experiment [1][2]. It was proposed in 2008 to determine the neutrino mass hierarchy (MH) by detecting reactor antineutrinos from nuclear power plants at a distance of 55 km. The JUNO detector consists of a central detector, a water Cherenkov detector and a muon tracker. The central detector (CD) is a LS detector of 20 kilotons fiducial mass with an energy resolution of $3\%/\sqrt{E(MeV)}$. The CD is submerged in a water pool (WP) to shield it from natural radioactivity from the surrounding rock and air. A diagram of the JUNO detector is shown in figure 1.

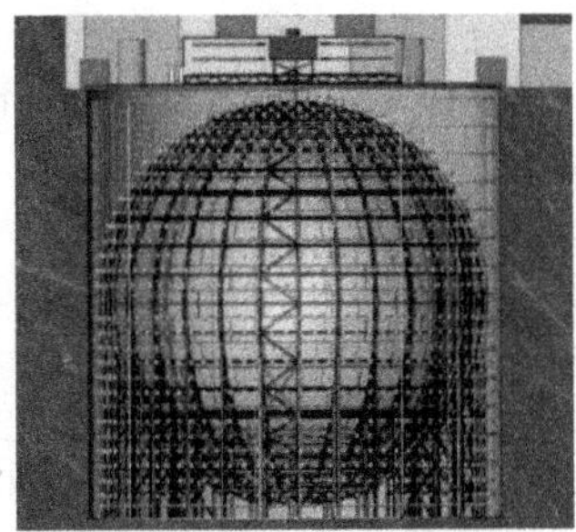

Fig. 1.   Overview of the JUNO detector.

The density of liquid scintillator (LS) is 0.859 g/ml. Twenty thousand LS is contained in a spherical container of radius of 17.7 m. The light emitted by the LS is detected by about 17000 20inch photomultiplier tubes (PMTs). PMTs are installed on a spherical structure of a radius of 19.5 m, and submerged in ultrapure water to protect the LS from the radioactivity of the PMT glass.

The mechanics of the central detector is very challenging. The CD is an acrylic sphere with 12 cm of its thickness, full of LS, which is submerged in a cylindrical WP. A chimney for calibration operation will connect the central detector to outside from the top. Special shielding against radioactivity and a muon detector will be designed for the chimney.

In section 2 we will discuss the design of filling and overflow system. Section 3 details the multiple liquid level measurements and emergency control. Section 4 will give a conclusion on the progress.

2. Overview of filling and overflow system design

2.1. *The layout of underground cavern*

A filling room is located above the WP and connects to the CD through the chimney. The size of the filling room is $20 \times 10 \times 10$ m^3. The whole setup is in 700 m underground. There is a vertical shaft and a sloped access tunnel connecting to the underground laboratory. It can transport a maximum device size of $4 \times 4 \times 9$ m^3.

2.2. *Overview of filling and overflow system requirements*

The purpose of filling system is the filling of the CD with pure LS and the WP with pure water [3]. Following the initial fill, the LS is circulated through a cleaning loop to ensure steady operation over 20 years. The overflow system ensures that the CD can run safely for about 20 years. Allowing for overflow is important because varying temperature will change the volume of the liquid.

The first step is simultaneous filling of the CD and WP with ultrapure water. This purges the air form the CD. Then we exchange the water in the CD with LS from the LS chamber. It is very important to control the levels of the liquids, especially in the CD. During the water filling process, the difference of levels between the inside and outside of the CD must be less than 2 m with an accuracy of 5 cm or less of adjustment. After LS exchanging, the LS level should be 4.5±0.2 m higher than the liquid level in the WP. The accuracy of this control should be less than 3 cm. The volume of the liquids changes with temperature. The LS will change 23 m^3 per degree Celsius. We can consider the height of water stable because of the large area at the surface of the WP.

JUNO's physics goals require a very low background. In the LS the singles rate from dust on the inner surface of the acrylic sphere should be less than 0.1 Hz. The mass of the dust should be less than 14 g. On the outer surface, the singles rate from dust should also be less than 0.1 Hz. The mass of that dust should be less than 25 g. We require less than 0.05 Bq/m^3 of Radon in the LS. The Radon source comes mostly from pure water and air.

The acrylic sphere must be sealed after cleaning. We require leak-tightness of 10^{-6} mbar • L/s for the whole system and 10^{-8} mbar • L/s for each single part.

2.3. *Filling scheme: Pure water filling*

We begin by simultaneously filling the CD and WP with pure water. Pure water will clean the surfaces of the acrylic sphere and purge air. The acrylic sphere might break under a large pressure difference between its inner and outer volume. We must minimize the pressure difference between the inside and outside of the CD. Because both are initially filled with water, this is equivalent to minimizing the liquid level difference.

This is achieved by adjusting a proportional valve to control the speed of CD filling and WP filling. On average, the speed of filling will be 80 m³/h for the WP and 20 m³/h for the CD. With the proportional valve, the total speed will always be 100 m³/h. Filling the water will take 2 months. As we can see in figure 1, the shapes of the WP and CD are different. We have to measure the liquid level carefully. We use pressure sensor, radar and other methods to get the correct liquid level information. Because the CD is narrow at the bottom and top, the rate at which the CD is filled must be controlled very carefully at the beginning and the end. We set a leak-gate in the CD to tell us when the process has concluded. The basic method is shown in figure 2.

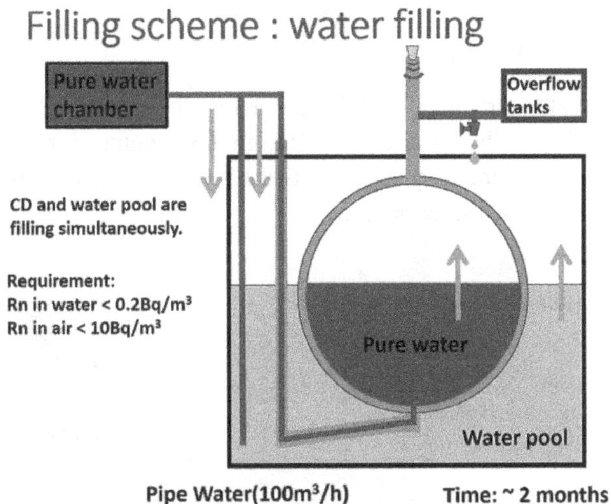

Fig. 2.   Water filling in the CD and WP.

2.4. *Filling scheme: LS filling*

The JUNO detector is a LS detector. After water filling, the water in the CD will be replaced by LS. At this step, the WP water level is stable. We utilize the lower density of LS and its poor solubility in water. Water will be pumped from the bottom of the CD and LS is filled through the chimney. The speed of LS filling is 6 m³/h, so this process will take

420

about 6 months. This speed is limited by the LS production. We pump LS from the purification chamber through the overflow tank in the filling room, and from there into the CD.

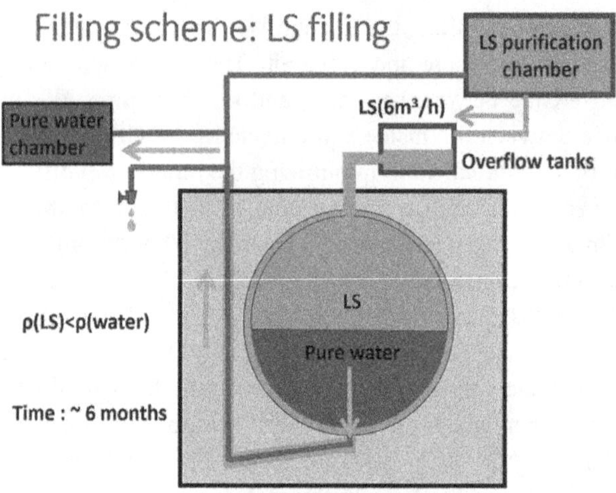

Fig. 3.   LS filling method, green represents on LS and blue represents on water (color online).

2.5. *Filling scheme: LS circulation*

JUNO will run for 20 years after LS filling. For long term running, LS will need to be circulated and purified. In the LS circulation design, LS is pumped from bottom of the CD and replaced through the chimney with purified LS from the purification chamber. This process is similar to the LS filling process.

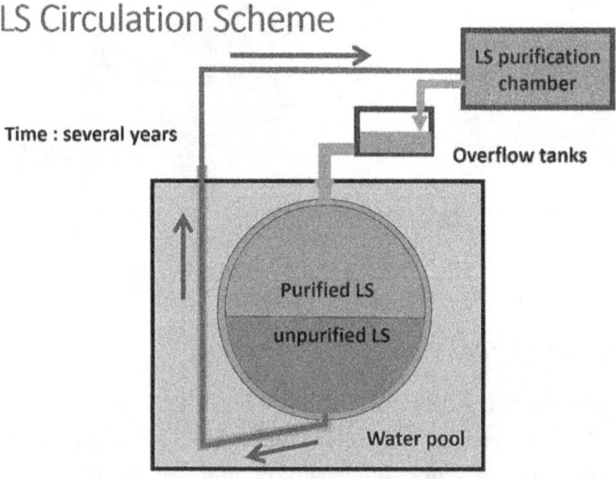

Fig. 4.   LS circulation method, the light green is new LS, the dark green is old LS (color online).

2.6. *LS overflow design*

Safety of the LS system during the running is important. The temperature is maintained by air conditioning. The liquid level in the WP is stable because of the large surface area. But the chimney is much thinner. If the temperature changes, the LS level changes a lot. We designed the overflow system to compensate for changes in the LS level. Ideally we would use a single high volume tank. But the width of the tunnel and size of the filling room make this impossible. Instead we use two tanks of 3 m diameter and 8 m height. They work together and serve as backups for each other. And one can be used as a reservoir if the other overflow tank needs to be cleaned or fixed. The layout of the filling room is shown in figure 5.

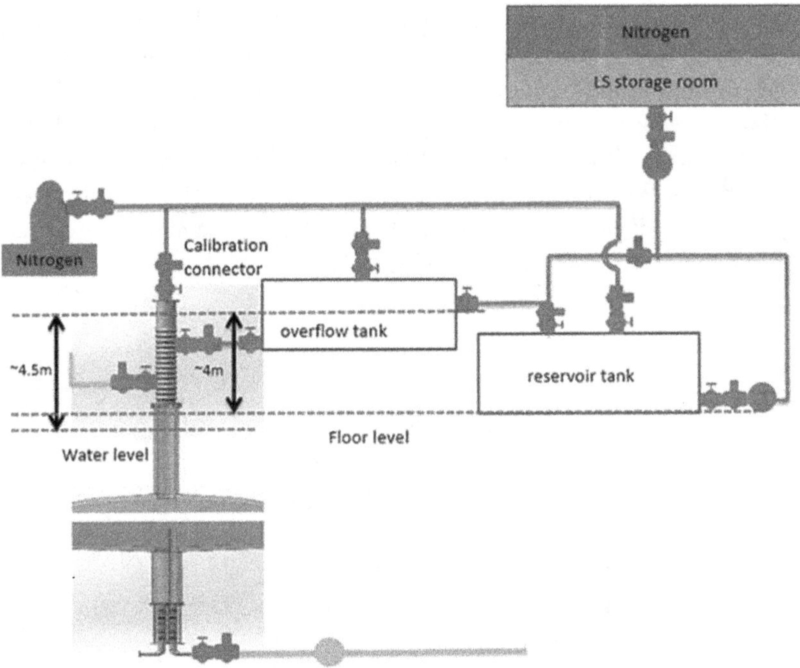

Fig. 5. Overflow design, the highest red dash line is safest liquid level (color online).

When the temperature rises, the volume rises, and the liquid level in the overflow tanks rises. The pressure which results from a level higher than 4.5 meters above the water level in the WP is unsafe for the CD. To avoid this an overflow gate is opened at this level. LS will drain. All the tanks are cylindrical. We want a large surface area for the liquid in the overflow tanks, so we place them on their sides with a stand to prevent them from rolling. The central height of the overflow tank will be 4.5 m above the WP surface to maximize the surface area. The overflow gate will be open all the time. When temperature rises, LS will flow into the reservoir tank through the overflow gate, so the LS level can't rise higher than it.

When the temperature falls, the volumes fall, and the liquid levels fall. We use a pump to pump LS from the reservoir into the overflow tank.

3. Multiple liquid level measurement and emergency control

As we introduced above, controlling the liquid level in the CD, WP, and overflow tank is important. For the CD and WP, the range is about 40 m. We need to use redundant methods to get the correct liquid levels to crosscheck one another. We use radar and hydraulic methods to measure the levels directly. Both have a long range and good precision. In addition, we can use an air pressure gauge to calculate the levels. The multiple liquid level measurement is shown in figure 6. In addition, we need to know where the interface between water and LS is during the LS filling process. Because of the low background requirements, we can't put any extra devices in the WP or CD. We can't see the interface directly. We use the difference in the liquid level in the communicating vessel and the LS in the chimney to calculate the interface position at pressure equilibrium.

For the emergency control, we design all devices including pipes and tanks with backups. The system will switch devices and instruments automatically and send warnings if an error happens. We set three states of emergency, minor problems that can be solved by the system or a shifter, serious problems that need remote administrator operation and fatal problems that need to shut down the system.

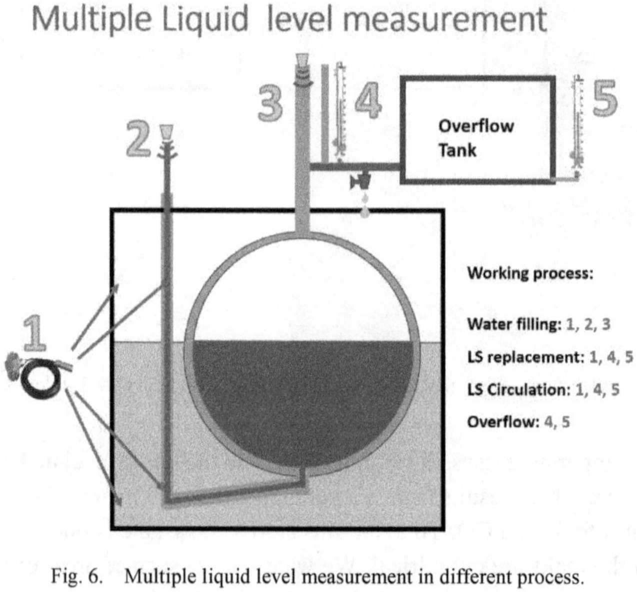

Fig. 6.   Multiple liquid level measurement in different process.

4. Conclusion

This system is designed to work automatically. The design accounts for many potential problems in different situations. We have done the primary survey on hardware and software. We built a test system in Sun Yat-Sen university to prove our system. We will finalize the filling design in the near future.

References

1. Fengpeng An *et al.*, J. Phys. G: Nucl. Part. Phys. **43** 030401 (2016).
2. Z. Djurcic *et al.* [JUNO Collaboration], arXiv:1508.07166 [physics.ins-det].
3. J. Benziger *et al.*, Nucl. Instr. and Meth. A 608 (2009) 464–474.

Recent Progress on
the Muon g-2 Experiment at Fermilab

Bingzhi Li

on behalf of the Muon g-2 Collaboration

Institute of Nuclear and Particle Physics, School of Physics and Astronomy,
Shanghai Jiao Tong University,
Shanghai Key Laboratory for Particle Physics and Cosmology,
800 Dongchuan Rd, Shanghai 200240, China
E-mail: bingzhili@sjtu.edu.cn

The Muon g-2 experiment measures the anomalous magnetic moment of muon. The previous measurement at Brookhaven National Laboratory, experiment E821, completed in 2001 found a greater than 3σ discrepancy from the predicted value. The new generation Muon g-2 experiment at Fermi National Laboratory is expected to accumulate 21 times more statistics and achieve a factor of four improvement in the precision with an upgraded apparatus and reduced systematic uncertainties. With both improvements from experiments and theoretical calculations, a more precise test of the Standard Model value will be made. The experiment construction was completed and the commissioning run successfully finished in Summer 2017. The physics data taking will begin in early 2018.

Keywords: Muon g-2; anomalous magnetic moment; procession frequency; beyond the SM.

1. Solving a Long Mystery

1.1. *Muon g-2 Theory*

For fermions, the magnetic moment is related to the spin:

$$\vec{\mu} = g\frac{Qe}{2m}\vec{s} \tag{1.1}$$

where $Q=\pm 1$ and e is the unit of elementary charge. The g-factor is given by Dirac's equation, which predicted for the electron to be 2, as shown in the non-relativistic form [1]:

$$\left(\frac{1}{2m_e}(P + qA)^2 + \frac{q}{2m_e}\sigma \cdot B - qA^0\right)\Psi_A = (E - m_e)\Psi_A \tag{1.2}$$

For particles such as the muon, this simple classical expectation differs from the observed value by a small fraction of a percent. The anomaly is denoted "a_μ" and defined as $a_\mu = (g - 2)/2$, called the anomalous magnetic moment of the muon. As for the electron, a measurement of the hyperfine structure of hydrogen in 1947 [2], showed that $a_e \sim 0.1\%$. Theoretical research indicates that the anomalous magnetic moment of the muon comes from quantum electrodynamics (QED), electroweak (EW) and quantum chromodynamics (QCD) hadronic corrections, as well as new physics (NP) beyond the Standard Model (BSM), as shown in Eq. (1.3). Feynman diagrams of contributions to a_μ are shown in

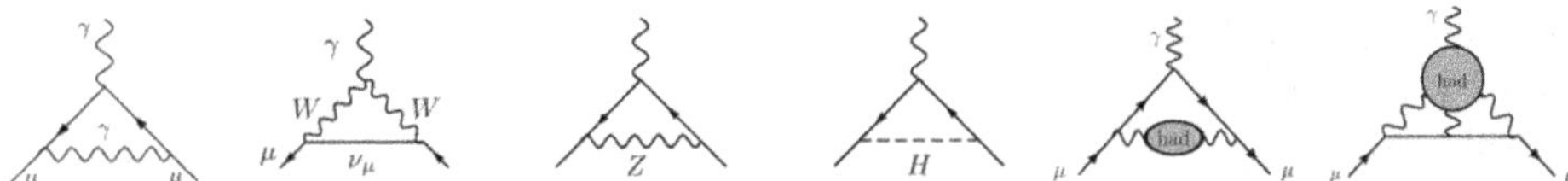

Fig. 1. Examples of contributions to a_μ. From left to right, the first plot shows a QED term. The middle 3 plots indicate the electroweak $W^\pm, Z, H$ exchange. The last 2 plots give the lowest-order hadronic vacuum polarization and hadronic light-by-light scattering.

Fig. 1. Because of the muon's larger mass, EW and hadronic contributions to the theoretical calculation of its anomalous magnetic moment are important at the current level of precision, whereas these effects are less important for the electron. The muon anomalous magnetic moment is also sensitive to contributions from BSM effects thus the muon g-2 experiment is a useful probe for new physics.

$$a_\mu = a_\mu^{QED} + a_\mu^{EW} + a_\mu^{Had} + (a_\mu^{NP}) \tag{1.3}$$

Calculated values for various components are summarized in Table 1. The QED component is the largest contribution to a_μ but has the smallest uncertainties. The EW component gives a small contribution and small uncertainty. The hadronic contribution contains two parts, the hadronic vacuum polarization (HVP) and hadronic light-by-light (HLbL) terms which give the largest uncertainties. HVP can be obtained from e^+e^- to hadronic measurements. HLbL must be estimated from a model and cannot be obtained directly from experiment data.

Table 1. Standard Model contributions to the muon anomalous magnetic moment, taken from [3]. Two values are quoted for lower order HVP process to reflect different estimates.

	Values in 10^{-11} units
QED	116584718.951$\pm$0.0009$\pm$0.019$\pm$0.007$\pm$0.077
HVP (lower order) [4]	6923$\pm$42
HVP (lower order) [5]	6949$\pm$43
HVP (higher order) [5]	-98.4$\pm$0.7
HLBL	105$\pm$26
EW	153.6$\pm$1.0
Total SM [4]	116591802$\pm$42$_{\text{H-LO}}\pm$26$_{\text{H-HO}}\pm$2$_{\text{other}}$ ($\pm$49$_{\text{total}}$)
Total SM [5]	116591828$\pm$43$_{\text{H-LO}}\pm$26$_{\text{H-HO}}\pm$2$_{\text{other}}$ ($\pm$50$_{\text{total}}$)

1.2. *The Muon g-2 Experiment*

The Muon g-2 experiment measures the anomalous magnetic moment of muon, $a_\mu \equiv (g-2)/2$, with a precision to 140 ppb.

The experiment uses Fermilab's powerful accelerator to explore the characteristics of the muon in "empty" space with a uniform and strong magnetic field. Even in the vacuum, the space is never "empty". According to the quantum physics, the invisible sea of virtual

426

particles pop in and out in a very short time. The Muon g-2 experiment uses muon beam stored in the storage ring with a uniform magnetic field to test the presence and nature of virtual particles.

The previous experiment at Brookhaven National Laboratory found a greater than 3σ (standard deviation) discrepancy between the observed value and theoretical prediction. With the expected four-fold increase precision, the new Muon g-2 experiment at Fermilab should be more sensitive to the new physics than any other previous similar experiments. Seven countries, 35 institutions and 190 collaborators are collaborating on the Muon g-2 experiment [6].

The comparison between experiments and SM value is shown in Fig. 2.

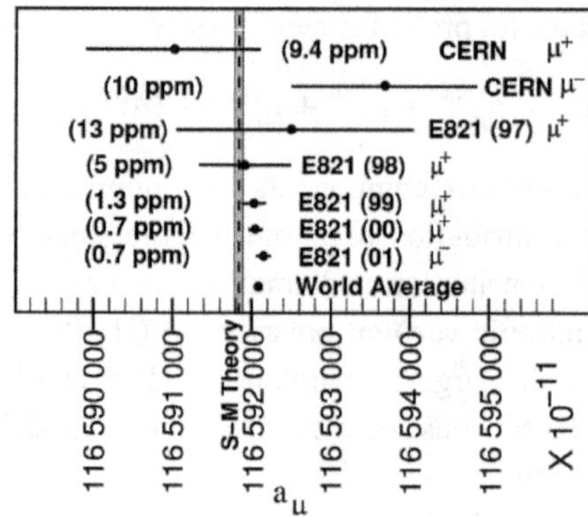

Fig. 2.   Measurements of a_μ from CERN and BNL E821 experiments, taken from [3]. The vertical band is the Standard Model prediction value using the hadronic contribution from [4].

2.   Experimental Setup

2.1.   *Generating and Delivering the Muon Beam*

The Muon g-2 experiment begins with a bunched proton beam from the 8 GeV Booster hitting a target and produces different kinds of particles. The emerging pions with energy of 3.11GeV/c$\pm$5% are selected. Pions then quickly decay to muons:

$$\pi^+ \rightarrow \nu_\mu + \mu^+$$

The pions and muons go through a triangular-shaped tunnel called the Muon Delivery Ring. The particle beam of pions and muons travels around the ring until pions all decay into muons. The muon beam with magic momentum of 3.094GeV/c is selected and injected into the Muon Storage Ring which has been used in BNL E821 experiment. The layout of the beamline of the Muon g-2 experiment at Fermilab is shown in Fig. 3.

2.2. *Storing Muon Beam with the Magic Momentum*

The muon beam with the magic momentum 3.094 GeV/c passes through a super-conducting inflector into the Muon Storage Ring, as shown in the Fig. 4. The muon beam is shifted onto orbit by three kickers.

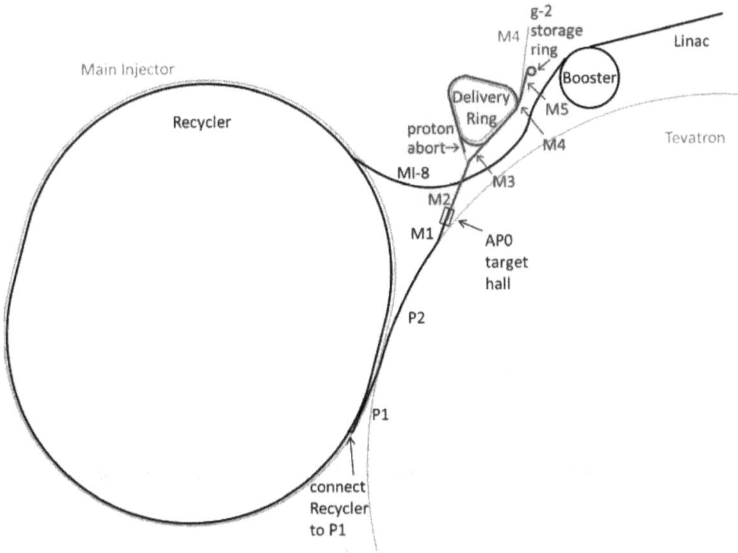

Fig. 3.   Path of the beam, taken from [3]. Protons (black) are accelerated in the Booster, are re-bunched in the Recycler, and then travel to the target. The pion and decayed muon beam (red) then travels through around the Delivery Ring, and the polarized muon then is injected into the muon storage ring (color online).

When muon beam circulates around the storage ring, there are two frequencies: the muon cyclotron frequency:

$$\omega_c = \frac{eB}{m\gamma} \tag{2.1}$$

and muon spin precession frequency:

$$\omega_s = \frac{eB}{m\gamma}(1 + \gamma a_\mu) \tag{2.2}$$

The difference between these two frequencies can be measured and it is directly related to muon anomalous magnetic moment a_μ:

$$\omega_a = \omega_s - \omega_c = a_\mu \frac{eB}{m} \tag{2.3}$$

To focus the beam, vertically electric quadrupole field is applied and the relationship between ω_a and magnetic field is:

428

$$\vec{\omega}_a = \frac{e}{m}\left[a_\mu \vec{B} - \left(a_\mu - \frac{1}{\gamma^2 - 1}\right)\left(\vec{\beta} \times \vec{E}\right)\right] \tag{2.4}$$

The electric term can be cancelled by choosing a magic $\gamma = 29.3$. Then the Eq. (2.4) has the same form as Eq. (2.3).

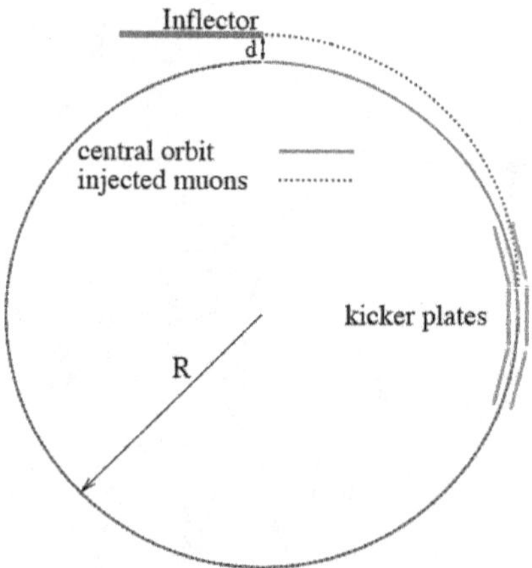

Fig. 4. Muon beam orbit in the storage ring, taken from [3]. The kickers shift the muon beam from the injected orbit to the storage orbit.

3. Analysis Method

3.1. *Muon Precession Analysis*

As the muons circulate around the storage ring, the direction between the muon spin and momentum changes due to the procession. When μ^+ decay into e^+

$$\mu^+ \rightarrow \nu_e + \bar{\nu}_\mu + e^+$$

the decayed positron carries the information from the muon's spin and momentum. By measuring the arrival time and energy of the positron precisely, ω_a can be obtained. Fig. 5 shows the particle decay and detection process together with the calorimeter and tracker detector layout. There are 24 calorimeters and 2 straw trackers around the storage ring.

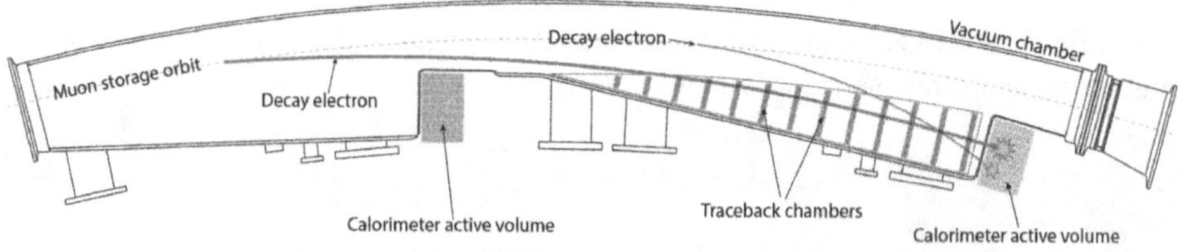

Fig. 5. Calorimeter and tracker in vacuum chamber and the detection process. Taken from [3]. The red line indicates the orbit of decay positron which hits the tracker and calorimeter (color online).

One method (called T-Method) to extract ω_a is to count the number of decay e^+ above an energy threshold (1.8GeV) and fit the positron's time distribution with a five-parameter fit. Another method (called Q-Method) to extract ω_a is to collect energy deposited (in calorimeter) and fit the energy time distribution instead. Both methods give a similar statistical uncertainty and comparable results. The two methods can be used to cross check each other because they are independent and expected to have different sensitivities to various systematic uncertainties.

3.2. *Magnetic Field Analysis*

The precise a_μ measurement could be extracted by measuring ω_a with a homogeneous magnetic field B (Eq. (2.3)). The magnetic field B is measured by Nuclear Magnetic Resonance (NMR) and can be expressed in term of the free proton precession frequency $\omega_p = 2\mu_p|\vec{B}|/\hbar$. The muon anomaly a_μ can be expressed to:

$$a_\mu = \frac{\omega_a}{\omega_p}\frac{2\mu_p}{\hbar}\frac{m_\mu}{e} = \frac{\omega_a}{\omega_p}\frac{\mu_p}{\mu_e}\frac{m_\mu}{m_e}\frac{g_e}{2} = \frac{\omega_a/\omega_p}{\mu_\mu/\mu_p - \omega_a/\omega_p} \tag{3.1}$$

with $\mu_e = g_e e\hbar/4m_e$ and $m_\mu/m_e = (g_e/g_\mu) \times (\mu_e/\mu_\mu)$. The BNL E821 experiment obtained an uncertainty $\delta\omega_p \approx 170$ parts per billion (ppb). The g-2 experiment at Fermilab updated the hardware apparatus and analysis techniques to reduce the uncertainty to reach the goal of $\delta\omega_p \approx 70$ppb.

Pulsed Nuclear Magnetic Resonance probe is the core of the magnetic field measurement. The NMR probes system mainly contains three parts:

- Fixed probes are located just above and below the muon storage ring
- Other sets of probes are pulled in a trolley used to determine the field seen by the muons in the storage ring
- A final set of probes is used for calibration

The fixed probes system is designed to measure the field continuously during the data taking. The trolley measures the magnetic field distribution around the storage ring when the muon beam is off. During a trolley run, about 6000 locations in azimuth data is taken by each of the 17 NMR probes (a total of 100,000 field points). The absolute calibration system is used to provide a chain relating field measurements to the equivalent Larmor frequency of a free proton.

4. Commissioning Run and Future Plan

First beam of muons was injected into the storage ring on May 31 and the end of run was on July 7. During the 38 days run about 10^{16} protons hit the target. The goals of commission run are to test the machine and exercise the entire parameter space. Figure 6 shows the result of the commissioning run. It shows two weeks' data accumulated in June 2017 with a about 700,000 positrons from muon decay. The statistics collected is already at the level of CERN II/III results.

The experiment is expected to take data through 2020 and accumulate 21 times more statistics than BNL E821 experiment. The increase of statistics combined with the improved systematic uncertainty is expected to give a factor of four improvement in the precision.

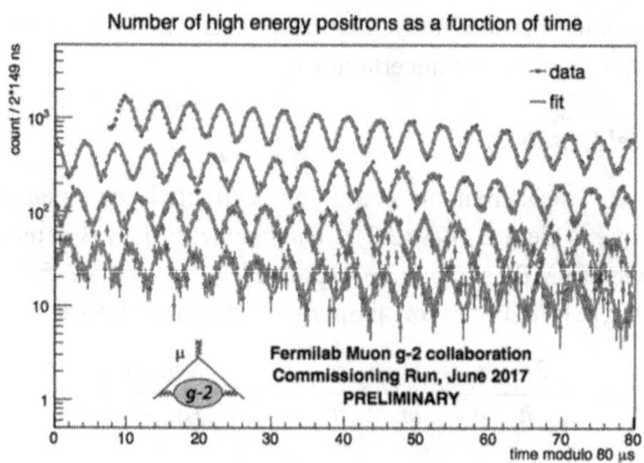

Fig. 6.   Preliminary result of the commissioning run, taken from [7].

Acknowledgment

This paper was supported by the National Natural Science Foundation of China (NSFC) under Grant 11375115. We also thank the staffs at Fermilab and collaborating institutions, and acknowledge support from the DOE and NSF (USA); STFC (UK); INFN (Italy).

References

1. F. Halzen and A. D. Martin, *Quarks and Leptons: An introductory course in modern particle physics* (John Wiley and Sons, New York, 1984).
2. J. E. Nafe, E. B. Nelson, and I. I. Rabi. The Hyperfine Structure of Atomic Hydrogen and Deuterium. *Phys. Rev.*, **71**:914–915, 1947.
3. J. Grange et al., *Muon (g-2) Technical Design Report 2015* (Preprint 1501.06858v1).
4. M. Davier, A. Hoecker, B. Malaescu and Z. Zhang, *Eur. Phys. J.* C71 (2011) 1515, Erratum-ibid. C72 (2012) 1874.
5. K. Hagiwara, R. Liao, A. D. Martin, D. Nomura and T. Teubner, *J. Phys.* G38 (2011) 085003.
6. http://muon-g-2.fnal.gov/collaboration.html.
7. https://cdcvs.fnal.gov/redmine/projects/g-2/wiki/ApprovedPlotsAndImages.

Physics Goals and R&D Progress of Jinping Neutrino Experiment*

ZHAO, Lin

on behalf of Jinping Neutrino Experiment research group

*Department of Engineering Physics, Tsinghua University,
Beijing, P. R. China
E-mail: l-zhao15@mails.tsinghua.edu.cn
https://jinping.hep.tsinghua.edu.cn*

Equipped in the China Jinping Underground Laboratory (CJPL), 2400 meters under Jinping Mountain, Sichuan province, P. R. China, the proposed Jinping Neutrino Experiment will be an unique observatory on low-energy neutrino physics researches, including solar neutrino, geo-neutrino and supernova relic neutrino. Jinping has a potential to significantly improve the measurements of neutrinos of a few MeV energy from the interior solar fusion processes, as well as geo-neutrinos with an unambiguous separation on U and Th cascade decays from the dominant crustal anti-electron neutrinos. With a slow liquid scintillator, Jinping is also very promising to make a discovery on supernova relic neutrino. Jinping one ton prototype detector is currently running at CJPL-I for several purposes, firstly, to understand the property of several important components of detector, such as acrylic, pure water and ultra-high molecular weight polyethylene rope, secondly, to study the technology of liquid scintillator and slow liquid scintillator in neutrino detection, thirdly, to measure the in-situ background like fast neutron of CJPL. Meanwhile, a package for Jinping neutrino experiment simulation and analysis has been developed for the study of detector R&D.

Keywords: CJPL; Jinping Neutrino Experiment; One-ton Detector Prototype; Liquid Scintillator and Slow Liquid Scintillator.

1. Introduction

In this paper, we present both the physics goals and the R&D progress of the proposed Jinping Neutrino Experiment. Section 2 gives a brief introduction to the experiment and the major physics goals. Section 3 reports the R&D study of the Jinping Neutrino Experiment. Section 4 gives a brief summary for this paper.

2. The Jinping Neutrino Experiment

2.1. *Location*

The Jinping Neutrino Experiment[1] is equipped in the China Jinping Underground Laboratory (CJPL)[2]. Located in Sichuan province, P. R. China, and 2400 meters under Jinping Mountain, CJPL has the lowest cosmic-ray muon flux (according

---

*This work is supported in part by, the National Natural Science Foundation of China (No. 11235006 and No. 11475093), the Tsinghua University Initiative Scientific Research Program (20121088035, 20131089288, and 20151080432), the Key Laboratory of Particle & Radiation Imaging (Tsinghua University), the CAS Center for Excellence in Particle Physics (CCEPP).

432

to the in-situ measurement[3], the cosmic-ray muon flux in CJPL is as low as $(2.0\pm0.4)\times10^{-10}/(\text{cm}^2\cdot\text{s}))$ and the lowest reactor neutrino flux (CJPL is far away from all the nuclear power plants[4] in operation and under construction), is one of the ideal sites to carry out low-energy neutrino experiments in the world.

2.2. *Conceptual detector design*

The conceptual design of Jinping Neutrino Experiment detector follows the principles adopted by the prevail underground neutrino experiments, with the special consideration on the unique feature of deep underground and the tunnel structure, as well as the major physics goals for low-energy neutrinos. The Jinping Neutrino Experiment plans to build two neutrino detectors with liquid scintillator or slow liquid scintillator as target material[1].

Fig. 1(a) or (b) shows a conceptual design of the detector for a cylinder or sphere inner vessel, respectively.

For the cylindrical scheme, the central vessel is made of acrylic, the height and diameter of the cylinder are both 14 meters. Filled with target material, the vessel is supported by an outer single-layer stainless steel truss, and both the truss diameter and height are 20 m. The acrylic vessel and the truss are connected by ropes made of synthetic fiber. On the stainless steel truss, the PMTs are uniformly distributed to detect the scintillation and cherenkov lights originated from neutrino interaction with the target material in the central region. The fiducial volume is defined by a cylinder of 11.2 m diameter and 11.2 m height.

In the spherical scheme, the stainless steel truss has a 20 m diameter, while the sphere has a diameter of 14 m. Because of the uniform stress distribution on the spherical vessel, the spherical scheme is more feasible for the structural design

(a: cylindric)

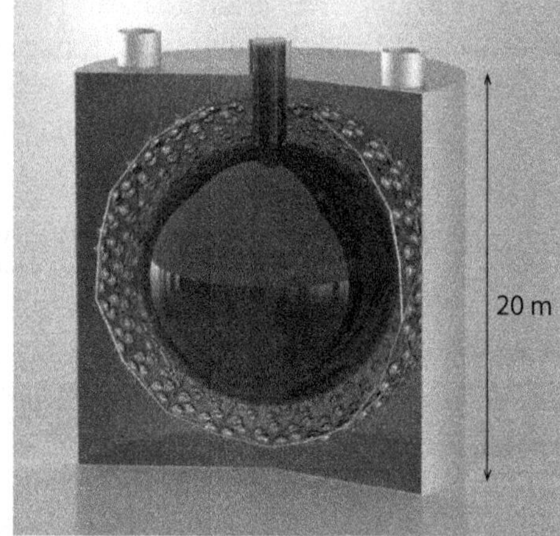

(b: spherical)

Fig. 1.   The conceptual design for Jinping Neutrino Experiment detector.

compared with the cylindrical scheme. But the cylindrical scheme has an advantage of making full use of the space and can contain more target materials[5].

2.3. *Physics goals*

Initial sensitivity studies have been conducted for the proposed Jinping Neutrino Experiment, assuming the lowest background levels either achieved as the other experiment or obtained due to special feature of CJPL.

2.3.1. *Solar neutrinos*

The proposed Jinping Neutrino Experiment has the potential to significantly improve the components, fluxes and spectra measurements of neutrinos with a few MeV energy from the interior solar fusion processes, see Fig. 2[1,6–8]. For the solar neutrino study, in which the detection process is neutrino-electron scattering, the total fiducial mass will be about 2 kiloton. The expected improvements are listed in the following items[1].

- Improvement on the known solar neutrino components, including pp neutrino, ^{7}Be neutrino, ^{8}B neutrino, and pep solar neutrino fluxes.
- Discover solar neutrinos from the carbon-nitrogen-oxygen (CNO) cycle, in order to discover the energy sources of massive stars, to find the solution of metal abundance of the solar core, and to attack the homogeneous chemical assumption of Carbon, Nitrogen, and Oxygen elements.
- Providing a critical test for the Mikheyev-Smirnov-Wolfenstein (MSW) theory in the high density environment, by measure the transition phase for the solar neutrino oscillation from vacuum to matter effect precisely.

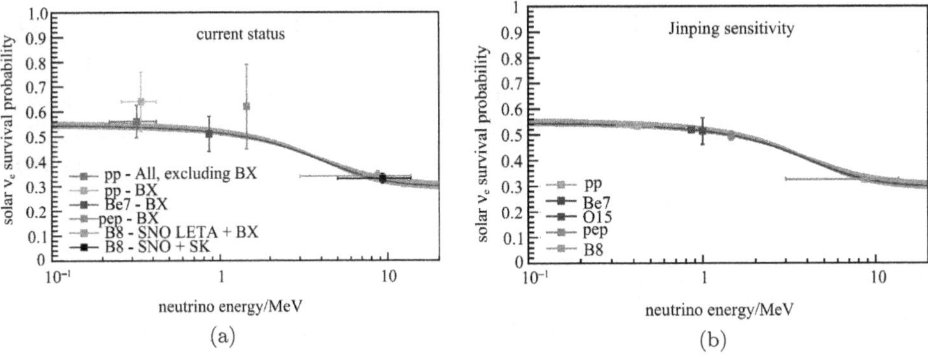

Fig. 2. The transition of oscillation probability from the vacuum to matter effect as a function of the solar neutrino energy. The present experimental status and Jinping sensitivity are shown in (a) and (b) plots, respectively. The horizontal error for ^{15}O bar is not shown for clarity.

2.3.2. *Geo-neutrino*

The Jinping Neutrino Experiment is an ideal site to precisely measure the geo-neutrino flux, because it's located far away from nuclear power plants[4]. The equivalent fiducial mass is 3 kiloton for the geo-neutrino research in the Jinping Neutrino Experiment[1], in which the neutrino interaction is the inverse beta decay (IBD) process.

The main purposes for geo-neutrino detection in the Jinping Neutrino Experiment are summarized below[9].

- Precision measurement of Th/U ratio
- Discriminate over the existing bulk silicate Earth (BSE) models.

2.3.3. *Supernova relic neutrinos*

Supernova relic neutrinos are unique tool for astronomy research because it will reveal the process of stellar evolution and the history of our universe. With the 3 kiloton fiducial mass of slow liquid scintillator detector and 10 years data, the Jinping Neutrino Experiment the potential to discover supernova relic neutrinos[11].

3. Detector R&D

3.1. *One-ton prototype detector*

An one-ton prototype detector of the Jinping Neutrino Experiment has been built to study the performance of the slow liquid scintillator, and to understand the background of the photomultiplier tubes (PMTs) and the radioactivity background originating from the detector materials.

Fig. 3 shows the structural scheme of the combination of the stainless steel tank, stainless steel truss, and acrylic vessel designed for the one-ton prototype detector. The installation of the prototype detector has finished in early 2017, as shown in Fig. 4(b). A total of 30 PMTs are uniformly installed in the one ton prototype detector.

The Prototype detector was running with pure water from May 8 2017 to July 27, 2017 and slow liquid scintillator from July 31, 2017 to now as the target. A gain study of the PMTs was also conducted. Fig. 5 shows the dark noise and gain change versus time. Each line in the plot represents a PMT status.

3.2. *The slow liquid scintillator*

The slow liquid scintillator used in the prototype detector is based on the relevant physical aspects of different combinations of linear alkylbenzene (LAB), with an solution of 0.07g/L 2,5-diphenyloxazole (PPO), and 13mg/L 1,4-bis (2-methylstyryl)-benzene (bis-MSB). The light yield, time profile, emission spectrum, attenuation length of scintillation emission and detected light yield of Cherenkov emission have been studied[12]. A acrylic container placed on the path of triggered muons to collect both scintillation and Cherenkov light in the liquid scintillator sample. The top and bottom PMTs[12], which were designed symmetrically aligned with the acrylic container, were used to collect the scintillation and Cherenkov light of the liquid scintillator Fig. 6 shows the bench test results for the sample of the slow liquid scintillator. Scintillation light is detected by the top PMT. In the first 20 ns of detection, the amplitude of the bottom PMT waveform is higher than that of the top PMT waveform because of Cherenkov light. A separation between Cherenkov and scintillation lights is seen. The Cherenkov light gives a prompt component and a narrow pulse shape, which can be used in the fit to the waveform of PMTs.

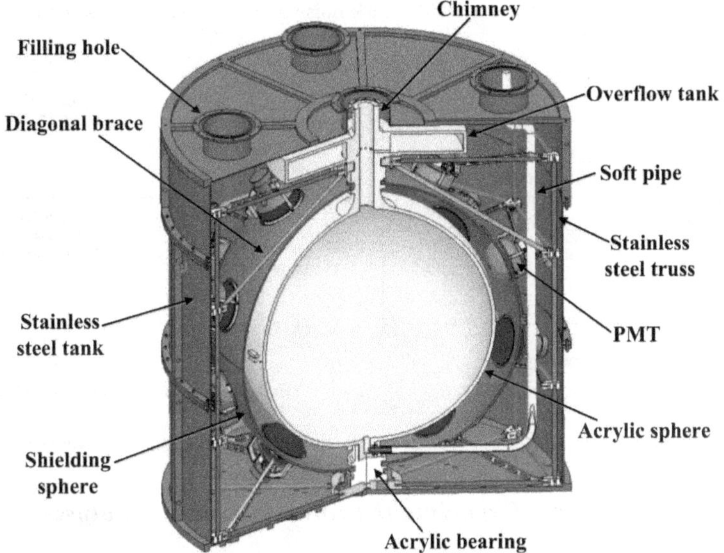

Fig. 3.   The structural scheme of the one ton prototype detector.

(a)   (b)

Fig. 4.   The prototype detector (a) has finished it's installation in early 2017 and taking data (b) now.

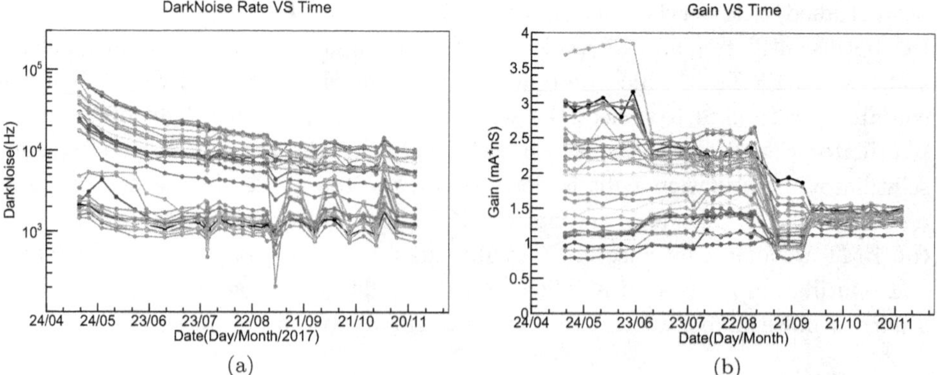

Fig. 5.   One ton prototype detector PMTs dark noise and gain change versus time are shown in (a) and (b) plots, respectively.

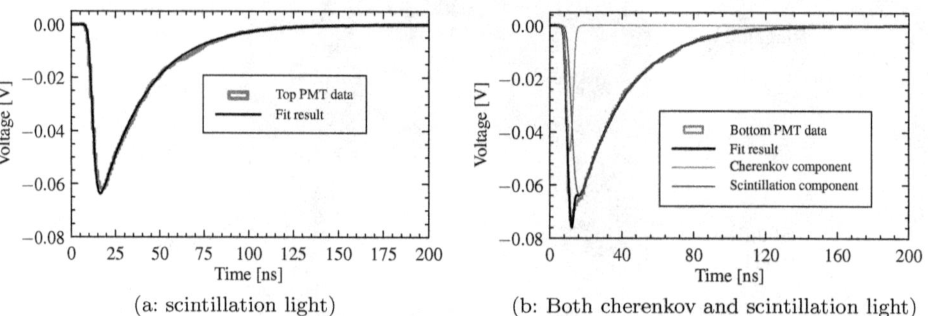

(a: scintillation light)   (b: Both cherenkov and scintillation light)

Fig. 6.   Scintillation light (a) and both cherenkov and scintillation light (b) waveform fitting result.

3.3. *Jinping Neutrino Experiment Simulation and Analysis Package*

Jinping Neutrino Experiment Simulation and Analysis Package (JSAP) is developed based on GEANT4[13] simulation package. JSAP now consists of 4 parts, namely simulation, calibration and reconstruction, and analysis tool. JSAP plays a very important role in the following items.

- Kiloton scale detector geometry simulation.
- Handle different geometry setup, including target material mass, PMT coverage, etc.
- Understand more about detector, like efficiency.
- Simulate natural radioactive elements and fast neutron background.

4. Summary

The proposed Jinping Neutrino Experiment, with its extremely low cosmic-ray muon flux, low reactor neutrino flux and a 2 kiloton fiducial mass for solar neutrino physics (equivalently, 3 kiloton for geo-neutrino and supernova relic neutrino physics), is expected to improve the present studies on solar neutrinos, geo-neutrinos and supernova neutrinos. Efforts on increasing the target mass with multi-modular neutrino detectors and developing the slow scintillator technique by one ton prototype detector will enable us to eventually enrich the Jinping Neutrino Experiment discovery potential[14–17].

References

1. John F. Beacom et al., "Physics prospects of the Jinping neutrino experiment." *Chinese physics C* 41.2 (2017): 023002.
2. Ke-Jun Kang et al., "Status and prospects of a deep underground laboratory in China." *Journal of Physics: Conference Series.* Vol. 203. No. 1. IOP Publishing, 2010.
3. Yu-Cheng Wu et al., "Measurement of cosmic ray flux in the China JinPing underground laboratory." *Chinese physics C* 37.8 (2013): 086001.
4. International Atomic Energy Agency, http://www.iaea.org/ (2015).
5. Zongyi Wang et al., "Design and analysis of a 1-ton prototype of the Jinping Neutrino Experiment." *Nuclear Instruments and Methods in Physics Research Section A: Accelerators, Spectrometers, Detectors and Associated Equipment* 855 (2017): 81-87.
6. Gianpaolo Bellini et al., "Precision measurement of the ^{7}Be solar neutrino interaction rate in Borexino." *Physical Review Letters* 107.14 (2011): 141302.
7. Gianpaolo Bellini et al., "First Evidence of Pep Solar Neutrinos by Direct Detection in Borexino." *Physical Review Letters* 108.5 (2012): 051302.
8. Gianpaolo Bellini et al., "Neutrinos from the primary proton-proton fusion process in the Sun." *Nature* 512.7515 (2014): 383-386.
9. Linyan Wan et al., "Geoneutrinos at Jinping: Flux prediction and oscillation analysis." *Physical Review D* 95.5 (2017): 053001.
10. Ondrej Sramek et al., "Revealing the Earth's mantle from the tallest mountains using the Jinping Neutrino Experiment." *Scientific reports* 6 (2016): 33034.
11. Hanyu Wei, Zhe Wang, and Shaomin Chen, "Discovery potential for supernova relic neutrinos with slow liquid scintillator detectors." *Physics Letters B* 769 (2017): 255-261.
12. Ziyi Guo et al., "Slow Liquid Scintillator Candidates for MeV-scale Neutrino Experiments." *arXiv preprint* arXiv:1708.07781 (2017).
13. Sea Agostinelli et al., "GEANT4-a simulation toolkit." *Nuclear instruments and methods in physics research section A: Accelerators, Spectrometers, Detectors and Associated Equipment* 506.3 (2003): 250-303.
14. Ghulam Hussain et al., "Assay of low-background stainless steel by smelting for the neutrino experiment at Jinping." *arXiv preprint* arXiv:1706.04506 (2017).
15. Jia-Hua Cheng et al., "Determination of the total absorption peak in an electromagnetic calorimeter." *Nuclear Instruments and Methods in Physics Research Section A: Accelerators, Spectrometers, Detectors and Associated Equipment* 827 (2016): 165-170.
16. Yu Zhi et al., "Wide field-of-view and high-efficiency light concentrator." *arXiv preprint* arXiv:1703.07527 (2017).
17. Zhe Wang and Shaomin Chen, "Reveal the Mantle and K-40 Components of Geoneutrinos with Liquid Scintillator Cherenkov Neutrino Detectors." *arXiv preprint* arXiv:1709.03743 (2017).

Electromagnetic Corrections to Pseudo-Scalar Meson Masses from the Perspective of Spontaneously Broken Chiral Symmetry

R. Ling

*Department of Mathematics and Physics, Shanghai Dianji University,
Shanghai 201306, China
E-mail: raimund.ling@gmail.com*

We are entering the age of precision tests of the standard model, when electromagnetic corrections cannot be ignored any more in the dynamics of strong interactions. A generalized axial-vector Ward identity has been demonstrated, where the Lagrangian conserves an approximate chiral symmetry, with effects of both quark masses and electromagnetism taken into account of. By analysing the pole structure of those matrix elements involved in the identity, we have got the mass formula of the pseudo-scalar octet, valid to all orders of perturbation theory. By clarify the connections between the 0^- meson decay constant and a new vertex with additional photon contribution, this mass formula may prove a classroom for realistic calculation on the highest level of precision.

Keywords: Electromagnetic corrections; pseudo-scalar meson masses; spontaneously broken chiral symmetry.

1. Introduction

It has been understood in history that weak interaction processes in which charge is exchanged between leptons and hadrons are well described at low energy by the effective Lagrangian

$$\frac{G_F}{\sqrt{2}} \left[\sum_l \bar{\nu}_l \gamma_\mu (1 + \gamma_5) l \right] J^\mu + \text{H.c.}, \tag{1}$$

where G_F is the conventional Fermi coupling constant; l runs over the renormalized fields of the three charged leptons e, μ and τ; $\bar{\nu}_l$ runs over the renormalized fields of the associated anti-neutrinos; and J^λ is a hadronic current. Within the quark model, the commutation and conservation properties of J^λ allow it to be identified with the quark current as

$$J^\mu = \overline{\begin{bmatrix} u \\ c \\ t \end{bmatrix}} \gamma^\mu (1 + \gamma_5) V \begin{bmatrix} d \\ s \\ b \end{bmatrix}, \tag{2}$$

where V is in the standard model a 3×3 unitary matrix known as the Kobayashi-Maskawa matrix.[1] As a matter of fact, contemporary experiments in flavour physics have reached a level of precision which makes it obligatory to consider electromagnetic corrections, in addition to explicit isospin symmetry breaking due to quark masses.[2] In the case of light quarks, important examples include the calculations of the leptonic decay constants F_π and F_K and of the form factor $F_+(0)$ for the semi-leptonic decay $K \to \pi + e/\mu + \nu$. These quantities are used to determine the Kobayashi-Maskawa matrix entries $|V_{us}|$ and $|V_{us}|/|V_{ud}|$ in high measure, which provides precise tests of the standard model.[3,4]

The low energy constants as F_π, and like them the hadron spectroscopy, are in general largely determined by quantum chromodynamics (QCD) in the low energy regime, where perturbation theory from first principles is inapplicable and non-perturbative methods have to be explored. One of the prevailing methods that satisfies the requirement is lattice gauge theory, the application of which in the examination of electromagnetic effects in the hadron spectrum and in the determination of quark masses dates back to decades ago.[5,6] Using various procedures to include electromagnetism in lattice QCD simulations, several collaborations have recently obtained strikingly accurate results for the mass differences between the charged members and their neutral counterparts in the systems of pseudo-scalar mesons and baryons.[a, 7-9]

Of all the problems of the mass splittings in hadron spectroscopy, those in the octet of 0^- mesons ($K^{+,0}$, $\pi^{\pm,0}$, η^0, $\bar{K}^{-,0}$) are of particular interest to both experimental and theoretical physicists. As is clear from the perspective of the modern theory–QCD, the strong interactions respect an approximate chiral $SU(3) \times SU(3)$ symmetry, because there are three fairly light quarks: the u, d and s, which have charges $+2/3$, $-1/3$ and $-1/3$ respectively. Since no parity doubling is observed in the hadron spectrum, we are forced to conclude that this chiral symmetry is spontaneously broken to its diagonal subgroup, the $SU(3)$ of Gell-Mann[10] and Ne'eman[11]. A broken approximate chiral symmetry entails the existence of an approximately massless Goldstone boson with the same quantum numbers as the generator of the broken symmetry:[12,13] it must be one of the states of negative parity, zero spin, and zero baryon number. In reality the lightest eight hadrons that have precisely these quantum numbers are the pseudo-scalar octet, so we are led to identify them as the Goldstone bosons associated with the spontaneous breaking of approximate chiral symmetry.

It is well known that the lowest order electromagnetic effect is the dominant contribution to the mass difference between the charged and neutral pions.[14] In the chiral limit where all the quarks are considered as massless, it is true for the kaons as well, and the electromagnetic corrections to the K^+ and π^+ masses are equal. This theorem, known as Dashen's theorem,[15-17] is broken by terms of order $O(\alpha m)$ away from the chiral limit. If we were to derive a mass formula for the pseudo-scalar octet, it should reproduce the results of Dashen's theorem in the leading order of $O(\alpha)$ for zero quark masses.

---

[a]The state of the art in lattice calculation is that sub-percent errors on low-lying hadron masses and other observable quantities are becoming the norm, while the uncertainties owing to explicit isospin symmetry breaking (of the order of $(m_u - m_d)/\Lambda_{QCD} \approx 0.01$) and to electromagnetic corrections (of the order of $\alpha \equiv e^2/4\pi \approx 1/137$) are comparable with, or even larger than, the quoted QCD errors.[2]

440

So far the most interesting consequences of the breaking of chiral symmetry have been derived by simply *assuming* that $SU(3) \times SU(3)$ is spontaneously broken to $SU(3)$ (or $SU(2) \times SU(2)$ is spontaneously broken to isotopic $SU(2)$, if merely two relatively light quarks u and d are considered), without any detailed understanding of the mechanism of this symmetry breaking. The reason behind this lies in the fact that it would involve all the complications of strong interaction dynamics to pursuit whether QCD actually exhibits such a pattern of symmetry breaking. We have made to derive an exact mass formula for the pseudo-scalar octet, by taking into consideration the effects of quantum electrodynamics (QED) and quark masses on the same footing.[18] By building the essential bridge between the approximate symmetry of the underlying theory (QCD+QED) and the realization of its physical states which manifest themselves in the 0^- meson spectrum, our work presents the platform of a potentially simple passage to tackle with the dynamics of strong interactions in the presence of radiative corrections.

2. Pseudo-scalar Meson Decay Constants

As we shall see in our proposed mass formula, the masses of the pseudo-scalar mesons are intrinsically connected with their corresponding leptonic decay constants, for which there is a major difference in the frameworks of pure QCD and QCD+QED that needs to be addressed separately.

Let us now focus on the leptonic decay processes of the charged pseudo-scalar meson. In strangeness-conserving semi-leptonic weak interactions like nuclear beta decay, it is the vector and axial-vector currents $\vec{V}^\mu$ and $\vec{A}^\mu$ (in the two-flavour version) of the strong interactions that happen to be the hadronic currents entering into the effective Lagrangian Eq. (1):

$$iJ^\mu = V_{ud}(V_+^\mu + A_+^\mu), \tag{3}$$

where $V_\pm^\mu$ and $A_\pm^\mu$ are the charge changing currents

$$V_\pm^\mu = V_1^\mu \pm iV_2^\mu, \qquad A_\pm^\mu = A_1^\mu \pm iA_2^\mu. \tag{4}$$

Here the (partially) conserved vector and axial-vector currents are written as

$$\vec{V}^\mu = i\bar{q}\gamma^\mu \vec{t}q, \qquad \vec{A}^\mu = i\bar{q}\gamma^\mu \gamma_5 \vec{t}q, \tag{5}$$

where q is the quark doublet $(u\ d)^{\mathrm{T}}$, and $\vec{t}$ is the three-vector of isospin matrices. For the process of pion decay $\pi^+ \to \mu^+ + \nu_\mu$ in the absence of electromagnetism, the only current matrix element we need is the matrix element of A_-^λ between a one-pion state and the vacuum:

$$\langle \mathrm{VAC}|A_i^\mu(x)|\pi_j\rangle = \frac{iF_\pi \delta_{ij}p_\pi^\mu e^{ip_\pi \cdot x}}{(2\pi)^{3/2}\sqrt{2p_\pi^0}}, \tag{6}$$

which is completely known except for the factor F_π. The rate for pion decay turns out to be

$$\Gamma(\pi \to \mu + \nu) = \frac{G_F^2|V_{ud}|^2 F_\pi^2 m_\mu^2 (m_\pi^2 - m_\mu^2)^2}{4\pi m_\pi^3}. \tag{7}$$

In the study of radiative corrections to the decay amplitudes, a special role is played by those corrections due to 'soft' photons, whose energy and momentum are much less than the mass and energy of the pseudo-scalar meson to decay characteristic of the process. The contributions of these photons of infinitely long wavelength are often so large that they take the form of divergent integrals. It is therefore impossible to give a well-defined expression of the decay constants like F_π as given in terms of the matrix element of the axial-vector current in Eq. (6) in the presence of electromagnetism, because of the contributions from the diagram in which photon is emitted by the hadron and absorbed by the charged lepton.[19,20]

These 'infra-red divergences' caused by soft photons all cancel in the standard way between diagrams containing different numbers of real and virtual photons.[21,22] Indeed in order to cancel the infra-red divergences as in Eq. (6) and obtain results for physical quantities like the decay rate, radiative corrections from virtual and real photons must be combined. At the order of $O(\alpha)$ the physical observable is the inclusive decay rate of the pseudo-scalar meson into a final state consisting of either $l^+ + \nu_l$ or $l^+ + \nu_l + \gamma$, with the energy of the emitted photon no greater than some small quantity ΔE.[23]

In our formula for the physical masses of the pseudo-scalar octet, the infra-red divergence of the pseudo-scalar meson decay constants, caused by internal exchange of virtual photons, is to be cancelled by another term in the same formula with an additional external photon line. The work to illustrate this point in detail is under progress.

3. Generalized Axial-Vector Ward Identities

In order to derive results valid to all orders in perturbation theory (and in some cases beyond perturbation theory), we have to exploit the field equations and commutation relations of the interacting fields in the Heisenberg picture. For our particular purpose, we are concerned with the Green's function for the axial-vector current $A_a^\mu(x)$ (in the three-flavour version with q being the quark triplet $(u\ d\ s)^{\mathrm{T}}$, and $\vec{t}$ replaced by λ_a, the complete set of 3×3 traceless Hermitian matrices named after Gell-Mann and Ne'eman, as compared with Eq. (5)), together with a Heisenberg picture quark field $q_n(y)$ of type n and the covariant adjoint quark field $\bar{q}_m(z)$ of flavour m:

$$\langle T\{A_a^\mu(x)q_n(y)\bar{q}_m(z)\}\rangle_{\mathrm{VAC}}. \tag{8}$$

Starting from the Lagrangian of QCD+QED, and assuming that the fields satisfy the corresponding Euler-Lagrange equations, we may get the divergence of the partially conserved axial-vector current as

$$\partial_\mu A_a^\mu = i\bar{q}\gamma_5\{M, \lambda_a\}q + eA_\mu\bar{q}\gamma^\mu\gamma_5[Q, \lambda_a]q, \tag{9}$$

where the quark mass matrix M is a diagonal matrix with non-zero elements m_u, m_d and m_s, the charge matrix Q is also a diagonal matrix but with elements $2/3$, $-1/3$

442

and $-1/3$, and A_μ represents the electromagnetic field. By drawing on this condition and the commutation relation between the time-component of the current A_a^0 and the quark fields q, it would yield a generalized Ward identity as[18]

$$\frac{\partial}{\partial x^\mu}\langle A_a^\mu(x)q_n(y)\bar{q}_m(z)\rangle_{\mathrm{VAC}}$$

$$= -ie\langle[Q, O_a(x)]q_n(y)\bar{q}_m(z)\rangle_{\mathrm{VAC}} + \langle\{M, \Phi_a(x)\}q_n(y)\bar{q}_m(z)\rangle_{\mathrm{VAC}} - \tag{10}$$

$$- \delta^4(x-z)\langle q_n(y)\bar{q}_{m'}(z)\rangle_{\mathrm{VAC}}(\gamma_5\lambda_a)_{m'm} - \delta^4(x-y)(\gamma_5\lambda_a)_{nn'}\langle q_{n'}(y)\bar{q}_m(z)\rangle_{\mathrm{VAC}}.$$

As a corollary of the conservation of the electric current, the (generalized) Ward identity in pure QED ensures that the radiative corrections to the electric coupling vertex function cancel the corrections due to radiation of the external fermion lines, leaving charge renormalization arising only from radiative corrections to the photon propagator.[24,25] In much the same way our generalized axial-vector Ward identity Eq. (10) in QCD+QED clarifies the detailed pattern of the cancellations among the great variety of the radiative corrections to the quark propagators and vertices induced by *three* kinds of currents all with the same quantum numbers – negative parity and zero baryon number. One of them is the axial-vector current A_a^μ, while the other two are the pseudoscalar current Φ_a and its electromagnetic counterpart O_a that are defined respectively by

$$\Phi_a \equiv i\bar{q}\gamma_5\lambda_a q, \tag{11}$$

and

$$O_a \equiv A_\mu A_a^\mu = iA_\mu\bar{q}\gamma^\mu\gamma_5\lambda_a q. \tag{12}$$

Lorentz invariance and isospin symmetry require the matrix elements of these currents between the vacuum and the one-particle state of the pseudo-Goldstone boson B of four-momentum q^μ to take the respective forms as[b]

$$\langle\mathrm{VAC}|a_a^\mu(0)|B_b\rangle = \frac{iF_a\delta_{ab}q^\mu}{(2\pi)^{3/2}Z_2\sqrt{2q^0}}, \tag{13}$$

$$\langle\mathrm{VAC}|\phi_a(0)|B_b\rangle = \frac{N_a\delta_{ab}}{(2\pi)^{3/2}Z_4\sqrt{2q^0}}, \tag{14}$$

and also

$$\langle\mathrm{VAC}|o_a(0)|B_b\rangle = \frac{R_a\delta_{ab}}{(2\pi)^{3/2}Z_A\sqrt{2q^0}}, \tag{15}$$

---

[b]Note that the symbols for the currents are written in their lower cases to emphasize that λ_a are now replaced by t_a which are suitable linear combinations of λ_a. We can write the pseudo-Goldstone boson fields in a real basis as π_a, with $B_{1/2} \equiv \pi^\pm = (\pi_1 \pm i\pi_2)/\sqrt{2}$, $B_3 \equiv \pi^0 = \pi_3$, $B_{4/5} \equiv K^\pm = (\pi_4 \pm i\pi_5)/\sqrt{2}$, $B_6 \equiv K^0 = (\pi_6 + i\pi_7)/\sqrt{2}$, $B_7 \equiv \bar{K}^0 = (\pi_6 - i\pi_7)/\sqrt{2}$, and $B_8 \equiv \eta^0 = \pi_8$. The generators t_a which represent the surviving isotopic $SU(3)$ symmetry are taken in parallelism as $t_{1/2} = (\lambda_1 \mp i\lambda_2)/\sqrt{2}$, $t_3 = \lambda_3$, $t_{4/5} = (\lambda_4 \mp i\lambda_5)/\sqrt{2}$, $t_{6/7} = (\lambda_6 \mp i\lambda_7)/\sqrt{2}$, and $t_8 = \lambda_8$.

where F_a, N_a and R_a are all non-zero constant coefficients to be determined, Z_2 is the renormalization constant of the quark fields, $Z_4 \equiv Z_2 Z_m$ with Z_m the renormalization constant of the quark mass, and $Z_A \equiv Z_2 Z_3^{-1/2}$ with Z_3 the renormalization constant of the electromagnetic field. Here all the renormalization constants are chosen so that renormalized quantities of the theory preserve the underlying symmetry as in the axial-vector Ward identity (10), which is equivalent to the statement that the manifestation of symmetries by the Ward identities shapes the behaviour of various radiative corrections of the theory which show themselves in the integrity of the renormalization constants.[18]

Aside from possible occurrence of infra-red divergences as discussed in the previous section, the author would like to remark on another important issue if one wanted to calculate the constants F, N and R using lattice methods in the framework of QCD+QED. On the lattice one first computes the matrix elements of the bare composite operators (as shown by Eqs. (5), (11) and (12)), which are dependent upon an ultra-violet cut-off provided by the inverse lattice spacing, and then multiplies the results with the corresponding renormalization factors. For the purpose of avoiding a generally significant discrepancy, one needs to develop feasible methods which allow a non-perturbative determination of the renormalization constants of composite operators, rather than to rely solely on perturbative lattice theory.[26]

4. The Mass Formula of the Pseudo-scalar Mesons

Let us consider the momentum-space amplitude

$$G_{lnm}(q_1, q_2, q_3) = \int \mathrm{d}^4x \mathrm{d}^4y \mathrm{d}^4z\, e^{-iq_1 \cdot x} e^{-iq_2 \cdot y} e^{+iq_3 \cdot z} \langle T\{\mathcal{O}_l(x) q_n(y)\bar{q}_m(z)\}\rangle_{\mathrm{VAC}}, \tag{16}$$

where $\mathcal{O}_l(x)$ can be any of the three currents appearing in the generalized axial-vector Ward identity (10). From Lorentz invariance and parity conservation we know that generally $\mathcal{O}_l$ has a non-zero matrix element between the vacuum and a one-pseudo-Goldstone-boson state $|B, \mathbf{q}_1\rangle$, which also has a non-vanishing matrix element with the state $q_n \bar{q}_m |\mathrm{VAC}\rangle$. Then according to the usual rules of polology, G has a pole at $q_1^2 = -m^2$, where m is the mass of the one-particle state, and the residue at this pole is given by

$$G_{lnm}(q_1, q_2, q_3) \to \frac{-2i\sqrt{\mathbf{q}_1^2 + m^2}}{q_1^2 + m^2 - i\epsilon}(2\pi)^3 \langle \mathrm{VAC}|\mathcal{O}_l(0)|B, \mathbf{q}_1\rangle$$

$$\times \int \mathrm{d}^4y \mathrm{d}^4z\, e^{-iq_2 \cdot y} e^{+iq_3 \cdot z} \langle B, \mathbf{q}_1|T\{q_n(y)\bar{q}_m(z)\}|\mathrm{VAC}\rangle. \tag{17}$$

Equating the pole terms in the Fourier transform of the generalized axial-vector Ward identity Eq. (10) entails the exact mass formula for each pseudo-Goldstone boson of type a:[18]

$$m_a^2 F_a = -\{M(\mu), N_a\} + ie(\mu)[Q, R_a], \tag{18}$$

where μ in parenthesis for M and e is understood to indicate the energy scale at which quark masses and electric charges are renormalized.

In the simplest case where the quarks are massless, F_a becomes a universal constant denoted by F_π, and

$$F_\pi N_\pi = 4v, \tag{19}$$

where in lowest order the vacuum expectation value of the quark bilinears are all equivalent to an unaltered value $\langle \bar{u}u \rangle_0 = \langle \bar{d}d \rangle_0 = \langle \bar{s}s \rangle_0 \equiv -v/Z_4$. In the limit where the symmetry-breaking contributions due to the quark mass terms are treated to first order in quark masses, and the corrections due to electromagnetism are treated to first order in the fine structure constant α with vanishing quark masses, the mass formula Eq. (18) of the pesudoscalar octet reproduces Dashen's results.

The truly new term in our mass formula is the one with the constant R_a defined by Eq. (15), which has never been obtained in the literature. Apart from its function of giving the apparent contribution to the electromagnetic masses of the charged pseudo-scalar mesons, so as to prepare the ground for the precise calculation of the mass splittings in the meson spectrum, it has theoretical significance in itself as well, say, to cancel the infra-red divergences shown in the definition of F_a as in Eq. (13), if electromagnetism is included.

5. Conclusions

With the electromagnetic interaction treated as an additive correction to the Lagrangian of QCD, we have presented a mass formula of the pseudo-scalar octet valid to all orders of perturbation theory. The generalized axial-vector Ward identity in this case entails a relationship between the newly discovered vertex with an explicit additional photon line and those from the matrix elements of the axial-vector and ordinary pseudo-scalar currents between one-0^--meson state and the vacuum. Our mass formula lays the foundations of the examination of the pseudo-scalar meson spectrum and of the determination of light quark masses to the highest extent of precision as one desires, when possible obstacles like infra-red divergences caused by soft photons are to be overcome.

Acknowledgments

We would like to thank Guido Martinelli and Davide Giusti for helpful discussions. This work is supported in part by the National Natural Science Foundation of China Grant No. 11647020, the Funding Programme on Training of Young Teachers from Universities and Colleges in Shanghai Grant No. ZZSDJ15033, and the Academic Discipline Project of Shanghai Dianji University Grant No. 16JCXK02.

References

1. M. Kobayashi and T. Maskawa, CP-violation in the renormalizable theory of weak interaction, *Prog. Theor. Phys.* **49**, 652 (1973).

2. S. Aoki *et al.*, Review of lattice results concerning low-energy particle physics, *Eur. Phys. J. C* **77**, 112 (2017).

3. K. Maltman, C. E. Wolfe, S. Banerjee, I. M. Nugent and J. M. Roney, Status of the hadronic τ decay determination of $|V_{us}|$, *Nucl. Phys. Proc. Suppl.* **189**, 175 (2009).

4. R. J. Hudspith, R. Lewis, K. Maltman, C. E. Wolfe and J. Zanotti, A resolution of the puzzle of low V_{us} values from inclusive flavor-breaking sum rule analyses of hadronic tau decay, in *Proc. 10th International Workshop on e+e- collisions from Phi to Psi (PHIPSI'15)*, (Hefei, Anhui, China, 2015).

5. A. Duncan, E. Eichten and H. Thacker, Electromagnetic splittings and light quark masses in lattice QCD, *Phys. Rev. Lett.* **76**, 3894 (1996).

6. T. Blum, R. Zhou, T. Doi, M. Hayakawa, T. Izubuchi, S. Uno and N. Yamada, Electromagnetic mass splittings of the low lying hadrons and quark masses from 2+1 flavor lattice QCD+QED, *Phys. Rev. D* **82**, 094508 (2010).

7. S. Borsanyi *et al.*, *Ab initio* calculation of the neutron-proton mass difference, *Science* **347**, 1452 (2015).

8. A. Patella, QED corrections to hadronic observables, in *Proc. 34th International Symposium on Lattice Field Theory (Lattice'16)*, (Southampton, UK, 2016).

9. D. Giusti *et al.*, Leading isospin-breaking corrections to pion, kaon, and charmed-meson masses with twisted-mass fermions, *Phys. Rev. D* **95**, 114504 (2017).

10. M. Gell-Mann and Y. Ne'eman, *The Eightfold Way* (Benjamin, New York, 1964).

11. Y. Ne'eman, Derivation of strong interactions from a gauge invariance, *Nucl. Phys.* **26**, 222 (1961).

12. J. Goldstone, A. Salam and S. Weinberg, Broken symmetries, *Phys. Rev.* **127**, 965 (1962).

13. S. Weinberg, Approximate symmetries and pseudo-Goldstone bosons, *Phys. Rev. Lett.* **29**, 1698 (1972).

14. T. Das, G. S. Guralnik, V. S. Mathur, F. E. Low and J. E. Young, Electromagnetic mass difference of pions, *Phys. Rev. Lett.* **18**, 759 (1967).

15. R. F. Dashen, Chiral $SU(3) \times SU(3)$ as a symmetry of the strong interactions, *Phys. Rev.* **183**, 1245 (1969).

16. P. Langacker and H. Pagels, Pion and kaon electromagnetic masses in chiral perturbation theory, *Phys. Rev. D* **8**, 4620 (1973).

17. S. Weinberg, The problem of mass, *Trans. N. Y. Acad. Sci.* **38**, 185 (1977).

18. R. Ling, B. L. Li and J. L. Ping, Exact formula of the spectrum of the pseudo-scalar octet, *Phys. Lett. B* **740**, 36 (2015).

19. J. Bijnens, Violations of Dashen's theorem, *Phys. Lett. B* **306**, 343 (1993).

20. J. Gasser and G. R. S. Zarnauskas, On the pion decay constant, *Phys. Lett. B* **693**, 122 (2010).

21. F. Bloch and A. Nordsieck, Note on the radiation field of the electron, *Phys. Rev.* **52**, 54 (1937).

22. D. R. Yennie, S. C. Frautschi and H. Suura, The infrared divergence phenomena and high-energy processes, *Ann. Phys.* **13**, 379 (1961).

23. N. Carrasco, V. Lubicz, G. Martinelli, C. T. Sachrajda, N. Tantalo, C. Tarantino and M. Testa, QED Corrections to Hadronic Processes in Lattice QCD, *Phys. Rev. D* **91**, 074506 (2015).

24. J. C. Ward, An identity in quantum electrodynamics, *Phys. Rev.* **78**, 182 (1950).

25. Y. Takahashi, On the generalized Ward identity, *Il Nuovo Cimento* **6**, 371 (1957).

26. M. Lüscher, Advanced lattice QCD, lectures given at *the Les Houches Summer School 'Probing the standard model of particle interactions'*, (DESY, Hamburg, Germany, 1997).

CP Violation Sensitivity at the Belle II Experiment

Tao Luo*

on behalf of the Belle II Collaboration

*Key Laboratory of Nuclear Physics and Ion-beam
Application (MOE) and Institute of Modern Physics,
Fudan University, Shanghai, China 200443
E-mail: luot@fudan.edu.cn*

The sensitivity on the measurements of the angles of the CKM unitarity triangle, i.e. ϕ_1, ϕ_2 and ϕ_3, for Belle II experiment is presented in this letter, the CKM mechanism is expected to be tested at 1% level on Belle II.

Keywords: *CP* violation; the angles of the CKM unitarity triangle; the Belle II experiment.

1. Introduction

The weak interactions of quarks are described by the Cabibbo-Kobayashi-Maskawa (CKM)[1,2] matrix, which needs to meet the unitary condition, so it has four degrees of freedom. If we select proper variables, the CKM matrix can be expressed by three rotation angles and one phase (A, λ, ρ, η). The unitary condition of CKM matrix can be converted to six free triangles in the complex plane, and one of the triangles is related to the B meson decays, which is shown in Fig. 1. In this triangle, the side lengths are related to the amount of branching fractions for the corresponding decays, and also the possible $b\bar{b}$ mixing, while the widths of the three angles are related to the amount of the *CP* Violation (*CPV*) of difference decay processes. These angles are just what we are interested in for the following parts of this letter. Due to history reasons, these three angles have two sets of names, α, β, γ, and ϕ_1, ϕ_2, ϕ_3. They are corresponding to each other like what are shown in Fig. 1. ϕ_1, ϕ_2, ϕ_3 are used in this letter.

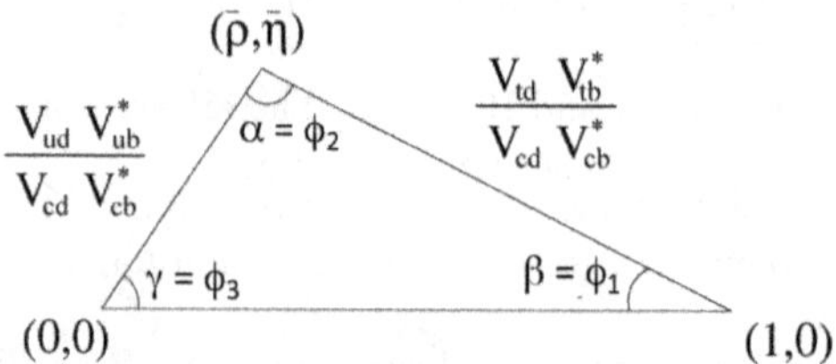

Fig. 1.   The B^0 Unitarity Triangle (UT).

*Supported by Fudan University Grant No. JIH5913023, No. IDH5913011/003.

The values of $\phi_1{}^{3-6}$ can be extracted from $b \to c\bar{c}s$ and $q\bar{q}s$ decay channels, such as: $B^0 \to J/\psi K_s^0$, $B^0 \to \phi K_S^0$, and $B^0 \to \eta' K_S^0$; the effective ϕ_2 can be extracted from $B \to u\bar{u}d$ processes, such as $B \to \pi\pi$, $\rho\rho$ and $\rho\pi$; the values of ϕ_3 can be extracted from $B \to c\bar{u}s$ processes, such as the golden mode $B^\pm \to DK^\pm$. The expected sensitivity of the measurement of these angles from part of these decay modes on Belle II will be introduced one by one in the following parts of this letter.

2. Time Dependent Measurements

One of the most important tasks on Belle II is to measure the time dependent CP violation of B meson decay, from this measurement, we can extract ϕ_1. This section just displays how to extract ϕ_1 from B^0 decay. With asymmetric energies, e^+ and e^- collide inside Belle II detector at $\Upsilon(4S)$ energy region, which is shown in Fig. 2.

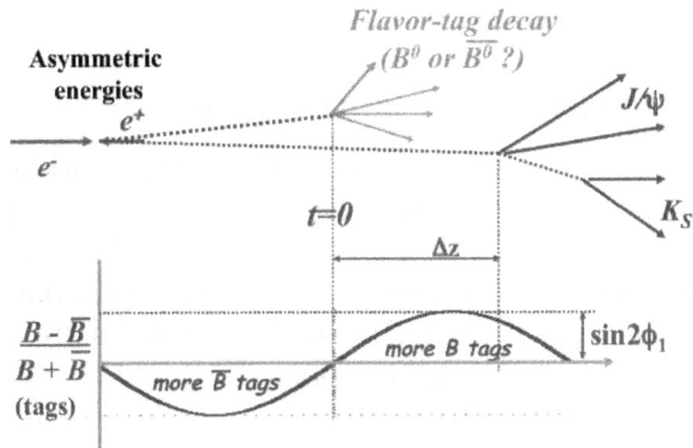

Fig. 2. Asymmetric B-factories at $\Upsilon(4S)$.

One B meson decays to J/ψ and K_S, which is taken as the signal CP side; the other B meson decays to other final states, which is taken as the flavor-tag side; the distance between this two B mesons can be determined by the detector, and then converted into time difference through the equation:

$$\Delta t = \Delta z / \beta\gamma c. \tag{1}$$

The difference of the time-dependent decay rates for the initial B meson, which decays to a CP eigenstate $J/\psi K_S$, can be expressed in Eq. (2); this CP asymmetry is related to S and C, which represent the indirect and direct CPV. By fitting to the B meson time dependent CP asymmetry ($\alpha_{f_{CP}}(\Delta t)$) distribution, we can extract S and ϕ_1 from the data.

$$\alpha_{f_{CP}}(\Delta t) \equiv \frac{\Gamma_{\bar{B} \to f_{CP}}(\Delta t) - \Gamma_{B \to f_{CP}}(\Delta t)}{\Gamma_{\bar{B} \to f_{CP}}(\Delta t) + \Gamma_{B \to f_{CP}}(\Delta t)} = S\sin(\Delta M \Delta t) - C\cos(\Delta M \Delta t), \tag{2}$$

where

$$S = -\xi_f \sin 2\phi_1, \, C \approx 0. \qquad (3)$$

3. Belle II Detector on SuperKEKB

KEKB has been upgraded to SuperKEKB, accordingly, the Belle detector has been upgraded to Belle II[7]. Two of the most important upgrades for the accelerator are: the beam size has been reduced to $\frac{1}{20}$ of that used for KEKB, and reaches nano meter level; the beam currents have been doubled. So totally the peaking luminosity of SuperKEKB will achieve 40 time of that for KEKB, reaching up to 8×10^{35} cm^{2}s^{-1}. The Belle II experiment expects to accumulate 50 ab^{-1} data[12] by 2024. There are many upgrades needed for the Belle II detector in order to increase the performance and cope with much more severe background conditions. The main improvement in performance is in two aspects: new tracking detector: Central Drift Chamber and new vertex determination detector; two new charged particle identification detectors[7].

4. ϕ_1 Determination from $b \to c\bar{c}s$ Decay Channels

As mentioned in previous section, $B^0 \to J/\psi K^0$ is the "golden mode" for extracting ϕ_1 for two reasons[3,4]: 1. The expected theoretical uncertainty is small; 2. The experimental signature is clean. This process happens mainly through the tree diagram process, but there is also penguin pollution. One good thing is that the theoretical estimates on penguin pollution have been significantly improved. Table 1 shows the current status of this angle determination from Belle. For S measurement, it is statistical uncertainty dominated. Table 2 shows the expected uncertainties on

Table 1. Current status of ϕ_1 determination from Belle[8].

PRL108 171802	-	Value	stat. (10^{-3})	syst. (10^{-3})
$J/\psi K^0$	S	+0.67	29	13
	$\mathcal{A} \equiv -C$	-0.015	21	$^{+45}_{-23}$
$c\bar{c}s$	S	+0.667	23	12
	$\mathcal{A} \equiv -C$	-0.006	16	12

Table 2. Belle2 expected uncertainties on ϕ_1 determination @ 50 ab^{-1}. Case1: irreducible syst. same as Belle; Case2: irreducible syst. (vertexing) reduced by a factor 2 due to the new Pixel Vertex detector and improved tracking and alignment algorithms[12].

Expected errors (10^{-3})	-	stat.	syst. reducible	syst. (case 1)	syst. (case 2)
$J/\psi K^0$	S	3.5	1.2	8.2	4.4
	$\mathcal{A} \equiv -C$	2.5	0.7	$^{+43}_{-22}$	$^{+42}_{-11}$
$c\bar{c}s$	S	2.7	2.6	7.0	3.6
	$\mathcal{A} \equiv -C$	1.9	1.4	10.6	8.7

Belle II at 50 ab^{-1} data, which will give results with precision better than 1% comparing to the current 5%. The results will become systematic uncertainty dominated by then.

5. ϕ_2 Determination on Belle II

The values of ϕ_2 angle are determined mainly through $b \to u\bar{u}d$ processes, such as $B \to \pi\pi$, $\rho\rho$, and $\rho\pi$. Because of the existence of non-negligible strong phase, we cannot extract ϕ_2 directly but determine the effective ϕ_2 like Eq. (4). C is no longer equal to 0, which means that there will be direct CP violation.

$$S = \sin(2\phi_2^{\text{eff}}), \phi_2^{\text{eff}} = \phi_2 + \Delta\phi_2; C \neq 0 \tag{4}$$

As an example, Table 3 shows the expected sensitivity of ϕ_2 determination from $B \to \pi\pi$ channels at 50 ab^{-1} Belle II data from the isospin analysis of MC simulation. Belle II will make big contributions on the research of these channels, especially on the branch fraction measurements.

Table 3. Branching fractions and CP asymmetry parameters entering in the isospin analysis of the $B \to \pi\pi$ system: Belle measurements at 0.8 ab^{-1} together with the expected Belle II sensitivity at 50 ab^{-1}.[12]

	Value	0.8 ab^{-1}	50 ab^{-1}
$\mathcal{B}_{\pi^+\pi^-}$ [10^{-6}]	5.04	$\pm 0.21 \pm 0.18$[10]	$\pm 0.03 \pm 0.08$
$\mathcal{B}_{\pi'\pi'}$ [10^{-6}]	1.31	$\pm 0.19 \pm 0.18$[9]	$\pm 0.04 \pm 0.04$
$\mathcal{B}_{\pi^+\pi'}$ [10^{-6}]	5.86	$\pm 0.26 \pm 0.38$[10]	$\pm 0.03 \pm 0.09$
$C_{\pi^+\pi^-}$	-0.33	$\pm 0.06 \pm 0.03$[11]	$\pm 0.01 \pm 0.03$
$S_{\pi^+\pi^-}$	-0.64	$\pm 0.08 \pm 0.03$[11]	$\pm 0.01 \pm 0.01$
$C_{\pi^0\pi^0}$	-0.14	$\pm 0.36 \pm 0.12$[9]	$\pm 0.03 \pm 0.01$
$S_{\pi^0\pi^0}$	-	-	$\pm 0.29 \pm 0.03$

6. Summary

Belle has been a successful B factory, especially for the research on CPV. Major upgrades of KEKB and Belle have been made to build SuperKEKB and Belle II. CKM mechanism will be tested at 1% level[12] on Belle II. Some flavor variables are still to be measured precisely, therefore a lot of room for discoveries at Belle II are expected!

References

1. N. Cabibbo, Phys. Rev. Lett. **10**, 531 (1963).
2. M. Kobayashi and T. Maskawa, Prog. Th. Phys. **49**, 652 (1973).
3. K. Abe et al. (Belle Collaboration) Phys. Rev. Lett. **87**, 091802 (2001).
4. B. Aubert et al. (BABAR Collaboration) Phys. Rev. Lett. **87**, 241801 (2001).

5. K. Abe et al. (Belle Collaboration) Phys. Rev. D **66**, 071102 (2002).
6. B. Aubert et al. (BABAR Collaboration) Phys. Rev. Lett. **89**, 201802 (2002).
7. T. Abe, Belle II Collaboration (2010), arXiv:1011.0352.
8. C. Patrignani et al. (Particle Data Group), Chin. Phys. C **40**, 100001 (2016) and 2017 update (2017).
9. arXiv:1705.02083.
10. Y.-T. Duh et al. (Belle Collaboration) Phys. Rev. D **87**, 031103 (2013).
11. J. Dalseno et al. (Belle Collaboration) Phys. Rev. D **88**, 092003 (2013).
12. The Belle II collaboration and B2TiP theory community, The Belle II Physics Book, to be published.

Status of the JUNO Detector[*]

Zhonghua Qin

on behalf of the JUNO Collaboration

*Institute of High Energy Physics, CAS,
Beijing, 100049, China
E-mail: qinzh@ihep.ac.cn*

The Jiangmen Underground Neutrino Observatory (JUNO) is a multi-purpose neutrino experiment. The primary goal of JUNO is to determine the neutrino mass hierarchy and measure the oscillation parameters in high precision by reactor anti-neutrinos. The JUNO detector is currently under R&D, which is defined as central detector, liquid scintillator (LS), PMT system, VETO detector, and calibration system. The central detector will build a large acrylic sphere of 35.4 m in diameter to contain the 20-kton LS, and the acrylic sphere is supported by a stainless-steel truss. Design and bidding of the acrylic sphere has been finished. The pilot plant of the LS has been constructed. The PMT system includes 20000 20-inch PMTs and 25000 3-inch PMTs. Mass production of the 20-inch PMTs has started with a few thousands of PMTs have been delivered. PMT testing and instrumentation is ongoing. The VETO detector is divided into a top tracker and a water Cherenkov detector. Calibration system with four complementary methods is considered. In this proceeding, the status of the JUNO detector will be addressed.

Keywords: JUNO; neutrino detector; PMT.

1. The JUNO experiment

The Jiangmen Underground Neutrino Observatory (JUNO) is a multipurpose neutrino experiment under construction[1,2]. It will be located 700 m deep underground in Jiangmen city, Guangdong province, with a distance of 53 km from Yangjiang and Taishan nuclear power plants. JUNO has a variety of physics programs as listed below, the main scientific goal is determination of the neutrino mass hierarchy (MH) by measurement of the reactor antineutrinos from the two nuclear power plants[2].

  (i) Mass hierarchy
 (ii) Precision measurement of mixing parameters
(iii) Supernova neutrinos
 (iv) Geo-neutrinos
  (v) Solar neutrinos
 (vi) Sterile neutrinos
(vii) Atmospheric neutrinos
(viii) Exotic searches

[*]This work is supported by the Strategic Priority Research Program of Chinese Academy of Sciences (XDA10010000).

2. Overview of the JUNO detector

The JUNO detector is defined as central detector (CD), liquid scintillator (LS), PMT system, VETO detector and calibration system, as shown in Figure 1. To reach $3\%/\sqrt{E(\text{MeV})}$ energy resolution, the central detector will build a super acrylic sphere and a stainless-steel truss, which will hold 20-kton liquid scintillator, 18000 20-inch PMTs and 25000 3-inch PMTs. The VETO detector will be divided into a top tracker and a water Cherenkov detector. The calibration system will provide different methods for JUNO calibration.

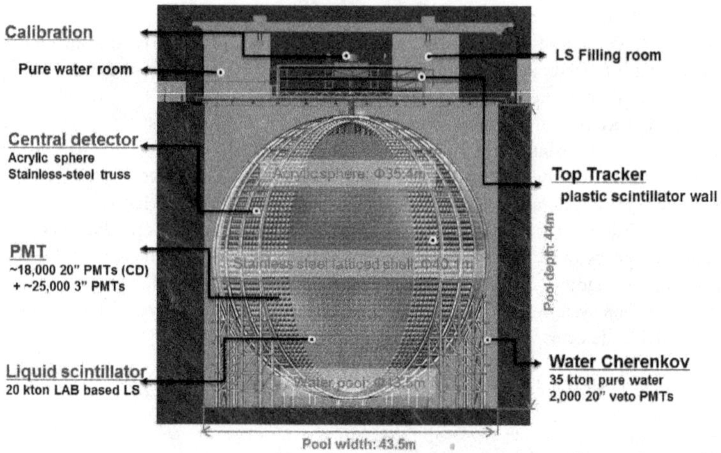

Fig. 1.   The JUNO detector.

2.1. *The central detector*

The central detector is composed of an acrylic sphere supported by a stainless-steel truss, as shown in Figure 2. The main parameters of the central detector are as follows:

- inner diameter of the acrylic sphere 35.4 m, thickness 120 mm;
- built from 265 acrylic sheets (max. sheet ~3 m x 8 m, ~3 tons in weight) by bulk polymerization;
- total weight 1200 tons (600 tons for acrylic sphere, 600 tons for steel truss);
- 590 nodes between acrylic sphere and steel truss;
- a chimney with diameter 80 cm;
- inner diameter of the steel truss 40.1 m;
- divided into 690 grids (23 in longitude x 30 in latitude);
- No. of supporting pillars 60, max. force of the pillar 58 ton.

Bidding of the acrylic sphere has been finished, and the winner is Donchamp acrylic company in China[4]. Now the mass production of the acrylic sheet is under preparation, including the raw material (MMA), polymerization water pool and the machining workshop. Samples of the spherical acrylic sheet and acrylic node have been produced.

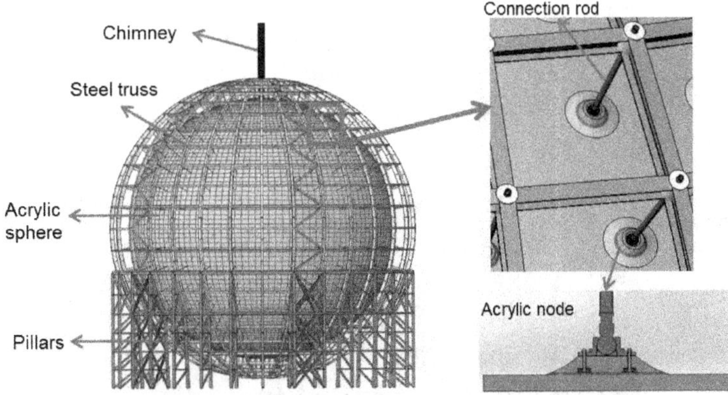

Fig. 2.   The central detector (with connection rod and acrylic node inserted).

2.2. *The liquid scintillator*

The main requirements of the JUNO LS are as follows:

(i)   Lower background: ^{238}U$<10^{-15}$g/g, ^{232}Th$<10^{-15}$g/g, ^{40}K$<10^{-17}$g/g;

(ii)   High light yield: ~10 k photons/MeV;

(iii)   Long attenuation length: >20m@430nm.

To reach such requirements, we need pure organic solvent with high concentration of the fluor to get high light-yield, and the solvent should be highly transparent to increase attenuation length. Base on Daya Bay experiment, a preliminary recipe is considered to add 3g/L PPO and 15mg/L bis-MSB in LAB. And there are four purification methods under development for JUNO, which are distillation, Al$_2$O$_3$ column purification, water extraction and gas stripping. Figure 3 shows a flow chart design of LS.

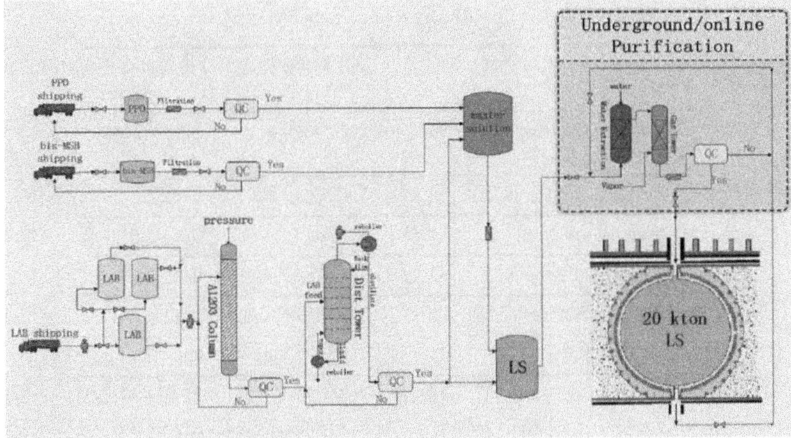

Fig. 3.   The flow chart design of LS.

454

To optimize the fluorescent component, check radioactive background, test the purification methods and pre-study the mass production, a pilot LS plant has been constructed, as shown in Figure 4. The pilot plant is located at the Daya Bay experiment hall, to do the study with one of the Daya Bay detectors. The final attenuation length of the LS in this plant reaches 23.8 m.

Fig. 4.   The pilot LS plant.

2.3. *The PMT sensors*

The JUNO PMT system includes totally 20000 20-inch PMTs and 25000 3-inch PMTs. For the 20-inch PMT, there are 15000 MCP-PMTs produced by a joint team from IHEP and NNVT[5]. The rest 5000 are from Hamamatsu company[6]. To get high quantum efficiency, a transmitted photocathode at top plus a reflective photocathode at the bottom is applied for the MCP-PMT. The MCP-PMT can have higher collection efficiency for the emitted photoelectrons than the dynode-PMT due to the less shadowing. Table 1 lists the main specifications of the two different types of PMT. Mass production of the 20-inch PMTs has started and a few thousands PMTs have been delivered to JUNO.

Table 1.   Specifications of 20-inch PMTs.

Characteristics	unit	MCP-PMT (NNVC)	R12860 (Hamamatsu)
Detection Efficiency (QE*CE*area)	%	27%, >24%	27%, >24%
P/V of SPE		3.5, >2.8	3, >2.5
TTS on the top point	ns	~12, <15	2.7, <3.5
Rise time/ Fall time	ns	R~2, F~12	R~5, F~9
Anode Dark Count	Hz	20K, <30K	10K, <50K
After Pulse Rate	%	1, <2	10, <15
Radioactivity of glass	ppb	238U:50 232Th:50 40K: 20	238U:400 232Th:400 40K: 40

The 25000 3-inch PMTs will be used for the double calorimetry by reducing the non-linearity from the 20-inch PMTs, they will be produced by HZC company in China[7]. The main specifications are listed in Table 2.

Table 2.   Specifications of 3-inch PMTs.

Parameters	HZC's response
QE×CE @ 420 nm	24% (>22%)
TTS(FWHM) of SPE	<5ns
P/V ratio of SPE	3 (>2)
SPE signal width (sigma)	35% (<45%)
Dark rate @ ¼ PE	1kHz (<1.8kHz)
QE uniformity	<30% in Φ60mm
Pre/after pulse ratio	<5%, <15%
Nonlinearity	<10%@1-100PE
Radioactivity	238U: <400ppb,232Th: <400ppb,40K: <200ppb

2.4. *Instrumentation of PMTs*

The goal of PMT instrumentation is to instrument both the 20-inch PMTs and 3-inch PMTs into the JUNO detector. The task includes testing, high voltage divider, potting, protection and installation.

PMT testing covers the acceptance test and parameter characterization. The acceptance test will be done after PMTs are delivered to JUNO to make sure they can meet JUNO specifications. Parameter characterization will be done after PMTs are potted. So far two test facilities, a 36-channel mass testing instrument for measurement of most of the parameters and a 1-channel scanning station for non-uniformity measurement, have been built and under commissioning in a warehouse close to the JUNO site.

PMT potting is to seal the PMT metal pins and the high voltage divider from water. Multiple waterproof layers will be applied to keep water leaking rate as low as possible.

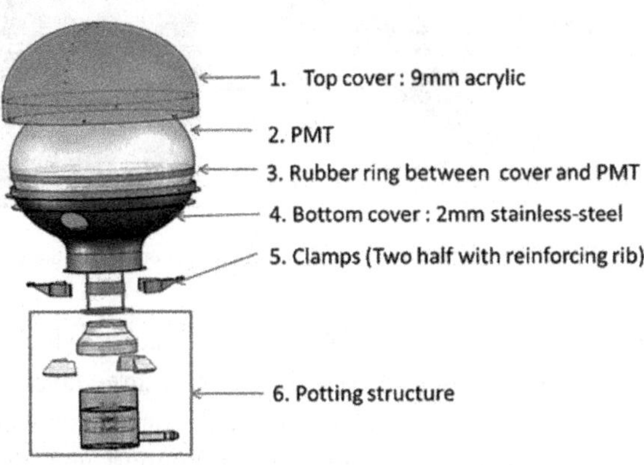

Fig. 5.   The schematic of single PMT assembly.

The target failure rate of PMT potting is less than 0.5% for the first 6 years and 5% for 20 years. Protection is to protect the PMTs from chain implosion which will damage all PMTs. The design of PMT protection will be an acrylic hemi-sphere covering the photocathode part plus a stainless-steel housing for the bottom part. Figure 5 shows an illustration after the PMT protection and potting are integrated.

PMT installation will consist of two steps, the first step is to mount the single PMT to a module, and the second step is to mount the module to the JUNO detector. To reach the required 75% coverage, there will be only a few-mm clearance between adjacent PMTs. A schematics of the PMT module design is shown in Figure 6.

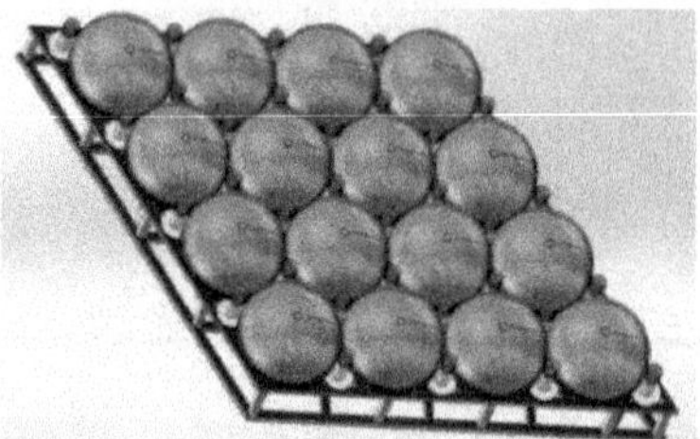

Fig. 6.   Illustration of a PMT module.

2.5. *The VETO detector*

A veto detector is needed to 1) reduce the cosmogenic isotope of $^9\mathrm{Li}/^8\mathrm{He}$ by precision reconstruction of muon track, 2) reject fast neutron background by passive shielding and possible tagging, and 3) shield the radioactivity from rock by water. So the veto detector need a top tracker and a water Cherenkov detector, as shown in Figure 7 and Figure 8.

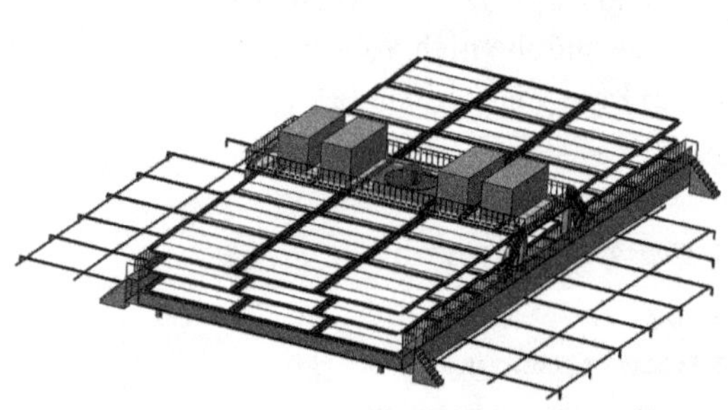
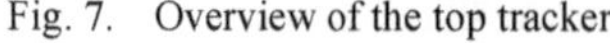

Fig. 7.   Overview of the top tracker.   Fig. 8.   Schematic of the water Cherenkov detector.

The top tracker will re-use the OPERA's Target Tracker which is made of plastic scintillators. Totally 62 scintillator walls will arranged in three layers spaced by 1 m and covering half of the top area of the water pool. The scintillator walls now have been shipped to near the JUNO site, and its mechanical support and electronics are under design.

For the water Cherenkov detector, there will be 2000 20-inch PMTs and 35-kton ultra-pure water. Its detection efficiency is expected to be large than 95%, reducing the fast neutron background to be about 0.1/day and radon less than 0.2 Bq/m^3. Also, coils used for the earth-magnetic-field compensation will be constructed in the water pool.

2.6. *The calibration system*

The challenge of the JUNO calibration system is to control the energy scale uncertainty less than 1% to reach the required energy resolution of 3%/Sqrt(E). There are four complementary methods under developing, which can perform 1D, 2D and 3D calibration for JUNO. An overview of the calibration system is shown in Figure 9.

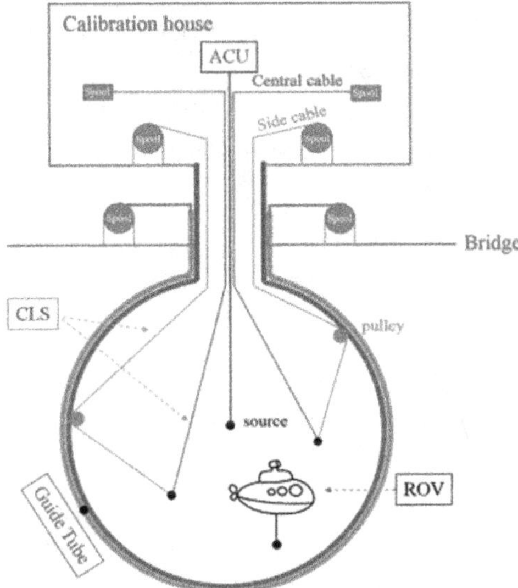

Fig. 9.   Overview of the calibration system.

The four calibration methods are as follows:

- ACU: Automatic Calibration Unit, one dimensional for the central axis scan;
- CLS: Cable Loop System, two dimensional which will scan the vertical planes inside of the acrylic sphere;
- GTCS: Guide Tube Calibration System, two dimensional for the acrylic outer surface scan;
- ROV: Remotely Operated under-LS Vehicle, three dimensional for the whole detector scan.

By prototyping, it shows that ACU is promising to reach sub-cm positioning. CLS test in large space can reconstruct the source in about 10 cm precision. Full-size track experiment of GTCS shows it can work well. Design of ROV is finished and a first prototype has been produced and tested in a water pool.

3. Conclusions

The JUNO experiment is under construction and R&D of the detector is ongoing. Design of the different sub-detector is close to final, with many important milestones have been reached which include PMT bidding and mass production, bidding of the acrylic sphere and steel truss, design and test of PMT protection and potting, and the LS pilot plant construction, etc. The JUNO experiment is expected to start running in 2020.

References

1. T. Adam et al. (JUNO Collaboration), JUNO Conceptual Design Report, arXiv: 1508.07166 (2015).
2. F. An et al. (JUNO Collaboration), J. Phys. G: Nucl. Part. Phys. 43 (2016) 030401.
3. L. Zhan et al., Phys. Rev. D **78**, 111103 (2008); Phys. Rev. D **79**, 073007 (2009).
4. Donchamp, http://www.donchamp.com.
5. North Night Vision of Technology, http://en.nvt.com.cn.
6. Hamamatsu Photonics, http://www.hamamatsu.com.
7. HZC, http://www.hzcphotonics.com.

Large Scale Lorentz Violation Gravity and Dark Energy

Jiayin Shen and Xun Xue

Department of Physics, East China Normal University,
Shanghai, 200241, China
E-mail: xxue@phy.ecnu.edu.cn

The accelerating expansion of universe can be described by the non-zero cosmological constant or the dark energy. However, the origin of the dark energy remains a mystery of modern physics. The local Lorentz invariance is the most exact symmetry of the Nature on the one hand, but all quantum gravity theories predict Lorentz violation on the other hand. The local Lorentz violation induced by the quantum gravity at the very early universe may be transformed into large scale by inflation. Combining the low-l anomalies of the CMB spectrum, we propose that the local Lorentz invariance may be broken at the large scale. We build the effective gravity theories by employing very special relativity symmetry groups and verify the self-consistency of these models. We construct the effective gravity at the cosmic scale with a local $SO(3)$ symmetry. The theory exhibits non-trivial contortion distribution even with scalar matter source. The FRW like solution of the theory is analyzed and the contortion distribution contributes a dark energy like effect which is responsible for the accelerating expansion of the universe. It reveals that the dark energy may be the remnants of quantum gravity in this sense.

Keywords: Lorentz violation; dark energy; accelerating expansion of universe; inflation.

1. Introduction

Though general relativity(GR) passes almost all observational examination, it is still controversial whether GR is the ultimate gravitational theory at all macroscopic scale even to cosmic one. There indeed observations deviating from predictions of GR at large scale. One is the galaxy rotation curve problem which can be explained with the existence of dark matter besides the luminous matter. The other is the discovery of accelerating expansion of the universe in 1998[1-3], which can be described by introducing a small positive cosmological constant Λ into the Einstein field equation or adjoining "dark energy" with $\rho \simeq -p$ to the energy-momentum tensor of cosmic media, known as ΛCDM model[4]. Although there are many dark energy models such as quintessence, phantom, etc, its origin is still a mystery of physics and there is no reasonable candidate for it. Therefore, while ΛCDM model has been a great success, dark energy's origin and essence is being the deep darkness, just as its name. On the other hand, all searches for the dark matter particles gives negative results until now.

It is reasonable to explore the deviation from GR at scale larger than the galaxy to the cosmic one with modified gravity rather than the assumption of dark matter or dark energy considering that the GR is only verified within the solar system or for the local phenomenon. Most of popular modified gravity models assume the local Lorentz invariance just as in the GR which can be regarded as a Lorentz gauge theory[5–10]. Though ΛCDM model can account for most of the CMB observation very well, one of cosmic anomalies is the equationment of low-l multipoles in the CMB angular power spectrum where ΛCDM's prediction do not fit the observation well. According to the recent Planck data, the normals of octopole plane, the quadrupole plane and the direction of dipole moment do not coincide evidently[11]. One cannot expect either the CMB spectrum obeys the cosmological principle in the CMB rest frame or the transformation from CMB rest frame to the peculiar motion frame is a simple Lorentz boost. There are many ways to compensate the low-l multipole anomalies, e.g. assuming the anisotropic dark energy etc. All these solutions either sacrifice the cosmological principle or assuming non-Riemann geometry or nontrivial topology of the space-time of the universe[11].

On the other hand, quantum gravity predicts the existence of minimum length scale L_p with respect to the maximum energy scale M_p and hence Lorentz invariance is lost at Planck scale, the main idea of doubly relativity[12]. There are many approaches in quantum gravity leading to Lorentz violation[13–15]. It is believed that our universe experiences a phase of inflation through which modern observable universe is expanded from a very tiny area of the primordial universe before inflation, when it is dominated by quantum gravity and therefore Lorentz violated, Fig. 1. The primordial Lorentz violation(LV) may be transformed into large scale one by inflation. The scale of LV may exceed the horizon after inflation and re-enter the horizon along with the expansion of the universe, Fig. 2. Therefore we expect there may be LV at large scale esp. at the cosmic scale. To account for the large scale Lorentz violation(LSLV), one natural way is employing gauge principle in the construction of large scale effective gravity(LSEG).

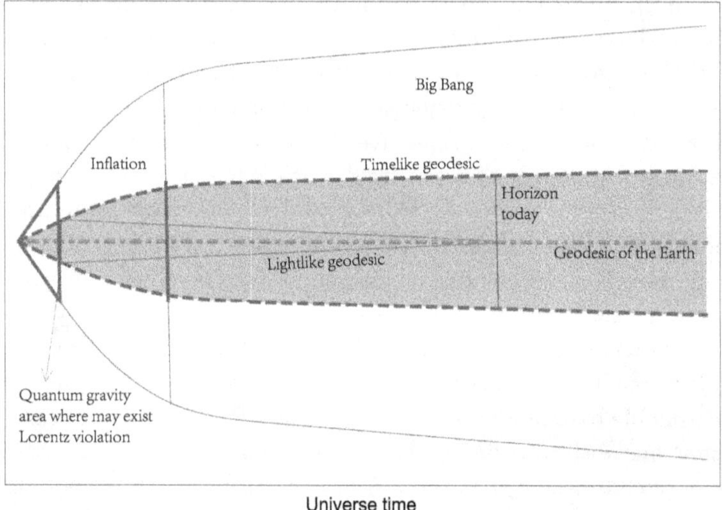

Fig. 1.   Illustration of the evolution of the universe and the primordial Lorentz violation.

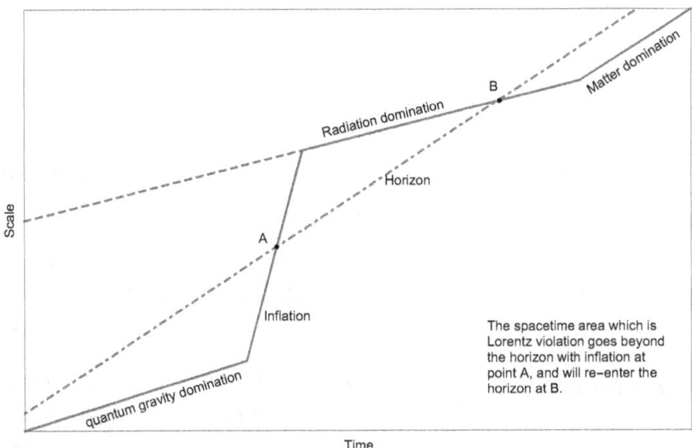

Fig. 2. The Lorentz violation at the quantum gravity dominating era may be transformed into large scale by inflation.

The idea of LSLV is first proposed in 2015 by Wu, Xue, Yang and Yuan (WXYY)[16]. The framework of constructing LSEG with LV is build up with gauge principle via the equivalence principle by utilizing the constrain dynamics and very special relativity(VSR)[17] symmetry as a modified gravity model[16,18–20]. The common feathers of these models are the non-trivial contortion distribution accompanied with the source matter energy-momentum distribution even with the scalar source matter, while contortion must be trivial and the space-time connection must be Levi-Civita one in GR in the case. The non-trivial contortion distribution contributes effectively a dark partner in addition to the matter source distribution. However, the VSR symmetry groups do not contain rotation symmetry totally, it is very hard to find a rotation symmetric solution in these models. The dark partner may contribute dark energy effect in the cosmic scale or the dark matter effect at scale larger than the galaxy. The scenario of the origin of dark partner comes from LSLV is very different from traditional dark energy and dark matter scenario which suggests microscopic origin. However, the Lorentz invariance at scale smaller than the cosmic one may be broken partly and its contribution to the dark partner may be only partly. It is most probably that the LSLV effect is at the cosmic scale, the dark energy effect. We therefore concentrate on the connection between the LSLV and the dark energy effect.

2. Gravity with Large Scale Lorentz Violation

To construct a gravitation theory with LSLV, one can learn from the tetrad formalism of general relativity[21,22]. The tetrad field h_a^μ is the coordinate transformation from anholonomic locally flat or free falling coordinates to a general holonomic coordinates. The commutator $[h_a, h_b] = f_{ab}{}^c h_c$ for tetrad basis $h_a = h_a^\mu \partial_\mu$ is non-trivial for anholonomic coordinates in general. The metric tensor can be decomposed into the tetrad fields in the way $g_{\mu\nu} = \eta_{ab}h^a_\mu h^b_\nu$. The local Lorentz invariance of matter part action is guaranteed by introducing the Lorentzian gauge field $A_\mu = \frac{1}{2}A^{ab}_{\ \ \mu}S_{ab}$, which behaves as the connection and has a corresponding linear connection $\Gamma^\rho_{\ \nu\mu} = h_a^{\ \rho}\left(\partial_\mu h^a_{\ \nu} + A^a_{\ b\mu}h^b_{\ \nu}\right)$, by the gauge principle with the substitution of ordinary derivative ∂_μ to the covariant derivative $\mathcal{D}_\mu = \partial_\mu - \frac{i}{2}A^{ab}_{\ \ \mu}S_{ab}$ in the matter part Lagrangian $\mathcal{L}_M$ in flat space-time background. Here S_{ab} stand for the generators of Lorentz algebra. The tetrad fields $h_a^{\ \mu}$ can also be regarded as gauge potential of the local translation of space-time in some sense. The Maurer-Cartan equation, $[D_a, D_b] = T_{ba}^{\ \ p}D_p + \frac{i}{2}R_{ba}^{\ \ pq}S_{pq}$, supplies the curvature and torsion as the field strengths for Lorentz connection and tetrad field respectively.

The gravity part of the action consists of the Lorentz gauge field $A^a_{\ b\mu}$ and tetrad field $h_a^{\ \mu}$ can only be of Hilbert-Einstein type,

$$S_E = \frac{1}{16\pi G}\int \mathrm{d}^4x\, hR \tag{1}$$

where $h = \det h^a_{\ \mu}$, because Yang-Mills one gives neither the correct dimension of the gravitational coupling constant nor the Newtonian gravity limit. Denoting the energy momentum tensor and the angular momentum tensor of source matter distribution by $(T_M)^a_{\ c}$ and $(C_M)_{ab}^{\ \ \mu}$ respectively, the equations of motion (EoM) for $A^a_{\ b\mu}$,

$$\mathcal{D}_\nu\left(hh_a^{[\nu}h_b^{\mu]}\right) = 8\pi G(C_M)_{ab}^{\ \ \mu} \tag{2}$$

constrains the connections to be the Levi-Civita one in the case $(C_M)_{ab}^{\ \ \mu} = 0$. The EoM for tetrad fields are the Einstein field equations

$$G^a_{\ b} = R^a_{\ b} - \frac{1}{2}R\delta^a_{\ b} = 8\pi G(T_M)^a_{\ b} \, . \tag{3}$$

To introduce Lorentz violation into the large scale effective gravitation theory, we take $Sim\,(2)$ symmetry as an illustration[16,18]. For a local $Sim\,(2)$ theory, one just need to restrict the local symmetry transformation on $Sim\,(2)$,

$$\psi \xrightarrow{\ h(x)\to h(x)\Lambda(x)\ } U\left(\Lambda\left(x\right)\right)\psi, \quad \Lambda\left(x\right) \in Sim\,(2) \, . \tag{4}$$

In the case of $Sim\,(2)$, the gauge potential takes the form

$$\begin{aligned}
A_\mu &= \frac{1}{2}A^{ab}_{\ \ \mu}S_{ab} \, , \\
&= \frac{1}{2}\left(A^{10}_{\ \ \mu} + A^{31}_{\ \ \mu}\right)T_1 + \frac{1}{2}\left(A^{20}_{\ \ \mu} - A^{23}_{\ \ \mu}\right)T_2 \\
&\quad + A^{30}_{\ \ \mu}K_3 + A^{12}_{\ \ \mu}J_3 \\
&\quad + \frac{1}{2}\left(A^{20}_{\ \ \mu} + A^{23}_{\ \ \mu}\right)\left(S_{20} + S_{23}\right) \\
&\quad + \frac{1}{2}\left(A^{10}_{\ \ \mu} - A^{31}_{\ \ \mu}\right)\left(S_{10} - S_{31}\right) \, .
\end{aligned} \tag{5}$$

The restriction on $Sim\,(2)$ implies

$$A^{10}{}_{\mu} - A^{31}{}_{\mu} = 0\,, \quad A^{20}{}_{\mu} + A^{23}{}_{\mu} = 0\,, \tag{6}$$

which can be realized by taking the action in the form

$$S_E = \frac{1}{16\pi G} \int d^4x\, h \left(R^{ab}{}_{ab} + \lambda_1^{\mu} \left(A^{10}{}_{\mu} - A^{31}{}_{\mu} \right) + \lambda_2^{\mu} \left(A^{20}{}_{\mu} + A^{23}{}_{\mu} \right) \right)\,, \tag{7}$$

where λ_1^{μ} and λ_2^{μ} are the Lagrange multipliers. The EoM for connection give constraints on tetrad and contorsion

$$\begin{aligned}
D_\nu &\left(h \left(h_a{}^\nu h_b{}^\mu - h_a{}^\mu h_b{}^\nu \right) \right) \\
&= \lambda_1^{\mu} h \left(\delta_a{}^1 \delta_b{}^0 - \delta_a{}^3 \delta_b{}^1 \right) + \lambda_2^{\mu} h \left(\delta_a{}^2 \delta_b{}^0 - \delta_a{}^2 \delta_b{}^3 \right)\,,
\end{aligned} \tag{8}$$

which can be reduced to

$$\mathcal{D}_\nu \left(h \left(h_1{}^\nu h_0{}^\mu - h_1{}^\mu h_0{}^\nu \right) \right) = -\mathcal{D}_\nu \left(h \left(h_3{}^\nu h_1{}^\mu - h_3{}^\mu h_1{}^\nu \right) \right)\,, \tag{9}$$

$$\mathcal{D}_\nu \left(h \left(h_2{}^\nu h_0{}^\mu - h_2{}^\mu h_0{}^\nu \right) \right) = \mathcal{D}_\nu \left(h \left(h_2{}^\nu h_3{}^\mu - h_2{}^\mu h_3{}^\nu \right) \right)\,, \tag{10}$$

$$\mathcal{D}_\nu \left(h \left(h_3{}^\nu h_0{}^\mu - h_3{}^\mu h_0{}^\nu \right) \right) = 0\,, \tag{11}$$

and

$$\mathcal{D}_\nu \left(h \left(h_1{}^\nu h_2{}^\mu - h_1{}^\mu h_2{}^\nu \right) \right) = 0\,. \tag{12}$$

The spin connection can be decomposed into torsionless part and contorsion $A^a{}_{bc} = \tilde{A}^a{}_{bc} + K^a{}_{bc}$, where $\tilde{A}^a{}_{bc} = \frac{1}{2}(f_b{}^a{}_c + f_c{}^a{}_b - f^a{}_{bc})$, $K^a{}_{bc} = \frac{1}{2}(T_b{}^a{}_c + T_c{}^a{}_b - T^a{}_{bc})$ and $T^a{}_{bc}$ is the torsion tensor. The constraint relations (9), (10), (11) and (12) can reduce the number of independent components of contorsion from 24 to 8, namely $K^{10}{}_0$, $K^{10}{}_1$, $K^{10}{}_2$, $K^{20}{}_0$, $K^{30}{}_0$, $K^{30}{}_1$, $K^{30}{}_2$, and $K^{12}{}_0$. The constraint (6) reduces to

$$\begin{aligned}
&2f^0{}_{10} + f^0{}_{31} - f^1{}_{30} + f^3{}_{10} = 0\,, \\
&f^1{}_{10} + f^1{}_{31} = 0\,, \\
&f^0{}_{12} - f^1{}_{20} + f^3{}_{12} - f^2{}_{10} + f^1{}_{23} - f^2{}_{31} = 0\,, \\
&f^1{}_{30} + f^0{}_{31} + f^3{}_{10} + 2f^3{}_{31} = 0\,, \\
&2f^0{}_{20} - f^2{}_{30} + f^3{}_{20} - f^0{}_{23} = 0\,, \\
&f^2{}_{10} + f^2{}_{31} + f^0{}_{12} + f^3{}_{12} + f^1{}_{20} - f^1{}_{23} = 0\,, \\
&f^2{}_{23} - f^2{}_{20} = 0\,, \\
&f^2{}_{03} + f^0{}_{23} + 2f^3{}_{23} - f^3{}_{20} = 0\,.
\end{aligned} \tag{13}$$

The tetrad EoM takes the form of (38). For only the symmetric part of connection can affect motion of particle through the geodesic equation, we decompose the curvature with the help of decomposition of spin connection as

$$R^{mn}{}_{ab} = \tilde{R}^{mn}{}_{ab} + R_K{}^{mn}{}_{ab} + R_{CK}{}^{mn}{}_{ab}\,, \tag{14}$$

464

where $\tilde{R}^{mn}{}_{ab}$ and $R_K{}^{mn}{}_{ab}$ are the curvatures composed of torsion-free connection and contorsion respectively, while $R_{CK}{}^{mn}{}_{ab}$ contains cross terms of them. We can rewrite (38) as

$$\tilde{R}^a_c - \frac{1}{2}\delta_c{}^a \tilde{R} = 8\pi G \left(T_{Sim(2)}\right)_c{}^a , \tag{15}$$

where

$$\left(T_{Sim(2)}\right)_c{}^a = \frac{1}{8\pi G}\left(\frac{1}{2}\delta_c{}^a\left(R_K + R_{CK}\right) - \left(R_{Kc}{}^a + R_{CKc}{}^a\right)\right) . \tag{16}$$

The force exerting on a particle moving in the gravitation field is supplied by the Levi-Civita connection for it moves along the geodesic curve, the curvature of which satisfies the Einstein field equation with the effective energy-momentum tensor $T_{Sim(2)}$ of Eq. (16) generated by an effective matter distribution formally. It is worthy to note that $T_{Sim(2)}$ will disappear if there is no matter distribution all over the space, *i.e.* Minkowski spacetime is still the vacuum solution. By including the source matter distribution, the complete gravitation equation will be

$$\tilde{R}_c{}^a - \frac{1}{2}\delta_c{}^a \tilde{R} = 8\pi G \left(T_{Sim(2)} + T_M\right)_c{}^a . \tag{17}$$

The effective energy-momentum tensor $T_{Sim(2)}$ contributes to the gravitation in addition to matter contribution T_M and appears as the dark partner of the matter distribution. Different source matter distribution is expected to give rise different dark distribution. At the cosmic scale, it is expected to lead to the possible contribution to dark energy effectively.

The complete constraints on the connection originated from (9), (10), (11) and (12) become

$$\mathcal{D}_\nu\left(hh_1{}^{[\nu}h_0{}^{\mu]}\right) + \mathcal{D}_\nu\left(hh_3{}^{[\nu}h_1{}^{\mu]}\right) = 8\pi G\left[(C_M)_{10}{}^\mu + (C_M)_{31}{}^\mu\right] , \tag{18}$$

$$\mathcal{D}_\nu\left(hh_2{}^{[\nu}h_0{}^{\mu]}\right) - \mathcal{D}_\nu\left(hh_2{}^{[\nu}h_3{}^{\mu]}\right) = 8\pi G\left[(C_M)_{20}{}^\mu - (C_M)_{23}{}^\mu\right] , \tag{19}$$

$$\mathcal{D}_\nu\left(hh_3{}^{[\nu}h_0{}^{\mu]}\right) = 8\pi G(C_M)_{30}{}^\mu , \tag{20}$$

and

$$\mathcal{D}_\nu\left(hh_1{}^{[\nu}h_2{}^{\mu]}\right) = 8\pi G(C_M)_{12}{}^\mu , \tag{21}$$

which lead to non-trivial torsion even in the scalar source-free case of $C_M = 0$ and hence the non-trivial effective energy-momentum tensor $T_{Sim(2)}$ can contribute at least partly to possible dark matter effect.

The discussions for other VSR symmetry groups are similar. The constraint conditions for local $Hom(2)$ symmetry gauge theories are

$$A^{10}{}_c - A^{31}{}_c = 0, \quad A^{20}{}_c + A^{23}{}_c = 0, \quad A^{12}{}_c = 0 , \tag{22}$$

while the 24 components of contorsion can be reduced to 12 independent ones, *e.g.*

$$K^{10}{}_{0},\ K^{10}{}_{1},\ K^{10}{}_{2},\ K^{20}{}_{0},\ K^{20}{}_{1},\ K^{30}{}_{0},$$

$$K^{30}{}_{1},\ K^{30}{}_{2},\ K^{12}{}_{0},\ K^{23}{}_{1},\ K^{23}{}_{3},\ K^{31}{}_{1}. \tag{23}$$

In the case of $C_M = 0$, the 12 constraint conditions (22) can be reduced to

$$\begin{aligned}
K^{10}{}_{0} - K^{31}{}_{3} &= 2f^{0}{}_{10} + f^{0}{}_{31} - f^{1}{}_{30} + f^{3}{}_{10}, \\
K^{20}{}_{1} + K^{23}{}_{1} &= \tfrac{1}{2}\left(f^{1}{}_{20} - f^{0}{}_{12} + f^{2}{}_{10} - f^{1}{}_{23} - f^{3}{}_{12} + f^{2}{}_{31}\right), \\
K^{20}{}_{0} + K^{23}{}_{3} &= 2f^{0}{}_{20} - f^{0}{}_{23} - f^{2}{}_{30} + f^{3}{}_{20}, \\
K^{12}{}_{0} &= \tfrac{1}{2}\left(-f^{0}{}_{12} - f^{1}{}_{20} + f^{2}{}_{10}\right), \\
K^{30}{}_{2} &= \tfrac{1}{2}\left(2f^{0}{}_{20} - f^{0}{}_{23} - f^{2}{}_{30} + f^{3}{}_{20} - 2f^{1}{}_{12}\right), \\
K^{30}{}_{1} &= \tfrac{1}{2}\left(2f^{0}{}_{10} + f^{0}{}_{31} - f^{1}{}_{30} + f^{3}{}_{10} + 2f^{2}{}_{12}\right),
\end{aligned} \tag{24}$$

and

$$\begin{aligned}
f^{0}{}_{10} + f^{0}{}_{31} + f^{3}{}_{10} + f^{3}{}_{31} &= 0, \\
f^{1}{}_{10} - f^{1}{}_{31} &= 0, \\
f^{1}{}_{20} - f^{1}{}_{23} + f^{2}{}_{10} + f^{2}{}_{31} &= 0, \\
f^{0}{}_{20} - f^{0}{}_{23} + f^{3}{}_{20} - f^{3}{}_{23} &= 0, \\
f^{2}{}_{23} - f^{2}{}_{20} &= 0, \\
3f^{1}{}_{23} - 2f^{1}{}_{20} + f^{2}{}_{31} - f^{3}{}_{12} &= 0.
\end{aligned} \tag{25}$$

For the $E(2)$ case, the constraint conditions (22) become

$$A^{10}{}_{c} - A^{31}{}_{c} = 0, \quad A^{20}{}_{c} + A^{23}{}_{c} = 0, \quad A^{30}{}_{c} = 0, \tag{26}$$

and 12 independent components of contorsion can be chosen as

$$K^{10}{}_{0},\ K^{10}{}_{1},\ K^{10}{}_{2},\ K^{10}{}_{3},\ K^{20}{}_{0},\ K^{20}{}_{2},$$

$$K^{20}{}_{3},\ K^{30}{}_{0},\ K^{30}{}_{1},\ K^{30}{}_{2},\ K^{12}{}_{0},\ K^{31}{}_{1}. \tag{27}$$

In the case of $C_M = 0$, the 12 constraint conditions (26) can be reduced to

$$\begin{aligned}
K^{10}{}_{0} - K^{10}{}_{3} &= \tfrac{1}{2}\left(-2f^{0}{}_{10} - f^{0}{}_{31} + f^{1}{}_{30} - f^{3}{}_{10}\right), \\
K^{20}{}_{3} - K^{20}{}_{0} &= \tfrac{1}{2}\left(2f^{0}{}_{20} - f^{0}{}_{23} - f^{2}{}_{30} + f^{3}{}_{20}\right), \\
K^{10}{}_{1} - K^{31}{}_{1} &= f^{2}{}_{23} - f^{2}{}_{20}, \\
K^{30}{}_{0} &= f^{0}{}_{30}, \\
K^{30}{}_{1} &= \tfrac{1}{2}\left(f^{0}{}_{31} - f^{1}{}_{30} - f^{3}{}_{10}\right), \\
K^{30}{}_{2} &= \tfrac{1}{2}\left(-f^{3}{}_{20} - f^{0}{}_{23} - f^{2}{}_{30}\right),
\end{aligned} \tag{28}$$

and

$$\begin{aligned}
f^{0}{}_{10} + f^{0}{}_{31} + f^{3}{}_{10} + f^{3}{}_{31} &= 0, \\
f^{1}{}_{10} + f^{1}{}_{31} - f^{2}{}_{20} + f^{2}{}_{23} &= 0, \\
f^{1}{}_{20} - f^{1}{}_{23} + f^{2}{}_{10} + f^{2}{}_{31} &= 0, \\
f^{0}{}_{20} - f^{0}{}_{23} + f^{3}{}_{20} - f^{3}{}_{23} &= 0, \\
f^{0}{}_{12} + f^{3}{}_{12} &= 0, \\
f^{0}{}_{30} + f^{1}{}_{10} + f^{1}{}_{31} + f^{3}{}_{30} &= 0.
\end{aligned} \tag{29}$$

466

We also take $T(2)$ symmetry into account, with 16 constraint equations and 16 independent components of contorsion. We can arrive at the conclusion that all VSR gauge theories are gravity theories with non-trivial torsion in general. We will be convinced that the dark energy effect can be emerged from the $SO(3)$ case. The self-consistency of effective gravity with VSR symmetry can be reached by verifying Maurer-Cartan equation and Bianchi Identity hold within $Sim(2)$ algebra, i.e. $Sim(2)$ is closed in the sense of VSR symmetric effective gravity. The curvature 2-form on $Sim(2)$ reads

$$\begin{aligned}
R^{pq}{}_{ab}S_{pq} = {} & 2\left(h_a{}^\nu h_b{}^\mu - h_a{}^\mu h_b{}^\nu\right)\left[\left(\partial_\mu A^{10}{}_\nu + A^{12}{}_\mu A^{20}{}_\nu - A^{10}{}_\mu A^{30}{}_\nu\right)T_1\right. \\
& \left. + \left(\partial_\mu A^{20}{}_\nu - A^{12}{}_\mu A^{10}{}_\nu - A^{20}{}_\mu A^{30}{}_\nu\right)T_2 + \partial_\mu A^{12}{}_\nu J_3 + \partial_\mu A^{30}{}_\nu K_3\right].
\end{aligned} \tag{30}$$

By including the contribution from the contortion, the Maurer-Cartan equation can be verified to hold within $Sim(2)$ algebra,

$$[D_a, D_b] = T^p{}_{ba}D_p + \frac{i}{2}R^{pq}{}_{ba}S_{pq} . \tag{31}$$

By the Jacobi Identity,

$$[D_m, [D_n, D_p]] + [D_p, [D_m, D_n]] + [D_n, [D_p, D_m]] = 0 \tag{32}$$

and Maurer-Cartan equation, one can reach the conclusion that the first Bianchi Identity

$$\begin{aligned}
& D_d T^a{}_{bc} + D_c T^a{}_{db} + D_b T^a{}_{cd} \\
& = R^a{}_{bcd} + R^a{}_{dbc} + R^a{}_{cdb} + T^e{}_{bd}T^a{}_{ec} + T^e{}_{dc}T^a{}_{eb} + T^e{}_{cb}T^a{}_{ed}
\end{aligned} \tag{33}$$

and the second Bianchi Identity

$$\mathcal{D}_\nu R^a{}_{b\rho\mu} + \mathcal{D}_\mu R^a{}_{b\nu\rho} + \mathcal{D}_\rho R^a{}_{b\mu\nu} = 0 \tag{34}$$

still hold within $Sim(2)$ algebra. Similar approach can be applied to other VSR symmetry groups.

3. Large Scale Lorentz Violation at the Cosmic Scale

To modify GR in the LSLV at cosmic scale case, one can restrict the components of Lorentz gauge field $A^a{}_{b\mu}$ nontrivial only on $SO(3)$ generators to incorporate the Lorentz boost violation just as the VSR symmetry case in [16,18–20]. However, the Lorentz boost is not violated totally at the large scale actually, the restriction of Lorentz gauge field's components on boost generators vanishing is so strong that to induce the dynamics degenerating, especially for the equations of the Robertson-Walker like solution. We turn to seek a moderate way of introducing the Lorentz boost violation in the gauge gravity scheme by observing that the Lorentz gauge potentials transform as

$$A'^a{}_{b\mu} = \Lambda^a{}_c(x)\,A^c{}_{d\mu}\Lambda_b{}^d(x) + \Lambda^a{}_c(x)\,\partial_\mu\Lambda_b{}^c(x) \tag{35}$$

under local Lorentz transformation $\Lambda(x)$. It is obvious that $A'^0{}_{i\mu} = \Lambda_i{}^j(x) A^0{}_{j\mu}$ for a rotation transformation $\Lambda \in SO(3)$. Hence the restriction to the Lorentz gauge potentials can proposed as

$$\left(A^0{}_{1\mu}\right)^2 + \left(A^0{}_{2\mu}\right)^2 + \left(A^0{}_{3\mu}\right)^2 = f_\mu^2 \tag{36}$$

where $f_\mu(x)$ are invariant under $SO(3)$ gauge transformation, which measures the magnitude of the boost violation in some sense.

The action of LSLV effective gravity can then be given by the constrain dynamics with Lagrange multiplier items adding to the Hilbert-Einstein part,

$$S_G = \frac{1}{16\pi G} \int \mathrm{d}^4 x \, h \left(R + \lambda^\mu \left(\left(A^0{}_{1\mu}\right)^2 + \left(A^0{}_{2\mu}\right)^2 + \left(A^0{}_{3\mu}\right)^2 - f_\mu^2 \right) \right). \tag{37}$$

The EoM of tetrad field keeps the form of Einstein equation,

$$G^a{}_b = R^a{}_b - \frac{1}{2} R \delta^a{}_b = 8\pi G (T_M)^a{}_b \,, \tag{38}$$

where $(T_M)^a{}_b$ is the energy-momentum tensor. The connection here is no longer the Levi-Civita one because the boost violation constraint (36) changes the EoM for gauge potential in a given basis of tetrad $h_a{}^\mu$ to

$$\mathcal{D}_\nu \left(h h_0{}^{[\nu} h_i{}^{\mu]} \right) + \frac{1}{2} \lambda^\mu h A^0{}_{i\mu} = 0$$
$$\mathcal{D}_\nu \left(h h_i{}^{[\nu} h_j{}^{\mu]} \right) = 0 \tag{39}$$

where $i, j = 1, 2, 3$ and the repeated μ does not mean summation, which lead to Levi-Civita connection in GR. With the decomposion $A^a{}_{b\mu}$ into Levi-Civita connection $\widetilde{A}^a{}_{b\mu}$ and contortion $K^a{}_{b\mu}$[22] by

$$A^a{}_{b\mu} = \widetilde{A}^a{}_{b\mu} + K^a{}_{b\mu}. \tag{40}$$

Eq. (39) can be expressed in detail

$$\begin{aligned}
&K^1{}_{23} = 0, K^2{}_{31} = 0, K^3{}_{12} = 0, \\
&K^0{}_{12} = K^0{}_{21}, K^0{}_{23} = K^0{}_{32}, K^0{}_{31} = K^0{}_{13}, \\
&K^2{}_{12} = -K^0{}_{10}, K^3{}_{23} = -K^0{}_{20}, K^1{}_{31} = -K^0{}_{30}, \\
&K^3{}_{13} = -K^0{}_{10}, K^1{}_{21} = -K^0{}_{20}, K^2{}_{32} = -K^0{}_{30}
\end{aligned} \tag{41}$$

and

$$\begin{aligned}
&2K^0{}_{10} h_0{}^\mu + \left(K^0{}_{22} + K^0{}_{33}\right) h_1{}^\mu - \left(K^1{}_{20} + K^0{}_{21}\right) h_2{}^\mu + \left(K^3{}_{10} - K^0{}_{31}\right) h_3{}^\mu \\
&\quad + \lambda^\mu \left(A^0{}_{10} h^0{}_\mu + A^0{}_{11} h^1{}_\mu + A^0{}_{12} h^2{}_\mu + A^0{}_{13} h^3{}_\mu\right) = 0 \\
&2K^0{}_{20} h_0{}^\mu + \left(K^1{}_{20} - K^0{}_{12}\right) h_1{}^\mu + \left(K^0{}_{11} + K^0{}_{33}\right) h_2{}^\mu - \left(K^2{}_{30} + K^0{}_{32}\right) h_3{}^\mu \\
&\quad + \lambda^\mu \left(A^0{}_{20} h^0{}_\mu + A^0{}_{21} h^1{}_\mu + A^0{}_{22} h^2{}_\mu + A^0{}_{23} h^3{}_\mu\right) = 0 \\
&2K^0{}_{30} h_0{}^\mu - \left(K^3{}_{10} + K^0{}_{13}\right) h_1{}^\mu + \left(K^2{}_{30} - K^0{}_{23}\right) h_2{}^\mu + \left(K^0{}_{11} + K^0{}_{22}\right) h_3{}^\mu \\
&\quad + \lambda^\mu \left(A^0{}_{30} h^0{}_\mu + A^0{}_{31} h^1{}_\mu + A^0{}_{32} h^2{}_\mu + A^0{}_{33} h^3{}_\mu\right) = 0
\end{aligned} \tag{42}$$

Since boost violation, the theory is not frame independent any more. One need to choose a frame to write down the EoM just like one does in a gauge theory where the EoM is gauge dependent and one needs to fixed a gauge. We choose the CMB rest frame as one in which the cosmology principle holds and suppose the theory stands in this frame. The cosmology principle guarantees FRW form of space-time metric $\mathrm{d}s^2 = \mathrm{d}t^2 - a(t)^2 \left(\dfrac{\mathrm{d}r^2}{1 - kr^2} + r^2\mathrm{d}\theta^2 + r^2\sin^2\theta\,\mathrm{d}\varphi^2 \right)$ from which it is easy to get the diagonal tetrad basis,

$$h_0 = \frac{\partial}{\partial t}, \ h_1 = \frac{\sqrt{1 - kr^2}}{a(t)}\frac{\partial}{\partial r},$$

$$h_2 = \frac{1}{a(t)r}\frac{\partial}{\partial \theta}, \ h_3 = \frac{1}{a(t)r\sin\theta}\frac{\partial}{\partial \varphi};$$

$$h^0 = \mathrm{d}t, \ h^1 = \frac{a(t)}{\sqrt{1 - kr^2}}\,\mathrm{d}r,$$

$$h^2 = a(t)r\,\mathrm{d}\theta, \ h^3 = a(t)r\sin\theta\,\mathrm{d}\varphi , \tag{43}$$

representing an isotropic observer in CMB Rest Frame. With (41), part of components of contortion K_μ is determined while part can be expressed in terms of $f_\mu(x)$ and vice versa. The cosmology principle also requires all cosmic physical quantities depend only on cosmic time t, and hence the condition

$$K^0{}_{11} = K^0{}_{22} = K^0{}_{33} = \mathscr{K}(t) \tag{44}$$

guarantees the requirement of $G^1{}_1 = G^2{}_2 = G^3{}_3$ from Eq. (38) by the ideal fluid assumption of the cosmic matter source. The requirement of $G^a{}_b = 0$, $\forall a \neq b$ reduces degree of freedom in $f_\mu(x)$ to $\mathscr{K}$. Finally, all the components of contortion can be determined to be trivial except $K^0{}_{11}, K^0{}_{22}, K^0{}_{33}$ (and their symmetric partners $K^1{}_{01}, K^2{}_{02}, K^3{}_{03}$ obviously) and $f_\mu(x)$ are determined as,

$$(f_t, f_r, f_\theta, f_\varphi) = (a(t)\mathscr{K}(t) + \dot{a}(t)) \cdot \left(0, \frac{1}{\sqrt{1 - kr^2}}, r, r\sin\theta \right) \tag{45}$$

4. Cosmology of Large Scale Lorentz Violation Gravity

Denote the covariant derivative and the Einstein tensor generated by Levi-Civita connection $\widetilde{A}_\mu$ as $\widetilde{\nabla}$ and $\widetilde{G}^a{}_c$ respectively, $G^a{}_c$ can be decomposed into,

$$G^a{}_c = \widetilde{G}^a{}_c + 2\left(\widetilde{\nabla}_{[c}K^{ab}{}_{b]} + K^a{}_{e[c}K^{eb}{}_{b]} - \frac{1}{2}\left(\widetilde{\nabla}_d K^{db}{}_b + K^d{}_{e[d}K^{eb}{}_{b]} \right)\delta^a{}_c \right) \tag{46}$$

by the decomposition (40). We can rewrite the tetrad field equation of the form (38) into

$$\widetilde{R}^a{}_c - \frac{1}{2}\widetilde{R}\delta^a{}_c = 8\pi G(T_M + T_\Lambda)^a{}_c \tag{47}$$

where $T_\Lambda{}^a{}_c = \dfrac{1}{8\pi G}\left(\widetilde{G}^a{}_c - G^a{}_c \right)$ can be regarded as the dark partner of the source matter distribution generated by the LSLV effect. If T_Λ can lead to the accelerating expansion of the universe, we can make the conclusion that the LSLV is the origin of the dark energy.

The effective dark partner for ideal fluid $[T_M]^a{}_c = Diag(\rho, -p, -p, -p)$ can be derived from (38) as

$$[T_\Lambda]^a{}_c = Diag(\rho_\Lambda, -p_\Lambda, -p_\Lambda, -p_\Lambda) \tag{48}$$

where $\rho_\Lambda = -\dfrac{1}{8\pi G}\left(3\mathscr{K}^2 + 6\mathscr{K}\dfrac{\dot{a}}{a}\right)$ and $p_\Lambda = \dfrac{1}{8\pi G}\left(\mathscr{K}^2 + 4\mathscr{K}\dfrac{\dot{a}}{a} + 2\dot{\mathscr{K}}\right)$ respectively. So is the modified Friedmann Equation,

$$\left(\frac{\dot{a}}{a}\right)^2 + \frac{k}{a^2} + 2\mathscr{K}\frac{\dot{a}}{a} + \mathscr{K}^2 = \frac{8\pi G}{3}\rho$$
$$\ddot{a} = -4\pi G \cdot a\left(p + \frac{\rho}{3}\right) - \frac{\mathrm{d}}{\mathrm{d}t}\left(a\mathscr{K}\right), \tag{49}$$

while one in ΛCDM model as a comparison,

$$\left(\frac{\dot{a}}{a}\right)^2 + \frac{k}{a^2} - \frac{1}{3}\Lambda = \frac{8\pi G}{3}\rho$$
$$\ddot{a} = -4\pi G \cdot a\left(p + \frac{\rho}{3}\right) + \frac{1}{3}a\Lambda. \tag{50}$$

By the modified Friedmann Equation (49), the accelerating expansion solution of the universe corresponds the condition $4\pi G \cdot a\left(p + \frac{\rho}{3}\right) + \frac{\mathrm{d}}{\mathrm{d}t}\left(a\mathscr{K}\right) < 0$. It should be noted that $\mathscr{K}$ in (49) cannot be determined by (49) even if the equation of state(EoS) of the cosmic media is given. From the discussion concerning (45), $\mathscr{K}$ is a phenomenological description of LSLV and should be given by more fundamental theory or as an observational input just like Λ in (50) of ΛCDM model. Since cosmology based on ΛCDM model is already very successful in many aspects, the LSLV modification should take ΛCDM model as a good approximation. However, the comparison of (49) and (50) reveals the modified Friedmann Equation cannot be equivalent to the one in ΛCDM model. We will hence make propositions in three cases to fix the evolution of $\mathscr{K}$ as close to ΛCDM model as possible.

Case A: By comparing Eq. (49) and Eq. (50), we can make the proposition

$$-\frac{\mathrm{d}}{\mathrm{d}t}\left(a\mathscr{K}\right) = \frac{1}{3}a\Lambda \tag{51}$$

and get the initial condition,

$$2\mathscr{K}(t_0)\frac{\dot{a}(t_0)}{a(t_0)} + \mathscr{K}(t_0)^2 = -\frac{1}{3}\Lambda \tag{52}$$

where $t_0 \simeq H_0^{-1}$ is the moment at present, or the age of universe now and H_0 is Hubble constant. From (52) one can get $\mathscr{K}(t_0) = H_0\left(\pm\sqrt{1 - \Omega_\Lambda} - 1\right) \simeq -0.465H_0$ or $-1.535H_0$ respectively, where $\Omega_\Lambda = \dfrac{\Lambda}{3H_0^2}$. The expressions of conservation law for cosmic media in ΛCDM and LSLV model can be derived from (49) and (50),

- ΛCDM:

$$\dot{\rho} + 3\left(\rho + p\right)\frac{\dot{a}}{a} = 0, \tag{53}$$

- LSLV:

$$\dot{\rho} + 3\left(\rho + p\right)\frac{\dot{a}}{a} + \left(3p + \rho\right)\mathscr{K} = 0 . \tag{54}$$

The general solution of Eq. (54) can be given by,

$$\frac{8\pi G}{3H_0^2}\rho = \Omega_m\left(\frac{a_0}{a}\right)^3 \exp\left(\int_t^{t_0}\mathscr{K}\,\mathrm{d}\tau\right) + \Omega_r\left(\frac{a_0}{a}\right)^4 \exp\left(2\int_t^{t_0}\mathscr{K}\,\mathrm{d}\tau\right) \tag{55}$$

where $a_0 = a(t_0)$ and Ω_m is density parameter for dust$(p = 0)$ while Ω_r is one for radiation$(p = \rho/3)$ respectively.

Case B: Suppose the EoS of cosmic media is $p(t) = w(t)\rho(t)$, (49) and (50) can be converted to,

- ΛCDM:

$$\frac{\ddot{a}}{a} + \frac{3w + 1}{2}\frac{\dot{a}^2 + k}{a^2} = \frac{w + 1}{2}\Lambda \tag{56}$$

- LSLV:

$$\frac{\ddot{a}}{a} + \frac{3w + 1}{2}\frac{\dot{a}^2 + k}{a^2} = -\dot{\mathscr{K}} - \frac{3w + 1}{2}\mathscr{K}^2 - (3w + 2)\frac{\dot{a}}{a}\mathscr{K} \tag{57}$$

We can make a proposition by requiring

$$\dot{\mathscr{K}} + \frac{3w + 1}{2}\mathscr{K}^2 + (3w + 2)\frac{\dot{a}}{a}\mathscr{K} = -\frac{w + 1}{2}\Lambda , \tag{58}$$

which relates contortion to $w(t)$. The initial conditions can be taken as the same as **Case A**, $\mathscr{K}(t_0) = -0.465H_0$ or $-1.535H_0$ respectively.

Case C: To mimic the contribution by cosmological constant to cosmology with one by contortion, we make an alternative proposition by requiring $T_\Lambda{}^a{}_b \propto \delta^a{}_b$, or $T_\Lambda{}^0{}_0 = T_\Lambda{}^1{}_1 = T_\Lambda{}^2{}_2 = T_\Lambda{}^3{}_3$, which leads to the equation satisfied by contortion,

$$\dot{\mathscr{K}} = \mathscr{K}^2 + \mathscr{K}\frac{\dot{a}}{a} . \tag{59}$$

The initial condition is obvious, $T_\Lambda{}^a{}_b(t_0) = \frac{1}{8\pi G}\Lambda\delta^a{}_b$, which is the same as one given by (52) exactly, i.e., $\mathscr{K}(t_0) = -0.465H_0$ or $-1.535H_0$.

We will concentrate on $k = 0$ case of the metric here for our universe is spatially flat by observation. The extension to $k = \pm 1$ is straightforward and will be presented elsewhere. The evolution of $H(t)$ and $\mathscr{K}(t)$ can be determined by the equations for Hubble parameters $H(t)$ derived from (49) together with Eq. (51), Eq. (58) and Eq. (59) and with the initial condition $H(t_0) = H_0$. Table 1 summarizes the cases discussed above.

The two initial values of $\mathscr{K}(t_0)$, $-0.465H_0$ and $-1.535H_0$ correspond to two different solutions of the modified Friedmann Equation (49), $H = \sqrt{\dfrac{8\pi G}{3}\rho} - \mathscr{K}$

Table 1. Proposed Models of LSLV Cosmology.

	Propositions on $\mathscr{K}(t)$	Values of $\mathscr{K}(t_0)$
Case A-1	$\dot{\mathscr{K}} + H\mathscr{K} = -\dfrac{1}{3}\Lambda$	$-0.465H_0$
Case A-2		$-1.535H_0$
Case B-1	$\dot{\mathscr{K}} + \dfrac{3w+1}{2}\mathscr{K}^2 + (3w+2)H\mathscr{K} = -\dfrac{w+1}{2}\Lambda$	$-0.465H_0$
Case B-2		$-1.535H_0$
Case C-1	$\dot{\mathscr{K}} = \mathscr{K}^2 + \mathscr{K}H$	$-0.465H_0$
Case C-2		$-1.535H_0$

and $H = -\sqrt{\dfrac{8\pi G}{3}\rho} - \mathscr{K}$ respectively. Obviously, the former is closer to result of ΛCDM model $H = \sqrt{\dfrac{8\pi G\rho + \Lambda}{3}}$. The evolution trend of $H(a)$ and $\mathscr{K}(a)$ versus scale factor a in these different models are presented in Fig. 3, Fig. 4 and Fig. 5. We can set $w(t) \simeq 0$ for the late universe which is dominated by cold matter with $\Omega_m \gg \Omega_r$ in Eq. (54).

The evolution curves of H in **Case B** and in ΛCDM are coincided basically, because **Case B** takes both matter and dark partner from LSLV into account. As can be observed, the evolution of all of these models is very close in the period near Hubble time, and **Case A-1** approaches exactly the same with ΛCDM model in far future. If $\mathscr{K}(t_0)$ takes value of $-1.535H_0$, the differences of $H(a)$ among these models become bigger than Fig. 3(a) near Hubble time. However trends of Fig. 4(b) and Fig. 3(b) are the same in the far future.

In the case of nontrivial contortion, the space-time is a Riemann-Cartan manifold on which the curve $x(\xi)$ satisfying autoparallel condition $\dfrac{\mathrm{d}^2 x^\rho}{\mathrm{d}\xi^2} + \Gamma^\rho{}_{(\mu\nu)}\dfrac{\mathrm{d}x^\mu}{\mathrm{d}\xi}\dfrac{\mathrm{d}x^\nu}{\mathrm{d}\xi} = 0$ is different from the geodesic one satisfying $\dfrac{\mathrm{d}^2 x^\rho}{\mathrm{d}\xi^2} + \tilde{\Gamma}^\rho{}_{\mu\nu}\dfrac{\mathrm{d}x^\mu}{\mathrm{d}\xi}\dfrac{\mathrm{d}x^\nu}{\mathrm{d}\xi} = 0$. The world line for a free falling particle can be determined by the Hamilton principle which

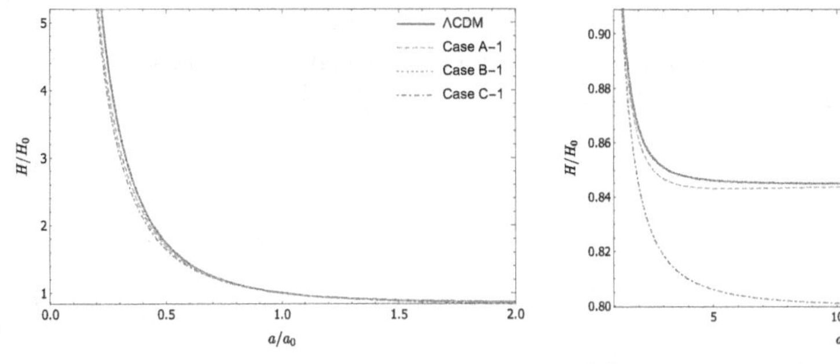

(a) H's evolution from $a \simeq 0$ to $a = 2a_0$.   (b) H's evolution from $a = a_0$ to $a = 20a_0$.

Fig. 3. Evolutions of H between LSLV model and ΛCDM model, with initial condition $\mathscr{K}(t_0) = -0.465H_0$.

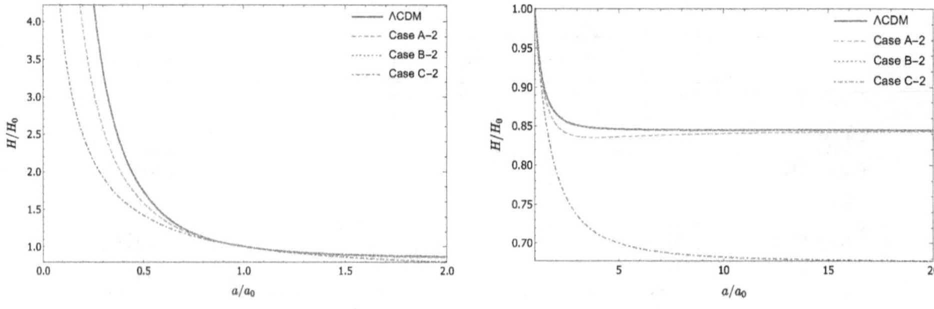

(a) H's evolution from $a \simeq 0$ to $a = 2a_0$.

(b) H's evolution from $a = a_0$ to $a = 20a_0$.

Fig. 4. Evolutions of H between LSLV model and ΛCDM model with initial conditions $\mathscr{K}(t_0) = -1.535 H_0$.

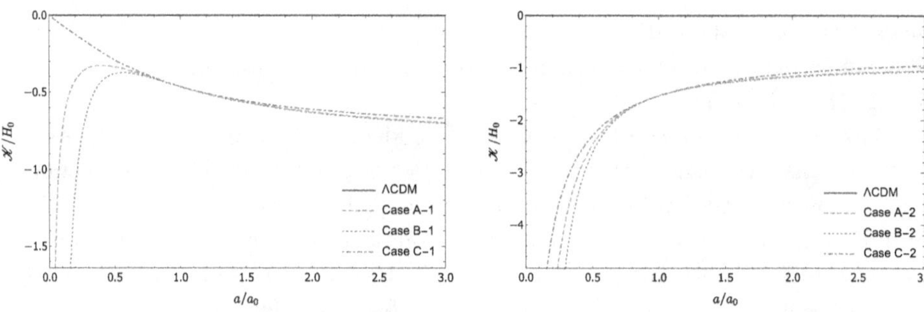

(a) Evolutions of $\mathscr{K}$ with initial conditions $\mathscr{K}(t_0) = -0.465 H_0$.

(b) Evolutions of $\mathscr{K}$ with initial conditions $\mathscr{K}(t_0) = -1.535 H_0$.

Fig. 5. $\mathscr{K}$'s evolution from $a \simeq 0$ to $a = 3a_0$.

gives exactly the geodesic curve. The world line of a photon along which the proper time vanishes identically is just the light-like geodesic curve rather than the auto-parallel one as in some literature[23]. A particle feels the gravity via the curvature of space-time generated by Levi-Civita connection. Contortion contributes to the gravity by the effective dark partner energy-momentum tensor as the source of gravity instead of by the Riemann-Cartan geometry directly as is shown in Eq. (47). Therefore the redshift formula is the same as in ΛCDM model,

$$1 + z = \frac{a_0}{a} \tag{60}$$

By definition of luminosity distance d_L (see[24]) we can get,

$$d_L(z) = (1 + z) \int_0^z \frac{1}{H(z')} \, \mathrm{d}z' \quad \text{(for } k = 0) \tag{61}$$

and

$$\frac{\mathrm{d}t}{\mathrm{d}z} = -\frac{1}{1+z} \frac{\mathrm{d}}{\mathrm{d}z} \left(\frac{d_L}{1+z} \right) \tag{62}$$

With Eq. (61) and Eq. (62), we can convert Eq. (49) and propositions in table.1 to equations for $d_L(z)$ and $\mathscr{K}(z)$ with redshift z as a variable.

Comparison of the results above among LSLV models and ΛCDM model as well as the measurement[25] are presented in Fig. 6 and Fig. 7. The distance modulus is defined as $\mu = 25 + 5\log_{10}(d_L/Mpc)$ (see[25]).

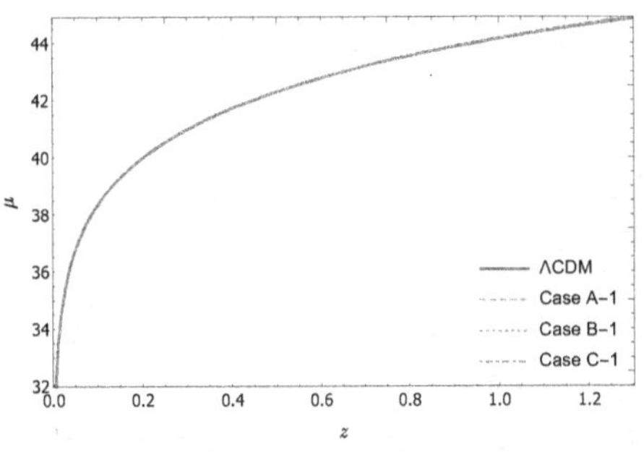

(a) Comparison of distance magnitudes. Curves of different models are hard to distinguish.

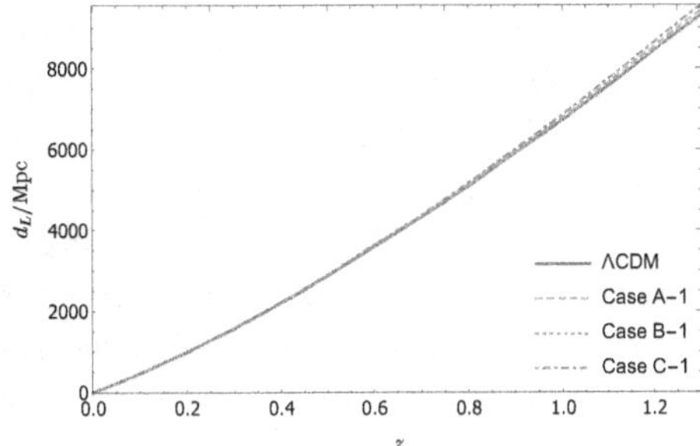

(b) Comparison of luminosity distance of these models.

Fig. 6. $\mathscr{K}(t_0) = -0.465H_0$ Case.

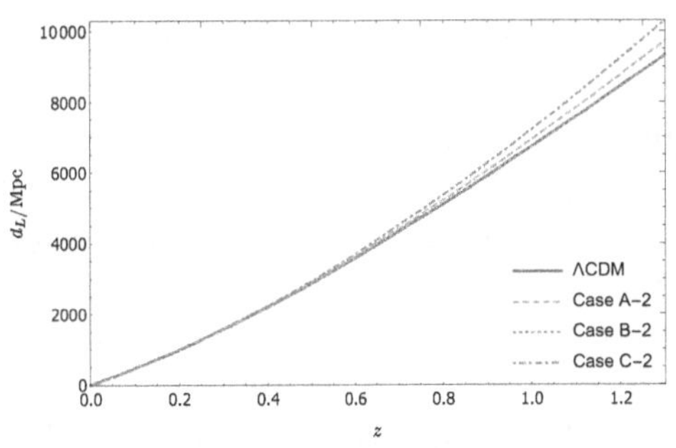

(a) Comparison of distance magnitudes. Divergences of different models are bigger than Fig. 6(a).

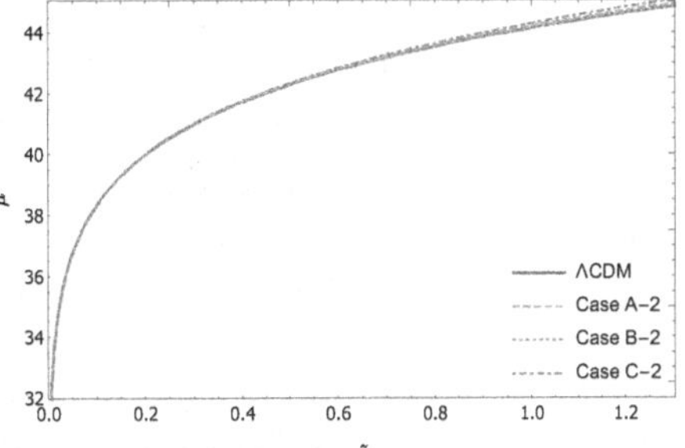

(b) Divergences of luminosity distance are obvious.

Fig. 7. $\mathscr{K}(t_0) = -1.535H_0$ Case.

5. Summary and Outlook

Because of the vacuum energy density is 54 to 112 order higher than the effective cosmological constant, one encounter the fine tuning problem in the approach that the value of cosmological constant and one of vacuum energy density cancels almost exactly and leaving a very tiny effective cosmological constant[26]. Local Lorentz symmetry is the most exact symmetry of the Nature so that the LV at the macroscopic scale, if there is, is inevitably very tiny for the Planck scale suppression. Our suggestion that the LSLV at the cosmic scale induced by quantum gravity through inflation is very different from the idea seeking LV in low energy physics by standard model extension etc. Our modified $SO(3)$ gauge gravity approach is an example which is successfully in assorting the LSLV to the dark energy like effect. Since it is a common feather of LV gauge gravity that the connection deviating from Levi-Civita one, it is reasonable to assert that the LSLV impact the evolution of the universe as the dark energy does or the LSLV may be regarded as the origin of the dark energy instead of quintessence or phantom particles or some strange scalar particle which gives negative pressure.

To make sure the assertion is model independent and universal, it is necessary to investigate various modified gravity with LV and compare the corresponding dark energy effects of different models especially the comparison between teleparallel gravity framework[22] and curvature approach one.

Donoghue[27] develops an idea recently that Lorentz connection may be part of the high energy gravity degree of freedom which confines or condensates to the Levi-Civita one at energy scale lower than the Planck scale similar to what happens in QCD. In this sense, the inflation may separates a phase of Lorentz connection plasma to the horizon scale almost in an instance to let the unconfined Lorentz connection phase lose interaction each other and may confined to other phase other than Levi-Civita one at large scale while confines to Levi-Civita one at short distance. It makes sense in this way our assumption on that the LV from quantum gravity can be transformed to the LSLV through inflation. Ideas from QCD can lead to the road how to realize LSLV from QGLV and it is worthy to investigate. We give here some predictions deviating from ΛCDM model both in the observation of the early universe as well as in the late universe. There is already some clue that the dark energy is time varying which cannot be explained by ΛCDM model[28]. This paper put a step toward a natural explanation about it.

It is widely believed that the physics approaching Planck scale will be quantum gravity. However the evidence of ever existence of quantum gravity is not obvious. Our approach indicates that the accelerating expansion of the universe or the dark energy may be regarded as the remnant of the existence of quantum gravity.

Acknowledgments

This work is supported by the National Natural Science Foundation of China, under Grant No. 11435005 and Grant No. 11775080.

References

1. A. V. Filippenko and A. G. Riess, Results from the high-z supernova search team, *Physics Reports* **307**, 31 (1998).
2. A. G. Riess *et al.*, Observational evidence from supernovae for an accelerating universe and a cosmological constant, *The Astronomical Journal* **116**, p. 1009 (1998).
3. S. Perlmutter *et al.*, Measurements of ω and λ from 42 high-redshift supernovae, *The Astrophysical Journal* **517**, p. 565 (1999).

4. Hinshaw Gary *et al.*, Nine-year wilkinson microwave anisotropy probe (WMAP) observations: Cosmological parameter results, *The Astrophysical Journal Supplement Series* **208**, p. 19 (2013).

5. R. Utiyama, Invariant theoretical interpretation of interaction, *Physical Review* **101**, p. 1597 (1956).

6. D. W. Sciama, On the analogy between charge and spin in general relativity, *Recent developments in general relativity*, p. 415 (1962).

7. T. W. Kibble, Lorentz invariance and the gravitational field, *Journal of Mathematical Physics* **2**, 212 (1961).

8. A. Trautman, Fiber bundles, gauge fields, and gravitation, in *General Relativity and Gravitation. Vol. 1. One hundred years after the birth of Albert Einstein*, 1980.

9. Y. Ne'eman, Gravity groups and gauges, in *General Relativity and Gravitation. Vol. 1. One hundred years after the birth of Albert Einstein*, 1980.

10. F. W. Hehl, P. Von der Heyde, G. D. Kerlick and J. M. Nester, General relativity with spin and torsion: Foundations and prospects, *Reviews of Modern Physics* **48**, p. 393 (1976).

11. L. Perivolaropoulos, Large scale cosmological anomalies and inhomogeneous dark energy, *Galaxies* **2**, 22 (2014).

12. G. Amelino-Camelia, Relativity: Special treatment, *Nature* **418**, 34 (2002).

13. V. A. Kostelecký and S. Samuel, Spontaneous breaking of lorentz symmetry in string theory, *Physical Review D* **39**, p. 683 (1989).

14. S. M. Carroll, J. A. Harvey *et al.*, Noncommutative field theory and lorentz violation, *Physical Review Letters* **87**, p. 141601 (2001).

15. J. Alfaro, H. A. Morales-Técotl, M. Reyes and L. Urrutia, Alternative approaches to lorentz violation invariance in loop quantum gravity inspired models, *Physical Review D* **70**, p. 084002 (2004).

16. Y. Wu, X. Xue, L. Yang and T.-C. Yuan, The effective gravitational theory at large scale with lorentz violation (2015).

17. A. G. Cohen and S. L. Glashow, Very special relativity, *Phys. Rev. Lett.* **97**, p. 021601 (2006).

18. Y. Wu and X. Xue, $SIM(2)$ gravity gauge theory, *Journal of East China Normal University (Natural Science)* **2016**, 76 (2016).

19. L. Yang, Y. Wu, W. Wei, X. Xue and T.-C. Yuan, The effective gravitation theory at large scale with lorentz violation, *Chinese Science Bulletin*, 944 (2017).

20. W. Wei, J. Shen, Y. Wu, L. Yang, X. Xue and T.-C. Yuan, $E(2)$ gauge theory model of effective gravitational theory at large scale, *Acta Phys. Sin.* **66**, 130301 (2017).

21. P. Ramond, *Field Theory: A Modern Primer* (Westview, 2001).

22. R. Aldrovandi and J. G. Pereira, *Teleparallel Gravity* (Springer, Dordrecht, 2013).

23. H. Zhang and L. Xu, Poincaré gauge gravity cosmology (2017).

24. S. Weinberg, *Cosmology* (2008).

25. J. T. Nielsen, A. Guffanti and S. Sarkar, Marginal evidence for cosmic acceleration from type Ia supernovae, *Scientific Reports* **6**, p. 35596 (2016).

26. J. Martin, Everything you always wanted to know about the cosmological constant problem (but were afraid to ask), *Comptes Rendus Physique* **13**, 566 (2012).

27. J. F. Donoghue, Is the spin connection confined or condensed?, *Phys. Rev.* **D96**, p. 044003 (2017).

28. G.-B. Zhao *et al.*, Dynamical dark energy in light of the latest observations, *Nat. Astron.* **1**, 627 (2017).

Lepton Flavor Violation and Leptoquark Decay in the Colored Zee-Babu Model

Fanrong Xu

*Department of Physics, Jinan University,
Guangzhou 510632, P. R. China
E-mail: fanrongxu@jnu.edu.cn*

We consider a neutrino mass generating model which employs a scalar leptoquark and a scalar diquark. The neutrino masses are generated at the two-loop level, as in the Zee-Babu model and the scalars play the role of the doubly and singly charged scalar in the Zee-Babu model. With a moderate working assumption that the magnitudes of the six Yukawa couplings between diquark and the down type quarks are of the same order. Strong connections are found between the neutrino masses and the charged lepton flavor violating processes. In particular, we study $\ell \to \ell'\gamma$ and $Z \to \ell\ell'$, and find that some portions of the parameter space of this model are within the reach of the planned charged lepton flavor violating experiments. Interesting lower bounds are predicted. The type of neutrino mass hierarchy could also be determined by measuring the charged lepton flavor violating double ratios. Moreover, definite leptoquark decay branching ratios are predicted when there is no Yukawa interaction between the right handed fermions and leptoquark, which could help refine the collider search limit on the scalar leptoquark mass.

Keywords: Neutrino mass; lepton flavor violation; Beyond Standard Model.

1. Introduction

It is now well established that at least two of the active neutrinos are massive. New physics beyond the Standard Model (SM) is required to give small but nonzero neutrino masses. It is still unclear whether neutrino mass is Dirac or Majorana type. In case of the Majorana mass, whatever the UV origin is, the key is to generate dimension-5 Weinberg effective operator[1], $(LH)^2$, which conserves the baryon number but violates lepton number by two units. If the new degrees of freedom at high energy also carry both nonzero lepton number and baryon number, the Weinberg operator will be generated at loop level leading to a naturally tiny Majorana neutrino mass.

Recently, a model to generate Weinberg operator and hence tiny neutrino masses, utilizing the scalar leptoquark and scalar di-quark, was discussed by[2], named as colored Zee-Babu Model (cZBM). In cZBM, one scalar leptoquark, Δ with SM $SU(3)_c \times SU(2)_L \times U(1)_Y$ quantum number $(3, 1, -1/3)$, and one scalar di-quark, S with SM quantum number $(6, 1, -2/3)$, were augmented to the SM particle content and the neutrino masses can be generated through the two-loop radiative corrections. The neutrino mass matrix pattern and the mixing angles are determined by Y_L and Y_S, the Yukawa couplings between leptoquark and di-quark and the SM fermions, see Eq. (1).

Generally speaking, Y_S and Y_L are arbitrary and a priori unknown. In this work, we perform a search for a realistic configurations of the Yukawa couplings to compatible with current neutrino experimental data. We find that sizable portion of the realistic solutions overlap with the designed sensitivities of the forthcoming lepton flavor violation experiments. Moreover, for the neutrino masses in both the normal hierarchy and inverted hierarchy, there are interesting and definite lower bounds on $B(Z \to \bar{l}l')$ and $B(l \to l'\gamma)$ which could be falsified in the future. Also, the type of neutrino mass hierarchy can be determined if the charged lepton flavor violating double ratios are measured to be within some specific ranges. If $Y_R = 0$, the model has concrete predictions for the scalar leptoquark decay branching ratios for both neutrino mass hierarchies. This will help refine the collider search limit on the scalar leptoquark mass for the $\beta = 1/2$ case.

In what follows, we briefly introduce the model and then study the connection between neutrino masses and Y_L as well as tree-level flavor violating processes. The loop-induced flavor violating processes are also discussed. Finally we give a numerical study the the conclusions are summarized.

2. Model and Neutrino Mass

As mentioned in the previous section, the SM is extended by adding S and Δ. After rotating the lepton fields into their weak basis and the quarks into their mass basis, the most general gauge invariant Yukawa interaction associated with S and Δ is

$$\mathscr{L}_Y = - \left[\overline{L_i^C}(Y_L)_{ij}i\sigma_2 Q_j + \overline{(\ell_{Ri})^C}(Y_R)_{ij}u_{Rj}\right]\Delta^* - \overline{(d_{Ri})^C}(Y_s)_{ij}d_{Rj}S^* + h.c. \quad (1)$$

where i, j are the flavor indices and the $SU(3)$ indices are suppressed. Apparently Y_S is symmetric in the flavor space while there is no such constraints on Y_L and Y_R. The neutrino masses will receive nonzero contributions through 2-loop quantum corrections if both Y_L and Y_S present. If one writes the effective Lagrangian for neutrino masses as $-\frac{1}{2}\overline{\nu_{Li}^C}(M_\nu)_{ij}\nu_{Lj}$, the neutrino mass matrix can be calculated to be

$$(M_\nu)_{ii'} = 24\mu(Y_L)_{ij}m_{dj}I_{jj'}(Y_s^\dagger)_{jj'}m_{dj'}(Y_L^T)_{j'i'}, \quad (2)$$

with the well-known two-loop integral $I_{jj'}$. For the later use, it is convenient to write the neutrino mass matrix in a compact form

$$M_\nu = Y_L\omega Y_L^T, \quad (3)$$

in which the matrix is defined as $\omega_{jj'} \equiv 24\mu I_\nu m_j m_{j'}(Y_s^\dagger)_{jj'}$. Qualitatively speaking, the resulting neutrino mass is about

$$m_\nu \sim \frac{\mu m_b^2 Y_L^2 Y_S}{32\pi^2 M^2} \sim 0.06\text{eV} \times \left(\frac{Y_L^2 Y_S}{10^{-6}}\right) \times \left(\frac{\text{TeV}}{M^2/\mu}\right). \quad (4)$$

One sees that, due to the 2-loop suppression, with a typical values $Y_L, Y_S \sim 0.01$ and $\mu, M \sim 1$ TeV, the sub-eV neutrino mass can be easily achieved without excessively fine tuning.

3. Neutrino Masses and the Tree-Level Flavor Violation

As discussed before, it is assumed that there is no hierarchy among the Y_S's. Since $m_b \gg m_s \gg m_d$, the matrix ω can be broken into the leading and sub-leading parts and $\omega = \omega^{(0)} + \omega^{(1)}$, where

$$\omega^{(0)} = 24\mu I_\nu \times \begin{pmatrix} 0 & 0 & 0 \\ 0 & 0 & m_b m_s (Y_S)^*_{23} \\ 0 & m_b m_s (Y_S)^*_{23} & m_b^2 (Y_S)^*_{33} \end{pmatrix}, \tag{5}$$

and $\mathcal{O}\left(\frac{\omega^{(1)}}{\omega^{(0)}}\right) \sim \mathcal{O}\left(\frac{m_d}{m_b}\right)$. It is easy to check that the leading neutrino mass matrix $M_\nu^{(0)} = Y_L \omega^{(0)} Y_L^T$ is of rank-2 and $\det M_\nu^{(0)} = 0$. Hence, at least one of the active neutrinos is nearly massless, $\sim (m_d/m_b) \times \max(m_\nu)$, and the scenario of quasi-degenerate neutrinos is disfavored in the cZBM. At leading order, $(Y_L)_{11,21,31}$ do not enter $M_\nu^{(0)}$ at all.

Once $m_{1,2,3}$ are fixed, the neutrino mass matrix can be worked out reversely by

$$M_\nu = U^*_{PMNS} \begin{pmatrix} m_1 & 0 & 0 \\ 0 & m_2 & 0 \\ 0 & 0 & m_3 \end{pmatrix} U^\dagger_{PMNS}. \tag{6}$$

where U_{PMNS} is the PMNS matrix in standard parametrization without two Majorana CP phases.

For simplicity, all Y_S's are assumed to be real. The leading order neutrino mass matrix has $5(= 6 - 1)$ independent entries[a]. For a given set of parameters, $\{\mu, m_S, m_\Delta, (Y_S)^{(0)}_{23}, (Y_S)^{(0)}_{33}, (Y_L)_{13}\}$, all the other 5 complex Yukawa couplings $(Y_L)_{ij}(j \neq 1)$ can be completely determined up to two signs by the leading $M_\nu^{(0)}$. For a real $(Y_L)_{13}$ case, one can write down explicitly the expressions for $(Y_L)^{(0)}_{23}, (Y_L)^{(0)}_{33}, (Y_L)^{(0)}_{12}, (Y_L)^{(0)}_{22}$ and $(Y_L)^{(0)}_{32}$ as well as the most important next to leading contribution to M_ν comes from $(Y_S)_{13}$. The exact expressions can be found in[3].

Without involving too many details of tree-level flavor violation, one may also obtain the range for combinations of Y_L and Y_s, which again are presented in[3].

Before ending this section, we recap the assumptions and discussion so far:

- Y_S's are assumed to be democratic and there is no outstanding hierarchy among these Yukawa couplings. This leads to one nearly massless active neutrino and $|Y_S| \lesssim 9 \times 10^{-3} \times (m_S/7\ \text{TeV})$ from the constrains of neutral meson mixings.
- The Yukawa couplings Y_R are turned off to minimize the TLFV.
- For a given set of $\{\mu, m_S, m_\Delta, (Y_S)_{13,23,33}\}$ and any one of the Y_L's, all the remaining 8 Y_L can be iteratively determined from the absolute neutrino masses and the U_{PMNS} matrix.

---

[a]The symmetric neutrino matrix $M_\nu^{(0)}$ has 6 elements minus 1 constraint that its determinant is zero.

4. Charged Lepton Flavor Violating Process at One-Loop

In this section, we shall study the charged lepton flavor violating (cLFV) processes $\ell \to \ell'\gamma$, $Z \to \ell'\bar{\ell}$ and the like which are induced at the 1-loop level with the leptoquark running in the loop, see Fig. 1.

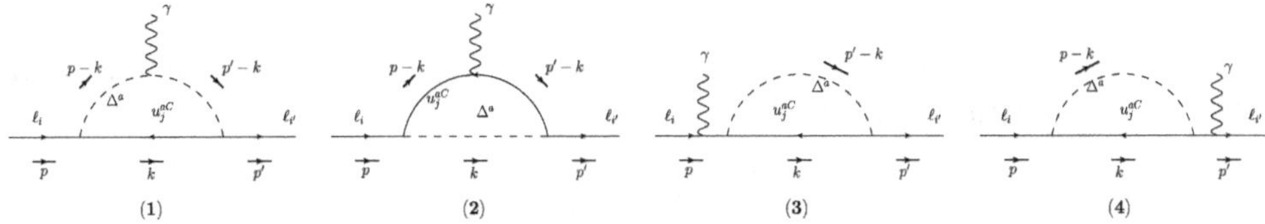

Fig. 1.   The Feynman diagrams for 1-loop LFV $\mu \to e\gamma$.

4.1. $\ell \to \ell'\gamma$

The effective Lagrangian responsible for the cLFV process $\ell \to \ell'\gamma$ is parameterized as

$$\mathscr{L} \supset \frac{1}{2}\bar{\ell}' \left(d_L^{ll'} \mathbb{P}_L + d_R^{ll'} \mathbb{P}_R \right) \sigma^{\mu\nu} \ell F_{\mu\nu} + h.c. \tag{7}$$

For $m'_\ell \ll m_\ell$, the partial decay width is given as

$$\Gamma(\ell \to \ell'\gamma) \simeq \frac{m_\ell^3}{16\pi}(|d_L^{ll'}|^2 + |d_R^{ll'}|^2). \tag{8}$$

A straightforward calculation yields

$$d_R^{ll'} = -\frac{N_c e}{16\pi^2 m_\Delta^2} \left[\left(m_{l'}(Y_R^*)_{l'q}(Y_R^T)_{ql} + m_l(Y_L^*)_{l'q}(Y_L^T)_{ql} \right) \mathcal{F}_1(r_q) \right.$$
$$\left. + m_q(Y_L^*)_{l'q}(Y_R^T)_{ql}\mathcal{F}_2(r_q) \right], \tag{9}$$

where the index q sums over $q = u, c, t$ and $r_q \equiv m_q^2/m_\Delta^2$. $d_L^{ll'}$ can be obtained by simply switching $Y_L \leftrightarrow Y_R$ in the above expression for $d_R^{ll'}$. The loop functions are

$$\mathcal{F}_1(x) = \frac{1 + 4x - 5x^2 + 2x(2 + x)\ln x}{12(1 - x)^4},$$

$$\mathcal{F}_2(x) = \frac{7 - 8x + x^2 + 2(2 + x)\ln x}{6(1 - x)^3}, \tag{10}$$

and they take the limits $\mathcal{F}_1 \to 1/12$ and $\mathcal{F}_2 \to 7/6 + (2\ln x)/3$ when $x \to 0$.

4.2. $Z \to \bar{\ell}\ell'$

The same Feynman diagrams in Fig. 1 with photon replaced by Z boson lead to cLFV $Z \to \bar{l}l'$ decays. Since Z is massive, it can also admit the vector or axial-vector

480

couplings other than the dipole transition couplings as in the $l \to l'\gamma$ cases. The most general gauge invariant $Z \to \bar{l}l'$ amplitude is parameterized as:

$$i\mathcal{M} = ie\overline{u}(p') \left[\left(c_R^Z \mathbb{P}_R + c_L^Z \mathbb{P}_L\right) \left(-g_{\mu\nu} + \frac{q_\mu q_\nu}{m_Z^2}\right)\gamma^\nu \right.$$
$$\left. + \frac{1}{m_Z}\left(d_L^Z \mathbb{P}_L + d_R^Z \mathbb{P}_R\right)\left(i\sigma_{\mu\nu}q^\nu\right)\right] v(-p)\epsilon^\mu(q), \tag{11}$$

where the 4-momentums are labeled as in Fig. 1. From the above parametrization, the branching ratio can be easily calculated to be

$$\mathcal{B}(Z \to \bar{\ell}\ell') = \frac{\alpha}{6}\frac{m_Z}{\Gamma_Z}\left[\left(|c_L^Z|^2 + |c_R^Z|^2\right) + \frac{1}{2}\left(|d_L^Z|^2 + |d_R^Z|^2\right)\right], \tag{12}$$

and the experimentally measured value $\Gamma_Z = 2.4952 \pm 0.0023\,\mathrm{GeV}$ is used in our study.

It is more useful to express the final result in the numerical form:

$$\mathcal{B}(Z \to \bar{\ell}\ell') \simeq 1.46 \times 10^{-7}\left|\sum_{q=u,c,t} a_q^Z (Y_L)^*_{\ell'q}(Y_L)_{\ell q}\right|^2 \times \left(\frac{\mathrm{TeV}}{m_\Delta}\right)^4, \tag{13}$$

where $a_u^Z = a_c^Z \simeq -0.125 - 0.077\mathrm{i} = -0.1468 e^{i31.63°}$ and $a_t^Z = 1$. The imaginary part of $a_{u,c}^Z$ comes from the pole of light-quark propagators in the loop when the light quarks are going on-shell in the Z decay.

The interference between the sub-diagrams with $u(c)$ and t running in the loop makes the relative phases between $a_{u,c}^Z$ and a_t^Z observable. This physical phase leads to CP violation and in general $\mathcal{B}(Z \to \bar{\ell}\ell') \neq \mathcal{B}(Z \to \bar{\ell}'\ell)$. Following the CP asymmetries are quantified as:

$$\eta_{\ell\ell'} \equiv \mathcal{B}(Z \to \bar{\ell}\ell') - \mathcal{B}(Z \to \bar{\ell}'\ell). \tag{14}$$

In this model, we have numerically

$$\eta_{\ell\ell'} \simeq (4.53 \times 10^{-8}) \times \mathbf{Im}\left[\left(Y_u^{\ell'\ell} + Y_c^{\ell'\ell}\right)(Y_t^{\ell'\ell})^*\right] \times \left(\frac{\mathrm{TeV}}{m_\Delta}\right)^4, \tag{15}$$

where the shorthand notation $Y_q^{\ell'\ell} \equiv (Y_L)^*_{\ell'q}(Y_L)_{\ell q}$. Interestingly, due to the sizable CP phase, the CP asymmetries and the cLFV decay branching ratios are of the same order. Also, for the later convenience, we define $\mathcal{B}_{\ell\ell'}^Z \equiv \mathcal{B}(Z \to \bar{\ell}\ell') + \mathcal{B}(Z \to \ell\bar{\ell}')$.

Now we have everything needed for the numerical and phenomenological study.

5. Numerical Study

5.1. $\ell \to \ell'\gamma$ and $Z \to \bar{\ell}\ell'$

The correlations between these cLFV processes are displayed in Fig. 2 for IH, as an illustration. For more explicitly, we also list the lower and upper bound of particular processes, including the CP asymmetries in $Z \to \bar{\ell}\ell'$.

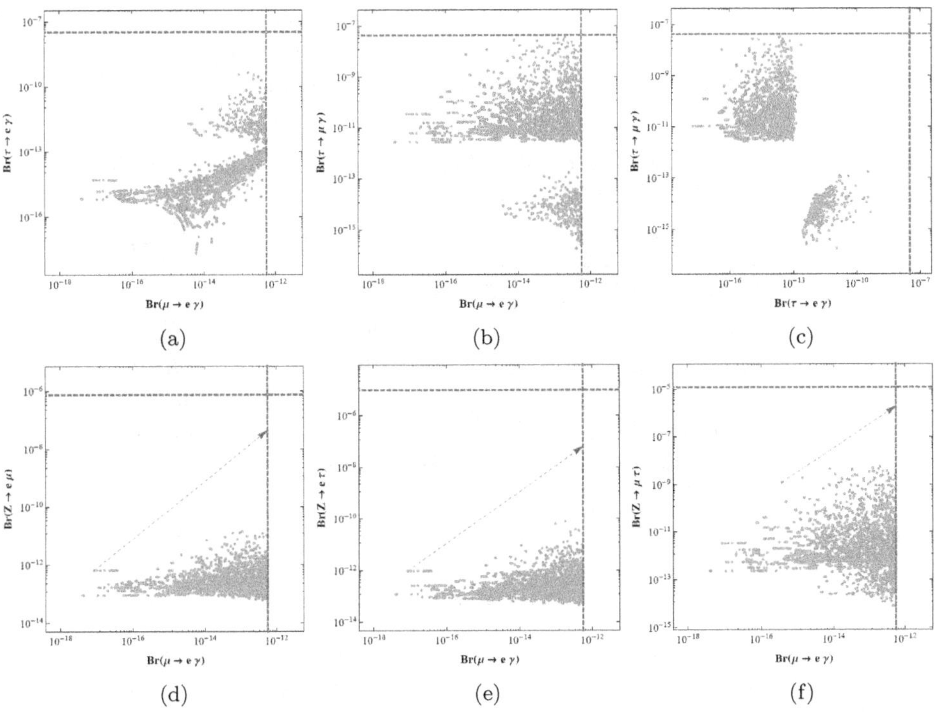

Fig. 2. Correlations among the charged lepton flavor violating branching ratios for neutrino masses are of the inverted hierarchy. In these plots, $m_\Delta = 1\,\text{TeV}$, $m_S = 7\,\text{TeV}$ and $|(Y_S)_{33}| = 0.0097$. The dashed lines represent the current experimental limits at 90%C.L.

Table 1. Range of $\mathcal{B}(\ell \to \ell'\gamma)$, $\mathcal{B}(Z \to \ell\ell')$, and $\eta_{\ell\ell'}$ for $m_\Delta = 1\,\text{TeV}$ and $m_S = 7\,\text{TeV}$. The numbers (in the parentheses) are for NH(IH) neutrino masses. The lower bounds are the lowest values found in the numerical search with $|(Y_S)_{33}| = 0.0097$. The upper bounds are found by rescaling Y_S, see text. Note that the sign for $\eta_{\ell\ell'}$ could be either ways. For a different leptoquark/di-quark mass, all the values should be multiplied by a factor of $(1\,\text{TeV} \cdot m_S/7m_\Delta^2)^2$.

	Lower Bounds	Upper Bounds (for $Y_R = 0$)
$\mathcal{B}(\mu \to e\gamma)$	$3.05 \times 10^{-16}\ (3.98 \times 10^{-18})$	$5.7(5.7) \times 10^{-13}$
$\mathcal{B}(\tau \to e\gamma)$	$3.16 \times 10^{-16}\ (2.03 \times 10^{-18})$	$2.3(0.51) \times 10^{-9}$
$\mathcal{B}(\tau \to \mu\gamma)$	$4.67 \times 10^{-17}\ (1.68 \times 10^{-16})$	$3.4(2.8) \times 10^{-8}$
$\mathcal{B}^Z_{e\mu}$	$2.5 \times 10^{-16}\ (4.9 \times 10^{-14})$	$2.2(8.7) \times 10^{-11}$
$\mathcal{B}^Z_{e\tau}$	$2.9 \times 10^{-16}\ (4.6 \times 10^{-14})$	$3.6(1.0) \times 10^{-10}$
$\mathcal{B}^Z_{\mu\tau}$	$2.5 \times 10^{-14}\ (7.8 \times 10^{-15})$	$5.5(4.5) \times 10^{-9}$
$\eta_{\mu e}$	$^{+.68}_{-.67}\left(^{+2.1}_{-.97}\right) \times 10^{-13}$	$^{+2.6}_{-5.4}\left(^{+9.3}_{-8.1}\right) \times 10^{-13}$
$\eta_{\tau e}$	$^{+2.4}_{-.20}\left(^{+.20}_{-1.2}\right) \times 10^{-12}$	$^{+2.3}_{-.56}\left(^{+.22}_{-.10}\right) \times 10^{-11}$
$\eta_{\tau\mu}$	$^{+2.3}_{-.78}\left(^{+1.3}_{-1.3}\right) \times 10^{-11}$	$^{+3.7}_{-8.1}\left(^{+3.0}_{-3.1}\right) \times 10^{-11}$

In particular, the neutrino mass hierarchy can be unambiguously determined if the measured values of the double ratio of any pair of cLFV process branching ratios fell into any of the decisive windows listed in the following Table 2.

Table 2. The definitions of the cLFV double ratios and the ranges which can be used to determine neutrino mass hierarchy.

Double Ratio	IH	NH
$R_1 \equiv \mathcal{B}^Z_{\mu\tau}/\mathcal{B}(\mu \to e\gamma)$	$R_1 > 10^4$ or $R_1 < 0.1$	N.A.
$R_2 \equiv \mathcal{B}^Z_{e\tau}/\mathcal{B}(\mu \to e\gamma)$	$R_2 > 10^3$	$R_2 < 0.1$
$R_3 \equiv \mathcal{B}^Z_{e\mu}/\mathcal{B}(\mu \to e\gamma)$	$R_3 > 10^2$	$R_3 < 0.1$
$R_4 \equiv \mathcal{B}(\tau \to \mu\gamma)/\mathcal{B}(\mu \to e\gamma)$	$R_4 > 10^6$	$R_4 < 0.003$
$R_5 \equiv \mathcal{B}(\tau \to \mu\gamma)/\mathcal{B}(\tau \to e\gamma)$	N.A.	$0.03 < R_5 < 30$
$R_6 \equiv \mathcal{B}(\tau \to e\gamma)/\mathcal{B}(\mu \to e\gamma)$	$R_6 < 0.03$	N.A.
$R_7 \equiv \mathcal{B}^Z_{\mu\tau}/\mathcal{B}^Z_{\mu e}$	$R_7 < 1.0$	$R_7 > 3 \times 10^4$
$R_8 \equiv \mathcal{B}^Z_{e\tau}/\mathcal{B}^Z_{e\mu}$	N.A.	$R_8 > 10^2$
$R_9 \equiv \mathcal{B}^Z_{\tau\mu}/\mathcal{B}^Z_{\tau e}$	$R_9 < 0.01$	$R_9 > 3 \times 10^4$

Most of these interesting double ratio windows are plagued by either small cLFV branching ratios or very limited parameter space. However, $R_5 \equiv \mathcal{B}(\tau \to \mu\gamma)/\mathcal{B}(\tau \to e\gamma)$ and $R_7 \equiv \mathcal{B}^Z_{\mu\tau}/\mathcal{B}^Z_{\mu e}$ look quite promising. For example, in the cZBM, if R_5 is measured in the future rare tau decay experiment to be within 0.03 and 30, the neutrino masses are of NH. The above discussion clearly demonstrates that the neutrino oscillation experiments and the cLFV measurements are complimentary to one another to better understand the origin of the neutrino masses.

5.2. *Leptoquark decay branching ratios*

The decay branching ratios for leptoquark from our numerical study are displayed in Fig. 3. Roughly speaking, for the IH case, the leptoquark decays are either (1) $B^\Delta_e \sim$ 1.0 or (2) $B^\Delta_\mu \sim 55\%$ and $B^\Delta_\tau \sim 45\%$. On the other hand, for the NH case, the B^Δ_μ and B^Δ_τ are concentrated in the region roughly enclosed by $0.7 \lesssim B^\Delta_\mu + B^\Delta_\tau \lesssim 1.0$ and $0.2 \lesssim B^\Delta_\tau \lesssim 0.8$. In other words, $B^\Delta_e \lesssim 0.3$ if the neutrino masses are in the NH.

Surprisingly, our numerical study has not found any configuration which has either $B^\Delta_\mu \sim 100\%$ or $B^\Delta_\tau \sim 100\%$. Dictated by the neutrino oscillation data, the model predicts that the leptoquark can NOT decays purely into the 2nd or the 3rd generation charged leptons. These concrete branching ratios could be used to provide the new benchmark leptoquark mass limits with a better motivation.

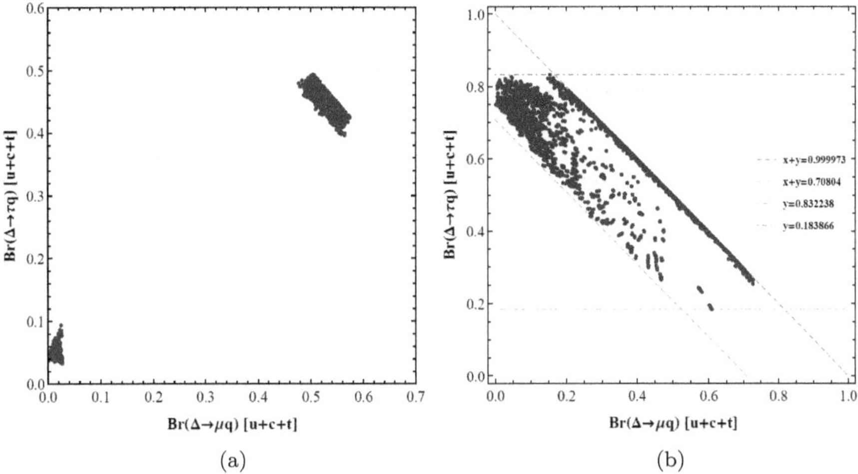

Fig. 3.   Leptoquark decay branching ratios for (a) IH, (b) NH.

6. Conclusion

Within the frame of cZBM, we have adopted a modest working assumption that all six $|Y_S|$ are of the same order. The tree-level flavor violating processes will be inevitably mediated by Δ or S with the realistic Y_L which accommodates the neutrino data. Due to the different chiral structures, the contributions to the flavor violating processes from Y_L and Y_R do not interfere with each other at the tree-level. We have considered the case that $Y_R = 0$ to minimized the tree-level flavor violating processes. A comprehensive numerical study has been performed to look for the realistic Y_L and Y_S configurations which pass all the known experimental constraints on the flavor violating processes. The viable configurations were collected and have been used to calculate the resulting 1-loop charged lepton flavor violating $Z \to \bar{l}l'$ and $l \to l'\gamma$. Some of the realistic configurations could be probed in the forthcoming cLFV experiments. Interesting and robust lower bounds have been found for these cLFV Moreover, the neutrino mass hierarchy could be determined if the measured cLFV double ratio(s) is/are in some specific range(s). For $Y_R = 0$, Δ has 50% of chance decaying into a charged lepton and an up-type quark. Specific ratios $\sum_j B(\Delta \to l_i u_j)$ for each generation charged lepton l_i have been predicted in this model. Given the potential link between the neutrino masses generation and Δ, it seems well-motivated using the predicted leptoquark branching ratios as a benchmark scenario for the future scalar leptoquark search limits.

Acknowledgments

FX is supported partially by NSFC (National Natural Science Foundation of China) under Grant No. 11605076, as well as the FRFCU (Fundamental Research Funds for the Central Universities in China) under the Grant No. 21616309.

References

1. S. Weinberg, Phys. Rev. Lett. **43**, 1566 (1979).
2. M. Kohda, H. Sugiyama, K. Tsumura, Phys. Lett. **B 718** (2013) 1436; arXiv: 1210.5622[hep-ph].
3. W. F. Chang, S. C. Liou, C. F. Wong and F. Xu, JHEP **1610**, 106 (2016) doi: 10.1007/JHEP10(2016)106 [arXiv:1608.05511[hep-ph]].

The Reactor Neutrino Energy Spectrum Measurement with a High Pressure Gas TPC Detector

Wenqi Yan[1,*], Jun Cao[1], Yufeng Li[1], Qian Liu[2], Zhe Ning[1], Xilei Sun[1],
Liangjian Wen[1], Wenhuan Wu[1], Liang Zhan[1]

[1]*Institute of High Energy Physics, IHEP*
[2]*University of Chinese Academy of Sciences, UCAS*
**E-mail: yanwq@ihep.ac.cn*

We are studying and designing a novel high pressure (~10bar, CF4 or Ne+TMA=95:5) gas TPC detector based on the thick-GEM or Micromegas to detect the neutrino energy spectrum of the reactor, which will provide an input for JUNO to determine the neutrino hierarchy. By detecting the energy and scattering angle of the elastic scattering electron, which happens between the reactor neutrino and gas, we will reconstruct the neutrino spectrum of the reactor. And the mainly physics motivation of the short baseline experiment is precisely measure the energy spectrum of reactor antineutrino with a 3% energy resolution. Providing the input for JUNO (Daya Bay ~8%; JUNO ~3%). So the high pressure gas TPC is a better option with a higher energy resolution and a better track reconstruction. We prepare design a 200kg's detector, which is placed at 20m away from power plant, to study the neutrino energy spectrum, mixing angle, sterile neutrino and abnormal magnetic moment.

Keywords: Neutrino energy spectrum; track and angle reconstruction; elastic scattering; energy resolution.

1. Introduction

With the micro pattern detector development, gas TPC detector with a better track and energy resolution has been used in the wider area. In the recent years, the dark matter and neutrinoless double beta decay detector has been a hotspot, such as panda-X and NEXT collaborators with a high pressure gas or liquid TPC. So we prepare to design a novel high pressure gas TPC detector to detect the neutrino energy spectrum of the reactor by detecting the scattering electron. As the following Figure 1, the concept structure of the neutrino detector, including the veto-Compton detector, outer SS chamber, inner chamber and shape

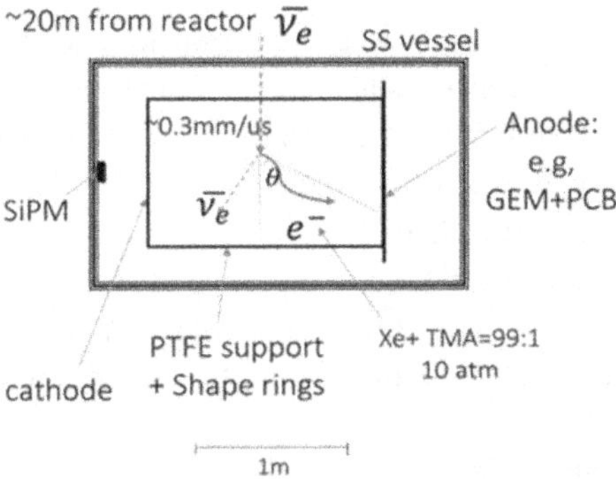

Figure 1.   The concept structure of the high pressure gas TPC detector.

rings and so on. As we know, the cross section of the elastic scattering with the reactor neutrino depends on the flux of the neutrino and electron density of the target. So we need to select a kind gas with a high electron density and the attachment coefficient is small enough with a long drift distance about 1m. The mainly motivation of the detector is to reconstruction the neutrino energy spectrum of the reactor, which is to provide an input for JUNO. The detector will have a deep study about fine structure of the neutrino energy spectrum with the less than 3% energy spectrum. In addition to this, the mixing angle, sterile neutrino and abnormal magnetic moment also is the study goals.

2. Physical requirement

The kinetic and angle of the elastic electron is the key parameters of the gas TPC detector, which determines the resolution of the neutrino spectrum. In the proceeding of the design, we prepare to place the detector at 20m away from reactor plant to increase the event rate. In addition to this, underground experiment is a key to shield the cosmic ray and other environment noise. In the calculation, we have raised a primary requirement about energy and angle resolution of the detector design based on the CF4 about 200kg with a pressure of 10atm as the target material.

2.1. *Event rate calculation*

As the following formula (1), this introduces the elastic scattering cross section between the neutrino and electron.

$$\frac{d\sigma}{dT} = \frac{G_F^2 m_e}{2\pi} \left[\left(g_V + g_A\right)^2 + \left(g_V - g_A\right)^2 \left(1 - \frac{T}{E_\nu}\right)^2 + \left(g_V^2 - g_A^2\right)\frac{m_e T}{E_\nu^2} \right] + \frac{\pi\alpha^2 \mu_\nu^2}{m_e^2} \frac{1 - T/E_\nu}{T} \quad (1)$$

When the neutrino is the electron anti-neutrino, $g_v = 2\sin^2\theta + 1/2$, $g_A = -1/2$.

$$T_e = \frac{2m_e E_\nu^2 \cos^2\theta}{(m_e + E_\nu)^2 - E_\nu^2 \cos^2\theta} \quad (2)$$

The m_e is the electron mass, T is the electron kinetic energy, E_ν is the neutrino energy, G_F is the weak interaction constant, μ_ν is the neutrino magnetic moment and θw is the mixing angle. The elastic scattering formula (2) is that electron scattering kinetic vary with the neutrino energy and scattering angle.

The event rate is calculated with the formula 1-2 and neutrino flux of the reactor. In the calculation, we select the 200kg CF4 as the working gas and the detector is 20m away from the reactor. As the following Figure 2 shown, the event rate per day varies with the electron kinetic, the lines with the different color show that the different value of the abnormal magnetic.

When the abnormal magnetic equals to zero, the event rate about 363 per day with the 400kg CF4 and 200keV's threshold based on the primary calculation result.

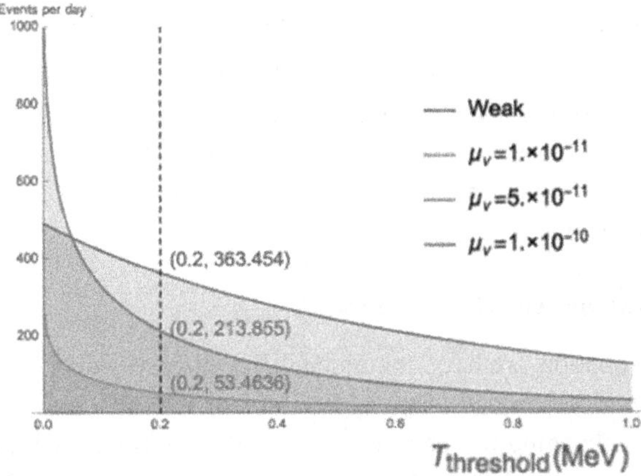

Figure 2.   The event rate with the gas TPC detector with 20m away from reactor.

2.2. *The gas TPC detector design requirement*

In order to providing the input for the Juno experiment, we need accurately measuring and studying whether has the secondary structure of the reactor neutrino energy spectrum. The 3% neutrino energy resolution is necessary, which also requires higher resolution of the electron kinetic and the scattering angle. For having a clear understanding, we have done some simulation of the neutrino energy reconstruction based on the experience. The

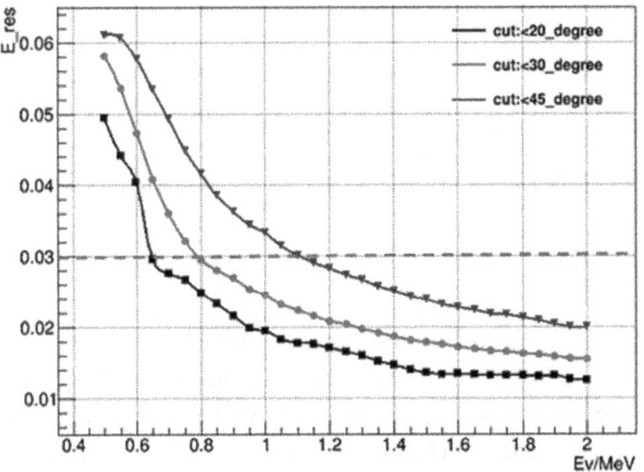

Figure 3.   The reconstruction of the neutrino energy resolution.

energy resolution is 1%@1MeV and the angle resolution is 3°@1MeV. As the above figure, the neutrino energy resolution varies with the neutrino energy. So we put forward the design parameters, including the detector energy resolution, angle resolution and scattering angle based on the primary simulation result.

3. Angle and energy reconstruction simulation of the detector

Angle resolution is the key parameter of the gas TPC detector. As we know, the commonly TPC detector is used as the high energy track detector in the accelerators and in the big experiments, but rarely is used as the low energy detector to measuring the energy and track. So we also do some simulation with geant4 and Garfield++, studying the electron track, energy and diffusion in the gas.

3.1. *Gases performance simulation and study*

In the angle simulation, we have compared different gases in common use, such as the neon, argon, CF4 and xenon with doping the TMA or iC4H10. And we ensure the same electron density by changing the pressure with a 10atm's Ne gas as the standard. In the simulation, we compare the electron drift diffusion and velocity with different pressure and different mixing gas component. On the other hand, we must consider the gas selection with the elastic scattering. As we know larger density of the electron, non-negative electronegative gases and less proton to avoid the inverse beta decay reactor and so on. We has done selected the optimal working gas by the simulation. Such as the Ne + TMA and the CF4 to consider the track reconstruction and the common factor.

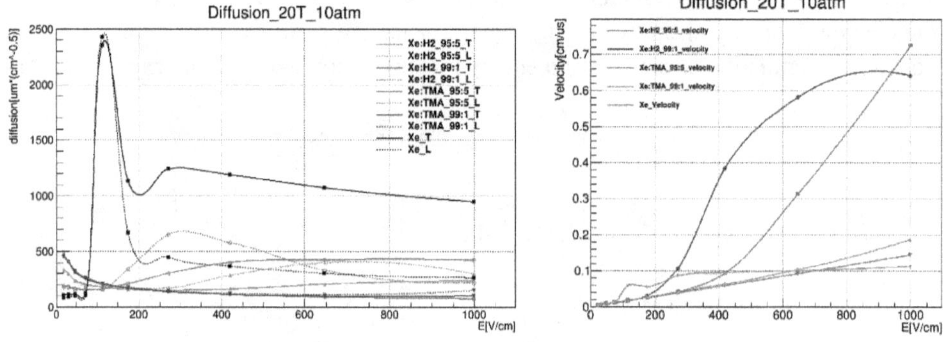

Figure 4.   The diffusion varies with the drift field in the xenon with different doping gases.

As shown Figure 4, the diffusion in the xenon with different doping gases, we can know that different doping gases can suppress the electron diffusion in the gas. But the TMA is optimal for the different polyatomic gas. So we can consider the optimal working gas then suppressing the diffusion by the TMA and the diffusion sigma is about 1mm when the drift distance is about 1m.

3.2. *Electron scattering angle reconstruction*

As Figure 5, we have done some simulation with the detector, which is filled with 10atm and 54atm xenon. We can know that the track bending and track length are different from different pressure gases.

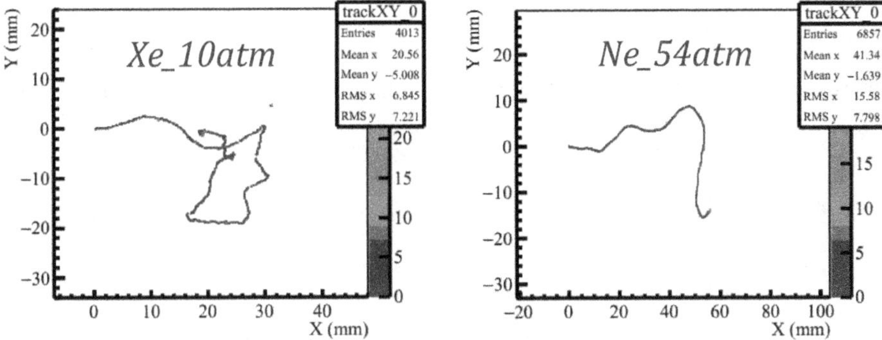

Figure 5.   The track in the xenon with the different pressure.

In the reconstruction, we assume that the track is liner at the start 10mm, so we can use the pol1 to fit the track and get the scattering angle. So when we select the gases we need to consider another problem whether the track is liner.

As the following Figure 6, the left figure shows that we do the linear fitting for the diffusion track and the right figure shows the angle reconstruction result with the 1000 tracks based on the 54atm Ne gas.

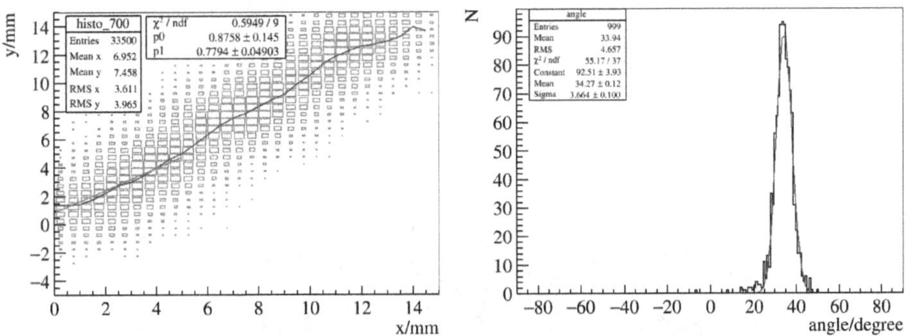

Figure 6.   Angle reconstruction method based on the 54atm neon gas.

We compared the different gases based on the method to study the optimal working gas. And we limit the angle about 3° to select the gas from the different pressure. As Figure 7 shows that working gas table.

Based on the candidate gases and reconstruction method, we study the scattering angle simulation result to select the optimal working gases.

From the Figure 8, we can know that the CF4, Neon and Argon are demandable and the CF4 is best with a less than 10atm pressure for Ne. But when the drift field is enough big, the CF4 will have electronegativity. So we select the Ne as the candidate working gas to study the neutrino energy spectrum reconstruction. The Figure 9 shows that the cosine RMS varies with the electron energy and input angle for the 10atm Neon gas.

Gas(Z,A)	$^{88}_{42}CF4$	$^{4}_{2}He$	$^{20}_{10}Ne$	$^{40}_{18}Ar$	$^{131}_{54}Xe$
Pressure	12.857atm	270atm	54atm	30atm	10atm@200kg
Pressure	7.143atm	150atm	30atm	16.667atm	5.556atm
Pressure	2.381atm	50atm	10atm	5.556atm	1.852atm
Pressure	1.19atm	25atm	5atm	2.778atm	0.926atm

Figure 7.   The candidate gases table with different pressure for the same electron density.

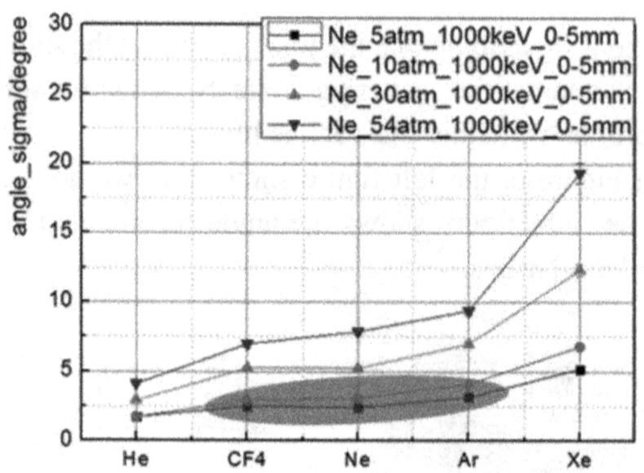

Figure 8.   The angle reconstruction with different pressure and gases.

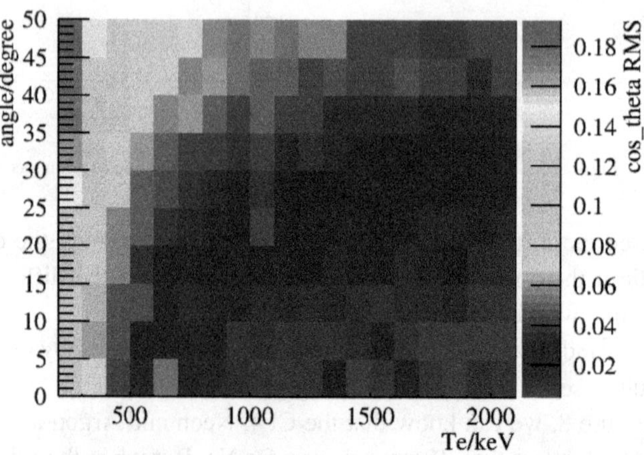

Figure 9.   The angle reconstruction result for Neon with 10atm.

3.3. *Neutrino energy spectrum reconstruction*

As Figure 3 shown, we reconstruct the neutrino energy spectrum with angle reconstruction result Figure 9, formula 2 and the electron energy resolution with the experience formula $1\%/\sqrt{E}$. And the electron energy resolution is an ideal approximate with the intrinsic energy resolution about 0.25% for Neon gas.

4. Conclusion

High pressure gas TPC has great potential to precisely measure reactor neutrino spectrum. And we have done some simulation for the angle resolution and electron energy resolution. We can get a 3% neutrino energy resolution based on the simulation result.

References

1. Escada J, Dias T H V T, Santos F P, et al. A Monte Carlo study of the fluctuations in Xe electroluminescence yield: pure Xe vs Xe doped with CH4 or CF4 and planar vs cylindrical geometries[J]. Physics, 2011, 6(8):1571-1574.
2. Daraktchieva Z, Amsler C, Avenier M, et al. Low energy tracking and particles identification in the MUNU Time Projection Chamber at 1 bar. Possible application in low energy solar neutrino spectroscopy[J]. Journal of Physics G Nuclear & Particle Physics, 2007, 35(12):1263-1269.
3. Daraktchieva Z. MUNU final results[J]. Nuclear Physics B, 2011, 221:62-66.
4. Tan A, Xiao M, Cui X, et al. Dark Matter Results from First 98.7-day Data of PandaX-II Experiment[J]. Physical Review Letters, 2016, 117(12):121303.
5. J Galan. Microbulk Micromegas for the search of $0\nu\beta\beta$ of 136Xe in the PandaX-III experiment[J]. Journal of Instrumentation, 2016, 11(04):P04024.
6. Ji X. Herman Feshbach Prize in Theoretical Nuclear Physics Xiangdong Ji, University of Maryland PandaX-III: high-pressure gas TPC for Xe136 neutrinoless double beta decay at CJPL[C]// APS April Meeting. APS April Meeting Abstracts, 2016.
7. Vuilleumier J L. The Gotthard and MUNU TPCs: From the past to the future?[C]// 2009:012013.

Research on Liquid Scintillator Energy Nonlinearity

Yuzi Yang* and Jiajie Ling

*Department of Physics, Sun Yat-sen University,
Guangzhou, Guangdong, China*
**E-mail: yangyz6@mail2.sysu.edu.cn*

Liquid scintillator(LS) calorimeter is a classical technology in the particle physics, especially for the reactor neutrino experiments, which is widely used for detecting electron anti-neutrinos though the inverse beta decay interaction channel. Because of the quenching effect, the scintillator detector has nonlinear energy response. It is critical to accurately measure the scintillator energy response for both the precision measurement of reactor antineutrino energy spectrum in Daya Bay Experiment and the neutrino mass hierarchy measurement determination in JUNO experiments. There are several bench measurements of the liquid scintillator energy nonlinearity response through the gamma-ray and the electron Compton scattering process. However, it is difficult to estimate the systematic uncertainties of those measurements are difficult to assess. In this paper, we used the Geant4 simulation package to study several systematic uncertainties, including the gamma-ray multiple scattering in the detector, the physical size of the detector and the edging effect. Our simulations shows that all these effects have marginal impact ($<1\%$) on the scintillator energy response measurement.

Keywords: Simulation; Geant4; scattering; reemission; edging effect.

1. Introduction

Liquid scintillator, which is an important material inside the neutrino detector, has certain unique features–outstanding low background level, large detector mass and low energy threshold. Thanks to the help of liquid scintillator, KamLAND managed to become the first experiment of observing the reactor anti-electron neutrino disappearance[4]. Daya Bay obtained the best constrain of θ_{13}. In the future, JUNO will also use LS to construct the centre detector, and major goal of JUNO is the determination of the neutrino mass hierarchy and the improvement of the precision of Δm_{21}^2, Δm_{32}^2 and $\sin^2\theta_{12}$. For JUNO, the crucial problem is that it requires a very high neutrino energy spectrum resolution. However, uncertainty of LS nonlinearity does not satisfy pivotal demand. Although the LS nonlinearity of JUNO had been measured before, disappointingly, the result was inconsistent with Daya Bay normal model (Fig. 1). At low energy range, the uncertainties are large, because of misalignment. In fact, the high pure germanium(HPGe) can cancel the misalignment uncertainty.

2. LS mechanism and IHEP measurement setup

Usually, people use LS to detect anti-electron neutrino via the observed positron energy in the inverse β-decay event, $\bar{\nu}+\mathrm{p}=\mathrm{e}^++\mathrm{n}$. Positron loss energy in ionization, and is finally annihilated with an electron into two $0.511\,\mathrm{MeV}$ photons. γ loss

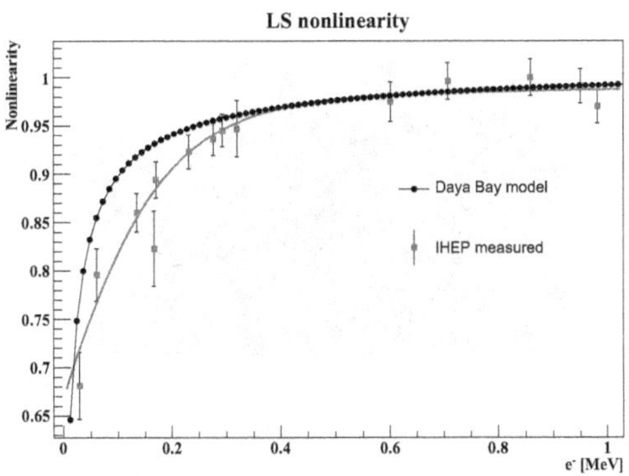

Fig. 1.   Daya Bay model and IHEP measured LS nonlinearity data.

energy in LS via Compton scattering. The scattered electron further loses energy via ionization again. The total deposited energy converts into photons in LS, which will be collected by PMTs. If the electron energy exceeds the threshold, Cherenkov light will also emit. In 1950s, J.B. Birks described the quenching mechanism with an empirical model-Birks Law. Later, KamLAND experiment found that the addition of Cherenkov radiation would optimize the scintillation mechanism[5]. Now we know that quenching energy is dependent on particle deposited energy in every interaction in LS, so we need to determine the every nonlinearity of the concerned particles. Here, this paper will focus on electron energy nonlinearity in LS. In this study, we research the nonlinearity of the Daya Bay LS for electron via the Compton scattering process in the laboratory. The LS consists of linear alkylbenzene(LAB) as the solvent, 3g/L 2,5-diphenyloxazole(PPO) as the primary fluorescence material, and 15mg/L p-bis-(o-methylstyryl)-benzene(bis-MSB) as the wavelength shifter. Through this study, it can also improve the understanding of energy response in LS detector.

The measurement was designed to use the Compton scattering of γ of known energy to produce mono-energetic electrons in the LS by tagging the scattered γ at certain angles. Fig. 2 shows the IHEP setup in Geant4 simulation. They used a 0.3mCi ^{22}Na source, it can produce γs of 0.511MeV and 1.275MeV. A 9mm diameter hole on the lead collimator, the inspired photons in LS will be collected by PMT below, seven coincidence detectors are placed in seven directions/($20°$, $30°$, $50°$, $60°$, $80°$, $100°$, $110°$) and 60cm away from the LS vessel. The energy of scattered electron in LS can be calculated with the Compton formula:

$$E_e = \frac{E_\gamma^2}{E_\gamma + \frac{m_e c^2}{1-\cos\theta}} \tag{1}$$

where E_e is the scattered electron energy, E_γ is the γ ray energy, m_e is the electron mass, and θ is the Compton scattering angle.

494

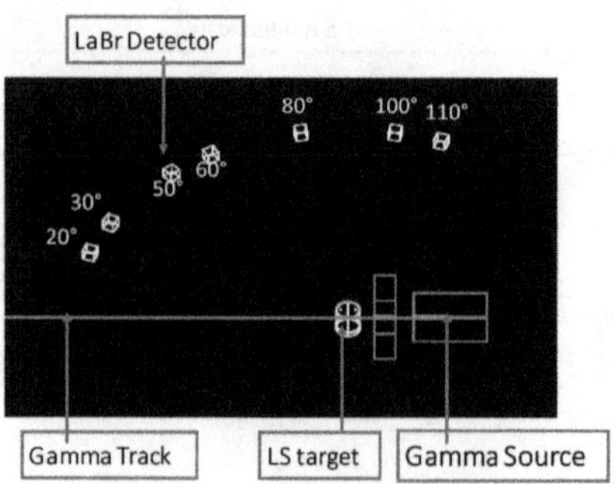

Fig. 2.   IHEP setup in Geant4 simulation.

3. Multiple scattering and energy loss

In the pre-measurement, an obvious uncertainty, γ multiple scattering in LS, was ignored. In our research process, we also found the recoil electron could lose in LS vessel, which is due to the hit on wall or the sub-recoil electron flee from LS volume. In Fig. 3, the red line corresponds to single gamma scattering and energy conservation event. If the electron leaks out LS volume, it is counted as an electron energy loss event (Green line). If gamma collide electron twice or more times, it is corresponding to multiple gamma scattering (Black). Gamma energy would lost in coincidence detector, we can use the full energy peak to determinate the energy conservation events (Blue). Considering the real experiment data, in Geant4

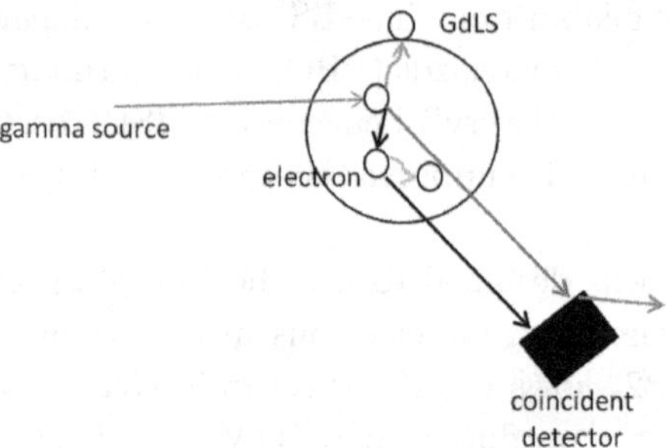

Fig. 3.   Multiple scattering and energy loss (color online).

simulation, we chose double gamma energy source 0.511MeV and 1.275MeV. Through the distribution of events at the different coincidence detector, these different kinds of events can be distinguished approximately (Fig. 4). Two bands of events are corresponding to different energy γs, we can image a hypothetical line, events on this line are energy conservation events, or they are energy loss events. Double enrichment areas are the concerned signal regions. Diffuse events on the energy conservation line correspond to multiple scattering events. In theory, we

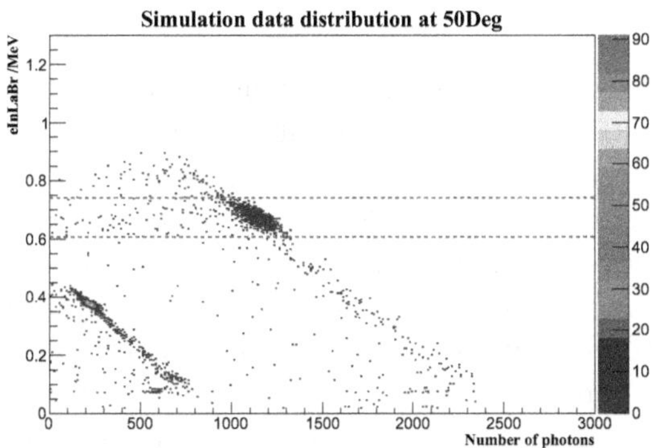

Fig. 4. Simulation data distribution at 50°. Photons are collected by PMT. For the good resolution of full energy peak, we ignored the energy loss events in coincidence detector. The red dotted line is corresponding to 1σ signal area of 1.275MeV γ (color online).

can refer to the attenuation length of γ ray in LS, the big value will suppress the times of γ scattering in LS, in my simulation data, conformably, the single Compton scattering events are dominant.

After some selection conditions, the raw data are used to analyze the nonlinearity. In the selection conditions, we hope to suppress the multiple scattering or energy loss events and increase the ratio of single scattering and energy conservation events. Similar to the pre–measurement, the coincidence detector(LaBr) energy cut is adopted(Red dotted line). It is incapable of selecting energy loss events in LS, but it can exclude most multiple scattering events. After the selection, we use Gauss function to fit the selection data, the fitted mean is the number of photons at this coincidence detector. Nonlinearity is the normalized ratio of number of photons and recoil electron energy. I determinate the effect via comparison the discrepancy of fitted mean between single scattering-energy conservation(SS-EC) events and all selected events(AS).

The slight discrepancy of double kinds of data indicates that the remnant multiple scattering or energy loss events have small effect to nonlinearity. It is believed that certain defective details of simulation code could make the simulated nonlinearity to do not match well with measured nonlinearity, however, the multiple scattering or energy loss events are rare in selected simulation data. Therefore, we believe that it would not cause significant change of the nonlinearity of LS.

Table 1. The fitted mean of 1.275MeV γ energy at different coincidence detector. At different coincidence detector, the fitted means are very close and their errors are also small. The 0.511MeV γ datas are similar, they are not shown here.

Angle	20°	30°	50°	60°	80°	100°	110°
SS-EC	287.95	570.32	1134.75	1355.59	1670.66	1870.90	1931.54
AS	287.23	569.94	1131.22	1353.17	1166.31	1867.90	1924.08

4. Target size

In the large LS detector experiments, it is important to confirm that when the LS volume changes, will the nonlinearity also change? Now, we will test it in our Geant4 simulation, comparing to the measurements in laboratory, the adjustment of LS volume in simulation code is sample and economical. In our Geant4 model, there are 5 physical processes relative to this research, they are scintillation, Cherenkov radiation, absorption, reemission and Rayleigh scattering. Scintillation is the primary source of photon yield, it is corresponding to the quenching process. If the velocity of charged particle exceed the threshold in LS, Cherenkov radiation arises. Absorption and reemission is the instinct of LS, photons would be absorbed and re-emit in LS in propagation, a crucial note is that when the wavelength of photon exceeds 345nm, the dominant absorption is bis-MSB absorption, or it is LAB and PPO absorption[6]. The efficiency of energy propagation from LAB to PPO is very high. Rayleigh scattering is dependent on wavelength of photons. In new Geant4 simulation, the physical geometry of LS vessel changes to a sphere from a cub, the incident particle is e$^-$ replacing γ, and the position of e$^-$ source is on the center of the sphere (Fig. 5).

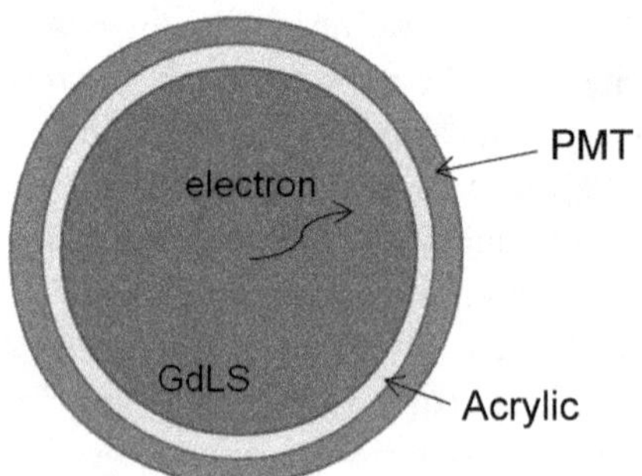

Fig. 5.　New setup in Geant4 code. The volume of LS could change, but the thickness of acrylic and PMT is stable. The cover rate of PMT is 1.

In our simulation, it follows some important terms:

- Extract energy from PPO spectrum as the photon energy of scintillator.
- Reemitted photon energy sample from PPO or bis-MSB spectrum.
- Cherenkov radiation is inclusive.
- Some vital parameters, such as attenuation length and reemission probability, are measured in pre-experiments.

A method to understand these physical processes' effect is through the superposition of processes one by one. In our simulation, at first, we run the code with only scintillator process, then the Cherenkov radiation, absorption, re-emission and Rayleigh scattering will be added one by one (Fig. 6).

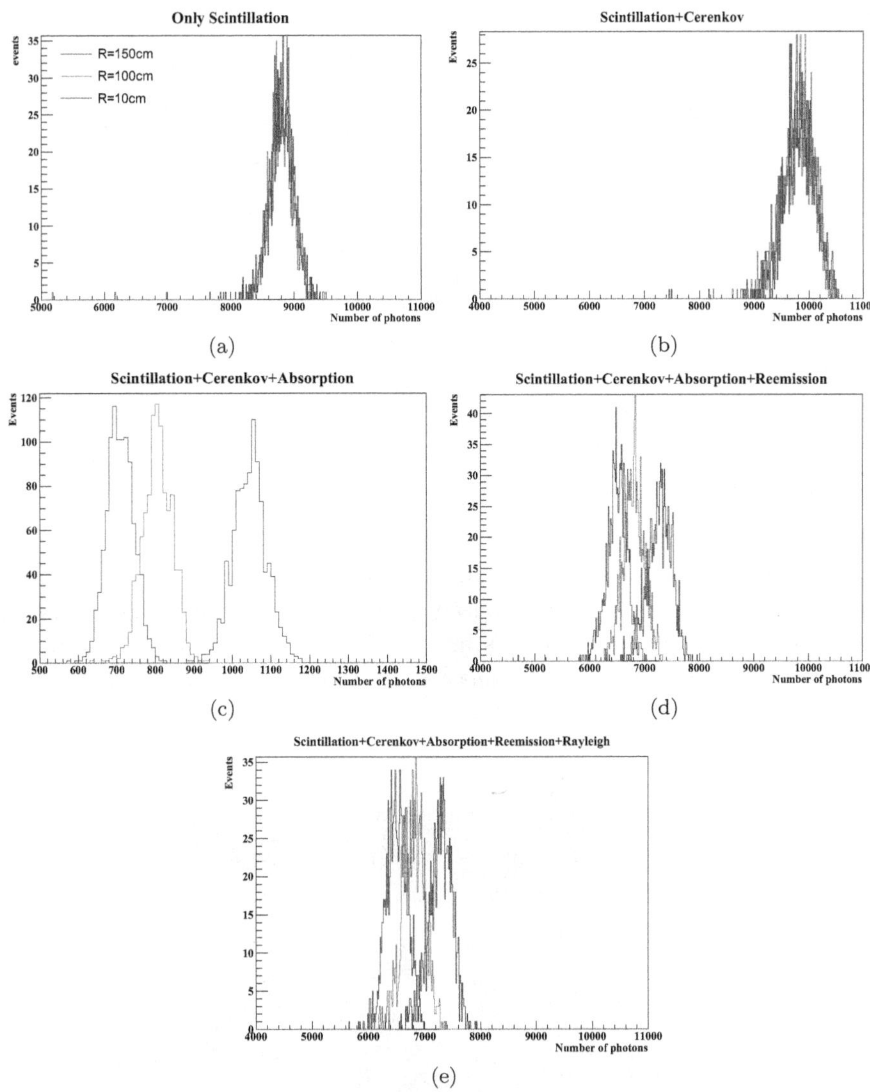

Fig. 6.   Events distribution at different runs. The difference arises when adding absorption and reemission process. For the especial wavelength, the Rayleigh scattering effect can be negligible.

We know that because of the effect of absorption and reemission, the absolute photon yield must be different in different LS volume. On the other hand, the relation between the nonlinearity curve and the LS volume also requires to be confirmed? In Fig. 7, the absolute photon yield curve is high when the LS volume is small, however, when we normalized these absolute curves to calculate the nonlinearity curves, they are overlapped. This indicates that the volume of LS can change the height of absolute photon yield, but it will not change the nonlinearity curve. Hereto, the effect of target size of LS is clearly understood.

498

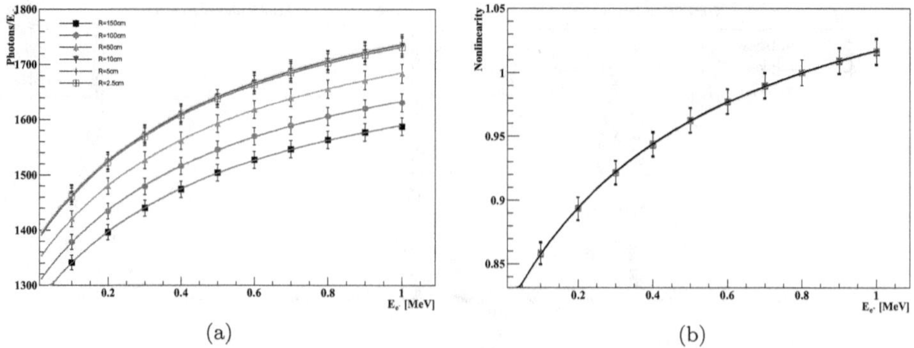

(a)   (b)

Fig. 7. (a) Absolute photon yield curves. (b) Nonlinearity. The nonlinearity curves are calculated by scaling photon yields of 0.8MeV to 1.

5. Edging effect

Edging effect is a complicated problem, it relates to not only to response of every area of LS, but also to the reflecting films. For this research, Geant4 simulation is a suitable method, the γs hit position and nonuniform collected photons by PMT of every area is clear at a glance. γs hit position is the position of causing recoil electrons. Generally, if the hit positions are close to the PMT below, more photons would be collected by PMT.

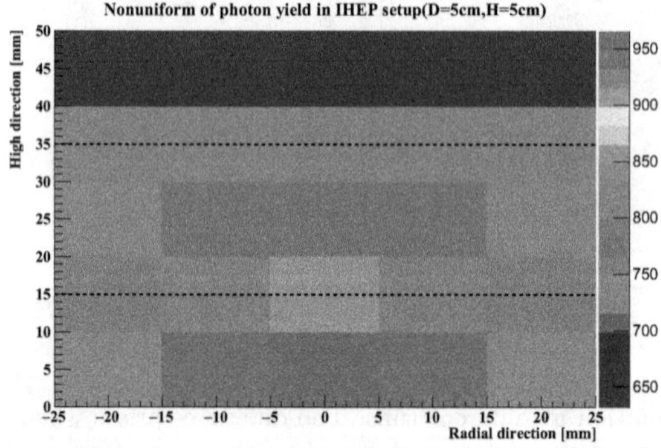

Fig. 8. Collected photons by PMT below nonuniform distribution. The area between two dash line is corresponding to the γs hit position in IHEP setup.

Moreover, we also notice that the recoil electron energies are diffuse. In our updated simulation code, we add this distribution, the information of diffusion parameters come from pre-simulation. After the data selection, we calculate the nonlinearity of electron incident position in the center with the IHEP setup and the nonlinearity of diffuse energy electron with inspired positions in the γs hit position range. Take notes that we also use the 1σ energy cut of electron energy in our

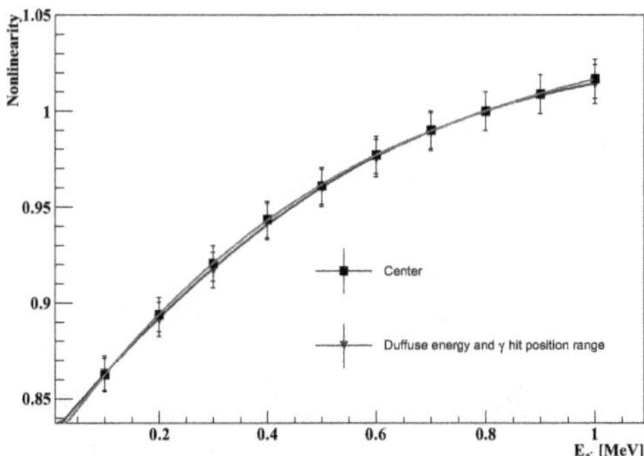

Fig. 9.   Compare to two different kinds of simulated datas. The high similarity implies the effect is negligible.

analysis to avoid the terrible fit status at low energy range (Fig. 9). Through our research, the edging effect has little impact on nonlinearity.

6. Summary

We use Geant4 simulation to estimate some equivocal system uncertainties in the measurement of LS nonlinearity, such as multiple scattering or energy loss effect, target size effect and edging effect. We compare the results through control different parameters in Geant4 to analyze uncertainties, finally, we come to the conclusion that these systematic uncertainties which have tiny impacts on LS nonlinearity (they are less than 1%). A more scientific and precise method should be explored for the LS measurements, for example replacing the coincidence detector with HPGe and purifying LS from O_2 before measurements.

Acknowledgments

This work was funded by the National Natural Science Foundation of China (NSFC) under No. 11775315.

References

1. AN F et al. (Daya Bay collaboration), Phys. Rev. D, 2000, 62: 072002
2. ZHANG Fei-Hong et al., Chinese Phys. C Vol. 39, No. 1 (2015) 016003
3. Liang Zhan, Phys. Rev. D 78, 111103 (2008)
4. Eguchi K et al. (KamLAND collabration), Phys. Rev. Lett., 2003, 90: 021802
5. Perevozchikov O, PhD Thesis, The University of Tennessee, 2009
6. Hu-Lin Xiao, Nation Natural Science Foundation of China (211202037)

Precision Measurement of $\sin^2 2\theta_{13}$ and $|\Delta m^2_{ee}|$ from Daya Bay Scintillator

Hongzhao Yu[*], Jiajie Ling

on behalf of the Daya Bay Collaboration

*Sun Yat-sen University,
Guangzhou, Guangdong Province, P. R. China*
[*]*E-mail: yuhzh3@mail2.sysu.edu.cn*

With eight functionally identical electron anti-neutrino detectors deployed at three experimental halls near three high-power nuclear reactor complexes, the Daya Bay Reactor Neutrino Experiment has achieved unprecedented precision in measuring the neutrino mixing angle θ_{13} via neutron captured on gadolinium (nGd). Combining 217 days of data collected using six antineutrinos detectors with 1013 days of data using eight detectors, a relative comparison of the rates and positron energy spectra of the detectors located far (~1500-1950m) relative to those near the reactors (~350-600m) gave $\sin^2 2\theta_{13} = 0.0841 \pm 0.0027(stat) \pm 0.0019(syst)$. In addition, an independent analysis using samples based on neutron captured on hydrogen (nH) is validated, which has yielded $\sin^2 2\theta_{13} = 0.071 \pm 0.011$. With the nGd result and the nH result combined, an improvement in precision can be reached. The technical details and the combination of nGd result and nH result will be presented in this proceeding.

Keywords: Daya Bay; neutrino; oscillation.

1. Introduction

Three-flavor neutrino oscillations are described by the unitary PMNS matrix [1,2], through which a neutrino with flavor α can be expressed as a linear combination of neutrino mass eigenstates.

$$\nu_\alpha = \Sigma_i U_{\alpha i} \nu_i \tag{1}$$

where $\alpha = e, \mu, \tau$ for flavor eigenstates, and i = 1,2,3 for mass eigenstates. The PMNS matrix are usually parameterized by the three angles θ_{12}, θ_{23}, θ_{13}, and the CP-violating phase δ_{CP}. Of these mixing angles, θ_{13} is the least known. Shortly after the evidence of non-zero θ_{13} noticed by the Double Chooz experiment [3], Daya Bay experiment first observed non-zero θ_{13} with 5.2σ significance in 2012 [4] and RENO collaboration confirmed the result with a significance of 4.9σ in the same year [5]. A non-zero value of θ_{13} makes it possible to measure the mass hierarchy of neutrino and the CP phase.

For reactor-based experiments, an unambiguous determination of θ_{13} can be extracted via the survival probability of the $\bar{\nu}_e$ at short distances from the reactors,

$$\begin{aligned}
P_{sur} &= 1 - \cos^4 \theta_{13} \sin^2 2\theta_{12} \sin^2 \Delta_{21} \\
&\quad - \sin^2 2\theta_{13} \left(\cos^2 \theta_{12} \sin^2 \Delta_{31} + \sin^2 \theta_{12} \sin^2 \Delta_{32}\right) \\
&\simeq 1 - \cos^4 \theta_{13} \sin^2 2\theta_{12} \sin^2 \Delta_{21} - \sin^2 2\theta_{13} \sin^2 \Delta_{ee}
\end{aligned} \tag{2}$$

where $\Delta_{ij} \simeq 1.267 \Delta m_{ij}^2 (eV^2) L(m)/E_\nu(MeV)$, and Δm_{ij}^2 is the square-mass difference between the mass eigenstates, E is the $\bar{\nu}_e$ energy in MeV and L is distance in meters between the $\bar{\nu}_e$ source and the detector (baseline). For short baseline reactor experiments, Δ_{31} and Δ_{32} are indistinguishable. Therefore, the expression can be approximated using an effective phase Δ_{ee}.

2. Daya Bay Experiment

The six pressurized water reactors are grouped into three pairs with each pair referred to as a nuclear power plant (NPP). The maximum thermal power of each reactor is 2.9 GW. Three underground experimental halls (EHs) are connected with horizontal tunnels. Two antineutrino detectors (ADs) are located in EH1 and EH2 (the near halls). Four ADs are positioned near the oscillation maximum in the far hall, EH3. The baselines have been surveyed with GPS and Total Station to a precision of 18 mm. The uncertainties in the baseline and the spatial distribution of the fission fractions in the core had a negligible effect to the results.

Each AD has three nested cylindrical volumes separated by concentric acrylic vessels. The outermost vessel is constructed of stainless steel (SSV), which holds 40 tons

Figure 1. Layout of the Daya Bay experiment. The Daya Bay and Ling Ao nuclear power plant (NPP) reactors (red circles) were situated on a narrow coastal shelf between the Daya Bay coastline and inland mountains. Two antineutrino detectors installed in each underground experimental hall near to the reactors (EH1 and EH2) measured the $\bar{\nu}_e$ flux emitted by the reactors, while four detectors in the far experimental hall (EH3) measured a deficit in the $\bar{\nu}_e$ flux due to oscillation. The plot is from [6] (color online).

502

of mineral oil (MO) to provide optical homogeneity and to shield the inner volumes from radioactivity. The innermost volume holds 20 tons of 0.1% by weight Gd-LS that serves as the antineutrino target. The middle volume is called gamma catcher is filled with 20 tons of un-doped liquid scintillator (LS) for detecting gamma-rays that escape the target volume. The gamma-catcher increases the containment of gamma energy thus improving the energy resolution and reducing the uncertainties of the antineutrino detection efficiency. Three automated calibration units (ACUs) mounted on the SSV lid allow for remote deployment of a light-emitting diode, a ^{68}Ge source, and a combined source of ^{241}Am-^{13}C and ^{60}Co into the Gd-LS and LS liquid volumes along three vertical axes.

The ADs in each EH are shielded with > 2.5 meter of high purity water against ambient radiation in all directions. Each water pool is segmented into inner and outer water shields (IWS and OWS) and instrumented with photomultiplier tubes (PMTs) to work as a veto system to remove the spallation neutrons and cosmogenic backgrounds. The water pool is covered with an array of RPC. Details of the detector system can be found in [7,8].

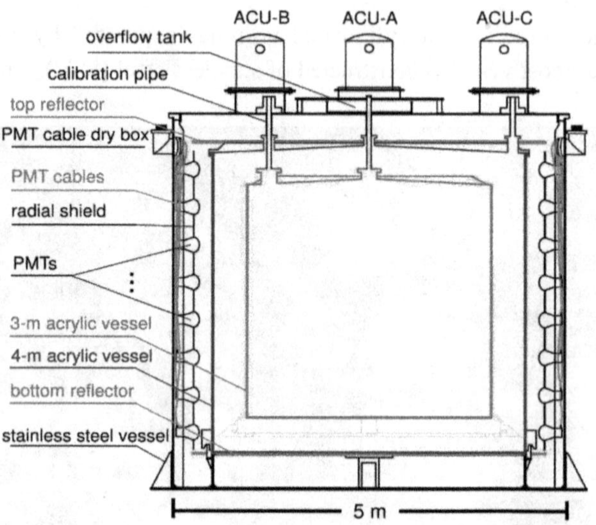

Figure 2.   Cross-sectional diagram of an AD. Scintillation light was produced when a reactor $\bar{\nu}_e$ interacted within the central 20-ton GdLS target, which was contained in a transparent acrylic vessel. The target was nested within an additional 20 tons of pure LS to increase efficiency for detection of gamma rays produced within the target. Scintillation light was detected by 192 photomultipliers mounted on the inner circumference of a 5 by 5 m stainless steel vessel, which was filled with MO. The plot is from [6].

3. Signal and Backgrounds

The $\bar{\nu}_e$ is detected via the inverse β-decay (IBD) reaction, $\bar{\nu}_e + p \rightarrow e^+ + n$, in a gadolinium-doped liquid scintillator (Gd-LS) [9,10]. The coincidence of the prompt scintillation from the e^+ and the delayed neutron capture on gadolinium (nGd) or on hydrogen (nH) provides a distinctive $\bar{\nu}_e$ signature. The prompt signals are between 0.7MeV and 12MeV. If neutron was captured on Gd, the capture time is about 30μs, and the excited

Gd will release ~8MeV of gamma rays. If it was captured on H, the capture time is about 200μs, and releasing ~2.2MeV gamma ray.

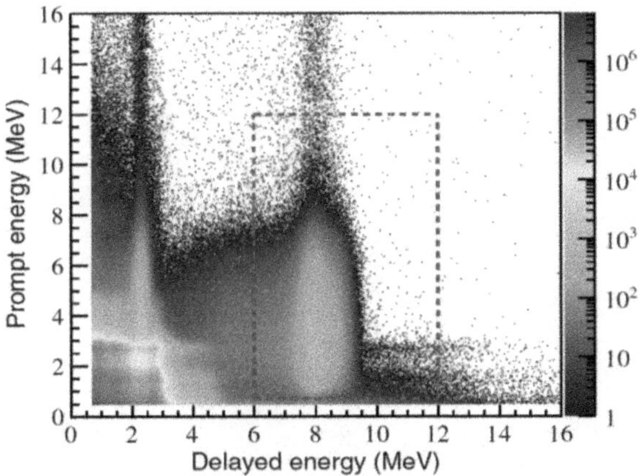

Figure 3. The distribution of prompt versus delayed energy for signal pairs which satisfied the $\bar{\nu}_e$ inverse beta-decay selection criteria. Interactions with prompt energy of $0.7\text{MeV} < E_p < 12\text{MeV}$ and delayed energy of $6\text{MeV} < E_d < 12\text{MeV}$ were used for the neutrino oscillation nGd analysis (red dashed box). A few-percent contamination from accidental backgrounds (symmetric under interchange of prompt and delayed energy) and ^9Li decay and fast neutron backgrounds (high prompt and $\sim 8MeV$ delayed energy) are visible within the selected region. Inverse beta-decay interactions where the neutron was captured on ^1H provided an additional signal region with $E_d \sim 2.2 MeV$, albeit with much higher background. The plot is from [6] (color online).

There are five main backgrounds including the accidents, the fast neutrons, the $^9Li/\,^8He$, $^{13}C(\alpha,n)^{16}O$, and the background from the calibration source ^{241}Am-^{13}C. In the antineutrino sample, the accidental background is the largest one. The rate and the spectrum of the accidental background have been determined by measuring the singles rates of prompt- and delayed-like signals. It gives 0.3% relative uncertainty. The other four are all significantly smaller, but they are correlated backgrounds. The fast neutrons and β-n decay of the cosmogenic $^9Li/\,^8He$ provides the prompt- and delayed-like signals. Besides, the gamma emission from the neutron capture also affects the IBD detections, which can come from the calibration source ^{241}Am-^{13}C and $^{13}C(\alpha,n)^{16}O$ background.

4. Detector Energy Calibration

A calibration process was implemented to reduce any potential differences in reactor $\bar{\nu}_e$ detection efficiency between ADs. To do the energy calibration, the first step is to do the PMT gain calibration based on the single photoelectron from the PMT dark noise or on the weekly deployment of LED. Then, the AD-to-AD differences were estimated with 13 difference calibration sources, including both deployed and naturally occurred sources. Figure 4 shows the comparison of the mean reconstructed energy between the ADs for 13 various sources. Relative energy scale uncertainty for all calibration sources were less than 0.2%.

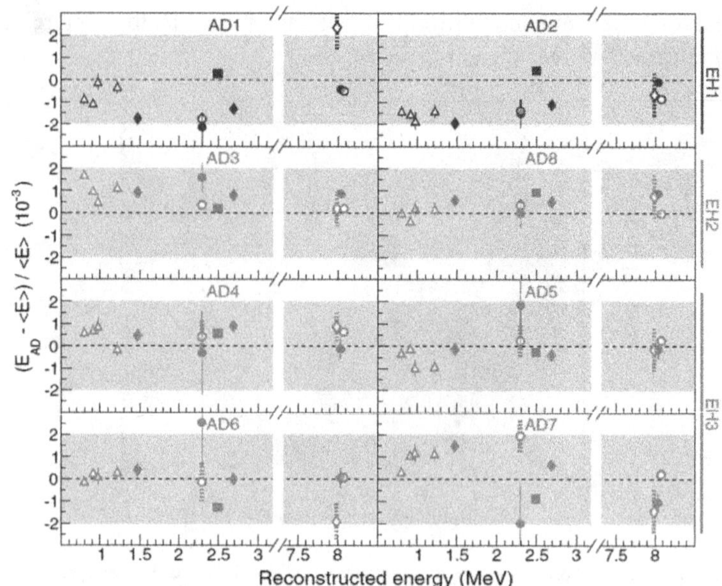

Figure 4. Comparison of the reconstructed energy between antineutrino detectors for a variety of calibration references. E_{AD} is the reconstructed energy determined using each AD, and $\langle E \rangle$ is the 8-detector average. The plot is from [6].

The characterization of the relationship between true energy and reconstructed IBD positron energy influences the measurement of the neutrino mass-square differences. The energy model of Daya Bay experiment includes the nonlinearity from liquid scintillator and the readout electronics. Twelve gamma lines and the continuous ^{12}B spectrum were utilized to calibrate the absolute energy response. Figure 5 shows the energy model, which the uncertainty band corresponds to the model consistent with the calibration data within 68.3% C.L.

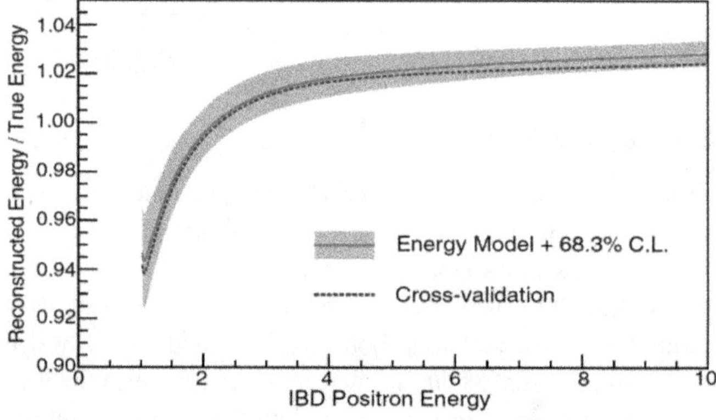

Figure 5. Estimated energy response of the detectors to positrons, including both kinetic and annihilation gamma energy (red solid curve). The plot is from [6] (color online).

5. Three-Flavor Neutrino Oscillation Analysis Using nGd

By selecting the nGd IBD candidates and by performing a relative measurement on the rate deficit and energy spectrum distortion through the detectors in the near halls and in the far hall, the best fit parameters in three-flavor scenario are

$$\sin^2 2\theta_{13} = 0.0841 \pm 0.0027(\text{stst}) \pm 0.0019(\text{syst})$$

$$|\Delta m_{ee}^2| = (2.50 \pm 0.06(stat) \pm 0.06(syst)) \times 10^{-3} eV^2$$

$$\Delta m_{32}^2(NH) = (2.45 \pm 0.06(stat) \pm 0.06(syst)) \times 10^{-3} eV^2$$

$$\Delta m_{32}^2(IH) = (-2.56 \pm 0.06(stat) \pm 0.06(syst)) \times 10^{-3} eV^2$$

Figure 6 shows the 2-D confidence and the $\Delta\chi^2$ versus $|\Delta m_{ee}^2|$ and $\sin^2 2\theta_{13}$. The upper panel shows the 1-D confidence of $\sin^2 2\theta_{13}$ when profiling $|\Delta m_{ee}^2|$, and the right panel gives the 1-D confidence of $|\Delta m_{ee}^2|$ when profiling $\sin^2 2\theta_{13}$. The electron antineutrino survival probability as a function of the ratio of the effective propagation distance L_{eff} over the average antineutrino energy $\langle E_\nu \rangle$ is also shown in Figure 6. More analysis details can be found in the publication [6].

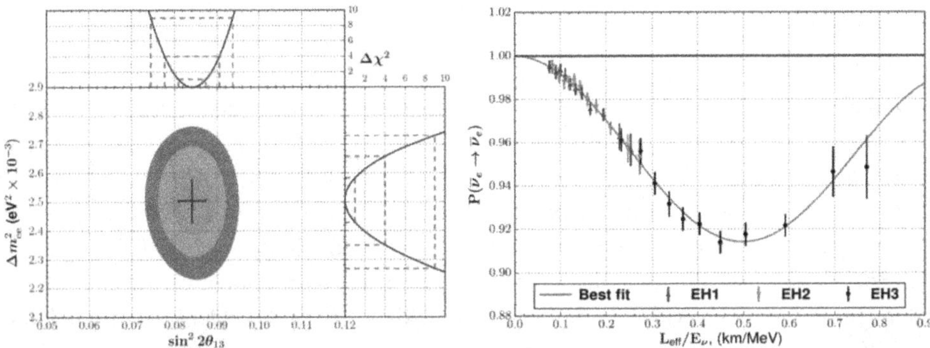

Figure 6.   Left: confidence interval for $|\Delta m_{ee}^2|$ and $\sin^2 2\theta_{13}$. Right: the measured electron antineutrino survival probability versus $L_{\text{eff}}/\langle E_\nu \rangle$, where L_{eff} is the effective baseline to reactors and $\langle E_\nu \rangle$ is the average antineutrino energy. The plots are from [6].

6. Independent $\sin^2 2\theta_{13}$ Measurement Using nH

An independent measurement of $\sin^2 2\theta_{13}$ has been measured via the detection of IBDs tagged by neutron capture on hydrogen (nH) by Daya Bay experiment. Based on the neutrino event rate deficit observed at the far hall from the 621 days of nH data, a value of $\sin^2 2\theta_{13} = 0.071 \pm 0.011$ is extracted in the three-flavor oscillation model. The result is shown in Figure 7. Figure 7 also shows a comparison of the prompt energy spectra of the IBD events of the far hall and the near halls weighted by the near-to-far baseline ratio, along with the ratio of the measured-to-predicted rates as a function of baseline. Clear evidence for electron antineutrino disappearance was observed.

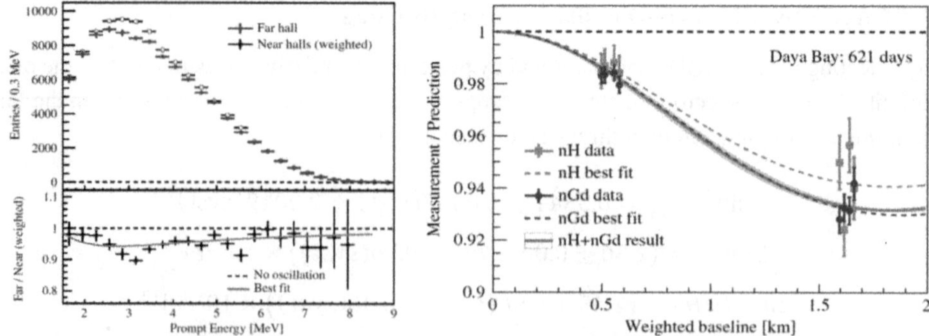

Figure 7.   Left: positron energy spectrum of the nH IBD sample, and the best-fit of neutrino oscillation. Right: the ratio of measured candidates to the predicted candidates assuming no oscillation in each detector versus the baseline. The nGd sample and the best-fit are also overlaid. The plots are from [11].

With the nGd-IBD result of $\sin^2 2\theta_{13} = 0.084 \pm 0.005$ and the nH-IBD result of $\sin^2 2\theta_{13} = 0.071 \pm 0.011$ from the same 621 days of data, both the analytical calculation and simultaneous fit resulted in $\sin^2 2\theta_{13} = 0.082 \pm 0.004$, which is an 8% improvement in precision. Details of the nH analysis can refer to [11].

7. Summary

With 1230 days of nGd data, a relative comparison of the rates and positron energy spectra of the detectors located far (~ 1500–1950 m) relative to those near the reactors (~ 350–600 m) gave $\sin^2 2\theta_{13} = 0.0841 \pm 0.0027(\text{stst}) \pm 0.0019(\text{syst})$ and the effective neutrino mass-squared difference of $|\Delta m_{ee}^2| = (2.50 \pm 0.06(stat) \pm 0.06(syst)) \times 10^{-3} eV^2$. This is equivalent to $\Delta m_{32}^2 = (2.45 \pm 0.06(stat) \pm 0.06(syst)) \times 10^{-3} eV^2$ assuming the normal mass hierarchy, or $\Delta m_{32}^2 = (-2.56 \pm 0.06(stat) \pm 0.06(syst)) \times 10^{-3} eV^2$ assuming the inverse hierarchy. Independent measurement using neutron captured on hydrogen with 621 days of data yields a new independent determination of $\sin^2 2\theta_{13} = 0.071 \pm 0.011$, which is consistent with that from the nGd analysis. Combining the nH- and nGd-IBD results from the same 621 days provides a new improved determination of $\sin^2 2\theta_{13} = 0.082 \pm 0.004$.

Acknowledgements

Daya Bay is supported in part by the Ministry of Science and Technology of China, the U.S. Department of Energy, the Chinese Academy of Sciences, the CAS Center for Excellence in Particle Physics, the National Natural Science Foundation of China, the Guangdong provincial government, the Shenzhen municipal government, the China General Nuclear Power Group, Key Laboratory of Particle and Radiation Imaging (Tsinghua University), the Ministry of Education, Key Laboratory of Particle Physics and Particle Irradiation (Shandong University), the Ministry of Education, Shanghai Laboratory for Particle Physics and Cosmology, the Research Grants Council of the

Hong Kong Special Administrative Region of China, the University Development Fund of The University of Hong Kong, the MOE program for Research of Excellence at National Taiwan University, National Chiao-Tung University, and NSC fund support from Taiwan, the U.S. National Science Foundation, the Alfred P. Sloan Foundation, the Ministry of Education, Youth, and Sports of the Czech Republic, the Joint Institute of Nuclear Research in Dubna, Russia, the CNFCRFBR joint research program, the National Commission of Scientific and Technological Research of Chile, and the Tsinghua University Initiative Scientific Research Program. We acknowledge Yellow River Engineering Consulting Co., Ltd., and China Railway 15th Bureau Group Co., Ltd., for building the underground laboratory. We are grateful for the ongoing cooperation from the China General Nuclear Power Group and China Light and Power Company.

This work was in part supported by the National Natural Science Foundation of China (NSFC) under Grant No. 11775315.

References

1. B. Pontecorvo, Sov. Phys. JETP 6, 429 (1957) and 26, 984 (1968).
2. Z. Maki, M. Nakagawa, and S. Sakata, Prog. Theor. Phys. 28, 870 (1962).
3. Y. Abe et al. (Double Chooz Collaboration), Phys. Rev. Lett. 108, 131801 (2012).
4. F. P. An et al. (Daya Bay Collaboration), Phys. Rev. Lett. 108, 191802 (2012).
5. J. K. Ahn et al. (RENO Collaboration), Phys. Rev. Lett. 108, 191802 (2012).
6. F. P. An et al. (Daya Bay Collaboration), Phys. Rev. D95, 072006 (2017).
7. F. P. An et al. (Daya Bay Collaboration), Nucl. Instrum. Meth. A811, 133 (2016).
8. F. P. An et al. (Daya Bay Collaboration), Nucl. Instrum. Meth. A733, 8 (2015).
9. Y. Y. Ding et al., Journal of Rare Earths 25, 310 (2007); Y. Y. Ding et al., Nucl. Instr. and Meth. 584, 38 (2008).
10. M. Yeh et al., Nucl. Instr. and Meth. A578, 329 (2007).
11. F. P. An et al. (Daya Bay Collaboration), Phys. Rev. D93, 072011 (2016).

Searching for Very Light Sterile Neutrino with Matter Effect Correction

Baobiao Yue* and Jiajie Ling[†]

*School of Physics, Sun Yat-sen University,
Guangzhou, Guangdong Province, China*
*E-mail: yuebb@mail2.sysu.edu.cn
[†]E-mail: lingjj5@mail.sysu.edu.cn
www.sysu.edu.cn

Wei Li[‡] and Fanrong Xu[§]

*Department of Physics, Jinan University,
Guangzhou, Guangdong Province, China*
[‡] E-mail: weili@stu2014.jnu.edu.cn
[§] E-mail: fanrongxu@jnu.edu.cn
www.jnu.edu.cn

The standard 3-flavor neutrino oscillation has been well established. However, some experimental hints suggest the existence of new types of neutrinos at mass of eV scale. Those sterile neutrinos will change the neutrino oscillation pattern for short-baseline neutrino experiments with the 3(active)+N(sterile) neutrino mixing framework. Recent proposed medium baseline reactor neutrino experiments are sensitive to those very light sterile neutrinos at 10^{-4} to 0.01 eV^2 mass-squared difference through electron antineutrino disappearance channel. In this paper, we will discuss the matter effect correction on those very light sterile neutrino oscillation. As shown in our study, the matter effect correction for sterile neutrino oscillation is quite small ($< 4\%$), and at the same scale as the 3-flavor active neutrino oscillation case.

Keywords: Sterile neutrino; the matter effect.

1. Introduction

Active neutrino has weak interaction, but sterile neutrino does not have. So the Hamiltonians are different for 3(active) neutrino and 3(active)+N(sterile) neutrino mixing framework. The difference will lead a diverse probability in each channel. Now, we want to study the probability for 3(active)+1(sterile) neutrino mixing framework in matter to make a contrast with 3(active) neutrino framework in JUNO [1-4] case. In this case, the baseline is about 53km, and we only care about $\bar{\nu}_e$ survival probability.

Firstly, we will talk about the neutrino oscillation to learn the difference between 3-Flavor and 4-Flavor neutrino oscillation. Secondly, we introduce the matter effect which shows the changes in oscillation probability. Finally, it is necessary to give a contrast to illustrate that the matter effect will affect 3-Flavor and 4-Flavor neutrino oscillation slightly and similarly in our case, and we will give the reason why the NC potential has almost no contribution to the $\bar{\nu}_e$ survival probability.

2. Neutrino oscillation

There are three known flavors of neutrinos: ν_e, ν_μ and ν_τ, and we suppose that there is an unknown flavor neutrino: ν_s. In the following, we will show the neutrino mixing matrix for both 3-Flavor and 4-Flavor cases.

2.1. *3-Flavor case*

U whose parameters come from [5] is the Pontecorvo-Maki-Nakagawa-Sakata Matrix:

$$U = \begin{bmatrix} 1 & 0 & 0 \\ 0 & c_{23} & s_{23} \\ 0 & -s_{23} & c_{23} \end{bmatrix} \begin{bmatrix} c_{13} & 0 & s_{13}e^{-i\delta} \\ 0 & 1 & 0 \\ -s_{13}e^{i\delta} & 0 & c_{13} \end{bmatrix} \begin{bmatrix} c_{12} & s_{12} & 0 \\ -s_{12} & c_{12} & 0 \\ 0 & 0 & 1 \end{bmatrix} = \begin{bmatrix} U_{e1} & U_{e2} & U_{e3} \\ U_{\mu1} & U_{\mu2} & U_{\mu3} \\ U_{\tau1} & U_{\tau2} & U_{\tau3} \end{bmatrix}, \quad (1)$$

where $c_{23} = \cos\theta_{23}$, $s_{23} = \sin\theta_{23}$, $c_{13} = \cos\theta_{13}$, $s_{13} = \sin\theta_{13}$, $c_{12} = \cos\theta_{12}$, $s_{12} = \sin\theta_{12}$, θ_{23}, θ_{13} and θ_{12} are mixing angles, and δ is the CP violation phase.

2.2. *4-Flavor case*

Here U is the mixing matrix for 4-Flavor case [6,7]:

$$U = \begin{bmatrix} 1 & 0 & 0 & 0 \\ 0 & 1 & 0 & 0 \\ 0 & 0 & c_{34} & s_{34}e^{-i\delta_{34}} \\ 0 & 0 & -s_{34}e^{i\delta_{34}} & c_{34} \end{bmatrix} \begin{bmatrix} 1 & 0 & 0 & 0 \\ 0 & c_{24} & 0 & s_{24} \\ 0 & 0 & 1 & 0 \\ 0 & -s_{24} & 0 & c_{24} \end{bmatrix} \begin{bmatrix} c_{14} & 0 & 0 & s_{14}e^{-i\delta_{14}} \\ 0 & 1 & 0 & 0 \\ 0 & 0 & 1 & 0 \\ -s_{14}e^{i\delta_{14}} & 0 & 0 & c_{14} \end{bmatrix}$$

$$\begin{bmatrix} 1 & 0 & 0 & 0 \\ 0 & c_{23} & s_{23} & 0 \\ 0 & -s_{23} & c_{23} & 0 \\ 0 & 0 & 0 & 1 \end{bmatrix} \begin{bmatrix} c_{13} & 0 & s_{13}e^{-i\delta} & 0 \\ 0 & 1 & 0 & 0 \\ -s_{13}e^{i\delta} & 0 & c_{13} & 0 \\ 0 & 0 & 0 & 1 \end{bmatrix} \begin{bmatrix} c_{12} & s_{12} & 0 & 0 \\ -s_{12} & c_{12} & 0 & 0 \\ 0 & 0 & 1 & 0 \\ 0 & 0 & 0 & 1 \end{bmatrix} \quad , \quad (2)$$

$$= \begin{bmatrix} U_{e1} & U_{e2} & U_{e3} & U_{e4} \\ U_{\mu1} & U_{\mu2} & U_{\mu3} & U_{\mu4} \\ U_{\tau1} & U_{\tau2} & U_{\tau3} & U_{\tau4} \\ U_{s1} & U_{s2} & U_{s3} & U_{s4} \end{bmatrix}$$

where $c_{34} = \cos\theta_{34}$, $s_{34} = \sin\theta_{34}$, $c_{24} = \cos\theta_{24}$, $s_{24} = \sin\theta_{24}$, $c_{14} = \cos\theta_{14}$, $s_{14} = \sin\theta_{14}$, θ_{34}, θ_{24} and θ_{14} are new mixing angles, and δ_{34} and δ_{14} are new CP phases from sterile neutrino.

2.3. *Neutrino oscillation probability*

We know U_{ei} elements affect $P_{\bar{\nu}_e \to \bar{\nu}_e}$ in vacuum, and we will show these parameters in the following. For 3-Flavor case:

$$U_{e1} = c_{12}c_{13} \quad (3)$$

$$U_{e2} = c_{13}s_{12} \quad (4)$$

510

$$U_{e3} = s_{13}e^{-i\delta_{13}} \tag{5}$$

For 4-Flavor case:

$$U_{e1} = c_{12}c_{13}c_{14} \tag{6}$$

$$U_{e2} = c_{13}c_{14}s_{12} \tag{7}$$

$$U_{e3} = c_{14}s_{13}e^{-i\delta_{13}} \tag{8}$$

$$U_{e4} = s_{14}e^{-i\delta_{14}} \tag{9}$$

It shows that $P_{\bar{\nu}_e \to \bar{\nu}_e}$ is affected only by s_{14} and δ_{14} which come from sterile neutrino parameters. Now we give an example to show the probability difference between 3-Flavor and 4-Flavor cases below.

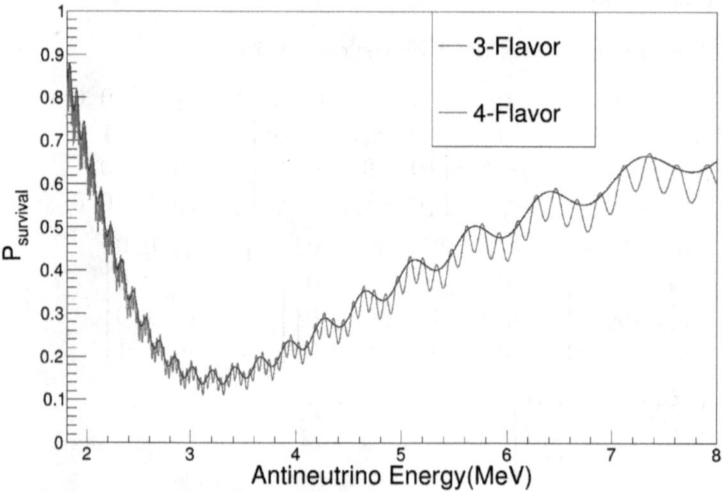

Fig. 1. For 3-Flavor case, it is normal hierarchy case. For 4-Flavor case, there is a high frequency oscillation with $\delta m^2_{41} = 0.01 eV^2$ and $\sin^2 2\theta_{14} = 0.1$.

3. Matter effect

When active neutrinos propagate through matter [8,9], the evolution equation of them is affected by effective potentials which are cased by coherent interactions with matter through coherent forward elastic weak CC (charged current) and NC (neutral current) scatterings. But sterile neutrino will not interact with matter, so there is no potential for it. In the following, 3-Flavor and 4-Flavor hamiltonians will be given separately.

3.1. *3-Flavor case*

The CC potential only affects the electron flavor neutrino. However, the NC potential affects all active flavor neutrino. If the matter density along the baseline is closely constant, the Hamiltonian in matter is given by

$$H = 2EH_{eff} = \begin{bmatrix} 0 & 0 & 0 \\ 0 & \delta m_{21}^2 & 0 \\ 0 & 0 & \delta m_{31}^2 \end{bmatrix} + U^\dagger \begin{bmatrix} A_{CC} & 0 & 0 \\ 0 & 0 & 0 \\ 0 & 0 & 0 \end{bmatrix} U, \tag{10}$$

where E is the neutrino energy, H_{eff} is the effective Hamiltonian.

For neutrino case:

$$A_{CC} = 2\sqrt{2}G_F N_e E = 7.63 \times 10^{-8}(eV^2)\left(\frac{\rho}{g/cm^3}\right)\left(\frac{E}{MeV}\right) \tag{11}$$

Here, G_F is the Fermi coupling constant, N_e is the electron number density, and ρ is the matter mass density along the baseline. A_{CC} has an opposite sign, and the mixing matrix U should be the conjugated matrix of Eq. (1) for antineutrino case.

3.2. *4-Flavor case*

Similarly, the CC potential only affects the electron flavor neutrino. However, the NC potential survives in this case. The Hamiltonian in matter is given by

$$H = 2EH_{eff} = \begin{bmatrix} 0 & 0 & 0 & 0 \\ 0 & \delta m_{21}^2 & 0 & 0 \\ 0 & 0 & \delta m_{31}^2 & 0 \\ 0 & 0 & 0 & \delta m_{41}^2 \end{bmatrix} + U^\dagger \begin{bmatrix} A_{CC} & 0 & 0 & 0 \\ 0 & 0 & 0 & 0 \\ 0 & 0 & 0 & 0 \\ 0 & 0 & 0 & A_{NC} \end{bmatrix} U, \tag{12}$$

where

$$A_{NC} = \sqrt{2}G_F N_n E = 3.815 \times 10^{-8}(eV^2)\left(\frac{\rho}{g/cm^3}\right)\left(\frac{E}{MeV}\right). \tag{13}$$

Here, N_n is the neutron number density. Similarly, A_{NC} has an opposite sign, and the mixing matrix U should be the conjugation of Eq. (2) for antineutrino case.

3.3. *Neutrino oscillation probability in matter*

We will give a contrast between 3-Flavor and 4-Flavor cases to show the similarities and differences. It is $\bar{\nu}_e$ survival probability for reactor antineutrino oscillation experiments. Here are two graphs which show the relative differences between P_{matter}(in matter) and P_{vacuum}(in vacuum) in Fig. 2.

3.4. *The influence of the matter effect in neutrino oscillation*

There is a table below to show the influence in Table 1.

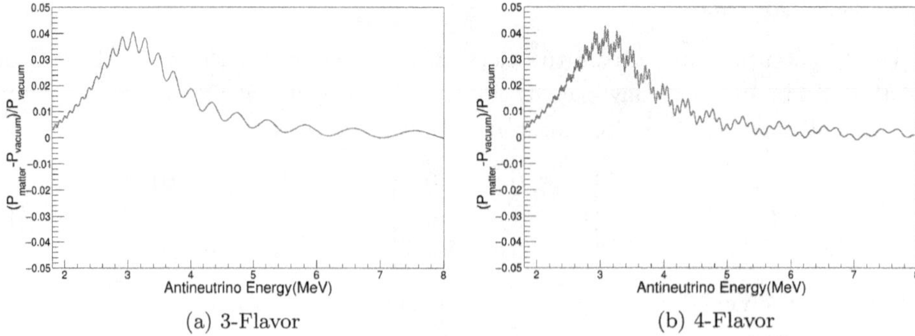

(a) 3-Flavor   (b) 4-Flavor

Fig. 2. (a) cited from [10] shows that the largest ratio is about 4% for 3-Flavor normal hierarchy case. However, (b) indicates the largest ratio is also about 4% in 4-Flavor case where $\delta m_{14}^2 = 0.01 eV^2$ and $\sin^2 2\theta_{14} = 0.1$ with normal hierarchy conditions. Inverted hierarchy case is the same.

Table 1. The influence of matter effect.

Normal hierarchy	$\delta m_{14}^2 (eV^2)$	$\sin^2 2\theta_{14}$	$\theta_{24}, \theta_{34}, \delta_{14}, \delta_{34}$	the largest $P_{relative}$
3-Flavor	-	-	-	$\sim 4\%$
4-Flavor	1	0.1	0	$\sim 4\%$
	0.1	0.1	0	$\sim 4\%$
	0.01	0.1	0	$\sim 4\%$
	0.001	0.1	0	$\sim 4\%$

The definition of $P_{relative}$ is given by

$$P_{relative} = \frac{P_{matter} - P_{vacuum}}{P_{vacuum}}. \tag{14}$$

It directly shows that the matter effect affects 3-Flavor and 4-Flavor cases similarly and slightly in Table 1, and inverted hierarchy case is the same.

3.5. *The main matter effect contribution in 4-Flavor case*

Here is a graph which shows the contribution to matter effect only from the NC potential in 4-Flavor case in Fig. 3.

Why is the NC potential contribution so small? Because, in Eq. (12), all factors of A_{CC} which mainly affect neutrino oscillation in matter are much larger than the corresponding factors of A_{NC} with small θ_{14}, θ_{24} and θ_{34} constrained in current experiments.

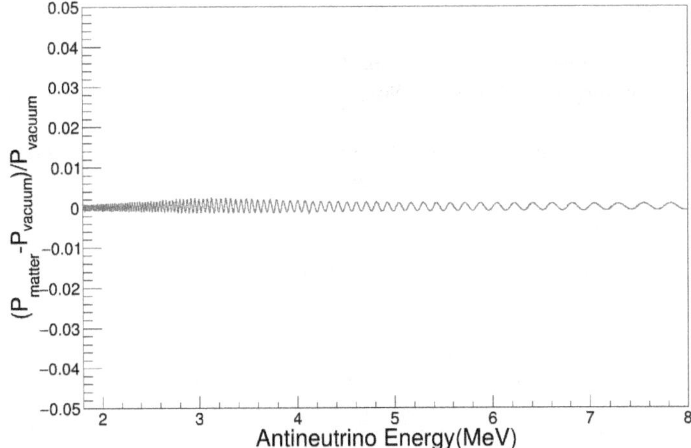

Fig. 3.   The contribution of Only NC potential matter effect is so small When $\delta m^2_{41} = 0.01 eV^2$ and $\sin^2 2\theta_{14} = 0.1$.

4. Conclusion

In summary, we have examined how slightly the matter effect affect neutrino oscillation in medium-baseline reactor experiments like JUNO ro RENO-50 for both 3-Flavor and 4-Flavor cases with possible sterile neutrino parameters constrained in current experiments. We also made a quick analysis about the reason of NC potential negligence, which shows why the matter effect is so similar for both 3-Flavor and 4-Flavor cases. We conclude that there is no need to take into account the NC potential in such experiments as a good approximation for 4-Flavor case. Due to negligible NC potential, the matter effect affects both 3-Flavor and 4-Flavor neutrino oscillation slightly and similarly with light sterile neutrino and small θ_{14} constrained by current experiments for JUNO case.

Acknowledgement

We thank the discussions with Yufeng Li, Zhenhui Huang and Steven Park. This work was in part supported by the National Natural Science Foundation of China (NSFC) under Grant No. 11775315.

References

1. F. An et al. (JUNO Collaboration), J. Phys. G, **43**, 030401 (2006).
2. Z. Djurcic et al. (JUNO Collaboration), arXiv:1508:07166.
3. Y. Wang and Z. Z. Xing, arXiv:1504.06155.
4. X. Qian and P. Vogel, Prog. Part. Nucl. Phys., **83**, 1 (2015).
5. C. Patrignani et al. (Particle Data Group), Chin. Phys. C, **40**, 100001 (2016).
6. C. Athanassopoulos et al. (LSND Collaboration), Phys. Rev. Lett. **75**, 2650 (1995); **77**, 3082 (1996); **81**, 1774.

7. A. A. Aguilar-Arevalo et al. (MiniBooNE Collaboration), Phys. Rev. Lett. **110**, 161801 (2013).
8. L. Wolfenstein, Phys. Rev. D, **17**, 2369 (1978).
9. S. P. Mikheev and A. Y. Smirnov, Sov. J. Nucl. Phys., **42**, 913 (1985) [Yad. Fiz., **42**, 1441 (1985)].
10. Y.-F. Li, Y. Wang and Z.-z. Xing, Terrestrial matter effects on reactor antineutrino oscillations at JUNO or RENO-50: how small is small?, arXiv:1605.00900.

Seasonal Muon Flux Modulation at the Daya Bay

Chengcai Zhang

on behalf of the Daya Bay Collaboration

*Institute of High Energy Physics,
University of Chinese Academy of Sciences, Beijing, China
E-mail: zhangcc85@ihep.ac.cn*

In the recent years, underground muon flux was observed to have a seasonal modulation, which was attributed to the atmosphere density seasonal modulations correlated by the temperature. The Daya Bay Experiment consists of eight identically designed detectors located in three underground experimental halls (named EH1, EH2, EH3), with respectively 250 m.w.e., 265 m.w.e. and 860 m.w.e. overburden. The underground muon rate is observed to be positively correlated with the effective atmospheric temperature, and to follow a seasonal modulation pattern. The correlation coefficient α, describing how a variation in the muon rate relates to a variation in the effective atmospheric temperature, is found to be $\alpha_{EH1} = 0.362 \pm 0.031$, $\alpha_{EH2} = 0.433 \pm 0.038$ and $\alpha_{EH3} = 0.641 \pm 0.057$ for each experimental hall.

1. Introduction

It is known that the muons result from the decay of charged mesons produced by interactions of primary cosmic rays with the upper atmosphere. Up till now many experiments have observed the underground muon intensity to be positively correlated with the atmospheric temperature [1–19]. Along with the temperature increases, the atmosphere becomes less dense, and the probability for a meson to interact with molecules in the atmosphere is reduced. The corresponding increase in meson decays yields a larger muon intensity over the summer months. The great majority of the experimental results are reported in terms of α, the correlation coefficient between the muon flux and the atmospheric temperature. This coefficient increases as a function of overburden, and hence Daya Bay Fig. 1, with three underground experimental halls at different depths, is an ideal setup to perform such a measurement.

2. Muon Data

2.1. *Muon Event Selection Criteria*

The muon event selection conditions is (i) the reconstructed energy in an AD is larger than 60 MeV, and (ii) more than 12 photomultipliers in the muon system (either IWS or OWS) produce a trigger within a 2 µs time window. We refer to (ii) as the muon tag. The only exception to the selection criteria is EH3 AD1. We know that there happened a tiny leak of liquid scintillator from the LS region to the buffer started in Summer 2012. The LS and MO levels stabilized in 2014 when an estimated 50 L of LS had leaked into the MO [20].

516

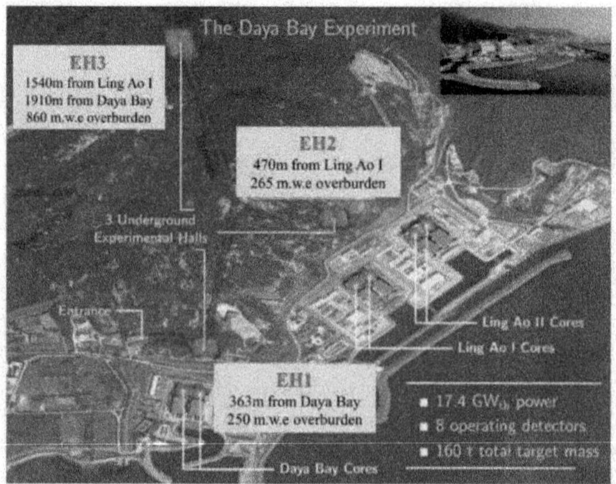

Figure 1.   Layout of the Daya Bay experiment. The Daya Bay and Ling Ao nuclear power plant (NPP) reactors (red circles) were situated on a narrow coastal shelf between the Daya Bay coastline and inland mountains. Two antineutrino detectors installed in each underground experimental hall near to the reactors (EH1 and EH2) measured the $\bar{\nu}_e$ flux emitted by the reactors, while four detectors in the far experimental hall (EH3) measured a deficit in the $\bar{\nu}_e$ flux due to oscillation (color online).

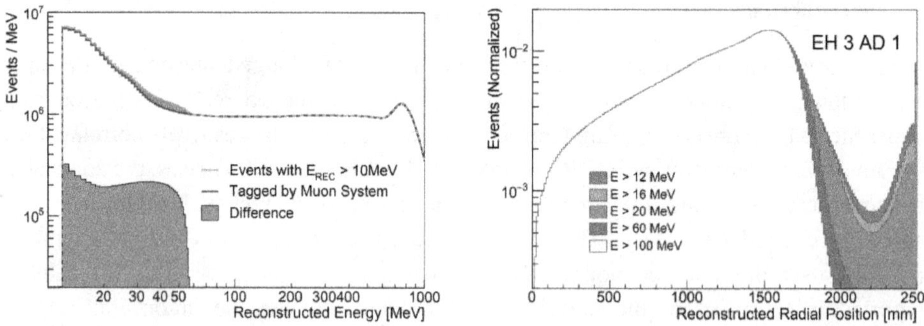

Figure 2.   (Left) plot is reconstructed energy spectrum of muon candidate events in EH1 AD1 with and without the muon tag. (Right) plot is the reconstructed radial distribution of muon candidate events in EH3 AD1. In the analysis, the MO region is treated as a single bin [2000 mm, 2500 mm]; here we provide finer binning for comparison only.

Hence the MO light yield increased, and the 60 MeV energy cut will be not enough to veto all events happened in MO region. So we raise the energy cut of EH3 AD1 to 100 MeV, and we will consider this AD separately from the others when computing systematic uncertainties.

2.2. *Muon Rate Variation Over Time*

We use a dataset collected from December 2011 to November 2013. The first 7 months of data only had 6 operating ADs [21], while the last 13 months changed to the full 8-AD configuration. The daily muon rate as a function of time in all the ADs is shown in Fig. 3.

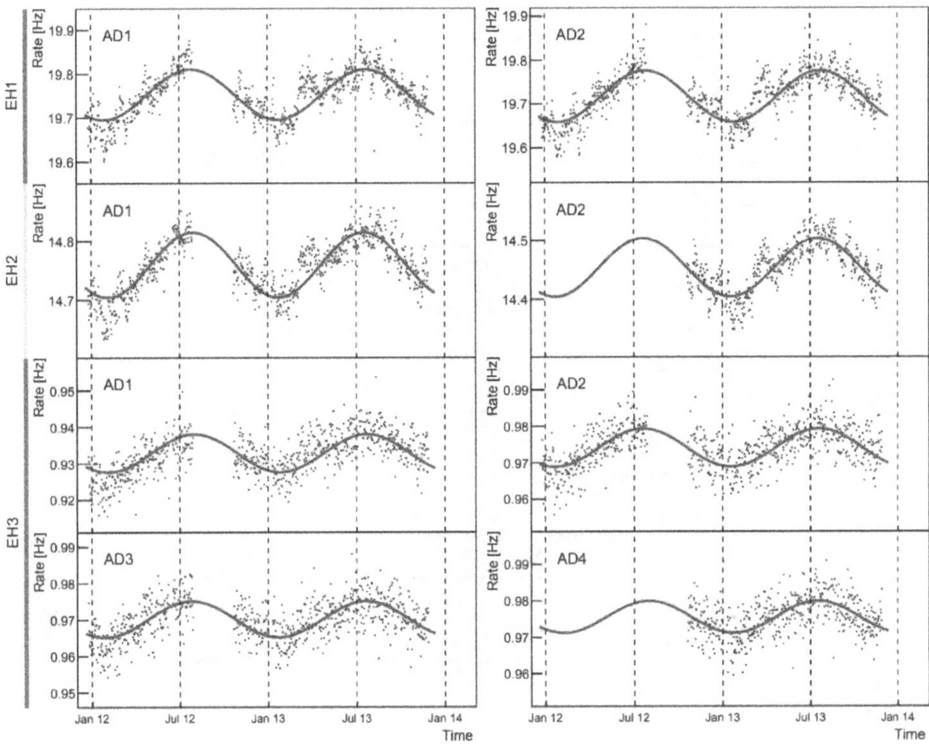

Figure 3.   Daily-binned muon rate as function of time in the eight ADs. The solid line shows the result of a sinusoidal fit to data. Fit parameters are shown in Table 1.

Table 1.   Parameters from fitting the muon modulation with a sinusoidal function.

Experimental Hall	Detector	Maximun	Period
EH1	AD1	21 Jul ± 2 days	361 ± 1 days
	AD2	21 Jul ± 2 days	361 ± 1 days
EH2	AD1	17 Jul ± 2 days	353 ± 1 days
	AD2	15 Jul ± 4 days	360 ± 3 days
EH3	AD1	22 Jul ± 4 days	356 ± 3 days
	AD2	18 Jul ± 5 days	363 ± 4 days
	AD3	20 Jul ± 5 days	360 ± 4 days
	AD4	15 Jul ± 10 days	348 ± 8 days

Fit each AD separately using a sinusoidal function, and the fit parameters are reported in Table 2. The oscillation period is one solar year, and the position of the oscillation maximum (i.e. the oscillation phase) occurs about the end of July. The oscillation amplitude depends on the average muon energy, and therefore on the overburden.

Table 2.   Central values and uncertainties of the parameters used in Eqs. (3.2)–(3.4).

Parameter	Value	Reference
A_π	1	[16]
A_K	0.38* $r_{K/\pi}$	[16]
$r_{K/\pi}$	0.149 ± 0.06	[25], [26]
B_π	1.460 ± 0.007	[16]
B_K	1.740 ± 0.028	[16]
Λ_N	120 g/cm^2	[25]
Λ_π	180 g/cm^2	[25]
Λ_K	160 g/cm^2	[25]
$<E_{thr}\cos\theta>_{EH1}$	37 ± 3 GeV	
$<E_{thr}\cos\theta>_{EH2}$	41 ± 3 GeV	
$<E_{thr}\cos\theta>_{EH3}$	143 ± 10 GeV	
γ	1.7 ± 0.1	[27]
ϵ_π	114 ± 3 GeV	[16]
ϵ_K	851 ± 14 GeV	[16]

3. Temperature Data

3.1. *Effective Atmospheric Temperature*

The atmospheric temperature data we used was European Centre for Medium-Range Weather Forecasts (ECMWF) [22]. The database contains different types of measurements (ground level, sounding balloon, satellite) at many locations over the world. Our analysis relies on the temperature values computed at the Daya Bay site (22.6°N, 114.5°E), which are provided four times a day (midnight, 6:00 am, noon, 6:00 pm) at 37 discrete pressure levels ranging from 1 hPa to 1000 hPa. We want to know the relationship between atmospheric temperature and muon production. However, nobody knows at which altitude a cosmic muon is produced, hence we follow [1, 11, 30] and regard the atmosphere as an isothermal body characterised by an effective temperature T_{eff}. T_{eff} is defined as the variable that would cause the observed muon flux changing, and it is computed as a weighted average of the temperature T over the atmospheric depth X, with weights W, as shown in Eq. (3.1).

$$T_{eff} = \frac{\int_0^\infty dX T(X) W(X)}{\int_0^\infty dX W(X)} \cong \frac{\sum_i \Delta X_i T(X_i) W(X_i)}{\sum_i \Delta X_i W(X_i)} \tag{3.1}$$

The left panel of Fig. 4 shows that pressure levels near the top of the atmosphere are weighted more heavily than the pressure levels at lower altitude. Both kaon and pion production and decay should be considered in the model [1]. However, because of limited sensitivity to the kaon contribution, previous experiments computed their weights using a pion-only model (e.g. Barrett [1], Sherman [3], Utah [17], Macro [11]). This approach changed with MINOS [16] which, for the first time, used a model put forward by Grashorn et al. [23] explicitly including kaons. Here we follow the latter approach, and we write $W(X_i) = W\pi(X_i) + W_K(X_i)$, where

$$W^{\pi,K}(X) \cong \frac{\left(1 - \frac{X}{\lambda_{\pi,K}}\right)^2 e^{\frac{-X}{\Lambda_{\pi,K}}} A_{\pi,K}}{\gamma + (1+\gamma)B_{\pi,K}K(X)\left(\frac{<E_{thr}\cos\theta>}{\epsilon_{\pi,K}}\right)^2} \tag{3.2}$$

$$K(X) \equiv \frac{X\left(1 - \frac{X}{\lambda_{\pi,K}}\right)^2}{\left(1 - e^{\frac{-X}{\lambda_{\pi,K}}}\right)\lambda_{\pi,K}} \tag{3.3}$$

$$\frac{1}{\lambda_{\pi,K}} = \frac{1}{\Lambda_N} - \frac{1}{\Lambda_{\pi,K}} \tag{3.4}$$

The parameters $A_{\pi,K}$ include the amount of inclusive meson production in the forward fragmentation region, the masses of mesons and muons, and the muon spectral index. The parameter $B_{\pi,K}$ reflect for the relative atmospheric attenuation of mesons. The parameter $\Lambda_{N,\pi,K}$ is the atmospheric attenuation length of the cosmic ray primaries, pions and kaons, respectively. The meson critical energy, π,K, is the meson energy for which decay and interaction have an equal probability. Finally, the muon spectral index is given by γ. The values of all the parameters are listed in Table 3 and are inherited from [16], with the exception of $<E_{thr}\cos\theta>$ calculated as described. The T_{eff} daily values computed at EH1 are shown in right panel of Fig. 4.

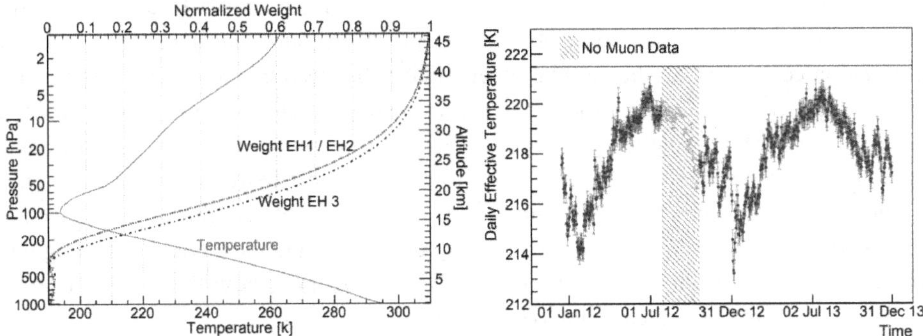

Figure 4. (Left) EH-dependent weights involved in the T_{eff} computation and time-averaged atmospheric temperature, both as a function of pressure and altitude. (Right) Daily T_{eff} values computed using EH1 weights. "No Muon Data" refers to the 2012 summer shutdown.

520

Table 3. Predicted and measured values of the correlation coefficient at the different experimental halls.

Exp. Hall	Prediction		This Work
	Including K and π [23]	Including π only [11]	
EH1	0.340 ± 0.019	0.362 ± 0.018	0.362 ± 0.031
EH2	0.362 ± 0.019	0.386 ± 0.018	0.433 ± 0.038
EH3	0.630 ± 0.019	0.687 ± 0.018	0.641 ± 0.057

3.2. *Temperature Uncertainty*

To estimate the uncertainty of the daily effective temperature $\sigma(T_{eff})$, we compute T_{eff} using a different temperature dataset. The new dataset is the Integrated Global Radiosonde Archive (IGRA) [24] provided by the US National Climatic Data Center. The temperature data is from sounding balloons launched from many meteorological stations around the world, and the closest station to Daya Bay is located in the city of Shantou, Guangdong (China). T_{eff} differences with inputs from two databases ECMWF and IGRA shows a spread of 0.4 K. We build a daily distribution of the difference between the smeared and the nominal T_{eff} values, and we find that over the whole data-taking period the spread induced by smearing the weights is 0.15 K, 0.15 K, 0.07 K, respectively for EH1, EH3 and EH3. Combining these uncertainties with the 0.4 K T_{eff} uncertainty common to all the experimental halls, we get a total $\sigma(T_{eff})$ of 0.43 K, 0.43 K and 0.41 K respectively.

4. Correlation Analysis

To study how a variation in the atmospheric temperature relates to a variation in the underground muon rate, we build a scatter plot where the x (y) axis represents the daily effective temperature (daily muon rate) relative variation with respect to its mean value. The y error bar on each data point represents the Poissonian uncertainty of muon counts, and the x error bar represents the temperature uncertainty mentioned before. A linear fitting is done to each scatter plot. We define the slope of the fitted linear function to be the correlation coefficient α. The correlation coefficient between the muon rate variation and the atmospheric temperature variation is known increasing along with the overburden, as a result of a harder muon energy spectrum. So we combine the results obtained from all ADs in same experimental hall and to perform a new linear regression on the combined scatter plots, as shown in Fig. 5. The aim is to provide our results as a function of $<E_{thr}\cos\theta>$ Fig. 6. We consider both the model accounting for π and K and the model accounting for π only, and we report our findings in Table 3. Figure 6 shows how the Daya Bay result compares to other experiments and to the model prediction.

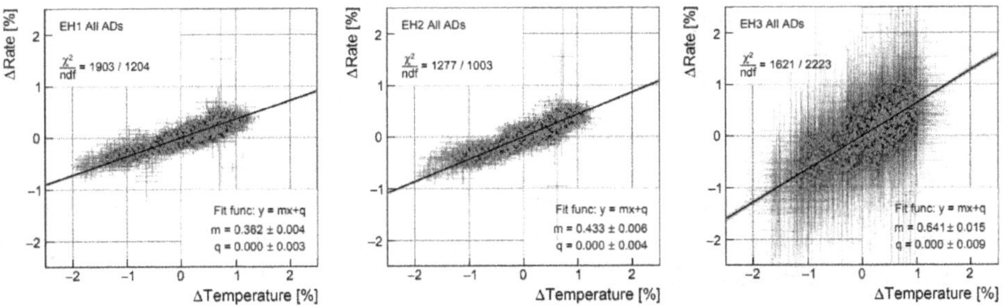

Figure 5. Relative muon rate variation vs relative effective temperature variation constructed by merging data from ADs belonging to the same experimental hall, together with the result of a linear regression accounting for uncertainties on both variables.

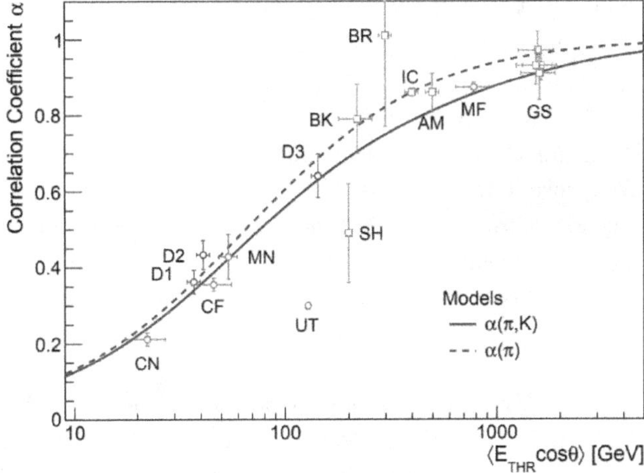

Figure 6. Comparison of the experimental α values with the model accounting for pions and kaons (solid line), and with the model accounting for pions only (dashed line). Values determined in this analysis are reported as D1, D2 and D3 respectively for the three experimental halls. Other experiments include Amanda (AM) [10], Baksan (BK) [9], Barrett (BR) [2], Icecube (IC) [14], MINOS Near (MN) [15] and far (MF) [16] detector, Double Chooz Near (CN) and Far (CF) detectors [19], Sherman (SH) [3], and Utah (UT) [17]. The four Gran Sasso (GS) based measurements are Macro [11], Borexino [12], and the two Gerda values [18]. Their $\langle E_{thr}\cos\theta \rangle$ are artifically displaced on the horizontal axis for the sake of visualization. The values and uncertainties of $\langle E_{thr}\cos\theta \rangle$ is shown in the figure for experiments tagged with a square marker have been estimated as described in the text.

5. Conclusion

Muon rate seasonal modulation variations over two years were measured in the Daya Bay experiment. Muon rates were correlated with the effective atmospheric temperature above the three experimental halls. The effective temperature is computed with a weighted average over all the pressure levels using the raw temperature dataset provided by the European Centre for Medium-Range Weather Forecasts. The correlation coefficient α at the three experimental halls was found to be:

$$\alpha_{EH1} = 0.362 \pm 0.031 \text{ at } < E_{thr}\cos\theta >_{EH1} 37 \pm 3 \text{ GeV}$$

$$\alpha_{EH2} = 0.433 \pm 0.038 \text{ at } < E_{thr}\cos\theta >_{EH2} 41 \pm 3 \text{ GeV}$$

$$\alpha_{EH3} = 0.641 \pm 0.057 \text{ at } < E_{thr}\cos\theta >_{EH3} 143 \pm 10 \text{ GeV}$$

The importance of this result lies in the fact that Daya Bay is able to probe the muon seasonal variation at different overburdens using identically designed detectors. Furthermore, the result of the three EHs offer an important validation of the model.

References

[1] P.H. Barrett et al., "Interpretation of Cosmic-Ray Measurements Far Underground", Rev. Mod. Phys. 24, 133 (1952).

[2] P.H. Barrett et al., "Atmospheric Temperature Effect for Mesons Far Underground", Rev. Mod. Phys. 95, 1573 (1954).

[3] N. Sherman, "Atmospheric Temperature Effect for μ Mesons Observed at a Depth of 846 m.w.e.", Phys. Rev. 93, 208 (1954).

[4] G. Cini Castagnoli et al., "Trains of diurnal waves in the muonic component at 70 mwe during cosmic-ray decreases", J. Geophys. Res. 74, 2414 (1969).

[5] A. Fenton et al. "Cosmic ray observations at 42 mwe underground at Hobart, Tasmania", Nuovo Cimento 22, 285 (1961).

[6] S. Yasue et al., "Observation of cosmic ray intensity variation with Matsushiro underground telescope", 17th International Cosmic Ray Conference, Vol. 4, p. 308–311 (1981).

[7] J.E. Humble et al., "Variations in Atmospheric Coefficients for Underground Cosmic-Ray Detectors", 16th International Cosmic Ray Conference, Vol. 4, p. 258 (1970),

[8] P.R.A. Lyons et al., "Further calculations of atmospheric coefficients for underground cosmic ray detectors", 17th International Cosmic Ray Conference, Vol. 4, p. 300–303 (1981).

[9] Y. Andreyev et al. (Baksan Collaboration), "Season Behaviour of the Amplitude of Daily Muon Intensity with Energy ≥ 220 GeV", in Proceedings of the 22th ICRC, 693 (1991).

[10] A. Bouchta (AMANDA Collaboration), "Seasonal Variation of the muon flux seen by Amanda", in Proceedings of the 26th ICRC [2, 108 (1999)].

[11] M. Ambrosio et al. [MACRO Collaboration], "Seasonal variations in the underground muon intensity as seen by MACRO", Astropart. Phys. 7, 109 (1997).

[12] G. Bellini et al. [Borexino Collaboration], "Cosmic-muon flux and annual modulation in Borexino at 3800 m water-equivalent depth", JCAP 1205, 015 (2012) [arXiv:1202.6403].

[13] M. Selvi (LVD Collaboration), in Proceedings of the 31st ICRC, Lodz, Poland, 2009.

[14] P. Desiati et al. [IceCube Collaboration], "Seasonal Variations of High Energy Cosmic Ray Muons Observed by the IceCube Observatory as a Probe of Kaon/Pion Ratio", Proceedings of the 32nd International Cosmic Ray Conference (ICRC 2011).

[15] P. Adamson et al., "Observation of muon intensity variations by season with the MINOS Near Detector", Phys. Rev. D 90, no. 1, 012010 (2014) [arXiv:1406.7019].

[16] P. Adamson et al. [MINOS Collaboration], "Observation of muon intensity variations by season with the MINOS far detector", Phys. Rev. D 81, 012001 (2010) [arXiv:0909.4012].

[17] D.J. Cutler and D.E. Groom, in Proceedings of the 17th ICRC, 290, Paris (1981), "Meteorological effects in cosmic ray muon production".

[18] M. Agostini et al. [GERDA Collaboration], "Flux Modulations seen by the Muon Veto of the GERDA Experiment", Astropart. Phys. 84 (2016) 29 [arXiv:1601.06007 [physics.ins-det]].

[19] T. Abrahao et al. [Double Chooz Collaboration], "Cosmic-muon characterization and annual modulation measurement with Double Chooz detectors", JCAP 1702 (2017) no. 02, 017 [arXiv:1611.07845 [hep-ex]].

[20] F.P. An et al. [Daya Bay Collaboration], "The Detector System of The Daya Bay Reactor Neutrino Experiment", Nucl. Instrum. Meth. A 811 (2016) 133 doi:10.1016/ j.nima.2015.11.144 [arXiv:1508.03943 [physics.ins-det]].

[21] F.P. An et al. [Daya Bay Collaboration], "Improved Measurement of Electron Antineutrino Disappearance at Daya Bay", Chin. Phys. C 37 (2013) 011001 doi: 10.1088/1674-1137/37/1/011001 [arXiv:1210.6327 [hep-ex]].

[22] The ERA-Interim database of the European Centre for Medium-Range Weather Forecasts.

[23] E.W. Grashorn et al. "The Atmospheric charged kaon/pion ratio using seasonal variation methods", Astropart. Phys. 33, 140 (2010) [arXiv:0909.5382].

[24] I. Durre et al., "Overview of the Integrated Global Radiosonde Archive".

[25] T.K. Gaisser, "Cosmic rays and particle physics", Cambridge, UK: Univ. Pr. (1990) 279 p.

[26] G.D. Barr et al., "Uncertainties in Atmospheric Neutrino Fluxes", Phys. Rev. D 74, 094009 (2006) [arXiv:astro-ph/0611266].

[27] P. Adamson et al. [MINOS Collaboration], "Measurement of the atmospheric muon charge ratio at TeV energies with MINOS", Phys. Rev. D 76, 052003 (2007) [arXiv:0705.3815].

Muon Reconstruction Performance of the ATLAS Detector at the LHC at $\sqrt{s} = 13$ TeV

Liqing Zhang

on behalf of the ATLAS Collaboration

University of Science and Technology of China,
No. 96, JinZhai Road Baohe District, Hefei, Anhui, 230026, P. R. China
E-mail: liqing.zhang@cern.ch

This article describes the performance of the ATLAS muon identification and reconstruction using the ATLAS dataset recorded at $\sqrt{s}$=13 TeV. Using a large sample of $J/\psi \to \mu\mu$ and $Z \to \mu\mu$ decays from pp collision data, measurements of the reconstruction efficiency, isolation efficiency, as well as of the momentum scale and resolution, are presented and compared to Monte Carlo simulations.

Keywords: Muon; reconstruction; identification; ATLAS.

1. Introduction

Excellent muon reconstruction is critical for many of the most important physics results published by the ATLAS experiment at the LHC. These results include the discovery of the Higgs boson in Ref. [1] and the measurement of its properties in Refs. [2, 3], the precise measurement of Standard Model processes in Ref. [4], and searches for physics beyond the Standard Model in Ref. [5]. This article describes performance of the ATLAS muon reconstruction with 2016 ATLAS data collected at $\sqrt{s}$=13 TeV. Measurements of muon reconstruction efficiency and calibrations of muon momentum scale and resolution are presented. The comparison between data and Monte Carlo (MC) simulation, and corrections to the simulation used in physics analyses are also discussed. The results are based on the analysis of a large sample of $J/\psi \to \mu\mu$ and $Z \to \mu\mu$ decays reconstructed in pp collisions recorded in ATLAS.

2. ATLAS detector

A detailed description of the ATLAS detector can be found in Ref. [6]. Information from the inner detector (ID) and the muon spectrometer (MS), supplemented by information from the calorimeters, is used to identify and precisely reconstruct muons produced in pp collisions.

The ID consists of three subdetectors: the silicon pixels (Pixel) and the semi-conductor tracker (SCT) with a pseudorapidity coverage up to $|\eta| = 2.5$, and the transition radiation tracker (TRT) with a pseudorapidity coverage up to $|\eta| = 2.0$. The ID reconstructs muon tracks close to the interaction point, providing accurate measurements of the track parameters inside an axial magnetic field of 2 T.

The MS is the outermost ATLAS subdetector. It is designed to detect muons in the pseudorapidity region up to $|\eta| = 2.7$, and to provide momentum measurements. The MS consists of one barrel ($|\eta| < 1.05$) and two endcap sections ($1.05 < |\eta| < 2.7$). A system of three large superconducting air-core toroidal magnets, each with eight coils, provides a magnetic field with a bending integral of about 2.5 Tm in the barrel and up to 6 Tm in the endcaps. Resistive plate chambers (RPC) with three doublet layers for $|\eta| < 1.05$ and thin gap chambers (TGC) with one triplet layer followed by two doublets for $1.0 < |\eta| < 2.4$ provide triggering capability to the detector, as well as (η, ϕ) position measurements with typical spatial resolution of 5–10 mm. A precise momentum measurement for muons with pseudorapidity up to $|\eta| = 2.7$ is provided by three layers of monitored drift tube chambers (MDT), with each chamber providing six to eight η measurements along the muon trajectory. For $|\eta| > 2$, the inner layer is instrumented with a quadruplet of cathode strip chambers (CSC) instead of MDTs. The single-hit resolution in the bending plane for the MDT and the CSC is about 80 μm and 60 μm, respectively.

3. Muon reconstruction

Muon reconstruction and identification are based on the information provided by the ID, MS, and calorimeters, described in Refs. [7, 8]. Four muon identification selections (*Medium*, *Loose*, *Tight*, and *High-p_{T}*) are provided to address specific needs of different physics analyses.

- **Loose** selection is designed to maximise the reconstruction efficiency while providing good-quality muon tracks.
- **Medium** selection provides the default selection for muons in ATLAS. This selection minimises the systematic uncertainties associated with muon reconstruction and calibration.
- **Tight** selection is selected to maximise the purity of muons.
- **High-p_{T}** selection aims to maximise the momentum resolution for tracks with transverse momentum above 100 GeV.

4. Reconstruction efficiency

Reconstruction efficiency is measured using the tag-and-probe method described in Ref. [8]. The tag muon has to be identified as a *Medium* muon that fires the trigger and the probe muon has to be reconstructed by a system independent of the one being studied. The method is based on the selection of pure muon sample from $J/\psi \to \mu\mu$ and $Z \to \mu\mu$ events. $J/\psi \to \mu\mu$ decays provide probes having 4 GeV $< p_{\mathrm{T}} < 15$ GeV. $Z \to \mu\mu$ decays provide a sample of probes with $p_{\mathrm{T}} > 10$ GeV.

Figure 1 shows the muon reconstruction efficiency as a function of p_{T} in different η regions, measured using probe muons from $J/\psi \to \mu\mu$ events. The efficiency is measured in data and MC, and the corresponding scale factors for the *Medium* selection are also shown. Figure 2 shows the muon reconstruction efficiency for the

526

Loose, *Medium* and *Tight* identification algorithms measured in $Z \to \mu\mu$ events as a function of the muon pseudorapidity for muons with $p_T > 10$ GeV. Figure 3 shows the reconstruction efficiencies for the *Medium* muon selection as a function of transverse momentum, including results from $Z \to \mu\mu$ and $J/\psi \to \mu\mu$, for muons with $0.1 < |\eta| < 2.5$. The efficiency is stable and slightly above 99% for $p_T > 5$ GeV. Values measured from $J/\psi \to \mu\mu$ and $Z \to \mu\mu$ events are in agreement in the overlap region between 10 and 20 GeV.

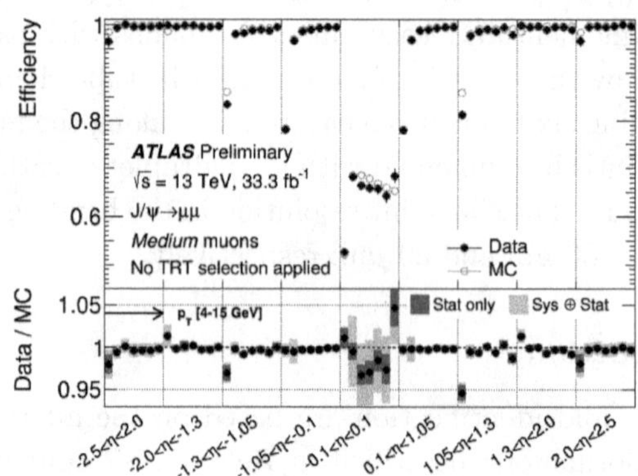

Fig. 1.   Muon reconstruction efficiency in different η regions measured in $J/\psi \to \mu\mu$ events for *Medium* muon selection. Within each η region, the efficiency is measured in seven p_T bins (4–5, 5–6, 6–7, 7–8, 8–10, 10–12, and 12–15 GeV). The resulting values are plotted as distinct measurements in each η bin with p_T increasing from 4 to 15 GeV going from left to right. The error bars on the efficiencies indicate the statistical uncertainty. The panel at the bottom shows the ratio of the measured to predicted efficiencies, with statistical and systematic uncertainties.

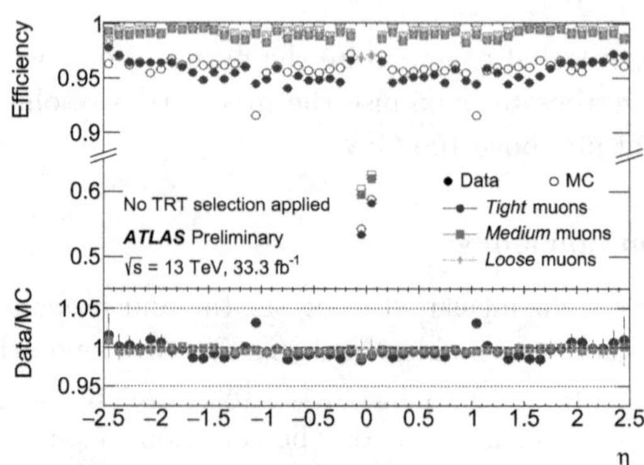

Fig. 2.   Muon reconstruction efficiencies for the *Loose*, *Medium*, *Tight* identification algorithms measured in $Z \to \mu\mu$ events as a function of the muon pseudorapidity for muons with $p_T > 10$ GeV. The prediction by the detector simulation is depicted as open circles, while filled dots indicate the observation in collision data with statistical errors. The bottom panel shows the ratio between expected and observed efficiencies, the efficiency scale factor. The errors in the bottom panel show the quadratic sum of statistical and systematic uncertainty.

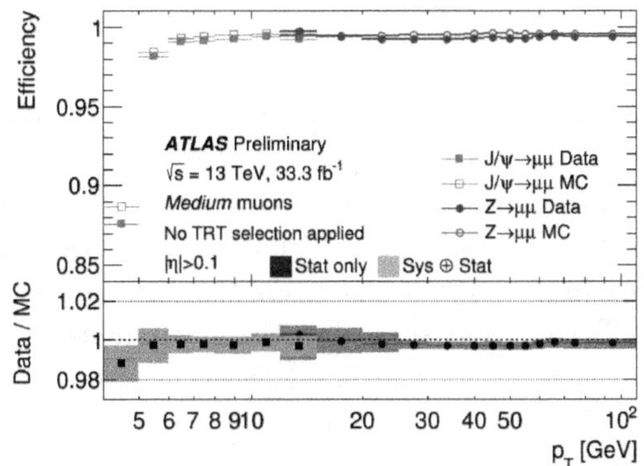

Fig. 3. Muon reconstruction efficiencies for the *Medium* identification algorithm measured in $J/\psi \rightarrow \mu\mu$ and $Z \rightarrow \mu\mu$ events as a function of the muon momentum. The prediction by the detector simulation is depicted as empty circles (squares), while the full circles (squares) indicate the observation in collision data for $Z \rightarrow \mu\mu$ ($J/\psi \rightarrow \mu\mu$) events. Only statistical errors are shown in the top panel. The bottom panel reports the efficiency scale factors. The darker error bands indicate the statistical uncertainty, while the lighter bands indicate the quadratic sum of statistical and systematic uncertainties.

5. Isolation

The measurement of the detector activity around a muon candidate is referred to as muon isolation. Two variables are defined to assess muon isolation: a track-based isolation variable and a calorimeter-based isolation variable. The isolation selection criteria are determined using the relative isolation variables, which are defined as the ratio of the track- or calorimeter-based isolation variables to the transverse momentum of the muon. The distribution of the relative isolation variables in muons from $Z \rightarrow \mu\mu$ events is shown in the top panels of Figure 4. Muons included in the plot satisfy the *Medium* identification criteria and are well separated from the other

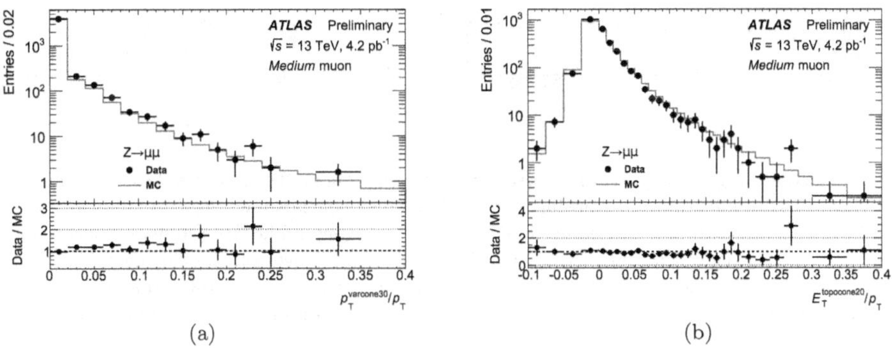

Fig. 4. Distributions of the track-based (a) and the calorimeter-based (b) relative isolation variables measured in $Z \rightarrow \mu\mu$ events. Muons are selected by the *Medium* identification algorithm. The dots show the distribution for data while the histograms show the distribution from simulation. The bottom panels show the ratio of data to simulation with the corresponding statistical uncertainty.

528

muon from the Z boson ($\Delta R_{\mu\mu} > 0.3$). The bottom panel shows the ratio of data to simulation.

Seven isolation working points *(Loose, Tight, Gradient, GradientLoose, Loose-TrackOnly, FixedCutTightTrackOnly, FixedCutLoose)* are defined using different selections on the track and calorimeter isolation variables, each optimised for different physics analyses. The efficiencies for the seven isolation working points are measured in data and simulation in $Z \to \mu\mu$ decays using the tag-and-probe method. Figure 5 shows the isolation efficiency measured for *Medium* muons in data and simulation as a function of the muon p_T for the *LooseTrackOnly, Loose, GradientLoose*, and *FixedCutLoose* working points, with the respective data/MC ratios included in the bottom panel.

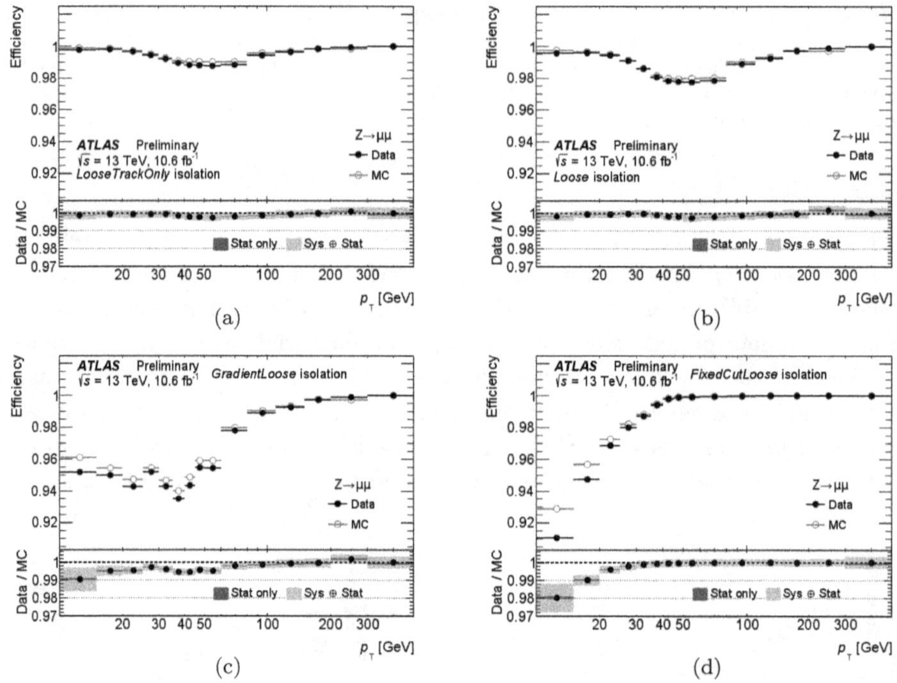

Fig. 5. Isolation efficiency for *the LooseTrackOnly* (a), *Loose* (b), *GradientLoose* (c), and *Fixed-CutLoose* (d) muon isolation working points. The efficiency is shown as a function of the muon transverse momentum p_T and is measured in $Z \to \mu\mu$ events. The full (empty) markers indicate the efficiency measured in data (MC) samples. The errors shown on the efficiency are statistical only. The bottom panel shows the ratio of the efficiency measured in data and simulation, as well as the statistical uncertainties and combination of statistical and systematic uncertainties.

6. Momentum scale and resolution

The accurate modelling of the ATLAS detector geometry and material distribution included in the simulation provides a good description of the p_T of the muons.

To account for second-order mis-modelling effects, a correction to the simulated muon momentum is derived separately for ID and MS tracks. The muon momentum scale and resolution are studied using $J/\psi \to \mu\mu$ and $Z \to \mu\mu$ decays. The corrected transverse momentum, $p_{\mathrm{T}}^{\mathrm{Cor,Det}}$ (Det=ID, MS), is described by the following equation:

$$p_{\mathrm{T}}^{\mathrm{Cor,Det}} = \frac{s_0 + (1 + s_1) \cdot p_{\mathrm{T}}^{\mathrm{MC,Det}}}{1 + \mathcal{N}(0,1) \cdot \sqrt{(\Delta r_0 / p_{\mathrm{T}}^{\mathrm{MC,Det}})^2 + \Delta r_1^2 + (\Delta r_2 \cdot p_{\mathrm{T}}^{\mathrm{MC,Det}})^2}} \tag{1}$$

where $p_{\mathrm{T}}^{\mathrm{MC,Det}}$ is the uncorrected transverse momentum in simulation, and the term $\mathcal{N}(0,1)$ is normally distributed random variables with zero mean and unit width.

The numerator of Eq. (1) describes the momentum scales. The s_1 term corrects for inaccuracy in the description of the magnetic field integral and the dimension of the detector in the direction perpendicular to the magnetic field. The s_0 term models the effect on the MS momentum from the inaccuracy in the simulation of the energy loss in the calorimeter and other materials between the interaction point and the MS. As the energy loss between the interaction point and the ID is negligible, s_0 for ID is set to zero.

The denominator of Eq. (1) describes the momentum smearing that broadens the relative p_{T} resolution in simulation to properly describe the data. The $\Delta r_0 / p_{\mathrm{T}}^{\mathrm{MC,Det}}$ term accounts mainly for fluctuations of the energy loss in the traversed material, the Δr_1 accounts mainly for multiple scattering, local magnetic field inhomogeneities and local radial displacements of the hits, and the $\Delta r_2 \cdot p_{\mathrm{T}}^{\mathrm{MC,Det}}$ term mainly describes intrinsic resolution effects caused by the spatial resolution of the hit measurements and by residual misalignment of the muon spectrometer.

The five correction parameters, s and Δr, are extracted from data using a binned maximum-likelihood fit with templates derived from simulation which compares the invariant mass distributions for $J/\psi \to \mu\mu$ and $Z \to \mu\mu$ candidates in data and simulation.

The invariant mass distributions for the $J/\psi \to \mu\mu$ and $Z \to \mu\mu$ candidates are shown in Figure 6. In the uncorrected simulation, it is noticeable that the signal distributions are narrower and slightly shifted with respect to data. After correction, the lineshapes of the two resonances in simulation agree with the data within the systematic uncertainties. Figure 7 shows the position of the peak of the invariant mass distribution $m_{\mu\mu}$ as a function of the pseudorapidity of the highest-p_{T} muon. The distributions are shown for data as well as corrected simulation. The corrected simulation is in very good agreement with the data. Figure 8 displays the dimuon mass resolution $\sigma(m_{\mu\mu})$ as a function of the leading-muon η for the two resonances. The dimuon mass resolution is about 1.2% and 1.6% at small η value, and increases to 1.6% and 1.9% in the endcaps for J/ψ and Z bosons respectively.

530

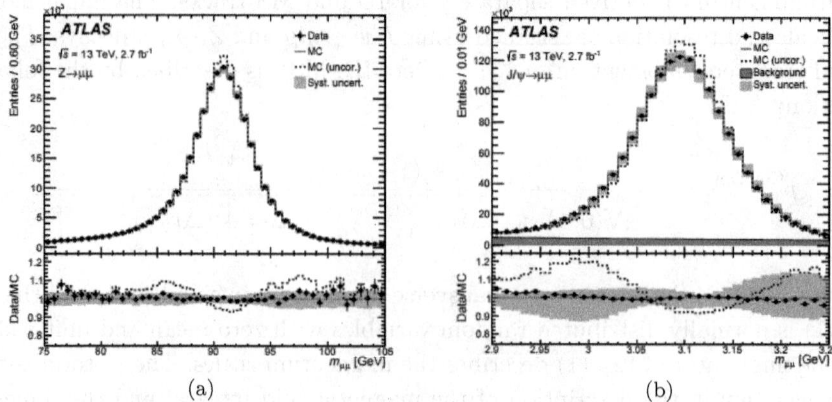

Fig. 6. Dimuon invariant mass distribution of $Z \to \mu\mu$ (a) and $J/\psi \to \mu\mu$ (b) candidate events reconstructed with combined muons. The upper panels show the invariant mass distribution for data and for the signal simulation plus the background estimate. The points show the data. The continuous line corresponds to the simulation with the MC momentum corrections applied while the dashed lines show the simulation when no correction is applied. Background estimates are added to the signal simulation. The band represents the effect of the systematic uncertainties on the MC momentum corrections. The lower panels show the data to MC ratios. In the Z sample, the MC background samples are added to the signal sample according to their expected cross sections. In the J/ψ sample, the background is estimated from a fit to the data as described in the text. The sum of background and signal MC distributions is normalised to the data.

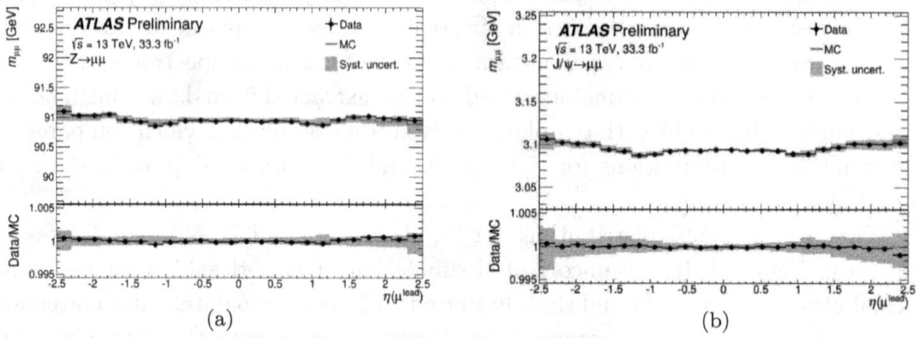

Fig. 7. Fitted mean mass of the dimuon system for combined muons for $Z \to \mu\mu$ (a) and $J/\psi \to \mu\mu$ (b) events for data and corrected simulation as a function of the pseudorapidity of the highest-p_T muon. The upper panels show the fitted mean mass value for data and corrected simulation. The small variations of the invariant mass estimator as a function of pseudorapidity are due to imperfect energy loss corrections and magnetic field description in the muon reconstruction. Both effects are well reproduced in the simulation. The lower panels show the data/MC ratio. The error bars represent the statistical uncertainty; the shaded bands represent the systematic uncertainty in the correction and the systematic uncertainty in the extraction method added in quadrature.

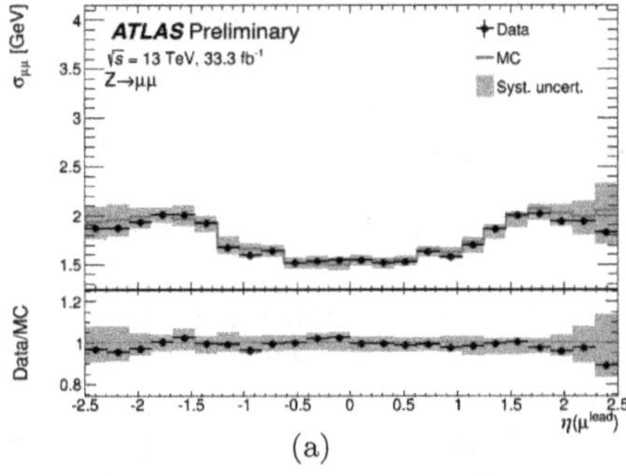 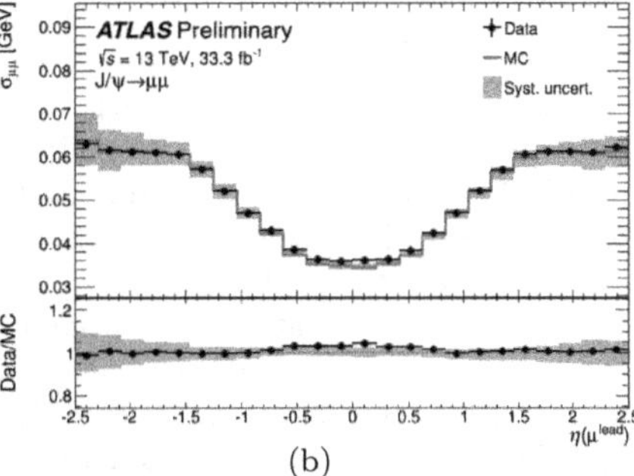

(a)   (b)

Fig. 8.　Dimuon invariant mass resolution for combined muons for $Z \to \mu\mu$ (a) and $J/\psi \to \mu\mu$ (b) events for data and corrected simulation as a function of the pseudorapidity of the highest-p_T muon. The upper panels show the fitted resolution value for data and corrected simulation. The lower panels show the data/MC ratio. The error bars represent the statistical uncertainty; the shaded bands represent the systematic uncertainty in the correction and the systematic uncertainty in the extraction method added in quadrature.

7. Conclusions

The performance of the ATLAS muon reconstruction has been measured using data from pp collisions at $\sqrt{s}$=13 TeV recorded at the LHC. A large calibration sample consisting of $Z \to \mu\mu$ decays and $J/\psi \to \mu\mu$ decays allows for a precise measurement of the reconstruction and isolation efficiency as well as of the momentum resolution and scale over a wide p_T range. The muon reconstruction efficiency measured using $J/\psi \to \mu\mu$ and $Z \to \mu\mu$ decays is close to 99% over most of the pseudorapidity range of $|\eta| < 2.5$ for $p_\mathrm{T} > 5$ GeV. The $Z \to \mu\mu$ sample is also used to measure the isolation efficiency for seven isolation working points. The isolation efficiency varies between 93% and 100% depending on the selection and on the momentum of the particle, and is well reproduced in the simulation. The muon momentum scale and resolution have been studied in detail using $J/\psi \to \mu\mu$ and $Z \to \mu\mu$ decays. These studies are used to correct the simulation to improve the agreement with data and to minimise the systematic uncertainties in physics analyses. After applying momentum corrections, the p_T resolution in data and simulation agree to better than 5% for most of the η range.

References

[1] ATLAS Collaboration, Observation of a new particle in the search for the Standard Model Higgs boson with the ATLAS detector at the LHC, *Phys. Lett. B* **716**, (2012) 1-29.

[2] ATLAS Collaboration, Measurement of the Higgs boson mass from the $H \to \gamma\gamma$ and $H \to ZZ^* \to 4l$ channel with the ATLAS detector using 25 fb^{-1} of pp collision data, *Phys. Rev. D* **90**, (2014) 052004.

[3] ATLAS Collaboration, Study of the spin and parity of the Higgs boson in diboson decays with the ATLAS detector, *Eur. Phys. J. C* **75**, (2015) 476.

[4] ATLAS Collaboration, Measurement of the ZZ Production Cross Section in pp Collisions at $\sqrt{s}$=13 TeV with the ATLAS Detector, *Phys. Rev. Lett.* **116**, (2016) 101801.

[5] ATLAS Collaboration, Search for new light gauge bosons in Higgs boson decays to four-lepton final states in pp collisions at $\sqrt{s}$=8 TeV with the ATLAS detector at the LHC, *Phys. Rev. D* **92**, (2015) 092001.

[6] ATLAS Collaboration, The ATLAS Experiment at the CERN Large Hadron Collider, *JINST* **3**, (2008) S08003.

[7] ATLAS Collaboration, Performance of the ATLAS Silicon Pattern Recognition Algorithm in Data and Simulation at $\sqrt{s}$=7 TeV, *ATLAS-CONF-2010-072*, (2010), http://cds.cern.ch/record/1281363.

[8] ATLAS Collaboration, Muon reconstruction performance of the ATLAS detector in proton-proton collision data at $\sqrt{s}$=13 TeV, *Eur. Phys. J. C* **76**, (2016) 292.

Double Calorimetry System in the JUNO Neutrino Experiment

Xuantong Zhang

on behalf of the JUNO Collaboration

Institution of High Energy Physics,
Beijing, China
E-mail: zhangxuantong@ihep.ac.cn

The Jiangmen Underground Neutrino Observatory (JUNO) is a multiple-purpose neutrino experiment with a 20 kiloton liquid scintillator (LS) detector at a depth of 700 meters. With about 18,000 20" photomultiplier tubes (PMTs) installed on the surface of a 35.4 meter diameter central detector, providing more than 75% coverage, JUNO aims for an unprecedented 3% energy resolution at 1 MeV. In order to achieve this energy resolution which needs a 1% precision in the absolute energy scale, the control of the energy response systematics is crucial. The single channel charge measurement is challenging in such a large detector, as the number of photoelectrons in each PMT can differ by two orders of magnitude depending on the event's position and energy. Second system of PMTs consisting of 25,000 3" PMTs placed in the space between the large PMTs and operating primarily in photon-counting mode will allow to calibrate the energy response of the large PMTs and to improve the physics of JUNO in several ways. The motivation for the double-calorimetry, as well as its benefits and implementation in JUNO will be discussed in this proceeding.

Keywords: JUNO; calorimetry; neutrino.

1. Introduction

JUNO[1] is a neutrino experiment under civil construction in Jiangmen, Guangdong province in southern China. The experiment is at the same distance (~53 km) from two powerful nuclear plants, Yangjiang and Taishan. The thickness of the overburden is 700 m to shield the cosmic ray. With 3% energy resolution (at 1 MeV), JUNO aims to determine the neutrino mass hierarchy using reactor antineutrinos by the interference between two oscillation frequency components driven by Δm^2_{31} and Δm^2_{32}, respectively. Moreover, due to the precise measurement of antineutrino spectra, JUNO is allowed to be the first experiment to measure solar and atmospheric mass splitting simultaneously[2].

The schematic design is shown in Fig. 1. There is a 35.4 m inner diameter acrylic sphere containing 20 kiloton of liquid scintillator. A 40.1 m-diameter stainless-steel truss supports the acrylic sphere. About 18,000 20" photomultiplier tubes (PMTs) and about 25,000 3" PMTs are installed on the stainless-steel shell watching inwards for the light generated by interactions of neutrinos and other events. A large pool filled with 35 kiloton pure water which immerses all the detectors serves as a Cherenkov detector. 2,000 20" PMTs are installed in the pool to receive Cherenkov photons tagging cosmic muons. On the top of the pool, plastic scintillators tiles are equipped to veto muon.

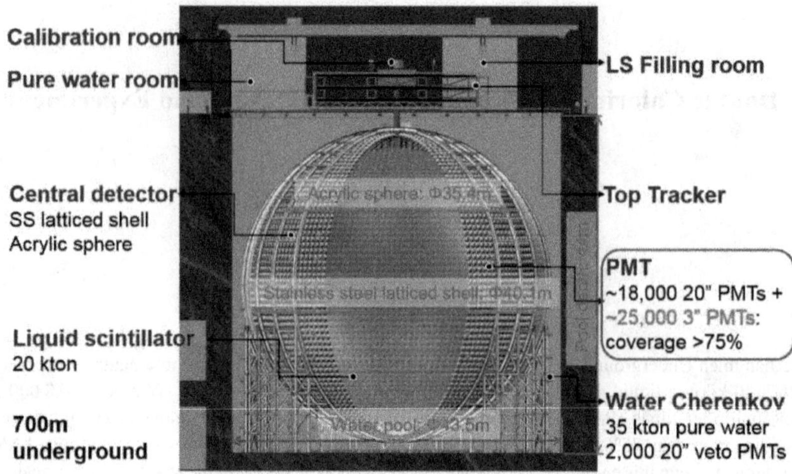

Fig. 1.   The JUNO schematic design.

2. Double Calorimetry

The main goal of JUNO is to determine the neutrino mass hierarchy by measuring the reactor antineutrino spectrum precisely to separate Δm_{31}^2 and Δm_{32}^2. Thus, JUNO aims for a 3% energy resolution which needs huge statistics of photons and strictly controlled systematic uncertainty.

2.1. *Detector Requirement*

To achieve the 3% energy resolution, the main requirements for JUNO's central detector (CD) are as follow:

- High light yield and high transparency liquid scintillator,
- High optical coverage and high quantum efficiency (QE) 20" PMTs,
- >1200 photoelectrons (p.e.) @1 MeV with <2.9% statistical fluctuation,
- Rooms for systematics uncertainty <1%.

2.2. *Calorimetry Systems in JUNO*

There are some challenges for the 20" PMT system. 20" PMT is so large that the detected number of photoelectrons varies by two orders of magnitude in the reactor antineutrino energy range in JUNO's huge central detector. Therefore, the nonlinear response of 20" PMTs is hard to be calibrated. Furthermore, because the charge range in a single PMT changes with the event vertex variety in different detector volume, the non-uniformity attributed to the nonlinearity of the single channel will deteriorate the energy resolution.

The 3" PMT system was firstly proposed to meet those challenges in 2014. As shown in Fig. 2, 25,000 3" PMTs are planned to be installed at the gaps between the 20" PMTs and increase of about 1/70 the photocathode area, compared to the 20" PMTs alone.

Because of the small photocathode area, 98% 3" PMTs only detect single p.e. in the reactor antineutrino energy range according a Monte Carlo simulation. As a result, the 3" PMT system works as a photon-counting system which almost has no charge nonlinearity. It greatly helps the calibration of 20" PMT system energy response. Combination of the 20" and 3" PMTs system composes the double calorimetry in JUNO. In addition, the 3" PMT system brings other advantages for JUNO:

- Some independent physics such as measurement of solar parameters.
- Help reconstruction for high energy physics such as muons and atmospheric neutrinos.
- Help measurement of supernova neutrino.

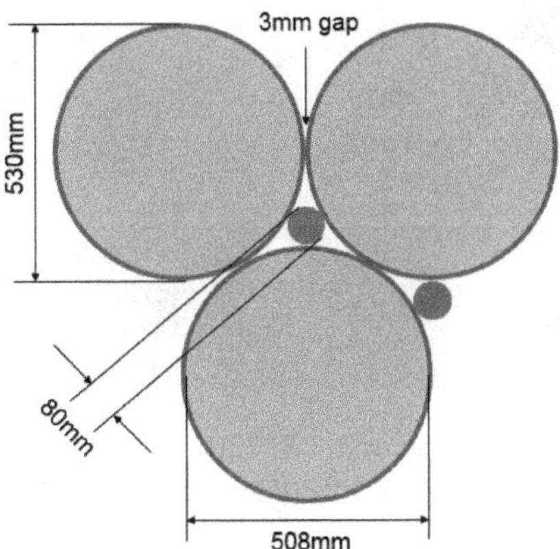

Fig. 2.   3" PMTs (red discs) are designed to be installed at the gaps between 20" PMTs (blue discs) and to increase the cover percentage of photocathode in central detector (color online).

3. 3" PMT System

As shown in Fig. 3, the 3" PMT system includes PMTs and readout electronics. A box containing the readout electronics system will be connected to 128 high voltage (HV) dividers with their corresponding 3" PMTs. Both the readout electronics system box and 3" PMTs are immersed in water. An about 100m long cable connects the underwater box and DAQ to transmit/receive power and data.

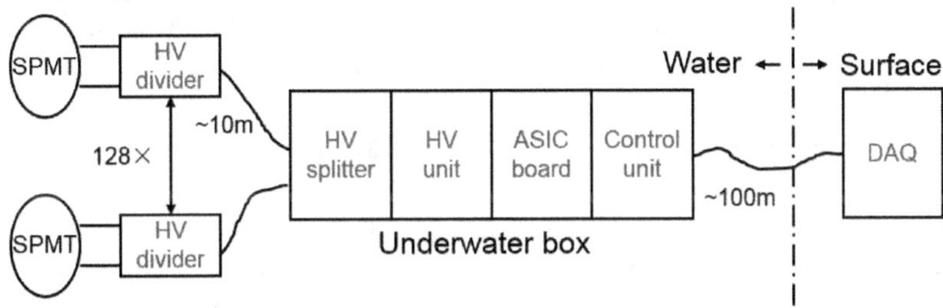

Fig. 3.   Schematic design of the 3" PMT system.

536

3.1. *PMT Tubes*

The XP72B22 3" PMTs from Hainan Zhanchuang (HZC) Photonics, a Chinese industry who introduced the technologies and producion line from PHOTONICS, won the international bidding of the 3" PMTs in May 2017. HZC plans to supply 25,000 XP72B22 3" PMTs in the next 2 years. In addition, HZC will also produce high voltage dividers and seal each PMT with water proofing in cooperation with the JUNO collaboration.

Compared to the previous XP72B20 in Fig. 4, XP72B22 is an improved model whose shape of the photocathode has been optimized to achieve better time performances.

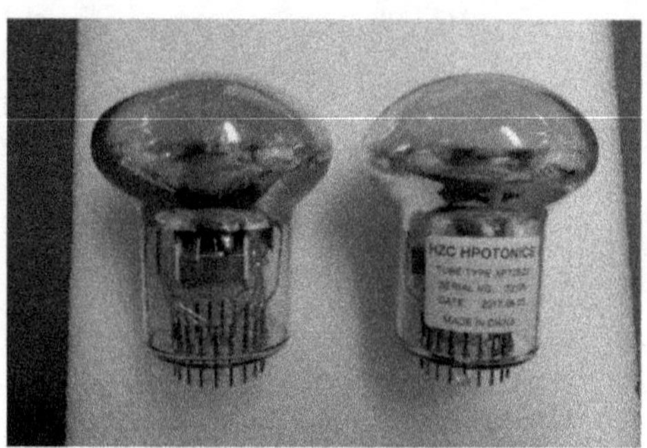

Fig. 4. The shape of the photocathode of XP72B22 has been optimized to get better time performances compared to the previous XP72B20. The XP72B20 is on the left and while the XP72B22 is on the right.

3.2. *PMT Performances*

HZC response performances of the XP72B22 3" PMT are listed in Table 1. All the response performances conform to the JUNO requirement. Considering that 3" PMTs are mostly working in the photon-counting mode, the requirement of the single p.e. resolution uniformity is particularly strict. 5 HZC 3" PMT samples' resolutions have been measured and turned to be 35%±2% showing an excellent uniformity.

Table 1. HZC response performances of XP72B22 and all of them meet JUNO's requirement. The parameters in brackets are the ranges of each parameter.

Parameters	HZC Response
Quantum Efficiency (QE) × Collection Efficiency (CE)	24% (>22%)
Transition Time Spread (TTS) (FWHM)	<5 ns
Peak/Valley ratio of single p.e. spectrum	3 (>2)
Single p.e. spectrum resolution	35% (<45%)
Dark rate @1/4 p.e.	1 kHz (<1.8 kHz)
Pre/after pulse ratio	<5% / <15%
Nonlinearity	<10% @1-100 p.e.
Radioactivity	^{238}U<400 ppb, ^{232}Th<400 ppb, ^{40}K<200 ppb

3.3. *PMT Electronic System*

The high voltage unit and the control unit of the 3" PMT system are determined to be the same as 20" PMT system. CATIROC[3], a multichannel front-end ASIC with almost no dead time, has been chosen to be the charge and time readout for 3" PMTs. 8 CATIROC ASICs are installed in one battery card and read out 128 PMTs. The preliminary design of the underwater box containing both high voltage splitter and readout ASICs includes a stainless-steel tube with two caps. A multichannel connector and cable between PMTs and the box is still in study and test.

4. Conclusion

JUNO aims to determine the neutrino mass hierarchy by a high precision measurement of reactor neutrino spectrum with 3% energy resolution. For the purpose of increasing statistics and controlling systematic, a 3" PMT system was proposed to work with the 20" PMT system as a double calorimetry to improve the energy resolution. 25,000 3" PMTs will be produced in HZC and installed in JUNO's central detector. PMT performances study and electronic design is ongoing.

Acknowledgments

This work is supported by the National Natural Science Foundation of China No. 11575226, and the Strategic Priority Research Program of the Chinese Academy of Sciences, Grant No. XDA10011200.

References

1. Adam, T. et al. [JUNO Collaboration], *JUNO conceptual design report.* arXiv: 1508.07166 (2015).
2. An, F. et al. [JUNO Collaboration], *Neutrino Physics with JUNO*, J. Phys. G 43, no. 3, 030401 (2016).
3. Blin, S. et al., Performance of CATIROC: ASIC for smart readout of large photo-multiplier arrays, Journal of Instrumentation 12.03 (2017): C03041.

Study of Invisible Neutrino Decays at Future Accelerator Neutrino Experiments Using Muon-Decay Beams

Yibing Zhang

School of Physics, Sun Yat-sen University, Guangzhou, 510275, China
E-mail: zhangyb27@mail2.sysu.edu.cn

In this paper we investigate the consequences of the invisible decays of the third neutrino mass eigenstate in future accelerator neutrino experiments using muon-decay beams. We find that the two experiments can improve the constraints on the lifetime of the third mass eigenstate τ_3/m_3. In addition, we discuss the correlation between θ_{23} and the neutrino lifetime in appearance and disappearance channels, respectively. A strong correlation between θ_{23} and τ_3/m_3 shows up in appearance channels. When all channels are considered, we find negligible impacts of neutrino decays on the precision measurements of θ_{23} and CP-violating phase δ in these two experiments.

Keywords: Neutrino oscillation; neutrino decay; precision measurement.

1. Introduction

The oscillation pattern of three-flavour neutrino mixing has been established through solar, atmospheric, accelerator and reactor neutrino experiments[1–4]. In the standard three-flavour paradigm, neutrino oscillations are dominated by two mass-squared splittings (i.e., Δm^2_{31}, Δm^2_{21}) and three mixing angles (i.e., θ_{12}, θ_{13}, θ_{23}). Up to now, most of the oscillation parameters have been measured well[5]. However, the CP-violating phase δ which characterizes the difference between matter and anti-matter as well as the mass hierarchy (Here, normal mass hierarchy: $\Delta m^2_{31} > 0$; inverted mass hierarchy: $\Delta m^2_{31} < 0$) has not been addressed. It is also not clear whether $\theta_{23} = 45°$ or not. The non-maximal value of θ_{23} is termed an octant degeneracy problem (i.e., lower octant: $\theta_{23} < 45°$ or higher octant: $\theta_{23} > 45°$). All these unknown parameters will be measured in the medium baseline reactor experiments: JUNO and RENO[6,7], and in the long-baseline accelerator neutrino experiments: T2K, NOνA, T2HK and DUNE[8–10]. Recent results from T2K and NOνA incline to a normal mass hierarchy and indicate a hint of $\delta_{CP} \approx 270°$ [11,12] only at a low confidence level. Therefore, we still require numerous data to attain a compelling conclusion. We probably need more data provided by the next-generation experiments such as muon-decay accelerator neutrino oscillation experiments.

Neutrino decays may be incorporated in neutrino oscillations[13–16]. As it depends on whether final states from the neutrino decays are detectable or not, the neutrino decays can be divided into visible and invisible decays[17,18]. Here we consider the third neutrino mass eigenstate decays into invisible products like active ν_1, ν_2 or even sterile neutrinos. In this situation, when the i-th mass eigenstate decays with lifetime τ_i, the energy of i-th mass eigenstate neutrino can be written as[19,20]:

$$E_i = \frac{m_i^2}{2E} - i\frac{\Gamma}{2},\tag{1}$$

with $\frac{1}{\Gamma} = \frac{E}{m_i}\tau_i$. We take the muon-decay accelerator neutrino experiments MuOn-decay MEdium baseline NeuTrino beam experiment (MOMENT)[21] and Low Energy Neutrino Factory Experiment (LENF)[22] as examples to interpret the possible effects generated by the neutrino decays.

This paper is organized as follows: we describe our implementations and simulation methods in Section 2. In Section 3, we present simulation results about the correlation between θ_{23} and the lifetime of ν_3, and the constraints on the ν_3 lifetime as well as the performance of precision measurements in MOMENT and LENF. Finally, it follows with the summary in Section 4.

2. Simulation details

The simulation details for MOMENT and LENF are shown in Table 1 with the neutrino sources, detector descriptions and running times[23,24]. MOMENT, as a medium muon decay accelerator neutrino experiment, is proposed as a future experiment to measure the leptonic CP-violating phase. The neutrino fluxes are offered by the MOMENT working group. Here we utilize eight oscillation channels: $\nu_e \to \nu_e$, $\nu_e \to \nu_\mu$, $\nu_\mu \to \nu_e$, $\nu_\mu \to \nu_\mu$ and their conjugate partners. We have to consider flavour and charge identifications to distinguish secondary particles by means of an advanced neutrino detector. The charged-current interactions are used to identify neutrino signals: $\nu_e + n \to p + e^-$, $\bar{\nu}_\mu + p \to n + \mu^+$, $\bar{\nu}_e + p \to n + e^+$, and $\nu_\mu + n \to p + \mu^-$. We consider the new technology using Gd-doping water to separate both Cherenkov and coincident signals from capture of thermal neu-

Table 1. Assumptions for the source, detector and the running time at MOMENT and LENF in the simulation.

Experiments	MOMENT	LENF
Fiducial mass	Gd-doping Water cherenkov(500 kton)	TASD (20 kton)
Channels	$\nu_e(\bar{\nu}_e) \to \nu_e(\bar{\nu}_e)$, $\nu_\mu(\bar{\nu}_\mu) \to \nu_\mu(\bar{\nu}_\mu)$, $\nu_e(\bar{\nu}_e) \to \nu_\mu(\bar{\nu}_\mu)$, $\nu_\mu(\bar{\nu}_\mu) \to \nu_e(\bar{\nu}_e)$	$\nu_\mu(\bar{\nu}_\mu)$ appearance, $\nu_\mu(\bar{\nu}_\mu)$ disappearance
Energy resolution	$12\%/E$	$15\%/\sqrt{E}$
Runtime	μ^- mode 5 yrs $+$ μ^+ mode 5 yrs	ν 4 yrs $+$ $\bar{\nu}$ 4 yrs
Baseline	150 km	1200 km
Energy range	100 MeV to 800 MeV	0.5 GeV to 5 GeV
Normalization (error on signal)	appearance channels: 2.5% disappearance channels: 5%	2% (all channels)
Normalization (error on background)	Neutral current, Atmospheric neutrinos Charge misidentification	NC events Charge misidentification

540

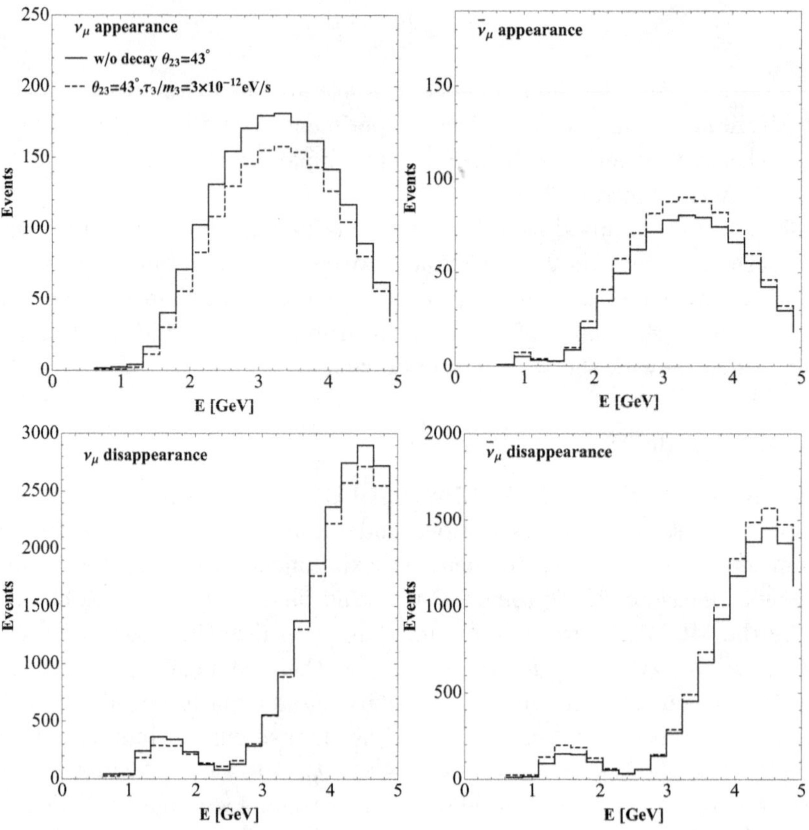

Fig. 1.   The event rates of appearance (upper panels) and disappearance channels (lower panels) in LENF.

trons. The major backgrounds are mostly from the atmospheric neutrinos, neutral current backgrounds and charge mis-identifications. They can be largely suppressed by the beam direction and proper modelling background spectra within the beam-off period, which is to be extensively studied in detector simulations. The LENF is a proposed long baseline muon-decay accelerator neutrino experiment. We assume the muon beam of 5 GeV with 10.6×10^{20} useful muon decays per year for each polarity. We take a baseline of $L = 1200$ km for the LENF and consider a magnetized Totally Active Scintillator Detector (TASD) with the fiducial mass of 20 kton and an energy resolution of $15\%/\sqrt{E}$. Our simulation is carried out by utilizing the GLOBES package[25,26]. The following central values and their uncertainties of the standard neutrino oscillation parameters are taken from the latest nu-fit results[5]: $\theta_{12} = 33.56°$ (2.3%), $\theta_{13} = 8.46°$ (1.8%), $\theta_{23} = 41.6°$ (5.8%), $\Delta m_{21}^2 = 7.5 \times 10^{-5} \text{eV}^2$ (2.4%), $\Delta m_{31}^2 = 2.524 \times 10^{-3} \text{eV}^2$ (1.6%). Unless otherwise mentioned, we assume the normal mass hierarchy in our simulation, i.e. $\Delta m_{31}^2 > 0$.

Based on the introduction to the characteristics and simulation details of MO-MENT and LENF, we further show the event rates in Fig. 1. Taking the LENF as an example, we present the impacts of neutrino decays on the event rates in ν_μ ($\bar{\nu}_\mu$) appearance channel and ν_μ ($\bar{\nu}_\mu$) disappearance channel. In both panels the solid black line shows the standard case without neutrino decays, the dashed black line shows case with $\tau_3/m_3 = 5 \times 10^{-12}$ eV/s. It is clear that when we turn on neutrino decays with $\tau_3/m_3 \neq 0$, a distinct difference between the standard case and the decay scenario can be easily read. Invisible decays can generate less neutrino events than the standard case, while for the antineutrinos there are more expected neutrinos than the standard case.

3. Results

In this section, we investigate the correlation between θ_{23} and δ_{cp} in appearance and disappearance channels, respectively. Then we will display the precise measurements of standard parameters with unstable neutrinos and constraints of the decay parameter τ_3/m_3 in MOMENT and LENF experiments.

3.1. *Correlation between θ_{23} and τ_3/m_3*

There are appearance and disappearance channels both in MOMENT and LENF. Here we present the numerical oscillation probabilities in LENF as examples as shown in Fig. 2[27]. In both panels, the variation of θ_{23} can give rise to the cyan bands, and the variation of τ_3/m_3 can give rise to the gray bands. In the appearance channel, the gray band and the cyan band overlap for some values of θ_{23} and τ_3/m_3 while they are distinct for other values, demonstrating the variation of τ_3/m_3 and

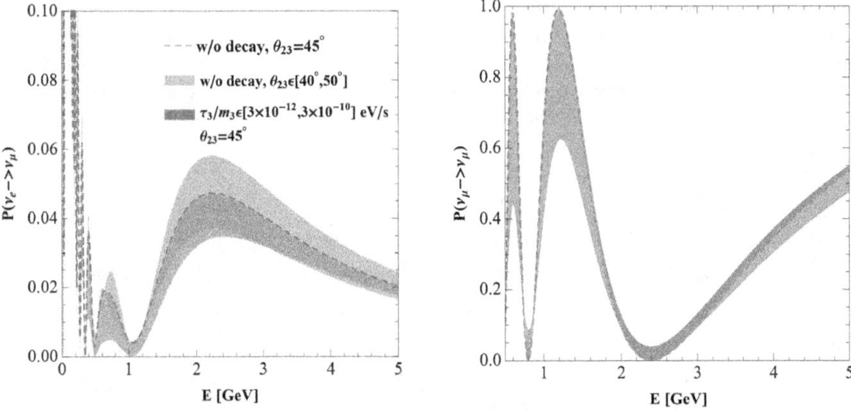

Fig. 2. The oscillation probabilities of ν_μ appearance channel (left panel) and ν_μ disappearance channel (right panel) in the LENF. The red line represents the 3 flavour case without neutrino decays. The gray regions shows the probability bands produced by variation of τ_3/m_3, and the cyan regions shows the probability bands produced by variation of θ_{23} (color online).

θ_{23} could interfere with the measurement of each other or even generate degeneracies between these two parameters. In the right panel, one can see the impacts of θ_{23} and τ_3/m_3 on the ν_μ disappearance probability. The value of τ_3/m_3 would influence the amplitude of the probability peak in low energy range and depress the probability in higher energy region, as shown in Fig. 1. However, the values of the θ_{23} would impact the depth of disappearance dip and adjust the curve in a high energy range. There are almost no overlapping areas between the gray region and the cyan region. Therefore, the correlation between the θ_{23} and τ_3/m_3 is not very strong for the ν_μ appearance channel.

In the next sections, we will carry out a discussion on the correlation between θ_{23} and τ_3/m_3 from the sensitivity level.

3.2. *Bound on the lifetime of* ν_3

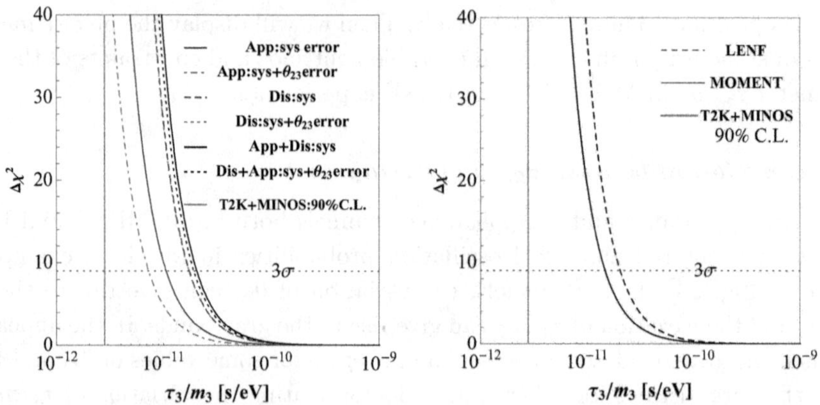

Fig. 3.   Both panels show the $\Delta\chi^2 = \chi^2 - \chi^2_{min}$ as a function of the test value of τ_3/m_3 where the input (true) value of τ_3/m_3 is assumed to be infinite. The left panel shows the constraints to τ_3/m_3 in LENF. Here "App" denotes appearance channels, "Dis" denotes disappearance channels and "sys" denotes systematic error are taken into account. The "θ_{23} error" refers to we imposed an 10% input error on central value of θ_{23}. The right panel shows the comparison between MOMENT and LENF with systematic errors and correlation errors of all standard parameters. Solid red line corresponds to the current bound obtained by T2K+MINOS [28] (color online).

We illustrate the impact of accuracy of θ_{23} on the constraint of τ_3/m_3 via the left panel of Fig. 3. It could be seen that the constraint to τ_3/m_3 would be much weaker when the uncertainty of θ_{23} is considered. However, the uncertainty of θ_{23} would not have a distinct influence on τ_3/m_3 for the disappearance channels. Consequently, there would be a very small difference between the case with systematic errors and the case with systematic errors and the error of θ_{23} after combining appearance and disappearance channels as the constraint to τ_3/m_3 is mainly from the disappearance channels.

In the right panel of Fig. 3, we show the results involving systematic errors and correlations of all standard parameters. The MOMENT is shown by the solid

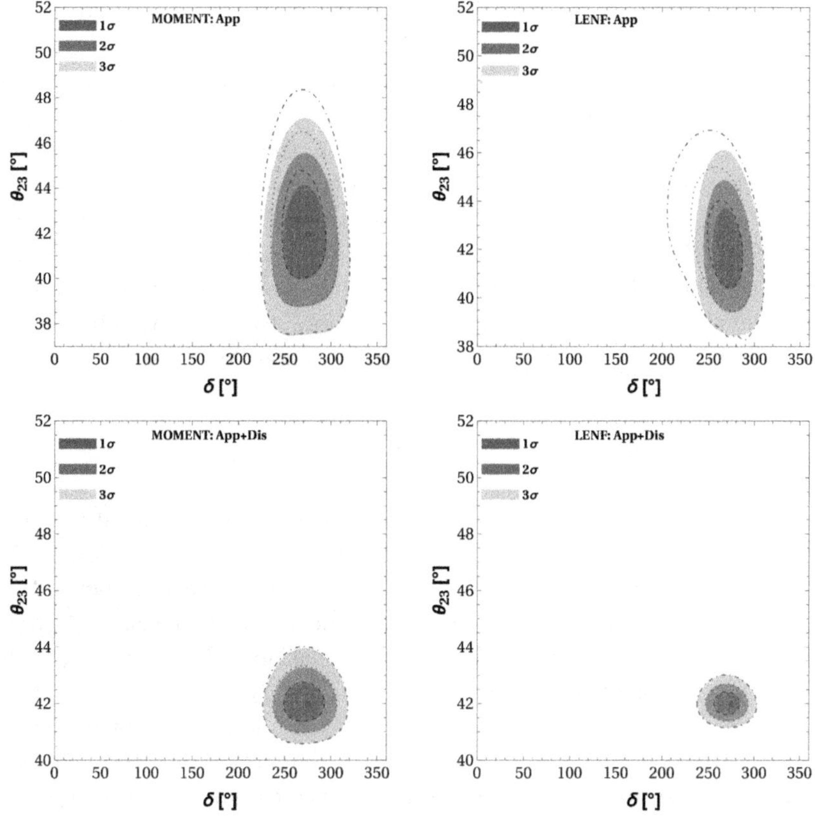

Fig. 4. The upper (lower) panels show the allowed region constrained by appearance channels (all channels). In both panels, the gray, magenta, and cyan regions show the 1σ, 2σ and 3σ C.L in standard case, respectively. The dashed, dotted and dash-dotted line show the allowed regions in 1σ, 2σ and 3σ C.L, respectively. The best fit values: $\theta_{23} = 42°$ and $\delta_{cp} = 270°$. In all panels the value of τ_3/m_3 vary within the range: $[3 \times 10^{-12}, 3 \times 10^{-10}]$ eV/s (color online).

black curve and the LENF is shown by the dashed black curve. We also include the current constraint to $\tau_3/m_3 : \tau_3/m_3 > 2.8 \times 10^{-12}$ at 90% C.L by combining T2K and MINOS using the solid red line. After a comparison we see the exclusion ranges in MOMENT: $\tau_3/m_3 < 1.47 \times 10^{-11}$ (3σ C.L), $\tau_3/m_3 < 2.81 \times 10^{-11}$ (90% C.L). Similarly, the LENF: $\tau_3/m_3 < 2 \times 10^{-11}$ (3σ C.L), $\tau_3/m_3 < 3.55 \times 10^{-11}$ (90% C.L). It is easy to see that MOMENT and LENF can significantly improve the bound on ν_3 lifetime by at least one order of magnitude, especially for the long baseline accelerator neutrino experiment LENF.

3.3. *Precise measurement*

In this section we display the impacts of ν_3 decay on determination of δ_{23} and CP-violating phase δ in MOMENT and LENF. We choose true values of θ_{23} and δ_{cp}

with $\delta_{23} = 42°$ and $\delta = 270°$ to simulate these two experiments and fit the spectra with/without ν_3 decay. The colorful regions show the allowed regions without decays in 1σ, 2σ and 3σ C.L and the lines show the allowed regions in the case with unstable ν_3. The allowed regions of upper panels are obtained only by the appearance channels, while the allowed regions in the lower panels are obtained by full channels. As we have discussed in Sec. 3.1, the appearance channels will suffer from τ_3/m_3 obviously. From Fig. 4 one can see that the existence of invisible decays would affect the determination of θ_{23} and δ from the appearance channels. However, the effects from the τ_3/m_3 are negligible for both experiments after taking into account the disappearance channels. Therefore, the correlation between the θ_{23} and τ_3 may be negligible after the combination regardless of some correlations in the appearance channel.

4. Conclusion

In this paper we have considered the third neutrino mass eigenstate ν_3 decays to invisible states in muon-decay accelerator neutrino experiments, MOMENT and LENF. Within the current allowed range of τ_3/m_3, the correlation between θ_{23} and τ_3 has been discussed in appearance and disappearance channels, respectively. For the appearance channels, there is an intense correlation of θ_{23} and τ_3/m_3 as they can both shift the height of the probability peak. Then θ_{23} and τ_3/m_3 would interfere with each other in appearance channels. However, there is no obvious correlation between the two parameters in disappearance channels. Moreover, the sensitivity of lifetime and θ_{23} are mainly from disappearance channels, so that the effects of τ_3/m_3 to precision measurements of θ_{23} and δ can be very tiny after combining all channels for neutrinos and antineutrinos. Finally, we show the constraint curves of τ_3/m_3 in the LENF and MOMENT, which demonstrates that these two muon-decay facilities can improve constraints on lifetime of ν_3 compared with the current bound.

Acknowledgement

This work is supported by the National Natural Science Foundation of China under Grant No. 11505301.

References

1. B. Aharmim *et al.*, Combined Analysis of all Three Phases of Solar Neutrino Data from the Sudbury Neutrino Observatory, *Phys. Rev.* **C88**, p. 025501 (2013).
2. R. Wendell *et al.*, Atmospheric neutrino oscillation analysis with sub-leading effects in Super-Kamiokande I, II, and III, *Phys. Rev.* **D81**, p. 092004 (2010).
3. S. Abe *et al.*, Precision Measurement of Neutrino Oscillation Parameters with Kam-LAND, *Phys. Rev. Lett.* **100**, p. 221803 (2008).
4. F. P. An *et al.*, Measurement of electron antineutrino oscillation based on 1230 days of operation of the Daya Bay experiment, *Phys. Rev.* **D95**, p. 072006 (2017).

5. I. Esteban, M. C. Gonzalez-Garcia, M. Maltoni, I. Martinez-Soler and T. Schwetz, Updated fit to three neutrino mixing: Exploring the accelerator-reactor complementarity, *JHEP* **01**, p. 087 (2017).

6. G. Ranucci, Status and prospects of the JUNO experiment, *J. Phys. Conf. Ser.* **888**, p. 012022 (2017).

7. H. Seo, Status of RENO-50, *PoS* **NEUTEL2015**, p. 083 (2015).

8. R. Acciarri *et al.*, Long-Baseline Neutrino Facility (LBNF) and Deep Underground Neutrino Experiment (DUNE) (2015).

9. P. Adamson *et al.*, First measurement of muon-neutrino disappearance in NOvA, *Phys. Rev.* **D93**, p. 051104 (2016).

10. K. Abe *et al.*, Neutrino oscillation physics potential of the T2K experiment, *PTEP* **2015**, p. 043C01 (2015).

11. K. Abe *et al.*, Combined Analysis of Neutrino and Antineutrino Oscillations at T2K, *Phys. Rev. Lett.* **118**, p. 151801 (2017).

12. P. Adamson *et al.*, Constraints on Oscillation Parameters from ν_e Appearance and ν_μ Disappearance in NOvA, *Phys. Rev. Lett.* **118**, p. 231801 (2017).

13. A. Acker, S. Pakvasa and J. T. Pantaleone, Decaying Dirac neutrinos, *Phys. Rev.* **D45**, 1 (1992).

14. A. Acker and S. Pakvasa, Solar neutrino decay, *Phys. Lett.* **B320**, 320 (1994).

15. G. B. Gelmini and M. Roncadelli, Left-Handed Neutrino Mass Scale and Spontaneously Broken Lepton Number, *Phys. Lett.* **99B**, 411 (1981).

16. S. Pakvasa, Do neutrinos decay?, *AIP Conf. Proc.* **542**, 99 (2000) [99(1999)].

17. A. M. Gago, R. A. Gomes, A. L. G. Gomes, J. Jones-Perez and O. L. G. Peres, Visible neutrino decay in the light of appearance and disappearance long baseline experiments, *JHEP* **11**, p. 022 (2017).

18. P. Coloma and O. L. G. Peres, Visible neutrino decay at DUNE (2017).

19. J. M. Berryman, A. de Gouvea, D. Hernandez and R. L. N. Oliveira, Non-Unitary Neutrino Propagation From Neutrino Decay, *Phys. Lett.* **B742**, 74 (2015).

20. J. A. Frieman, H. E. Haber and K. Freese, Neutrino Mixing, Decays and Supernova Sn1987a, *Phys. Lett.* **B200**, 115 (1988).

21. J. Cao *et al.*, Muon-decay medium-baseline neutrino beam facility, *Phys. Rev. ST Accel. Beams* **17**, p. 090101 (2014).

22. E. Fernandez Martinez, T. Li, S. Pascoli and O. Mena, Improvement of the low energy neutrino factory, *Phys. Rev.* **D81**, p. 073010 (2010).

23. J. Tang and Y. Zhang, Study of Non-Standard Charged-Current Interactions at the MOMENT experiment (2017).

24. J. Tang, Y. Zhang and Y.-F. Li, Probing Direct and Indirect Unitarity Violation in Future Accelerator Neutrino Facilities, *Phys. Lett.* **B774**, 217 (2017).

25. P. Huber, M. Lindner and W. Winter, Simulation of long-baseline neutrino oscillation experiments with GLoBES (General Long Baseline Experiment Simulator), *Comput. Phys. Commun.* **167**, p. 195 (2005).

26. P. Huber, J. Kopp, M. Lindner, M. Rolinec and W. Winter, New features in the simulation of neutrino oscillation experiments with GLoBES 3.0: General Long Baseline Experiment Simulator, *Comput. Phys. Commun.* **177**, 432 (2007).

27. T. Hahn, Routines for the diagonalization of complex matrices (2006).

28. R. A. Gomes, A. L. G. Gomes and O. L. G. Peres, Constraints on neutrino decay lifetime using long-baseline charged and neutral current data, *Phys. Lett.* **B740**, 345 (2015).

546

Commissioning and Validation of the ATLAS Level-1 Topological Trigger in Run 2

Daniel Zheng

*Department of Electrical and Computer Engineering, University of Pittsburgh,
Pittsburgh, PA 15213, USA
E-mail: daniel.zheng@pitt.edu*

Andrew Aukerman

*Department of Physics and Astronomy, University of Pittsburgh,
Pittsburgh, PA 15213, USA
E-mail: ataukerman@pitt.edu*

Tae Min Hong

*Department of Physics and Astronomy, University of Pittsburgh,
Pittsburgh, PA 15213, USA
E-mail: tmhong@pitt.edu*

On behalf of the ATLAS Collaboration

The ATLAS experiment has introduced and recently commissioned a completely new
hardware sub-system of its first-level trigger: the topological processor (L1Topo). L1Topo
consist of two AdvancedTCA blades mounting state-of-the-art FPGA processors, providing high input bandwidth (up to 4 Gb/s) and low latency data processing (200 ns).
L1Topo is able to select collision events by applying kinematic and topological requirements on candidate objects (energy clusters, jets, and muons) measured by calorimeters
and muon sub-detectors. Results from data recorded using the L1Topo trigger will be
presented. These results demonstrate a significantly improved background event rejection, thus allowing for rate reduction with minimal efficiency loss. This improvement has
been shown for several physics processes leading to low-p_T leptons, including $H \to \tau\tau$
and $J/\psi \to \mu\mu$. In addition to describing the L1Topo trigger system, we will discuss the
use of an accurate L1Topo simulation as a powerful tool to validate and optimize the
performance of this new system. To reach the required accuracy, the simulation must
mimic the approximations applied in firmware to execute the kinematic calculations.

1. Introduction

The ATLAS Experiment at the European Organization for Nuclear Research
(CERN) is one of two general purpose experiments at the Large Hadron Collider
(LHC) in Meyrin, Switzerland. The LHC collides proton bunches with a frequency
of 40 MHz.[1] The ATLAS Trigger and Data Acquisition (TDAQ) system discards
collision events that are not of physics interest.[2] As the LHC operates at increasing
instantaneous luminosity, the need for further rate reduction arises.

1.1. *TDAQ*

The TDAQ system is comprised of two levels, the first being a hardware based, low-granularity Level-1 (L1) system with a fixed latency of 2.5 μs.[3] This system constructs Regions-of-Interest seeding the software based algorithms used in the subsequent High-Level Trigger (HLT) system, reconstructing the event with full detector read-out granularity. The L1 system is further divided into three subcomponents: the Calorimeter Trigger (L1Calo), the Muon Trigger (L1Muon) and the Central Trigger Processor (CTP). The L1 system reduces the 40 MHz collision rate to 100 kHz, which is further reduced by the HLT trigger to 3-4 kHz.[3] The Level-1 Topological Trigger (L1Topo) system is a new component of the ATLAS TDAQ system that makes it possible to impose topological constraints to L1 triggers.[4] The goal of this system is to improve L1 event selection by imposing kinematic requirements based on the event topology of certain processes. Its electronic boards compute angular and kinematic quantities between various L1 Trigger Objects (TOBs).[4] Its role in the L1 trigger is shown in the flowchart included in Fig. 1. The system is designed to receive and process up to 6Tb/s of real time data and was commissioned and validated in the 2017 LHC run at $\sqrt{s} = 7\ TeV$.[4]

2. Motivation

As higher luminosities are reached at the LHC, the production rate for physics signatures increases. Traditionally, this is dealt with using prescales, which save only 1 of every n events that pass the trigger, or by tightening selection criteria, e.g., by increasing the transverse momentum threshold required.[5] Neither of these approaches is optimal, however, as they both cause data loss that affects particular studies involving high-p_T objects. There are a variety of physics analyses that benefit from the introduction of topological cuts. For example, L1Topo provides rate reduction for the $B^0 \to J/\psi\phi$ analysis, which was reaching unsustainable rates for low p_T muon triggers.[6]

3. L1Topo Algorithm Execution

3.1. *L1Muon*

In Run 1, the Muon CTP Interface (MUCTPI) extracted only muon multiplicity and momentum quantities for the Central Trigger Processor (CTP) to use in the L1Muon decision.[2] With the introduction of L1Topo, additional kinematic and angular cuts can be performed.[4] Thus, for Run 2, the MUCTPI was upgraded to also extract coarse-grained muon candidates to send to the L1Topo module.[7] The MUCTPI system includes 16 Muon Octant (MIOCT) modules, one for each octant in each half of the detector. Up to 32 muon candidates are sent to L1Topo. A feasibility study in which a high resolution phase scan was performed on data received from the overclocked MIOCT modules showed that there exists an optimal sampling point for DDR data transmission, with a bit error rate lower than 10-15 at 95%

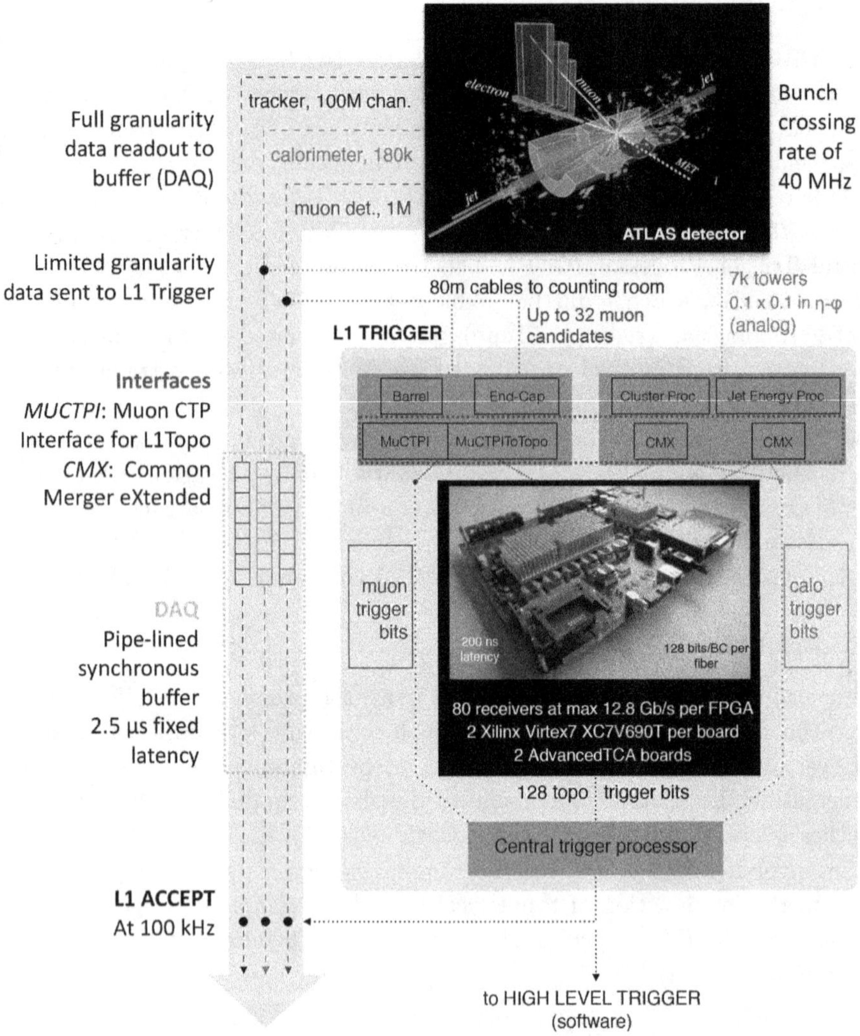

Fig. 1. Data flow through the L1 Trigger System with emphasis on L1Topo's role.

confidence level.[7] This demonstrates that the MUCTPI trigger outputs can reliably transmit data to L1Topo. The total aggregated data rate of the MUCTPI system is 10.24 Gb/s.[7] Commissioning tests of this system specifically include use of a high-level software simulation to generate input vectors and calculation of expected results.[7] No errors appeared in a sample of snapshot memory from 1010 bunches. The parallel output from the MUCTPI must also be encoded into optical output for the L1Topo modules. This is implemented in a dedicated MUCTPIToTopo board, which includes a Xilinx VC707 development kit and two FMC cards. Integration tests showed zero errors out of $2.6 * 10^6$ test bunches.[7]

3.2. *L1Calo*

The L1Calo Common Merger Modules (CMMs), which interface with the CTP, had to be replaced to run at higher bandwidth and to function with the proposed L1Topo module. The CMM modules, in the same vein as the MIOCT modules, collect information from the calorimeter specific to their crate numbers.[8] The upgrade of these crate modules to Common Merger eXtended (CMX) models allows transmission of 120 e/γ candidates, 120 τ candidates, 64 jet candidates, and MET information to L1Topo.[8]

3.3. *L1Topo*

The L1Topo system is a crate equipped with two AdvancedTCA L1Topo modules. In each module, TOB data is received from the L1Calo and L1Muon systems through optical ribbon fibers in the backplane and front panel. These optical signals are then converted to electrical ones and transmitted to two Virtex-7 FPGAs. Algorithms are executed within 75 ns, corresponding to three LHC bunch crossings. The first two ticks are used for data reduction, consisting of sorting TOBs by E_t value or a selection algorithm providing an E_t cut. Decision algorithms are subsequently applied to the reduced input lists and output bits are sent to the CTP for use in the L1 trigger decision. The output consists of a decision and an overflow bit. The overflow sets the decision to true so the HLT can examine the event with more granular information and complex algorithms. The module can handle a maximum of 128 trigger algorithms (32 per FPGA).[9]

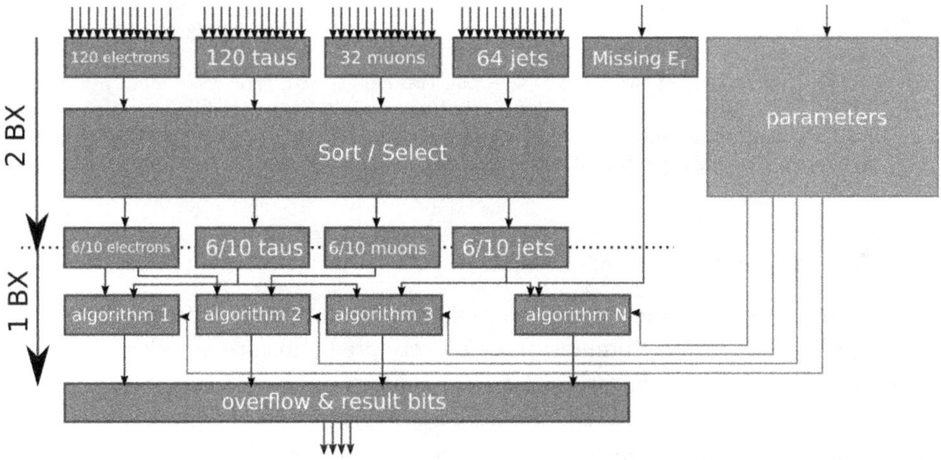

Fig. 2.   Diagram of the L1Topo algorithm execution.[9]

4. Commissioning

The goal L1Topo is to improve the purity of L1 trigger selection by imposing kinematic and angular constraints. This results in a rate reduction when the rate for a L1 trigger including a topological requirement is compared with the one for the same multiplicity and energy thresholds, but without the topological requirement. An example of this rate reduction is for the case of a trigger targeting low-mass dimuon resonances. In this case the topological requirements are that the two muons are collinear, $0.2 < \Delta R(\mu, \mu) < 1.5$, and the muons have an invariant mass consistent with the resonance, $2\ GeV < m(\mu, \mu) < 9\ GeV$ (Fig. 3). This rate reduction is achieved without significant loss in the selection efficiency with respect to the L1 trigger requiring only two muons with $p_T > 6\ GeV$.[4]

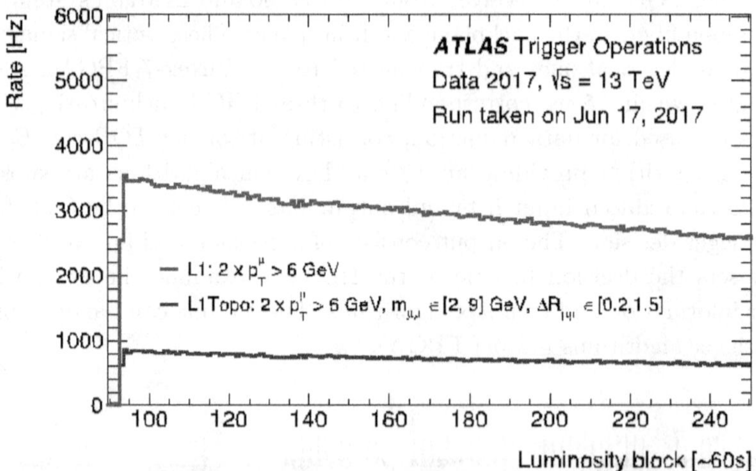

Fig. 3.   The rate vs. time of two triggers selecting two muons with transverse momentum larger than 6 GeV. The L1Topo trigger rate (blue) includes the additional topological requirement that the two L1 muons form an invariant mass $2\ GeV < M_{\mu\mu} < 9\ GeV$ and have angular separation $0.2 < \Delta R < 1.5$, showing a significant rate reduction.[5] (color online)

4.1. *Xmon Online Rate Monitoring*

The Xmon rate monitoring system is a tool available to trigger shifters and experts. Xmon provides real-time, luminosity scaled rate predictions using pileup linear regressions unique to each trigger. Each trigger can be written in terms of cross-section by considering the current rate of the trigger and the instantaneous luminosity and adjusting for number of bunches and prescale. The detected pileup is used to perform a linear regression to estimate this cross section. The estimated parameters are then used in real time and multiplied by luminosity to predict trigger rates. For L1Topo, an additional functionality was added to the Xmon tool to monitor the ratios between trigger rates. Specifically, the ratio of L1Topo items with their

L1 equivalent without topological requirements. This in turn allows for the fast detection of potential issues arising with the MUCTPI, CMX modules, CTP link, and the L1Topo modules themselves. Issues involving L1Topo trigger rates would immediately be visible if they involved the new L1Topo components. An example showing the ratio between the L1Topo rate and L1 rate for muons is shown in Fig. 4.

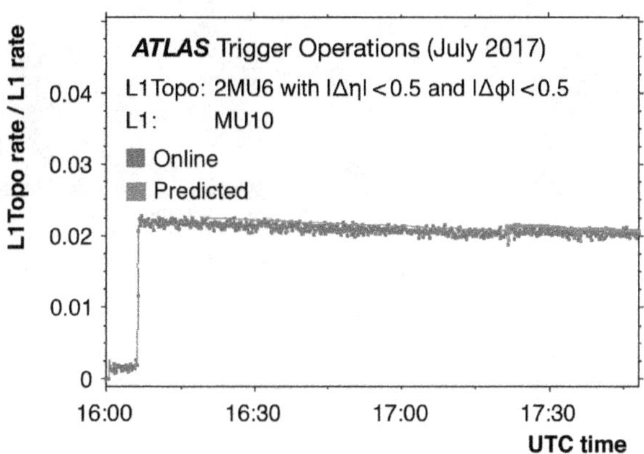

Fig. 4. L1Topo rate ratio monitoring. The rate of the L1Topo trigger requiring two muons with $p_T > 6~GeV$ and $|\Delta\eta|, |\Delta\phi| > 0.5$ is divided by the rate of an L1 muon trigger requiring $p_T > 10~GeV$.[5]

5. Validation

5.1. *Hardware and Firmware Validation*

The execution of the L1Topo algorithms was tested in several ways. One such test was to compare the hardware output with software simulations, which were accurate to O(1%) as shown in Fig. 5. These results were further improved with bitwise simulations that account for the fact that the simulation runs on a floating point architecture, while the algorithms are executed on FPGAs capable of only integer arithmetic.

6. Conclusions

The demands placed on the L1Topo subsystem, such as low latency, high-throughput and computational accuracy, have all been met. The commissioning and validation of L1Topo has successfully demonstrated that despite the stringent bandwidth, latency, and hardware constraints, topological information can be extracted and utilized to improve the capabilities of the ATLAS trigger system. With the introduction of the Level-1 Topological Trigger, the ATLAS trigger system is now capable of taking data at high luminosity with manageable rates while maintaining efficiency for several signatures relevant to important physics analyses.

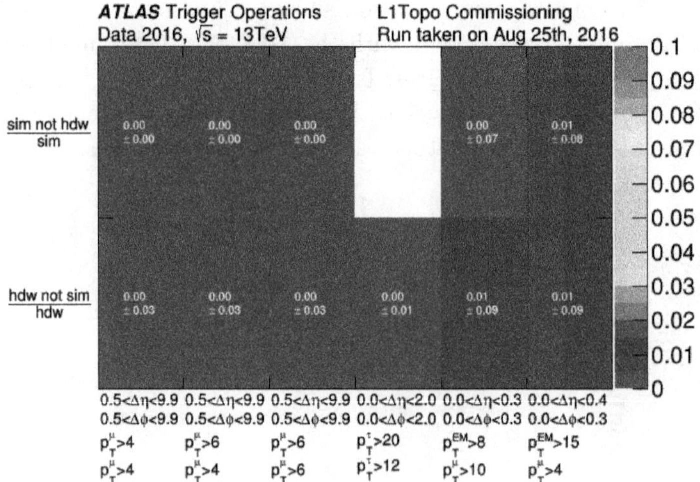

Fig. 5. Detailed comparison of event mismatches in simulation and hardware. O(1%) error are shown for false positives (top) and false negatives (bottom) for various L1Topo algorithms.[5]

References

1. ATLAS Collaboration, The ATLAS Experiment at the CERN Large Hadron Collider (2008, JINST 3 S08003).
2. ATLAS Collaboration, Performance of the ATLAS Trigger System in 2010 (2011, arXiv:1110.1530).
3. ATLAS TDAQ Collaboration, The ATLAS Data Acquisition and High Level Trigger system (2008, JINST 11 P06008).
4. E. Simioni, The Topological Processor for the future ATLAS Level-1 Trigger: From design to commissioning (2014, arXiv:1406.4316).
5. ATLAS Collaboration, Trigger Operation Public Results (2017) https://twiki.cern.ch/twiki/bin/view/AtlasPublic/TriggerOperationPublicResults
6. ATLAS Collaboration, ATLAS B-physics studies at increased LHC luminosity, potential for CP-violation measurement in the $B_s^0 \to J/\psi\phi$ decay (2013, ATL-PHYS-PUB-2013-010).
7. F. Anulli et al., The Level-1 Trigger Muon Barrel System of the ATLAS experiment at CERN (2009, JINST 4 P0401).
8. W.T. Fedorak and P.P. Plucinski, CMX Base Function FPGA Firmware Functionality (2015).
9. S. Artz, Physics performance with the new ATLAS Level-1 Topological trigger in Run 2 (2016, Proc. of the Fourth Annual Large Hadron Collider Physics, ATL-DAQ-PROC-2016-012).

Detector Description and Event Display for the JUNO Experiment[*]

Jiang Zhu, Kaijie Li and Zhengyun You

*School of Physics, Sun Yat-sen University,
Guangzhou, 510275, China*

Yumei Zhang[†]

*Sino-French Institute of Nuclear Engineering and Technology, Sun Yat-sen University,
Guangzhou, 510275, China
E-mail: zhangym26@mail.sysu.edu.cn*

On behalf of the JUNO Collaboration

Offline software system plays an important role in improving the efficiency and quality of physics analysis. In offline software of Jiangmen Underground Neutrino Observatory Experiment (JUNO), the detector geometry management system describes the detector data and its structure, and provides the interfaces of detector data for applications such as simulation, calibration, reconstruction, analysis and event display. To guarantee the consistency of geometry data in each application, a unified detector description is provided. In JUNO offline software, an event display system based on ROOT has been developed. We also use Unity, a renowned game engine, to improve its performance and make it available in different platforms. Compared with ROOT, Unity can give a more vivid demonstration of high energy physics experiments and the application built by Unity can be easily transplanted into different platforms. A new event display system with Unity is under development for JUNO. It provides an intuitive way to observe the detector structure, the particle trajectory and the hit time distribution.

Keywords: Detector description; event display; JUNO; ROOT; Unity.

1. Introduction

The next generation reactor neutrino detector, Jiangmen Underground Neutrino Observatory Experiment (JUNO), is under construction in Jiangmen city, Guangdong province, China, which is in a distance of 53km from both of Yangjiang and Taishan Nuclear Power Plants (NNP)[1]. It will begin to take data in 2020 if the construction and installation make progress successfully as the schedule planned.

The offline software system is very important in High Energy Physics as the experiment in HEP will generate a great amount of data which consumes a lot of computing resource and storage volume when doing physics analysis, so physicists will benefit a lot from the improvement of the offline software as it will speed up the analysis and get better data. The experiment in HEP like JUNO, will have a software system as an integrity to do the simulation, calibration, reconstruction, event display and other parts of physics analysis. The event data[2] will be used in different applications as mentioned above. Besides the event

[*]This work is supported by National Natural Science Foundation of China (11405279, 11675275) and the Strategic Priority Research Program of Chinese Academy of Sciences (XDA10010900).
[†]Corresponding author

data, the detector information is also necessary for different applications in the offline software to do analysis. To keep the consistency of the geometry data between applications, a unified detector description needs to be developed. When doing the physics analysis, event display is a very useful tool for physicists to have an institute way to observe rare event. The visualization of the experiment makes it possible to show the detector structure clearly. Two event display systems have been built for JUNO to meet different requirements of the users.

2. JUNO Experiment

The main scientific purpose of JUNO experiment is to measure the mass hierarchy of the three generations of neutrinos with Inverse Beta Decay (IBD)[3]. As a reactor neutrino experiment, JUNO will capture the neutrinos generated by the two NPPs in a distance of 53km. The neutrinos will react in the JUNO detector and then the liquid scintillator will glow as it absorbs the energy from the IBD. The light generated by the liquid scintillator will be captured by the photomultiplier tube (PMT), very sensitive light detectors in the ultraviolet, visible and close infrared electromagnetic spectrum[4]. With the read-out of the PMTs, we can run the reconstruction algorithm to learn about the information of neutrinos that the detector captured.

The detector structure of JUNO is shown in Fig. 1. The detector will be constructed at a shallow depth, with an average overburden of 700m to reduce the backgrounds induced by cosmogenic muons. The detector consists of a central detector, a muon tracker and a Cherenkov detector. The central detector is an acrylic ball with 35.4-meter diameter, which contains 20000-ton liquid scintillator inside and there are 18000 20-inch PMTs surrounding the sphere. The outer layer of the central detector is the water pool to support the weight of the acrylic ball with about 2000 veto PMTs in it. When we are doing the simulation, reconstruction and some other physics analysis, the users may want to use the detail information of the detector, like the position of each PMT, the direction of support

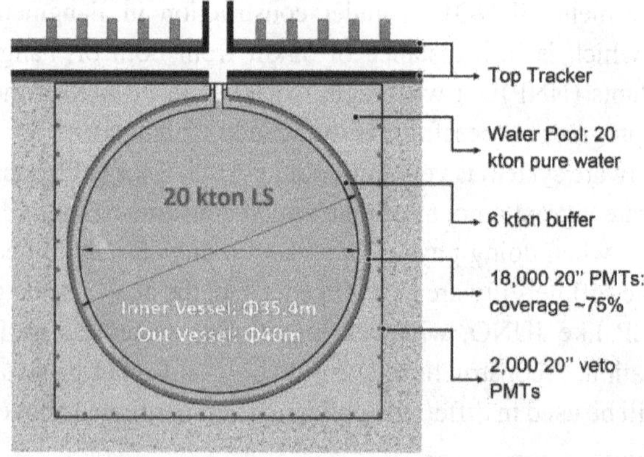

Fig. 1.   A schematic view of the JUNO detector.

steel, the detector hierarchy and so on. A unified detector description is developed in the geometry management service to provide a lot of useful functions for users.

3. Detector Description

The geometry service is designed to describe the details of the detector and to provide the consistent data for all the applications in JUNO offline software system. There are some important criterions when developing the detector description. First, it need to be accurate as we want to get all information of the detector. Second, a consistent description is necessary. All the applications in JUNO offline software will share a single source of detector description. Third, it should be flexible, which means different versions of geometry can coexist in the software, detector geometry can be changed dynamically over time and the detector units can be turned on or off. Last but not least, it should be user-friendly.

In JUNO offline software system, each unit of the detector has a unique ID, which means each specific unit in any sub-detector corresponds to an identifier. The identifier in the offline is numbered with 32 bits unsigned integer. What geometry service does in the offline software is to construct every unique detector unit, map the units with its identifier, organize the detector hierarchy and provide application interface for users.

With the geometry service, users can access the detail information of JUNO detector, such as the information of central detector and the PMTs on it. Some useful functions are provided to get the parameters and sub structure of the central detector. What's more, by building the physics node of each units, the hierarchy of JUNO detector is constructed from the top word volume to the level of each PMT, as Fig. 2 shows. Then the users can get all the information of each PMT, such as its material, shape, rotation angles, positons, id of layer and azimuth.

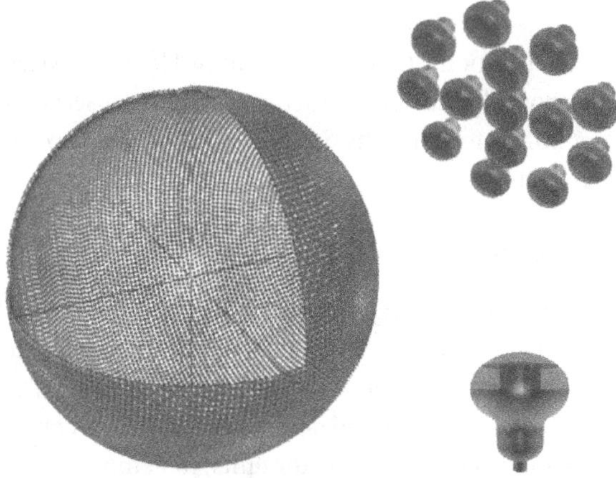

Fig. 2.   Central detector and PMTs.

Fig. 3 shows the current scheme of geometry service in JUNO offline software. The original geometry information is input into the offline software in text format. Then Geant4[5] will run the detector simulation and generate the GDML files[6] which contain the detector data. Next, the GDML files are transformed into the ROOT[7] files automatically[8,9] and are saved together with the event data if users choose to save the geometry information when running the simulation. In the following steps of analysis like calibration, reconstruction and event display, after initialization with the detector description ROOT files, the geometry service provides the interface to applications so that they can access the geometry data.

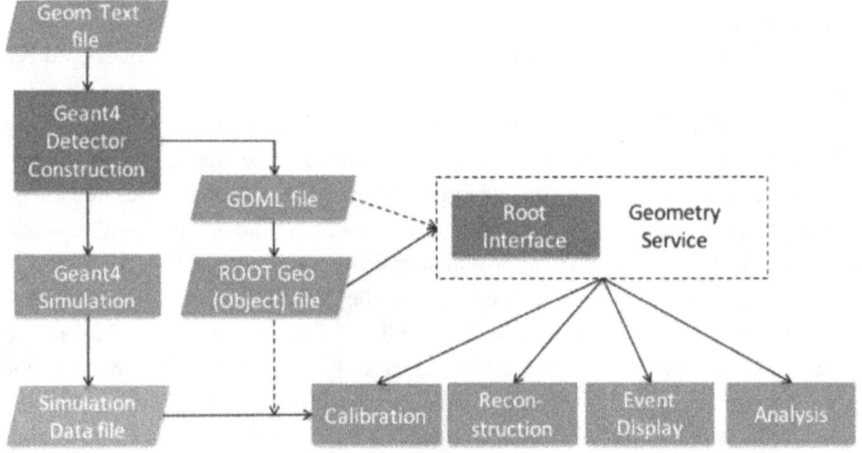

Fig. 3.   The geometry management scheme in JUNO offline software.

4. Event Display

Event display is the visualization of the experiment in HEP. The users can use the event display tools to learn about the detector structure easily to observe the rare events in an intuitive way. It is not only a tool for physics analysis but can also be used to improve of the reconstruction algorithm.

4.1. *Event display based on ROOT*

An event display tool is developed in JUNO offline software called SERENA, which stands for Software for Event Display with Root Eve in Neutrino Analysis. The main graphic user interface of SERENA is shown in Fig. 4. As an event display system based on ROOT, it is widely used in JUNO for detector optimization, simulation, reconstruction and physics study. SERENA is able to visualize the detector units, to demonstrate the hits distributions on PMTs, to compare the associations between reconstruction and MC truth and to show the 2D projection view.

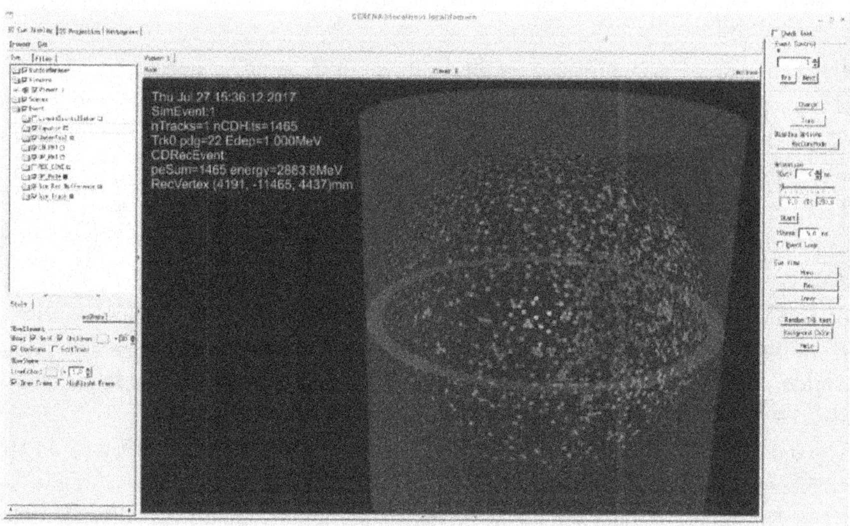

Fig. 4.   The Graphical User Interface of SERENA.

Fig. 5 shows the scheme of event display in JUNO Offline. SERENA is integrated in the JUNO offline software and is based on ROOT. It gets the detector data from the geometry service and builds the geometry object with the ROOT Event Visualization Environment(EVE) package[10]. Then it reads the event data generated by the JUNO offline software and receives the settings changed by users with the GUI components to change the visual effect of the corresponding geometry object.

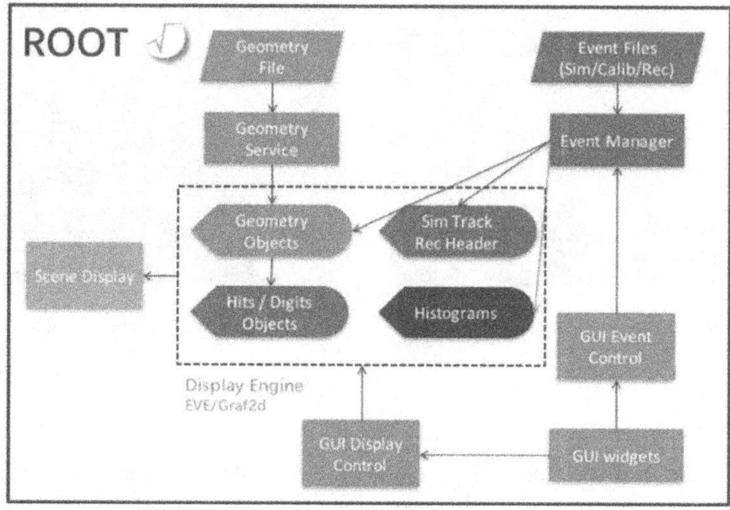

Fig. 5. Current event display scheme in JUNO offline.

4.2. *Event display based on Unity*

A new event display system based on the Unity engine[11] is under development for JUNO. We are trying to build a new event display as a client to make it less dependent on the JUNO offline software so that users can run the event display on their own PCs without setting up the whole offline software.

The Unity engine is a renowned game engine. Many companies and independent developers use it to build their own games, especially the ones in mobile device, because an obvious advantage of Unity is that it allows developers to target many devices very easily. It is convenient for developers to transplant their applications into other platforms. What's more, Unity is not just for game development, and it can also be used in education, simulation, visualization and so on. There are some successful projects in HEP, like the CAMELIA[12].

The data flow of the new event display system is similar to the one based on ROOT. The difference is that the new event display system is independent with the offline software. The geometry data and the event data generated by the offline software in ROOT format will be transformed and then be read into the event display system based on Unity, so the event display and offline software are separated. Users just need a client in their own PCs. For now, the new event display system has realized some basic functions, like the visualization of the detector, the animation of hits distribution and demonstration of the detail information, as Fig. 6 and Fig. 7 show. In comparison to the one based on ROOT a great advantage of the event display system based on Unity is that it allows users to run the event display system on their laptops without setting up the offline software, and it is easy to realize fancier visual effect as a game engine.

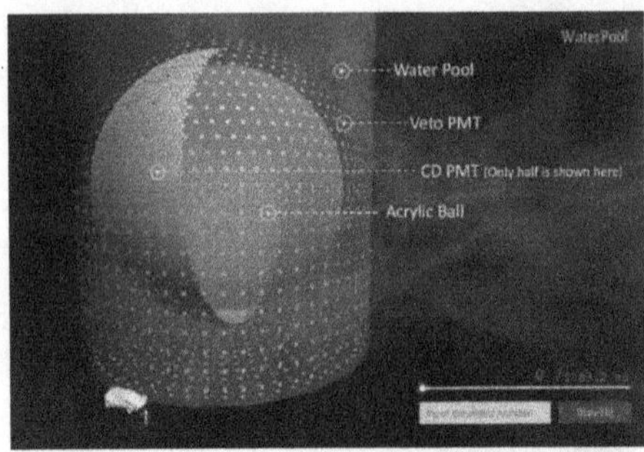

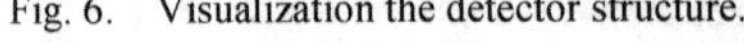

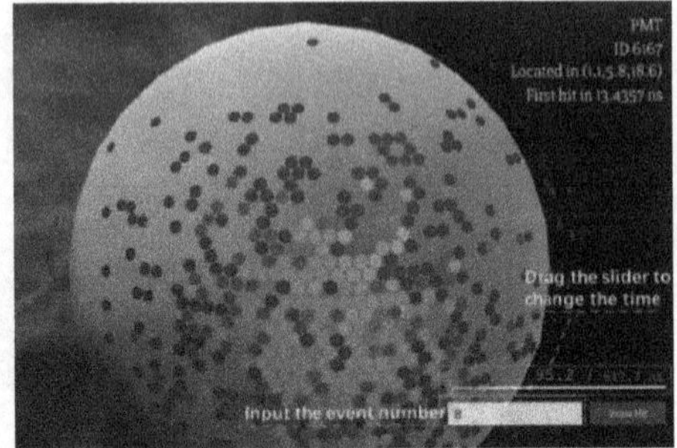

Fig. 6.   Visualization the detector structure.

Fig. 7.   Demonstration the event hit and reconstruction vertex.

5. Conclusion

A consistent detector description is provided for the JUNO experiment. Two event display systems have been developed to help users tuning reconstruction algorithms and analyzing simulation data. The one based on ROOT is approaching full functionality and the other one based on Unity is basically available.

Acknowledgments

The authors would like to thank the members of JUNO for their valuable discussions and suggestions, and thank the support from National Natural Science Foundation of China (11405279, 11675275) and the Strategic Priority Research Program of Chinese Academy of Sciences (XDA10010900).

References

1. Z. Djurcic, X. Li, W. Hu et al., *JUNO conceptual design report*, arXiv: 1508.07166.
2. T. Li, X. Xia, X. T. Huang et al., *Design and development of JUNO event data model*, *Chinese physics C*, **41**(6), 066201 (2017).
3. F. An, G. An, Q. An et al., *Neutrino physics with JUNO*, *Journal of Physics G: Nuclear and Particle Physics* **43**(3), 030401 (2016).
4. A. G. Wright, *The Photomultiplier Handbook, Oxford University Press* (2017).
5. S. Agostinelli et al., *GEANT4 — A simulation toolkit, Nuclear instruments and methods in physics research section A: Accelerators, Spectrometers, Detectors and Associated Equipment* **506**(3), 250–303 (2003).
6. R. Chytracek, J. McCormick, W. Pokorski and G. Santin, *Geometry description markup language for physics simulation and analysis applications, IEEE Trans. Nucl. Sci.* **53**, 2892 (2006).
7. R. Brun, F. Rademakers, *ROOT — An object oriented data analysis framework, Nucl. Inst. Meth. A* **389**, 81–86 (1997).
8. Z. Y. You, Y. T. Liang, Y. J. Mao, *A method for detector description exchange among ROOT GEANT4 and GEANT3, Chin. Phys. C* **32**(7), 572–575 (2008).
9. Y. T. Liang, B. Zhu, Z. Y. You et al., *A uniform geometry description for simulation, reconstruction and visualization in the BESIII experiment, Nucl. Inst. Meth. A* **603**, 325–327 (2009).
10. M. Tadel, *Overview of EVE — The event visualization environment of ROOT, Journal of Physics: Conference Series*, Vol. **219**. No. 4. IOP Publishing (2010).
11. https://unity3d.com.
12. http://medialab.web.cern.ch/content/camelia.

CPSIA information can be obtained
at www.ICGtesting.com
Printed in the USA
FSHW021608080320
67827FS